T0180328

Linearization of Nonlinear Control Systems

Hong-Gi Lee

Linearization of Nonlinear Control Systems

 Springer

Hong-Gi Lee
Chung-Ang University
Seoul, Korea (Republic of)

ISBN 978-981-19-3645-6   ISBN 978-981-19-3643-2   (eBook)
https://doi.org/10.1007/978-981-19-3643-2

This Springer imprint is published by the registered company Springer Nature Singapore Pte Ltd.
The registered company address is: 152 Beach Road, #21-01/04 Gateway East, Singapore 189721,
Singapore

For Yeonhee, Jaekwon, and Jung-Gyu Sung

Preface

The theory of nonlinear control systems attracts much attention with the recent rapid development of digital equipment. The linearization problem of a nonlinear control system is to find the state transformation and feedback such that a nonlinear system satisfies, in the new state, a linear system equation. It is well known as one of the effective control techniques if applicable. In addition to linear algebra, differential geometry is essential to understanding the linearization problems of the control nonlinear systems. However, according to the author's experience, the basics of differential geometry are hard to learn for engineering students. In this book, the basics of differential geometry and Lie algebra formulas needed in linearization are explained for the students who are not accustomed to differential geometry. Standard definitions in differential geometry are also found in the Appendix.

The conditions in the linearization problems are complicated to check because the Lie bracket calculation of vector fields by hand needs much attention. This book provides the MATLAB programs for most of the theorems. The MATLAB programs in this book might be helpful for further research in nonlinear control problems.

The book's contents are organized as follows: Chap. 2 gives the mathematical background for understanding the linearization problem. Conditions of linearization problems cannot be understood without differential geometry. Chapters 3–6 consider the continuous-time systems. State equivalence to a linear system (LS) and feedback linearization are discussed in Chaps. 3 and 4, respectively. In Chap. 5, the state equation and the output equation are considered for linearization. It is shown, in Chap. 6, that we can enlarge the class of linearizable systems by using dynamic feedback instead of static feedback. Chapter 7 deals with the discrete version of Chaps. 3–5. The conditions for linearization of discrete-time systems are quite different from those for continuous-time cases. The discrete version of Chap. 6 can also be found, even though it is omitted in this book. Chapter 8 deals with the observer linearization problem. If we find a state transformation that transforms the nonlinear system into a nonlinear observer canonical form, then a Luenberger-like observer design is possible. Finally, input-output decoupling is explained in Chap. 9. MATLAB programs are provided for examples and problems at the end of every chapter. I like to refer to the book [A3] by A. Isidori and the book [A5] by H. Nijmeijer and

A. van der Schaft for more advanced topics of nonlinear control systems. They have been beneficial for me to understand the nonlinear control theory.

I wish to thank Prof. Hyung-Jong Ko, Prof. Yong-Min Kim, and Prof. Ho-Jae Lee for their comments and encouragement. Also, I thank Ms. Hyeran Hong for her contribution to drawing the figures. Finally, I would like to thank Prof. Steven I. Marcus and Prof. Ari Arapostathis for their help.

Seoul, Korea (Republic of)   Hong-Gi Lee
January 2022

Acknowledgements This book was supported by Basic Science Research Program through the National Research Foundation of Korea (NRF) funded by the Ministry of Education (NRF-2019R1F1A1057480).

Contents

Acronyms

DOEL	Dynamic observer error linearizable
GNOCF	Generalized nonlinear observer canonical form
LOCF	Linear observer canonical form
LS	Linear system
MIMO	Multi input multi output
NLS	Nonlinear system
NOCF	Nonlinear observer canonical form
OT	Output transformation
RDOEL	Restricted dynamic observer error linearizable
SISO	Single input single output

Chapter 1
Introduction

For the control of complex nonlinear systems such as robots and aircraft, more advanced control techniques are required. Thus, the theory of nonlinear control systems has developed rapidly over the past several decades. This book deals with linearization, one of the significant trends in modern nonlinear control system theory.

1.1 Trends of Nonlinear Control System Theory

Consider the following state equation and output equation of the control system:

$$\dot{x} = F(x, u) ; \quad y = h(x, u).$$

If $F(x, u)$ and $h(x, u)$ are linear functions of state x and control input u, then it is a linear control system. Otherwise, it is a nonlinear system. For example, system (1.2) and system (1.3) are nonlinear control systems.

$$\dot{x} = Ax + Bu ; \quad y = Cx + Du \tag{1.1}$$

$$\begin{bmatrix} \dot{x}_1 \\ \dot{x}_2 \end{bmatrix} = \begin{bmatrix} 2x_1 x_2 - 2x_2^3 + (1 + 2x_2)u \\ x_1 - x_2^2 + u \end{bmatrix} ; \quad y = x_1 - x_2^2 + x_2 \tag{1.2}$$

$$\begin{bmatrix} \dot{x}_1 \\ \dot{x}_2 \end{bmatrix} = \begin{bmatrix} x_2 \\ -2x_1 - 2x_2 + x_1^2 + x_2^2 + (1 + x_1^2)u \end{bmatrix} ; \quad y = x_1 \tag{1.3}$$

© The Author(s), under exclusive license to Springer Nature Singapore Pte Ltd. 2022
H.-G. Lee, *Linearization of Nonlinear Control Systems*,
https://doi.org/10.1007/978-981-19-3643-2_1

For linear control system (1.1), it is clear that $t \geq 0$,

$$y(t) = Ce^{At}x(0) + \int_0^t Ce^{A(t-\tau)}Bu(\tau)d\tau + Du(t) \tag{1.4}$$

and the transfer function $G(s)$ can be defined such that $Y(s) = G(s)U(s)$. For a nonlinear system, $y(t)$ cannot be expressed in the closed-form, such as (1.4). Instead, it can be expressed as a series, called the Volterra series (See (5.66)). The transfer function cannot even be defined. Thus, the nonlinear systems are more complicated to analyze and control than the linear systems. The classical method of controlling a nonlinear system is the first-order approximate linearization (See Sect. 1.2.) In other words, the linear system can be obtained by ignoring all terms above the second order in the Taylor series of $F(x, u)$ about the nominal trajectory. In other words, if the nominal trajectory of system (1.2) is the origin, the system can be linearized, by ignoring $2x_1x_2$, $-2x_2^3$, $2x_2u$, $-x_2^2$, and $-x_2^2$, as follows:

$$\begin{bmatrix} \dot{x}_1 \\ \dot{x}_2 \end{bmatrix} = \begin{bmatrix} 0 & 0 \\ 1 & 0 \end{bmatrix} \begin{bmatrix} x_1 \\ x_2 \end{bmatrix} + \begin{bmatrix} 1 \\ 1 \end{bmatrix} u$$

$$y = \begin{bmatrix} 1 & 1 \end{bmatrix} \begin{bmatrix} x_1 \\ x_2 \end{bmatrix}$$

If the nominal trajectory of system (1.3) is the origin, the system (1.3) can also be linearized approximately, by ignoring x_1^2, x_2^2, and x_1^2u, as follows:

$$\begin{bmatrix} \dot{x}_1 \\ \dot{x}_2 \end{bmatrix} = \begin{bmatrix} 0 & 1 \\ -2 & -2 \end{bmatrix} \begin{bmatrix} x_1 \\ x_2 \end{bmatrix} + \begin{bmatrix} 0 \\ 1 \end{bmatrix} u$$

$$y = \begin{bmatrix} 1 & 0 \end{bmatrix} \begin{bmatrix} x_1 \\ x_2 \end{bmatrix}$$

One of the nonlinear control research trends is to extend the results for the linear systems to the nonlinear systems. For example, the input-output decoupling (or diagonalization) problem has been considered for the linear system. Suppose that the number of the outputs and the inputs are the same. If the transfer function is a diagonal matrix, each output can be controlled independently.

Example 1.1.1 Consider the following MIMO linear system:

$$\begin{bmatrix} \dot{x}_1 \\ \dot{x}_2 \\ \dot{x}_3 \end{bmatrix} = \begin{bmatrix} 0 & 1 & 0 \\ -2 & -3 & -4 \\ 0 & 0 & -3 \end{bmatrix} \begin{bmatrix} x_1 \\ x_2 \\ x_3 \end{bmatrix} + \begin{bmatrix} 0 & 0 \\ 1 & 0 \\ 0 & 1 \end{bmatrix} \begin{bmatrix} u_1 \\ u_2 \end{bmatrix} = Ax + Bu$$

$$\begin{bmatrix} y_1 \\ y_2 \end{bmatrix} = \begin{bmatrix} 1 & 0 & 0 \\ 0 & 0 & 1 \end{bmatrix} \begin{bmatrix} x_1 \\ x_2 \\ x_3 \end{bmatrix} = Cx$$

Find out a nonsingular linear feedback such that the transfer function matrix of the closed-loop system is a diagonal matrix.

Solution   It is easy to see that $\dot{y}_1 = x_2$

$$\begin{bmatrix} \ddot{y}_1 \\ \dot{y}_2 \end{bmatrix} = \begin{bmatrix} -2 & -3 & -4 \\ 0 & 0 & -3 \end{bmatrix} \begin{bmatrix} x_1 \\ x_2 \\ x_3 \end{bmatrix} + \begin{bmatrix} 1 & 0 \\ 0 & 1 \end{bmatrix} \begin{bmatrix} u_1 \\ u_2 \end{bmatrix}$$

and

$$G(s) = C(sI - A)^{-1} B = \begin{bmatrix} \frac{1}{s^2+3s+2} & \frac{-4}{(s^2+3s+2)(s+3)} \\ 0 & \frac{1}{s+3} \end{bmatrix}.$$

If we let

$$\begin{bmatrix} u_1 \\ u_2 \end{bmatrix} = \begin{bmatrix} 2 & 3 & 4 \\ 0 & 0 & 3 \end{bmatrix} \begin{bmatrix} x_1 \\ x_2 \\ x_3 \end{bmatrix} + \begin{bmatrix} v_1 \\ v_2 \end{bmatrix} = Fx + Gv$$

then we have that

$$\begin{bmatrix} \ddot{y}_1 \\ \dot{y}_2 \end{bmatrix} = \begin{bmatrix} v_1 \\ v_2 \end{bmatrix}$$

or

$$Y(s) = G_c(s)V(s) = \begin{bmatrix} \frac{1}{s^2} & 0 \\ 0 & \frac{1}{s} \end{bmatrix} V(s).$$

□

Example 1.1.2   Consider the following MIMO nonlinear system:

$$\begin{bmatrix} \dot{x}_1 \\ \dot{x}_2 \\ \dot{x}_3 \end{bmatrix} = \begin{bmatrix} x_2 \\ x_1^2 + u_1 + u_2 \\ (1 + x_2^2)u_2 \end{bmatrix} ; \quad \begin{bmatrix} y_1 \\ y_2 \end{bmatrix} = \begin{bmatrix} x_1 \\ x_3 \end{bmatrix}.$$

Find out the nonsingular feedback $u = \gamma(x, v)$, such that

$$\begin{bmatrix} \ddot{y}_1 \\ \dot{y}_2 \end{bmatrix} = \begin{bmatrix} v_1 \\ v_2 \end{bmatrix}.$$

Solution   It is easy to see that $\dot{y}_1 = x_2$ and

$$\begin{bmatrix} \ddot{y}_1 \\ \ddot{y}_2 \end{bmatrix} = \begin{bmatrix} x_1^2 \\ 0 \end{bmatrix} + \begin{bmatrix} 1 & 1 \\ 0 & 1 + x_2^2 \end{bmatrix} \begin{bmatrix} u_1 \\ u_2 \end{bmatrix} = \begin{bmatrix} v_1 \\ v_2 \end{bmatrix}.$$

Therefore, it is clear that

$$\begin{bmatrix} u_1 \\ u_2 \end{bmatrix} = - \begin{bmatrix} 1 & 1 \\ 0 & 1 + x_2^2 \end{bmatrix}^{-1} \begin{bmatrix} x_1^2 \\ 0 \end{bmatrix} + \begin{bmatrix} 1 & 1 \\ 0 & 1 + x_2^2 \end{bmatrix}^{-1} \begin{bmatrix} v_1 \\ v_2 \end{bmatrix}$$

$$= \begin{bmatrix} -x_1^2 \\ 0 \end{bmatrix} + \begin{bmatrix} 1 & -\frac{1}{1+x_2^2} \\ 0 & \frac{1}{1+x_2^2} \end{bmatrix} \begin{bmatrix} v_1 \\ v_2 \end{bmatrix} = \gamma(x, v).$$

$\square$

Example 1.1.2 is the nonlinear version of Example 1.1.1. In other words, the nonlinear version of the input-output decoupling problem can be defined. (See Chap. 9.) Similarly, many researchers have studied the nonlinear version of controllability, observability, noninteracting control, disturbance decoupling, controlled invariant distribution, adaptive control, optimal control, etc.

Another research trend is feedback linearization, which transforms a nonlinear system into a linear system using nonlinear state transformation and feedback. If a given nonlinear system is feedback linearizable, it is possible to use a well-developed linear system theory to control the nonlinear system. Thus, the feedback linearization problem has attracted tremendous interest from many researchers. In Sect. 1.3, the linearization problems are briefly introduced.

1.2   Approximate Linearization of the Nonlinear Systems

In this section, the classical approximate linearization method is introduced. Consider the following nonlinear control system:

$$\frac{dx(t)}{dt} = f(x(t), u(t)), \quad x \in \mathbb{R}^n, u \in \mathbb{R}^m \tag{1.5}$$

Suppose that the nominal trajectory $x^0(t)$ and nominal input $u^0(t)$ satisfies

$$\dot{x}^0(t) = f(x^0(t), u^0(t)) ; \quad x^0(0) = x(0). \tag{1.6}$$

If we expand the right-hand side of (1.5) into a Taylor series about $(x^0(t), u^0(t))$, then we have

$$\frac{dx(t)}{dt} = f(x^0(t), u^0(t)) + \left.\frac{\partial f}{\partial x}\right|_{\substack{x=x^0(t)\\u=u^0(t)}} (x(t) - x^0(t))$$

$$+ \left.\frac{\partial f}{\partial u}\right|_{\substack{x=x^0(t)\\u=u^0(t)}} (u(t) - u^0(t)) + \cdots \tag{1.7}$$

Let

$$\Delta x(t) \triangleq x(t) - x^0(t) \; ; \quad \Delta u(t) \triangleq u(t) - u^0(t). \tag{1.8}$$

If $\|\Delta x(t)\|$ and $\|\Delta u(t)\|$ are very small, then it is clear, by Taylor Theorem, that

$$\frac{d}{dt}(\Delta x(t)) = \dot{x}(t) - \dot{x}^0(t)$$

$$\cong \left.\frac{\partial f}{\partial x}\right|_{\substack{x=x^0(t)\\u=u^0(t)}} \Delta x(t) + \left.\frac{\partial f}{\partial u}\right|_{\substack{x=x^0(t)\\u=u^0(t)}} \Delta u(t) \tag{1.9}$$

$$\triangleq A(t)\Delta x(t) + B(t)\Delta u(t).$$

Example 1.2.1 Find a nominal trajectory $x^0(t)$, which is the solution to system (1.3) with initial conditions $x^0(0) = \begin{bmatrix} 0 \\ 0 \end{bmatrix}$ and input $u^0(t) = 0$, $t \geq 0$. Also, linearize system (1.3) about the nominal trajectory $x^0(t)$ and nominal input $u^0(t)$.

Solution Omitted. □

Example 1.2.2 Find a nominal trajectory $x^0(t)$ and input $u^0(t)$, which is the solution to system (1.3) with initial conditions $x^0(0) = \begin{bmatrix} 1 \\ 1 \end{bmatrix}$. Also, linearize system (1.3) about the nominal trajectory $x^0(t)$ and nominal input $u^0(t)$.

Solution Omitted. □

In Example 1.2.2, it is not easy to find the nominal trajectory for the first-order approximation linearization. Besides, if the state is far from the nominal trajectory, the approximation becomes inaccurate, and the new approximation equation about the new nominal trajectory must be obtained for accurate control.

1.3 Exact Linearization of the Nonlinear Systems

In the previous section, we studied the approximate linearization. This section introduces the exact linearization problem. This method has been defined in the early 80s and has attracted much attention. For example, if we consider nonlinear control system (1.3), then the closed-loop system with nonlinear state feedback $u = -\frac{x_1^2 + x_2^2}{1 + x_1^2} + \frac{1}{1 + x_1^2} v$ satisfies the following linear system equation:

$$\begin{bmatrix} \dot{x}_1 \\ \dot{x}_2 \end{bmatrix} = \begin{bmatrix} 0 & 1 \\ -2 & -2 \end{bmatrix} \begin{bmatrix} x_1 \\ x_2 \end{bmatrix} + \begin{bmatrix} 0 \\ 1 \end{bmatrix} v.$$

The above equation is not an approximation, but an exact one. In other words, nonlinear feedback could eliminate some nonlinear terms of the state equation. However, for nonlinear control system (1.2), there is no feedback to remove all the nonlinear terms. Another way to linearize a nonlinear system is to use a nonlinear coordinate transformation. To understand this, consider the following example.

Example 1.3.1 Consider the following linear system:

$$\begin{bmatrix} \dot{z}_1 \\ \dot{z}_2 \end{bmatrix} = \begin{bmatrix} 0 & 0 \\ 1 & 0 \end{bmatrix} \begin{bmatrix} z_1 \\ z_2 \end{bmatrix} + \begin{bmatrix} 1 \\ 1 \end{bmatrix} u \tag{1.10}$$

Let us define nonlinear state transform $x = S(z)$ by

$$\begin{bmatrix} x_1 \\ x_2 \end{bmatrix} = S(z) = \begin{bmatrix} z_1 + z_2^2 \\ z_2 \end{bmatrix} \quad \text{or} \quad z = S^{-1}(x) = \begin{bmatrix} x_1 - x_2^2 \\ x_2 \end{bmatrix}. \tag{1.11}$$

Show that the new state x satisfies Eq. (1.2).

Solution It is easy to see that

$$\dot{x}_1 = \dot{z}_1 + 2z_2 \dot{z}_2 = u + 2z_2(z_1 + u) = 2x_1 x_2 - 2x_2^3 + (1 + 2x_2)u$$

and

$$\dot{x}_2 = \dot{z}_2 = z_1 + u = x_1 - x_2^2 + u.$$

Thus, we have

$$\begin{bmatrix} \dot{x}_1 \\ \dot{x}_2 \end{bmatrix} = \begin{bmatrix} 2x_1 x_2 - 2x_2^3 \\ x_1 - x_2^2 \end{bmatrix} + \begin{bmatrix} 1 + 2x_2 \\ 1 \end{bmatrix} u.$$

$\square$

As shown in Example 1.3.1, when the nonlinear coordinate transformation of Eq. (1.11) is applied to linear system (1.10), the system becomes nonlinear system (1.2) in the new coordinate system. In other words, when the nonlinear coordinate transformation of Eq. (1.11) is applied to nonlinear system (1.2), the system becomes linear system (1.10) in the new coordinate system. Thus, nonlinear system (1.2) is linearizable by a state transformation.

So far, it has been shown that nonlinear feedback and nonlinear state transformation can be used to linearize nonlinear systems. Linearization techniques, if applicable, are known to be very powerful techniques for developing effective control laws for nonlinear systems. Several linearization problems can be defined. These are discussed in turn, starting with the linearization by state transformation. The next chapter introduces basic mathematics necessary to understand linearization problems and conditions. Chapter 2 would be useful not only for linearization theory, but also for other fields of nonlinear control system theory.

Chapter 2
Basic Mathematics for Linearization

2.1 Vector Calculus

We define the partial derivative of scalar function $h(x) = h(x_1, \ldots, x_n)$ with respect to vector variable $x = \begin{bmatrix} x_1 \\ \vdots \\ x_n \end{bmatrix}$ by

$$\frac{\partial h(x)}{\partial x} \triangleq \begin{bmatrix} \frac{\partial h(x)}{\partial x_1} & \cdots & \frac{\partial h(x)}{\partial x_n} \end{bmatrix}. \tag{2.1}$$

In other words, $\frac{\partial h(x)}{\partial x}$ is a $1 \times n$ matrix. Then it is easy to see that for scalar functions $h(x)$ and $\eta(x)$

$$\frac{\partial \{h(x)\eta(x)\}}{\partial x} = \eta(x)\frac{\partial h(x)}{\partial x} + h(x)\frac{\partial \eta(x)}{\partial x}. \tag{2.2}$$

Example 2.1.1 Let $x = \begin{bmatrix} x_1 & \cdots & x_n \end{bmatrix}^{\mathsf{T}}$. Suppose that C and A are $1 \times n$ constant matrix and $n \times n$ constant matrix, respectively. Show that

(a) $\frac{\partial}{\partial x}(Cx) = C$
(b) $\frac{\partial}{\partial x}(x^{\mathsf{T}}C^{\mathsf{T}}) = C$
(c) $\frac{\partial}{\partial x}\left(x^{\mathsf{T}}Ax\right) = x^{\mathsf{T}}(A^{\mathsf{T}} + A)$.

Solution Omitted. (Problem 2-1, 4.) □

With (2.1), the derivative with respect to a vector variable can be expressed as simple as the derivative with respect to a scalar variable, as shown in (a) of the above example. However, (b) and (c) require some care. We can also define the partial

© The Author(s), under exclusive license to Springer Nature Singapore Pte Ltd. 2022
H.-G. Lee, *Linearization of Nonlinear Control Systems*,
https://doi.org/10.1007/978-981-19-3643-2_2

derivative of $m \times 1$ column vector function $h(x) = \begin{bmatrix} h_1(x) \\ \vdots \\ h_m(x) \end{bmatrix}$ with respect to vector

variable $x = \begin{bmatrix} x_1 & \cdots & x_n \end{bmatrix}^\mathsf{T}$ by

$$
\frac{\partial h(x)}{\partial x} \triangleq \begin{bmatrix} \frac{\partial h_1(x)}{\partial x} \\ \vdots \\ \frac{\partial h_m(x)}{\partial x} \end{bmatrix} = \begin{bmatrix} \frac{\partial h_1(x)}{\partial x_1} & \cdots & \frac{\partial h_1(x)}{\partial x_n} \\ \vdots & \ddots & \vdots \\ \frac{\partial h_m(x)}{\partial x_1} & \cdots & \frac{\partial h_m(x)}{\partial x_n} \end{bmatrix}.
$$

Example 2.1.2 Let $x = \begin{bmatrix} x_1 & \cdots & x_n \end{bmatrix}^\mathsf{T}$. Suppose that $h(x)$ is a scalar matrix. Show that $\frac{\partial}{\partial x}\left(\frac{\partial h(x)}{\partial x}\right)^\mathsf{T}$ is a symmetric $n \times n$ matrix.

Solution Omitted. (Problem 2-2.) $\qquad\qquad\qquad\qquad\qquad\qquad\qquad\qquad$ □

Example 2.1.3 Let $x = \begin{bmatrix} x_1 & \cdots & x_n \end{bmatrix}^\mathsf{T}$. Suppose that $b(x)$ and $c(x)$ are $1 \times m$ matrix and $m \times 1$ matrix, respectively. Show that

$$
\frac{\partial}{\partial x}\{b(x)c(x)\} = c(x)^\mathsf{T}\frac{\partial b(x)^\mathsf{T}}{\partial x} + b(x)\frac{\partial c(x)}{\partial x}. \tag{2.3}
$$

Solution Omitted. (Problem 2-3.) $\qquad\qquad\qquad\qquad\qquad\qquad\qquad\qquad$ □

Example 2.1.4 Let $x = \begin{bmatrix} x_1 & \cdots & x_n \end{bmatrix}^\mathsf{T}$. Suppose that $A(x)$ and $b(x)$ are $q \times m$ matrix and $m \times 1$ matrix, respectively. Show that

$$
\frac{\partial}{\partial x}(A(x)b(x)) = \begin{bmatrix} b(x)^\mathsf{T}\frac{\partial A_1(x)^\mathsf{T}}{\partial x} \\ \vdots \\ b(x)^\mathsf{T}\frac{\partial A_q(x)^\mathsf{T}}{\partial x} \end{bmatrix} + A(x)\frac{\partial b(x)}{\partial x} \tag{2.4}
$$

where $A_i(x)$ is the ith row of $A(x)$.

Solution Omitted. (Problem 2-5.) $\qquad\qquad\qquad\qquad\qquad\qquad\qquad\qquad$ □

To define the partial differentiation of a matrix function with respect to a vector variable, it is difficult to effectively arrange it on two-dimensional paper. In this case, the Kronecker product $\otimes$ can be used. Let

$$
A \otimes B = \begin{bmatrix} a_{11}B & a_{12}B & \cdots & a_{1q}B \\ a_{21}B & a_{22}B & \cdots & a_{2q}B \\ \vdots & \vdots & \cdots & \vdots \\ a_{p1}B & a_{p2}B & \cdots & a_{pq}B \end{bmatrix}
$$

where a_{ij} is (i, j) element of $p \times q$ matrix A and B is a $m \times n$ matrix. Similarly, we can define $\frac{\partial B}{\partial A}$ by

$$
\frac{\partial B}{\partial A} =
\begin{bmatrix}
\frac{\partial}{\partial a_{11}} B & \frac{\partial}{\partial a_{12}} B & \cdots & \frac{\partial}{\partial a_{1q}} B \\
\frac{\partial}{\partial a_{21}} B & \frac{\partial}{\partial a_{22}} B & \cdots & \frac{\partial}{\partial a_{2q}} B \\
\vdots & \vdots & \cdots & \vdots \\
\frac{\partial}{\partial a_{p1}} B & \frac{\partial}{\partial a_{p2}} B & \cdots & \frac{\partial}{\partial a_{pq}} B
\end{bmatrix}_{(mp) \times (nq)}.
$$

Theorem 2.1 (Chain rule) *Suppose that* $f(x) : \mathbb{R}^n \to \mathbb{R}^m$ *and* $g(y) : \mathbb{R}^m \to \mathbb{R}^\ell$ *are smooth functions. Then the derivative of the composite function satisfies*

$$
\frac{\partial (g \circ f)(x)}{\partial x} = \frac{\partial g(f(x))}{\partial x} = \frac{\partial g(y)}{\partial y}\bigg|_{y=f(x)} \frac{\partial f(x)}{\partial x}.
$$

For example, if $S(x) : \mathbb{R}^n \to \mathbb{R}^n$ and $x(t) : \mathbb{R} \to \mathbb{R}^n$, then it is clear, by chain rule, that

$$
\frac{d S(x(t))}{dt} = \frac{\partial S(x)}{\partial x} \frac{dx(t)}{dt} = \sum_{i=1}^{n} \frac{\partial S(x)}{\partial x_i} \frac{dx_i(t)}{dt}.
$$

Example 2.1.5 (a) Find out $\frac{d}{dx}(e^{x^2})$.

(b) Find out $\frac{\partial g(f(x))}{\partial x}$, where $g(y) = \begin{bmatrix} e^{y_1} \cos y_3 \\ y_2 \sin y_1 \end{bmatrix}$ and $f(x) = \begin{bmatrix} x_1^2 \\ x_1 + x_2 \\ e^{x_3} \end{bmatrix}$.

Solution (a) $\frac{d}{dx} e^{x^2} = \frac{de^y}{dy}\big|_{y=x^2} \frac{d}{dx} x^2 = 2x e^{x^2}$.

(b) Since

$$
\frac{\partial g(y)}{\partial y} = \begin{bmatrix} e^{y_1} \cos y_3 & 0 & -e^{y_1} \sin y_3 \\ y_2 \cos y_1 & \sin y_1 & 0 \end{bmatrix} ; \quad \frac{\partial f(x)}{\partial x} = \begin{bmatrix} 2x_1 & 0 & 0 \\ 1 & 1 & 0 \\ 0 & 0 & e^{x_3} \end{bmatrix}
$$

we have, by chain rule, that

$$
\frac{\partial g(f(x))}{\partial x} = \frac{\partial g(y)}{\partial y}\bigg|_{y=f(x)} \frac{\partial f(x)}{\partial x}
$$

$$
= \begin{bmatrix} 2x_1 e^{x_1^2} \cos(e^{x_3}) & 0 & -e^{x_3 + x_1^2} \sin(e^{x_3}) \\ 2x_1(x_1 + x_2) \cos(x_1^2) + \sin(x_1^2) & \sin(x_1^2) & 0 \end{bmatrix}.
$$

$\square$

2.2 State Transformation

In short, a state coordinate change is differentiable bijective (1-1 and onto) function $z = S(x) : \mathbb{R}^n \to \mathbb{R}^n$. Thus, inverse function $x = S^{-1}(z)$ exists. We can assume without loss of generality that $S(0) = 0$, if necessary. The precise definition of state transformation is given in Definition 2.4.

Definition 2.1 (C^r and C^∞)

(a) A function $S(x) : U \subset \mathbb{R}^n \to \mathbb{R}$ defined on an open set U of $\mathbb{R}^n$ is said to be of class C^0 if it is continuous on U. A function $S(x) : U \subset \mathbb{R}^n \to \mathbb{R}^m$ is said to be of class C^0 if $S_i(x)$ is of class C^0 for $1 \leq i \leq m$.

(b) Let r be a positive integer. A function $S(x) : U \subset \mathbb{R}^n \to \mathbb{R}$ defined on an open set U of $\mathbb{R}^n$ is said to be of class C^r (or r-times continuously differentiable) on U if all partial derivatives

$$\frac{\partial^\lambda S(x)}{\partial x_1^{\lambda_1} \partial x_2^{\lambda_2} \cdots \partial x_n^{\lambda_n}}$$

exist and are continuous on U, for every $\lambda_1, \lambda_2, \ldots, \lambda_n$ nonnegative integers, such that $\lambda = \lambda_1 + \lambda_2 + \cdots + \lambda_n \leq r$. A function $S(x) : U \subset \mathbb{R}^n \to \mathbb{R}^m$ is said to be of class C^r if $S_i(x)$ is of class C^r for $1 \leq i \leq m$.

(c) A function $S(x) : U \subset \mathbb{R}^n \to \mathbb{R}^m$ defined on an open set U of $\mathbb{R}^n$ is said to be of class C^∞, or smooth, on U if $S(x)$ is of class C^r on U for all positive integer r.

Definition 2.2 (*homeomorphism*) A function $S(x) : U \subset \mathbb{R}^n \to \mathbb{R}^n$ defined on an open set U of $\mathbb{R}^n$ is said to be a homeomorphism if $S(x)$ is bijective (or 1-1 and onto) and functions $S(x)$ and $S^{-1}(z) : S(U) \to \mathbb{R}^n$ are continuous (or of class C^0).

Definition 2.3 (*diffeomorphism*) A function $S(x) : U \subset \mathbb{R}^n \to \mathbb{R}^n$ defined on an open set U of $\mathbb{R}^n$ is said to be a diffeomorphism if $S(x)$ is bijective (or 1-1 and onto) and functions $S(x)$ and $S^{-1}(z) : S(U) \to \mathbb{R}^n$ are smooth (or of class C^∞).

Definition 2.4 (*state transformation*) A function $S(x) : U \subset \mathbb{R}^n \to \mathbb{R}^n$ defined on an open set U of $\mathbb{R}^n$ is said to be a state transformation on $U(\subset \mathbb{R}^n)$ if $S(x)$ is a diffeomorphism.

For example $\begin{bmatrix} z_1 \\ z_2 \end{bmatrix} = S(x) = \begin{bmatrix} x_1 + x_2 \\ x_2 \end{bmatrix} = \begin{bmatrix} 1 & 1 \\ 0 & 1 \end{bmatrix} \begin{bmatrix} x_1 \\ x_2 \end{bmatrix}$ and $\begin{bmatrix} z_1 \\ z_2 \end{bmatrix} = S(x) = \begin{bmatrix} x_1 + x_2^2 \\ x_2 \end{bmatrix}$ are state transformations on $\mathbb{R}^2$.

Example 2.2.1 (*Similarity Transformation*) Consider the following linear system:

$$\dot{x} = Ax + Bu ; \quad y = Cx \tag{2.5}$$

Suppose that state transformation $z = S(x) = P^{-1}x$ is linear. Show that system (2.5) satisfies, in z-coordinates, a new linear system equation.

Solution It is easy to see that

$$\dot{z} = P^{-1}\dot{x} = P^{-1}(Ax + Bu) = P^{-1}APz + P^{-1}Bu$$
$$\triangleq \tilde{A}z + \tilde{B}u.$$

and

$$y = Cx = CPz = \tilde{C}z.$$

□

In the linear system theory, the linear state transformation (or similarity transformation) of Example 2.2.1 is used to transform system (2.5) into various canonical forms such as controllable canonical form (CCF), observable canonical form (OCF), Jordan canonical form (JCF), etc.

An open set $U(\subset \mathbb{R}^n)$ is said to be a neighborhood of a point $p(\in \mathbb{R}^n)$, if $p \in U$. It is easy to see that $z = S(x) = \begin{bmatrix} x_1 + x_2^2 & x_2 \end{bmatrix}^T$ is invertible and thus it is a state transformation. But, it is not easy to see whether $z = S(x) = \begin{bmatrix} x_1 + x_2^2 & x_2 + x_1^2 \end{bmatrix}^T$ is an invertible function (or local state transformation) or not. The following theorem gives the condition for a smooth function to be invertible on a neighborhood of a point.

Theorem 2.2 (inverse function theorem) *Suppose that $S(x) : \mathbb{R}^n \to \mathbb{R}^n$ is a smooth function. If $\frac{\partial S(x)}{\partial x}\Big|_{x=a}$ is a nonsingular matrix, then there exists a neighborhood U of a such that $S(x) : U \to S(U)$ is a diffeomorphism.*

Theorem 2.2 means that if $\det\left(\frac{\partial S(x)}{\partial x}\Big|_{x=a}\right) \neq 0$, smooth function $z = S(x)$ is a local state transformation (or diffeomorphism) on a neighborhood of $x = a$.

Example 2.2.2 Show that $z = S(x) = \begin{bmatrix} x_1 + \frac{1}{2}x_2^2 \\ x_2 + \frac{1}{2}x_1^2 \end{bmatrix}$ is a local state transformation on a neighborhood of the origin.

Solution Note that

$$\det\left(\frac{\partial S(x)}{\partial x}\Big|_{x=0}\right) = \det\left(\begin{bmatrix} 1 & 0 \\ 0 & 1 \end{bmatrix}\right) = 1$$

which implies, by inverse function theorem, that $z = S(x)$ is a local state transformation (or diffeomorphism) on a neighborhood of the origin. □

Example 2.2.3 Show that $\begin{bmatrix} z_1 \\ z_2 \end{bmatrix} = S(x) = \begin{bmatrix} e^{x_1}\cos x_2 \\ e^{x_1}\sin x_2 \end{bmatrix}$ is a local state transformation. Is it a global state transformation?

Solution Note that

$$\det\left(\frac{\partial S(x)}{\partial x}\right) = \det\left(\begin{bmatrix} e^{x_1}\cos x_2 & -e^{x_1}\sin x_2 \\ e^{x_1}\sin x_2 & e^{x_1}\cos x_2 \end{bmatrix}\right) = e^{2x_1} \neq 0$$

which implies, by inverse function theorem, that $z = S(x)$ is a local state transformation. But, since $z = S(x)$ is not injective in the entire region $(S([0\,0]^{\mathsf{T}}) = S([0\,2\pi]^{\mathsf{T}}))$, it is not a global state transformation. ▢

Theorem 2.3 (implicit function theorem) *Suppose that $f(x, y) : \mathbb{R}^{n+m} \to \mathbb{R}^m$ is a smooth function with $f(x_0, y_0) = 0$. If $\frac{\partial f(x,y)}{\partial y}\Big|_{(x,y)=(x_0,y_0)}$ is a $m \times m$ nonsingular matrix, then there exist a neighborhood $V(\subset \mathbb{R}^n)$ and a unique smooth function $g(x) : V \to \mathbb{R}^m$ such that $g(x_0) = y_0$ and*

$$f(x, g(x)) = 0, \text{ for all } x \in V.$$

Given implicit equation $f(x, y) = O_{m \times 1}$, $y \in \mathbb{R}^m$, implicit function theorem gives the condition for the existence of explicit function $y = g(x)$.

2.3   Nonsingular State Feedback

In this book, we consider the following nonlinear systems:

$$\dot{x} = F(x, u) ; \quad y = h(x) \tag{2.6}$$

and

$$\dot{x} = f(x) + g(x)u ; \quad y = h(x) \tag{2.7}$$

where $x \in \mathbb{R}^n$, $u \in \mathbb{R}^m$, and $y \in \mathbb{R}^q$. Also, we assume that $F(x, u)$, $f(x)$, $g(x)$, and $h(x)$ are smooth functions with $F(0, 0) = 0$, $f(0) = 0$, and $h(0) = 0$. In this book, we assume that $(0, 0)$ is the equilibrium point of the system. System (2.7) is said to be an affine system. $F(x, u)$, $f(x)$, and $g(x)$ are said to be vector fields. (See the next section.) The solution of differential equation (2.6) depends on the vector field $F(x, u)$ and the initial state $x(0)$. Vector field $F(x, u)$ can be changed by state feedback $u = \gamma(x, v)$, where $v(\in \mathbb{R}^m)$ is the new input. We assume that $\gamma(0, 0) = 0$, so that $(0, 0)$ is the equilibrium point of the closed-loop system. With state feedback $u = \gamma(x, v)$, we have the closed-loop system

$$\dot{x} = F(x, \gamma(x, v)) \triangleq \bar{F}(x, v).$$

State feedback

$$u = \alpha(x) + \beta(x)v; \quad \alpha(0) = 0$$

is said to be an affine feedback. For affine system (2.7), we have, with affine state feedback $u = \alpha(x) + \beta(x)v$, the affine closed-loop system

$$\dot{x} = f(x) + g(x)\alpha(x) + g(x)\beta(x)v \triangleq \bar{f}(x) + \bar{g}(x)v.$$

Consider

$$\begin{bmatrix} \dot{x}_1 \\ \dot{x}_2 \\ \dot{x}_3 \end{bmatrix} = \begin{bmatrix} x_2 \\ x_3 \\ 0 \end{bmatrix} + \begin{bmatrix} 1 + x_1^2 \\ 1 + x_2^2 \\ 1 + x_3^2 \end{bmatrix} u$$

If we let state feedback $u = \alpha(x) + \beta(x)v = 0$, then it is clear that the closed-loop system is linear. However, we cannot control the closed-loop system. Therefore, the nonsingular (or regular) state feedback is used in this book.

Definition 2.5 (*nonsingular state feedback*) A state feedback $u = \alpha(x) + \beta(x)v$ (or $u = \gamma(x, v)$) is said to be nonsingular if

$$\text{rank}\,(\beta(0)) = \text{rank}\left(\beta(0)^{-1}\right) = m$$

$$\left(\text{or rank}\left(\frac{\partial \gamma(x, v)}{\partial v}\bigg|_{(0,0)}\right) = \text{rank}\left(\frac{\partial \gamma(x, v)}{\partial v}\bigg|_{(0,0)}^{-1}\right) = m\right).$$

For system (2.7), if we consider the dynamic feedback

$$\begin{aligned} u &= c(x, z) + d(x, z)v \\ \dot{z} &= a(x, z) + b(x, z)v \end{aligned} \tag{2.8}$$

then we have the extended system

$$\dot{x}_E = \begin{bmatrix} \dot{x} \\ \dot{z} \end{bmatrix} = \begin{bmatrix} f(x) + g(x)c(x, z) \\ a(x, z) \end{bmatrix} + \begin{bmatrix} g(x)d(x, z) \\ b(x, z) \end{bmatrix} v \tag{2.9}$$

$$= f_E(x_E) + g_E(x_E)v.$$

where $z \in \mathbb{R}^d$ and $x_E = \begin{bmatrix} x \\ z \end{bmatrix}$.

Definition 2.6 (*regular dynamic feedback*) A dynamic feedback (2.8) is said to be regular if the extended system (2.9) with output $u = c(x, z) + d(x, z)v$ is dynamic input-output decouplable (Refer to Chap. 9 for dynamic i-o decoupling).

The regular dynamic state feedback is considered in Chap. 6 for dynamic feedback linearization.

2.4 Vector Field and Tangent Vector

In this section, vector field and tangent vector on subsets of Euclidean space will be studied. Vector field and tangent vector on manifolds can be found in Appendix. The right-hand side of the state equation in (2.7) is called a vector field on $\mathbb{R}^n$. Suppose that U be an open subset of $\mathbb{R}^n$. A function $f : U(\subset \mathbb{R}^n) \to \mathbb{R}$ is said to belong to $C^\infty(U)$, if f is C^∞ (or smooth). In other words, $C^\infty(U)$ is the set of all smooth scalar functions on U.

Definition 2.7 (*smooth vector field on Euclidean space*) A vector-valued function $f : U(\subset \mathbb{R}^n) \to \mathbb{R}^n$ is said to be a smooth vector field on U, if $f(x) = \begin{bmatrix} f_1(x) \\ \vdots \\ f_n(x) \end{bmatrix}$ and $f_i \in C^\infty(U)$ for $1 \le i \le n$.

Suppose that $x \triangleq [x_1 \ x_2 \ \cdots \ x_n]^\mathsf{T}$ is a Cartesian coordinate system of $\mathbb{R}^n$. Then a vector field $f(x)$ can be expressed by

$$f(x) = \begin{bmatrix} f_1(x) \\ f_2(x) \\ \vdots \\ f_n(x) \end{bmatrix} = f_1(x)\frac{\partial}{\partial x_1} + f_2(x)\frac{\partial}{\partial x_2} + \cdots + f_n(x)\frac{\partial}{\partial x_n}$$

where

$$\frac{\partial}{\partial x_1} \triangleq \begin{bmatrix} 1 \\ 0 \\ \vdots \\ 0 \end{bmatrix}, \quad \frac{\partial}{\partial x_2} \triangleq \begin{bmatrix} 0 \\ 1 \\ \vdots \\ 0 \end{bmatrix}, \quad \cdots, \text{ and } \frac{\partial}{\partial x_n} \triangleq \begin{bmatrix} 0 \\ \vdots \\ 0 \\ 1 \end{bmatrix}.$$

For system (2.7), $f(x)$, $g(x)$, and $f(x) + g(x)u$ are smooth vector fields on $\mathbb{R}^n$, if $f(x)$ and $g(x)$ are smooth functions on $\mathbb{R}^n$. Addition of vector fields and scalar multiplication are defined by

$$\begin{bmatrix} f_1(x) \\ f_2(x) \\ \vdots \\ f_n(x) \end{bmatrix} + \begin{bmatrix} g_1(x) \\ g_2(x) \\ \vdots \\ g_n(x) \end{bmatrix} \triangleq \begin{bmatrix} f_1(x) + g_1(x) \\ f_2(x) + g_2(x) \\ \vdots \\ f_n(x) + g_n(x) \end{bmatrix} \tag{2.10}$$

and

$$r(x) \begin{bmatrix} f_1(x) \\ f_2(x) \\ \vdots \\ f_n(x) \end{bmatrix} \triangleq \begin{bmatrix} r(x)f_1(x) \\ r(x)f_2(x) \\ \vdots \\ r(x)f_n(x) \end{bmatrix}, \quad \forall r(x) \in C^\infty(\mathbb{R}^n). \tag{2.11}$$

Example 2.4.1 Show that the set of all smooth vector fields on $\mathbb{R}^n$ is a vector space over field $\mathbb{R}$.

Solution Omitted. (Problem 2-9.) □

Example 2.4.2 Consider the following control system:

$$\begin{bmatrix} \dot{x}_1 \\ \dot{x}_2 \end{bmatrix} = \begin{bmatrix} -x_2 \\ x_1 \end{bmatrix} = f(x)$$

Let $x(0) = \begin{bmatrix} 1 \\ 0 \end{bmatrix}$. Then it is easy to see that $x(t) = \begin{bmatrix} \cos t \\ \sin t \end{bmatrix}$ and $\frac{dx(t)}{dt} = \begin{bmatrix} -\sin t \\ \cos t \end{bmatrix}$.

Note that $x(\frac{\pi}{4}) = \begin{bmatrix} \frac{1}{\sqrt{2}} \\ \frac{1}{\sqrt{2}} \end{bmatrix}$ and

$$\left. \frac{dx(t)}{dt} \right|_{t=0} = \begin{bmatrix} 0 \\ 1 \end{bmatrix} = f\left(\begin{bmatrix} 1 \\ 0 \end{bmatrix}\right) ; \quad \left. \frac{dx(t)}{dt} \right|_{t=\frac{\pi}{4}} = \begin{bmatrix} -\frac{1}{\sqrt{2}} \\ \frac{1}{\sqrt{2}} \end{bmatrix} = f\left(\begin{bmatrix} \frac{1}{\sqrt{2}} \\ \frac{1}{\sqrt{2}} \end{bmatrix}\right).$$

Thus, $f(\bar{x})$ is a tangent vector of solution curve $x(t)$ at point $x = \bar{x}(\in \mathbb{R}^2)$ (See Fig. 2.1).

Given smooth vector field $f(x)$ on $\mathbb{R}^n$

$$f(\bar{x}) = [f_1(\bar{x}) \ f_2(\bar{x}) \ \cdots \ f_n(\bar{x})]^\mathsf{T} = \sum_{i=1}^n f_i(\bar{x}) \left. \frac{\partial}{\partial x_i} \right|_{\bar{x}}$$

is said to be a tangent vector at point $\bar{x}(\in \mathbb{R}^n)$ that is a vector starting at point p. The vector at $\mathbb{R}^n$ has a magnitude and direction. The tangent vector at point $\bar{x}(\in \mathbb{R}^n)$ is a vector starting at point $\bar{x}$. If $\bar{x} \neq \hat{x}$, then $f(\bar{x}) + f(\hat{x})$ can not be defined or required.

Fig. 2.1 Tangent vectors of
Example 2.4.2

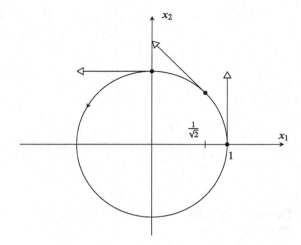

Suppose that $f(x)$ and $g(x)$ are smooth vector fields on $\mathbb{R}^n$. Also, let $h(x) \in C^\infty(\mathbb{R}^n)$. The following two operations (Lie bracket and Lie derivative) will be used very often in this book.

Definition 2.8 (*Lie bracket*) The Lie bracket of vector field $f(x)$ and vector field $g(x)$ is defined by

$$[f(x), g(x)] \triangleq \frac{\partial g(x)}{\partial x} f(x) - \frac{\partial f(x)}{\partial x} g(x). \qquad (2.12)$$

Definition 2.9 (*Lie derivative*) The Lie derivative of scalar function $h(x)$ with respect to vector field $f(x)$ is defined by

$$L_{f(x)}h(x) \triangleq \frac{\partial h(x)}{\partial x} f(x) = \sum_{i=1}^{n} f_i(x) \frac{\partial h(x)}{\partial x_i}. \qquad (2.13)$$

In engineering mathematics, it is learned that the directional derivative of $h(x)$ at $x = p$ in the direction of $f(p)$ is given by $\left\{ \frac{1}{\|f(x)\|} L_f h(x) \right\}\Big|_{x=p}$. It is clear, by Definition 2.8, that

$$[f(x), f(x)] = 0 \text{ and } [f(x), 0] = 0.$$

For simplicity, we use 0 instead of $O_{n \times 1}$. Also, if b_1 and b_2 are constant vector fields on $\mathbb{R}^n$, it is clear that

$$[b_1, b_2] = 0. \qquad (2.14)$$

For $\forall h(x), \beta(x) \in C^\infty(\mathbb{R}^n)$, it is easy to see that

$$L_{f(x)+g(x)}h(x) = L_{f(x)}h(x) + L_{g(x)}h(x) \tag{2.15}$$

$$L_{\beta(x)g(x)}h(x) = \beta(x)L_{g(x)}h(x). \tag{2.16}$$

$\frac{\partial f(x)}{\partial x}$ can be calculated via **jacobian**(f,x) or **diff**($\cdot,\cdot$) of Matlab program. Thus, Lie bracket $[f(x), g(x)]$ and Lie derivative $L_f h(x)$ can also be easily calculated by Matlab program (See **adfg**(f,g,x) and **Lfh**(f,h,x) in Appendix C).

Example 2.4.3 Let $h(x) = x_1 x_2$, $f(x) = \begin{bmatrix} x_2 \\ 1 \end{bmatrix}$, and $g(x) = \begin{bmatrix} 1 \\ x_1 \end{bmatrix}$. Find out $[f(x), g(x)]$ and $L_f h(x)$.

Solution

$$[f(x), g(x)] = \frac{\partial g(x)}{\partial x} f(x) - \frac{\partial f(x)}{\partial x} g(x)$$
$$= \begin{bmatrix} 0 & 0 \\ 1 & 0 \end{bmatrix} \begin{bmatrix} x_2 \\ 1 \end{bmatrix} - \begin{bmatrix} 0 & 1 \\ 0 & 0 \end{bmatrix} \begin{bmatrix} 1 \\ x_1 \end{bmatrix} = \begin{bmatrix} -x_1 \\ x_2 \end{bmatrix}$$

and

$$L_f h(x) = \frac{\partial h(x)}{\partial x} f(x) = \begin{bmatrix} x_2 & x_1 \end{bmatrix} \begin{bmatrix} x_2 \\ 1 \end{bmatrix} = x_2^2 + x_1.$$

$\square$

Example 2.4.4 Let $h_1(x), h_2(x) \in C^\infty(\mathbb{R}^n)$. Show the following:

(a) linearity

$$L_f \{r_1 h_1(x) + r_2 h_2(x)\} = r_1 L_f h_1(x) + r_2 L_f h_1(x), \quad \forall r_1, r_2 \in \mathbb{R}$$

(b) Leibniz rule

$$L_f \{h_1(x) h_2(x)\} = h_2(x) L_f h_1(x) + h_1(x) L_f h_2(x).$$

Solution It is clear, by (2.2), that

$$L_f \{h_1(x) h_2(x)\} = \frac{\partial \{h_1(x) h_2(x)\}}{\partial x} f = \left(h_2(x) \frac{\partial h_1(x)}{\partial x} + h_1(x) \frac{\partial h_2(x)}{\partial x} \right) f(x)$$
$$= h_2(x) L_f h_1(x) + h_1(x) L_f h_2(x).$$

$\square$

Example 2.4.5 Use Example 2.1.3 to show that for all $h(x) \in C^\infty(\mathbb{R}^n)$

$$L_{[f,g]}h(x) = L_f L_g h(x) - L_g L_f h(x). \tag{2.17}$$

Solution

$$
\begin{aligned}
L_f L_g h(x) - L_g L_f h(x) &= L_f\left(\frac{\partial h}{\partial x}g\right) - L_g\left(\frac{\partial h}{\partial x}f\right) \\
&= \frac{\partial}{\partial x}\left(\frac{\partial h}{\partial x}g\right)f - \frac{\partial}{\partial x}\left(\frac{\partial h}{\partial x}f\right)g \\
&= \left(g^{\mathsf{T}}\frac{\partial}{\partial x}\left(\frac{\partial h}{\partial x}\right)^{\mathsf{T}} + \frac{\partial h}{\partial x}\frac{\partial g}{\partial x}\right)f - \left(f^{\mathsf{T}}\frac{\partial}{\partial x}\left(\frac{\partial h}{\partial x}\right)^{\mathsf{T}} + \frac{\partial h}{\partial x}\frac{\partial f}{\partial x}\right)g \\
&= (g^{\mathsf{T}}h_{xx}f - f^{\mathsf{T}}h_{xx}g) + \frac{\partial h}{\partial x}\left(\frac{\partial g}{\partial x}f - \frac{\partial f}{\partial x}g\right) \\
&= \frac{\partial h(x)}{\partial x}[f(x), g(x)] = L_{[f,g]}h(x)
\end{aligned}
$$

where $h_{xx} \triangleq \frac{\partial}{\partial x}\left(\frac{\partial h}{\partial x}\right)^{\mathsf{T}} = h_{xx}^{\mathsf{T}}$.   □

The relation in (2.17) is used very often in this book. In fact, it is used as the definition of Lie bracket $[f(x), g(x)]$ for the vector fields on manifolds. (See (B.1) in Appendix.)

Example 2.4.6 Suppose that $f(x)$, $g(x)$, and $\tau(x)$ are smooth vector fields on $\mathbb{R}^n$. Show the following:

(a) bilinear

$$[r_1 f(x) + r_2 g(x), \tau(x)] = r_1[f(x), \tau(x)] + r_2[g(x), \tau(x)], \quad \forall r_1, r_2 \in \mathbb{R}$$

$$[\tau(x), r_1 f(x) + r_2 g(x)] = r_1[\tau(x), f(x)] + r_2[\tau(x), g(x)], \quad \forall r_1, r_2 \in \mathbb{R}$$

(b) anticommutativity or skew-commutative

$$[f(x), g(x)] = -[g(x), f(x)]$$

(c) Jacobi identity

$$[f(x), [g(x), \tau(x)]] + [g(x), [\tau(x), f(x)]] + [\tau(x), [f(x), g(x)]] = 0 \tag{2.18}$$

Solution It is obvious that (a) and (b) are satisfied. Note that if $L_f h(x) = 0$ for $\forall h(x) \in C^\infty(\mathbb{R}^n)$, then $f_i(x) = L_f x_i = 0$ for $1 \le i \le n$ or $f(x) = 0$. It is easy, by (2.15) and (2.17), to show that for $\forall h(x) \in C^\infty(\mathbb{R}^n)$

$$
\begin{aligned}
L_{[f,[g,\tau]]+[g,[\tau,f]]+[\tau,[f,g]]}h &= L_{[f,[g,\tau]]}h + L_{[g,[\tau,f]]}h + L_{[\tau,[f,g]]}h \\
&= L_f L_{[g,\tau]}h - L_{[g,\tau]}L_f h + L_g L_{[\tau,f]}h - L_{[\tau,f]}L_g h + L_\tau L_{[f,g]}h - L_{[f,g]}L_\tau h \\
&= L_f L_g L_\tau h - L_f L_\tau L_g h - L_g L_\tau L_f h + L_\tau L_g L_f h + L_g L_\tau L_f h - L_g L_f L_\tau h \\
&\quad - L_\tau L_f L_g h + L_f L_\tau L_g h + L_\tau L_f L_g h - L_\tau L_g L_f h - L_f L_g L_\tau h + L_g L_f L_\tau h \\
&= 0.
\end{aligned}
$$

Therefore, (2.18) is satisfied. Of course, (2.18) can also be shown directly by using (2.4) and (2.12) (See Problem 2-11). For example

$$
\begin{aligned}
[\tau, [f, g]] &= \left[\tau, \left(\frac{\partial g}{\partial x}f - \frac{\partial f}{\partial x}g\right)\right] = \frac{\partial}{\partial x}\left(\frac{\partial g}{\partial x}f - \frac{\partial f}{\partial x}g\right)\tau - \frac{\partial \tau}{\partial x}\left(\frac{\partial g}{\partial x}f - \frac{\partial f}{\partial x}g\right) \\
&= \begin{bmatrix} f^{\mathsf{T}}(g_1)_{xx}\tau \\ \vdots \\ f^{\mathsf{T}}(g_n)_{xx}\tau \end{bmatrix} + \frac{\partial g}{\partial x}\frac{\partial f}{\partial x}\tau - \begin{bmatrix} g^{\mathsf{T}}(f_1)_{xx}\tau \\ \vdots \\ g^{\mathsf{T}}(f_n)_{xx}\tau \end{bmatrix} - \frac{\partial f}{\partial x}\frac{\partial g}{\partial x}\tau - \frac{\partial \tau}{\partial x}\left(\frac{\partial g}{\partial x}f - \frac{\partial f}{\partial x}g\right).
\end{aligned}
$$

$\square$

It is easy, by Examples 2.4.1 and 2.4.6, to show that the set of all smooth vector fields on $\mathbb{R}^n$ is a Lie algebra over field $\mathbb{R}$. The set of all $n \times n$ real matrices is a (linear) algebra over field $\mathbb{R}$. Lie algebra can be used in the nonlinear system theory, whereas linear algebra can be used in the linear system theory.

Definition 2.10 (*differential map of tangent vector*) Suppose that $z = S(x) : \mathbb{R}^n \to \mathbb{R}^m$ is a smooth function. Differential map S_* of $z = S(x)$ is a linear map from the set of tangent vectors at x^0 in $\mathbb{R}^n$ to the set of tangent vectors at $z^0(= S(x^0))$ in $\mathbb{R}^m$. The differential map $S_*(f(x^0))$ of tangent vector $f(x^0)$ is defined by

$$
S_*(f(x^0)) = \left.\frac{\partial S}{\partial x}\right|_{x^0} f(x^0).
$$

Example 2.4.7 Let $z = S(x) = \begin{bmatrix} x_1 \\ (1+x_1)x_2 + 2x_3 \end{bmatrix}$, $f(x) = \begin{bmatrix} 1 \\ 0 \\ x_2 \end{bmatrix}$, and $g(x) = \begin{bmatrix} 0 \\ 1 \\ x_1 \end{bmatrix}$. Find the tangent vectors $S_*(f(x^0))$, $S_*(f(x^1))$, $S_*(g(x^0))$, and $S_*(g(x^1))$, where $x^0 = \begin{bmatrix} 1 \\ 1 \\ 0 \end{bmatrix}$ and $x^1 = \begin{bmatrix} 1 \\ 0 \\ 1 \end{bmatrix}$.

Solution Note that $z^0 = S(x^0) = S(x^1) = \begin{bmatrix} 1 \\ 2 \end{bmatrix}$. Thus, $S_*(f(x^0))$, $S_*(f(x^1))$,

$S_*(g(x^0))$, and $S_*(g(x^1))$ are tangent vectors at $z = \begin{bmatrix} 1 \\ 2 \end{bmatrix} \in \mathbb{R}^2$.

$$S_*(f(x^0)) = \left.\frac{\partial S}{\partial x}\right|_{x=x^0} f(x^0) = \begin{bmatrix} 1 & 0 & 0 \\ 1 & 2 & 2 \end{bmatrix} \begin{bmatrix} 1 \\ 0 \\ 1 \end{bmatrix} = \begin{bmatrix} 1 \\ 3 \end{bmatrix}$$

$$= \left.\frac{\partial}{\partial z_1}\right|_{\left[\begin{smallmatrix}1\\2\end{smallmatrix}\right]} + 3 \left.\frac{\partial}{\partial z_2}\right|_{\left[\begin{smallmatrix}1\\2\end{smallmatrix}\right]}$$

Similarly, it is easy to see that

$$S_*(f(x^1)) = \begin{bmatrix} 1 \\ 0 \end{bmatrix}, \ S_*(g(x^0)) = \begin{bmatrix} 0 \\ 4 \end{bmatrix}, \text{ and } S_*(g(x^1)) = \begin{bmatrix} 0 \\ 4 \end{bmatrix}.$$

$\square$

Definition 2.11 (*well-defined vector field of differential map*) Suppose that smooth function $z = S(x) : \mathbb{R}^n \to \mathbb{R}^m$ is surjective (or onto). Let $f(x)$ is a vector field on $\mathbb{R}^n$. $S_*(f(x))$ is said to be a well-defined vector field on $\mathbb{R}^m$, if $S_*(f(x)) = S_*(f(\bar{x}))$ whenever $S(x) = S(\bar{x})$.

It is easy to see that if $z = S(x) : \mathbb{R}^n \to \mathbb{R}^n$ is a state transformation (or diffeomorphism), then $\tilde{f}(z) \triangleq S_*(f(x))$ is a well-defined vector field on $\mathbb{R}^n$. By Definition 2.11, $S_*(f(x))$ in Example 2.4.7 is not a well-defined vector field on $\mathbb{R}^2$, whereas $S_*(g(x))$ in Example 2.4.7 might be a well-defined vector field on $\mathbb{R}^2$. (See Fig. 2.2.) If $S(x)$ is surjective and $m < n$, there exists a constant $(n - m) \times n$ matrix A such that $\begin{bmatrix} \left.\frac{\partial S}{\partial x}\right|_{x=0} \\ A \end{bmatrix}$ is an invertible matrix. Thus, if we let

$$\bar{z} \triangleq \begin{bmatrix} z \\ \tilde{z} \end{bmatrix} = \begin{bmatrix} S(x) \\ \tilde{S}(x) \end{bmatrix} \triangleq \begin{bmatrix} S(x) \\ Ax \end{bmatrix} \triangleq \bar{S}(x) \tag{2.19}$$

then it is clear, by Theorem 2.2, that $\bar{z} = \bar{S}(x)$ has an inverse function $x = \bar{S}^{-1}(\bar{z})$ locally. In Example 2.4.7, if we let $\bar{z} \triangleq \begin{bmatrix} z \\ \tilde{z}_1 \end{bmatrix} = \begin{bmatrix} S(x) \\ x_3 \end{bmatrix} = \bar{S}(x)$, then we have that

$$x = \bar{S}^{-1}(\bar{z}) = \begin{bmatrix} z_1 & \frac{z_2 - 2\tilde{z}_1}{1 + z_1} & \tilde{z}_1 \end{bmatrix}^{\mathsf{T}}$$

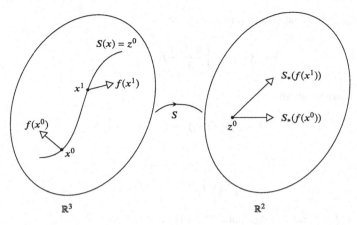

Fig. 2.2 $S_*(f(x))$ of Example 2.4.7

$$\frac{\partial S}{\partial x} f(x)\Big|_{x=\bar{S}^{-1}(\bar{z})} = \begin{bmatrix} 1 & 0 & 0 \\ x_2 & 1+x_1 & 2 \end{bmatrix} \begin{bmatrix} 1 \\ 0 \\ x_2 \end{bmatrix}\Big|_{x=\bar{S}^{-1}(\bar{z})} = \begin{bmatrix} 1 \\ 3x_2 \end{bmatrix}\Big|_{x=\bar{S}^{-1}(\bar{z})}$$

$$= \begin{bmatrix} 1 \\ \frac{3z_2-6z_1}{1+z_1} \end{bmatrix}$$

and

$$\frac{\partial S}{\partial x} g(x)\Big|_{x=\bar{S}^{-1}(\bar{z})} = \begin{bmatrix} 0 \\ 1+3x_1 \end{bmatrix}\Big|_{x=\bar{S}^{-1}(\bar{z})} = \begin{bmatrix} 0 \\ 1+3z_1 \end{bmatrix}. \tag{2.20}$$

Since $\frac{\partial S}{\partial x} g(x)\big|_{x=\bar{S}^{-1}(\bar{z})}$ depends on z only, $S_*(g(x))$ is a well-defined vector field on $\mathbb{R}^m$. But, since $\frac{\partial S}{\partial x} f(x)\big|_{x=\bar{S}^{-1}(\bar{z})}$ does not depend on z only, $S_*(f(x))$ is not a well-defined vector field on $\mathbb{R}^m$. Geometric condition for well-defined vector field can be also found in Theorem 2.6.

Definition 2.12 (*differential map of vector field*) Suppose that $f(x)$ is a smooth vector field on $\mathbb{R}^n$ and smooth function $z = S(x) : \mathbb{R}^n \to \mathbb{R}^m$ is surjective. Also, suppose that $S_*(f(x))$ is a well-defined vector field in $\mathbb{R}^m$. The differential map $S_*(f(x))$ of vector field $f(x)$ under smooth function $z = S(x)$ is defined by

$$S_*(f(x)) = \left\{ \frac{\partial S(x)}{\partial x} f(x) \right\}\Big|_{x=\bar{S}^{-1}(\bar{z})}$$

where $\bar{z} = \bar{S}(x)$ is defined in (2.19).

If $z = S(x)$ is a state transformation, then

$$S_*(f(x)) = \left\{ \frac{\partial S(x)}{\partial x} f(x) \right\} \bigg|_{x = S^{-1}(z)}. \tag{2.21}$$

Also, it is easy to see that

$$(T \circ S)_*(f(x)) = T_* \circ S_*(f(x)) \tag{2.22}$$

and

$$S_*^{-1}(S_*(f(x))) = f(x) \tag{2.23}$$

where $w = T(z)$ and $z = S(x)$ are state transformations.

Example 2.4.8 Consider the following control system:

$$\dot{x} = f(x) + \sum_{i=1}^{m} u_i g_i(x) = f(x) + g(x)u \tag{2.24}$$

$$y = h(x)$$

where $x \in \mathbb{R}^n$, $u \in \mathbb{R}^m$, and $y \in \mathbb{R}^q$. Suppose that $z = S(x) : \mathbb{R}^n \to \mathbb{R}^n$ is a state transformation and system (2.24) satisfies, in z-coordinates,

$$\dot{z} = \tilde{f}(z) + \sum_{i=1}^{m} u_i \tilde{g}_i(z) = \tilde{f}(z) + \tilde{g}(z)u$$

$$y = \tilde{h}(z).$$

Show that for $1 \leq i \leq m$

$$\tilde{f}(z) = S_*(f(x)) ; \quad \tilde{g}_i(z) = S_*(g_i(x))$$

$$\tilde{h}(z) = h \circ S^{-1}(z).$$

In other words, $f(x)$ and $\tilde{f}(z)(= S_*(f(x)))$ are the same vector fields expressed in x-coordinates and z-coordinates, respectively.

Solution It is easy to see that

$$\dot{z} = \frac{d}{dt}S(x(t)) = \frac{\partial S(x)}{\partial x}\dot{x} = \frac{\partial S(x)}{\partial x}\left\{f(x) + \sum_{i=1}^{m} u_i g_i(x)\right\}$$

$$= \frac{\partial S(x)}{\partial x}f(x)\bigg|_{x=S^{-1}(z)} + \sum_{i=1}^{m} u_i \frac{\partial S(x)}{\partial x}g_i(x)\bigg|_{x=S^{-1}(z)}$$

$$= S_*(f(x)) + \sum_{i=1}^{m} u_i S_*(g_i(x))$$

and

$$y = h(x) = h \circ S^{-1}(z).$$

$\square$

In Example 2.2.1, since $z = S(x) = P^{-1}x$, then it is easy to see that $\tilde{f}(z) = S_*(Ax) = P^{-1}APz$, $\tilde{g}(z) = S_*(B) = P^{-1}B$, and $\tilde{h}(z) = Cx|_{x=S^{-1}(z)} = CPz$.

Example 2.4.9 Suppose that

$$\begin{bmatrix} z_1 \\ z_2 \end{bmatrix} = S(x) = \begin{bmatrix} x_1 \\ x_1 + x_2 \end{bmatrix}.$$

Find out $S_*\left(\begin{bmatrix} 1 \\ 0 \end{bmatrix}\right)$ and $S_*\left(\begin{bmatrix} 0 \\ 1 \end{bmatrix}\right)$.

Solution It is clear that

$$S_*\left(\begin{bmatrix} 1 \\ 0 \end{bmatrix}\right) = \begin{bmatrix} 1 & 0 \\ 1 & 1 \end{bmatrix}\begin{bmatrix} 1 \\ 0 \end{bmatrix}\bigg|_{x=S^{-1}(z)} = \begin{bmatrix} 1 \\ 1 \end{bmatrix}; \quad S_*\left(\begin{bmatrix} 0 \\ 1 \end{bmatrix}\right) = \begin{bmatrix} 0 \\ 1 \end{bmatrix}$$

or

$$\begin{bmatrix} 1 \\ 0 \end{bmatrix}\bigg|_x = \frac{\partial}{\partial x_1} = \frac{\partial}{\partial z_1} + \frac{\partial}{\partial z_2} = \begin{bmatrix} 1 \\ 1 \end{bmatrix}\bigg|_z \tag{2.25}$$

and

$$\begin{bmatrix} 0 \\ 1 \end{bmatrix}\bigg|_x = \frac{\partial}{\partial x_2} = \frac{\partial}{\partial z_2} = \begin{bmatrix} 0 \\ 1 \end{bmatrix}\bigg|_z. \tag{2.26}$$

See Fig. 2.3.

$\square$

Suppose that $z = S(x)$ is a state transformation. Then, by chain rule (Theorem 2.1), we have with a slight abuse of notation that

Fig. 2.3 Unit vectors of
Example 2.4.9

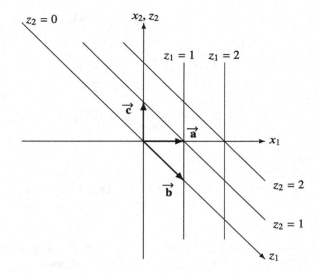

$$\frac{\partial}{\partial x_i} h(z) = \frac{\partial}{\partial x_i} h(z(x)) = \sum_{j=1}^{n} \frac{\partial h}{\partial z_j} \frac{\partial z_j}{\partial x_i} = \sum_{j=1}^{n} \frac{\partial z_j}{\partial x_i} \frac{\partial}{\partial z_j} h(z)$$

or

$$\frac{\partial}{\partial x_i} = \sum_{j=1}^{n} \frac{\partial z_j}{\partial x_i} \frac{\partial}{\partial z_j}. \tag{2.27}$$

We can use (2.27) to show that (2.25) and (2.26) are satisfied. Vector fields in (2.21)
can be written in the operator form as follows:

$$\sum_{i=1}^{n} f_i(x) \frac{\partial}{\partial x_i} = \sum_{i=1}^{n} \sum_{j=1}^{n} f_i(x) \frac{\partial z_j}{\partial x_i} \frac{\partial}{\partial z_j}.$$

Example 2.4.10 Use Example 2.4.8 to solve Example 1.3.1.

Solution Note that $\tilde{f}(z) = \begin{bmatrix} 0 \\ z_1 \end{bmatrix}$ and $\tilde{g}(z) = \begin{bmatrix} 1 \\ 1 \end{bmatrix}$.

$$f(x) = S_*(\tilde{f}(z)) = \frac{\partial S(z)}{\partial z} \tilde{f}(z) \Big|_{z=S^{-1}(x)} = \begin{bmatrix} 1 & 2z_2 \\ 0 & 1 \end{bmatrix} \begin{bmatrix} 0 \\ z_1 \end{bmatrix} \Big|_{z=S^{-1}(x)}$$

$$= \begin{bmatrix} 2x_1 x_2 - 2x_2^3 \\ x_1 - x_2^2 \end{bmatrix}$$

and

$$g(x) = S_*(\tilde{g}(z)) = \left.\frac{\partial S(z)}{\partial z}\tilde{g}(z)\right|_{z=S^{-1}(x)} = \left.\begin{bmatrix} 1 & 2z_2 \\ 0 & 1 \end{bmatrix}\begin{bmatrix} 1 \\ 1 \end{bmatrix}\right|_{z=S^{-1}(x)}$$

$$= \begin{bmatrix} 1+2x_2 \\ 1 \end{bmatrix}.$$

□

The following two theorems show that Lie bracket and Lie derivative defined in Definitions 2.8 and 2.9 are coordinate free. They are very important and are used very often in the rest of this book.

Theorem 2.4 *Suppose that $f(x)$ and $g(x)$ are smooth vector fields on $\mathbb{R}^n$ and smooth function $z = S(x) : \mathbb{R}^n \to \mathbb{R}^m$ is surjective (or onto). Also, suppose that $(S_*(f(x))$ and $S_*(g(x))$ are well-defined vector fields on $\mathbb{R}^m$. Then the following is satisfied:*

$$S_*([f(x), g(x)]) = [(S_*(f(x)), S_*(g(x))]. \tag{2.28}$$

Proof If $\frac{\partial}{\partial z}\left\{\left.\Phi(x)\right|_{x=\bar{S}^{-1}(\bar{z})}\right\} = O_{n\times(n-m)}$, then we have that

$$\frac{\partial}{\partial z}\left\{\left.\Phi(x)\right|_{x=\bar{S}^{-1}(\bar{z})}\right\} \left.S_x(x)\right|_{x=\bar{S}^{-1}(\bar{z})}$$

$$= \left[\frac{\partial}{\partial z}\left\{\left.\Phi(x)\right|_{x=\bar{S}^{-1}(\bar{z})}\right\} \quad O_{n\times(n-m)}\right] \left.\begin{bmatrix} S_x(x) \\ \bar{S}_x(x) \end{bmatrix}\right|_{x=\bar{S}^{-1}(\bar{z})}$$

$$= \frac{\partial}{\partial \bar{z}}\left\{\left.\Phi(x)\right|_{x=\bar{S}^{-1}(\bar{z})}\right\} \left.\bar{S}_x(x)\right|_{x=\bar{S}^{-1}(\bar{z})} \tag{2.29}$$

$$= \left.\frac{\partial \Phi(x)}{\partial x}\right|_{x=\bar{S}^{-1}(\bar{z})} \frac{\partial \bar{S}^{-1}(\bar{z})}{\partial \bar{z}} \left.\frac{\partial \bar{S}(x)}{\partial x}\right|_{x=\bar{S}^{-1}(\bar{z})}$$

$$= \left.\frac{\partial \Phi(x)}{\partial x}\right|_{x=\bar{S}^{-1}(\bar{z})} \left.\frac{\partial\left\{\bar{S}^{-1}\circ\bar{S}(x)\right\}}{\partial x}\right|_{x=\bar{S}^{-1}(\bar{z})} = \left.\frac{\partial \Phi(x)}{\partial x}\right|_{x=\bar{S}^{-1}(\bar{z})}.$$

Thus, if we denote $\frac{\partial S(x)}{\partial x} = S_x(x)$, then we have, by (2.29), that

$$\left[(S_*f)(z), (S_*g)(z)\right] = \left[\left.(S_x(x)f(x))\right|_{x=\bar{S}^{-1}(\bar{z})}, \left.(S_x(x)g(x))\right|_{x=\bar{S}^{-1}(\bar{z})}\right]$$

$$= \frac{\partial}{\partial z}\left\{\left.(S_x(x)g(x))\right|_{x=\bar{S}^{-1}(\bar{z})}\right\} \left.S_x(x)\right|_{x=\bar{S}^{-1}(\bar{z})} \left.f(x)\right|_{x=\bar{S}^{-1}(\bar{z})}$$

$$- \frac{\partial}{\partial z}\left\{\left.(S_x(x)f(x))\right|_{x=\bar{S}^{-1}(\bar{z})}\right\} \left.S_x(x)\right|_{x=\bar{S}^{-1}(\bar{z})} \left.g(x)\right|_{x=\bar{S}^{-1}(\bar{z})}$$

$$= \left\{\frac{\partial (S_x(x)g(x))}{\partial x}f(x) - \frac{\partial (S_x(x)f(x))}{\partial x}g(x)\right\}\Bigg|_{x=\bar{S}^{-1}(\bar{z})}.$$

Thus, it is easy to see, by (2.4), that

$$
\begin{aligned}
&\left[(S_* f)(z),\, (S_* g)(z)\right] \\
&= \left\{ \begin{bmatrix} g(x)^{\mathsf T}\frac{\partial}{\partial x}\left(\frac{\partial S_1}{\partial x}\right)^{\mathsf T} \\ \vdots \\ g(x)^{\mathsf T}\frac{\partial}{\partial x}\left(\frac{\partial S_n}{\partial x}\right)^{\mathsf T} \end{bmatrix} f(x) + \frac{\partial S(x)}{\partial x}\frac{\partial g(x)}{\partial x} f(x) \right\} \Bigg|_{x=\bar S^{-1}(\bar z)} \\
&\quad - \left\{ \begin{bmatrix} f(x)^{\mathsf T}\frac{\partial}{\partial x}\left(\frac{\partial S_1}{\partial x}\right)^{\mathsf T} \\ \vdots \\ f(x)^{\mathsf T}\frac{\partial}{\partial x}\left(\frac{\partial S_n}{\partial x}\right)^{\mathsf T} \end{bmatrix} g(x) - \frac{\partial S(x)}{\partial x}\frac{\partial f(x)}{\partial x} g(x) \right\} \Bigg|_{x=\bar S^{-1}(\bar z)} \\
&= \left\{ \frac{\partial S(x)}{\partial x}\frac{\partial g(x)}{\partial x} f(x) - \frac{\partial S(x)}{\partial x}\frac{\partial f(x)}{\partial x} g(x) \right\} \Bigg|_{x=\bar S^{-1}(\bar z)} \\
&= \left\{ \frac{\partial S(x)}{\partial x}[f(x),\, g(x)] \right\} \Bigg|_{x=\bar S^{-1}(\bar z)} = S_*\left([f(x),\, g(x)]\right)
\end{aligned}
$$

since $\left(\frac{\partial}{\partial x}\left(\frac{\partial S_1}{\partial x}\right)^{\mathsf T}\right)^{\mathsf T} = \frac{\partial}{\partial x}\left(\frac{\partial S_1}{\partial x}\right)^{\mathsf T}$ and $g^{\mathsf T}\frac{\partial}{\partial x}\left(\frac{\partial S_1}{\partial x}\right)^{\mathsf T} f = f^{\mathsf T}\frac{\partial}{\partial x}\left(\frac{\partial S_1}{\partial x}\right)^{\mathsf T} g$.   □

Theorem 2.5 *Suppose that $f(x)$ is a smooth vector field on $\mathbb{R}^n$ and $z = S(x)$ is a state transformation. Then for $\forall h(x) \in C^\infty(\mathbb{R}^n)$,*

$$
L_{\tilde f(z)}\tilde h(z) = L_f h(x)\big|_{x=S^{-1}(z)} \tag{2.30}
$$

where $\tilde f(z) = S_*(f(x))$ and $\tilde h(z) = h \circ S^{-1}(z)$.

Proof Note, by chain rule, that

$$
\begin{aligned}
L_{\tilde f(z)}\tilde h(z)\Big|_{z=S(x)} &= \frac{\partial\left(h \circ S^{-1}(z)\right)}{\partial z}\bigg|_{z=S(x)} \tilde f(z)\big|_{z=S(x)} \\
&= \frac{\partial h(x)}{\partial x}\frac{\partial S^{-1}(z)}{\partial z}\bigg|_{z=S(x)} \frac{\partial S(x)}{\partial x} f(x) \\
&= \frac{\partial h(x)}{\partial x}\frac{\partial\left(S^{-1} \circ S(x)\right)}{\partial x} f(x) = L_f h(x)
\end{aligned}
$$

which implies that (2.30) is satisfied.   □

Consider state transformation $z = S(x)$ and vector fields $f(x)$, $\tilde f(z)$, $g(x)$, and $\tilde g(z)$, where $\tilde f(z) = S_*(f(x))$ and $\tilde g(z) = S_*(g(x))$. Thus, $f(x)$ and $\tilde f(z)$ are the same vector fields expressed in the different coordinates. Theorem 2.4 means that $[f(x), g(x)]$ and $[\tilde f(z), \tilde g(z)]$ are also the same vector fields expressed in the different coordinates. In other words, Lie bracket operation of vector fields is coordinate free. Also, since $h(x)$ and $\tilde h(z) \left(= h \circ S^{-1}(z)\right)$ are the same scalar functions expressed in the different coordinates, Theorem 2.5 means that $L_f h(x)$ and $L_{\tilde{f}(z)}\tilde{h}(z)$ are also the same scalar functions expressed in the different coordinates. In other words, Lie derivative operation of scalar function with respect to vector field is also coordinate free.

Example 2.4.11 Consider the scalar function $h(x) = x_1 x_2$ and vector fields $f(x) = \begin{bmatrix} x_2 \\ 1 \end{bmatrix}$ and $g(x) = \begin{bmatrix} 1 \\ x_1 \end{bmatrix}$ given in Example 2.4.3. Let $z = S(x) = \begin{bmatrix} x_1 + x_2^2 \\ x_2 \end{bmatrix}$ be a state transformation. Find out $\tilde{h}(z) \left(\triangleq h \circ S^{-1}(z) \right)$, $\tilde{f}(z) \left(\triangleq S_*(f(x)) \right)$, $\tilde{g}(z) \left(\triangleq S_*(g(x)) \right)$, $[\tilde{f}(z), \tilde{g}(z)]$, and $L_{\tilde{f}(z)}\tilde{h}(z)$. Also, show that $S_*([f(x), g(x)]) = [\tilde{f}(z), \tilde{g}(z)]$ and $L_{\tilde{f}(z)}\tilde{h}(z) = L_f h(x)\big|_{x=S^{-1}(z)}$.

Solution Note, by Example 2.4.3, that $[f(x), g(x)] = \begin{bmatrix} -x_1 \\ x_2 \end{bmatrix}$ and $L_f h(x) = x_1 + x_2^2$. Also, it is easy to see that

$$x = S^{-1}(z) = \begin{bmatrix} z_1 - z_2^2 \\ z_2 \end{bmatrix}; \quad \tilde{h}(z) = h \circ S^{-1}(z) = z_2(z_1 - z_2^2)$$

$$\tilde{f}(z) = S_*(f(x)) = \frac{\partial S}{\partial x} f(x)\bigg|_{x=S^{-1}(z)} = \begin{bmatrix} 1 & 2x_2 \\ 0 & 1 \end{bmatrix} \begin{bmatrix} x_2 \\ 1 \end{bmatrix}\bigg|_{x=S^{-1}(z)}$$

$$= \begin{bmatrix} 3z_2 \\ 1 \end{bmatrix} = 3z_2 \frac{\partial}{\partial z_1} + \frac{\partial}{\partial z_2}$$

$$\tilde{g}(z) = S_*(g(x)) = \begin{bmatrix} 1 + 2x_1 x_2 \\ x_1 \end{bmatrix}\bigg|_{x=S^{-1}(z)} = \begin{bmatrix} 1 + 2z_2(z_1 - z_2^2) \\ z_1 - z_2^2 \end{bmatrix}$$

$$[\tilde{f}(z), \tilde{g}(z)] = \frac{\partial \tilde{g}(z)}{\partial z} \tilde{f}(z) - \frac{\partial \tilde{f}(z)}{\partial z} \tilde{g}(z) = \begin{bmatrix} -z_1 + 3z_2^2 \\ z_2 \end{bmatrix}$$

$$S_*([f(x), g(x)]) = \begin{bmatrix} -x_1 + 2x_2^2 \\ x_2 \end{bmatrix}\bigg|_{x=S^{-1}(z)} = \begin{bmatrix} -z_1 + 3z_2^2 \\ z_2 \end{bmatrix}$$

$$L_{\tilde{f}}\tilde{h}(z) = \frac{\partial \tilde{h}(z)}{\partial z} \tilde{f}(z) = \begin{bmatrix} z_2 & z_1 - 3z_2^2 \end{bmatrix} \begin{bmatrix} 3z_2 \\ 1 \end{bmatrix} = z_1$$

and

$$L_f h(x)\big|_{x=S^{-1}(z)} = z_1.$$

□

Theorem 2.6 gives geometric necessary and sufficient conditions for a well-defined vector field.

Theorem 2.6 (geometric condition for a well-defined vector field) *Suppose that smooth function $y = S(x) : \mathbb{R}^n \to \mathbb{R}^m$ is surjective and $f(x)$ is a vector field on $\mathbb{R}^n$. $S_*(f(x))$ is a well-defined vector field on $\mathbb{R}^m$, if and only if*

$$[f(x), \ker S_*] \subset \ker S_* \tag{2.31}$$

where $\ker S_* \triangleq \{\tau(x) \mid S_*(\tau(x)) = 0\}$.

Proof Necessity. Suppose that $S_*(f(x))$ is a well-defined vector field on $\mathbb{R}^m$. Let $\tau(x) \in \ker S_*$. Then it is clear, by Theorem 2.4, that

$$S_*([f(x), \tau(x)]) = [S_*(f(x)), S_*(\tau(x))] = [S_*(f(x)), 0] = 0$$

which implies that (2.31) is satisfied.

Sufficiency. Suppose that (2.31) is satisfied. Let $S(\bar{x}) = S(x)$. Let $\gamma_t(x)$ be a smooth parameterized curve such that $\gamma_0(x) = x$, $\gamma_1(x) = \bar{x}$, and

$$S(\gamma_t(x)) = S(x), \ 0 \leq t \leq 1. \tag{2.32}$$

(For example, in Example 2.4.7, $\gamma_t(x) = [x_1 \ \ x_2 + t(\bar{x}_2 - x_2) \ \ x_3 - \frac{t}{2}(1 + x_1)(\bar{x}_2 - x_2)]^\mathsf{T}$.) If we can show that for $0 \leq t \leq 1$

$$\left\{ \frac{\partial S(x)}{\partial x} f(x) \right\}\Bigg|_{x=\gamma_t(x)} = \frac{\partial S(x)}{\partial x} f(x)$$

or

$$\frac{d}{dt}\left(\left\{ \frac{\partial S(x)}{\partial x} f(x) \right\}\Bigg|_{x=\gamma_t(x)} \right)\Bigg|_{t=0} = 0$$

then $S_*(f(x))$ is a well-defined vector field on $\mathbb{R}^m$. Note, by (2.32), that

$$\frac{dS(\gamma_t(x))}{dt}\Bigg|_{t=0} = \frac{\partial S(x)}{\partial x} \frac{d\gamma_t(x)}{dt}\Bigg|_{t=0} = 0$$

or

$$b(x) \triangleq \left. \frac{d\gamma_t(x)}{dt} \right|_{t=0} \in \ker S_*. \tag{2.33}$$

Thus, it is clear, by (2.31) and (2.33), that

$$\frac{\partial S(x)}{\partial x} \frac{\partial b(x)}{\partial x} f(x) - \frac{\partial S(x)}{\partial x} \frac{\partial f(x)}{\partial x} b(x) = \frac{\partial S(x)}{\partial x} [f(x), b(x)] = 0$$

which implies, together with (2.4), that

$$\frac{d}{dt}\left(\left\{ \frac{\partial S(x)}{\partial x} f(x) \right\} \bigg|_{x=\gamma_t(x)} \right) \bigg|_{t=0} = \frac{\partial}{\partial x} \left\{ \frac{\partial S(x)}{\partial x} f(x) \right\} \frac{d\gamma_t(x)}{dt} \bigg|_{t=0}$$

$$= \begin{bmatrix} f(x)^{\mathsf{T}} \frac{\partial}{\partial x} \left(\frac{\partial S_1(x)}{\partial x} \right)^{\mathsf{T}} b(x) \\ \vdots \\ f(x)^{\mathsf{T}} \frac{\partial}{\partial x} \left(\frac{\partial S_m(x)}{\partial x} \right)^{\mathsf{T}} b(x) \end{bmatrix} + \frac{\partial S(x)}{\partial x} \frac{\partial f(x)}{\partial x} b(x)$$

$$= \begin{bmatrix} b(x)^{\mathsf{T}} \frac{\partial}{\partial x} \left(\frac{\partial S_1(x)}{\partial x} \right)^{\mathsf{T}} f(x) \\ \vdots \\ b(x)^{\mathsf{T}} \frac{\partial}{\partial x} \left(\frac{\partial S_m(x)}{\partial x} \right)^{\mathsf{T}} f(x) \end{bmatrix} + \frac{\partial S(x)}{\partial x} \frac{\partial b(x)}{\partial x} f(x)$$

$$= \frac{\partial}{\partial x} \left\{ \frac{\partial S(x)}{\partial x} b(x) \right\} f(x) = 0.$$

$\square$

If $z = S(x)$ is a diffeomorphism on a neighborhood U of the origin, then $\ker S_* = \mathrm{span}\{0\}$. Thus, $S_*(f(x))$ is, by Theorem 2.6, a well-defined vector field as (2.21), for any smooth vector field $f(x)$.

Example 2.4.12   Consider Example 2.4.7 again. Let

$$f(x) = \begin{bmatrix} 1 \\ 0 \\ x_2 \end{bmatrix}, g(x) = \begin{bmatrix} 0 \\ 1 \\ x_1 \end{bmatrix}, \quad \text{and } z = S(x) = \begin{bmatrix} x_1 \\ (1 + x_1)x_2 + 2x_3 \end{bmatrix}.$$

Use Theorem 2.6 to show that $S_*(f(x))$ is not a well-defined vector field on $\mathbb{R}^2$ and $S_*(g(x))$ is a well-defined vector field on $\mathbb{R}^2$.

Solution Note that

$$\frac{\partial S(x)}{\partial x} = \begin{bmatrix} 1 & 0 & 0 \\ x_2 & 1+x_1 & 2 \end{bmatrix} \text{ and } \ker S_* = \text{span} \left\{ \begin{bmatrix} 0 \\ -2 \\ 1+x_1 \end{bmatrix} \right\}.$$

Thus, we have that

$$\left[f(x), \begin{bmatrix} 0 \\ -2 \\ 1+x_1 \end{bmatrix} \right] = \begin{bmatrix} 0 \\ 0 \\ 3 \end{bmatrix} \notin \ker S_* \text{ and } \left[g(x), \begin{bmatrix} 0 \\ -2 \\ 1+x_1 \end{bmatrix} \right] = \begin{bmatrix} 0 \\ 0 \\ 0 \end{bmatrix} \in \ker S_*.$$

Hence, it is clear, by Theorem 2.6, that $S_*(f(x))$ is not a well-defined vector field on $\mathbb{R}^2$ and $S_*(g(x))$ is a well-defined vector field on $\mathbb{R}^2$. Vector field $S_*(g(x))$ is given by (2.20).   □

Let us define

$$L_f^0 h(x) \triangleq h(x); \quad L_f^{k+1} h(x) \triangleq L_f\left(L_f^k h(x)\right), \text{ for } k \geq 0 \qquad (2.34)$$

and

$$\text{ad}_f^0 g(x) \triangleq g(x); \quad \text{ad}_f^{k+1} g(x) \triangleq \left[f(x), \text{ad}_f^k g(x)\right], \text{ for } k \geq 0. \qquad (2.35)$$

For example, we write $L_f^4 h(x)$ and $\text{ad}_{f(x)}^3 g(x)$ instead of $L_f(L_f(L_f(L_f h(x))))$ and $[f(x), [f(x), [f(x), g(x)]]]$, respectively.

Example 2.4.13 (a) Consider the following linear system:

$$\dot{x}(t) = Ax(t); \quad y(t) = Cx(t), \quad x \in \mathbb{R}^n, \ y \in \mathbb{R}.$$

Use chain rule to show that for $k \geq 0$

$$y^{(k)}(t) \triangleq \frac{d^k}{dt^k} y(t) = CA^k x(t).$$

(b) Consider the following nonlinear system:

$$\dot{x}(t) = f(x(t)); \quad y(t) = h(x(t)), \quad x \in \mathbb{R}^n, \ y \in \mathbb{R}.$$

Use chain rule to show that for $k \geq 0$

$$y^{(k)}(t) \triangleq \frac{d^k}{dt^k} y(t) = L_f^k h(x(t)). \qquad (2.36)$$

Solution (a) Omitted.
(b) Note that

$$\dot{y}(t) = \frac{\partial h(x)}{\partial x}\dot{x} = \frac{\partial h(x)}{\partial x}f(x) = L_f h(x(t))$$

$$\ddot{y}(t) = \frac{\partial \left(L_f h(x)\right)}{\partial x}\dot{x} = \frac{\partial L_f h(x)}{\partial x}f(x) = L_f L_f h(x) = L_f^2 h(x(t)).$$

It is easy to show, by mathematical induction, that (2.36) is satisfied.   □

Example 2.4.14 Suppose that

$$\tilde{f}(z) = S_*(f(x)), \quad \tilde{g}(z) = S_*(g(x)), \quad \text{and} \quad \tilde{h}(z) = h \circ S^{-1}(z)$$

where $z = S(x)$ is a state transformation.

(a) Show that for $k \geq 0$

$$\text{ad}_{\tilde{f}}^k \tilde{g}(z) = S_* \left(\text{ad}_f^k g(x)\right) \quad \text{or} \quad \text{ad}_f^k g(x) = S_*^{-1}\left(\text{ad}_{\tilde{f}}^k \tilde{g}(z)\right) \qquad (2.37)$$

(b) Show that if $\tilde{f}(z) = Az$ and $\tilde{g}(z) = b$, then

$$\text{ad}_{\tilde{f}}^k \tilde{g}(z) = (-1)^k A^k b, \quad k \geq 0 \qquad (2.38)$$

and

$$[\text{ad}_f^i g(x), \text{ad}_f^j g(x)] = 0, \quad i \geq 0, \ j \geq 0.$$

(c) Show that for $k \geq 0$

$$L_{\tilde{f}}^k \tilde{h}(z) = L_f^k h(x)\big|_{x=S^{-1}(z)} \quad \text{or} \quad L_f^k h(x) = L_{\tilde{f}}^k \tilde{h}(z)\big|_{z=S(x)} \qquad (2.39)$$

(d) Show that if $\tilde{f}(z) = Az$, $\tilde{g}(z) = b$, and $\tilde{h}(z) = cz$, then for $k \geq 0$

$$L_{\tilde{f}}^k \tilde{h}(z) = cA^k z \quad \text{or} \quad L_f^k h(x) = cA^k S(x) \qquad (2.40)$$

and

$$L_g L_f^k h(x) = cA^k b. \qquad (2.41)$$

Solution Since $\tilde{f}(z) = S_*(f(x))$ and $\tilde{g}(z) = S_*(g(x))$, it is easy to show, by (2.23), (2.28), (2.35), and mathematical induction, that (2.37) is satisfied. It is obvious that (2.38) is satisfied for $k = 0$. Assume that (2.38) is satisfied for $k \le i$ and $i \ge 0$. Then we have that

$$\text{ad}_{\tilde{f}}^{i+1}\tilde{g}(z) = \left[\tilde{f}(z), \text{ad}_{\tilde{f}}^{i}\tilde{g}(z)\right] = (-1)^i \left\{\frac{\partial\left(A^i b\right)}{\partial z} Az - \frac{\partial(Az)}{\partial z} A^i b\right\}$$
$$= (-1)^{i+1} A^{i+1} b$$

which implies, by mathematical induction, that (2.38) is satisfied for $k \ge 0$. Therefore, it is easy to see, by (2.14), (2.28), (2.37), and (2.38), that for $i \ge 0$ and $j \ge 0$

$$\left[\text{ad}_f^i g(x), \text{ad}_f^j g(x)\right] = \left[S_*^{-1}\left(\text{ad}_{\tilde{f}}^i \tilde{g}(z)\right), S_*^{-1}\left(\text{ad}_{\tilde{f}}^j \tilde{g}(z)\right)\right]$$
$$= S_*^{-1}\left(\left[\text{ad}_{\tilde{f}}^i \tilde{g}(z), \text{ad}_{\tilde{f}}^j \tilde{g}(z)\right]\right)$$
$$= S_*^{-1}\left(\left[(-1)^i A^i b, (-1)^j A^j b\right]\right) = S_*^{-1}(0) = 0.$$

Since $\tilde{f}(z) = S_*(f(x))$ and $\tilde{h}(z) = h \circ S^{-1}(z)$, it is easy to show, by (2.30) and mathematical induction, that (2.39) is satisfied (See Problem 2-20). It is also clear, by Example 2.4.13 and (2.39), that (2.40) is satisfied. Finally, we have, by (2.30) and (2.40), that

$$L_g L_f^k h(x) = L_g\left(cA^k S(x)\right) = L_{\tilde{g}}\left(cA^k S \circ S^{-1}(z)\right)\big|_{z=S(x)}$$
$$= cA^k b\big|_{z=S(x)} = cA^k b.$$

$\square$

Example 2.4.15 Show the following useful properties for any scalar functions $\lambda(x)$ and $a(x)$ and vector fields $f(x)$ and $g(x)$.

(a)

$$[f(x), \lambda(x)g(x)] = \lambda(x)[f(x), g(x)] + (L_f\lambda(x))g(x) \qquad (2.42)$$

(b)

$$[a(x)f(x), \lambda(x)g(x)] = a(x)\lambda(x)[f(x), g(x)] + a(x)(L_f\lambda(x))g(x)$$
$$- \lambda(x)(L_g a(x))f(x) \qquad (2.43)$$

(c)

$$\text{ad}^i_{f(x)}\{\lambda(x)g(x)\} = \sum_{k=0}^{i} \binom{i}{k} L^k_f \lambda(x) \, \text{ad}^{i-k}_f g(x), \ \ i \geq 0 \qquad (2.44)$$

(d)

$$L_{\text{ad}^i_f g} h(x) = \sum_{k=0}^{i} (-1)^k \binom{i}{k} L^{i-k}_f L_g L^k_f h(x), \ \ i \geq 0 \qquad (2.45)$$

where $\binom{i}{k} \triangleq \frac{i!}{k!(i-k)!}$ and $\binom{i}{k-1} + \binom{i}{k} = \binom{i+1}{k}$.

Solution (a)

$$\begin{aligned}
[f(x), \lambda(x)g(x)] &= \frac{\partial(\lambda g)}{\partial x} f - \frac{\partial f}{\partial x} \lambda g = g \frac{\partial \lambda}{\partial x} f + \lambda \frac{\partial g}{\partial x} f - \lambda \frac{\partial f}{\partial x} g \\
&= (L_f \lambda)g + \lambda[f, g]
\end{aligned}$$

(b)

$$\begin{aligned}
[af, \lambda g] &= (L_{af}\lambda)g + \lambda[af, g] = a(L_f\lambda)g - \lambda[g, af] \\
&= a(L_f\lambda)g - \lambda \left(a[g, f] + (L_g a)f \right) \\
&= a\lambda[f, g] + a(L_f\lambda)g - \lambda(L_g a)f
\end{aligned}$$

(c)

$$\begin{aligned}
\text{ad}^{i+1}_f(\lambda g) &= \left[f, \text{ad}^i_f(\lambda g) \right] = \sum_{k=0}^{i} \binom{i}{k} \left[f, L^k_f \lambda \, \text{ad}^{i-k}_f g \right] \\
&= \sum_{k=0}^{i} \binom{i}{k} L^k_f \lambda \, \text{ad}^{i-k+1}_f g + \sum_{k=0}^{i} \binom{i}{k} L^{k+1}_f \lambda \, \text{ad}^{i-k}_f g \\
&= \text{ad}^{i+1}_f g + \sum_{k=1}^{i} \binom{i}{k} L^k_f \lambda \, \text{ad}^{i-k+1}_f g \\
&\quad + \sum_{k=1}^{i} \binom{i}{k-1} L^k_f \lambda \, \text{ad}^{i-k+1}_f g + L^{i+1}_f \lambda \, g \\
&= \sum_{k=0}^{i+1} \binom{i+1}{k} L^k_f \lambda \, \text{ad}^{i+1-k}_f g
\end{aligned}$$

(d)

$$L_{\mathrm{ad}_f^{i+1}g}h = L_{[f,\mathrm{ad}_f^i g]}h = L_f L_{\mathrm{ad}_f^i g}h - L_{\mathrm{ad}_f^i g}L_f h$$

$$= \sum_{k=0}^{i}(-1)^k \binom{i}{k} L_f^{i-k+1} L_g L_f^k h - \sum_{k=0}^{i}(-1)^k \binom{i}{k} L_f^{i-k} L_g L_f^{k+1} h$$

$$= L_f^{i+1} L_g + \sum_{k=1}^{i}(-1)^k \binom{i}{k} L_f^{i-k+1} L_g L_f^k h$$

$$+ \sum_{k=1}^{i}(-1)^k \binom{i}{k-1} L_f^{i-k+1} L_g L_f^k h - (-1)^i L_g L_f^{i+1} h$$

$$= \sum_{k=0}^{i+1}(-1)^k \binom{i+1}{k} L_f^{i+1-k} L_g L_f^k h.$$

$\square$

Example 2.4.16 By using (2.45), show that the following statements are equivalent.

(a)

$$L_g L_f^i h(x) = \begin{cases} a_i, & 0 \le i < N-1 \\ c(x), & i = N-1 \end{cases} \tag{2.46}$$

(b)

$$L_{\mathrm{ad}_f^i g} L_f^k h(x) = \begin{cases} (-1)^i a_{i+k}, & i+k < N-1 \\ (-1)^i c(x), & i+k = N-1 \end{cases} \tag{2.47}$$

(c)

$$L_{\mathrm{ad}_f^i g} h(x) = \begin{cases} (-1)^i a_i, & i < N-1 \\ (-1)^i c(x), & i = N-1 \end{cases} \tag{2.48}$$

where $a_i, 0 \le i \le N-1$ are constants.

Solution Suppose that (2.46) is satisfied. Then we have, by (2.45), that

$$L_{\mathrm{ad}_f^i g} L_f^k h(x) = \sum_{j=0}^{i}(-1)^j \binom{i}{j} L_f^{i-j} L_g L_f^{k+j} h(x)$$

$$= \begin{cases} 0, & k+i < N-1 \\ (-1)^i L_g L_f^{k+i} h(x), & k+i = N-1 \end{cases}$$

which implies that (2.47) is satisfied. Suppose that (2.47) is satisfied. Then it is obvious, with $k = 0$, that (2.48) is satisfied. Suppose that (2.48) is satisfied. Then it is obvious that (2.46) is satisfied when $i = 0$. Assume that (2.46) is satisfied for $i \leq k$ and $0 \leq k \leq N - 2$. Then we have, by (2.45), that

$$L_{\text{ad}_f^{k+1} g} h(x) = \sum_{j=0}^{k+1} (-1)^j \binom{k+1}{j} L_f^{k+1-j} L_g L_f^j h(x) = (-1)^{k+1} L_g L_f^{k+1} h(x)$$

which implies that (2.46) is satisfied for $i \leq k + 1$. Therefore, by mathematical induction, (2.46) is satisfied.   □

Example 2.4.17 Let $z = S(x)$ be a state coordinates change. Show the following useful property for any scalar function $\lambda(x)$ and vector field $g(x)$.

$$S_* (\lambda(x) g(x)) = \lambda \circ S^{-1}(z) \, S_* (g(x)) \tag{2.49}$$

Solution

$$S_* (\lambda(x) g(x)) = \left. \frac{\partial S(x)}{\partial x} \lambda(x) g(x) \right|_{x = S^{-1}(z)} = \left. \lambda(x) \frac{\partial S(x)}{\partial x} g(x) \right|_{x = S^{-1}(z)}$$
$$= \lambda \left(S^{-1}(z) \right) \, S_* (g(x)).$$

□

Example 2.4.18 By using Jacobi identity, show that the following statements are equivalent:

(a)

$$[\text{ad}_f^i g(x), \text{ad}_f^k g(x)] = 0, \ 0 \leq i \leq s_1 \text{ and } 0 \leq k \leq s_2 \tag{2.50}$$

(b)

$$[\text{ad}_f^i g(x), \text{ad}_f^k g(x)] = 0, \ 0 \leq i + k \leq s_1 + s_2. \tag{2.51}$$

Solution If (2.51) holds, then (2.50) is obviously satisfied. Suppose that (2.50) is satisfied. Then, it is easy to see, by (2.18), that for $0 \leq i \leq s_1$ and $0 \leq k \leq s_2$

$$[\text{ad}_f^i g, \text{ad}_f^k g] = \left[[f, \text{ad}_f^{i-1} g], \text{ad}_f^k g \right] = -\left[[\text{ad}_f^{i-1} g, \text{ad}_f^k g], f \right]$$
$$- \left[[\text{ad}_f^k g, f], \text{ad}_f^{i-1} g \right] = 0 - [\text{ad}_f^{i-1} g, \text{ad}_f^{k+1} g].$$

In this manner, it is easy to see that for $0 \le i \le s_1, 0 \le k \le s_2$, and $-k \le j \le i$

$$[\mathrm{ad}_f^{i-j} g, \mathrm{ad}_f^{k+j} g] = (-1)^j [\mathrm{ad}_f^i g, \mathrm{ad}_f^k g] = 0$$

which implies that (2.51) is satisfied. $\square$

Example 2.4.19 Let $m(\ge 3)$ be odd. Suppose that for $2 \le i + k \le m$

$$\left[\mathrm{ad}_f^{i-1} g(x), \mathrm{ad}_f^{k-1} g(x)\right] = 0. \tag{2.52}$$

Show that (2.52) is satisfied for $2 \le i + k \le m + 1$.

Solution Suppose that (2.52) is satisfied for $2 \le i + k \le m$. Then, by Example 2.4.18, it is clear that (2.52) is satisfied for $1 \le i \le \frac{m-1}{2}$ and $1 \le k \le \frac{m+1}{2}$. Since $\left[\mathrm{ad}_f^{\frac{m+1}{2}-1} g(x), \mathrm{ad}_f^{\frac{m+1}{2}-1} g(x)\right] = 0$, (2.52) is satisfied for $1 \le i \le \frac{m+1}{2}$ and $1 \le k \le \frac{m+1}{2}$. Hence, it is clear, by Example 2.4.18, that (2.52) is satisfied for $2 \le i + k \le m + 1$. $\square$

Example 2.4.20 Suppose that $\{Y^1(x), Y^2(x), \dots, Y^n(x)\}$ is a set of linearly independent vector fields on a neighborhood of $0 \in \mathbb{R}^n$. Let

$$Y^{n+1}(x) = \sum_{i=1}^{n} a_i(x) Y^i(x).$$

Show that if for $1 \le i \le n+1$ and $1 \le j \le n+1$

$$[Y^i(x), Y^j(x)] = 0,$$

then

$$Y^{n+1}(x) = \sum_{i=1}^{n} a_i Y^i(x)$$

for some constants $a_i \in \mathbb{R}$, $1 \le i \le n$.

Solution Let $Y^{n+1}(x) = \sum_{i=1}^{n} a_i(x) Y^i(x)$. Then we have that for $1 \le j \le n$

$$0 = \left[Y^j(x), Y^{n+1}(x)\right] = \sum_{i=1}^{n} \left[Y^j(x), a_i(x) Y^i(x)\right]$$

$$= \sum_{i=1}^{n} a_i(x) \left[Y^j(x), Y^i(x)\right] + \sum_{i=1}^{n} L_{Y^j} a_i(x) Y^i(x) = \sum_{i=1}^{n} L_{Y^j} a_i(x) Y^i(x)$$

which implies that for $1 \leq i \leq n$

$$O_{1 \times n} = \left[L_{Y^1} a_i(x) \cdots L_{Y^n} a_i(x) \right] = \frac{\partial a_i(x)}{\partial x} \left[Y^1(x) \cdots Y^n(x) \right].$$

Since $\frac{\partial a_i(x)}{\partial x} = O_{1 \times n}$ for $1 \leq i \leq n$, it is clear that $a_i(x)$ is a constant for $1 \leq i \leq n$.

$\square$

2.5   Covector Field and One Form

A covector field on Euclidean space is the transpose of a vector field. Suppose that U be an open subset of $\mathbb{R}^n$.

Definition 2.13 (*smooth covector field on Euclidean space*) A vector-valued function $w : U(\subset \mathbb{R}^n) \to \mathbb{R}^n$ is said to be a smooth covector field on U, if $w = [w_1 \ w_2 \ \cdots \ w_n]$ and $w_i \in C^\infty(U)$ for $1 \leq i \leq n$.

Suppose that $x \triangleq [x_1 \ x_2 \ \cdots \ x_n]^\mathsf{T}$ is a Cartesian coordinate system of $\mathbb{R}^n$. Then a covector field $w(x)$ can be expressed by

$$\begin{aligned} w(x) &= \left[w_1(x) \ w_2(x) \ \cdots \ w_n(x) \right] \\ &= w_1(x) dx_1 + w_2(x) dx_2 + \cdots + w_n(x) dx_n \end{aligned}$$

where

$$dx_1 \triangleq [1 \ 0 \ \cdots \ 0], \ dx_2 \triangleq [0 \ 1 \ 0 \ \cdots \ 0], \ \cdots, \ \text{and } dx_n \triangleq [0 \ \cdots \ 0 \ 1].$$

Addition of covector fields and scalar multiplication are defined by the transpose of (2.10) and (2.11).

Example 2.5.1 Show that the set of all smooth covector fields on $\mathbb{R}^n$ is a vector space over field $\mathbb{R}$.

Solution Omitted. (Problem 2-15.) $\square$

Let us define $\langle w(x), f(x) \rangle$ by

$$\left\langle \sum_{i=1}^n w_i(x) dx_i, \sum_{j=1}^n f_j(x) \frac{\partial}{\partial x_j} \right\rangle \triangleq \sum_{i=1}^n w_i(x) f_i(x) = w(x) f(x).$$

With the operator $\langle w(x), \cdot \rangle$, a smooth covector field $w(x)$ can be thought of a function from the set of smooth vector fields to $C^\infty(\mathbb{R}^n)$. For example, dx_i is a linear function such that

$$\left\langle dx_i, \frac{\partial}{\partial x_j} \right\rangle = \begin{cases} 1, & \text{if } j = i \\ 0, & \text{if } j \neq i. \end{cases}$$

The differential (or total derivative) $dh(x)$ of $h(x) \in C^\infty(\mathbb{R}^n)$ is defined by

$$dh(x) \triangleq \left[\frac{\partial h(x)}{\partial x_1} \quad \frac{\partial h(x)}{\partial x_2} \quad \cdots \quad \frac{\partial h(x)}{\partial x_n} \right] = \sum_{i=1}^{n} \frac{\partial h}{\partial x_i} dx_i.$$

A smooth covector field $w(x)$ is obtained when a scalar function is differentiated once, so it is also called a differential one form, or simply a one form. The Lie derivative of $h(x)$ with respect to $f(x)$ can also be written by

$$L_f h(x) = \frac{\partial h(x)}{\partial x} f(x) = \langle dh(x), f(x) \rangle.$$

Definition 2.14 (*exact one form*) One form $w(x)$ is said to be an exact one form, if there exists a scalar function $h(x)$ such that $w(x) = \frac{\partial h}{\partial x}$ or $w(x) = dh(x)$.

Note that $\frac{\partial^2 h(x)}{\partial x_i \partial x_j} = \frac{\partial^2 h(x)}{\partial x_j \partial x_i}$, for $1 \leq i \leq n$ and $1 \leq j \leq n$. If $w(x)$ is an exact one form, then $w_i(x) = \frac{\partial h(x)}{\partial x_i}$, $1 \leq i \leq n$ for some scalar function $h(x)$. Thus, it is clear that

$$\frac{\partial w_j(x)}{\partial x_i} = \frac{\partial w_i(x)}{\partial x_j}, \quad 1 \leq i \leq n, \ 1 \leq j \leq n \tag{2.53}$$

or

$$\frac{\partial w(x)^\mathsf{T}}{\partial x} = \left(\frac{\partial w(x)^\mathsf{T}}{\partial x} \right)^\mathsf{T}.$$

Conversely, if (2.53) is satisfied, then $w(x)$ is an exact one form (See Lemma 2.1).

Lemma 2.1 *Let* $1 \leq k \leq n$, $x = \begin{bmatrix} x^1 \\ x^2 \end{bmatrix}$, $x^1 = \begin{bmatrix} x_1 \\ \vdots \\ x_k \end{bmatrix}$, *and* $x^2 = \begin{bmatrix} x_{k+1} \\ \vdots \\ x_n \end{bmatrix}$. *Suppose that* $w_i(x) \in C^\infty(\mathbb{R}^n)$ *for* $1 \leq i \leq k$. *There exists a function* $h(x) \in C^\infty(\mathbb{R}^n)$ *such that* $h(0, x^2) = 0$ *and* $\frac{\partial h(x)}{\partial x^1} = [w_1(x) \ \cdots \ w_k(x)] \triangleq w^1(x)$, *if and only if for* $1 \leq i \leq k$ *and* $1 \leq j \leq k$

$$\frac{\partial w_j(x)}{\partial x_i} = \frac{\partial w_i(x)}{\partial x_j} \tag{2.54}$$

or

$$\frac{\partial w^1(x)^\mathsf{T}}{\partial x^1} = \left(\frac{\partial w^1(x)^\mathsf{T}}{\partial x^1} \right)^\mathsf{T}.$$

We denote

$$h(x) = \int [w_1(x) \ \cdots \ w_k(x)] d(x_1 \ \cdots \ x_k)$$

$$\triangleq \int w^1(x) dx^1.$$

Proof Necessity. Obvious.

Sufficiency. Suppose that (2.54) is satisfied. Let

$$h(x) = \sum_{j=1}^{k} Q_j(x) \tag{2.55}$$

where

$$Q_1(x) = \int w_1(x) dx_1$$

$$Q_i(x) = \int w_i(x) dx_i - \sum_{j=1}^{i-1} \int \frac{\partial Q_j(x)}{\partial x_i} dx_i, \quad 2 \le i \le k. \tag{2.56}$$

Then it is easy to see, by (2.56), that

$$\frac{\partial}{\partial x_i} \left(\sum_{j=1}^{i} Q_j(x) \right) = w_i(x), \quad 1 \le i \le k. \tag{2.57}$$

Now we will show, by mathematical induction, that for $2 \le i \le k$

$$\frac{\partial Q_i(x)}{\partial x_\ell} = 0, \quad 1 \le \ell \le i - 1. \tag{2.58}$$

Since $Q_2(x) = \int w_2(x) dx_2 - \int \int \frac{\partial w_1(x)}{\partial x_2} dx_1 dx_2$, we have, by (2.54), that

$$\frac{\partial Q_2(x)}{\partial x_1} = \int \frac{\partial w_2(x)}{\partial x_1} dx_2 - \int \frac{\partial w_1(x)}{\partial x_2} dx_2 = \int \left(\frac{\partial w_2(x)}{\partial x_1} - \frac{\partial w_1(x)}{\partial x_2} \right) dx_2 = 0$$

which implies that (2.58) is satisfied when $i = 2$. Assume that (2.58) is satisfied for $2 \le i \le p$ and $2 \le p \le k - 1$. Let $1 \le q \le p$. Then we have, by (2.54), (2.56), and (2.57), that

$$\frac{\partial Q_{p+1}(x)}{\partial x_q} = \int \frac{\partial w_{p+1}(x)}{\partial x_q} dx_{p+1} - \int \frac{\partial^2 \left(\sum_{j=1}^{p} Q_j(x)\right)}{\partial x_{p+1}\partial x_q} dx_{p+1} \quad \text{(by (2.56))}$$

$$= \int \frac{\partial w_{p+1}(x)}{\partial x_q} dx_{p+1} - \int \frac{\partial^2 \left(\sum_{j=1}^{q} Q_j(x)\right)}{\partial x_{p+1}\partial x_q} dx_{p+1} \quad \text{(by assumption)}$$

$$= \int \frac{\partial w_{p+1}(x)}{\partial x_q} dx_{p+1} - \int \frac{\partial w_q(x)}{\partial x_{p+1}} dx_{p+1} \quad \text{(by (2.57))}$$

$$= \int \left(\frac{\partial w_{p+1}(x)}{\partial x_q} - \frac{\partial w_q(x)}{\partial x_{p+1}}\right) dx_{p+1} = 0 \quad \text{(by (2.54))}$$

which implies that (2.58) is satisfied for $i = p + 1$. Therefore, (2.58) is, by mathematical induction, satisfied for $2 \le i \le k$. Hence, it is easy to see, by (2.55), (2.57), and (2.58), that for $1 \le i \le k$

$$\frac{\partial h(x)}{\partial x_i} = \frac{\partial}{\partial x_i} \left(\sum_{j=1}^{k} Q_j(x)\right) = \frac{\partial}{\partial x_i} \left(\sum_{j=1}^{i} Q_j(x)\right) = w_i(x).$$

$\square$

Example 2.5.2 Show that one form $w(x) = \begin{bmatrix} 1 & x_1 \end{bmatrix}$ is not exact.

Solution Since

$$\frac{\partial w_1(x)}{\partial x_2} = 0 \ne 1 = \frac{\partial w_2(x)}{\partial x_1} \quad \text{or} \quad \frac{\partial w(x)^\mathsf{T}}{\partial x} = \begin{bmatrix} 0 & 0 \\ 1 & 0 \end{bmatrix} \ne \left(\frac{\partial w(x)^\mathsf{T}}{\partial x}\right)^\mathsf{T}$$

(2.53) is not satisfied. Thus, $w(x) = \begin{bmatrix} 1 & x_1 \end{bmatrix}$ is not an exact one form. $\square$

Example 2.5.3 Show that $w(x) = \begin{bmatrix} x_2 & x_1 + x_3 & x_2 + 2x_3 \end{bmatrix}$ is an exact one form. Find out scalar function $h(x)$ such that $w(x) = dh(x)$ and $h(0) = 0$.

Solution Since

$$\frac{\partial w_1}{\partial x_2} = 1 = \frac{\partial w_2}{\partial x_1} \ ; \quad \frac{\partial w_1}{\partial x_3} = 0 = \frac{\partial w_3}{\partial x_1} \ ; \quad \frac{\partial w_2}{\partial x_3} = 1 = \frac{\partial w_3}{\partial x_2}$$

or

$$\frac{\partial w(x)^\mathsf{T}}{\partial x} = \begin{bmatrix} 0 & 1 & 0 \\ 1 & 0 & 1 \\ 0 & 1 & 2 \end{bmatrix} = \left(\frac{\partial w(x)^\mathsf{T}}{\partial x}\right)^\mathsf{T}$$

(2.53) is satisfied. Therefore, one form $w(x) = \begin{bmatrix} x_2 & x_1 & x_3 \end{bmatrix}$ is exact. We have $h(x) = x_1 x_2 + R_1(x_2, x_3)$ from $\frac{\partial h(x)}{\partial x_1} = w_1(x) = x_2$. (or $Q_1(x) = x_1 x_2$.) Also, since $\frac{\partial h(x)}{\partial x_2} = w_2(x) = x_1 + x_3$, we have $\frac{\partial R_1(x_2, x_3)}{\partial x_2} = x_3$ and $R_1(x_2, x_3) = x_2 x_3 + R_2(x_3)$.

(or $Q_2(x) = x_2 x_3$.) Finally, since $\frac{\partial h(x)}{\partial x_3} = w_3(x) = x_2 + 2x_3$, we have $\frac{\partial R_2(x_3)}{\partial x_3} = 2x_3$ and $R_2(x_3) = x_3^2 + \text{const.}$ (or $Q_3(x) = x_3^2$.) Hence, we have $h(x) = x_1 x_2 + x_2 x_3 + x_3^2 (= Q_1(x) + Q_2(x) + Q_3(x))$. □

2.6 Distribution and Frobenius Theorem

When vector field $f(x)$ and a state transformation $z = S(x)$ are given, vector field $\tilde{f}(z)(= S_*(f(x)))$, that is the same vector field expressed in z-coordinates, can be found as in Example 2.4.11. In this section, we first try to find a state transformation $z = S(x)$ such that for $1 \le i \le n$

$$S_*(f_i(x)) = \frac{\partial}{\partial z_i} \tag{2.59}$$

when $\{f_1(x), \cdots, f_n(x)\}$ are a set of linearly independent vector fields.

Example 2.6.1 Consider vector fields $f(x) = \begin{bmatrix} x_2 \\ 1 \end{bmatrix}$ and $\tau(x) = \begin{bmatrix} 1 \\ 0 \end{bmatrix}$. Find a state transformation $z = S(x)$ such that $S_*(f(x)) = \frac{\partial}{\partial z_1} = \begin{bmatrix} 1 \\ 0 \end{bmatrix}$ and $S_*(\tau(x)) = \frac{\partial}{\partial z_2} = \begin{bmatrix} 0 \\ 1 \end{bmatrix}$.

Solution We need to find a state transformation $z = S(x)$ such that

$$\left[S_*(f(x)) \ S_*(\tau(x)) \right] = \left[\frac{\partial S(x)}{\partial x} f(x) \ \frac{\partial S(x)}{\partial x} \tau(x) \right]\Big|_{x = S^{-1}(z)} = \begin{bmatrix} 1 & 0 \\ 0 & 1 \end{bmatrix}.$$

Since

$$\frac{\partial S(x)}{\partial x} \left[f(x) \ \tau(x) \right] = \frac{\partial S(x)}{\partial x} \begin{bmatrix} x_2 & 1 \\ 1 & 0 \end{bmatrix} = I$$

we have that

$$\begin{bmatrix} \frac{\partial S_1(x)}{\partial x} \\ \frac{\partial S_2(x)}{\partial x} \end{bmatrix} = \frac{\partial S(x)}{\partial x} = \begin{bmatrix} x_2 & 1 \\ 1 & 0 \end{bmatrix}^{-1} = \begin{bmatrix} 0 & 1 \\ 1 & -x_2 \end{bmatrix}.$$

Since one forms $\begin{bmatrix} 0 & 1 \end{bmatrix}$ and $\begin{bmatrix} 1 & -x_2 \end{bmatrix}$ are exact, there exist scalar functions $S_1(x)$ and $S_2(x)$ such that $\frac{\partial S_1(x)}{\partial x} = \begin{bmatrix} 0 & 1 \end{bmatrix}$ and $\frac{\partial S_2(x)}{\partial x} = \begin{bmatrix} 1 & -x_2 \end{bmatrix}$. By easy calculation, we have $S(x) = \begin{bmatrix} x_2 \\ x_1 - \frac{1}{2}x_2^2 \end{bmatrix}$. □

Example 2.6.2 Consider vector fields $f(x) = \begin{bmatrix} x_2 \\ 1 \end{bmatrix}$ and $g(x) = \begin{bmatrix} e^{x_1} \\ 0 \end{bmatrix}$. Can we find

a state transformation $z = S(x)$ such that $S_*(f(x)) = \frac{\partial}{\partial z_1} = \begin{bmatrix} 1 \\ 0 \end{bmatrix}$ and $S_*(g(x)) =$

$\frac{\partial}{\partial z_2} = \begin{bmatrix} 0 \\ 1 \end{bmatrix}$?

Solution A state transformation $z = S(x)$ should satisfy

$$\begin{bmatrix} \frac{\partial S_1(x)}{\partial x} \\ \frac{\partial S_2(x)}{\partial x} \end{bmatrix} = \begin{bmatrix} f(x) & g(x) \end{bmatrix}^{-1} = \begin{bmatrix} 0 & 1 \\ e^{-x_1} & -x_2 e^{-x_1} \end{bmatrix}.$$

Since one form $\begin{bmatrix} e^{-x_1} & -x_2 e^{-x_1} \end{bmatrix}$ is not exact, there does not exist a scalar function
$S_2(x)$ such that $\frac{\partial S_2(x)}{\partial x} = \begin{bmatrix} e^{-x_1} & -x_2 e^{-x_1} \end{bmatrix}$. □

In the above Examples, it can be easily shown that $[f(x), \tau(x)] = 0$ and
$[f(x), g(x)] \neq 0$. Theorem 2.7 gives the conditions for the existence of a state transformation $z = S(x)$ such that (2.59) holds. The following Theorem is often used in
this book.

Theorem 2.7 *Suppose that* $\{f_1(x), \ldots, f_n(x)\}$ *is a set of linearly independent
smooth vector fields on open set* U *of* $\mathbb{R}^n$. *There exists a state transformation*
$z = S(x) : U \to \mathbb{R}^n$ *such that for* $1 \leq i \leq n$

$$S_*(f_i(x)) = \frac{\partial}{\partial z_i} \tag{2.60}$$

if and only if for $1 \leq i \leq n$ *and* $1 \leq j \leq n$

$$[f_i(x), f_j(x)] = 0. \tag{2.61}$$

Furthermore, state transformation $z = S(x)$ *satisfies*

$$\frac{\partial S(x)}{\partial x} = \begin{bmatrix} f_1(x) & \cdots & f_n(x) \end{bmatrix}^{-1}. \tag{2.62}$$

Proof Theorem 2.7 is a special case of Theorem 2.9 with $k = n$. Since Frobenius
Theorem (Theorem 2.8) is needed to prove the sufficiency part of Theorem 2.9 when
$k \leq n - 1$, Theorem 2.9 is considered after Theorem 2.8. However, Theorem 2.7 can
be proven without Frobenius Theorem. The proof is omitted, since it is very similar to that of Theorem 2.9 with $k = n$. It is easy to see, by (2.71), that (2.62) is
satisfied. □

From now on, we try to find a state transformation $z = S(x)$ such that for $k + 1 \leq p \leq n$

$$L_{f_i(x)} S_p(x) = 0, \ 1 \leq i \leq k$$

or

$$\frac{\partial S_p(x)}{\partial x} \left[f_1(x) \ \cdots \ f_k(x) \right] = \left[0 \ \cdots \ 0 \right]$$

when $k \leq n - 1$ and $\{ f_1(x), \ \cdots, \ f_k(x) \}$ are a set of linearly independent vector fields.

Example 2.6.3 Let $f_1(x) = \begin{bmatrix} 1 \\ 0 \\ 0 \end{bmatrix}$ and $f_2(x) = \begin{bmatrix} x_1^2 \\ 1 \\ x_2 \end{bmatrix}$. Find a state transformation $z = S(x)$ such that $L_{f_1} S_3(x) = 0$ and $L_{f_2} S_3(x) = 0$.

Solution Since

$$\begin{bmatrix} 0 & 0 \end{bmatrix} = \begin{bmatrix} L_{f_1} S_3(x) & L_{f_2} S_3(x) \end{bmatrix} = \begin{bmatrix} \frac{\partial S_3(x)}{\partial x} f_1(x) & \frac{\partial S_3(x)}{\partial x} f_2(x) \end{bmatrix}$$

$$= \frac{\partial S_3(x)}{\partial x} \begin{bmatrix} f_1(x) & f_2(x) \end{bmatrix} = \frac{\partial S_3(x)}{\partial x} \begin{bmatrix} 1 & x_1^2 \\ 0 & 1 \\ 0 & x_2 \end{bmatrix}$$

we have that $\frac{\partial S_3(x)}{\partial x} = \begin{bmatrix} 0 & -x_2 a(x) & a(x) \end{bmatrix}$, where $a(x)$ is a smooth nonzero function. If we let $a(x) = 1$, then one form $\begin{bmatrix} 0 & -x_2 a(x) & a(x) \end{bmatrix}$ is exact and it is easy to see that $S_3(x) = x_3 - \frac{1}{2} x_2^2$. We can choose any smooth functions $S_1(x)$ and $S_2(x)$ such that $\{ dS_1(x), dS_2(x), dS_3(x) \}$ are linearly independent. For example, let $S_1(x) = x_1$ and $S_2(x) = x_2$. Then it is clear that $z = S(x) = \begin{bmatrix} x_1 \\ x_2 \\ x_3 - \frac{1}{2} x_2^2 \end{bmatrix}$ is a state transformation such that $L_{f_1} S_3(x) = 0$ and $L_{f_2} S_3(x) = 0$. □

Example 2.6.4 Let $g_1(x) = \begin{bmatrix} 1 \\ 0 \\ 0 \end{bmatrix}$ and $g_2(x) = \begin{bmatrix} 0 \\ 1 \\ x_1 \end{bmatrix}$. Show that there does not exist a state transformation $z = S(x)$ such that $L_{g_1} S_3(x) = 0$ and $L_{g_2} S_3(x) = 0$.

Solution Since

$$\begin{bmatrix} 0 & 0 \end{bmatrix} = \begin{bmatrix} L_{g_1} S_3(x) & L_{g_2} S_3(x) \end{bmatrix} = \begin{bmatrix} \frac{\partial S_3(x)}{\partial x} g_1(x) & \frac{\partial S_3(x)}{\partial x} g_2(x) \end{bmatrix}$$

$$= \frac{\partial S_3(x)}{\partial x} \begin{bmatrix} g_1(x) & g_2(x) \end{bmatrix} = \frac{\partial S_3(x)}{\partial x} \begin{bmatrix} 1 & 0 \\ 0 & 1 \\ 0 & x_1 \end{bmatrix}$$

we have that $\frac{\partial S_3(x)}{\partial x} = \begin{bmatrix} 0 & -x_1 a(x) & a(x) \end{bmatrix} \triangleq \begin{bmatrix} w_1(x) & w_2(x) & w_3(x) \end{bmatrix}$, where $a(x)$ is a smooth nonzero function. Note that $\frac{\partial w_3(x)}{\partial x_1} = \frac{\partial a(x)}{\partial x_1}$, $\frac{\partial w_2(x)}{\partial x_1} = -a(x) - x_1 \frac{\partial a(x)}{\partial x_1}$, and $\frac{\partial w_1(x)}{\partial x_3} = \frac{\partial w_1(x)}{\partial x_2} = 0$. In order for one form $\begin{bmatrix} w_1(x) & w_2(x) & w_3(x) \end{bmatrix}$ to be exact, it is clear that $\frac{\partial w_3(x)}{\partial x_1} = \frac{\partial w_1(x)}{\partial x_3}$ and $\frac{\partial w_2(x)}{\partial x_1} = \frac{\partial w_1(x)}{\partial x_2}$. Thus we have that $a(x) = 0$ and $S_3(x) = 0$. Therefore, there does not exist a state transformation $z = S(x)$ such that $L_{g_1} S_3(x) = 0$ and $L_{g_2} S_3(x) = 0$.   □

In Examples 2.6.3 and 2.6.4, note that

$$[f_1, f_2] = \begin{bmatrix} 2x_1 \\ 0 \\ 0 \end{bmatrix} \in \{c_1(x)f_1 + c_2(x)f_2 \mid c_1(x), c_2(x) \in C^\infty(\mathbb{R}^3)\}$$

$$[g_1, g_2] = \begin{bmatrix} 0 \\ 0 \\ 1 \end{bmatrix} \notin \{c_1(x)g_1 + c_2(x)g_2 \mid c_1(x), c_2(x) \in C^\infty(\mathbb{R}^3)\}.$$

For simplicity, we let

$$\mathrm{span}\{f_1(x), \ldots, f_k(x)\} \triangleq \left\{ \sum_{i=1}^k c_i(x)f_i(x) \,\middle|\, c_i(x) \in C^\infty(\mathbb{R}^n),\ 1 \le i \le k \right\}.$$

Definition 2.15 (*distribution*) $D(x)$ is said to be a k-dimensional distribution on open set U of $\mathbb{R}^n$, if $D(x)$ is k-dimensional subspace of $\mathbb{R}^n$ for any $x \in U$.

Let $p_1, p_2 \in U \subset \mathbb{R}^n$ and $p_1 \ne p_2$. If $\dim D(p_1) \ne \dim D(p_2)$, then $D(x)$ is not a distribution. For example, let

$$D(x) = \mathrm{span}\left\{ \begin{bmatrix} 1 + x_2 \\ 0 \end{bmatrix}, \begin{bmatrix} 0 \\ 1 \end{bmatrix} \right\}.$$

Since $D(\begin{bmatrix} 0 \\ -1 \end{bmatrix}) = \mathrm{span}\left\{ \begin{bmatrix} 0 \\ 0 \end{bmatrix}, \begin{bmatrix} 0 \\ 1 \end{bmatrix} \right\}$, we have that $\dim(D(\begin{bmatrix} 0 \\ -1 \end{bmatrix})) = 1$. Thus, $D(x)$ is a 2-dimensional distribution not on $\mathbb{R}^2$ but on open set $U \triangleq \{(x_1, x_2) \in \mathbb{R}^2 \mid x_2 \ne -1\}$ of $\mathbb{R}^2$.

Definition 2.16 (*smooth distribution*) $D(x)$ is said to be a k-dimensional smooth distribution on open set U of $\mathbb{R}^n$, if there exists a set of smooth vector fields $\{f_1(x), \ldots, f_k(x)\}$, defined on a neighborhood $\bar{U}$ of $p \in U$, such that for any $x \in \bar{U}$

$$D(x) = \mathrm{span}\{f_1(x), \ldots, f_k(x)\}.$$

Here $\{f_1(x), \ldots, f_k(x)\}$ is called a local basis of distribution $D(x)$.

Definition 2.17 (*involutive distribution*) Smooth distribution $D(x)$ is said to be involutive, if for any smooth vector fields $f(x) \in D(x)$ and $g(x) \in D(x)$

$$[f(x), g(x)] \in D(x).$$

In other words, smooth distribution $D(x)$ is said to be involutive, if $D(x)$ is closed under bracket operation.

Example 2.6.5 Show that smooth distribution $D(x) = \text{span}\{g_1(x), g_2(x)\}$ is not involutive, where $g_1(x) = \frac{\partial}{\partial x_1} = \begin{bmatrix} 1 \\ 0 \\ 0 \end{bmatrix}$ and $g_2(x) = \frac{\partial}{\partial x_2} + x_1 \frac{\partial}{\partial x_3} = \begin{bmatrix} 0 \\ 1 \\ x_1 \end{bmatrix}$.

Solution $D(x) = \text{span}\{g_1(x), g_2(x)\}$ is not involutive, since

$$[g_1(x), g_2(x)] = \frac{\partial}{\partial x_3} = \begin{bmatrix} 0 \\ 0 \\ 1 \end{bmatrix} \notin D(x).$$

$\square$

Example 2.6.6 Let $D(x) = \text{span}\{f_1(x), \ldots, f_k(x)\}$, where $\{f_1(x), \ldots, f_k(x)\}$ is a set of linearly independent smooth vector fields. Show that distribution $D(x)$ is involutive, if and only if for $1 \le i \le k$ and $1 \le j \le k$

$$\left[f_i(x), f_j(x)\right] \in D(x).$$

Solution Omitted. (Problem 2-22.) $\square$

By Example 2.6.6, only a finite number of brackets of vector fields belonging to the basis are needed to check in order to know whether the distribution is involutive or not.

Example 2.6.7 Show that smooth distribution $D(x) = \text{span}\{f_1(x), f_2(x)\}$ is involutive, where $f_1(x) = \begin{bmatrix} 1 \\ x_1 x_2 \\ x_1 x_3 \end{bmatrix}$ and $f_2(x) = \begin{bmatrix} 0 \\ 1 \\ 1 \end{bmatrix}$.

Solution It is easy to see that

$$[f_1(x), f_2(x)] = \begin{bmatrix} 0 \\ -x_1 \\ -x_1 \end{bmatrix} = -x_1 f_2(x) \in D(x).$$

Thus, by Example 2.6.6, $D(x)$ is an involutive distribution. $\square$

Definition 2.18 (*completely integrable distribution*) Suppose that $D(x) = \text{span}\{f_1(x), \ldots, f_k(x)\}$ is a k-dimensional smooth distribution on open set U of $\mathbb{R}^n$. Distribution $D(x)$ is said to be completely integrable, if there exists a state transformation $z = S(x) : U \to \mathbb{R}^n$ such that

$$\tilde{D}(z) = S_*(D(x)) \triangleq \text{span}\,\{S_*(f_1(x)), \ldots, S_*(f_k(x))\}$$

$$= \text{span}\left\{\frac{\partial}{\partial z_1}, \ldots, \frac{\partial}{\partial z_k}\right\} \tag{2.63}$$

or for $k + 1 \leq j \leq n$

$$L_{f(x)}S_j(x) = 0, \quad \forall f(x) \in D(x). \tag{2.64}$$

Theorem 2.8 (Frobenius Theorem) *A distribution $D(x)$ is completely integrable, if and only if $D(x)$ is involutive.*

Proof Omitted. (Refer to [A1].) ☐

Suppose that $\{S_{k+1}(x), \ldots, S_n(x)\}$ is a set of scalar functions such that (2.64) is satisfied. Let $\{S_1(x), \ldots, S_k(x)\}$ be any set of scalar functions such that $z = S(x)$ is a state transformation or $\{dS_1(x), \ldots, dS_n(x)\}$ are linearly independent. Then (2.63) is also satisfied, even though it may not be true that $S_*(f_i(x)) = \frac{\partial}{\partial z_i}$, $1 \leq i \leq k$.

Example 2.6.8 For involutive distribution $D(x)$ in Example 2.6.7, find out a scalar function $h(x)$ (or $S_3(x)$) such that $h(0) = 0$, $\left.\frac{\partial h(x)}{\partial x}\right|_{x=0} \neq 0$, and

$$L_{f(x)}h(x) = 0, \quad \forall f(x) \in D(x).$$

Also, find out a state transformation $z = S(x)$ such that

$$\tilde{D}(z) \triangleq \text{span}\,\{S_*(f_1(x)), S_*(f_2(x))\} = \text{span}\left\{\frac{\partial}{\partial z_1}, \frac{\partial}{\partial z_2}\right\}. \tag{2.65}$$

Solution Note that

$$[0\ 0] = \frac{\partial h(x)}{\partial x}\left[f_1(x)\ f_2(x)\right] = \frac{\partial h(x)}{\partial x}\begin{bmatrix} 1 & 0 \\ x_1x_2 & 1 \\ x_1x_3 & 1 \end{bmatrix}$$

which implies that $a(0) \neq 0$ and

$$\frac{\partial h(x)}{\partial x} = \begin{bmatrix} \frac{\partial h(x)}{\partial x_1} & \frac{\partial h(x)}{\partial x_2} & \frac{\partial h(x)}{\partial x_3} \end{bmatrix} = \begin{bmatrix} x_1(x_2 - x_3)a(x) & -a(x) & a(x) \end{bmatrix} \triangleq \omega(x).$$

Thus, we need to find $a(x)$ such that $\omega(x)$ is an exact one form (or $\frac{\partial \omega_i(x)}{\partial x_j} = \frac{\partial \omega_j(x)}{\partial x_i}$, $i \neq j$). Since distribution $D(x)$ is involutive, there exists, by Theorem 2.8, a scalar function $a(x)$ such that $\omega(x)$ is an exact one form. In general, a scalar function $a(x)$ is complicated to find.

$$\frac{\partial \omega_1(x)}{\partial x_2} = x_1 a(x) + x_1(x_2 - x_3)\frac{\partial a(x)}{\partial x_2} \; ; \quad \frac{\partial \omega_2(x)}{\partial x_1} = -\frac{\partial a(x)}{\partial x_1}$$

$$\frac{\partial \omega_1(x)}{\partial x_3} = -x_1 a(x) + x_1(x_2 - x_3)\frac{\partial a(x)}{\partial x_3} \; ; \quad \frac{\partial \omega_3(x)}{\partial x_1} = \frac{\partial a(x)}{\partial x_1}$$

$$\frac{\partial \omega_2(x)}{\partial x_3} = -\frac{\partial a(x)}{\partial x_3} \; ; \quad \frac{\partial \omega_3(x)}{\partial x_2} = \frac{\partial a(x)}{\partial x_2}$$

If we let $\frac{\partial a(x)}{\partial x_2} = \frac{\partial a(x)}{\partial x_3} = 0$, we can obtain scalar functions $a(x) = e^{-\frac{1}{2}x_1^2}$ and $h(x) = e^{-\frac{1}{2}x_1^2}(-x_2 + x_3)$. Of course, we may be able to obtain a different $a(x)$ without $\frac{\partial a(x)}{\partial x_2} = \frac{\partial a(x)}{\partial x_3} = 0$. We can choose any smooth functions $S_1(x)$ and $S_2(x)$ such that $\{dS_1(x), dS_2(x), dS_3(x)\}$ are linearly independent. For example, let $S_1(x) = x_1$ and $S_2(x) = x_2$. Then it is clear that $z = S(x) = \begin{bmatrix} x_1 \\ x_2 \\ e^{-\frac{1}{2}x_1^2}(-x_2 + x_3) \end{bmatrix}$ is a state transfor-

mation. Since $x = S^{-1}(z) = \begin{bmatrix} z_1 \\ z_2 \\ z_3 e^{\frac{1}{2}z_1^2} + z_2 \end{bmatrix}$, we have

$$S_*(f_1(x)) = \left\{ \frac{\partial S(x)}{\partial x} f_1(x) \right\} \Big|_{x = S^{-1}(z)} = \begin{bmatrix} 1 \\ z_1 z_2 \\ 0 \end{bmatrix} = \frac{\partial}{\partial z_1} + z_1 z_2 \frac{\partial}{\partial z_2}$$

$$S_*(f_2(x)) = \left\{ \begin{bmatrix} 1 & 0 & 0 \\ 0 & 1 & 0 \\ -x_1 e^{-\frac{1}{2}x_1^2}(-x_2 + x_3) & -e^{-\frac{1}{2}x_1^2} & e^{-\frac{1}{2}x_1^2} \end{bmatrix} \begin{bmatrix} 0 \\ 1 \\ 1 \end{bmatrix} \right\} \Big|_{x = S^{-1}(z)}$$

$$= \begin{bmatrix} 0 \\ 1 \\ 0 \end{bmatrix} = \frac{\partial}{\partial z_2}$$

which implies that (2.65) is satisfied. $\square$

Theorem 2.9 *Suppose that $\{f_1(x), \ldots, f_k(x)\}$ is a set of linearly independent smooth vector fields on open set U of $\mathbb{R}^n$. There exists a state transformation $z = S(x) : U \to \mathbb{R}^n$ such that for $1 \le i \le k$,*

$$S_*(f_i(x)) = \frac{\partial}{\partial z_i} \tag{2.66}$$

if and only if for $1 \le i \le k$ and $1 \le j \le k$

$$[f_i(x), f_j(x)] = 0. \tag{2.67}$$

Proof Necessity. Suppose that there exists a state transformation $z = S(x)$ such that (2.66) is satisfied. Then, it is clear, by Theorem 2.4, that for $0 \le i \le k$ and $0 \le j \le k$

$$\left[f_i(x), f_j(x) \right] = \left[S_*^{-1} \left(\frac{\partial}{\partial z_i} \right), S_*^{-1} \left(\frac{\partial}{\partial z_j} \right) \right] = S_*^{-1} \left(\left[\frac{\partial}{\partial z_i}, \frac{\partial}{\partial z_j} \right] \right) = 0.$$

Sufficiency. Suppose that (2.67) is satisfied. Then, distribution $D(x)$ is involutive, where

$$D(x) \triangleq \mathrm{span}\{f_1(x), \, \cdots, \, f_k(x)\}.$$

Thus, it is clear, by Frobenius Theorem (Theorem 2.8), that there exists a state transformation $\tilde{x} = T(x)$ such that

$$\tilde{D}(\tilde{x}) \triangleq \mathrm{span}\left\{ \tilde{f}_1(\tilde{x}), \ldots, \tilde{f}_k(\tilde{x}) \right\} = \mathrm{span}\left\{ \frac{\partial}{\partial \tilde{x}_1}, \ldots, \frac{\partial}{\partial \tilde{x}_k} \right\} \tag{2.68}$$

where $\tilde{f}_i(\tilde{x}) \triangleq T_*(f_i(x))$ for $1 \le i \le k$ (If $k = n$, we can let $\tilde{x} = T(x) = x$). Thus, we can let

$$\tilde{f}_i(\tilde{x}) \triangleq \begin{bmatrix} \hat{f}_i(\tilde{x}) \\ O_{(n-k) \times 1} \end{bmatrix}, \quad 1 \le i \le k. \tag{2.69}$$

Also, it is clear, by Theorem 2.4, that for $1 \le i \le k$ and $1 \le j \le k$

$$[\tilde{f}_i(\tilde{x}), \tilde{f}_j(\tilde{x})] = 0 \ \text{ or } \ \frac{\partial \hat{f}_i(\tilde{x})}{\partial \tilde{x}^1} \hat{f}_j(\tilde{x}) - \frac{\partial \hat{f}_j(\tilde{x})}{\partial \tilde{x}^1} \hat{f}_i(\tilde{x}) = O_{k \times 1}. \tag{2.70}$$

Let $\tilde{x} = \begin{bmatrix} \tilde{x}^1 \\ \tilde{x}^2 \end{bmatrix}, \tilde{x}^1 = \begin{bmatrix} \tilde{x}_1 \\ \vdots \\ \tilde{x}_k \end{bmatrix}$, and $\tilde{x}^2 = \begin{bmatrix} \tilde{x}_{k+1} \\ \vdots \\ \tilde{x}_n \end{bmatrix}$. Also, define $1 \times k$ matrix $w_i(\tilde{x})$, $1 \le i \le k$ by

$$\begin{bmatrix} w_1(\tilde{x}) \\ \vdots \\ w_k(\tilde{x}) \end{bmatrix} \triangleq \left[\hat{f}_1(\tilde{x}) \cdots \hat{f}_k(\tilde{x}) \right]^{-1} \tag{2.71}$$

or

$$\begin{bmatrix} w_1(\tilde{x}) \\ \vdots \\ w_k(\tilde{x}) \end{bmatrix} \left[\hat{f}_1(\tilde{x}) \cdots \hat{f}_k(\tilde{x}) \right] = \begin{bmatrix} 1 & \cdots & 0 \\ \vdots & & \vdots \\ 0 & \cdots & 1 \end{bmatrix}.$$

In other words, $w_p(\tilde{x})\hat{f}_i(\tilde{x}) = \delta_{p,i}$ for $1 \leq p \leq k$ and $1 \leq i \leq k$, where $\delta_{p,i}$ is the Kronecker delta function. Thus, we have, by (2.3), that for $1 \leq p \leq k$ and $1 \leq i \leq k$

$$\frac{\partial\left(w_p(\tilde{x})\hat{f}_i(\tilde{x})\right)}{\partial\tilde{x}^1} = w_p(\tilde{x})\frac{\partial\hat{f}_i(\tilde{x})}{\partial\tilde{x}^1} + \hat{f}_i(\tilde{x})^\mathsf{T}\frac{\partial\left(w_p(\tilde{x})^\mathsf{T}\right)}{\partial\tilde{x}^1} = O_{1\times k}.$$

Therefore, we have that for $1 \leq p \leq k$, $1 \leq i \leq k$, and $1 \leq j \leq k$

$$w_p(\tilde{x})\frac{\partial\hat{f}_i(\tilde{x})}{\partial\tilde{x}^1}\hat{f}_j(\tilde{x}) + \hat{f}_i(\tilde{x})^\mathsf{T}\frac{\partial\left(w_p(\tilde{x})^\mathsf{T}\right)}{\partial\tilde{x}^1}\hat{f}_j(\tilde{x}) = 0$$

and

$$w_p(\tilde{x})\frac{\partial\hat{f}_j(\tilde{x})}{\partial\tilde{x}^1}\hat{f}_i(\tilde{x}) + \hat{f}_j(\tilde{x})^\mathsf{T}\frac{\partial\left(w_p(\tilde{x})^\mathsf{T}\right)}{\partial\tilde{x}^1}\hat{f}_i(\tilde{x}) = 0$$

which imply that

$$w_p\left(\frac{\partial\hat{f}_i}{\partial\tilde{x}^1}\hat{f}_j - \frac{\partial\hat{f}_j}{\partial\tilde{x}^1}\hat{f}_i\right) + \hat{f}_i^\mathsf{T}\left\{\frac{\partial\left(w_p^\mathsf{T}\right)}{\partial\tilde{x}^1} - \left(\frac{\partial\left(w_p^\mathsf{T}\right)}{\partial\tilde{x}^1}\right)^\mathsf{T}\right\}\hat{f}_j = 0.$$

Therefore, it is easy to see, by (2.70), that for $1 \leq p \leq k$

$$\begin{bmatrix}\hat{f}_1(\tilde{x})^\mathsf{T}\\\vdots\\\hat{f}_k(\tilde{x})^\mathsf{T}\end{bmatrix}\left\{\frac{\partial\left(w_p(\tilde{x})^\mathsf{T}\right)}{\partial\tilde{x}^1} - \left(\frac{\partial\left(w_p(\tilde{x})^\mathsf{T}\right)}{\partial\tilde{x}^1}\right)^\mathsf{T}\right\}[\hat{f}_1(\tilde{x}) \cdots \hat{f}_k(\tilde{x})] = O_{k\times k}.$$

Since $[\hat{f}_1(\tilde{x}) \cdots \hat{f}_k(\tilde{x})]$ has rank k, we have $\frac{\partial(w_p(\tilde{x})^\mathsf{T})}{\partial\tilde{x}^1} = \left(\frac{\partial(w_p(\tilde{x})^\mathsf{T})}{\partial\tilde{x}^1}\right)^\mathsf{T}$ for $1 \leq p \leq k$ and thus there exists, by Lemma 2.1, a scalar function $\tilde{S}_p(\tilde{x})$ for $1 \leq p \leq k$ such that $\frac{\partial\tilde{S}_p(\tilde{x})}{\partial\tilde{x}^1} = w_p(\tilde{x})$. Let $\tilde{S}(\tilde{x}) \triangleq [\tilde{S}_1(\tilde{x}) \cdots \tilde{S}_k(\tilde{x}) \ \tilde{x}_{k+1} \cdots \tilde{x}_n]^\mathsf{T} \triangleq \begin{bmatrix}\tilde{S}^1(\tilde{x})\\\tilde{x}^2\end{bmatrix}$. Since

$\frac{\partial\tilde{S}^1(\tilde{x})}{\partial\tilde{x}^1} = \begin{bmatrix}w_1(\tilde{x})\\\vdots\\w_k(\tilde{x})\end{bmatrix}$, it is easy to see, by (2.71), that $z = \tilde{S}(\tilde{x})$ is a state transformation.

Therefore, we have, by (2.69) and (2.71), that

$$
\left[\tilde{S}_*(\tilde{f}_1(\tilde{x})) \cdots \tilde{S}_*(\tilde{f}_k(\tilde{x}))\right] = \left[\frac{\partial \tilde{S}(\tilde{x})}{\partial \tilde{x}} \tilde{f}_1(\tilde{x}) \cdots \frac{\partial \tilde{S}(\tilde{x})}{\partial \tilde{x}} \tilde{f}_k(\tilde{x})\right]\Bigg|_{\tilde{x}=\tilde{S}^{-1}(z)}
$$

$$
= \left\{\frac{\partial \tilde{S}(\tilde{x})}{\partial \tilde{x}} \left[\tilde{f}_1(\tilde{x}) \cdots \tilde{f}_k(\tilde{x})\right]\right\}\Bigg|_{\tilde{x}=\tilde{S}^{-1}(z)}
$$

$$
= \left\{\left[\begin{matrix} \frac{\partial \tilde{S}^1(\tilde{x})}{\partial \tilde{x}^1} & \frac{\partial \tilde{S}^1(\tilde{x})}{\partial \tilde{x}^2} \\ O_{(n-k)\times k} & I_{n-k} \end{matrix}\right] \left[\begin{matrix} \hat{f}_1(\tilde{x}) & \cdots & \hat{f}_k(\tilde{x}) \\ O_{(n-k)\times 1} & \cdots & O_{(n-k)\times 1} \end{matrix}\right]\right\}\Bigg|_{\tilde{x}=\tilde{S}^{-1}(z)}
$$

$$
= \left[\begin{matrix} I_k \\ O_{(n-k)\times k} \end{matrix}\right]
$$

which implies that $\tilde{S}_*(\tilde{f}_i(\tilde{x})) = \frac{\partial}{\partial z_i}$ for $1 \leq i \leq k$. Hence, it is clear that for $1 \leq i \leq k$

$$
S_*(f_i(x)) = \tilde{S}_* \circ T_*(f_i(x)) = \tilde{S}_*(\tilde{f}_i(\tilde{x})) = \frac{\partial}{\partial z_i}
$$

where $z = S(x) \triangleq \tilde{S} \circ T(x)$.   □

Corollary 2.1 *Let* $x = \left[\begin{matrix} x^1 \\ x^2 \end{matrix}\right]$, $x^1 = \left[\begin{matrix} x_1 \\ \vdots \\ x_k \end{matrix}\right]$, *and* $x^2 = \left[\begin{matrix} x_{k+1} \\ \vdots \\ x_n \end{matrix}\right]$. *Suppose that* $\{f_1(x),$
..., $f_k(x)\}$ is a set of linearly independent smooth vector fields on open set U of $\mathbb{R}^n$
and that

$$
\text{span}\{f_1(x), \cdots, f_k(x)\} = \text{span}\left\{\frac{\partial}{\partial x_1}, \cdots, \frac{\partial}{\partial x_k}\right\}. \tag{2.72}
$$

There exists a state transformation $z = S(x) : U \to \mathbb{R}^n$ *such that for* $1 \leq i \leq k$

$$
S_*(f_i(x)) = \frac{\partial}{\partial z_i} \tag{2.73}
$$

if and only if for $1 \leq i \leq k$ *and* $1 \leq j \leq k$

$$
\left[f_i(x), f_j(x)\right] = 0. \tag{2.74}
$$

Furthermore, a state transformation $z = S(x) \triangleq \left[\begin{matrix} S^1(x) \\ x^2 \end{matrix}\right]$ *satisfies*

$$
\frac{\partial S^1(x)}{\partial x} = \left[\hat{f}_1(x) \cdots \hat{f}_k(x)\right]^{-1} \tag{2.75}
$$

where $f_i(x) \triangleq \left[\begin{matrix} \hat{f}_i(x) \\ O_{(n-k)\times 1} \end{matrix}\right]$ *for* $1 \leq i \leq k$.

Proof If (2.72) is satisfied, Frobenius Theorem is not needed. In other words, the proof of Corollary 2.1 is the same as the proof of Theorem 2.9 with $\tilde{x} = T(x) = x$.   □

The annihilator $D(x)^\perp$ of distribution $D(x)$ in open set U of $\mathbb{R}^n$ is defined by

$$D(x)^\perp \triangleq \{w(x) \mid \langle w(x), X \rangle = 0, \ \forall X \in D(x)\}.$$

In other words, $D(x)^\perp$ is the set of all one forms which are perpendicular to all vector fields in $D(x)$. Thus, it is easy to see that $\dim(D(x)^\perp) = n - \dim(D(x))$.

Example 2.6.9 Let $D(x)$ be the smooth distribution in Example 2.6.7. Find the annihilator $D(x)^\perp$ of $D(x)$.

Solution It is easy to see that

$$D(x)^\perp = \text{span}\left\{ \begin{bmatrix} x_1(x_2 - x_3) & -1 & 1 \end{bmatrix} \right\}.$$

Since $D(x)$ is involutive, there exists a scalar function $h(x) = e^{-\frac{1}{2}x_1^2}(-x_2 + x_3)$ such that $D(x)^\perp = \text{span}\{dh(x)\}$ (Refer to Example 2.6.8)   □.

Suppose that $S : \mathbb{R}^n \to \mathbb{R}^m$ is a smooth surjective function and $D(x) = \text{span}\{f_1(x), \ldots, f_k(x)\}$ is a k-dimensional smooth distribution on $\mathbb{R}^n$. Define the set $S_*(D(x))$ of tangent vectors on $\mathbb{R}^m$ by

$$S_*(D(x)) \triangleq \left\{ S_* \left(\sum_{i=1}^{k} a_i(x) f_i(x) \right) \ \middle| \ a_i(x) \in C^\infty(\mathbb{R}^n) \right\}.$$

Example 2.6.10 Suppose that $x \in \mathbb{R}^3$ and $z = S(x) = \begin{bmatrix} x_1 \\ (1 + x_1)x_2 + 2x_3 \end{bmatrix}$. Consider the following smooth distributions:

$$D_1(x) = \text{span}\left\{ \begin{bmatrix} 0 \\ 1 \\ x_1 \end{bmatrix} \right\} = \text{span}\{f_1(x)\}$$

$$D_2(x) = \text{span}\left\{ \begin{bmatrix} 0 \\ e^{x_3} \\ x_1 e^{x_3} \end{bmatrix} \right\} = \text{span}\{f_2(x)\}$$

$$D_3(x) = \text{span}\left\{ \begin{bmatrix} 1 \\ 0 \\ x_2 \end{bmatrix} \right\} = \text{span}\{f_3(x)\}$$

$$D_4(x) = \text{span}\left\{ \begin{bmatrix} 0 \\ 1 \\ x_1 \end{bmatrix}, \begin{bmatrix} 0 \\ -2 \\ 1 + x_1 \end{bmatrix} \right\} = \text{span}\{f_1(x), f_4(x)\}.$$

Let $\bar{x} = \begin{bmatrix} x_1 \\ a \\ x_3 + \frac{1}{2}(1+x_1)x_2 - \frac{1}{2}a(1+x_1) \end{bmatrix}$. Then it is clear that $S(\bar{x}) = S(x)$ for all $a \in \mathbb{R}$. Find $S_*(D_i(x))$, $1 \leq i \leq 4$ and $S_*(D_i(\bar{x}))$, $1 \leq i \leq 4$.

Solution Note that $S_*(D_i(x))$, $1 \leq i \leq 4$ and $S_*(D_i(\bar{x}))$, $1 \leq i \leq 4$ are the sets of tangent vectors at $z = \begin{bmatrix} x_1 \\ (1+x_1)x_2 + 2x_3 \end{bmatrix} \in \mathbb{R}^2$. It is clear that

$$S_*(f_1(x)) = \frac{\partial S}{\partial x} f_1(x) = \begin{bmatrix} 1 & 0 & 0 \\ x_2 & 1+x_1 & 2 \end{bmatrix} \begin{bmatrix} 0 \\ 1 \\ x_1 \end{bmatrix} = \begin{bmatrix} 0 \\ 1+3x_1 \end{bmatrix}$$

$$S_*(f_2(x)) = \begin{bmatrix} 0 \\ (1+3x_1)e^{x_3} \end{bmatrix} ; \quad S_*(f_3(x)) = \begin{bmatrix} 1 \\ 3x_2 \end{bmatrix}$$

$$S_*(f_4(x)) = \begin{bmatrix} 0 \\ 0 \end{bmatrix}$$

Also, we have that

$$S_*(f_1(\bar{x})) = \frac{\partial S}{\partial x} f_1(x) \bigg|_{x=\bar{x}} = \begin{bmatrix} 0 \\ 1+3x_1 \end{bmatrix}$$

$$S_*(f_2(\bar{x})) = \begin{bmatrix} 0 \\ (1+3x_1)e^{x_3 + \frac{1}{2}(1+x_1)x_2 - \frac{1}{2}a(1+x_1)} \end{bmatrix}$$

$$S_*(f_3(\bar{x})) = \begin{bmatrix} 1 \\ 3a \end{bmatrix} ; \quad S_*(f_4(\bar{x})) = \begin{bmatrix} 0 \\ 0 \end{bmatrix}.$$

Thus, it is easy to see that

$$S_*(D_1(\bar{x})) = \text{span} \left\{ \begin{bmatrix} 0 \\ 1+3x_1 \end{bmatrix} \right\} = S_*(D_1(x))$$

$$S_*(D_2(\bar{x})) = \text{span} \left\{ \begin{bmatrix} 0 \\ (1+3x_1)e^{x_3 + \frac{1}{2}(1+x_1)x_2 - \frac{1}{2}a(1+x_1)} \end{bmatrix} \right\}$$

$$= \text{span} \left\{ \begin{bmatrix} 0 \\ 1+3x_1 \end{bmatrix} \right\} = S_*(D_2(x))$$

$$S_*(D_3(\bar{x})) = \text{span} \left\{ \begin{bmatrix} 1 \\ 3a \end{bmatrix} \right\} \neq \text{span} \left\{ \begin{bmatrix} 1 \\ 3x_2 \end{bmatrix} \right\} = S_*(D_3(x))$$

and

$$S_*(D_4(\bar{x})) = \text{span}\left\{\begin{bmatrix} 0 \\ 1 + 3x_1 \end{bmatrix}\right\} = \text{span}\left\{\begin{bmatrix} 0 \\ 1 + 3x_1 \end{bmatrix}\right\} = S_*(D_4(x)).$$

$\square$

In the above Example, since $S_*(f_1(x))$ is a well-defined vector field $\tilde{f}_1(z) \triangleq$ $S_*(f_1(x)) = \begin{bmatrix} 0 \\ 1 + 3z_1 \end{bmatrix} = (1 + 3z_1)\frac{\partial}{\partial z_2}$, it is clear that $\tilde{D}_1(z) \triangleq S_*(D_1(x)) =$ span $\left\{\begin{bmatrix} 0 \\ 1 + 3z_1 \end{bmatrix}\right\}$ is a distribution on a neighborhood $U = \{\|z\| < \frac{1}{3}\}$ of the origin $\begin{bmatrix} 0 \\ 0 \end{bmatrix} \in \mathbb{R}^2$. Even though $S_*(f_2(x))$ is not a well-defined vector field, $S_*(e^{-x_3}f_2(x))$ is a well-defined vector field $S_*(e^{-x_3}f_2(x)) = \begin{bmatrix} 0 \\ 1 + 3z_1 \end{bmatrix} = (1 + 3z_1)\frac{\partial}{\partial z_2}$ and $D_2(x) =$ span $\{f_2(x)\} = \text{span}\left\{e^{-x_3}f_2(x)\right\} = D_1(x)$. Thus, $\tilde{D}_2(z) \triangleq S_*(D_2(x)) = \tilde{D}_1(z) =$ span $\left\{\begin{bmatrix} 0 \\ 1 + 3z_1 \end{bmatrix}\right\}$ is also a distribution on a neighborhood U of the origin $\begin{bmatrix} 0 \\ 0 \end{bmatrix} \in \mathbb{R}^2$. Both of $S_*(f_3(x))$ and $S_*(f_3(\bar{x}))$ are the set of tangent vectors on $z = S(x) = S(\bar{x})$. However, since $S_*(D_3(x)) \neq S_*(D_3(\bar{x}))$ unless $a = x_2$, it is clear that $S_*(D_3(x))$ is not a distribution on $\mathbb{R}^2$. Since $S_*(D_4(x)) = S_*(D_4(\bar{x}))$, it is clear that $S_*(D_4(x))$ is a distribution on $\mathbb{R}^2$.

Suppose that $D(x) = \text{span}\{f_1(x), \ldots, f_k(x)\}$ is a k-dimensional smooth distribution on $\mathbb{R}^n$. If $S : \mathbb{R}^n \to \mathbb{R}^n$ is a state transformation (or diffeomophism), then $S_*(f_i(x))$, $1 \leq i \leq k$ are well-defined vector fields and

$$\tilde{D}(z) \triangleq S_*(D(x)) = \text{span}\{S_*(f_1(x)), \ldots, S_*(f_k(x))\}$$
$$\triangleq \text{span}\left\{\tilde{f}_1(z), \ldots, \tilde{f}_k(z)\right\}$$

is a smooth distribution.

Definition 2.19 (*well-defined smooth distribution*) Suppose that $S : \mathbb{R}^n \to \mathbb{R}^m$ is a smooth surjective function and $D(x)$ is a smooth distribution on $\mathbb{R}^n$. $S_*(D(x))$ is said to be a well-defined smooth distribution on $\mathbb{R}^m$, if $S_*(D(\bar{x})) = S_*(D(x))$ whenever $S(\bar{x}) = S(x)$.

A geometric condition for the differential map of the distribution to be a well-defined distribution is given in the following Theorem.

Theorem 2.10 (geometric condition for well-defined distribution) *Suppose that $S : \mathbb{R}^n \to \mathbb{R}^m$ is a smooth surjective function. Also, suppose that $D(x) = \text{span}\{f_1(x), \ldots, f_k(x)\}$ is a k-dimensional smooth distribution on $\mathbb{R}^n$. Then*

$S_*(D(x))$ *is a well-defined distribution on a neighborhood of* $z(= S(x))$ *in* $\mathbb{R}^m$, *if and only if*

$$[f_i(x), \ \ker S_*] \subset D(x) + \ker S_*, \ 1 \le i \le k. \tag{2.76}$$

Proof Necessity. Let $z = S(x)$. Suppose that $S_*(D(x))$ is a well-defined $\bar{k}$-dimensional distribution on a neighborhood of $z(= S(x))$ in $\mathbb{R}^m$. Without loss of generality, assume that for $\bar{k} + 1 \le i \le k$

$$f_i(x) \in \ker S_* \ \text{ or } \ S_*(f_i(x)) = 0. \tag{2.77}$$

Let $z = S(x)$. Then we have that

$$\tilde{D}(z) = S_*(D(x)) = S_* \left(\text{span} \left\{ f_1(x), \ldots, f_{\bar{k}}(x)\right\}\right)$$
$$= \text{span} \left\{ \bar{f}_1(z), \ldots, \bar{f}_{\bar{k}}(z)\right\}$$

and for $1 \le j \le \bar{k}$

$$\bar{f}_j(z) = S_* \left(\sum_{i=1}^{\bar{k}} a_{ji}(x) f_i(x) \right)$$

for some smooth functions $a_{ji}(x)$. Let $\tau(x) \in \ker S_*$. Thus, it is clear, by Theorem 2.6 and (2.42), that for $1 \le j \le \bar{k}$

$$\left[\sum_{i=1}^{k} a_{ji}(x) f_i(x), \ \tau(x) \right] \in \ker S_*$$

or

$$\sum_{i=1}^{k} a_{ji}(x) [f_i(x), \ \tau(x)] - \sum_{i=1}^{k} L_\tau a_{ji}(x) f_i(x) \in \ker S_*$$

which implies that for $1 \le j \le \bar{k}$

$$\sum_{i=1}^{k} a_{ji}(x) [f_i(x), \ \tau(x)] \in D(x) + \ker S_*. \tag{2.78}$$

Since rank $\left(\begin{bmatrix} a_{11} & \cdots & a_{1\bar{k}} \\ \vdots & & \vdots \\ a_{\bar{k}1} & \cdots & a_{\bar{k}\bar{k}} \end{bmatrix} \right) = \bar{k}$, it is clear, by (2.78), that (2.76) is satisfied for

$1 \leq i \leq \bar{k}$. Since ker S_* is involutive, (2.76) is, by (2.77), satisfied for $\bar{k} + 1 \leq i \leq k$.

Sufficiency. Suppose that (2.76) is satisfied. Let $S(\bar{x}) = S(x)$. Let $\gamma_t(x)$ be a smooth parameterized curve such that $\gamma_0(x) = x$, $\gamma_1(x) = \bar{x}$, and

$$S(\gamma_t(x)) = S(x), \quad 0 \leq t \leq 1. \tag{2.79}$$

(For example, in Example 2.4.7, $\gamma_t(x) = \begin{bmatrix} x_1 \\ x_2 + t(\bar{x}_2 - x_2) \\ x_3 - \frac{t}{2}(1 + x_1)(\bar{x}_2 - x_2) \end{bmatrix}$.) Note that

$$S_*(D(\gamma_t(x))) = \text{span} \left\{ \left\{ \frac{\partial S(x)}{\partial x} f_1(x) \right\} \bigg|_{x = \gamma_t(x)}, \ldots, \left\{ \frac{\partial S(x)}{\partial x} f_k(x) \right\} \bigg|_{x = \gamma_t(x)} \right\}.$$

If we can show that for $0 \leq t \leq 1$

$$S_*(D(\gamma_t(x))) = S_*(D(x))$$
$$= \text{span} \left\{ \frac{\partial S(x)}{\partial x} f_1(x), \ldots, \frac{\partial S(x)}{\partial x} f_k(x) \right\}$$

or for $1 \leq i \leq k$

$$\frac{d}{dt} \left(\left\{ \frac{\partial S(x)}{\partial x} f_i(x) \right\} \bigg|_{x = \gamma_t(x)} \right) \bigg|_{t=0} \in S_*(D(x)) \tag{2.80}$$

then $S_*(D(x))$ is a well-defined distribution on a neighborhood of $z(= S(x))$ in $\mathbb{R}^m$. Note, by (2.79), that

$$\frac{dS(\gamma_t(x))}{dt} \bigg|_{t=0} = \frac{\partial S(x)}{\partial x} \frac{d\gamma_t(x)}{dt} \bigg|_{t=0} = 0$$

or

$$b(x) \triangleq \frac{d\gamma_t(x)}{dt} \bigg|_{t=0} \in \ker S_*. \tag{2.81}$$

Thus, it is easy to see, by (2.76) and (2.81), that there exists $\tilde{f}_i(x) \in D(x)$ such that for $1 \leq i \leq k$

$$\frac{\partial S(x)}{\partial x}([f_i(x), b(x)]) = \frac{\partial S(x)}{\partial x}\frac{\partial b(x)}{\partial x}f_i(x) - \frac{\partial S(x)}{\partial x}\frac{\partial f_i(x)}{\partial x}b(x)$$

$$= \frac{\partial S(x)}{\partial x}\tilde{f}_i(x)$$

which implies, together with (2.4) and (2.81), that for $1 \leq i \leq k$,

$$\frac{d}{dt}\left(\left\{\frac{\partial S(x)}{\partial x}f_i(x)\right\}\bigg|_{x=\gamma_t(x)}\right)\bigg|_{t=0} = \frac{\partial}{\partial x}\left\{\frac{\partial S(x)}{\partial x}f_i(x)\right\}\frac{d\gamma_t(x)}{dt}\bigg|_{t=0}$$

$$= \begin{bmatrix} f_i(x)^\mathsf{T}\frac{\partial}{\partial x}\left(\frac{\partial S_1(x)}{\partial x}\right)^\mathsf{T}b(x) \\ \vdots \\ f_i(x)^\mathsf{T}\frac{\partial}{\partial x}\left(\frac{\partial S_m(x)}{\partial x}\right)^\mathsf{T}b(x) \end{bmatrix} + \frac{\partial S(x)}{\partial x}\frac{\partial f_i(x)}{\partial x}b(x)$$

$$= \begin{bmatrix} b(x)^\mathsf{T}\frac{\partial}{\partial x}\left(\frac{\partial S_1(x)}{\partial x}\right)^\mathsf{T}f_i(x) \\ \vdots \\ b(x)^\mathsf{T}\frac{\partial}{\partial x}\left(\frac{\partial S_m(x)}{\partial x}\right)^\mathsf{T}f_i(x) \end{bmatrix} + \frac{\partial S(x)}{\partial x}\frac{\partial b(x)}{\partial x}f_i(x) - \frac{\partial S(x)}{\partial x}\tilde{f}_i(x)$$

$$= \frac{\partial}{\partial x}\left\{\frac{\partial S(x)}{\partial x}b(x)\right\}f_i(x) - \frac{\partial S(x)}{\partial x}\tilde{f}_i(x)$$

$$= -\frac{\partial S(x)}{\partial x}\tilde{f}_i(x) \in S_*(D(x)).$$

In other words, (2.80) is satisfied. Hence, $S_*(D(x))$ is a well-defined distribution on a neighborhood of $z(= S(x))$ in $\mathbb{R}^m$. $\square$

Example 2.6.11 Suppose that $D(x)$ is involutive distribution on $\mathbb{R}^n$ and $S_*(D(x))$ is a well-defined smooth distribution on $\mathbb{R}^m$. Show that $S_*(D(x))$ is also an involutive distribution.

Solution Omitted. (Problem 2-24.) $\square$

Example 2.6.12 Suppose that $x \in \mathbb{R}^3$ and $z = S(x) = \begin{bmatrix} x_1 \\ (1+x_1)x_2 + 2x_3 \end{bmatrix}$. Consider the following smooth distributions:

$$D_1(x) = \text{span}\left\{\begin{bmatrix} 0 \\ 1 \\ x_1 \end{bmatrix}\right\} = \text{span}\{f_1(x)\}$$

$$D_2(x) = \text{span}\left\{\begin{bmatrix} 0 \\ e^{x_3} \\ x_1 e^{x_3} \end{bmatrix}\right\} = \text{span}\{f_2(x)\}$$

$$D_3(x) = \text{span} \left\{ \begin{bmatrix} 1 \\ 0 \\ x_2 \end{bmatrix} \right\} = \text{span}\{f_3(x)\}$$

$$D_4(x) = \text{span} \left\{ \begin{bmatrix} 0 \\ 1 \\ x_1 \end{bmatrix}, \begin{bmatrix} 0 \\ -2 \\ 1+x_1 \end{bmatrix} \right\} = \text{span}\{f_1(x), f_4(x)\}$$

$$D_5(x) = \text{span} \left\{ \begin{bmatrix} 1 \\ 0 \\ 0 \end{bmatrix}, \begin{bmatrix} 0 \\ 1 \\ 0 \end{bmatrix} \right\} = \text{span}\{f_5(x), f_6(x)\}.$$

Use Theorem 2.10 to find whether $S_*(D(x))$ is a well-defined smooth distribution on a neighborhood of the origin in $\mathbb{R}^2$ or not. If it is a well-defined smooth distribution, then express it in terms of z-coordinates.

Solution Since $\frac{\partial S(x)}{\partial x} = \begin{bmatrix} 1 & 0 & 0 \\ x_2 & 1+x_1 & 2 \end{bmatrix}$, it is clear that

$$\ker S_* = \text{span} \left\{ -2\frac{\partial}{\partial x_2} + (1+x_1)\frac{\partial}{\partial x_3} \right\} = \text{span} \left\{ \begin{bmatrix} 0 \\ -2 \\ 1+x_1 \end{bmatrix} \right\} \triangleq \text{span}\{\tau(x)\}.$$

Thus, it is easy to see that

$$[f_1(x), \tau(x)] = 0 \in \ker S_* \subset D_1(x) + \ker S_*$$

$$[f_2(x), \tau(x)] = -(1+x_1)\begin{bmatrix} 0 \\ e^{x_3} \\ x_1 e^{x_3} \end{bmatrix} \in D_2(x) + \ker S_*$$

$$[f_3(x), \tau(x)] = \begin{bmatrix} 0 \\ 0 \\ 3 \end{bmatrix} \notin D_3(x) + \ker S_*$$

$$[f_4(x), \tau(x)] = 0 \in \ker S_* \subset D_4(x) + \ker S_*$$

$$[f_5(x), \tau(x)] = \begin{bmatrix} 0 \\ 0 \\ 1 \end{bmatrix} \in D_5(x) + \ker S_*$$

$$[f_6(x), \tau(x)] = 0 \in \ker S_* \subset D_5(x) + \ker S_*$$

$$\left[f_5(x) - \frac{x_2}{1+x_1}f_6(x), \tau(x) \right] = \frac{1}{1+x_1}\tau(x) \in \ker S_*$$

which imply that $S_*(D_3(x))$ is not a well-defined smooth distribution on a neighborhood of the origin in $\mathbb{R}^2$ and

$$S_*(D_1(x)) = S_*(D_2(x)) = S_*(D_4(x)) = \text{span}\left\{\begin{bmatrix} 0 \\ 1 + 3z_1 \end{bmatrix}\right\}$$

$$S_*(D_5(x)) = \text{span}\left\{\begin{bmatrix} 1 \\ 0 \end{bmatrix}, \begin{bmatrix} 0 \\ 1 + 3z_1 \end{bmatrix}\right\} = \text{span}\left\{\begin{bmatrix} 1 \\ 0 \end{bmatrix}, \begin{bmatrix} 0 \\ 1 \end{bmatrix}\right\}.$$

$\square$

If $\Phi(x)$ is a distribution and $f(x) - g(x) \in \Phi(x)$, we denote

$$f(x) \equiv g(x) \mod \Phi(x).$$

For example, suppose that

$$f(x) = \begin{bmatrix} 0 \\ 3 \\ 1 + x_1 \end{bmatrix}, \ g(x) = \begin{bmatrix} 0 \\ 1 \\ x_3 \end{bmatrix}, \text{ and } \Phi(x) = \begin{bmatrix} 0 \\ 0 \\ 2 \end{bmatrix}.$$

Then we have that

$$f(x) \equiv g(x) \mod \Phi(x).$$

2.7 State Equivalence and Feedback Equivalence

The concept of an equivalence relation is important in mathematics and will be used in the form of state equivalence or feedback equivalence. In mathematics, an equivalence relation is a binary relation that is reflexive, symmetric, and transitive. So the equivalence relationship divides the set into separate equivalence classes.

Definition 2.20 (*equivalence relation*) Binary relation $\sim$ on a set A is said to be an equivalence relation, if for all elements a, b, c of A

(a) $a \sim a$ (reflexivity)
(b) $a \sim b \implies b \sim a$ (symmetry)
(c) $a \sim b$ and $b \sim c \implies a \sim c$ (transitivity).

Example 2.7.1 Suppose that $a \sim b$ if $a - b$ is even for $a \in \mathbb{Z}$ and $b \in \mathbb{Z}$. Show that binary relation $\sim$ on a set $\mathbb{Z}$ is an equivalence relation.

Solution Omitted. (Problem 2-25.) $\square$

Definition 2.21 (*equivalence class*) Suppose that a binary relation $\sim$ on a set A is an equivalence relation. The equivalence class of an element x in A is defined to be the set $[x] \triangleq \{a \in A \mid a \sim x\}$.

Example 2.7.2 For the equivalence relation defined in Example 2.7.1, show that $[1] = [-3]$ and $[0] = [6]$.

Solution Omitted. (Problem 2-27.)   □

Consider the following systems:

$$\Sigma_1 : \dot{x} = f(x) + g(x)u = f(x) + \sum_{i=1}^{m} u_i g_i(x) \qquad (2.82)$$

$$\Sigma_2 : \dot{z} = \tilde{f}(z) + \tilde{g}(z)u = \tilde{f}(z) + \sum_{i=1}^{m} u_i \tilde{g}_i(z) \qquad (2.83)$$

where $x \in \mathbb{R}^n$, $z \in \mathbb{R}^n$, $u \in \mathbb{R}^m$, and $f(0) = \tilde{f}(0) = 0$.

Definition 2.22 (*state equivalence of the systems*) System (2.82) is said to be state equivalent to system (2.83), if there exists a state transformation $z = S(x)$ such that system (2.82) satisfies, in z-coordinates, the state equation of system (2.83). In other words, for all $u(\in \mathbb{R}^m)$

$$S_*\left(f(x) + g(x)u\right) = \tilde{f}(z) + \tilde{g}(z)u.$$

Example 2.7.3 Show that the relation of Definition 2.22 is equivalence relation.

Solution We need to prove that the conditions of Definition 2.20 are satisfied.

(a) Reflexivity is obviously satisfied with $S(x) = x$.
(b) Suppose that $\Sigma_1 \sim \Sigma_2$. Then there exists a state transformation $z = S(x)$ such that $S_*(f(x) + g(x)u) = \tilde{f}(z) + \tilde{g}(z)u$. Since $S_*^{-1}\left(\tilde{f}(z) + \tilde{g}(z)u\right) = f(x) + g(x)u$, it is clear that $\Sigma_2 \sim \Sigma_1$.
(c) Suppose that $\Sigma_1 \sim \Sigma_2$ and $\Sigma_2 \sim \Sigma_3$, where

$$\Sigma_3 : \dot{\xi} = \bar{f}(\xi) + \bar{g}(\xi)u.$$

Then there exist state transformations $z = S^1(x)$ and $\xi = S^2(z)$ such that $S_*^1(f(x) + g(x)u) = \tilde{f}(z) + \tilde{g}(z)u$ and $S_*^2(\tilde{f}(z) + \tilde{g}(z)u) = \bar{f}(\xi) + \bar{g}(\xi)u$. Since $(S^2 \circ S^1)_*(f(x) + g(x)u) = S_*^2 \circ S_*^1(f(x) + g(x)u) = \bar{f}(\xi) + \bar{g}(\xi)u$ and $\xi = S^2 \circ S^1(x)$ is a state transformation, it is clear that $\Sigma_1 \sim \Sigma_3$.   □

By Example 2.7.3, the binary relationship of Definition 2.22 can be called the state equivalence.

Example 2.7.4 Show that if system Σ_1 and system Σ_2 are state equivalent, then the eigenvalues of $\left.\frac{\partial f(x)}{\partial x}\right|_{x=0}$ are the same as the eigenvalues of $\left.\frac{\partial \tilde{f}(z)}{\partial z}\right|_{z=0}$.

Solution Since

$$\tilde{f}(z) = S_*(f(x))(z) = \left.\frac{\partial S(x)}{\partial x}f(x)\right|_{x=S^{-1}(z)}$$

we have, by chain rule, that

$$\frac{\partial \tilde{f}(z)}{\partial z} = \frac{\partial}{\partial x} \left(\frac{\partial S(x)}{\partial x} f(x) \right) \Big|_{x=S^{-1}(z)} \frac{\partial S^{-1}(z)}{\partial z}$$

and

$$\frac{\partial \tilde{f}(z)}{\partial z} \Big|_{z=0} = \frac{\partial}{\partial x} \left(\frac{\partial S(x)}{\partial x} f(x) \right) \Big|_{x=0} \frac{\partial S^{-1}(z)}{\partial z} \Big|_{z=0}.$$

If we let the ith row of $\frac{\partial S(x)}{\partial x}$ by $A_i(x)$, then it is easy, by (2.4), to see that

$$\frac{\partial}{\partial x} \left(\frac{\partial S(x)}{\partial x} f(x) \right) \Big|_{x=0} = \begin{bmatrix} f(0)^{\mathsf{T}} \frac{\partial A_1(x)^{\mathsf{T}}}{\partial x} \\ \vdots \\ f(0)^{\mathsf{T}} \frac{\partial A_n(x)^{\mathsf{T}}}{\partial x} \end{bmatrix}_{x=0} + \frac{\partial S(x)}{\partial x} \Big|_{x=0} \frac{\partial f(x)}{\partial x} \Big|_{x=0}$$

$$= \frac{\partial S(x)}{\partial x} \Big|_{x=0} \frac{\partial f(x)}{\partial x} \Big|_{x=0}.$$

Therefore, we have

$$\frac{\partial \tilde{f}(z)}{\partial z} \Big|_{z=0} = \frac{\partial S(x)}{\partial x} \Big|_{x=0} \frac{\partial f(x)}{\partial x} \Big|_{x=0} \frac{\partial S^{-1}(z)}{\partial z} \Big|_{z=0}$$

$$= P \frac{\partial f(x)}{\partial x} \Big|_{x=0} P^{-1}$$

where $P = \frac{\partial S(x)}{\partial x} \Big|_{x=0}$. Since $\frac{\partial f(x)}{\partial x} \Big|_{x=0}$ and $\frac{\partial \tilde{f}(z)}{\partial z} \Big|_{z=0}$ are similar, it is clear that the eigenvalues of them are the same. □

Example 2.7.5 Use Example 2.7.4 to show that system (1.2) is not state equivalent to the following system:

$$\begin{bmatrix} \dot{z}_1 \\ \dot{z}_2 \end{bmatrix} = \begin{bmatrix} 0 & 0 \\ 1 & 1 \end{bmatrix} \begin{bmatrix} z_1 \\ z_2 \end{bmatrix} + \begin{bmatrix} 1 \\ 1 \end{bmatrix} u$$

Solution Omitted. (See Problem 2-28.) □

Example 2.7.6 Show that the following two systems are state equivalent with $z = S(x) = \begin{bmatrix} x_2 + \frac{1}{2}x_1^2 \\ -x_1 \end{bmatrix}$:

$$\begin{bmatrix} \dot{x}_1 \\ \dot{x}_2 \end{bmatrix} = \begin{bmatrix} x_2 + \frac{1}{2}x_1^2 \\ -x_1x_2 - \frac{1}{2}x_1^3 \end{bmatrix} + \begin{bmatrix} 0 \\ 1 \end{bmatrix} v \tag{2.84}$$

$$\begin{bmatrix} \dot{z}_1 \\ \dot{z}_2 \end{bmatrix} = \begin{bmatrix} 0 & 0 \\ -1 & 0 \end{bmatrix} \begin{bmatrix} z_1 \\ z_2 \end{bmatrix} + \begin{bmatrix} 1 \\ 0 \end{bmatrix} v \tag{2.85}$$

Solution It is clear that $x = S^{-1}(z) = \begin{bmatrix} -z_2 \\ z_1 - \frac{1}{2}z_2^2 \end{bmatrix}$ and

$$S_* \left(\begin{bmatrix} x_2 + \frac{1}{2}x_1^2 \\ -x_1x_2 - \frac{1}{2}x_1^3 + v \end{bmatrix} \right) = \begin{bmatrix} x_1 & 1 \\ -1 & 0 \end{bmatrix} \begin{bmatrix} x_2 + \frac{1}{2}x_1^2 \\ -x_1x_2 - \frac{1}{2}x_1^3 + v \end{bmatrix} \Bigg|_{x=S^{-1}(z)}$$

$$= \begin{bmatrix} v \\ -x_2 - \frac{1}{2}x_1^2 \end{bmatrix} \Bigg|_{x=S^{-1}(z)} = \begin{bmatrix} v \\ -z_1 \end{bmatrix}.$$

Therefore, system (2.84) is state equivalent to system (2.85) via $z = S(x) = \begin{bmatrix} x_2 + \frac{1}{2}x_1^2 \\ -x_1 \end{bmatrix}$. $\qquad\square$

Consider the following systems:

$$\Sigma_1 : \dot{x} = f(x) + g(x)u = f(x) + \sum_{i=1}^{m} u_i g_i(x) \tag{2.86}$$

and

$$\Sigma_2 : \dot{z} = \tilde{f}(z) + \tilde{g}(z)v = \tilde{f}(z) + \sum_{i=1}^{m} v_i \tilde{g}_i(z) \tag{2.87}$$

where $x \in \mathbb{R}^n$, $z \in \mathbb{R}^n$, and $f(0) = \tilde{f}(0) = 0$.

Definition 2.23 (*feedback equivalence of the systems*) System (2.86) is said to be feedback equivalent to system (2.87), if there exist a nonsingular feedback $u = \alpha(x) + \beta(x)v$ and a state transformation $z = S(x)$ such that the closed-loop system of (2.86) satisfies, in z-coordinates, the state equation of system (2.87) or

$$S_* (f(x) + g(x)\alpha(x) + g(x)\beta(x)v) = \tilde{f}(z) + \tilde{g}(z)v.$$

In other words, system (2.86) is said to be feedback equivalent to system (2.87) if there exists a nonsingular feedback $u = \alpha(x) + \beta(x)v$ such that the closed-loop system of (2.86) is state equivalent to system (2.87).

Example 2.7.7 Show that the relation of Definition 2.23 is equivalence relation.

Solution We need to prove that the conditions of Definition 2.20 are satisfied.

(a) Reflexivity is obviously satisfied with state transformation $z = S(x) = x$ and feedback $u = v$.

(b) Suppose that $\Sigma_1 \sim \Sigma_2$. Then there exist a state transformation $z = S(x)$ and feedback $u = \alpha(x) + \beta(x)v$ such that $S_*\left(f(x) + g(x)\alpha(x) + g(x)\beta(x)v\right) = \tilde{f}(z) + \tilde{g}(z)v$. Since $\tilde{g}(z) = S_*(g(x)\beta(x)) = S_*(g(x))\left(\beta \circ S^{-1}(z)\right)$, we have that $\tilde{g}(z)\left(\beta \circ S^{-1}(z)\right)^{-1} = S_*(g(x))$ and $S_*^{-1}\left(\tilde{g}(z)\left(\beta \circ S^{-1}(z)\right)^{-1}\right) = g(x)$. Since $\tilde{f}(z) = S_*(f(x) + g(x)\alpha(x)) = S_*(f(x)) + S_*(g(x))\left(\alpha \circ S^{-1}(z)\right)$, it is easy to see that

$$S_*(f(x)) = \tilde{f}(z) - S_*(g(x))\left(\alpha \circ S^{-1}(z)\right)$$
$$= \tilde{f}(z) - \tilde{g}(z)\left(\beta \circ S^{-1}(z)\right)^{-1}\left(\alpha \circ S^{-1}(z)\right)$$

Therefore, it is clear that

$$S_*(f(x) + g(x)u) = \tilde{f}(z) - \tilde{g}(z)\left(\beta \circ S^{-1}(z)\right)^{-1}\left(\alpha \circ S^{-1}(z)\right)$$
$$+ \tilde{g}(z)\left(\beta \circ S^{-1}(z)\right)^{-1} u$$

or

$$S_*^{-1}\left(\tilde{f} + \tilde{g}\left(\beta \circ S^{-1}(z)\right)^{-1}\left(-\alpha \circ S^{-1}(z) + u\right)\right) = f(x) + g(x)u.$$

Hence, $\Sigma_2 \sim \Sigma_1$ with state transformation $x = S^{-1}(z)$ and feedback $v = \left(\beta \circ S^{-1}(z)\right)^{-1}\left(-\alpha \circ S^{-1}(z) + u\right)$.

(c) Suppose that $\Sigma_1 \sim \Sigma_2$ with state transformations $z = S^1(x)$ and feedback $u = \alpha_1(x) + \beta_1(x)v$ and $\Sigma_2 \sim \Sigma_3$ with state transformations $\xi = S^2(z)$ and feedback $v = \alpha_2(z) + \beta_2(z)w$, where

$$\Sigma_3 : \dot{\xi} = \bar{f}(\xi) + \bar{g}(\xi)w.$$

In other words,

$$S_*^1\left(f(x) + g(x)\alpha_1(x) + g(x)\beta_1(x)v\right) = \tilde{f}(z) + \tilde{g}(z)u$$

and

$$S_*^2\left(\tilde{f}(z) + \tilde{g}(z)\alpha_2(z) + \tilde{g}(z)\beta_2(z)w\right) = \bar{f}(\xi) + \bar{g}(\xi)w.$$

Thus, it is easy to see that

$$(S^2 \circ S^1)_* \left(f(x) + g(x)\{\alpha_1(x) + \beta_1(x)(\alpha_2 \circ S_1(x) + \beta_2 \circ S_1(x)w)\} \right)$$
$$= S_*^2 \left(S_*^1 \left(\{f(x) + g(x)\alpha_1(x)\} + g(x)\beta_1(x) \{\alpha_2 \circ S_1(x) + \beta_2 \circ S_1(x)w\} \right) \right)$$
$$= S_*^2 \left(\tilde{f}(z) + \tilde{g}(z)\{\alpha_2(z) + \beta_2(z)w\} \right)$$
$$= \bar{f}(\xi) + \bar{g}(\xi)w$$

Since $\xi = S^2 \circ S^1(x)$ is a state transformation, it is clear that $\Sigma_1 \sim \Sigma_3$ with state transformations $\xi = S^2 \circ S^1(x)$ and feedback

$$u = \alpha_1(x) + \beta_1(x) \{\alpha_2 \circ S_1(x) + \beta_2 \circ S_1(x)w\}.$$

$\square$

By Example 2.7.7, the binary relationship of Definition 2.23 can be called the feedback equivalence.

Definition 2.24 (*feedback equivalent to a linear system*) System (2.86) is said to be feedback equivalent to a linear system, if there exist a nonsingular feedback $u = \alpha(x) + \beta(x)v$ and a state transformation $z = S(x)$ such that for all $v(\in \mathbb{R}^m)$,

$$S_* \left(f(x) + g(x)\alpha(x) + g(x)\beta(x)v \right) = Az + Bv$$

for some constant $n \times n$ matrix A and $n \times m$ matrix B.

The feedback may change the eigenvalues of $\left. \frac{\partial f(x)}{\partial x} \right|_{x=0}$. Thus, the eigenvalues of $\left. \frac{\partial f(x)}{\partial x} \right|_{x=0}$ and $\left. \frac{\partial \tilde{f}(z)}{\partial z} \right|_{z=0}$ may not be the same when system Σ_1 and system Σ_2 are feedback equivalent.

Example 2.7.8 Suppose that system Σ_1 and system Σ_2 are feedback equivalent. Show that for $k \geq 1$,

$$\text{rank} \left(\left[\tilde{g}(0) \quad \left. \frac{\partial \tilde{f}(z)}{\partial z} \right|_{z=0} \tilde{g}(0) \cdots \left(\left. \frac{\partial \tilde{f}(z)}{\partial z} \right|_{z=0} \right)^{k-1} \tilde{g}(0) \right] \right)$$
$$= \text{rank} \left(\left[g(0) \quad \left. \frac{\partial f(x)}{\partial x} \right|_{x=0} g(0) \cdots \left(\left. \frac{\partial f(x)}{\partial x} \right|_{x=0} \right)^{k-1} g(0) \right] \right).$$

Solution Omitted. (See Problem 2-29.) $\square$

Example 2.7.9 Show that a single input controllable linear system is feedback equivalent to the following Brunovsky canonical form:

$$\dot{z} = \begin{bmatrix} 0 & 1 & 0 & \cdots & 0 & 0 \\ 0 & 0 & 1 & \cdots & 0 & 0 \\ \vdots & \vdots & \vdots & & \vdots & \vdots \\ 0 & 0 & 0 & \cdots & 0 & 1 \\ 0 & 0 & 0 & \cdots & 0 & 0 \end{bmatrix} z + \begin{bmatrix} 0 \\ 0 \\ \vdots \\ 0 \\ 1 \end{bmatrix} v = A_0 z + b_0 v \qquad (2.88)$$

Solution Consider the following single input controllable linear system:

$$\dot{\zeta} = A\zeta + bw \qquad (2.89)$$

where rank $\left(\begin{bmatrix} b & Ab & \cdots & A^{n-1}b \end{bmatrix} \right) = n$ and

$$A^n b = \sum_{i=1}^{n} a_i A^{i-1} b.$$

Let $z = P^{-1}\zeta$, where

$$P = \begin{bmatrix} b & Ab & \cdots & A^{n-1}b \end{bmatrix} \begin{bmatrix} -a_2 & -a_3 & \cdots & -a_n & 1 \\ -a_3 & -a_4 & \cdots & 1 & 0 \\ \vdots & \vdots & & \vdots & \vdots \\ -a_{n-1} & -a_n & \cdots & 0 & 0 \\ -a_n & 1 & \cdots & 0 & 0 \\ 1 & 0 & \cdots & 0 & 0 \end{bmatrix}.$$

Then it is easy to see that

$$\dot{z} = P^{-1}APz + P^{-1}bw$$

$$= \begin{bmatrix} 0 & 1 & 0 & \cdots & 0 & 0 \\ 0 & 0 & 1 & \cdots & 0 & 0 \\ \vdots & \vdots & \vdots & & \vdots & \vdots \\ 0 & 0 & 0 & \cdots & 0 & 1 \\ a_1 & a_2 & a_3 & \cdots & a_{n-1} & a_n \end{bmatrix} z + \begin{bmatrix} 0 \\ 0 \\ \vdots \\ 0 \\ 1 \end{bmatrix} w = \bar{A}z + b_0 w. \qquad (2.90)$$

Therefore, system (2.89) is feedback equivalent to system (2.90) with state transformation $z = P^{-1}\zeta$ and nonsingular feedback $w = w$. Also, system (2.90) is feedback equivalent to system (2.88) with state transformation $z = z$ and nonsingular feedback $w = - \begin{bmatrix} a_1 & a_2 & \cdots & a_n \end{bmatrix} z + v$. Hence, system (2.89) is feedback equivalent to system (2.88) with state transformation $z = P^{-1}\zeta$ and nonsingular feedback $w = - \begin{bmatrix} a_1 & a_2 & \cdots & a_n \end{bmatrix} P^{-1}\zeta + v$. $\qquad \square$

By Example 2.7.9, it is clear that if a single input system is feedback equivalent to a controllable linear system, then it is feedback equivalent to the Brunovsky canonical form. It is also true for the multi-input system.

Example 2.7.10 Show that a multi-input controllable linear system is feedback equivalent to the following Brunovsky canonical form:

$$
\dot{z} = \begin{bmatrix} A_{11} & O & \cdots & O \\ O & A_{22} & \cdots & O \\ \vdots & \vdots & \ddots & \vdots \\ O & O & \cdots & A_{mm} \end{bmatrix} z + \begin{bmatrix} B_{11} & O & \cdots & O \\ O & B_{22} & \cdots & O \\ \vdots & \vdots & \ddots & \vdots \\ O & O & \cdots & B_{mm} \end{bmatrix} v
$$

$$
= Az + Bv
$$

where $\sum_{i=1}^{m} \kappa_i = n$ and

$$
A_{ii} = \begin{bmatrix} 0 & 1 & 0 & \cdots & 0 & 0 \\ 0 & 0 & 1 & \cdots & 0 & 0 \\ \vdots & \vdots & \vdots & & \vdots & \vdots \\ 0 & 0 & 0 & \cdots & 0 & 1 \\ 0 & 0 & 0 & \cdots & 0 & 0 \end{bmatrix} (\kappa_i \times \kappa_i) ; \quad B_{ii} = \begin{bmatrix} 0 \\ 0 \\ \vdots \\ 0 \\ 1 \end{bmatrix} (\kappa_i \times 1).
$$

Solution Omitted. (Problem 2-31.) ☐

2.8 MATLAB Programs

In this section, the following subfunctions in Appendix C are needed:
    adfg, ChZero, ChCommute, ChExact, ChInverseF,
    Codi, Lfh, sstarmap

The following MATLAB program is for Examples 2.4.3, 2.4.7, 2.4.10, and 2.4.11.

```
clear all
syms x1 x2 x3 x4 x5 x6 x7 x8 x9 real
syms z1 z2 z3 z4 z5 z6 z7 z8 z9 real

EX=2403
% EX=2407
% EX=2410
% EX=2411

if EX==2403
  f=[x2; 1]; g=[1; x1]; h=x1*x2
  n=size(f,1); x=sym('x',[n,1]);
  out1=adfg(f,g,x)
  out2=Lfh(f,h,x)
end
```

```
if EX==2407
  S=[x1; (1+x1)*x2+2*x3]
  f=[1; 0; x2]; g=[0; 1; x1]
  p0=[1; 1; 0]; p1=[1; 0; 1]
  n=size(f,1); x=sym('x',[n,1]);
  dS=jacobian(S,x)
  Sf=dS*f
  Sg=dS*g
  ans1=subs(Sf,x,p0)
  ans2=subs(Sf,x,p1)
  ans3=subs(Sg,x,p0)
  ans4=subs(Sg,x,p1)
end

if EX==2410
  S=[z1+z2^2; z2]
  iS=[x1-x2^2; x2]
  fz=[0; z1]; gz=[1; 1]
  n=size(fz,1); x=sym('x',[n,1]); z=sym('z',[n,1]);
  dS=jacobian(S,z)
  Tfz=dS*fz
  fx=subs(Tfz,z,iS)
  Tgz=dS*gz
  gx=subs(Tgz,z,iS)
end

if EX==2411
  S=[x1+x2^2; x2]; iS=[z1-z2^2; z2]
  fx=[x2; 1]; gx=[1; x1]; h=x1*x2;
  n=size(fx,1); x=sym('x',[n,1]); z=sym('z',[n,1]);
  hz=subs(h,x,iS)
  dS=jacobian(S,x);
  Tfx=dS*fx;
  fz=subs(Tfx,x,iS)
  Tgx=dS*gx;
  gz=subs(Tgx,x,iS)
  adfzgz=adfg(fz,gz,z)
  adfxgx=adfg(fx,gx,x)
  Sadfxgx=subs(dS*adfg(fx,gx,x),x,iS)
  Lfzhz=Lfh(fz,hz,z)
  LfxhxiS=subs(Lfh(fx,h,x),x,iS)
end

return
```

The following MATLAB program is for Examples 2.6.1, 2.6.5, 2.6.7, and 2.6.8.

```
clear all
syms x1 x2 x3 x4 x5 x6 x7 x8 x9 real
syms z1 z2 z3 z4 z5 z6 z7 z8 z9 real

EX=2601
% EX=2605
% EX=2607
% EX=2608

if EX==2601
  f=[x2; 1]; tau=[1; 0];
  n=size(f,1); x=sym('x',[n,1]);
  idS=[f tau]
  if ChCommute(idS,x)==0
    return
  end
  dS=simplify(inv(idS))
  S=Codi(dS,x)
end

if or(EX==2605,EX==2607)
  g1=[1; 0; 0]; g2=[0; 1; x1];
  f1=[1; x1*x2; x1*x3]; f2=[0; 1; 1];
  n=length(g1); x=sym('x',[n,1]);
  if EX==265
    D=[g1 g2]
  else
    D=[f1 f2]
  end
  T12=adfg(D(:,1),D(:,2),x)
  if rank([T12 D]) > rank(D)
    display('NOT Involutive.')
    return
  end
  display('Involutive.')
end

if EX==2608
  f1=[1; x1*x2; x1*x3]; f2=[0; 1; 1];
  n=length(f1); x=sym('x',[n,1]); z=sym('z',[n,1]);
  D=[f1 f2]
  e3=[0; 0; 1]
  bD=[D e3]
  Tomega=e3'*inv(bD)
  if ChExact(Tomega,x)==0
    display('Find out a(x) without MATLAB.')
    ax=exp(-x1^2/2)
  else
    ax=1
  end
  omega=ax*Tomega
```

```
  if ChExact(omega,x)==0
    display('a(x) is not correct.')
    return
  end
  S3=Codi(omega,x)
  S=[x1; x2; S3]
  iS=[x1; x2; x3*exp(x1^2/2)+x2]
  if ChInverseF(S,iS,x)==0
    display('Inverse function iS is not correct.')
    return
  end
  Tf1z=sstarmap(S,iS,f1,x)
  f1z=subs(Tf1z,x,z)
  Tf2z=sstarmap(S,iS,f2,x)
  f2z=subs(Tf2z,x,z)
  Dz=[f1z f2z]
  Dz2=[[1; 0; 0] [0; 1; 0]]
  r1=rank(Dz)
  r2=rank(Dz2)
  r3=rank([Dz Dz2])
end

return
```

The following MATLAB program is for Problem 2-12, 16, 17, 19, and 21.

```
clear all
syms x1 x2 x3 x4 x5 x6 x7 x8 x9 real
syms z1 z2 z3 z4 z5 z6 z7 z8 z9 real

EX=2912
% EX=2916
% EX=2917
% EX=2919
% EX=2921

if EX==2912
  x=sym('x',[2,1]); z=sym('z',[2,1]);
  S=[x1; x2+x1^2]; iS=[z1; z2-z1^2]
  f1=[1; 0]; f2=[x2; 1]
  Tf1x=jacobian(S,x)*f1
  f1z=subs(Tf1x,x,iS)
  Tf2x=jacobian(S,x)*f2
  f2z=subs(Tf2x,x,iS)
end

if EX==2916
  x=sym('x',[2,1]); z=sym('z',[2,1]);
  f=[2*x1*x2-2*x2^3; x1-x2^2]
  g=[1+2*x2; 1]
```

```
   X2=adfg(f,g,x)
   ANSa=adfg(g,X2,x)
   idS=[g X2]
   dS=inv(idS)
   S=Codi(dS,x)
   Tfz=simplify(dS*f)
   Tgz=simplify(dS*g)
end

if EX==2917
   x=sym('x',[3,1]); z=sym('z',[3,1]);
   f=[-2*x2*(x1+x2+x2^2); x1+x2+x2^2; -2*x2*(x1+x2+x2^2)]
   g=[1 0; 0 0; 0 1]
   X2=adfg(f,g(:,1),x)
   idS=[g(:,1) X2 g(:,2)]
   r1=rank(idS)
   ANSa=ChCommute(idS,x)
   dS=inv(idS)
   S=Codi(dS,x)
   Tfz=simplify(dS*f)
   Tgz=simplify(dS*g)
end

if EX==2919
   syms a real
   x=sym('x',[3,1]); z=sym('z',[3,1]);

   Sab=[x2-x1^2; x1]; dSab=jacobian(Sab,x)
   kerSab=[0; 0; 1]
   fa=[1; 0; 0]
   Ta=adfg(fa,kerSab,x)
   if rank([Ta kerSab])>rank(kerSab)
     display('S_*(fa) is NOT a well-defined vector field.')
   end
   Tfax=dSab*fa
   iSab=[z2; z1+z2^2; a]
   faz=subs(Tfax,x,iSab)

   fb=[0; 0; 1]
   Tb=adfg(fb,kerSab,x)
   if rank([Tb kerSab])>rank(kerSab)
     display('S_*(fb) is NOT a well-defined vector field.')
   end
   Tfbx=dSab*fb
   fbz=subs(Tfbx,x,iSab)

   Sc=[x2-x1^2; x3*(1+x1)]; dSc=jacobian(Sc,x)
   kerSc=[1; 2*x1; -x3/(1+x1)]; fc=[0; 0; 1]
   Tc=adfg(fc,kerSc,x)
```

```
  if rank([Tc kerSc])>rank(kerSc)
    display('S_*(fc) is NOT a well-defined vector field.')
  end

  Sde=[x2-x1*(x2^2+x3); x2^2+x3]; dSde=jacobian(Sde,x)
  kerSde=[1; x2^2+x3; -2*x2*(x2^2+x3)];
  fd=[1; 0; 0]
  Td=adfg(fd,kerSde,x)
  if rank([Td kerSde])>rank(kerSde)
    display('S_*(fd) is NOT a well-defined vector field.')
  end
  Tfdx=dSde*fd
  iSde=[a; z1+a*z2; z2-(z1+a*z2)^2]
  fbz=subs(Tfdx,x,iSde)

  fe=[0; 1; 0]
  Te=adfg(fe,kerSde,x)
  if rank([Te kerSde])>rank(kerSde)
    display('S_*(fe) is NOT a well-defined vector field.')
  end
end

if EX==2921
  omegaA=[1 -2*x2]
  x=sym('x',[2,1]);
  Ta=jacobian(omegaA',x)
  ha=Codi(omegaA,x)
  omegaB=[x2 x1 x3 1]
  x=sym('x',[4,1]);
  Tb=jacobian(omegaB',x)
  hb=Codi(omegaB,x)
end

return
```

2.9 Problems

2-1. Solve Example 2.1.1.
2-2. Solve Example 2.1.2.
2-3. Solve Example 2.1.3.
2-4. By using Example 2.1.3, solve Example 2.1.1(c).
2-5. By using Example 2.1.3, solve Example 2.1.4.

2-6. Prove that (2.15) and (2.16) are satisfied.

2-7. Prove the following:

(a) $L_{f+gu}h(x) = L_f h(x) + L_g h(x)\, u$

(b) $L^2_{f+gu}h(x) = L^2_f h(x) + (L_g L_f h(x) + L_f L_g h(x))u + L^2_g h(x)\, u^2$

2-8. Consider the following nonlinear control system:

$$\dot{x}(t) = f(x(t)) + g(x(t))u(t), \quad u \in \mathbb{R}$$
$$y(t) = h(x(t))$$

(a) Show that

$$y^{(2)}(t) \triangleq \frac{d^2 y(t)}{dt^2} = L^2_f h + (L_g L_f h + L_f L_g h)u + L^2_g h\, u^2 + L_g h\, \dot{u}(t)$$

(b) Find out $y^{(3)}(t)$.

(c) Define natural number ρ by $L_g L^\ell_f h(x) = 0, \ell \le \rho - 2$ and $L_g L^{\rho-1}_f h(x) \ne 0$. Show that

$$y^{(i)}(t) = L^i_f h(x), \quad 0 \le i \le \rho - 1$$
$$y^{(\rho)}(t) = L^\rho_f h(x) + L_g L^{\rho-1}_f h(x)\, u.$$

2-9. Solve Example 2.4.1.

2-10. Find $f(x), g(x), h_1(x)$, and $h_2(x)$ such that

$$L_f L_g (h_1(x)h_2(x)) \ne h_2(x)L_f L_g(h_1(x)) + h_1(x)L_f L_g(h_2(x)).$$

In other words, $L_f L_g(\cdot)$ does not satisfy Leibniz rule in Example 2.4.4(b).

2-11. Use (2.4) and (2.12) to show that (2.18) is satisfied.

2-12. Define state transformation $z = S(x)$ by

$$z = \begin{bmatrix} z_1 \\ z_2 \end{bmatrix} = S(x) = \begin{bmatrix} x_1 \\ x_2 + x_1^2 \end{bmatrix}.$$

Find out $S_*(\frac{\partial}{\partial x_1})$ and $S_*\left(x_2\frac{\partial}{\partial x_1} + \frac{\partial}{\partial x_2}\right)$.

2-13. Let

$$
\hat{f}(\xi) = \begin{bmatrix} A_1\xi^1 \\ \hat{\Phi}(\xi^1, \xi^2) \end{bmatrix}, \quad \hat{g}(\xi) = \begin{bmatrix} b_1 \\ \hat{\Psi}(\xi^1, \xi^2) \end{bmatrix}, \quad \xi = \begin{bmatrix} \xi^1 \\ \xi^2 \end{bmatrix}
$$

$$
A_1 = \begin{bmatrix} 0 & 1 & 0 & \cdots & 0 & 0 \\ 0 & 0 & 1 & \cdots & 0 & 0 \\ \vdots & \vdots & \vdots & & \vdots & \vdots \\ 0 & 0 & 0 & \cdots & 0 & 1 \\ 0 & 0 & 0 & \cdots & 0 & 0 \end{bmatrix} (\rho \times \rho), \quad b_1 = \begin{bmatrix} 0 \\ 0 \\ \vdots \\ 0 \\ 1 \end{bmatrix} (\rho \times 1)
$$

$$
\xi^1 = \begin{bmatrix} \xi^1_1 \\ \vdots \\ \xi^1_\rho \end{bmatrix}, \quad \xi^2 = \begin{bmatrix} \xi^2_1 \\ \vdots \\ \xi^2_{n-\rho} \end{bmatrix}.
$$

Show that for $k \geq 0$

$$
\mathrm{ad}^k_{\hat{f}}\hat{g}(\xi) = \begin{bmatrix} (-1)^k A_1^k b_1 \\ * \end{bmatrix}
$$

or

$$
\mathrm{ad}^k_{\hat{f}}\hat{g}(\xi) \equiv \begin{bmatrix} (-1)^k A_1^k b_1 \\ O_{(n-\rho)\times 1} \end{bmatrix} \quad \mathrm{mod} \ \mathrm{span} \left\{ \begin{bmatrix} O_{\rho\times 1} \\ 1 \\ 0 \\ \vdots \\ 0 \end{bmatrix}, \ldots, \begin{bmatrix} O_{\rho\times 1} \\ 0 \\ \vdots \\ 0 \\ 1 \end{bmatrix} \right\}.
$$

2-14. Suppose that $\{g(x), \mathrm{ad}_f g(x), \ldots, \mathrm{ad}^{n-1}_f g(x)\}$ is a set of linearly independent vector fields on a neighborhood of $0 \in \mathbb{R}^n$. Let $z = S(x)$ be a state transformation such that

$$
S_* \left(\mathrm{ad}^{i-1}_f g(x) \right) = \frac{\partial}{\partial z_i}.
$$

Show that if for $1 \leq i \leq n$ and $1 \leq j \leq n$

$$
\left[\mathrm{ad}^{i-1}_f g(x), \mathrm{ad}^{j-1}_f g(x) \right] = 0
$$

then

$$
S_* \left(\mathrm{ad}^n_f g(x) \right) = \sum_{i=1}^n a_i(z_n) S_* \left(\mathrm{ad}^{i-1}_f g(x) \right) \tag{2.91}
$$

for some scalar functions $a_i(z_n)$, $1 \leq i \leq n$.

2-15. Solve Example 2.5.1.

2-16. Consider the following nonlinear control system:

$$\begin{bmatrix} \dot{x}_1 \\ \dot{x}_2 \end{bmatrix} = \begin{bmatrix} 2x_1 x_2 - 2x_2^3 \\ x_1 - x_2^2 \end{bmatrix} + \begin{bmatrix} 1 + 2x_2 \\ 1 \end{bmatrix} u = f(x) + g(x)u.$$

(a) Show that $[g(x), \text{ad}_f g(x)] = 0$.

(b) Find out state coordinates change $z = S(x)$ such that $S_*(g(x)) = \frac{\partial}{\partial z_1}$ and $S_* \left(\text{ad}_f g(x) \right) = \frac{\partial}{\partial z_2}$.

(c) Find out the state equation that the new state z in (b) satisfies.

2-17. Consider the following nonlinear control system:

$$\begin{bmatrix} \dot{x}_1 \\ \dot{x}_2 \\ \dot{x}_3 \end{bmatrix} = \begin{bmatrix} -2x_2(x_1 + x_2 + x_2^2) \\ x_1 + x_2 + x_2^2 \\ -2x_2(x_1 + x_2 + x_2^2) \end{bmatrix} + \begin{bmatrix} 1 & 0 \\ 0 & 0 \\ 0 & 1 \end{bmatrix} \begin{bmatrix} u_1 \\ u_2 \end{bmatrix}$$

$$= f(x) + g_1(x)u_1 + g_2(x)u_2$$

(a) Let $X_1(x) = g_1(x)$, $X_2(x) = \text{ad}_f g_1(x)$, and $X_3(x) = g_2(x)$. Show that $\{X_1(x), X_2(x), X_3(x)\}$ satisfies (2.61) of Theorem 2.7.

(b) Find out state coordinates change $z = S(x)$ such that $S_* (X_i(x)) = \frac{\partial}{\partial z_i}$, $1 \leq i \leq 3$.

(c) Find out the state equation for the new state z in (b).

2-18. Suppose that $z = S(x)$ is a state coordinates change. By using that $\frac{\partial S(x)}{\partial x}$ is a nonsingular matrix, show that if $\{f_1(x), \ldots, f_k(x)\}$ is a set of linearly independent vector fields, then $\{S_*(f_1(x)), \ldots, S_*(f_k(x))\}$ is also a set of linearly independent vector fields.

2-19. Consider the following smooth functions $z = S(x) : \mathbb{R}^3 \to \mathbb{R}^2$. Use Theorem 2.6 to determine whether $S_*(f(x))$ is a well-defined vector field on a neighborhood of $0 \in \mathbb{R}^2$ or not. If it is a well-defined vector field, then find $S_*(f(x))$.

(a) $S(x) = \begin{bmatrix} x_2 - x_1^2 \\ x_1 \end{bmatrix}$, $f(x) = \frac{\partial}{\partial x_1} = \begin{bmatrix} 1 \\ 0 \\ 0 \end{bmatrix}$

(b) $S(x) = \begin{bmatrix} x_2 - x_1^2 \\ x_1 \end{bmatrix}$, $f(x) = \frac{\partial}{\partial x_3}$

(c) $S(x) = \begin{bmatrix} x_2 - x_1^2 \\ x_3 + x_1 x_3 \end{bmatrix}$, $f(x) = \frac{\partial}{\partial x_3}$

(d) $S(x) = \begin{bmatrix} x_2 - x_1(x_2^2 + x_3) \\ x_2^2 + x_3 \end{bmatrix}$, $f(x) = \frac{\partial}{\partial x_1}$

(e) $S(x) = \begin{bmatrix} x_2 - x_1(x_2^2 + x_3) \\ x_2^2 + x_3 \end{bmatrix}$, $f(x) = \frac{\partial}{\partial x_2}$

2-20. Suppose that

$$\tilde{f}(z) = S_*(f(x)) \text{ and } \tilde{h}(z) = h \circ S^{-1}(z)$$

where $z = S(x)$ is a state transformation.

(a) Show that for $k \geq 0$,

$$L_{\tilde{f}}^k \tilde{h}(z) = L_f^k h(x)\big|_{x = S^{-1}(z)}$$

(b) Show that if $\tilde{f}(z) = Az$ and $\tilde{h}(z) = cz$, then

$$L_f^k h(x) = cA^k S(x), \quad k \geq 0.$$

2-21. Show that one form $w(x)$ is exact and find the scalar function $h(x)$ such that $dh(x) = w(x)$ and $h(0) = 0$.

(a) $w(x) = \begin{bmatrix} 1 & -2x_2 \end{bmatrix}$
(b) $w(x) = \begin{bmatrix} x_2 & x_1 & x_3 & 1 \end{bmatrix}$.

2-22. Solve Example 2.6.6 by using (2.43).
2-23. Find out annihilator $D(x)^{\perp}$ of distribution

$$D(x) = \text{span} \left\{ x_3 \frac{\partial}{\partial x_1} + \frac{\partial}{\partial x_3}, \frac{\partial}{\partial x_2} + \frac{\partial}{\partial x_3} \right\} = \text{span} \left\{ \begin{bmatrix} x_3 \\ 0 \\ 1 \end{bmatrix}, \begin{bmatrix} 0 \\ 1 \\ 1 \end{bmatrix} \right\}.$$

2-24. Solve Example 2.6.11.
2-25. Solve Example 2.7.1.
2-26. Suppose that $a \sim b$ if $a - b$ is odd for $a \in \mathbb{Z}$ and $b \in \mathbb{Z}$. Show that binary relation $\sim$ on $\mathbb{Z}$ is not an equivalence relation.
2-27. Solve Example 2.7.2.
2-28. Solve Example 2.7.5.
2-29. Solve Example 2.7.8.
2-30. Suppose that $\Phi_t^{f(x)}(x_0)$ is the solution of the following differential equation:

$$\dot{x} = f(x); \; x(0) = x_0.$$

In other words,

$$\frac{d}{dt}\Phi_t^{f(x)}(x_0) = f(\Phi_t^{f(x)}(x_0)); \quad \Phi_0^{f(x)}(x_0) = x_0.$$

Show that

$$L_f h(x)\big|_{x=x_0} = \frac{d}{dt}h(\Phi_t^f(x_0))\big|_{t=0}.$$

2-31. Solve Example 2.7.10.

Chapter 3
Linearization by State Transformation

3.1 Introduction

In Example 1.3.1, we have obtained a nonlinear state equation by a nonlinear state transformation from a linear state equation. Conversely, we could have a linear state equation by a nonlinear state transformation from a nonlinear state equation. It motivates the linearization problem by state transformation. Consider the following nonlinear control system:

$$\dot{x} = f(x) + g(x)u = f(x) + \sum_{i=1}^{m} u_i g_i(x) \tag{3.1}$$

where $x \in \mathbb{R}^n$, $u \in \mathbb{R}^m$, and $f(0) = 0$.

Definition 3.1 (*state equivalence to a linear system*) System (3.1) is said to be (locally) state equivalent to a linear system (or linearizable by state transformation), if there exist a neighborhood U of origin and a state transformation $z = S(x) : U \to \mathbb{R}^n$ such that system (3.1) satisfies, in z-coordinates, the following linear controllable system:

$$\dot{z} = Az + Bu, \quad z \in \mathbb{R}^n, \quad u \in \mathbb{R}^m. \tag{3.2}$$

In other words, system (3.1) is said to be state equivalent to a linear system (or linearizable by state transformation), if there exists a controllable linear system that is state equivalent to system (3.1). Thus, linearization problem by state transformation is to find the equivalence class of the set of all controllable linear systems. If $U = \mathbb{R}^n$ in the above definition, we call it the global linearization problem. Throughout the book, we consider the local linearization problems only. In almost all references, the (A, B) matrix of the linear system (3.2) is assumed to be a controllable pair, because

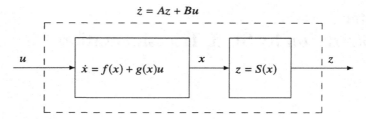

Fig. 3.1 state equivalence to a linear system

it is difficult to solve the above problem without this assumption. Therefore, in this book, the linear system (3.2) in the above definition is assumed to be controllable. The state equivalence to a linear system is shown, in Fig. 3.1, as a block diagram. If nonlinear system (3.1) is linearizable by state transformation, then system (3.1) can be controlled as easily as linear system (3.2). For example, if we want to find a control input $u(t)$ which steers the state $x(t)$ from the initial state x^0 at $t = 0$ to the final state x^f at $t = t_f$, it is enough to find a control input $u(t)$ for linear system (3.2) which steers the state $z(t)$ from the initial state $z^0 = S(x^0)$ to the final state $z^f = S(x^f)$. Also, note that $S(0) = 0$. Thus, if $A - BK$ is asymptotically stable (or Hurwitz) and feedback control $u(t) = \alpha(x) = -KS(x)$ is applied to system (3.1), then we obtain that $\lim_{t \to \infty} x(t) = 0$.

3.2 Single Input Nonlinear Systems

This section deals with the linearization problem of the following smooth single input nonlinear system:

$$\dot{x} = f(x) + g(x)u, \quad x \in \mathbb{R}^n, \ u \in \mathbb{R}. \tag{3.3}$$

We can assume, without loss of generality, that $f(0) = 0$. Let $z = S(x)$ be a state transformation. Then system (3.3) satisfies, in z-coordinates, that

$$\begin{aligned}
\dot{z} &= \frac{\partial S(x)}{\partial x}\dot{x} = \frac{\partial S(x)}{\partial x}(f(x) + g(x)u) \\
&= \left\{\frac{\partial S(x)}{\partial x}f(x)\right\}\bigg|_{x=S^{-1}(z)} + \left\{\frac{\partial S(x)}{\partial x}g(x)\right\}\bigg|_{x=S^{-1}(z)} u \\
&= S_*(f(x)) + S_*(g(x))u = \tilde{f}(z) + \tilde{g}(z)u.
\end{aligned} \tag{3.4}$$

Therefore, the linearization by state transformation is to find a state transformation $z = S(x)$ such that

$$\tilde{f}(z) = S_*(f(x)) = Az \text{ and } \tilde{g}(z) = S_*(g(x)) = b \tag{3.5}$$

where

$$\text{rank}\left(\begin{bmatrix} b & Ab & \cdots & A^{n-1}b \end{bmatrix}\right) = n. \tag{3.6}$$

Theorem 3.1 (necessary and sufficient condition) *System* (3.3) *is state equivalent to a linear system with state transformation* $z = S(x)$, *if and only if*

(i) $\text{rank}\left(\begin{bmatrix} g(x) & \text{ad}_f g(x) & \cdots & \text{ad}_f^{n-1} g(x) \end{bmatrix}\Big|_{x=0}\right) = n$

(ii) $\begin{bmatrix} \text{ad}_f^{i-1} g(x), \text{ad}_f^{j-1} g(x) \end{bmatrix} = 0, \quad 1 \le i \le n+1, \ 1 \le j \le n+1.$

Furthermore, a state transformation $z = S(x)$ *can be obtained by*

$$\frac{\partial S(x)}{\partial x} = \begin{bmatrix} g(x) & \text{ad}_f g(x) & \cdots & \text{ad}_f^{n-1} g(x) \end{bmatrix}^{-1}. \tag{3.7}$$

Proof Necessity. Suppose that system (3.3) is state equivalent to a linear system. Then there exists a state transformation $z = S(x)$ such that (3.5) is satisfied. It is easy to see, by Example 2.4.14, that for $i \ge 0$,

$$S_*\left(\text{ad}_f^i g(x)\right) = (-1)^i A^i b \tag{3.8}$$

and for $1 \le i \le n+1$ and $1 \le j \le n+1$,

$$\begin{bmatrix} \text{ad}_f^{i-1} g(x), \text{ad}_f^{j-1} g(x) \end{bmatrix} = 0 \tag{3.9}$$

which implies that condition (ii) is satisfied. Also, we have, by (3.8), that

$$\begin{bmatrix} b & -Ab & \cdots & (-1)^{n-1} A^{n-1} b \end{bmatrix}$$
$$= \left\{ \frac{\partial S(x)}{\partial x} \begin{bmatrix} g(x) & \text{ad}_f g(x) & \cdots & \text{ad}_f^{n-1} g(x) \end{bmatrix} \right\}\Big|_{x=S^{-1}(z)} \tag{3.10}$$
$$= \frac{\partial S(x)}{\partial x}\Big|_{x=0} \begin{bmatrix} g(x) & \text{ad}_f g(x) & \cdots & \text{ad}_f^{n-1} g(x) \end{bmatrix}\Big|_{x=0}$$

which implies, together with (3.6), that condition (i) is satisfied.

Sufficiency. Suppose that condition (i) and (ii) are satisfied. Then, by Theorem 2.7, there exists a state transformation $z = S(x)$ such that for $1 \le i \le n$,

$$S_*\left(\text{ad}_f^{i-1} g(x)\right) = \frac{\partial}{\partial z_i} \tag{3.11}$$

or

$$\frac{\partial S(x)}{\partial x} \left[g(x) \ \text{ad}_f g(x) \ \cdots \ \text{ad}_f^{n-1} g(x) \right] = I.$$

Thus, it is clear that $\tilde{g}(z) \triangleq S_*(g(x)) = \frac{\partial}{\partial z_1} = \begin{bmatrix} 1 & 0 & \cdots & 0 \end{bmatrix}^\mathsf{T} \triangleq b$. We will show that $\tilde{f}(z) \triangleq S_*(f(x)) = Az$ for some constant matrix A. It is easy to see, by (2.28) and (3.11), that for $1 \leq i \leq n-1$,

$$-\frac{\partial \tilde{f}(z)}{\partial z_i} = \left[\tilde{f}(z), \frac{\partial}{\partial z_i} \right] = \left[S_*(f(x)), S_* \left(\text{ad}_f^{i-1} g(x) \right) \right]$$

$$= S_* \left(\text{ad}_f^i g(x) \right) = \frac{\partial}{\partial z_{i+1}}$$

which implies that for $1 \leq j \leq n$ and $1 \leq i \leq n-1$,

$$\frac{\partial \tilde{f}_j(z)}{\partial z_i} = \begin{cases} -1, & \text{if } j = i+1 \\ 0. & \text{otherwise.} \end{cases}$$

Also, it is clear, by Example 2.4.20 and condition (i) and (ii), that there exist some constants $a_i \in \mathbb{R}$, $1 \leq i \leq n$ such that

$$\text{ad}_f^n g(x) = \sum_{i=1}^n a_i \text{ad}_f^{i-1} g(x).$$

Thus, we have, by (2.28) and (3.11), that

$$-\frac{\partial \tilde{f}(z)}{\partial z_n} = \left[\tilde{f}(z), \frac{\partial}{\partial z_n} \right] = \left[S_*(f(x)), S_* \left(\text{ad}_f^{n-1} g(x) \right) \right] = S_* \left(\text{ad}_f^n g(x) \right)$$

$$= \sum_{i=1}^n a_i S_* \left(\text{ad}_f^{i-1} g(x) \right) = \sum_{i=1}^n a_i \frac{\partial}{\partial z_i} = \begin{bmatrix} a_1 & \cdots & a_n \end{bmatrix}^\mathsf{T}.$$

Since $\tilde{f}(0) = 0$, it is clear that

$$\frac{\partial \tilde{f}(z)}{\partial z} = - \begin{bmatrix} 0 & 0 & 0 & \cdots & 0 & a_1 \\ 1 & 0 & 0 & \cdots & 0 & a_2 \\ 0 & 1 & 0 & \cdots & 0 & a_3 \\ \vdots & \vdots & \vdots & & \vdots & \vdots \\ 0 & 0 & 0 & \cdots & 1 & a_n \end{bmatrix} = A \tag{3.12}$$

and $\tilde{f}(z) \triangleq S_*(f(x)) = Az$.   □

We say that vector field $f(x)$ and vector field $g(x)$ commute if $[f(x), g(x)] = 0$. Condition (ii) of Theorem 3.1 is that $\left\{ g, \text{ad}_f g, \ldots, \text{ad}_f^{n-1} g, \text{ad}_f^n g \right\}$ is a set of commuting vector fields. If we replace condition (ii) by that $\left\{ g, \mathrm{ad}_f g, \ldots, \mathrm{ad}_f^{n-1} g \right\}$ is a set of commuting vector fields, then the state transformation $z = S(x)$ in (3.7) still exists. However, vector field $S_*(f(x))$ may not be linear in z.

Example 3.2.1 Suppose that $\left\{ g(x), \mathrm{ad}_f g(x), \ldots, \mathrm{ad}_f^{n-1} g(x) \right\}$ is a set of commuting linearly independent vector fields. Let $z = S(x)$ be the state transformation in (3.7). Use (2.91) to show that

$$\tilde{f}(z) \triangleq S_*(f(x)) = - \begin{bmatrix} 0\,0\,0 \cdots 0\,0 \\ 1\,0\,0 \cdots 0\,0 \\ 0\,1\,0 \cdots 0\,0 \\ \vdots\,\vdots\,\vdots \quad \vdots\,\vdots \\ 0\,0\,0 \cdots 1\,0 \end{bmatrix} z + \gamma(z_n)$$

for some vector function $\gamma : \mathbb{R} \to \mathbb{R}^n$.

Solution Omitted. (Problem 3-1.) $\qquad\square$

Since $f(0) = 0$, it is clear that

$$(\mathrm{ad}_f g)(0) = \left.\frac{\partial g}{\partial x}\right|_{x=0} f(0) - \left.\frac{\partial f}{\partial x}\right|_{x=0} g(0) = - \left.\frac{\partial f}{\partial x}\right|_{x=0} g(0) \qquad (3.13)$$

and

$$(\mathrm{ad}_f^i g)(0) = (-1)^i \left(\left.\frac{\partial f}{\partial x}\right|_{x=0} \right)^i g(0), \quad i \geq 0. \qquad (3.14)$$

Therefore, condition (i) of Theorem 3.1 is satisfied on a neighborhood of the origin, if and only if

$$(\mathrm{i})' \quad \mathrm{rank}\left(\left[g(0) \;\; \left.\frac{\partial f}{\partial x}\right|_{x=0} g(0) \;\; \cdots \;\; \left(\left.\frac{\partial f}{\partial x}\right|_{x=0} \right)^{n-1} g(0) \right] \right) = n. \qquad (3.15)$$

Example 3.2.2 Show, by using the Jacobi identity of vector fields, that condition (ii) of Theorem 3.1 can also be expressed as follows.

$$(\mathrm{ii})' \quad \mathrm{ad}_g \mathrm{ad}_f^k g(x) = 0, \quad 0 \leq k \leq 2n - 1. \qquad (3.16)$$

Solution By Examples 2.4.18 and 2.4.19, it is easy to show. (Problem 3-2.) $\qquad\square$

Example 3.2.3 Consider the nonlinear system (1.2).

$$\begin{bmatrix} \dot{x}_1 \\ \dot{x}_2 \end{bmatrix} = \begin{bmatrix} 2x_1 x_2 - 2x_2^3 \\ x_1 - x_2^2 \end{bmatrix} + \begin{bmatrix} 1 + 2x_2 \\ 1 \end{bmatrix} u = f(x) + g(x)u. \qquad (3.17)$$

In Example 1.3.1, we have obtained system (1.2) from linear system

$$\begin{bmatrix} \dot{z}_1 \\ \dot{z}_2 \end{bmatrix} = \begin{bmatrix} 0 & 0 \\ 1 & 0 \end{bmatrix} \begin{bmatrix} z_1 \\ z_2 \end{bmatrix} + \begin{bmatrix} 1 \\ 1 \end{bmatrix} u$$

by nonlinear state transformation

$$\begin{bmatrix} z_1 \\ z_2 \end{bmatrix} = S(x) = \begin{bmatrix} x_1 - x_2^2 \\ x_2 \end{bmatrix}.$$

Show that system (1.2) satisfies the conditions of Theorem 3.1. Also, find the state transformation in (3.7).

Solution Since

$$\begin{aligned} \mathrm{ad}_f g(x) &= \frac{\partial g(x)}{\partial x} f(x) - \frac{\partial f(x)}{\partial x} g(x) \\ &= \begin{bmatrix} 0 & 2 \\ 0 & 0 \end{bmatrix} \begin{bmatrix} 2x_1 x_2 - 2x_2^3 \\ x_1 - x_2^2 \end{bmatrix} - \begin{bmatrix} 2x_2 & 2x_1 - 6x_2^2 \\ 1 & -2x_2 \end{bmatrix} \begin{bmatrix} 1 + 2x_2 \\ 1 \end{bmatrix} \\ &= \begin{bmatrix} -2x_2 \\ -1 \end{bmatrix} \end{aligned}$$

and

$$\mathrm{ad}_f^2 g(x) = \frac{\partial \left(\mathrm{ad}_f g(x) \right)}{\partial x} f(x) - \frac{\partial f(x)}{\partial x} \mathrm{ad}_f g(x) = \begin{bmatrix} 0 \\ 0 \end{bmatrix}$$

it is easy to see that condition (i) and (ii) of Theorem 3.1 are satisfied. Therefore, system (1.2) is state equivalent to a linear system. It is clear, by (3.7), that

$$\frac{\partial S(x)}{\partial x} = \begin{bmatrix} 1 + 2x_2 & -2x_2 \\ 1 & -1 \end{bmatrix}^{-1} = \begin{bmatrix} 1 & -2x_2 \\ 1 & -1 - 2x_2 \end{bmatrix}$$

and

$$\begin{bmatrix} z_1 \\ z_2 \end{bmatrix} = S(x) = \begin{bmatrix} x_1 - x_2^2 \\ x_1 - x_2 - x_2^2 \end{bmatrix}. \qquad (3.18)$$

It is easy to see that

$$\begin{bmatrix} \dot{z}_1 \\ \dot{z}_2 \end{bmatrix} = S_*(f(x)) + S_*(g(x))u = \begin{bmatrix} 0 & 0 \\ -1 & 0 \end{bmatrix} \begin{bmatrix} z_1 \\ z_2 \end{bmatrix} + \begin{bmatrix} 1 \\ 0 \end{bmatrix} u. \qquad (3.19)$$

Note that system (3.19) is state equivalent to system (1.10) with a linear state transformation (i.e., similarity transformation). □

Example 3.2.4 Consider the nonlinear system (1.2).

(a) Let $x(0) = \begin{bmatrix} 1 \\ 1 \end{bmatrix}$. Find an input $u(t)$, $0 \le t \le t_f$ such that $t_f = 2$ and $x(t_f) = \begin{bmatrix} 0 \\ 0 \end{bmatrix}$.

(b) Find a nonlinear feedback $u = \alpha(x)$ such that $\lim_{t \to \infty} x(t) = 0$ for the nonlinear system (1.2).

Solution The controllability Gramian of linear system (3.19) can be calculated as follows:

$$W(0, t) \triangleq \int_0^t e^{-A\tau} bb^\mathsf{T} (e^{-A\tau})^\mathsf{T} d\tau$$

$$= \int_0^t \begin{bmatrix} 1 & 0 \\ \tau & 1 \end{bmatrix} \begin{bmatrix} 1 \\ 0 \end{bmatrix} \begin{bmatrix} 1 & 0 \end{bmatrix} \begin{bmatrix} 1 & \tau \\ 0 & 1 \end{bmatrix} d\tau = \begin{bmatrix} t & \frac{1}{2}t^2 \\ \frac{1}{2}t^2 & \frac{1}{3}t^3 \end{bmatrix}.$$

Note that $z(0) = S(x(0)) = \begin{bmatrix} 0 \\ -1 \end{bmatrix}$ and $z(t_f) = S(x(t_f)) = \begin{bmatrix} 0 \\ 0 \end{bmatrix}$, where $z = S(x)$ is given in (3.18). Thus,

$$u(t) = -b^\mathsf{T} (e^{-At})^\mathsf{T} W(0, t_f)^{-1} z(0) = -\begin{bmatrix} 1 & 0 \end{bmatrix} \begin{bmatrix} 1 & t \\ 0 & 1 \end{bmatrix} \begin{bmatrix} 2 & 2 \\ 2 & \frac{8}{3} \end{bmatrix}^{-1} \begin{bmatrix} 0 \\ -1 \end{bmatrix}$$

$$= \frac{3}{2}(t - 1), \quad 0 \le t \le 2$$

is an input such that $z(2) = \begin{bmatrix} 0 \\ 0 \end{bmatrix}$ and $x(2) = \begin{bmatrix} 0 \\ 0 \end{bmatrix}$. If we let $u = \tilde{\alpha}(z) = -2z_1 + 2z_2$, then the closed-loop system of system (3.19) satisfies the following asymptotically stable linear system:

$$\begin{bmatrix} \dot{z}_1 \\ \dot{z}_2 \end{bmatrix} = \begin{bmatrix} -2 & 2 \\ -1 & 0 \end{bmatrix} \begin{bmatrix} z_1 \\ z_2 \end{bmatrix}$$

and $\lim_{t \to \infty} z(t) = 0$. Thus,

$$u = \tilde{\alpha} \circ S(x) = -2(x_1 - x_2^2) + 2(x_1 - x_2 - x_2^2) = -2x_2$$

is a nonlinear feedback such that $\lim_{t \to \infty} x(t) = 0$. □

Example 3.2.5 Show that the following nonlinear system is (locally) state equivalent to a linear system.

$$\begin{bmatrix} \dot{x}_1 \\ \dot{x}_2 \end{bmatrix} = \begin{bmatrix} 0 \\ x_1 \cos^2 x_2 \end{bmatrix} + \begin{bmatrix} 1 \\ 0 \end{bmatrix} u = f(x) + g(x)u. \tag{3.20}$$

Solution It is easy to see that

$$\text{ad}_f g(x) = \frac{\partial g(x)}{\partial x} f(x) - \frac{\partial f(x)}{\partial x} g(x) = \begin{bmatrix} 0 \\ -\cos^2 x_2 \end{bmatrix}$$

and

$$\text{ad}_f^2 g(x) = \frac{\partial \text{ad}_f g(x)}{\partial x} f(x) - \frac{\partial f(x)}{\partial x} \text{ad}_f g(x) = \begin{bmatrix} 0 \\ 0 \end{bmatrix}.$$

Therefore, condition (i) and (ii) of Theorem 3.1 are satisfied and thus system (3.20) is state equivalent to a linear system. It is clear, by (3.7), that

$$\frac{\partial S(x)}{\partial x} = \begin{bmatrix} 1 & 0 \\ 0 & -\cos^2 x_2 \end{bmatrix}^{-1} = \begin{bmatrix} 1 & 0 \\ 0 & -\sec^2 x_2 \end{bmatrix}$$

and

$$\begin{bmatrix} z_1 \\ z_2 \end{bmatrix} = S(x) = \begin{bmatrix} x_1 \\ -\tan x_2 \end{bmatrix}.$$

Then we have that

$$\begin{bmatrix} \dot{z}_1 \\ \dot{z}_2 \end{bmatrix} = S_*(f(x)) + S_*(g(x))u = \begin{bmatrix} 0 & 0 \\ -1 & 0 \end{bmatrix} \begin{bmatrix} z_1 \\ z_2 \end{bmatrix} + \begin{bmatrix} 1 \\ 0 \end{bmatrix} u.$$

Note that $\left(\frac{\partial S(x)}{\partial x} \right)^{-1}$ is not nonsingular when $x_2 = \frac{\pi}{2}$. Thus, $S(x)$ is a state transformation on $\{x \in \mathbb{R}^2 \mid |x_2| < \frac{\pi}{2}\}$. In other words, the state equivalence does not work on the entire state space $\mathbb{R}^2$. It is called the local linearization. □

Example 3.2.6 Show that the following nonlinear system is not state equivalent to a linear system.

$$\begin{bmatrix} \dot{x}_1 \\ \dot{x}_2 \end{bmatrix} = \begin{bmatrix} x_2 \\ x_1^2 \end{bmatrix} + \begin{bmatrix} 0 \\ 1 \end{bmatrix} u = f(x) + g(x)u. \tag{3.21}$$

Solution It is easy to see that

$$g(x) = \begin{bmatrix} 0 \\ 1 \end{bmatrix}, \quad \mathrm{ad}_f g(x) = \begin{bmatrix} -1 \\ 0 \end{bmatrix}, \quad \mathrm{ad}_f^2 g(x) = \begin{bmatrix} 0 \\ 2x_1 \end{bmatrix}$$

and

$$\left[\mathrm{ad}_f g(x), \mathrm{ad}_f^2 g(x) \right] = \begin{bmatrix} 0 \\ -2 \end{bmatrix} \neq \begin{bmatrix} 0 \\ 0 \end{bmatrix}.$$

Therefore, condition (ii) of Theorem 3.1 is not satisfied and thus system (3.21) is not state equivalent to a linear system. □

If we use feedback $u = -x_1^2 + v$ for system (3.21), we have the following linear closed-loop system

$$\begin{bmatrix} \dot{x}_1 \\ \dot{x}_2 \end{bmatrix} = \begin{bmatrix} x_2 \\ 0 \end{bmatrix} + \begin{bmatrix} 0 \\ 1 \end{bmatrix} v = f_c(x) + g_c(x)v. \tag{3.22}$$

In other words, system (3.21) can be linearized by using feedback $u = -x_1^2 + v$.

Example 3.2.7 Show that the following nonlinear system is not state equivalent to a linear system.

$$\begin{bmatrix} \dot{x}_1 \\ \dot{x}_2 \end{bmatrix} = \begin{bmatrix} 2x_1 x_2 - x_2^3 - x_1(1 + 2x_2) \\ -x_2^2 \end{bmatrix} + \begin{bmatrix} 1 + 2x_2 \\ 1 \end{bmatrix} u = f(x) + g(x)u. \tag{3.23}$$

Solution It is easy to see that

$$g(x) = \begin{bmatrix} 1 + 2x_2 \\ 1 \end{bmatrix}, \quad \mathrm{ad}_f g(x) = \begin{bmatrix} 1 + 2x_2 + 4x_2^2 \\ 2x_2 \end{bmatrix}$$

and

$$\left[g(x), \mathrm{ad}_f g(x) \right] = \begin{bmatrix} 2 + 4x_2 \\ 2 \end{bmatrix} \neq \begin{bmatrix} 0 \\ 0 \end{bmatrix}.$$

Therefore, condition (ii) of Theorem 3.1 is not satisfied and thus system (3.23) is not state equivalent to a linear system. □

If we use feedback $u = x_1 + v$ for system (3.23), we have

$$\begin{bmatrix} \dot{x}_1 \\ \dot{x}_2 \end{bmatrix} = \begin{bmatrix} 2x_1 x_2 - 2x_2^3 \\ x_1 - x_2^2 \end{bmatrix} + \begin{bmatrix} 1 + 2x_2 \\ 1 \end{bmatrix} v = f_c(x) + g_c(x)v \tag{3.24}$$

that is state equivalent to a linear system. (Refer to Example 3.2.3) In other words, system (3.23) can be linearized by using state transformation (3.18) and feedback $u = x_1 + v$. The linearization by using both state transformation and feedback will be discussed in the next chapter.

3.3 Multi Input Nonlinear Systems

In this section, we extend the single input results of the previous section to multi-input systems. Consider the following smooth multi-input control systems:

$$\dot{x} = f(x) + \sum_{i=1}^{m} g_i(x)u_i = f(x) + g(x)u \tag{3.25}$$

where $x \in \mathbb{R}^n$, $u \in \mathbb{R}^m$, and $f(0) = 0$. We want to find a state transformation $z = S(x)$ such that

$$\dot{z} = Az + \sum_{i=1}^{m} b_i u_i = Az + Bu \tag{3.26}$$

or

$$S_*(f(x)) = Az \quad \text{and} \quad S_*(g_i(x)) = b_i, \ 1 \le i \le m \tag{3.27}$$

where

$$\text{rank}\left(\begin{bmatrix} b_1 \ Ab_1 \ \cdots \ A^{\kappa_1-1}b_1 \ \cdots \ b_m \ \cdots \ A^{\kappa_m-1}b_m \end{bmatrix}\right) = n. \tag{3.28}$$

Definition 3.2 (*Kronecker indices*) For the list of mn constant vector fields of the form

$$\left(g_1, \ldots, g_m, \text{ad}_f g_1, \ldots, \text{ad}_f g_m, \ldots, \text{ad}_f^{n-1} g_1, \ldots, \text{ad}_f^{n-1} g_m \right)\Big|_{x=0}$$

delete all vector fields that are linearly dependent on the set of preceding vector fields and obtain the unique set of linearly independent constant vector fields

$$\left\{ g_1, \text{ad}_f g_1, \ldots, \text{ad}_f^{\kappa_1-1} g_1, \ldots, g_m, \text{ad}_f g_m, \ldots, \text{ad}_f^{\kappa_m-1} g_m \right\}\Big|_{x=0}.$$

$(\kappa_1, \ldots, \kappa_m)$ are said to be the Kronecker indices of system (3.25).

In other words, κ_i is the smallest nonnegative integer such that

$$\begin{aligned}
\text{ad}_f^{\kappa_i} g_i(x)\Big|_{x=0} &\in \text{span}\{ \text{ad}_f^{\ell-1} g_j(x)\Big|_{x=0} \ \Big| \ 1 \le j \le m, \quad 1 \le \ell \le \kappa_i \} \\
&+ \text{span}\{ \text{ad}_f^{\kappa_i} g_j(x)\Big|_{x=0} \ \Big| \ 1 \le j \le i-1 \}.
\end{aligned} \tag{3.29}$$

If $\sum_{i=1}^{m} \kappa_i = n$, system (3.25) is said to be reachable on a neighborhood of the origin.

Example 3.3.1 Show that the Kronecker indices are invariant under state transformation. In other words, the Kronecker indices of system (3.25) are the same as the Kronecker indices of system (3.26).

Solution Suppose that $\tilde{f}(z) = S_*(f(x))$ and $\tilde{g}(z) = S_*(g(x))$, where $z = S(x)$ is a state transformation. Since rank $\left(\frac{\partial S(x)}{\partial x} \Big|_{x=0} \right) = n$ and

$$\left[\tilde{g}_1 \cdots, \tilde{g}_m \text{ ad}_{\tilde{f}} \tilde{g}_1 \cdots \text{ ad}_{\tilde{f}} \tilde{g}_m \cdots \text{ad}_{\tilde{f}}^{n-1} \tilde{g}_1 \cdots \text{ad}_{\tilde{f}}^{n-1} \tilde{g}_m \right]\Big|_{z=0}$$

$$= \frac{\partial S(x)}{\partial x}\Big|_{x=0} \left[g_1 \cdots g_m \text{ ad}_f g_1 \cdots, \text{ad}_f g_m \cdots \text{ad}_f^{n-1} g_1 \cdots \text{ad}_f^{n-1} g_m \right]\Big|_{x=0}$$

we obtain, after deletion of Definition 3.2, the unique set of linearly independent constant vector fields

$$\left\{ \tilde{g}_1, \text{ad}_{\tilde{f}} \tilde{g}_1, \ldots, \text{ad}_{\tilde{f}}^{\kappa_1-1} \tilde{g}_1, \ldots, \tilde{g}_m, \text{ad}_{\tilde{f}} \tilde{g}_m, \ldots, \text{ad}_{\tilde{f}}^{\kappa_m-1} \tilde{g}_m \right\}\Big|_{z=0} .$$

Therefore, the Kronecker indices of system (3.26) are $(\kappa_1, \ldots, \kappa_m)$. $\qquad\square$

Theorem 3.2 (necessary and sufficient condition) *System* (3.25) *is state equivalent to a linear system, if and only if*

(i) $\sum_{i=1}^{m} \kappa_i = n.$

(ii) *for* $1 \leq i \leq m, 1 \leq j \leq m, 1 \leq \ell_i \leq \kappa_i + 1,$ *and* $1 \leq \ell_j \leq \kappa_j + 1,$

$$[\text{ad}_f^{\ell_i-1} g_i(x), \text{ad}_f^{\ell_j-1} g_j(x)] = 0. \tag{3.30}$$

Furthermore, a state transformation $z = S(x)$ *can be obtained by*

$$\frac{\partial S(x)}{\partial x} = \left[g_1 \text{ ad}_f g_1 \cdots \text{ad}_f^{\kappa_1-1} g_1 \cdots g_m \cdots \text{ad}_f^{\kappa_m-1} g_m \right]^{-1}. \tag{3.31}$$

Proof Necessity. Suppose that system (3.25) is state equivalent to a linear system. Then there exists a state transformation $z = S(x)$ such that (3.27) is satisfied. It is easy to see, by Example 2.4.14, that for $1 \leq i \leq m$ and $k \geq 0,$

$$S_*(\text{ad}_f^k g_i(x)) = (-1)^k A^k b_i. \tag{3.32}$$

Thus, it is easy to see that for $1 \leq i \leq m, 1 \leq j \leq m, 1 \leq \ell_i \leq \kappa_i + 1$, and $1 \leq \ell_j \leq \kappa_j + 1$,

$$
\begin{aligned}
\left[\mathrm{ad}_f^{\ell_i-1} g_i(x), \mathrm{ad}_f^{\ell_j-1} g_j(x)\right] &= [S_*^{-1}\{(-1)^{\ell_i-1} A^{\ell_i-1} b_i\}, S_*^{-1}\{(-1)^{\ell_j-1} A^{\ell_j-1} b_i\}] \\
&= S_*^{-1}\{[(-1)^{\ell_i-1} A^{\ell_i-1} b_i, (-1)^{\ell_j-1} A^{\ell_j-1} b_i]\} = 0
\end{aligned}
$$

and condition (ii) is satisfied. Also, we have, by (3.32), that

$$
\begin{aligned}
&\left[b_1 \ -Ab_1 \ \cdots \ (-1)^{\kappa_1-1} A^{\kappa_1-1} b_1 \ \cdots \ b_m \ \cdots \ (-1)^{\kappa_m-1} A^{\kappa_m-1} b_m\right] \\
&= \left\{\frac{\partial S(x)}{\partial x}\left[g_1 \ \cdots \ \mathrm{ad}_f^{\kappa_1-1} g_1 \ \cdots \ g_m \ \cdots \ \mathrm{ad}_f^{\kappa_m-1} g_m\right]\right\}\bigg|_{x=S^{-1}(z)} \\
&= \frac{\partial S(x)}{\partial x}\left[g_1 \ \cdots \ \mathrm{ad}_f^{\kappa_1-1} g_1 \ \cdots \ g_m \ \cdots \ \mathrm{ad}_f^{\kappa_m-1} g_m\right]
\end{aligned}
$$

which implies, together with (3.28), that condition (i) is satisfied.

Sufficiency. Suppose that condition (i) and (ii) are satisfied. Then, by Theorem 2.7, there exists a state transformation $z = S(x)$ such that for $1 \leq i \leq m$ and $1 \leq \ell \leq \kappa_i$,

$$
S_*\left(\mathrm{ad}_f^{\ell-1} g_i(x)\right) = \frac{\partial}{\partial z_\ell^i}
$$

$$
z = \begin{bmatrix} z^1 \\ \vdots \\ z^m \end{bmatrix}, \quad z^i = \begin{bmatrix} z_1^i \\ \vdots \\ z_{\kappa_i}^i \end{bmatrix} \tag{3.33}
$$

or

$$
\frac{\partial S(x)}{\partial x}\left[g_1 \ \mathrm{ad}_f g_1 \ \cdots \ \mathrm{ad}_f^{\kappa_1-1} g_1 \ \cdots \ g_m \ \cdots \ \mathrm{ad}_f^{\kappa_m-1} g_m\right] = I.
$$

Thus, it is clear that $S_*(g_i(x)) = \frac{\partial}{\partial z_1^i} \triangleq b_i$ for $1 \leq i \leq m$. We will show that $S_*(f(x)) = Az$ for some constant matrix A. Let

$$
S_*(f(x)) = \sum_{i=1}^m \sum_{j=1}^{\kappa_i} \tilde{f}_j^i(z) \frac{\partial}{\partial z_j^i}. \tag{3.34}
$$

Then, it is easy to see, by (3.33), (3.34), and condition (ii), that for $1 \leq k_1 \leq m$, $1 \leq k_2 \leq m, 1 \leq \ell_1 \leq \kappa_{k_1}, 1 \leq \ell_2 \leq \kappa_{k_2}$,

$$0 = S_* \left(\left[[f(x), \mathrm{ad}_f^{\ell_1-1} g_{k_1}(x)], \mathrm{ad}_f^{\ell_2-1} g_{k_2}(x) \right] \right)$$

$$= \left[\left[S_*(f(x)), S_*(\mathrm{ad}_f^{\ell_1-1} g_{k_1}(x)) \right], S_*(\mathrm{ad}_f^{\ell_2-1} g_{k_2}(x)) \right]$$

$$= \left[\left[\sum_{i=1}^{m} \sum_{j=1}^{\kappa_j} \tilde{f}_j^i(z) \frac{\partial}{\partial z_j^i}, \frac{\partial}{\partial z_{\ell_1}^{k_1}} \right], \frac{\partial}{\partial z_{\ell_2}^{k_2}} \right] \qquad (3.35)$$

$$= \sum_{i=1}^{m} \sum_{j=1}^{\kappa_j} \frac{\partial^2 \tilde{f}_j^i(z)}{\partial z_{\ell_1}^{k_1} \partial z_{\ell_2}^{k_2}} \frac{\partial}{\partial z_j^i}.$$

Since $\frac{\partial}{\partial z}(\frac{\partial \tilde{f}_j^i(z)}{\partial z})^\mathsf{T} = 0$ for $1 \leq i \leq m$, $1 \leq j \leq \kappa_i$, it is clear that

$$\tilde{f}_i(z) = \tilde{f}_i(0) + A_i z, \quad 1 \leq i \leq n.$$

Since $f(0) = 0$ and $S(0) = 0$, it is clear that $\tilde{f}(0) = S_*(f(0)) = 0$. Therefore, we have that $\tilde{f}_i(z) = A_j^i z$ and

$$S_*(f(x)) = \sum_{i=1}^{m} \sum_{i=1}^{\kappa_j} A_j^i z \frac{\partial}{\partial z_j^i} = Az \qquad (3.36)$$

where

$$A = \begin{bmatrix} A^1 \\ \vdots \\ A^m \end{bmatrix}, \quad A^i = \begin{bmatrix} A_1^i \\ \vdots \\ A_{\kappa_i}^i \end{bmatrix}, \quad 1 \leq i \leq m.$$

$\square$

Example 3.3.2 Show, by using the Jacobi identity of vector fields, that condition (ii) of Theorem 3.2 can also be expressed as follows.

$$(\text{ii})' \qquad \mathrm{ad}_{g_i} \mathrm{ad}_f^\ell g_j(x) = 0 \quad \text{for} \quad 0 \leq \ell \leq \kappa_i + \kappa_j. \qquad (3.37)$$

Solution By Example 2.4.18, it is easy to show. (See Problem 3-8.) $\square$

Example 3.3.3 Show that the following nonlinear system is state equivalent to a linear system. Also, find out the state transformation $z = S(x)$ in (3.31).

$$\begin{bmatrix} \dot{x}_1 \\ \dot{x}_2 \\ \dot{x}_3 \end{bmatrix} = \begin{bmatrix} -2x_2(x_1 + x_2 + x_2^2) \\ x_1 + x_2 + x_2^2 \\ -2x_2(x_1 + x_2 + x_2^2) \end{bmatrix} + \begin{bmatrix} 1 & 0 \\ 0 & 0 \\ 0 & 1 \end{bmatrix} \begin{bmatrix} u_1 \\ u_2 \end{bmatrix} \qquad (3.38)$$

$$= f(x) + g_1(x)u_1 + g_2(x)u_2.$$

Solution By simple calculation, we have that $(\kappa_1, \kappa_2) = (2, 1)$ and

$$\mathrm{ad}_f g_1(x) = \begin{bmatrix} 2x_2 \\ -1 \\ 2x_2 \end{bmatrix}, \quad \mathrm{ad}_f g_2(x) = \begin{bmatrix} 0 \\ 0 \\ 0 \end{bmatrix}, \quad \mathrm{ad}_f^2 g_1(x) = \begin{bmatrix} -2x_2 \\ 1 \\ -2x_2 \end{bmatrix}.$$

It is easy to see that condition (i) and (ii) of Theorem 3.2 are satisfied. Therefore, system (3.38) is state equivalent to a linear system. It is clear, by (3.31), that

$$\frac{\partial S(x)}{\partial x} = \begin{bmatrix} 1 & 2x_2 & 0 \\ 0 & -1 & 0 \\ 0 & 2x_2 & 1 \end{bmatrix}^{-1} = \begin{bmatrix} 1 & 2x_2 & 0 \\ 0 & -1 & 0 \\ 0 & 2x_2 & 1 \end{bmatrix}$$

and

$$\begin{bmatrix} z_1 \\ z_2 \\ z_3 \end{bmatrix} = S(x) = \begin{bmatrix} x_1 + x_2^2 \\ -x_2 \\ x_3 + x_2^2 \end{bmatrix}.$$

Then we have that

$$\begin{bmatrix} \dot{z}_1 \\ \dot{z}_2 \end{bmatrix} = S_*(f(x)) + S_*(g_1(x))u_1 + S_*(g_2(x))u_2$$

$$= \begin{bmatrix} 0 & 0 & 0 \\ -1 & 1 & 0 \\ 0 & 0 & 0 \end{bmatrix} \begin{bmatrix} z_1 \\ z_2 \\ z_3 \end{bmatrix} + \begin{bmatrix} 1 & 0 \\ 0 & 0 \\ 0 & 1 \end{bmatrix} \begin{bmatrix} u_1 \\ u_2 \end{bmatrix}.$$

□

Example 3.3.4 Show that the following nonlinear system is not state equivalent to a linear system.

$$\begin{bmatrix} \dot{x}_1 \\ \dot{x}_2 \\ \dot{x}_3 \end{bmatrix} = \begin{bmatrix} x_2 \\ -x_1 + x_2^2 \\ x_3^2 \end{bmatrix} + \begin{bmatrix} 0 & 0 \\ 1 + x_1^2 & 0 \\ 0 & 1 \end{bmatrix} \begin{bmatrix} u_1 \\ u_2 \end{bmatrix} \tag{3.39}$$

$$= f(x) + g_1(x)u_1 + g_2(x)u_2$$

Solution By simple calculation, we have that $(\kappa_1, \kappa_2) = (2, 1)$ and

$$\mathrm{ad}_f g_1(x) = \begin{bmatrix} -1 - x_1^2 \\ 2x_2(x_1 - 1 - x_1^2) \\ 0 \end{bmatrix}.$$

Since $[g_1(x), \mathrm{ad}_f g_1(x)] \neq 0$, condition (ii) of Theorem 3.2 is not satisfied. Therefore, system (3.39) is not state equivalent to a linear system.   □

3.4 MATLAB Programs

In this section, the following subfunctions in Appendix C are needed:
**adfg, adfgk, adfgM, adfgkM, ChCommute, ChZero,
Codi, Delta, Kindex0, TauFG**

MATLAB program for Theorem 3.1:

```
clear all
syms x1 x2 x3 x4 x5 x6 x7 x8 x9 real

f=[2*x1*x2-2*x2^3; x1-x2^2]; g=[1+2*x2 ; 1]; %Ex:3.2.3

% f=[0; x1*cos(x2)^2]; g=[1; x1-x1]; %Ex:3.2.5

% f=[x2; x1^2]; g=[x1-x1; 1]; %Ex:3.2.6

% f=[2*x1*x2-2*x2^3-x1*(1+2*x2); -x2^2];
% g=[1+2*x2; 1]; %Ex:3.2.7

% f=[x1*x2+x2-x2^3; 0]; g=[2*x2; 1]; %P:3-3

% f=[2*x2-2*x1*x2+2*x2^2+2*x2^3; -x1+x2+x2^2];
% g=[1; x1-x1]; %P:3-4

% f=[x2+(1/2)*x1^2; x1^2]; g=[x1-x1; 1]; %P:3-5

% f=[x2+x3^2; x3; 0]; g=[2*x3; -2*x3; 1]; %P:3-6

% f=[x1-x1]; g=[1+x1]; %P:3-7

f=simplify(f)
g=simplify(g)

[n,m]=size(g);
x=sym('x',[n,1]);

T(:,1)=g;
for k=2:n+1
  T(:,k)=adfg(f,T(:,k-1),x);
end

T=simplify(T)
BD=T(:,1:n)
BD0=subs(BD,x,x-x);

if rank(BD0) < n
  display('condition (i) of Thm 3.1 is not satisfied.')
  display('System is NOT state equivalent to a LS.')
  return
```

```
end

if ChCommute(T,x) == 0
  display('condition (ii) of Thm 3.1 is not satisfied.')
  display('System is NOT state equivalent to a LS.')
  return
end

display('System is state equivalent to a LS with')

dS=simplify(inv(BD))
S=Codi(dS,x)

AS=simplify(dS*f);
dAS=simplify(jacobian(AS,x));
A=simplify(dAS*BD)
B=simplify(dS*g)

return
```

MATLAB program for Theorem 3.2:

```
clear all
syms x1 x2 x3 x4 x5 x6 x7 x8 x9 real

f=[-2*x2*(x1+x2+x2^2); x1+x2+x2^2; -2*x2*(x1+x2+x2^2)];
g=[1  x1-x1; 0  0; 0  1]; %Ex:3.3.3

% f=[x2; -x1+x2^2; x3^2];
% g=[0  x1-x1; 1+x1^2  0; 0  1]; %Ex:3.3.4

% f=[x2; x4; x4+3*x2^2*x4; 0];
% g=[0 2*x4; 1 0; 3*x2^2 0; 0 1]; %P:3-9

% f=[-x1+x2^2; -2*x2+sin(x2)];
% g=[1 x1-x1; 0 1]; %P:3-10

f=simplify(f)
g=simplify(g)
[n,m]=size(g);
x=sym('x',[n,1]);

[kappa,D]=Kindex0(f,g,x)

if sum(kappa)<n
  display('condition (i) of Thm 3.2 is not satisfied.')
  return
end

BDD=TauFG(f,g,x,kappa+1)
```

```
if ChCommute(BDD,x) == 0
  display('condition (ii) of Thm 3.2 is not satisfied.')
  return
end

display('System is state equivalent to a LS with')

BD=TauFG(f,g,x,kappa)
dS=simplify(inv(BD))
S=Codi(dS,x)

AS=simplify(dS*f)
dAS=simplify(jacobian(AS,x));
A=simplify(dAS*BD)
B=simplify(dS*g)

return
```

3.5 Problems

3-1. Solve Example 3.2.1.

3-2. Solve Example 3.2.2.

3-3. Show that the following nonlinear system is state equivalent to a linear system. Also, find the state transformation in (3.7).

$$\dot{x} = \begin{bmatrix} x_1 x_2 + x_2 - x_2^3 \\ 0 \end{bmatrix} + \begin{bmatrix} 2x_2 \\ 1 \end{bmatrix} u.$$

3-4. Show that the following nonlinear system is state equivalent to a linear system. Also, find the state transformation $z = S(x)$ and the linear system.

$$\dot{x} = \begin{bmatrix} 2x_2 - 2x_1 x_2 + 2x_2^2 + 2x_2^3 \\ -x_1 + x_2 + x_2^2 \end{bmatrix} + \begin{bmatrix} 1 \\ 0 \end{bmatrix} u.$$

3-5. Show that the following nonlinear system is not state equivalent to a linear system.

$$\begin{bmatrix} \dot{x}_1 \\ \dot{x}_2 \end{bmatrix} = \begin{bmatrix} x_2 + \frac{1}{2}x_1^2 \\ x_1^2 \end{bmatrix} + \begin{bmatrix} 0 \\ 1 \end{bmatrix} u. \tag{3.40}$$

3-6. Find a nonlinear feedback $u = \alpha(x)$ such that $\lim\limits_{t \to \infty} x(t) = 0$ for the following nonlinear control system.

$$\dot{x} = \begin{bmatrix} x_2 + x_3^2 \\ x_3 \\ 0 \end{bmatrix} + \begin{bmatrix} 2x_3 \\ -2x_3 \\ 1 \end{bmatrix} u.$$

3-7. Linearize the following nonlinear control system by state transformation.

$$\dot{x} = (1 + x)u$$

Find the subset (containing the origin of the state) where this linearization technique is effective. Also, linearize the above nonlinear control system by using feedback.

3-8. Solve Example 3.3.2.

3-9. Linearize the following nonlinear system by state transformation.

$$\dot{x} = \begin{bmatrix} x_2 \\ x_4 \\ x_4 + 3x_2^2 x_4 \\ 0 \end{bmatrix} + \begin{bmatrix} 0 & 2x_4 \\ 1 & 0 \\ 3x_2^2 & 0 \\ 0 & 1 \end{bmatrix} u.$$

3-10. Show that the following nonlinear system is not state equivalent to a linear system.

$$\dot{x} = \begin{bmatrix} -x_1 + x_2^2 \\ -2x_2 + \sin x_2 \end{bmatrix} + \begin{bmatrix} 1 \\ 0 \end{bmatrix} u_1 + \begin{bmatrix} 0 \\ 1 \end{bmatrix} u_2.$$

3-11. Consider the nonlinear system in Example 3.2.3.

$$\begin{bmatrix} \dot{x}_1 \\ \dot{x}_2 \end{bmatrix} = \begin{bmatrix} 2x_1 x_2 - 2x_2^3 \\ x_1 - x_2^2 \end{bmatrix} + \begin{bmatrix} 1 + 2x_2 \\ 1 \end{bmatrix} u.$$

Find the state transformation $z = S(x)$ such that

$$\begin{bmatrix} \dot{z}_1 \\ \dot{z}_2 \end{bmatrix} = \begin{bmatrix} 0 & 1 \\ 0 & 0 \end{bmatrix} \begin{bmatrix} z_1 \\ z_2 \end{bmatrix} + \begin{bmatrix} 0 \\ 1 \end{bmatrix} u.$$

3-12. Suppose that system (3.3) is state equivalent to a linear system.

(a) Show that

$$\text{ad}_f^n g(x) = \sum_{i=1}^{n} (-1)^i c_{i-1} \text{ad}_f^{i-1} g(x)$$

for some constants $c_0, c_1, \ldots, c_{n-1}$.

(b) Find out the state transformation $z = S(x)$ such that the system satisfies, in z-coordinates, the following controllable canonical form:

$$\dot{x} = \begin{bmatrix} 0 & 1 & 0 & \cdots & 0 & 0 \\ 0 & 0 & 1 & \cdots & 0 & 0 \\ \vdots & \vdots & \vdots & & \vdots & \vdots \\ 0 & 0 & 0 & \cdots & 0 & 1 \\ -c_0 & -c_1 & -c_2 & \cdots & -c_{n-2} & -c_{n-1} \end{bmatrix} + \begin{bmatrix} 0 \\ 0 \\ \vdots \\ 0 \\ 1 \end{bmatrix} u.$$

3-13. Suppose that condition (i) of Theorem 3.1 is satisfied. Show that condition (ii) of Theorem 3.1 is equivalent to the following condition.

(ii)'' $\left[\text{ad}_f^{i-1} g(x), \text{ad}_f^{j-1} g(x) \right] = 0, \quad i \geq 1, \ j \geq 1.$

Chapter 4
Feedback Linearization

4.1 Introduction

In Chap. 3, we considered the linearization by state transformation only. This chapter deals with the linearization problems by both state transformation and feedback. In Example 3.2.6, we have shown that a nonlinear system (3.21), that is not state equivalent to a linear system, can be linearized by using nonlinear feedback $u = -x_1^2 + v$. Consider

$$\begin{bmatrix} \dot{x}_1 \\ \dot{x}_2 \end{bmatrix} = \begin{bmatrix} x_2 + \frac{1}{2}x_1^2 \\ x_1^2 \end{bmatrix} + \begin{bmatrix} 0 \\ 1 \end{bmatrix} u = f(x) + g(x)u. \tag{4.1}$$

that is not linearizable by state transformation (See Problem 3-3.5). We cannot eliminate the nonlinear term $\frac{1}{2}x_1^2$ by feedback. If we let

$$u = -x_1^2 - x_1 x_2 - \frac{1}{2}x_1^3 + v \tag{4.2}$$

then we have the following closed-loop system:

$$\begin{bmatrix} \dot{x}_1 \\ \dot{x}_2 \end{bmatrix} = \begin{bmatrix} x_2 + \frac{1}{2}x_1^2 \\ -x_1 x_2 - \frac{1}{2}x_1^3 \end{bmatrix} + \begin{bmatrix} 0 \\ 1 \end{bmatrix} v = f_c(x) + g_c(x)v. \tag{4.3}$$

If we let $z = S(x) = \begin{bmatrix} x_2 + \frac{1}{2}x_1^2 \\ -x_1 \end{bmatrix}$, then we have

$$\begin{bmatrix} \dot{z}_1 \\ \dot{z}_2 \end{bmatrix} = \begin{bmatrix} 0 & 0 \\ -1 & 0 \end{bmatrix} \begin{bmatrix} z_1 \\ z_2 \end{bmatrix} + \begin{bmatrix} 1 \\ 0 \end{bmatrix} v. \tag{4.4}$$

(See Example 2.7.6). System (4.1) is not linearizable by state transformation. However, we can transform system (4.1) into a linear system by using both feedback (4.2) and state transformation $z = S(x) = \begin{bmatrix} x_2 + \frac{1}{2}x_1^2 \\ -x_1 \end{bmatrix}$. In other words, the larger class of the nonlinear systems can be linearized by using both feedback and state transformation. It motivates the linearization problem by state transformation and feedback (or simply feedback linearization problem). Consider the following nonlinear control system:

$$\dot{x} = f(x) + g(x)u = f(x) + \sum_{i=1}^{m} u_i g_i(x) \tag{4.5}$$

where $x \in \mathbb{R}^n$, $u \in \mathbb{R}^m$, and $f(0) = 0$.

Definition 4.1 (*Feedback linearization*) System (4.5) is said to be feedback linearizable. If there exist a state transformation $z = S(x)$ and a nonsingular feedback $u = \alpha(x) + \beta(x)v$ such that the closed-loop system satisfies, in z-coordinates, the following Brunovsky canonical form:

$$\dot{z} = \begin{bmatrix} A_{11} & O & \cdots & O \\ O & A_{22} & \cdots & O \\ \vdots & \vdots & \ddots & \vdots \\ O & O & \cdots & A_{mm} \end{bmatrix} z + \begin{bmatrix} B_{11} & O & \cdots & O \\ O & B_{22} & \cdots & O \\ \vdots & \vdots & \ddots & \vdots \\ O & O & \cdots & B_{mm} \end{bmatrix} v$$

$$= Az + Bv$$

or

$$S_* \left(f(x) + g(x)\alpha(x) + g(x)\beta(x)v \right) = Az + Bv \tag{4.6}$$

where $\sum_{i=1}^{m} \kappa_i = n$, $z = \begin{bmatrix} z^1 \\ \vdots \\ z^m \end{bmatrix}$, $z^i = \begin{bmatrix} z_1^i \\ \vdots \\ z_{\kappa_i}^i \end{bmatrix}$, and

$$A_{ii} = \begin{bmatrix} 0 & 1 & 0 & \cdots & 0 & 0 \\ 0 & 0 & 1 & \cdots & 0 & 0 \\ \vdots & \vdots & \vdots & & \vdots & \vdots \\ 0 & 0 & 0 & \cdots & 0 & 1 \\ 0 & 0 & 0 & \cdots & 0 & 0 \end{bmatrix} \ (\kappa_i \times \kappa_i), \quad B_{ii} = \begin{bmatrix} 0 \\ 0 \\ \vdots \\ 0 \\ 1 \end{bmatrix} \ (\kappa_i \times 1).$$

For the same reason as in Chap. 3, we assume that $\sum_{i=1}^{m} \kappa_i = n$ or

$$\dim \left(\begin{bmatrix} B & AB & \cdots & A^{n-1}B \end{bmatrix} \right) = n. \tag{4.7}$$

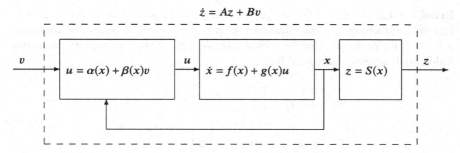

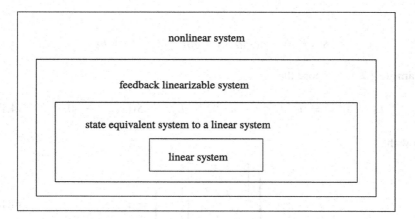

Fig. 4.1 Feedback linearization

Fig. 4.2 Relation of feedback linearization and state equivalence to a linear system

Figure 4.1 gives the block diagram of feedback linearization. If a system is state equivalent to a linear system, it is also feedback linearizable with $u = v$. Figure 4.2 shows the relationship between linearization by state transformation and feedback linearization.

4.2 Single Input Nonlinear Systems

This section deals with the feedback linearization problem of the following smooth single input nonlinear system:

$$\dot{x} = f(x) + g(x)u, \quad x \in \mathbb{R}^n,\ u \in \mathbb{R} \tag{4.8}$$

We can assume, without loss of generality, that $f(0) = 0$.

Definition 4.2 (*Feedback linearization*) System (4.8) is said to be feedback linearizable. if there exist a state transformation $z = S(x)$ and a nonsingular feedback $u = \alpha(x) + \beta(x)v$ such that the closed-loop system satisfies, in z-coordinates, the following Brunovsky canonical form:

$$\dot{z} = \begin{bmatrix} 0 & 1 & 0 & \cdots & 0 & 0 \\ 0 & 0 & 1 & \cdots & 0 & 0 \\ \vdots & \vdots & \vdots & & \vdots & \vdots \\ 0 & 0 & 0 & \cdots & 0 & 1 \\ 0 & 0 & 0 & \cdots & 0 & 0 \end{bmatrix} z + \begin{bmatrix} 0 \\ 0 \\ \vdots \\ 0 \\ 1 \end{bmatrix} v = Az + bv \qquad (4.9)$$

or

$$S_* \left(f(x) + g(x)\alpha(x) + g\beta(x)v \right) = Az + bv. \qquad (4.10)$$

Example 4.2.1 Suppose that

$$L_g L_f^i S_1(x) = 0, \quad 0 \le i \le n - 2 ; \quad L_g L_f^{n-1} S_1(x)\big|_{x=0} \ne 0. \qquad (4.11)$$

Show that

$$\text{rank} \left(\begin{bmatrix} \frac{\partial S_1(x)}{\partial x} \\ \frac{\partial L_f S_1(x)}{\partial x} \\ \vdots \\ \frac{\partial L_f^{n-1} S_1(x)}{\partial x} \end{bmatrix} \Bigg|_{x=0} \right) = n \qquad (4.12)$$

and

$$\text{rank} \left(\begin{bmatrix} g(x) & \text{ad}_f g(x) & \cdots & \text{ad}_f^{n-1} g(x) \end{bmatrix} \big|_{x=0} \right) = n. \qquad (4.13)$$

Solution It is easy to see, by Example 2.4.16, that

$$\begin{bmatrix} \frac{\partial S_1(x)}{\partial x} \\ \frac{\partial L_f S_1(x)}{\partial x} \\ \vdots \\ \frac{\partial L_f^{n-1} S_1(x)}{\partial x} \end{bmatrix} \begin{bmatrix} g(x) & \text{ad}_f g(x) & \cdots & \text{ad}_f^{n-1} g(x) \end{bmatrix}$$

$$
= \begin{bmatrix} L_g S_1(x) & L_{\mathrm{ad}_f g} S_1(x) & \cdots & L_{\mathrm{ad}_f^{n-1} g} S_1(x) \\ \vdots & \vdots & & \vdots \\ L_g L_f^{n-2} S_1(x) & L_{\mathrm{ad}_f g} L_f^{n-2} S_1(x) & \cdots & L_{\mathrm{ad}_f^{n-1} g} L_f^{n-2} S_1(x) \\ L_g L_f^{n-1} S_1(x) & L_{\mathrm{ad}_f g} L_f^{n-1} S_1(x) & \cdots & L_{\mathrm{ad}_f^{n-1} g} L_f^{n-1} S_1(x) \end{bmatrix}
$$

$$
= \begin{bmatrix} 0 & 0 & \cdots & (-1)^{n-1} L_g L_f^{n-1} S_1(x) \\ \vdots & \vdots & & * \\ 0 & -L_g L_f^{n-1} S_1(x) \cdots & & * \\ L_g L_f^{n-1} S_1(x) & * & \cdots & * \end{bmatrix}.
$$

Since the matrix of the right-hand side has rank n, it is clear that (4.12) and (4.13) are satisfied. □

Lemma 4.1 *System (4.8) is feedback linearizable with state transformation $z = S(x) = [S_1(x) \ \cdots \ S_n(x)]^\mathsf{T}$ and feedback $u = \alpha(x) + \beta(x)v$, if and only if there exists a scalar function $S_1(x)$ such that*

(i) $L_g L_f^i S_1(x) = 0$, $0 \le i \le n-2$

(ii) $L_g L_f^{n-1} S_1(x)\big|_{x=0} \ne 0$.

Furthermore, state transformation $z = S(x)$ and feedback $u = \alpha(x) + \beta(x)v$ satisfy

$$
z = S(x) = \left[S_1(x) \ L_f S_1(x) \ \cdots \ L_f^{n-1} S_1(x) \right]^\mathsf{T}. \tag{4.14}
$$

and

$$
\alpha(x) = -\frac{L_f^n S_1(x)}{L_g L_f^{n-1} S_1(x)} \ ; \quad \beta(x) = \frac{1}{L_g L_f^{n-1} S_1(x)}. \tag{4.15}
$$

Proof Necessity. Suppose that system (4.8) is feedback linearizable. Then, there exist a state transformation $z = S(x)$ and a nonsingular feedback $u = \alpha(x) + \beta(x)v$ ($\beta(x) \ne 0$) such that (4.10) is satisfied. Thus, we have that

$$
\begin{bmatrix} \frac{\partial S_1(x)}{\partial x} \\ \vdots \\ \frac{\partial S_{n-1}(x)}{\partial x} \\ \frac{\partial S_n(x)}{\partial x} \end{bmatrix} \{ f(x) + g(x)\alpha(x) + g(x)\beta(x)v \} = AS(x) + bv = \begin{bmatrix} S_2(x) \\ \vdots \\ S_n(x) \\ v \end{bmatrix}.
$$

In other words, for $1 \le i \le n-1$

$$
S_{i+1}(x) = L_{f+g(\alpha+\beta v)} S_i(x) = L_f S_i(x) + L_g S_i(x)\{\alpha(x) + \beta(x)v\}
$$

and

$$v = L_{f+g(\alpha+\beta v)} S_n(x) = L_f S_n(x) + L_g S_n(x)\{\alpha(x) + \beta(x)v\}.$$

Since $\beta(0) \neq 0$, it is easy to see that for $1 \leq i \leq n-1$

$$S_{i+1}(x) = L_f S_i(x) ; \quad L_g S_i(x) = 0 \tag{4.16}$$

and

$$L_f S_n(x) + L_g S_n(x)\alpha(x) = 0 ; \quad L_g S_n(x)\beta(x) = 1 \tag{4.17}$$

which imply that (4.14) is satisfied. Therefore, it is easy to see, by (4.16) and (4.17), that condition (i), (ii) and (4.15) are satisfied.

Sufficiency. Suppose that there exists a scalar function $S_1(x)$ such that condition (i)–(ii) are satisfied. Let us define $z = S(x) = [S_1(x) \cdots S_n(x)]^\top$ and feedback $u = \alpha(x) + \beta(x)v$ as (4.14) and (4.15), respectively. Then it is clear, from Example 4.2.1, that $z = S(x)$ is a state transformation. Also, it is easy to see, by condition (i), (4.14), and (4.15), that

$$S_* (f(x) + g(x)(\alpha(x) + \beta(x)v)) = \left. \frac{\partial S(x)}{\partial x}(f + g(\alpha + \beta v)) \right|_{x=S^{-1}(z)}$$

$$= \begin{bmatrix} \frac{\partial S_1(x)}{\partial x} \\ \frac{\partial L_f S_1(x)}{\partial x} \\ \vdots \\ \frac{\partial L_f^{n-1} S_1(x)}{\partial x} \end{bmatrix} \left. \{f + g(\alpha + \beta v)\} \right|_{x=S^{-1}(z)}$$

$$= \begin{bmatrix} L_f S_1(x) + L_g S_1(x)(\alpha + \beta v) \\ \vdots \\ L_f^{n-1} S_1(x) + L_g L_f^{n-2} S_1(x)(\alpha + \beta v) \\ L_f^n S_1(x) + L_g L_f^{n-1} S_1(x)(\alpha + \beta v) \end{bmatrix}_{x=S^{-1}(z)}$$

$$= \begin{bmatrix} L_f S_1(x) \\ \vdots \\ L_f^{n-1} S_1(x) \\ v \end{bmatrix}_{x=S^{-1}(z)} = \begin{bmatrix} z_2 \\ \vdots \\ z_n \\ v \end{bmatrix}.$$

$\square$

By using Lemma 4.1, the verifiable necessary and sufficient conditions can be obtained as follows.

Theorem 4.1 (Conditions for feedback linearization) *System (4.8) is feedback linearizable, if and only if*

(i) $\operatorname{rank}\left(\left[g(x) \ \operatorname{ad}_f g(x) \ \cdots \ \operatorname{ad}_f^{n-1} g(x)\right]\Big|_{x=0}\right) = n.$

(ii) Distribution $\Delta_{n-2}(x)$ is involutive, where

$$\Delta_{n-2}(x) \triangleq \text{span} \left\{ g(x), \text{ad}_f g(x), \ldots, \text{ad}_f^{n-2} g(x) \right\}.$$

Proof Necessity. Suppose that system (4.8) is feedback linearizable. Then, by Lemma 4.1, there exists a smooth function $S_1(x)$ such that

$$L_g L_f^i S_1(x) = 0, \quad 0 \le i \le n-2; \quad L_g L_f^{n-1} S_1(x) \Big|_{x=0} \ne 0.$$

Thus, by Example 4.2.1, condition (i) is satisfied. Also, it is clear, by Example 2.4.16, that

$$L_{\text{ad}_f^i g} S_1(x) = 0, \quad 0 \le i \le n-2; \quad L_{\text{ad}_f^{n-1} g} S_1(x) \Big|_{x=0} \ne 0.$$

Therefore, by Frobenius Theorem (or Theorem 2.8), distribution $\Delta_{n-2}(x)$ is involutive and condition (ii) is satisfied.

Sufficiency. Suppose that condition (i) and (ii) are satisfied. Then, there exists, by Frobenius Theorem (or Theorem 2.8), a smooth function $S_1(x)$ such that $S_1(0) = 0$ and

$$L_{\text{ad}_f^i g} S_1(x) = 0, \quad 0 \le i \le n-2; \quad L_{\text{ad}_f^{n-1} g} S_1(x) \Big|_{x=0} \ne 0. \tag{4.18}$$

Then, it is clear, from Example 2.4.16, that

$$L_g L_f^i S_1(x) = 0, \quad 0 \le i \le n-2; \quad L_g L_f^{n-1} S_1(x) \Big|_{x=0} \ne 0. \tag{4.19}$$

Therefore, it is clear that $S_1(x)$ satisfies condition (i) and (ii) of Lemma 4.1. Hence, by Lemma 4.1, system (4.8) is feedback linearizable. $\square$

(i) of Theorem 4.1 is called the controllability (or more precisely, reachability) condition, and (ii) is called the involutivity condition. If $n = 2$, then $\Delta_0 = \text{span}\{g\}$ and condition (ii) of Theorem 4.1 is obviously satisfied. Thus, when $n = 2$, we need to check the controllability condition only for feedback linearizability.

Suppose that conditions of Theorem 4.1 are satisfied. Then we need to find $S_1(x)$, that satisfies (4.18), in order to find a state transformation $z = S(x)$ and a nonsingular feedback $u = \alpha(x) + \beta(x)v$. In other words

$$\frac{\partial S_1(x)}{\partial x} \left[g(x) \, \text{ad}_f g(x) \, \cdots \, \text{ad}_f^{n-2} g(x) \, \text{ad}_f^{n-1} g(x) \Big|_{x=0} \right]$$
$$= c(x) \left[0 \, 0 \, \cdots \, 0 \, (-1)^{n-1} \right] \tag{4.20}$$

where $c(0) \neq 0$ (For example, $c(0) = 1$). Thus, we have

$$\frac{\partial S_1(x)}{\partial x} = c(x) \left[0 \;\cdots\; 0 \; (-1)^{n-1} \right] \left[g(x) \;\cdots\; \mathrm{ad}_f^{n-2} g(x) \;\; \mathrm{ad}_f^{n-1} g(x) \Big|_{x=0} \right]^{-1}$$

$$\triangleq c(x)\omega(x).$$

If one form $\omega(x) = \left[\omega_1(x) \;\cdots\; \omega_n(x) \right]$ is exact, then we can let $c(x) = 1$. Otherwise, we need to find a scalar function $c(x)$ such that $c(x)\omega(x)$ is exact. By Theorem 2.8 (Frobenius Theorem), we know the existence of such $c(x)$. Thus, we have, by Lemma 2.1, that

$$\frac{\partial \left(c(x)\omega(x)^{\mathsf{T}} \right)}{\partial x} = \left(\frac{\partial \left(c(x)\omega(x)^{\mathsf{T}} \right)}{\partial x} \right)^{\mathsf{T}}$$

or

$$c(x)\frac{\partial \omega(x)^{\mathsf{T}}}{\partial x} + \omega(x)^{\mathsf{T}}\frac{\partial c(x)}{\partial x} = c(x) \left(\frac{\partial \omega(x)^{\mathsf{T}}}{\partial x} \right)^{\mathsf{T}} + \left(\frac{\partial c(x)}{\partial x} \right)^{\mathsf{T}} \omega(x)$$

which implies that

$$\left(\frac{\partial \ln c(x)}{\partial x} \right)^{\mathsf{T}} \omega(x) - \omega(x)^{\mathsf{T}}\frac{\partial \ln c(x)}{\partial x} = \frac{\partial \omega(x)^{\mathsf{T}}}{\partial x} - \left(\frac{\partial \omega(x)^{\mathsf{T}}}{\partial x} \right)^{\mathsf{T}} \tag{4.21}$$

$$\triangleq Q(x).$$

Since the both sides of (4.21) are skew-symmetric matrix, we have the following $\binom{n}{2}$ equations:

$$\frac{\partial \ln c(x)}{\partial x} \left[W_1(x) \; W_2(x) \;\cdots\; W_{n-1}(x) \right] \triangleq \frac{\partial \ln c(x)}{\partial x} \bar{W}(x) = \bar{Q}(x) \tag{4.22}$$

where $Q(x) = \{q_{ij}(x)\}$ and for $1 \leq i \leq n-1$,

$$W_i(x) = \begin{bmatrix} O_{(i-1)\times 1} & O_{(i-1)\times 1} & \cdots & O_{(i-1)\times 1} \\ \omega_{i+1}(x) & \omega_{i+2}(x) & \cdots & \omega_n(x) \\ -\omega_i(x) & 0 & \cdots & 0 \\ 0 & -\omega_i(x) & \cdots & 0 \\ \vdots & \vdots & & \vdots \\ 0 & 0 & \cdots & -\omega_i(x) \end{bmatrix} \quad (n \times (n-i) \text{ matrix})$$

and

$$\bar{Q}(x) = \left[q_{12}(x) \;\cdots\; q_{1n}(x) \; q_{23}(x) \;\cdots\; q_{2n}(x) \;\cdots\; q_{(n-1)n}(x) \right].$$

For example, if $n = 4$, then we have

$$\frac{\partial \ln c(x)}{\partial x} \begin{bmatrix} \omega_2(x) & \omega_3(x) & \omega_4(x) & 0 & 0 & 0 \\ -\omega_1(x) & 0 & 0 & \omega_3(x) & \omega_4(x) & 0 \\ 0 & -\omega_1(x) & 0 & -\omega_2(x) & 0 & \omega_4(x) \\ 0 & 0 & -\omega_1(x) & 0 & -\omega_2(x) & -\omega_3(x) \end{bmatrix}$$
$$= \begin{bmatrix} q_{12}(x) & q_{13}(x) & q_{14}(x) & q_{23}(x) & q_{24}(x) & q_{34}(x) \end{bmatrix}.$$

Then it is easy to see that

$$\omega(x)\bar{W}(x) = O.$$

Since $\omega(0) \neq 0$, it is easy to see that rank $(\bar{W}(x)) = $ rank $(\bar{W}(0)) = n - 1$ and $\frac{\partial \ln c(x)}{\partial x} = \frac{-1}{\omega_K(x)} \begin{bmatrix} q_{K1}(x) & q_{K2}(x) & \cdots & q_{Kn}(x) \end{bmatrix} = \frac{-1}{\omega_K(x)} \mathbf{q}_K(x)$ is a particular solution of linear algebraic equation (4.22), where $\omega_K(0) \neq 0$ and $\mathbf{q}_K(x)$ is the Kth row of $Q(x)$. Therefore, the general solution of linear equation (4.22) is

$$\frac{\partial \ln c(x)}{\partial x} = \frac{-1}{\omega_K(x)} \mathbf{q}_K(x) + d(x)\omega(x) \tag{4.23}$$

where $d(x)$ is a smooth function on a neighborhood of the origin. If one form $\frac{-1}{\omega_K(x)}\mathbf{q}_K(x)$ is exact, then we can find easily $\ln c(x)$, $c(x)$, and $S_1(x)$ such that

$$\frac{\partial \ln c(x)}{\partial x} = \frac{-1}{\omega_K(x)} \mathbf{q}_K(x) \tag{4.24}$$

and

$$\frac{\partial S_1(x)}{\partial x} = c(x)\omega(x) \tag{4.25}$$

(See MATLAB function $\mathbf{S1}(f, g, x)$ and $\mathbf{CXexact}(\omega, x)$ in Appendix C.) However, if one form $\frac{-1}{\omega_K(x)}\mathbf{q}_K(x)$ is not exact, we need to find $c(x)$ without MATLAB program such that $c(x)\omega(x)$ is exact.

Example 4.2.2 In Example 3.2.6, it is shown that system (3.21) is not state equivalent to a linear system. Show that system (3.21) is feedback linearizable.

$$\begin{bmatrix} \dot{x}_1 \\ \dot{x}_2 \end{bmatrix} = \begin{bmatrix} x_2 \\ x_1^2 \end{bmatrix} + \begin{bmatrix} 0 \\ 1 \end{bmatrix} u = f(x) + g(x)u \tag{4.26}$$

Solution Since

$$g(x) = \begin{bmatrix} 0 \\ 1 \end{bmatrix} \quad \text{and} \quad \text{ad}_f g(x) = \begin{bmatrix} -1 \\ 0 \end{bmatrix}$$

condition (i) and (ii) of Theorem 4.1 are satisfied. Therefore, system (3.21) is feedback linearizable. A state transformation (4.14) and feedback (4.15) can be found as follows. By (4.20), we have

$$c(x) \begin{bmatrix} 0 & -1 \end{bmatrix} = \begin{bmatrix} \frac{\partial S_1(x)}{\partial x_1} & \frac{\partial S_1(x)}{\partial x_2} \end{bmatrix} \begin{bmatrix} 0 & -1 \\ 1 & 0 \end{bmatrix}$$

which implies that $\begin{bmatrix} \frac{\partial S_1(x)}{\partial x_1} & \frac{\partial S_1(x)}{\partial x_2} \end{bmatrix} = c(x) \begin{bmatrix} 1 & 0 \end{bmatrix}$. We need to find a scalar function $c(x)(\neq 0)$ such that $\frac{\partial}{\partial x_2} \left(\frac{\partial S_1(x)}{\partial x_1} \right) = \frac{\partial}{\partial x_1} \left(\frac{\partial S_1(x)}{\partial x_2} \right)$ or $\frac{\partial c(x)}{\partial x_2} = 0$. Since one form $\begin{bmatrix} 1 & 0 \end{bmatrix}$ is exact, we have that $c(x) = 1$ and $S_1(x) = x_1$. ($c(x)$ is not unique. $c(x) = 1 + 2x_1$ and $S_1(x) = x_1 + x_1^2$ also work.) Thus, it is easy to see that

$$\begin{bmatrix} z_1 \\ z_2 \end{bmatrix} = \begin{bmatrix} S_1(x) \\ L_f S_1(x) \end{bmatrix} = \begin{bmatrix} x_1 \\ x_2 \end{bmatrix}$$

and

$$u = -\frac{L_f^2 S_1(x)}{L_g L_f S_1(x)} + \frac{1}{L_g L_f S_1(x)} v = -x_1^2 + v.$$

It is easy to see that

$$\begin{bmatrix} \dot{z}_1 \\ \dot{z}_2 \end{bmatrix} = \begin{bmatrix} 0 & 1 \\ 0 & 0 \end{bmatrix} \begin{bmatrix} z_1 \\ z_2 \end{bmatrix} + \begin{bmatrix} 0 \\ 1 \end{bmatrix} v.$$

$\square$

Example 4.2.3 Find out a state transformation $z = S(x)$ and a feedback $u = \alpha(x) + \beta(x)v$ such that the closed-loop system of system (3.23) in Example 3.2.7 satisfies the Brunovsky canonical form

$$\begin{bmatrix} \dot{x}_1 \\ \dot{x}_2 \end{bmatrix} = \begin{bmatrix} -2x_2^3 - x_1 \\ -x_2^2 \end{bmatrix} + \begin{bmatrix} 1 + 2x_2 \\ 1 \end{bmatrix} u = f(x) + g(x)u$$

Solution Since

$$\det \left(\begin{bmatrix} g(x) & \mathrm{ad}_f g(x) \end{bmatrix} \right) = \det \left(\begin{bmatrix} 1 + 2x_2 & 1 + 2x_2 + 4x_2^2 \\ 1 & 2x_2 \end{bmatrix} \right) = 1$$

controllability condition (i) of Theorem 4.1 is satisfied. Since $n = 2$, involutivity condition (ii) of Theorem 4.1 is trivially satisfied. Therefore, by Theorem 4.1, system (3.23) is feedback linearizable. By (4.20), we have

$$c(x)\begin{bmatrix} 0 & -1 \end{bmatrix} = \frac{\partial S_1(x)}{\partial x} \left[g(x) \ \mathrm{ad}_f g(x)\big|_{x=0} \right]$$

$$= \begin{bmatrix} \frac{\partial S_1(x)}{\partial x_1} & \frac{\partial S_1(x)}{\partial x_2} \end{bmatrix} \begin{bmatrix} 1 + 2x_2 & 1 \\ 1 & 0 \end{bmatrix}$$

which implies that $\begin{bmatrix} \frac{\partial S_1(x)}{\partial x_1} & \frac{\partial S_1(x)}{\partial x_2} \end{bmatrix} = c(x)\begin{bmatrix} -1 & 1 + 2x_2 \end{bmatrix}$. We need to find a scalar function $c(x)(\neq 0)$ such that $\frac{\partial}{\partial x_2}\left(\frac{\partial S_1(x)}{\partial x_1} \right) = \frac{\partial}{\partial x_1}\left(\frac{\partial S_1(x)}{\partial x_2} \right)$ or $-\frac{\partial c(x)}{\partial x_2} = (1 + 2x_2)\frac{\partial c(x)}{\partial x_1}$. Since one form $\begin{bmatrix} -1 & 1 + 2x_2 \end{bmatrix}$ is exact, we have that $c(x) = 1$ and $S_1(x) = -x_1 + x_2 + x_2^2$. Thus, we have that

$$\begin{bmatrix} z_1 \\ z_2 \end{bmatrix} = \begin{bmatrix} S_1(x) \\ L_f S_1(x) \end{bmatrix} = \begin{bmatrix} -x_1 + x_2 + x_2^2 \\ x_1 - x_2^2 \end{bmatrix}$$

and

$$u = -\frac{L_f^2 S_1(x)}{L_g L_f S_1(x)} + \frac{1}{L_g L_f S_1(x)} v = x_1 + v.$$

Then, it is easy to see that

$$\begin{bmatrix} \dot{z}_1 \\ \dot{z}_2 \end{bmatrix} = \begin{bmatrix} 0 & 1 \\ 0 & 0 \end{bmatrix} \begin{bmatrix} z_1 \\ z_2 \end{bmatrix} + \begin{bmatrix} 0 \\ 1 \end{bmatrix} v.$$

$\square$

Example 4.2.4 Show that the following nonlinear control system is feedback linearizable. Also, find a linearizing state transformation and feedback.

$$\begin{bmatrix} \dot{x}_1 \\ \dot{x}_2 \\ \dot{x}_3 \end{bmatrix} = \begin{bmatrix} x_2 \\ x_1 \\ x_2 + x_1 x_3 \end{bmatrix} + \begin{bmatrix} 1 \\ 0 \\ 0 \end{bmatrix} u = f(x) + g(x)u \qquad (4.27)$$

Solution By simple calculation, we have

$$\mathrm{ad}_f g(x) = \begin{bmatrix} 0 \\ -1 \\ -x_3 \end{bmatrix} \quad \text{and} \quad \mathrm{ad}_f^2 g(x) = \begin{bmatrix} 1 \\ 0 \\ 1 - x_2 \end{bmatrix}$$

which implies that condition (i) of Theorem 4.1 is satisfied. Since $[g(x), \mathrm{ad}_f g(x)] = 0$, distribution $\Delta_1(x) = \mathrm{span}\{g(x), \mathrm{ad}_f g(x)\}$ is involutive and condition (ii) is also satisfied. Therefore, by Theorem 4.1, system (4.27) is feedback linearizable. By (4.20), we have

$$c(x)\begin{bmatrix}0 & 0 & 1\end{bmatrix} = \begin{bmatrix}L_g S_1(x) & L_{\mathrm{ad}_{fg}} S_1(x) & L_{\mathrm{ad}_f^2 g}\big|_{x=0} S_1(x)\end{bmatrix}$$

$$= \begin{bmatrix}\frac{\partial S_1(x)}{\partial x_1} & \frac{\partial S_1(x)}{\partial x_2} & \frac{\partial S_1(x)}{\partial x_3}\end{bmatrix}\begin{bmatrix}1 & 0 & 1\\ 0 & -1 & 0\\ 0 & -x_3 & 1\end{bmatrix}$$

which implies that $\begin{bmatrix}\frac{\partial S_1(x)}{\partial x_1} & \frac{\partial S_1(x)}{\partial x_2} & \frac{\partial S_1(x)}{\partial x_3}\end{bmatrix} = c(x)\begin{bmatrix}0 & -x_3 & 1\end{bmatrix} \triangleq c(x)\omega(x)$. Note that one form $\omega(x)$ is not exact. We need to find a scalar function $c(x)$ $(c(0) = 1)$ such that $\frac{\partial}{\partial x_j}\left(\frac{\partial S_1(x)}{\partial x_i}\right) = \frac{\partial}{\partial x_i}\left(\frac{\partial S_1(x)}{\partial x_j}\right)$ for $i \neq j$ or

$$\frac{\partial c(x)}{\partial x_1} = 0$$

$$\frac{\partial c(x)}{\partial x_2} = \frac{\partial (-c(x)x_3)}{\partial x_3} = -c(x) - \frac{\partial c(x)}{\partial x_3}x_3.$$

We have, by (4.24) and (4.25), that $\omega_3(0) = 1 \neq 0$

$$Q(x) \triangleq \frac{\partial \omega(x)^\mathsf{T}}{\partial x} - \left(\frac{\partial \omega(x)^\mathsf{T}}{\partial x}\right)^\mathsf{T} = \begin{bmatrix}0 & 0 & 0\\ 0 & 0 & -1\\ 0 & 1 & 0\end{bmatrix}$$

$$\frac{\partial \ln c(x)}{\partial x} = \frac{-1}{\omega_3(0)}\mathbf{q}_3(x) = \begin{bmatrix}0 & -1 & 0\end{bmatrix}.$$

Since one form $\frac{-1}{\omega_3(0)}\mathbf{q}_3(x)$ is exact, we have $\ln c(x) = -x_2$, $c(x) = e^{-x_2}$, and $S_1(x) = x_3 e^{-x_2}$. Thus, we have that

$$\begin{bmatrix}z_1\\ z_2\\ z_3\end{bmatrix} = \begin{bmatrix}S_1(x)\\ L_f S_1(x)\\ L_f^2 S_1(x)\end{bmatrix} = \begin{bmatrix}x_3 e^{-x_2}\\ x_2 e^{-x_2}\\ x_1(1 - x_2)e^{-x_2}\end{bmatrix}$$

and

$$u = -\frac{L_f^3 S_1(x)}{L_g L_f^2 S_1(x)} + \frac{1}{L_g L_f^2 S_1(x)}v = \frac{x_1^2 x_2 - 2x_1^2 - x_2^2 + x_2}{x_2 - 1} + \frac{e^{x_2}}{1 - x_2}v.$$

Then, it is easy to see that

$$\begin{bmatrix}\dot{z}_1\\ \dot{z}_2\\ \dot{z}_3\end{bmatrix} = \begin{bmatrix}0 & 1 & 0\\ 0 & 0 & 1\\ 0 & 0 & 0\end{bmatrix}\begin{bmatrix}z_1\\ z_2\\ z_3\end{bmatrix} + \begin{bmatrix}0\\ 0\\ 1\end{bmatrix}v.$$

$\square$

Example 4.2.5 Feedback linearize the following nonlinear system:

$$
\begin{bmatrix} \dot{x}_1 \\ \dot{x}_2 \\ \dot{x}_3 \end{bmatrix} = \begin{bmatrix} x_2 - x_1^2 \\ x_3 + 2x_1x_2 - x_1^3 \\ x_1^2 - 3x_1^2x_2 + 3x_1^4 \end{bmatrix} + \begin{bmatrix} 0 \\ 0 \\ 1 + x_1 \end{bmatrix} u = f(x) + g(x)u \tag{4.28}
$$

Solution By simple calculation, we have

$$
\mathrm{ad}_f g(x) = \begin{bmatrix} 0 \\ -1 - x_1 \\ x_2 - x_1^2 \end{bmatrix} \quad \text{and} \quad \mathrm{ad}_f^2 g(x) = \begin{bmatrix} 1 + x_1 \\ 4x_1^2 + 2x_1 - 2x_2 \\ -2x_1^3 - 3x_1^2 + x_3 \end{bmatrix}
$$

which implies that condition (i) of Theorem 4.1 is satisfied. Since $[g(x), \mathrm{ad}_f g(x)] = 0$, distribution $\Delta_1(x) = \mathrm{span}\{g(x), \mathrm{ad}_f g(x)\}$ is involutive and condition (ii) is also satisfied. Therefore, by Theorem 4.1, system (4.28) is feedback linearizable. By (4.20), we have

$$
c(x)\begin{bmatrix} 0 & 0 & 1 \end{bmatrix} = \begin{bmatrix} \frac{\partial S_1(x)}{\partial x_1} & \frac{\partial S_1(x)}{\partial x_2} & \frac{\partial S_1(x)}{\partial x_3} \end{bmatrix} \begin{bmatrix} 0 & 0 & 1 \\ 0 & -1 - x_1 & 0 \\ 1 + x_1 & x_2 - x_1^2 & 0 \end{bmatrix}
$$

which implies that $\begin{bmatrix} \frac{\partial S_1(x)}{\partial x_1} & \frac{\partial S_1(x)}{\partial x_2} & \frac{\partial S_1(x)}{\partial x_3} \end{bmatrix} = c(x)\begin{bmatrix} 1 & 0 & 0 \end{bmatrix}$. Since one form $\begin{bmatrix} 1 & 0 & 0 \end{bmatrix}$ is exact, we can let $c(x) = 1$ and $S_1(x) = x_1$. Thus, we have

$$
\begin{bmatrix} z_1 \\ z_2 \\ z_3 \end{bmatrix} = \begin{bmatrix} S_1(x) \\ L_f S_1(x) \\ L_f^2 S_1(x) \end{bmatrix} = \begin{bmatrix} x_1 \\ x_2 - x_1^2 \\ x_3 + x_1^3 \end{bmatrix} \tag{4.29}
$$

and

$$
u = -\frac{L_f^3 S_1(x)}{L_g L_f^2 S_1(x)} + \frac{1}{L_g L_f^2 S_1(x)} v = -\frac{x_1^2}{1 + x_1} + \frac{1}{1 + x_1} v. \tag{4.30}
$$

Then, it is easy to see that

$$
\begin{bmatrix} \dot{z}_1 \\ \dot{z}_2 \\ \dot{z}_3 \end{bmatrix} = \begin{bmatrix} 0 & 1 & 0 \\ 0 & 0 & 1 \\ 0 & 0 & 0 \end{bmatrix} \begin{bmatrix} z_1 \\ z_2 \\ z_3 \end{bmatrix} + \begin{bmatrix} 0 \\ 0 \\ 1 \end{bmatrix} v. \tag{4.31}
$$

$\square$

Brunovsky canonical form (4.31) is not asymptotically stable, because all eigenvalues are zero. In Example 4.2.5, if we use the feedback

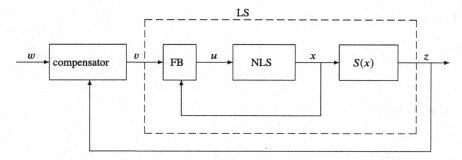

Fig. 4.3 Feedback linearization with the compensator

$$u = -\frac{x_1^2}{1 + x_1} + \frac{1}{1 + x_1}\{-a_0 z_1 - a_1 z_2 - a_2 z_3 + w\}$$

$$= -\frac{x_1^2}{1 + x_1} + \frac{1}{1 + x_1}\{-a_0 S_1(x) - a_1 S_2(x) - a_2 S_3(x) + w\}$$

instead of (4.30), then we have

$$\begin{bmatrix} \dot{z}_1 \\ \dot{z}_2 \\ \dot{z}_3 \end{bmatrix} = \begin{bmatrix} 0 & 1 & 0 \\ 0 & 0 & 1 \\ -a_0 & -a_1 & -a_2 \end{bmatrix} \begin{bmatrix} z_1 \\ z_2 \\ z_3 \end{bmatrix} + \begin{bmatrix} 0 \\ 0 \\ 1 \end{bmatrix} w$$

whose characteristic equation is $s^3 + a_2 s^2 + a_1 s + a_0 = 0$. In other words, once the nonlinear system is feedback linearized, we can use the well-known linear system theory to control it. Figure 4.3 shows the block diagram of feedback linearization with the compensation.

In Example 4.2.5, state transformation (4.29) is valid for all x, but feedback (4.30) works only when $x_1 \neq -1$. Therefore, the nonlinear system (4.28) is not globally feedback linearizable. System (4.28) is locally feedback linearizable (on $\{x \in \mathbb{R}^3 \mid x_1 > -1\}$). The neighborhood of the origin, for local linearization, depends on how far the state transformation and feedback are valid.

For system (4.8), let us define the following distributions:

$$\Delta_0(x) = \text{span}\{g(x)\}$$
$$\Delta_i(x) = \Delta_{i-1}(x) + [f(x), \Delta_{i-1}(x)], \quad i \geq 1 \tag{4.32}$$

or for $i \geq 0$,

$$\Delta_i(x) \triangleq \text{span}\left\{ \text{ad}_f^k g(x) \mid 0 \leq k \leq i \right\}$$
$$= \text{span}\left\{ g(x), \text{ad}_f g(x), \ldots, \text{ad}_f^i g(x) \right\}. \tag{4.33}$$

With nonsingular feedback $u = \alpha(x) + \beta(x)v$, we have the following closed-loop system:

$$\dot{x} = f(x) + g(x)\alpha(x) + g(x)\beta(x)v$$
$$\triangleq \hat{f}(x) + \hat{g}(x)v. \tag{4.34}$$

It is clear that

$$f(x) = \hat{f}(x) + \hat{g}(x)\hat{\alpha}(x) \; ; \; g(x) = \hat{g}(x)\hat{\beta}(x) \tag{4.35}$$

where $\hat{\beta}(x) = \beta(x)^{-1}$ and $\hat{\alpha}(x) = -\hat{\beta}(x)\alpha(x)$. For the closed-loop system (4.47), we can also define the following distributions:

$$\hat{\Delta}_0(x) = \text{span}\{\hat{g}(x)\}$$
$$\hat{\Delta}_i(x) = \hat{\Delta}_{i-1}(x) + \text{ad}_{\hat{f}}\hat{\Delta}_{i-1}(x), \quad i \geq 1 \tag{4.36}$$

or for $i \geq 0$,

$$\hat{\Delta}_i(x) \triangleq \text{span}\left\{ \text{ad}_{\hat{f}}^k \hat{g}(x) \mid 0 \leq k \leq i \right\}$$
$$= \text{span}\left\{ \hat{g}(x), \text{ad}_{\hat{f}}\hat{g}(x), \ldots, \text{ad}_{\hat{f}}^i \hat{g}(x) \right\}. \tag{4.37}$$

Example 4.2.6 Suppose that $\dim(\Delta_{n-1}(x)) = n$ and $0 \leq k \leq n - 2$. Show that if distribution $\Delta_k(x)$ is involutive, then $\Delta_{k-1}(x)$ is also involutive.

Solution Suppose that $\Delta_k(x)$ is involutive. Assume that $\Delta_{k-1}(x)$ is not involutive. Then, there exists i and j such that $j < i \leq k - 1$ and

$$\left[\text{ad}_f^i g(x), \text{ad}_f^j g(x) \right] = c(x)\text{ad}_f^k g(x) + Y(x)$$

where $c(x) \neq 0$ and $Y(x) \in \Delta_{k-1}(x)$. By Jacobi identity (or (2.18)), it is clear that

$$\left[f, \left[\text{ad}_f^i g, \text{ad}_f^j g \right] \right] = \left[\text{ad}_f^i g, \text{ad}_f^{j+1} g \right] - \left[\text{ad}_f^j g, \text{ad}_f^{i+1} g \right].$$

Thus, we have, by (2.42), that

$$\left[\text{ad}_f^j g, \text{ad}_f^{i+1} g \right] = \left[\text{ad}_f^i g, \text{ad}_f^{j+1} g \right] - \left[f, \left[\text{ad}_f^i g, \text{ad}_f^j g \right] \right]$$
$$= \left[\text{ad}_f^i g, \text{ad}_f^{j+1} g \right] - c(x)\text{ad}_f^{k+1} g - L_f c(x)\text{ad}_f^k g - [f, Y(x)]$$
$$\triangleq -c(x)\text{ad}_f^{k+1} g + Z(x)$$

where $Z(x) = \left[\text{ad}_f^i g, \text{ad}_f^{j+1} g\right] - L_f c(x) \text{ad}_f^k g - [f, Y(x)] \in \Delta_k(x)$. Since $\text{ad}_f^{k+1} g$
$\notin \Delta_k(x)$, $c(x) \neq 0$, and $j < i + 1 \leq k$, $\left[\text{ad}_f^j g, \text{ad}_f^{i+1} g\right] \notin \Delta_k(x)$ and thus $\Delta_k(x)$ is
not involutive. It contradicts. Hence, $\Delta_{k-1}(x)$ is involutive. $\square$

Theorem 4.2 (Conditions for feedback linearization) *System* (4.8) *is feedback linearizable, if and only if*

(i) $\dim(\Delta_{n-1}(x)) = n$
(ii) $\Delta_k(x)$, $0 \leq k \leq n - 2$ *are constant dimensional involutive distributions.*

Proof Obvious by Theorem 4.1 and Example 4.2.6. $\square$

Example 4.2.7 Show that the following nonlinear system is not feedback linearizable.

$$\begin{bmatrix} \dot{x}_1 \\ \dot{x}_2 \\ \dot{x}_3 \end{bmatrix} = \begin{bmatrix} x_2 \\ x_3 + x_2^2 \\ 0 \end{bmatrix} + \begin{bmatrix} 0 \\ 1 \\ 1 \end{bmatrix} u = f(x) + g(x)u \qquad (4.38)$$

Solution By simple calculation, we have

$$\text{ad}_f g(x) = \begin{bmatrix} -1 \\ -1 - 2x_2 \\ 0 \end{bmatrix} \quad \text{and} \quad \text{ad}_f^2 g(x) = \begin{bmatrix} 1 + 2x_2 \\ 2x_2^2 + 2x_2 - 2x_3 \\ 0 \end{bmatrix}$$

which implies that condition (i) of Theorem 4.1 is satisfied. However, since
$[g(x), \text{ad}_f g(x)] = \begin{bmatrix} 0 & -2 & 0 \end{bmatrix}^{\mathsf{T}} \notin \Delta_1(x)$, distribution $\Delta_1(x) = \text{span}\{g(x), \text{ad}_f g(x)\}$
is not involutive and condition (ii) is not satisfied. Therefore, by Theorem 4.1, system (4.28) is not feedback linearizable. $\square$

4.3   Multi-input Nonlinear Systems

In this section, we extend the single input results of the previous section to multi-input systems. Consider the following smooth multi-input control systems:

$$\dot{x} = f(x) + \sum_{i=1}^{m} g_i(x)u_i = f(x) + g(x)u \qquad (4.39)$$

where $x \in \mathbb{R}^n$, $u \in \mathbb{R}^m$, and $f(0) = 0$. Let us define the following set of vector fields:

$$\begin{aligned} \Delta_0(x) &= \text{span}\{ g_i(x) \mid 1 \leq i \leq m\} \\ \Delta_i(x) &= \Delta_{i-1} + \left[f(x), \Delta_{i-1}(x)\right], \quad i \geq 1 \end{aligned} \qquad (4.40)$$

or

$$\Delta_i(x) \triangleq \text{span} \left\{ \text{ad}_f^k g_j(x) \big| 1 \leq j \leq m, \ 0 \leq k \leq i \right\}, \quad i \geq 0. \tag{4.41}$$

Then it is clear that

$$\Delta_0(x) \subset \Delta_1(x) \subset \Delta_2(x) \subset \cdots \subset \Delta_{n-1}(x) = \Delta_n(x) = \cdots \tag{4.42}$$

Example 4.3.1 Suppose that $\dim(\Delta_i(x)) = \dim(\Delta_i(0))$, $i \geq 0$ on a neighborhood U of $0 \in \mathbb{R}^n$. In other words, $\Delta_i(x)$, $i \geq 0$ are distributions on a neighborhood U of $0 \in \mathbb{R}^n$. Show that for $i \geq 0$,

$$\Delta_i(x) = \text{span} \left\{ \text{ad}_f^k g_j(x) \big| 1 \leq j \leq m, \ 0 \leq k \leq \min(i, \kappa_j - 1) \right\} \tag{4.43}$$

and

$$\dim(\Delta_i(x)) = \sum_{j=1}^{m} \min(i + 1, \kappa_j). \tag{4.44}$$

Solution Suppose that $1 \leq p \leq m$, $\ell \geq \kappa_p$, and $i \leq \kappa_p - 1$. If

$$\text{ad}_f^\ell g_p(x) \notin \text{span} \left\{ \text{ad}_f^k g_j(x) \big| 1 \leq j \leq m, \ 0 \leq k \leq \min(i, \kappa_j - 1) \right\}$$

then we have that

$$\text{ad}_f^\ell g_p(x) \big|_{x=0} \notin \text{span} \left\{ \text{ad}_f^k g_j(x) \big|_{x=0} \Big| 1 \leq j \leq m, \ 0 \leq k \leq \min(i, \kappa_j - 1) \right\}$$

which contradicts the definition of the Kronecker indices. Thus, it is clear that for $1 \leq p \leq m$ and $\ell \geq \kappa_p$

$$\text{ad}_f^\ell g_p(x) \in \text{span} \left\{ \text{ad}_f^k g_j(x) \big| 1 \leq j \leq m, \ 0 \leq k \leq \min(i, \kappa_j - 1) \right\}.$$

Therefore, it is easy to see that (4.43) and (4.44) are satisfied.   □

Definition 4.3 (*g-invariant distribution*) For system (4.39), distribution $D(x)$ is said to be g-invariant, if for $1 \leq i \leq m$

$$[g_i(x), D(x)] \subset D(x). \tag{4.45}$$

Definition 4.4 ((*f, g*)-*invariant distribution*) For system (4.39), distribution $D(x)$ is said to be (*f, g*)-invariant, if for $1 \leq i \leq m$

$$[f(x), D(x)] \subset D(x) \quad \text{and} \quad [g_i(x), D(x)] \subset D(x). \tag{4.46}$$

With nonsingular feedback $u = \alpha(x) + \beta(x)v$, we have the following closed-loop system:

$$\dot{x} = f(x) + g(x)\alpha(x) + g(x)\beta(x)v$$
$$\triangleq \hat{f}(x) + \hat{g}(x)v. \tag{4.47}$$

It is clear that

$$f(x) = \hat{f}(x) + \hat{g}(x)\hat{\alpha}(x) ; \quad g(x) = \hat{g}(x)\hat{\beta}(x) \tag{4.48}$$

where $\hat{\beta}(x) = \beta(x)^{-1}$ and $\hat{\alpha}(x) = -\hat{\beta}(x)\alpha(x)$. For the closed-loop system (4.47), we can also define the following distributions:

$$\hat{\Delta}_0(x) = \text{span}\{\hat{g}_j(x) \mid 1 \le j \le m\}$$
$$\hat{\Delta}_i(x) = \hat{\Delta}_{i-1}(x) + \left[\hat{f}(x), \hat{\Delta}_{i-1}(x)\right], \quad i \ge 1 \tag{4.49}$$

or

$$\hat{\Delta}_i(x) \triangleq \text{span}\left\{\text{ad}^k_{\hat{f}}\hat{g}_j(x)\middle| 1 \le j \le m, \ 0 \le k \le i\right\}, \quad i \ge 0. \tag{4.50}$$

Example 4.3.2 Show that if $\Delta_i(x)$, $i \ge 0$ are g-invariant, then

$$\hat{\Delta}_i(x) = \Delta_i(x), \quad i \ge 0. \tag{4.51}$$

Solution Suppose that for $i \ge 0$,

$$[g_j(x), \Delta_i(x)] \subset \Delta_i(x), \quad 1 \le j \le m. \tag{4.52}$$

Let $\beta_{kj}(x)$ and $\hat{\beta}_{kj}(x)$ be the kj-element of $\beta(x)$ and $\hat{\beta}(x)$, respectively. Since $\hat{g}_j(x) = \sum_{k=1}^{m}\beta_{kj}(x)g_k(x)$ and $g_j(x) = \sum_{k=1}^{m}\hat{\beta}_{kj}(x)\hat{g}_k(x)$, we have that $\hat{\Delta}_0(x) \subset \Delta_0(x)$ and $\Delta_0(x) \subset \hat{\Delta}_0(x)$, respectively. Thus, it is clear that $\hat{\Delta}_0(x) = \Delta_0(x)$. Assume that $i \ge 1$ and $\hat{\Delta}_{i-1}(x) = \Delta_{i-1}(x)$. Then it is easy to see, by (2.42), (4.40), (4.42), and (4.52), that

$$\hat{\Delta}_i(x) = \hat{\Delta}_{i-1}(x) + \left[\hat{f}(x), \hat{\Delta}_{i-1}(x)\right]$$

$$= \Delta_{i-1}(x) + \left[f(x) + \sum_{k=1}^{m} \alpha_k(x)g_k(x), \Delta_{i-1}(x)\right]$$

$$\subset \Delta_{i-1}(x) + [f, \Delta_{i-1}] + \sum_{k=1}^{m} \alpha_k(x)[g_k, \Delta_{i-1}] + \Delta_0$$

$$\subset \Delta_i(x).$$

Similarly, we can show that $\Delta_i(x) \subset \hat{\Delta}_i(x)$. Thus, we have $\hat{\Delta}_i(x) = \Delta_i(x)$. Therefore, by mathematical induction, (4.51) is satisfied. $\qquad\square$

Lemma 4.2 *If system* (4.39) *is feedback linearizable, then*

$$\dim(\Delta_{n-1}(x)) = n.$$

Proof Suppose that system (4.39) is feedback linearizable with state transformation $z = S(x)$ and nonsingular feedback $u = \alpha(x) + \beta(x)v$. Then it is clear, by (4.6), that

$$S_*(\hat{f}(x)) = Az \; ; \quad S_*(\hat{g}_j(x)) = b_j$$

where $\hat{f}(x) = f(x) + g(x)\alpha(x)$, $\hat{g}(x) = g(x)\beta(x)$, $f(x) = \hat{f}(x) + \hat{g}(x)\hat{\alpha}(x)$, and $g(x) = \hat{g}(x)\hat{\beta}(x)$. It is easy to see, by (2.28) and Example 2.4.14, that for $i \geq 0$, $1 \leq j \leq m$, $1 \leq \ell \leq m$, and $0 \leq k \leq i$

$$\left[\hat{g}_j(x), \mathrm{ad}_{\hat{f}}^k \hat{g}_\ell(x)\right] = [S_*^{-1}(b_j), S_*^{-1}(A^k b_\ell)] = S_*^{-1}\left([b_j, A^k b_\ell]\right)$$

$$= 0 \in \hat{\Delta}_i(x)$$

which implies that $\hat{\Delta}_i(x)$, $i \geq 0$ are g-invariant. Therefore, it is clear, from Example 4.3.2, that $\hat{\Delta}_i(x) = \Delta_i(x)$, $i \geq 0$. Since for $i \geq 0$,

$$\hat{\Delta}_i(x) \triangleq \mathrm{span}\left\{\mathrm{ad}_{\hat{f}}^k \hat{g}_j(x) \,\middle|\, 1 \leq j \leq m, \; 0 \leq k \leq i\right\}$$

$$= \mathrm{span}\left\{S_*^{-1}(A^k b_\ell) \,\middle|\, 1 \leq j \leq m, \; 0 \leq k \leq i\right\}$$

it is easy to see that $\dim(\Delta_{n-1}(x)) = \dim(\hat{\Delta}_{n-1}(x)) = \sum_{i=1}^{m} \kappa_i = n$. $\qquad\square$

Example 4.3.3 Let $1 \leq j \leq m$. Suppose that for $1 \leq i \leq j$

$$L_g L_f^k S_{i1}(x) = 0, \quad 0 \leq k \leq \kappa_i - 2 \tag{4.53}$$

and

$$\text{rank} \left(\begin{bmatrix} L_g L_f^{\kappa_1 - 1} S_{11}(x) \\ L_g L_f^{\kappa_2 - 1} S_{21}(x) \\ \vdots \\ L_g L_f^{\kappa_j - 1} S_{j1}(x) \end{bmatrix} \Bigg|_{x=0} \right) = j \tag{4.54}$$

Show that

$$\left\{ d \left(L_f^\ell S_{i1}(x) \right) \Big|_{x=0} \mid 1 \le i \le j, \ 0 \le \ell \le \kappa_i - 1 \right\}$$

is a set of linearly independent 1-forms.

Solution We can assume, without loss of generality, that $\kappa_1 \ge \kappa_2 \ge \cdots \ge \kappa_j$. Let

$$\kappa_1 = \cdots = \kappa_{m_1} > \kappa_{m_1+1} = \cdots = \kappa_{m_2} > \cdots > \kappa_{m_{p-1}+1} = \cdots = \kappa_{m_p}$$

and for $1 \le q \le p$

$$S^q(x) \triangleq \begin{bmatrix} S_{(m_{q-1}+1)1}(x) \\ \vdots \\ S_{m_q 1}(x) \end{bmatrix}$$

where $m_p = j$ and $m_0 \triangleq 0$. Suppose that

$$\sum_{i=1}^{p-1} \sum_{\ell=0}^{\kappa_{m_i} - \kappa_{m_{i+1}} - 1} c_\ell^i \begin{bmatrix} d \left(L_f^{\ell + \kappa_{m_1} - \kappa_{m_i}} S^1(x) \right) \\ d \left(L_f^{\ell + \kappa_{m_2} - \kappa_{m_i}} S^2(x) \right) \\ \vdots \\ d \left(L_f^\ell S^i(x) \right) \end{bmatrix} \Bigg|_{x=0}$$

$$+ \sum_{\ell=0}^{\kappa_{m_p} - 1} c_\ell^p \begin{bmatrix} d \left(L_f^{\ell + \kappa_{m_1} - \kappa_{m_p}} S^1(x) \right) \\ d \left(L_f^{\ell + \kappa_{m_2} - \kappa_{m_p}} S^2(x) \right) \\ \vdots \\ d \left(L_f^\ell S^p(x) \right) \end{bmatrix} \Bigg|_{x=0} = O_{1 \times n} \tag{4.55}$$

where c_ℓ^i, $1 \le i \le p$ are $1 \times m_i$ vectors for all ℓ. If we postmultiply (4.55) by $\begin{bmatrix} g_1(x) \cdots g_m(x) \end{bmatrix}\big|_{x=0}$, then we have, by (4.53), that

$$O_{1 \times m_p} = c^p_{\kappa_{m_p}-1} \left. \begin{bmatrix} L_g L_f^{\kappa_{m_1}-1} S^1(x) \\ L_g L_f^{\kappa_{m_2}-1} S^2(x) \\ \vdots \\ L_g L_f^{\kappa_{m_p}-1} S^p(x) \end{bmatrix} \right|_{x=0}$$

which implies, together with (4.54), that $c^p_{\kappa_{m_p}-1} = O_{1 \times m_p}$. If $\kappa_{m_p} \geq 2$, then we can show that $c^q_{\kappa_{m_p}-2} = O_{1 \times m_p}$ by post-multiplying (4.55) by $[\mathrm{ad}_f g_1(x) \cdots \mathrm{ad}_f g_m(x)]\big|_{x=0}$. In this manner, it can be easily shown that $c^p_\ell = O_{1 \times m_p}$ for $0 \leq \ell \leq \kappa_{m_p} - 1$. Similarly, we can show that $c^{p-1}_{\kappa_{m_{p-1}}-\kappa_{m_p}-1} = O_{1 \times m_{p-1}}$ by post-multiplying (4.55) by $\left[\mathrm{ad}_f^{\kappa_{m_p}} g_1(x) \cdots \mathrm{ad}_f^{\kappa_{m_p}} g_m(x) \right]\big|_{x=0}$. In this manner, it can be easily shown that $c^i_\ell = O_{1 \times m_i}$ for $1 \leq i \leq p - 1$ and $0 \leq \ell \leq \kappa_{m_i} - \kappa_{m_{i+1}} - 1$. Therefore, $\left\{ d\left(L_f^\ell S_{i1}(x) \right)\big|_{x=0} \mid 1 \leq i \leq j, \ 0 \leq \ell \leq \kappa_i - 1 \right\}$ is a set of linearly independent 1-forms. $\qquad \square$

The following Lemma is the multi-input version of Lemma 4.1.

Lemma 4.3 *System* (4.39) *is feedback linearizable with state transformation* $z = S(x) = [S_{11}(x) \cdots S_{1\kappa_1}(x) \cdots S_{m1}(x) \cdots S_{m\kappa_m}(x)]^\mathsf{T}$ *and feedback* $u = \alpha(x) + \beta(x)v$, *if and only if for* $1 \leq i \leq m$

(i) $L_g L_f^k S_{i1}(x) = 0, \ 0 \leq k \leq \kappa_i - 2$

(ii) rank $\left(\left. \begin{bmatrix} L_g L_f^{\kappa_1-1} S_{11}(x) \\ \vdots \\ L_g L_f^{\kappa_m-1} S_{m1}(x) \end{bmatrix} \right|_{x=0} \right) = m.$

Furthermore, state transformation $z = S(x) = [S_{11}(x) \cdots S_{1\kappa_1}(x) \cdots S_{m1}(x) \cdots S_{m\kappa_m}(x)]^\mathsf{T}$ *and feedback* $u = \alpha(x) + \beta(x)v$ *satisfy that for* $1 \leq i \leq m$,

$$S_{ik}(x) = L_f^{k-1} S_{i1}(x), \ 2 \leq k \leq \kappa_i \tag{4.56}$$

and

$$\beta(x) = \begin{bmatrix} L_g L_f^{\kappa_1-1} S_{11}(x) \\ \vdots \\ L_g L_f^{\kappa_m-1} S_{m1}(x) \end{bmatrix}^{-1} ; \quad \alpha(x) = -\beta(x) \begin{bmatrix} L_f^{\kappa_1} S_{11}(x) \\ \vdots \\ L_f^{\kappa_m} S_{m1}(x) \end{bmatrix}. \tag{4.57}$$

Proof Necessity. Suppose that system (4.39) is feedback linearizable with state transformation $z = S(x) = [S_{11}(x) \cdots S_{1\kappa_1}(x) \cdots S_{m1}(x) \cdots S_{m\kappa_m}(x)]^{\mathsf{T}}$ and feedback $u = \alpha(x) + \beta(x)v$. Then, we have, by (4.6), that for $1 \le i \le m$ and $1 \le k \le \kappa_i - 1$

$$
\begin{aligned}
S_{i(k+1)}(x) &= L_{f+g(\alpha+\beta v)} S_{ik}(x) \\
&= L_f S_{ik}(x) + L_g S_{ik}(x)\{\alpha(x) + \beta(x)v\}
\end{aligned}
$$

and

$$
\begin{bmatrix} v_1 \\ \vdots \\ v_m \end{bmatrix} = \begin{bmatrix} L_{f+g(\alpha+\beta v)} S_{1\kappa_1}(x) \\ \vdots \\ L_{f+g(\alpha+\beta v)} S_{m\kappa_m}(x) \end{bmatrix}
$$

$$
= \begin{bmatrix} L_f S_{1\kappa_1}(x) \\ \vdots \\ L_f S_{m\kappa_m}(x) \end{bmatrix} + \begin{bmatrix} L_g S_{1\kappa_1}(x) \\ \vdots \\ L_g S_{m\kappa_m}(x) \end{bmatrix} \{\alpha(x) + \beta(x)v\}.
$$

Since $\det(\beta(0)) \ne 0$, it is easy to see that for $1 \le i \le m$ and $1 \le k \le \kappa_i - 1$

$$
S_{i(k+1)}(x) = L_f S_{ik}(x) ; \quad L_g S_{ik}(x) = O_{1 \times m} \tag{4.58}
$$

and

$$
\begin{bmatrix} L_f S_{1\kappa_1}(x) \\ \vdots \\ L_f S_{m\kappa_m}(x) \end{bmatrix} + \begin{bmatrix} L_g S_{1\kappa_1}(x) \\ \vdots \\ L_g S_{m\kappa_m}(x) \end{bmatrix} \alpha(x) = 0 ; \quad \begin{bmatrix} L_g S_{1\kappa_1}(x) \\ \vdots \\ L_g S_{m\kappa_m}(x) \end{bmatrix} \beta(x) = I_m \tag{4.59}
$$

which imply that (4.56) is satisfied. Therefore, it is easy to see, by (4.58) and (4.59), that condition (i), condition (ii) and (4.57) are satisfied.

Sufficiency. Suppose that there exist scalar functions $S_{i1}(x)$, $1 \le i \le m$ such that condition (i) and condition (ii) are satisfied. Let us define

$$
\begin{aligned}
z &\triangleq [z_{11} \cdots z_{1\kappa_1} \cdots z_{m1} \cdots z_{m\kappa_m}]^{\mathsf{T}} \\
&= S(x) = [S_{11}(x) \cdots S_{1\kappa_1}(x) \cdots S_{m1}(x) \cdots S_{m\kappa_m}(x)]^{\mathsf{T}}
\end{aligned}
$$

and feedback $u = \alpha(x) + \beta(x)v$ as (4.56) and (4.57), respectively. Then it is clear, by Example 4.3.3, that $z = S(x)$ is a state transformation. It is easy to see, by condition (i) and (4.56), that

$$S_* \left(f + g(\alpha + \beta v) \right) = \frac{\partial S(x)}{\partial x} \left(f + g(\alpha + \beta v) \right) \Big|_{x=S^{-1}(z)}$$

$$= \begin{bmatrix} \dfrac{\partial S_{11}(x)}{\partial x} \\[4pt] \dfrac{\partial L_f S_{11}(x)}{\partial x} \\[4pt] \vdots \\[4pt] \dfrac{\partial L_f^{\kappa_1 - 1} S_{11}(x)}{\partial x} \\[4pt] \vdots \\[4pt] \dfrac{\partial S_{m1}(x)}{\partial x} \\[4pt] \vdots \\[4pt] \dfrac{\partial L_f^{\kappa_m - 1} S_{m1}(x)}{\partial x} \end{bmatrix} \{ f + g(\alpha + \beta v) \} \Bigg|_{x=S^{-1}(z)}$$

$$= \begin{bmatrix} L_f S_{11}(x) + L_g S_{11}(x)(\alpha + \beta v) \\[4pt] \vdots \\[4pt] L_f^{\kappa_1 - 1} S_{11}(x) + L_g L_f^{\kappa_1 - 2} S_{11}(x)(\alpha + \beta v) \\[4pt] L_f^{\kappa_1} S_{11}(x) + L_g L_f^{\kappa_1 - 1} S_{11}(x)(\alpha + \beta v) \\[4pt] \vdots \\[4pt] L_f S_{11}(x) + L_g S_{m1}(x)(\alpha + \beta v) \\[4pt] \vdots \\[4pt] L_f^{\kappa_m} S_{m1}(x) + L_g L_f^{\kappa_m - 1} S_{m1}(x)(\alpha + \beta v) \end{bmatrix} \Bigg|_{x=S^{-1}(z)}$$

$$= \begin{bmatrix} L_f S_{11}(x) \\[4pt] \vdots \\[4pt] L_f^{\kappa_1 - 1} S_{11}(x) \\[4pt] v_1 \\[4pt] \vdots \\[4pt] L_f S_{m1}(x) \\[4pt] \vdots \\[4pt] L_f^{\kappa_m - 1} S_{m1}(x) \\[4pt] v_m \end{bmatrix} \Bigg|_{x=S^{-1}(z)} = \begin{bmatrix} z_{12} \\[4pt] \vdots \\[4pt] z_{1\kappa_1} \\[4pt] v_1 \\[4pt] \vdots \\[4pt] z_{m2} \\[4pt] \vdots \\[4pt] z_{m\kappa_m} \\[4pt] v_m \end{bmatrix}.$$

□

Suppose that $(\kappa_1, \kappa_2, \ldots, \kappa_m)$ is the Kronecker indices of system (4.39). If we let $\kappa_{\max} \triangleq \max\{\kappa_i,\ 1 \le i \le m\}$, it is clear, by the definition of the Kronecker indices, that $\Delta_{\kappa_{\max} - 1}(x) = \Delta_{n-1}(x)$.

Lemma 4.4 *Suppose that system* (4.39) *satisfies*

(i) $\dim(\Delta_{\kappa_{\max}-1}(x)) = n$

(ii) $\Delta_i(x)$, $0 \le i \le \kappa_{\max} - 2$ *are involutive distributions on a neighborhood of* $0 \in$ $\mathbb{R}^n$.

Then there exist scalar functions $\{S_{11}(x), \dots, S_{m1}(x)\}$ *such that for* $1 \le i \le m$ *and* $1 \le j \le m$, $S_{i1}(0) = 0$

$$\frac{\partial S_{i1}(x)}{\partial x} \mathrm{ad}_f^{k-1} g_j(x) = 0, \ 1 \le k \le \kappa_i - 1 \tag{4.60}$$

$$\left.\frac{\partial S_{i1}(x)}{\partial x}\right|_{x=0} \mathrm{ad}_f^{\kappa_i - 1} g_j(x)\Big|_{x=0} = (-1)^{\kappa_i - 1}\delta_{i,j}, \ \text{if } \kappa_j \ge \kappa_i \tag{4.61}$$

and

$$\Delta_i(x)^\perp \triangleq \mathrm{span}\left\{d\left(L_f^k S_{j1}(x)\right) \mid 1 \le j \le m, \ 0 \le k \le \kappa_j - i - 2\right\}.$$

In other words, there exist scalar functions $\{S_{11}(x), \dots, S_{m1}(x)\}$ *such that condition (i) and condition (ii) of Lemma 4.3 are satisfied.*

Proof Suppose that condition (i) and condition (ii) of Lemma 4.4 are satisfied. We can assume, without loss of generality, that $\kappa_1 \ge \kappa_2 \ge \cdots \ge \kappa_m$. Let

$$\kappa_1 = \cdots = \kappa_{m_1} > \kappa_{m_1+1} = \cdots = \kappa_{m_2} > \cdots > \kappa_{m_{p-1}+1} = \cdots = \kappa_{m_p}$$

and

$$g^i(x) \triangleq \left[g_{m_{i-1}+1}(x) \cdots g_{m_i}(x)\right]$$

where $m_p = m$ and $m_0 \triangleq 0$. Note, by Example 4.3.1, that

$$\dim(\Delta_{\kappa_{m_1}-2}(x)) = \sum_{j=1}^m \min(\kappa_{m_1} - 1, \kappa_j) = \sum_{j=1}^{m_1}(\kappa_{m_1} - 1) + \sum_{j=m_1+1}^m \kappa_j$$

$$= \sum_{j=1}^m \kappa_j - m_1 = n - m_1. \tag{4.62}$$

Thus, there exist, by Frobenius Theorem (or Theorem 2.8), smooth functions $S_{i1}(x)$, $1 \le i \le m_1$ such that $S_{i1}(0) = 0$ and

$$\Delta_{\kappa_{m_1}-2}(x)^\perp = \mathrm{span}\{dS_{i1}(x) \mid 1 \le i \le m_1\} \tag{4.63}$$

or for $1 \le i \le m_1$ and $1 \le j \le m$

$$L_{\mathrm{ad}_f^k g_j} S_{i1}(x) = 0, \quad 0 \le k \le \kappa_{m_1} - 2$$

$$L_{g_1} L_f^{\kappa_{m_1}-1} S^1(x)\Big|_{x=0} = I_{m_1} \tag{4.64}$$

where $S^1(x) \triangleq \begin{bmatrix} S_{11}(x) \\ \vdots \\ S_{m_11}(x) \end{bmatrix}$. It is clear, by (4.64) and Example 2.4.16, that for $1 \le i \le m_1$ and $1 \le j \le m$

$$L_{\mathrm{ad}_f^k g_j} L_f^\ell S_{i1}(x) = 0, \quad 0 \le k + \ell \le \kappa_{m_1} - 2. \tag{4.65}$$

which implies, together with (4.41), that for $1 \le i \le m_1$ and $0 \le \ell \le \kappa_{m_1} - \kappa_{m_2}$

$$d\left(L_f^\ell S_{i1}(x)\right) \in \Delta_{\kappa_{m_2}-2}(x)^\perp. \tag{4.66}$$

Also, it is easy to show, by (4.64) and Example 4.3.3, that

$$\dim\left(\mathrm{span}\left\{d\left(L_f^\ell S_{i1}(x)\right) \mid 1 \le i \le m_1, \ 0 \le \ell \le \kappa_i - 1\right\}\right) = m_1\kappa_{m_1}. \tag{4.67}$$

Similarly, we have, by Example 4.3.1, that

$$\begin{aligned}
\dim(\Delta_{\kappa_{m_2}-2}(x)) &= \sum_{j=1}^m \min(\kappa_{m_2}-1, \kappa_j) = \sum_{j=1}^{m_2}(\kappa_{m_2}-1) + \sum_{j=m_2+1}^m \kappa_j \\
&= \sum_{j=1}^m \kappa_j - \sum_{j=1}^{m_1}(\kappa_{m_1}-\kappa_{m_2}) - m_2 \\
&= n - m_1(\kappa_{m_1}-\kappa_{m_2}) - m_2.
\end{aligned} \tag{4.68}$$

Thus, there exist, by Frobenius Theorem (or Theorem 2.8), smooth functions $h_i(x)$, $1 \le i \le m_1(\kappa_{m_1}-\kappa_{m_2}) + m_2$ such that $h_i(0) = 0$ and

$$\Delta_{\kappa_{m_2}-2}(x)^\perp = \mathrm{span}\left\{dh_i(x) \mid 1 \le i \le m_1(\kappa_{m_1}-\kappa_{m_2}) + m_2\right\}. \tag{4.69}$$

By (4.66) and (4.69), there exist at least $(m_2 - m_1)$ functions $h_{s_j}(x)$, $1 \le j \le m_2 - m_1$ such that

$$dh_{s_j}(x) \notin \mathrm{span}\left\{d\left(L_f^\ell S_{i1}(x)\right) \mid 1 \le i \le m_1, \ 0 \le \ell \le \kappa_{m_1} - \kappa_{m_2}\right\}.$$

Let $S_{m_1+j}(x) = h_{s_j}(x)$ for $1 \le j \le m_2 - m_1$. Then, it is easy to see that

$$\Delta_{\kappa_{m_2}-2}(x)^\perp = \mathrm{span}\left\{d\left(L_f^\ell S_{i1}(x)\right) \mid 1 \le i \le m_2, \ 0 \le \ell \le \kappa_i - \kappa_{m_2}\right\}.$$

In other words, there exist smooth functions $S_{i1}(x)$, $m_1 + 1 \le i \le m_2$ such that for $m_1 + 1 \le i \le m_2$ and $1 \le j \le m$

$$L_{\mathrm{ad}_f^k g_j} S_{i1}(x) = 0, \quad 0 \le k \le \kappa_i - 2 \tag{4.70}$$

and

$$\left[\begin{array}{cc} L_{g^1} L_f^{\kappa_{m_2}-1} L_f^{\kappa_{m_1}-\kappa_{m_2}} S^1(x) & L_{g^2} L_f^{\kappa_{m_2}-1} L_f^{\kappa_{m_1}-\kappa_{m_2}} S^1(x) \\ L_{g^1} L_f^{\kappa_{m_2}-1} S^2(x) & L_{g^2} L_f^{\kappa_{m_2}-1} S^2(x) \end{array} \right]\Bigg|_{x=0}$$
$$= \left[\begin{array}{cc} L_{g^1} L_f^{\kappa_{m_1}-1} S^1(x) & L_{g^2} L_f^{\kappa_{m_1}-1} S^1(x) \\ L_{g^1} L_f^{\kappa_{m_2}-1} S^2(x) & L_{g^2} L_f^{\kappa_{m_2}-1} S^2(x) \end{array} \right]\Bigg|_{x=0} = C = \left[\begin{array}{cc} I_{m_1} & C_{12} \\ O & I_{m_2-m_1} \end{array} \right] \tag{4.71}$$

where

$$S^1(x) \triangleq \left[\begin{array}{c} S_{11}(x) \\ \vdots \\ S_{m_1 1}(x) \end{array} \right] \quad \text{and} \quad S^2(x) \triangleq \left[\begin{array}{c} S_{(m_1+1)1}(x) \\ \vdots \\ S_{m_2 1}(x) \end{array} \right].$$

$$\left(\text{If } C = \left[\begin{array}{cc} I_{m_1} & C_{12} \\ C_{21} & C_{22} \end{array} \right] \ne \left[\begin{array}{cc} I_{m_1} & C_{12} \\ O & I_{m_2-m_1} \end{array} \right], \text{ then use new scalar functions} \right.$$

$$\bar{S}^2(x) = (C_{22} - C_{21}C_{12})^{-1} \left\{ S^2(x) - C_{21} L_f^{\kappa_{m_1}-\kappa_{m_2}} S^1(x) \right\}$$

instead of $S^2(x)$.$\Big)$ Also, it is easy to show, by (4.64), (4.70), (4.71), and Example 4.3.3, that

$$\dim \left(\mathrm{span} \left\{ d \left(L_f^\ell S_{i1}(x) \right) \mid 1 \le i \le m_2, \, 0 \le \ell \le \kappa_i - 1 \right\} \right)$$
$$= m_1 \kappa_{m_1} + (m_2 - m_1)\kappa_{m_2} = \sum_{i=1}^{m_2} \kappa_i.$$

Let $\Delta_{-1}(x) \triangleq \mathrm{span}\{O_{n \times 1}\}$. In this manner, we can find smooth functions $S_{i1}(x)$, $1 \le i \le m_p = m$ such that $S_{i1}(0) = 0$ and for $1 \le q \le p$

$$\Delta_{\kappa_{m_q}-2}(x)^\perp = \mathrm{span} \left\{ d \left(L_f^\ell S_{i1}(x) \right) \mid 1 \le i \le m_q. \, 0 \le \ell \le \kappa_i - \kappa_{m_q} \right\}.$$

In other words, there exist smooth functions $S_{i1}(x)$, $m_{q-1} + 1 \le i \le m_q$ such that for $m_{q-1} + 1 \le i \le m_q$ and $1 \le j \le m$

$$L_{\mathrm{ad}_f^k g_j} S_{i1}(x) = 0, \quad 0 \le k \le \kappa_i - 2 \tag{4.72}$$

and

$$
\begin{bmatrix}
L_{g^1}L_f^{\kappa_{m_1}-1}S^1(x) & L_{g^2}L_f^{\kappa_{m_1}-1}S^1(x) & \cdots & L_{g^q}L_f^{\kappa_{m_1}-1}S^1(x) \\
L_{g^1}L_f^{\kappa_{m_2}-1}S^2(x) & L_{g^2}L_f^{\kappa_{m_2}-1}S^2(x) & \cdots & L_{g^q}L_f^{\kappa_{m_2}-1}S^2(x) \\
\vdots & \vdots & & \vdots \\
L_{g^1}L_f^{\kappa_{m_q}-1}S^q(x) & L_{g^2}L_f^{\kappa_{m_q}-1}S^q(x) & \cdots & L_{g^q}L_f^{\kappa_{m_q}-1}S^q(x)
\end{bmatrix}\Bigg|_{x=0}
\tag{4.73}
$$

$$
=
\begin{bmatrix}
I_{m_1} & C_{12} & \cdots & C_{1q} \\
O & I_{m_2-m_1} & \cdots & C_{2q} \\
\vdots & \vdots & & \vdots \\
O & O & \cdots & I_{m_q-m_{q-1}}
\end{bmatrix}
$$

where $m_0 = 0$ and for $1 \le q \le p$

$$
S^q(x) \triangleq
\begin{bmatrix}
S_{(m_{q-1}+1)1}(x) \\
\vdots \\
S_{m_q 1}(x)
\end{bmatrix}.
$$

Also, it is easy to show, by (4.72), (4.73), and Example 4.3.3, that

$$
\dim\left(\operatorname{span}\left\{d\left(L_f^\ell S_{i1}(x)\right) \mid 1 \le i \le m_q,\ 0 \le \ell \le \kappa_i - 1\right\}\right) = \sum_{i=1}^{m_q} \kappa_i.
\tag{4.74}
$$

In this manner, we can find smooth functions $S_{i1}(x)$, $1 \le i \le m_p = m$ such that $S_{i1}(0) = 0$

$$
L_{\operatorname{ad}_f^k g}S_{i1}(x) = 0, \quad 0 \le k \le \kappa_i - 2
\tag{4.75}
$$

$$
\operatorname{rank}\left(
\begin{bmatrix}
L_{\operatorname{ad}_f^{\kappa_1-1}g_1}S_{11}(x) & \cdots & L_{\operatorname{ad}_f^{\kappa_1-1}g_m}S_{11}(x) \\
\vdots & & \vdots \\
L_{\operatorname{ad}_f^{\kappa_m-1}g_1}S_{m1}(x) & \cdots & L_{\operatorname{ad}_f^{\kappa_m-1}g_m}S_{m1}(x)
\end{bmatrix}
\right)
$$
$$
= \operatorname{rank}\left(
\begin{bmatrix}
L_g L_f^{\kappa_1-1}S_{11}(x) \\
\vdots \\
L_g L_f^{\kappa_m-1}S_{m1}(x)
\end{bmatrix}
\right) = m
\tag{4.76}
$$

and

$$\dim\left(\operatorname{span}\left\{d\left(L_f^\ell S_{i1}(x)\right) \mid 1 \le i \le m,\ 0 \le \ell \le \kappa_i - 1\right\}\right) = \sum_{i=1}^m \kappa_i = n.$$

□

The proof of Lemma 4.4 seems more complicated than the sufficiency proof of Theorem 4.1. For example, let $n = 7$, $m = 3$, and $(\kappa_1, \kappa_2, \kappa_3) = (4, 2, 1)$. Then we can find a scalar function $S_{11}(x)$ such that

$$dS_{11}(x) \in \left(\operatorname{span}\left\{g_1, g_2, g_3, \operatorname{ad}_f g_1, \operatorname{ad}_f g_2, \operatorname{ad}_f^2 g_1\right\}\right)^\perp \quad \text{and}$$

$$L_{\operatorname{ad}_f^3 g_1} S_{11}(x)\Big|_{x=0} = -1 \quad \left(\text{or } L_{g_1} L_f^3 S_{11}(x)\big|_{x=0} = 1\right).$$

Also, we can find a scalar function $S_{21}(x)$ such that

$$dS_{21}(x) \in (\operatorname{span}\{g_1(x), g_2(x), g_3(x)\})^\perp \quad \text{and}$$

$$\left[L_{\operatorname{ad}_f g_1} S_{21}(x)\ \ L_{\operatorname{ad}_f g_2} S_{21}(x)\right]\big|_{x=0} = \begin{bmatrix} 0 & -1 \end{bmatrix}.$$

Finally, we can find a scalar function $S_{31}(x)$ such that

$$\left[L_{g_1} S_{31}(x)\ \ L_{g_2} S_{31}(x)\ \ L_{g_3} S_{31}(x)\right]\big|_{x=0} = \begin{bmatrix} 0 & 0 & 1 \end{bmatrix}.$$

Then it is easy to see that $S_{i1}(x)$, $1 \le i \le 3$ satisfy

$$L_g L_f^k S_{i1}(x) = 0,\ \ 1 \le i \le 3 \text{ and } 0 \le k \le \kappa_i - 2$$

and

$$\begin{bmatrix} L_g L_f^3 S_{11}(x) \\ L_g L_f S_{21}(x) \\ L_g S_{31}(x) \end{bmatrix}\Bigg|_{x=0} = \begin{bmatrix} 1 & * & * \\ 0 & 1 & * \\ 0 & 0 & 1 \end{bmatrix}.$$

Theorem 4.3 (Conditions for feedback linearization) *System* (4.39) *is locally feedback linearizable, if and only if*

(i) $\dim\left(\Delta_{\kappa_{\max}-1}(0)\right) = n$
(ii) $\Delta_i(x),\ 0 \le i \le \kappa_{\max} - 2$ *are involutive distributions on a neighborhood of* $0 \in \mathbb{R}^n$.

Proof Necessity. Suppose that system (4.39) is feedback linearizable. Then, by Lemma 4.2, condition (i) of Theorem 4.3 is satisfied. Also, by Lemma 4.3, there exist smooth functions $S_{i1}(x)$, $1 \le i \le m$ such that condition (i) and (ii) of Lemma 4.3 are satisfied. Thus, we have that for $1 \le i \le m$

$$L_g L_f^k S_{i1}(x) = 0, \ 0 \le k \le \kappa_i - 2$$

and

$$\text{rank}\left(\left[\begin{array}{c} L_g L_f^{\kappa_1 - 1} S_{11}(x) \\ \vdots \\ L_g L_f^{\kappa_m - 1} S_{m1}(x) \end{array}\right]\Bigg|_{x=0}\right) = m.$$

Then, it is clear, by Example 2.4.16, that for $1 \le i \le m$ and $1 \le j \le m$

$$L_{\text{ad}_f^k g_j} S_{i1}(x) = 0, \quad 0 \le k \le \kappa_i - 2$$

and

$$\text{rank}\left(\left[\begin{array}{ccc} L_{\text{ad}_f^{\kappa_1-1} g_1} S_{11}(x) & \cdots & L_{\text{ad}_f^{\kappa_1-1} g_m} S_{11}(x) \\ \vdots & & \vdots \\ L_{\text{ad}_f^{\kappa_m-1} g_1} S_{m1}(x) & \cdots & L_{\text{ad}_f^{\kappa_m-1} g_m} S_{m1}(x) \end{array}\right]\Bigg|_{x=0}\right) = m.$$

Thus, it is easy to see, by (4.41), Examples 4.3.3, and 2.4.16, that for $i \ge 0$

$$\Delta_i(x)^{\perp} = \text{span}\left\{ dL_f^k S_{j1}(x) \big| 1 \le j \le m, \ 0 \le k \le \kappa_j - 2 - i \right\}$$

and

$$\dim(\Delta_i(x)^{\perp}) = \sum_{j=1}^{m} \max(\kappa_j - 1 - i, 0) = \sum_{j=1}^{m} \{\kappa_j - \min(i+1, \kappa_j)\}$$

$$= n - \sum_{j=1}^{m} \min(i+1, \kappa_j).$$

Therefore, by Frobenius Theorem (or Theorem 2.8), $\Delta_i(x)$, $0 \le i \le \kappa_{\max} - 2$ are involutive distributions with dimension $\sum_{j=1}^{m} \min(i+1, \kappa_j)$ and condition (ii) of Theorem 4.3 is satisfied.

Sufficiency. Suppose that conditions (i) and (ii) of Theorem 4.3 are satisfied. Then, by Lemma 4.4, there exist scalar functions $\{S_{11}(x), \ldots, S_{m1}(x)\}$ such that conditions (i) and (ii) of Lemma 4.3 are satisfied. Therefore, by Lemma 4.3, system (4.39) is feedback linearizable. □

Example 4.3.4 In Example 3.3.4, it is shown that system (3.39) is not state equivalent to a linear system. Show that system (3.39) is feedback linearizable.

$$\begin{bmatrix} \dot{x}_1 \\ \dot{x}_2 \\ \dot{x}_3 \end{bmatrix} = \begin{bmatrix} x_2 \\ -x_1 + x_2^2 \\ x_3^2 \end{bmatrix} + \begin{bmatrix} 0 & 0 \\ 1 + x_1^2 & 0 \\ 0 & 1 \end{bmatrix} \begin{bmatrix} u_1 \\ u_2 \end{bmatrix}$$

$$= f(x) + g_1(x)u_1 + g_2(x)u_2$$

Solution By simple calculation, we have that $(\kappa_1, \kappa_2) = (2, 1)$ and

$$\mathrm{ad}_f g_1(x) = \begin{bmatrix} -1 - x_1^2 \\ 2x_2(x_1 - 1 - x_1^2) \\ 0 \end{bmatrix}.$$

Since $\kappa_1 + \kappa_2 = 3$, condition (i) of Theorem 4.3 is satisfied. Since $\dim(\Delta_0(0)) = \dim(\Delta_0(x)) = \dim(\mathrm{span}\{g_1(x), g_2(x)\}) = 2$, $\Delta_0(x)$ is a distribution. Also, it is easy to see that distribution $\Delta_0(x) = \mathrm{span}\{g_1(x), g_2(x)\}$ is involutive and condition (ii) of Theorem 4.3 is satisfied. Therefore, by Theorem 4.3, system (3.39) is feedback linearizable. We need to find scalar functions $S_{11}(x)$ and $S_{21}(x)$ such that

$$\mathrm{span}\{dS_{11}(x)\} = \Delta_0(x)^\perp = \mathrm{span}\{g_1(x), g_2(x)\}^\perp$$
$$\mathrm{span}\{dS_{11}(x), dL_f S_{11}(x), dS_{21}(x)\} = \Delta_{-1}(x)^\perp \triangleq \mathbb{R}^3.$$

In other words, we have, by (4.60) and (4.61), that $c_1(0) = 1$ and

$$c_1(x)\begin{bmatrix} 0 & 0 & -1 \end{bmatrix} = \frac{\partial S_{11}(x)}{\partial x}\begin{bmatrix} g_1(x) & g_2(x) & \mathrm{ad}_f g_1(x)\big|_{x=0} \end{bmatrix}$$

$$= \begin{bmatrix} \frac{\partial S_{11}(x)}{\partial x_1} & \frac{\partial S_{11}(x)}{\partial x_2} & \frac{\partial S_{11}(x)}{\partial x_3} \end{bmatrix}\begin{bmatrix} 0 & 0 & -1 \\ 1 + x_1^2 & 0 & 0 \\ 0 & 1 & 0 \end{bmatrix}$$

which implies that $\frac{\partial S_{11}(x)}{\partial x} = c_1(x)\begin{bmatrix} 1 & 0 & 0 \end{bmatrix}$. Since $\begin{bmatrix} 1 & 0 & 0 \end{bmatrix}$ is exact, we have $c_1(x) = 1$ and $S_{11}(x) = x_1$. ($S_{11}(x)$ is not unique.) Also, we have, by (4.60) and (4.61), that $c_2(0) = 1$ and

$$c_2(x)\begin{bmatrix} 0 & 1 \end{bmatrix} = \frac{\partial S_{21}(x)}{\partial x}\begin{bmatrix} g_1(0) & g_2(0) \end{bmatrix}$$

$$= \begin{bmatrix} \frac{\partial S_{21}(x)}{\partial x_1} & \frac{\partial S_{21}(x)}{\partial x_2} & \frac{\partial S_{21}(x)}{\partial x_3} \end{bmatrix}\begin{bmatrix} 0 & 0 \\ 1 & 0 \\ 0 & 1 \end{bmatrix}$$

which implies that $\frac{\partial S_{21}(x)}{\partial x} = c_2(x)\begin{bmatrix} d(x) & 0 & 1 \end{bmatrix}$. ($d(x)$ is a smooth function.) Since $\begin{bmatrix} 0 & 0 & 1 \end{bmatrix}$ is exact, we have $d(x) = 0$, $c_2(x) = 1$, and $S_{21}(x) = x_3$. ($S_{21}(x)$ is not unique.) Then, it is clear, by (4.56) and (4.57), that

$$\begin{bmatrix} z_1 \\ z_2 \\ z_3 \end{bmatrix} = S(x) = \begin{bmatrix} S_{11}(x) \\ L_f S_{11}(x) \\ S_{21}(x) \end{bmatrix} = \begin{bmatrix} x_1 \\ x_2 \\ x_3 \end{bmatrix}$$

and

$$\begin{bmatrix} u_1 \\ u_2 \end{bmatrix} = \begin{bmatrix} L_g L_f S_{11}(x) \\ L_g S_{21}(x) \end{bmatrix}^{-1} \left(-\begin{bmatrix} L_f^2 S_{11}(x) \\ L_f S_{21}(x) \end{bmatrix} + \begin{bmatrix} v_1 \\ v_2 \end{bmatrix} \right)$$

$$= \begin{bmatrix} 1 + x_1^2 & 0 \\ 0 & 1 \end{bmatrix}^{-1} \left(-\begin{bmatrix} -x_1 + x_2^2 \\ x_3^2 \end{bmatrix} + \begin{bmatrix} v_1 \\ v_2 \end{bmatrix} \right)$$

$$= \begin{bmatrix} \frac{x_1 - x_2^2}{1 + x_1^2} \\ -x_3^2 \end{bmatrix} + \begin{bmatrix} \frac{1}{1+x_1^2} & 0 \\ 0 & 1 \end{bmatrix} \begin{bmatrix} v_1 \\ v_2 \end{bmatrix} = \alpha(x) + \beta(x)v.$$

It is easy to see that

$$\begin{bmatrix} \dot{z}_1 \\ \dot{z}_2 \\ \dot{z}_3 \end{bmatrix} = S_* \left(f(x) + g(x)\alpha(x) + g(x)\beta(x)v \right)$$

$$= \begin{bmatrix} 0 & 1 & 0 \\ 0 & 0 & 0 \\ 0 & 0 & 0 \end{bmatrix} \begin{bmatrix} z_1 \\ z_2 \\ z_3 \end{bmatrix} + \begin{bmatrix} 0 & 0 \\ 1 & 0 \\ 0 & 1 \end{bmatrix} \begin{bmatrix} v_1 \\ v_2 \end{bmatrix}.$$

□

Example 4.3.5 Show that the following nonlinear system is feedback linearizable.

$$\dot{x} = \begin{bmatrix} x_2 + 2x_4 x_5 \\ x_3 \\ 0 \\ x_5 \\ 0 \end{bmatrix} + \begin{bmatrix} 0 & 0 \\ 0 & 0 \\ 1 & x_1 \\ 0 & 0 \\ 0 & 1 \end{bmatrix} \begin{bmatrix} u_1 \\ u_2 \end{bmatrix} \tag{4.77}$$

$$= f(x) + g_1(x)u_1 + g_2(x)u_2$$

Also, find a state transformation and feedback.

Solution By simple calculation, we have

$$[\text{ad}_f g_1(x) \ \text{ad}_f g_2(x)] = \begin{bmatrix} 0 & -2x_4 \\ -1 & -x_1 \\ 0 & x_2 + 2x_4 x_5 \\ 0 & -1 \\ 0 & 0 \end{bmatrix} \quad \text{and} \quad \text{ad}_f^2 g_1(x) = \begin{bmatrix} 1 \\ 0 \\ 0 \\ 0 \\ 0 \end{bmatrix}.$$

It is clear that $\dim(\Delta_0(0)) = \dim(\Delta_0(x)) = \dim(\text{span}\{g_1(x), g_2(x)\}) = 2$, $\dim(\Delta_1(0)) = \dim(\Delta_1(x)) = \dim\left(\text{span}\{g_1(x), g_2(x), \text{ad}_f g_1(x), \text{ad}_f g_2(x)\}\right) = 4$, and $(\kappa_1, \kappa_2) = (3, 2)$. Since $\kappa_1 + \kappa_2 = 5$, condition (i) of Theorem 4.3 is satisfied. Also, it is easy to see that $\Delta_0(x)$ and $\Delta_1(x)$ are involutive distributions and thus condition (ii) of Theorem 4.3 is satisfied. Therefore, by Theorem 4.3, system (4.77) is feedback linearizable. We need to find scalar functions $S_{11}(x)$ and $S_{21}(x)$ such that

$$\text{span}\{dS_{11}(x)\} = \Delta_1(x)^{\perp}$$
$$\text{span}\{dS_{11}(x), \, dL_f S_{11}(x), \, dS_{21}(x)\} = \Delta_0(x)^{\perp}.$$

In other words, we have, by (4.60) and (4.61), that $c_1(0) = 1$, $c_2(0) = 1$

$$c_1(x)\begin{bmatrix} 0 & 0 & 0 & 0 & 1 \end{bmatrix} = \frac{\partial S_{11}(x)}{\partial x}\begin{bmatrix} 0 & 0 & 0 & -2x_4 & 1 \\ 0 & 0 & -1 & -x_1 & 0 \\ 1 & x_1 & 0 & x_2 + 2x_4 x_5 & 0 \\ 0 & 0 & 0 & -1 & 0 \\ 0 & 1 & 0 & 0 & 0 \end{bmatrix}$$

and

$$c_2(x)\begin{bmatrix} 0 & 0 & 0 & -1 \end{bmatrix} = \frac{\partial S_{21}(x)}{\partial x}\begin{bmatrix} 0 & 0 & 0 & 0 \\ 0 & 0 & -1 & 0 \\ 1 & x_1 & 0 & 0 \\ 0 & 0 & 0 & -1 \\ 0 & 1 & 0 & 0 \end{bmatrix}$$

which imply that

$$\frac{\partial S_{11}(x)}{\partial x} = c_1(x)\begin{bmatrix} 1 & 0 & 0 & -2x_4 & 0 \end{bmatrix}$$

and

$$\frac{\partial S_{21}(x)}{\partial x} = c_2(x)\begin{bmatrix} d_2(x) & 0 & 0 & 1 & 0 \end{bmatrix}.$$

Since $\begin{bmatrix} 1 & 0 & 0 & -2x_4 & 0 \end{bmatrix}$ is exact, we have $c_1(x) = 1$ and $S_{11}(x) = x_1 - x_4^2$. ($S_{11}(x)$ is not unique.) Since $\begin{bmatrix} d_2(x) & 0 & 0 & 1 & 0 \end{bmatrix}$ is exact with $d_2(x) = 0$, we have $c_2(x) = 1$ and $S_{21}(x) = x_4$. ($S_{21}(x)$ is not unique.) Then, it is clear, by (4.56) and (4.57), that

$$\begin{bmatrix} z_1 \\ z_2 \\ z_3 \\ z_4 \\ z_5 \end{bmatrix} = S(x) = \begin{bmatrix} S_{11}(x) \\ L_f S_{11}(x) \\ L_f^2 S_{11}(x) \\ S_{21}(x) \\ L_f S_{21}(x) \end{bmatrix} = \begin{bmatrix} x_1 - x_4^2 \\ x_2 \\ x_3 \\ x_4 \\ x_5 \end{bmatrix}$$

and

$$\begin{bmatrix} u_1 \\ u_2 \end{bmatrix} = \begin{bmatrix} L_g L_f^2 S_{11}(x) \\ L_g L_f S_{21}(x) \end{bmatrix}^{-1} \left(-\begin{bmatrix} L_f^3 S_{11}(x) \\ L_f^2 S_{21}(x) \end{bmatrix} + \begin{bmatrix} v_1 \\ v_2 \end{bmatrix} \right)$$

$$= \begin{bmatrix} 1 & x_1 \\ 0 & 1 \end{bmatrix}^{-1} \left(-\begin{bmatrix} 0 \\ 0 \end{bmatrix} + \begin{bmatrix} v_1 \\ v_2 \end{bmatrix} \right)$$

$$= \begin{bmatrix} 1 & -x_1 \\ 0 & 1 \end{bmatrix} \begin{bmatrix} v_1 \\ v_2 \end{bmatrix} = \alpha(x) + \beta(x)v.$$

It is easy to see that

$$\begin{bmatrix} \dot{z}_1 \\ \dot{z}_2 \\ \dot{z}_3 \\ \dot{z}_4 \\ \dot{z}_5 \end{bmatrix} = \begin{bmatrix} 0 & 1 & 0 & 0 & 0 \\ 0 & 0 & 1 & 0 & 0 \\ 0 & 0 & 0 & 0 & 0 \\ 0 & 0 & 0 & 0 & 1 \\ 0 & 0 & 0 & 0 & 0 \end{bmatrix} \begin{bmatrix} z_1 \\ z_2 \\ z_3 \\ z_4 \\ z_5 \end{bmatrix} + \begin{bmatrix} 0 & 0 \\ 0 & 0 \\ 1 & 0 \\ 0 & 0 \\ 0 & 1 \end{bmatrix} \begin{bmatrix} v_1 \\ v_2 \end{bmatrix}$$

$$= S_*(f(x) + g(x)\alpha(x) + g(x)\beta(x)v).$$

□

Example 4.3.6 Show that the following nonlinear system is feedback linearizable:

$$\dot{x} = \begin{bmatrix} x_2 - x_4(x_3 + x_4) \\ 0 \\ x_4 \\ 0 \end{bmatrix} + \begin{bmatrix} 0 & x_1 \\ 0 & 1 \\ -1 & -x_1 \\ 1 & x_1 \end{bmatrix} \begin{bmatrix} u_1 \\ u_2 \end{bmatrix} \qquad (4.78)$$

$$= f(x) + g_1(x)u_1 + g_2(x)u_2$$

Solution By simple calculation, we have that $(\kappa_1, \kappa_2) = (2, 2)$ and

$$[\mathrm{ad}_f g_1(x) \ \mathrm{ad}_f g_2(x)] = \begin{bmatrix} x_3 + x_4 & x_2 + (x_1 - x_4)(x_3 + x_4) - 1 \\ 0 & 0 \\ -1 & x_4(x_3 + x_4) - x_2 - x_1 \\ 0 & x_2 - x_4(x_3 + x_4) \end{bmatrix}.$$

Since $\kappa_1 + \kappa_2 = 4$, condition (i) of Theorem 4.3 is satisfied. It is clear that $\dim(\Delta_0(0)) = \dim(\Delta_0(x)) = \dim(\mathrm{span}\{g_1(x), g_2(x)\}) = 2$. Also, it is easy to see that distribution $\Delta_0(x)$ is involutive and thus condition (ii) of Theorem 4.3 is

satisfied. Therefore, by Theorem 4.3, system (4.78) is feedback linearizable. We
need to find scalar functions $S_{11}(x)$ and $S_{21}(x)$ such that

$$
\begin{aligned}
\mathrm{span}\{dS_{11}(x), dS_{21}(x)\} &= \Delta_0(x)^\perp \\
&= \mathrm{span}\left\{\begin{bmatrix}1 & -x_1 & 0 & 0\end{bmatrix}, \begin{bmatrix}0 & 0 & 1 & 1\end{bmatrix}\right\} \\
&= \mathrm{span}\{dx_1 - x_1 dx_2, \ dx_3 + dx_4\}
\end{aligned}
$$

In other words, we have, by (4.60) and (4.61), that $c_1(0) = 1$, $c_2(0) = 1$

$$
c_1(x)\begin{bmatrix}0 & 0 & -1 & 0\end{bmatrix} = \frac{\partial S_{11}(x)}{\partial x}\begin{bmatrix}0 & x_1 & 0 & -1 \\ 0 & 1 & 0 & 0 \\ -1 & -x_1 & -1 & 0 \\ 1 & x_1 & 0 & 0\end{bmatrix}
$$

and

$$
c_2(x)\begin{bmatrix}0 & 0 & 0 & -1\end{bmatrix} = \frac{\partial S_{21}(x)}{\partial x}\begin{bmatrix}0 & x_1 & 0 & -1 \\ 0 & 1 & 0 & 0 \\ -1 & -x_1 & -1 & 0 \\ 1 & x_1 & 0 & 0\end{bmatrix}
$$

which imply that

$$
\frac{\partial S_{11}(x)}{\partial x} = c_1(x)\begin{bmatrix}0 & 0 & 1 & 1\end{bmatrix}
$$

and

$$
\frac{\partial S_{21}(x)}{\partial x} = c_2(x)\begin{bmatrix}1 & -x_1 & 0 & 0\end{bmatrix}.
$$

Since $\begin{bmatrix}0 & 0 & 1 & 1\end{bmatrix}$ is exact, we have $c_1(x) = 1$ and $S_{11}(x) = x_3 + x_4$. ($S_{11}(x)$ is
not unique.) Since $\begin{bmatrix}1 & -x_1 & 0 & 0\end{bmatrix}$ is not exact, we need to find $c_2(x)$ such that
$c_2(x)\begin{bmatrix}1 & -x_1 & 0 & 0\end{bmatrix}$ is exact. It is easy to see, by (4.24) and (4.25), that $c_2(x) = e^{-x_2}$
and $S_{21}(x) = x_1 e^{-x_2}$ work. ($S_{21}(x)$ is not unique.) Then, it is clear, by (4.56) and
(4.57), that

$$
\begin{bmatrix}z_1 \\ z_2 \\ z_3 \\ z_4\end{bmatrix} = S(x) = \begin{bmatrix}S_{11}(x) \\ L_f S_{11}(x) \\ S_{21}(x) \\ L_f S_{21}(x)\end{bmatrix} = \begin{bmatrix}x_3 + x_4 \\ x_4 \\ x_1 e^{-x_2} \\ (x_2 - x_4(x_3 + x_4))\,e^{-x_2}\end{bmatrix}
$$

and

$$\begin{bmatrix} u_1 \\ u_2 \end{bmatrix} = \begin{bmatrix} L_g L_f S_{11}(x) \\ L_g L_f S_{21}(x) \end{bmatrix}^{-1} \left(-\begin{bmatrix} L_f^2 S_{11}(x) \\ L_f^2 S_{21}(x) \end{bmatrix} + \begin{bmatrix} v_1 \\ v_2 \end{bmatrix} \right)$$

$$= \begin{bmatrix} 1 & x_1 \\ -(x_3 + x_4)e^{-x_2} & (1 - x_2 - (x_1 - x_4)(x_3 + x_4)) e^{-x_2} \end{bmatrix}^{-1}$$

$$\cdot \left(\begin{bmatrix} 0 \\ x_4^2 e^{-x_2} \end{bmatrix} + \begin{bmatrix} v_1 \\ v_2 \end{bmatrix} \right)$$

$$= \begin{bmatrix} -\frac{x_1 x_4^2}{1 - x_2 + x_4(x_3 + x_4)} \\ \frac{x_4^2}{1 - x_2 + x_4(x_3 + x_4)} \end{bmatrix} + \begin{bmatrix} \frac{1 - x_2 - (x_1 - x_4)(x_3 + x_4)}{1 - x_2 + x_4(x_3 + x_4)} & -\frac{x_1 e^{x_2}}{1 - x_2 + x_4(x_3 + x_4)} \\ \frac{x_3 + x_4}{1 - x_2 + x_4(x_3 + x_4)} & \frac{e^{x_2}}{1 - x_2 + x_4(x_3 + x_4)} \end{bmatrix} \begin{bmatrix} v_1 \\ v_2 \end{bmatrix}$$

$$= \alpha(x) + \beta(x)v.$$

It is easy to see that

$$\begin{bmatrix} \dot{z}_1 \\ \dot{z}_2 \\ \dot{z}_3 \\ \dot{z}_4 \end{bmatrix} = \begin{bmatrix} 0 & 1 & 0 & 0 \\ 0 & 0 & 0 & 0 \\ 0 & 0 & 0 & 1 \\ 0 & 0 & 0 & 0 \end{bmatrix} \begin{bmatrix} z_1 \\ z_2 \\ z_3 \\ z_4 \end{bmatrix} + \begin{bmatrix} 0 & 0 \\ 1 & 0 \\ 0 & 0 \\ 0 & 1 \end{bmatrix} \begin{bmatrix} v_1 \\ v_2 \end{bmatrix}$$

$$= S_* \left(f(x) + g(x)\alpha(x) + g(x)\beta(x)v \right).$$

□

Example 4.3.7 Show that the following nonlinear system is not feedback linearizable.

$$\dot{x} = \begin{bmatrix} x_2 \\ x_3 \\ x_4 \\ 0 \end{bmatrix} + \begin{bmatrix} 0 \\ 0 \\ 1 \\ 0 \end{bmatrix} u_1 + \begin{bmatrix} x_1 \\ 0 \\ 0 \\ 1 \end{bmatrix} u_2 = f(x) + g_1(x)u_1 + g_2(x)u_2 \qquad (4.79)$$

Solution By simple calculation, we have that $(\kappa_1, \kappa_2) = (3, 1)$ and

$$\begin{bmatrix} \mathrm{ad}_f g_1(x) & \mathrm{ad}_f g_2(x) & \mathrm{ad}_f^2 g_1(x) & \mathrm{ad}_f^2 g_2(x) \end{bmatrix} = \begin{bmatrix} 0 & x_2 & 1 & x_3 \\ -1 & 0 & 0 & 1 \\ 0 & -1 & 0 & 0 \\ 0 & 0 & 0 & 0 \end{bmatrix}.$$

Since $\kappa_1 + \kappa_2 = 4$, condition (i) of Theorem 4.3 is satisfied. Since $4 = \dim(\Delta_1(x)) \neq \dim(\Delta_1(0)) = 3$, it is clear that

$$\Delta_1(x) \left(\triangleq \mathrm{span}\{g_1(x), g_2(x), \mathrm{ad}_f g_1(x), \mathrm{ad}_f g_2(x)\} \right)$$

is not a distribution on a neighborhood of the origin and condition (ii) of Theorem 4.3 is not satisfied. Therefore, by Theorem 4.3, system (4.79) is not feedback linearizable. If we consider the local linearization on a neighborhood of x^0 ($x_2^0 \neq 0$) instead of a neighborhood of 0, we have that $(\kappa_1, \kappa_2) = (2, 2)$ on a neighborhood of x^0 ($x_2^0 \neq 0$) and system (4.79) is feedback linearizable with

$$
\begin{bmatrix} z_1 \\ z_2 \\ z_3 \\ z_4 \end{bmatrix} = S(x) = \begin{bmatrix} S_{11}(x) \\ L_f S_{11}(x) \\ S_{21}(x) \\ L_f S_{21}(x) \end{bmatrix} = \begin{bmatrix} x_2 \\ x_3 \\ x_1 e^{-x_4} \\ x_2 e^{-x_4} \end{bmatrix}
$$

and

$$
\begin{bmatrix} u_1 \\ u_2 \end{bmatrix} = \begin{bmatrix} L_g L_f S_{11}(x) \\ L_g L_f S_{21}(x) \end{bmatrix}^{-1} \left(- \begin{bmatrix} L_f^2 S_{11}(x) \\ L_f^2 S_{21}(x) \end{bmatrix} + \begin{bmatrix} v_1 \\ v_2 \end{bmatrix} \right)
$$

$$
= \begin{bmatrix} 1 & 0 \\ 0 & -x_2 e^{-x_4} \end{bmatrix}^{-1} \left(- \begin{bmatrix} x_4 \\ x_3 e^{-x_4} \end{bmatrix} + \begin{bmatrix} v_1 \\ v_2 \end{bmatrix} \right)
$$

$$
= \begin{bmatrix} -x_4 \\ \frac{x_3}{x_2} \end{bmatrix} + \begin{bmatrix} 1 & 0 \\ 0 & -\frac{e^{x_4}}{x_2} \end{bmatrix} \begin{bmatrix} v_1 \\ v_2 \end{bmatrix} = \alpha(x) + \beta(x) v.
$$

$\qquad\qquad\qquad\qquad\qquad\qquad\qquad\qquad\qquad\qquad\qquad\qquad\qquad\qquad\qquad$ □

Example 4.3.8 Show that the following nonlinear system is not feedback linearizable:

$$
\dot{x} = \begin{bmatrix} x_2 \\ 0 \\ 0 \end{bmatrix} + \begin{bmatrix} 0 \\ 1 \\ 0 \end{bmatrix} u_1 + \begin{bmatrix} x_2^2 \\ 0 \\ 1 \end{bmatrix} u_2 = f(x) + g_1(x) u_1 + g_2(x) u_2 \qquad (4.80)
$$

Solution By simple calculation, we have that $(\kappa_1, \kappa_2) = (2, 1)$ and

$$
\begin{bmatrix} \mathrm{ad}_f g_1(x) \ \mathrm{ad}_f g_2(x) \end{bmatrix} = \begin{bmatrix} -1 & 0 \\ 0 & 0 \\ 0 & 0 \end{bmatrix}.
$$

Since $\kappa_1 + \kappa_2 = 3$, condition (i) of Theorem 4.3 is satisfied. Since $\dim(\Delta_0(x)) = \dim(\Delta_0(0)) = 2$, it is clear that $\Delta_0(x)$ is a distribution on a neighborhood of the origin. However, since

$$
[g_1(x), g_2(x)] = \begin{bmatrix} 2x_2 \\ 0 \\ 0 \end{bmatrix} \notin \Delta_0(x) = \mathrm{span}\{g_1(x), g_2(x)\}
$$

it is clear that distribution $\Delta_0(x) = \mathrm{span}\{g_1(x), g_2(x)\}$ is not involutive and condition (ii) of Theorem 4.3 is not satisfied. Therefore, by Theorem 4.3, system (4.80) is not feedback linearizable. $\qquad\qquad\qquad\qquad\qquad\qquad\qquad\qquad\qquad\qquad\qquad\qquad\qquad$ □

4.4 Applications of Feedback Linearization

As seen in Example 4.3.8, even a simple nonlinear system may not be feedback linearizable. That is, the class of control systems that is feedback linearizable is relatively small. However, many control systems, including robots, aircraft, and AC motors, belong to this class. For this reason, the problem of feedback linearization has attracted considerable attention.

Example 4.4.1 (*magnetic-ball-suspension system*) The dynamic equations of the magnetic-ball-suspension system in Fig. 4.4 are

$$M\frac{d^2 y(t)}{dt^2} = Mg - \frac{i(t)^2}{y(t)}$$

$$L\frac{di(t)}{dt} + Ri(t) = e(t)$$

where $y(t)$, M, g, R, L, $i(t)$, and $e(t)$ are steel ball position, steel ball mass, gravitational acceleration, winding resistance, winding inductance, winding current, and input voltage, respectively. With the state variables $x_1(t) = y(t)$, $x_2(t) = \dot{y}(t)$, and $x_3(t) = i(t)$, we can obtain the following state equation:

$$\begin{bmatrix} \dot{x}_1 \\ \dot{x}_2 \\ \dot{x}_3 \end{bmatrix} = \begin{bmatrix} x_2 \\ g - \frac{1}{M}\frac{x_3^2}{x_1} \\ -\frac{R}{L}x_3 \end{bmatrix} + \begin{bmatrix} 0 \\ 0 \\ \frac{1}{L} \end{bmatrix} e \triangleq F(x, e) \tag{4.81}$$

Fig. 4.4 Magnetic-ball-suspension system of Example 4.4.1

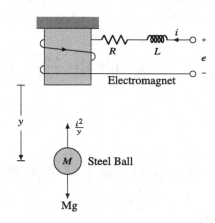

Note that $F(0, 0) \neq 0$. Let $x^0 = \begin{bmatrix} y_0 & 0 & \sqrt{Mgy_0} \end{bmatrix}^{\mathsf{T}}$ and $e^0 = R\sqrt{Mgy_0}$, where $y_0 > 0$. Then $F(x^0, e^0) = 0$ and (x^0, e^0) is an equilibrium point of system (4.81). Since $\dot{\xi} = F(\xi + x^0, u + e^0)$, we have

$$
\begin{bmatrix} \dot{\xi}_1 \\ \dot{\xi}_2 \\ \dot{\xi}_3 \end{bmatrix} = \begin{bmatrix} \xi_2 \\ g - \dfrac{1}{M}\dfrac{(\xi_3 + x_3^0)^2}{\xi_1 + y_0} \\ -\dfrac{R}{L}\xi_3 \end{bmatrix} + \begin{bmatrix} 0 \\ 0 \\ \dfrac{1}{L} \end{bmatrix} u \triangleq \bar{f}(\xi) + \bar{g}(\xi)u \tag{4.82}
$$

where $x_3^0 \triangleq \sqrt{Mgy_0}$,

$$
\begin{bmatrix} \xi_1 \\ \xi_2 \\ \xi_3 \end{bmatrix} \triangleq x - x^0 = \begin{bmatrix} x_1 - y_0 \\ x_2 \\ x_3 - \sqrt{Mgy_0} \end{bmatrix} \text{ and } u \triangleq e - e^0 = e - R\sqrt{Mgy_0}. \tag{4.83}
$$

Note that $\bar{f}(0) = 0$. By simple calculation, we have

$$
\begin{bmatrix} \mathrm{ad}_{\bar{f}}\bar{g}(\xi) & \mathrm{ad}_{\bar{f}}^2\bar{g}(\xi) \end{bmatrix} = \begin{bmatrix} 0 & -\dfrac{2(\xi_3 + x_3^0)}{LM(\xi_1 + y_0)} \\ \dfrac{2(\xi_3 + x_3^0)}{LM(\xi_1 + y_0)} & \dfrac{2x_3^0(R\xi_1 + Ry_0 - L\xi_2) - 2L\xi_2\xi_3}{L^2M(\xi_1 + y_0)^2} \\ \dfrac{R}{L^2} & \dfrac{R^2}{L^3} \end{bmatrix}
$$

which implies that condition (i) of Theorem 4.1 is satisfied. Since

$$
[\bar{g}(\xi), \mathrm{ad}_{\bar{f}}\bar{g}(\xi)] = \begin{bmatrix} 0 \\ \dfrac{2}{L^2M(\xi_1 + y_0)} \\ 0 \end{bmatrix} = \dfrac{1}{L(\xi_3 + x_3^0)}\left(\mathrm{ad}_{\bar{f}}\bar{g}(\xi) - \dfrac{R}{L}\bar{g}(\xi) \right)
$$

distribution $\Delta_1(x) = \mathrm{span}\{\bar{g}(\xi), \mathrm{ad}_{\bar{f}}\bar{g}(\xi)\}$ is involutive and condition (ii) is satisfied. Therefore, by Theorem 4.1, system (4.82) is feedback linearizable. By (4.20), we have

$$
c(\xi)\begin{bmatrix} 0 & 0 & 1 \end{bmatrix} = \dfrac{\partial S_1(\xi)}{\partial \xi}\left[\bar{g}(\xi)\ \ \mathrm{ad}_{\bar{f}}\bar{g}(\xi)\ \ \mathrm{ad}_{\bar{f}}^2\bar{g}(\xi) \Big|_{\xi=0} \right]
$$

$$
= \begin{bmatrix} \dfrac{\partial S_1(\xi)}{\partial \xi_1} & \dfrac{\partial S_1(\xi)}{\partial \xi_2} & \dfrac{\partial S_1(\xi)}{\partial \xi_3} \end{bmatrix} \begin{bmatrix} 0 & 0 & -\dfrac{2x_3^0}{LMy_0} \\ 0 & \dfrac{2(\xi_3 + x_3^0)}{LM(\xi_1 + y_0)} & \dfrac{2Rx_3^0}{L^2My_0} \\ \dfrac{1}{L} & \dfrac{R}{L^2} & \dfrac{R^2}{L^3} \end{bmatrix}
$$

which implies that $\begin{bmatrix} \dfrac{\partial S_1(\xi)}{\partial \xi_1} & \dfrac{\partial S_1(\xi)}{\partial \xi_2} & \dfrac{\partial S_1(\xi)}{\partial \xi_3} \end{bmatrix} = c(\xi)\begin{bmatrix} -\dfrac{LMy_0}{2x_3^0} & 0 & 0 \end{bmatrix}$. Since one form $\begin{bmatrix} -\dfrac{LMy_0}{2x_3^0} & 0 & 0 \end{bmatrix}$ is exact, we can let $c(\xi) = -\dfrac{2x_3^0}{LMy_0}$ and $S_1(\xi) = \xi_1$. Thus, we have that

$$\begin{bmatrix} z_1 \\ z_2 \\ z_3 \end{bmatrix} = \begin{bmatrix} S_1(\xi) \\ L_{\tilde{f}}S_1(\xi) \\ L_{\tilde{f}}^2 S_1(\xi) \end{bmatrix} = \begin{bmatrix} \xi_1 \\ \xi_2 \\ g - \frac{1}{M}\frac{(\xi_3 + x_3^0)^2}{\xi_1 + y_0} \end{bmatrix}$$

and

$$u = -\frac{L_{\tilde{f}}^3 S_1(\xi)}{L_g L_{\tilde{f}}^2 S_1(\xi)} + \frac{1}{L_g L_{\tilde{f}}^2 S_1(\xi)} v$$

$$= R\xi_3 + \frac{L\xi_2(\xi_3 + x_3^0)}{2(\xi_1 + y_0)} - \frac{LM(\xi_1 + y_0)}{2(\xi_3 + x_3^0)} v.$$

Then, it is easy to see that

$$\begin{bmatrix} \dot{z}_1 \\ \dot{z}_2 \\ \dot{z}_3 \end{bmatrix} = \begin{bmatrix} 0 & 1 & 0 \\ 0 & 0 & 1 \\ 0 & 0 & 0 \end{bmatrix} \begin{bmatrix} z_1 \\ z_2 \\ z_3 \end{bmatrix} + \begin{bmatrix} 0 \\ 0 \\ 1 \end{bmatrix} v.$$

Refer to the MATLAB program in Sect. 4.5.   □

Example 4.4.2 (*robot arm*) The dynamic equation of the robot arm with p degrees of freedom (or p joints) is

$$M(q)\ddot{q} + B(q, \dot{q})\dot{q} + G(q) + F(q, \dot{q}) = \tau \tag{4.84}$$

where q is a $p \times 1$ vector representing the position (distance or angle) of each joint. Here, the input τ is a $p \times 1$ vector representing the force or the torque. With the state variables $x^1 \triangleq q$ and $x^2 \triangleq \dot{q}$, we can obtain the following state equation.

$$\begin{bmatrix} \dot{x}^1 \\ \dot{x}^2 \end{bmatrix} = \begin{bmatrix} x^2 \\ -M(x^1)^{-1}\{B(x^1, x^2)x^2 + G(x^1) + F(x^1, x^2)\} \end{bmatrix} + \begin{bmatrix} 0 \\ M(x^1)^{-1} \end{bmatrix} \tau \tag{4.85}$$

If we consider nonlinear feedback

$$\tau = B(x^1, x^2)x^2 + G(x^1) + F(x^1, x^2) + M(x^1)v \tag{4.86}$$
$$= \alpha(x) + \beta(x)v$$

then we have the linear closed-loop system

$$\dot{x} = \begin{bmatrix} \dot{x}^1 \\ \dot{x}^2 \end{bmatrix} = \begin{bmatrix} O & I_p \\ O & O \end{bmatrix} \begin{bmatrix} x^1 \\ x^2 \end{bmatrix} + \begin{bmatrix} O \\ I_p \end{bmatrix} v$$

where I_p is the $p \times p$ identity matrix. Therefore, system (4.85) is linearizable with feedback (4.86).

Example 4.4.3 (*induction motor*) In the $d - q$ coordinate frame rotating synchronously with an angular speed w_s, the dynamic equations of a p-pole pair induction motor are

$$\dot{i}_{ds} = -a_1 i_{ds} + w_s i_{qs} + a_2 \Phi_{dr} + pa_3 w_r \Phi_{qr} + c V_{ds}$$

$$\dot{i}_{qs} = -w_s i_{ds} - a_1 i_{qs} - pa_3 w_r \Phi_{dr} + a_2 \Phi_{qr} + c V_{qs}$$

$$\dot{\Phi}_{dr} = a_5 i_{ds} - a_4 \Phi_{dr} + (w_s - pw_r)\Phi_{qr}$$

$$\dot{\Phi}_{qr} = a_5 i_{qs} - (w_s - pw_r)\Phi_{dr} - a_4 \Phi_{qr}$$

$$\dot{w}_r = \frac{-Dw_r + K_T(\Phi_{dr} i_{qs} - \Phi_{qr} i_{ds}) - T_L}{J}$$

Here, V_{ds}, V_{qs}, and w_s are the control inputs. The constants c, D, J, K_T and a_i, $i = 1, \ldots, 5$ are the parameters of the induction motor.

V_a, V_b, V_c stator phase voltages
i_a, i_b, i_c stator phase currents
$V_{ds}(V_{qs})$ d-axis (q-axis) stator voltage
$i_{ds}(i_{qs})$ d-axis (q-axis) stator current
$\Phi_{dr}(\Phi_{qr})$ d-axis (q-axis) rotor flux
w_r rotor angular speed
w_s slip angular speed
$R_s(R_r)$ stator (rotor) resistance
$L_s(L_r)$ stator (rotor) self-inductance
M stator/rotor mutual inductance
p number of pole pairs
σ leakage coefficient $(= 1 - \frac{M^2}{L_s L_r})$
$c = \frac{1}{\sigma L_s}$
$a_1 = c\left(R_s + \frac{M^2 R_r}{L_r^2}\right)$; $a_2 = \frac{cM R_r}{L_r^2}$; $a_3 = \frac{cM}{L_r}$; $a_4 = \frac{R_r}{L_r}$; $a_5 = \frac{M R_r}{L_r}$
J rotor inertia of the MG set
D damping coefficient of the MG set
K_T torque constant $(= \frac{3pM}{2L_r})$
$T_e = K_T(\Phi_{dr} i_{qs} - \Phi_{qr} i_{ds})$ generated torque
T_L disturbance torque

Refer to (F6) for the symbol meaning. With the state variables

$$x \triangleq \begin{bmatrix} i_{ds} & i_{qs} & \Phi_{dr} & \Phi_{qr} & w_r \end{bmatrix}^\mathsf{T}$$

we obtain the following state equation.

$$
\begin{bmatrix} \dot{x}_1 \\ \dot{x}_2 \\ \dot{x}_3 \\ \dot{x}_4 \\ \dot{x}_5 \end{bmatrix} = \begin{bmatrix} a_2 x_3 - a_1 x_1 + p a_3 x_4 x_5 \\ a_2 x_4 - a_1 x_2 - p a_3 x_3 x_5 \\ a_5 x_1 - a_4 x_3 - p x_4 x_5 \\ a_5 x_2 - a_4 x_4 + p x_3 x_5 \\ \frac{K_T(x_2 x_3 - x_1 x_4) - T_L - D x_5}{J} \end{bmatrix} + \begin{bmatrix} c & 0 & x_2 \\ 0 & c & -x_1 \\ 0 & 0 & x_4 \\ 0 & 0 & -x_3 \\ 0 & 0 & 0 \end{bmatrix} \begin{bmatrix} V_{ds} \\ V_{qs} \\ w_s \end{bmatrix} \tag{4.87}
$$

$$
\triangleq f(x) + g(x)u
$$

Since

$$
[\mathrm{ad}_f g_1(x)\ \mathrm{ad}_f g_2(x)\ \mathrm{ad}_f g_3(x)] = \begin{bmatrix} a_1 c & 0 & 0 \\ 0 & a_1 c & 0 \\ -a_5 c & 0 & 0 \\ 0 & -a_5 c & 0 \\ \frac{K_T c x_4}{J} & \frac{-K_T c x_3}{J} & 0 \end{bmatrix}
$$

we have that $(\kappa_1, \kappa_2, \kappa_3) = (2, 2, 1)$ on a neighborhood of $x = x_0(\neq 0)$. Since $3 = \dim(\Delta_0(x)) \neq \dim(\Delta_0(0)) = 2$ for $x \neq 0$, $\Delta_0(x) \left(\triangleq \mathrm{span}\{g_1(x), g_2(x), g_3(x)\} \right)$ is not a distribution on a neighborhood of $x = 0$ and condition (ii) of Theorem 4.3 is not satisfied. Therefore, by Theorem 4.3, system (4.87) is not feedback linearizable on a neighborhood of $x = 0$. If we consider the local linearization on a neighborhood of $x = x_0(\neq 0)$ instead of a neighborhood of 0, we have that $\kappa_1 + \kappa_2 + \kappa_3 = 5$ on a neighborhood of $x = x_0$ and condition (i) of Theorem 4.3 is satisfied. Also, it is clear that $\Delta_0(x) \left(\triangleq \mathrm{span}\{g_1(x), g_2(x), g_3(x)\} \right)$ is a involutive distribution on a neighborhood of $x = x_0$ and thus condition (ii) of Theorem 4.3 is satisfied. Hence, system (4.87) is feedback linearizable with

$$
S = \begin{bmatrix} x_3^2 + x_4^2 \\ -2a_4 x_3^2 + 2a_5 x_1 x_3 - 2a_4 x_4^2 + 2a_5 x_2 x_4 \\ x_5 \\ \frac{K_T(x_2 x_3 - x_1 x_4) - T_L - D x_5}{J} \\ x_1 \end{bmatrix}
$$

and

$$
\begin{bmatrix} V_{ds} \\ V_{qs} \\ w_s \end{bmatrix} = \begin{bmatrix} L_g L_f S_{11}(x) \\ L_g L_f S_{21}(x) \\ L_g S_{31}(x) \end{bmatrix}^{-1} \left(- \begin{bmatrix} L_f^2 S_{11}(x) \\ L_f^2 S_{21}(x) \\ L_f S_{31}(x) \end{bmatrix} + \begin{bmatrix} v_1 \\ v_2 \\ v_3 \end{bmatrix} \right)
$$

$$
= \begin{bmatrix} 2a_5 c x_3 & 2a_5 c x_4 & 0 \\ \frac{-K_T c x_4}{J} & \frac{K_T c x_3}{J} & 0 \\ c & 0 & x_2 \end{bmatrix}^{-1} \left(- \begin{bmatrix} L_f^2 S_{11}(x) \\ L_f^2 S_{21}(x) \\ L_f S_{31}(x) \end{bmatrix} + \begin{bmatrix} v_1 \\ v_2 \\ v_3 \end{bmatrix} \right)
$$

$$
= \begin{bmatrix} \alpha_1(x) \\ \alpha_2(x) \\ \alpha_3(x) \end{bmatrix} + \begin{bmatrix} \frac{x_3}{2a_5 c(x_3^2 + x_4^2)} & \frac{-J x_4}{K_T c(x_3^2 + x_4^2)} & 0 \\ \frac{x_4}{2a_5 c(x_3^2 + x_4^2)} & \frac{J x_3}{K_T c(x_3^2 + x_4^2)} & 0 \\ \frac{-x_3}{2a_5 x_2(x_3^2 + x_4^2)} & \frac{J x_4}{K_T x_2(x_3^2 + x_4^2)} & \frac{1}{x_2} \end{bmatrix} \begin{bmatrix} v_1 \\ v_2 \\ v_3 \end{bmatrix} = \alpha(x) + \beta(x)v
$$

where

$$L_f^2 S_{11}(x) = 2a_5 x_3 f_1(x) + 2a_5 x_4 f_2(x) + (2a_5 x_1 - 4a_4 x_3) f_3(x)$$
$$+ (2a_5 x_2 - 4a_4 x_4) f_4(x)$$
$$L_f^2 S_{21}(x) = \frac{K_T(-x_4 f_1(x) + x_3 f_2(x) + x_2 f_3(x) - x_1 f_4(x))}{J} - \frac{D f_5(x)}{J}$$
$$L_f S_{31}(x) = a_2 x_3 - a_1 x_1 + a_3 p x_4 x_5$$

and

$$\alpha(x) = -\begin{bmatrix} \frac{x_3}{2a_5 c(x_3^2+x_4^2)} & \frac{-J x_4}{K_T c(x_3^2+x_4^2)} & 0 \\ \frac{x_4}{2a_5 c(x_3^2+x_4^2)} & \frac{J x_3}{K_T c(x_3^2+x_4^2)} & 0 \\ \frac{-x_3}{2a_5 x_2(x_3^2+x_4^2)} & \frac{J x_4}{K_T x_2(x_3^2+x_4^2)} & \frac{1}{x_2} \end{bmatrix} \begin{bmatrix} L_f^2 S_{11}(x) \\ L_f^2 S_{21}(x) \\ L_f S_{31}(x) \end{bmatrix}$$

Then it is easy to see that

$$\begin{bmatrix} \dot{z}_1 \\ \dot{z}_2 \\ \dot{z}_3 \\ \dot{z}_4 \\ \dot{z}_5 \end{bmatrix} = \begin{bmatrix} 0 & 1 & 0 & 0 & 0 \\ 0 & 0 & 0 & 0 & 0 \\ 0 & 0 & 0 & 1 & 0 \\ 0 & 0 & 0 & 0 & 0 \\ 0 & 0 & 0 & 0 & 0 \end{bmatrix} \begin{bmatrix} z_1 \\ z_2 \\ z_3 \\ z_4 \\ z_5 \end{bmatrix} + \begin{bmatrix} 0 & 0 & 0 \\ 1 & 0 & 0 \\ 0 & 0 & 0 \\ 0 & 1 & 0 \\ 0 & 0 & 1 \end{bmatrix} \begin{bmatrix} v_1 \\ v_2 \\ v_3 \end{bmatrix}$$
$$= Az + Bv.$$

Refer to the MATLAB program in Sect. 4.5.

Example 4.4.4 (*the aircraft*) The dynamic equations for the aircraft model can be written as follows: (Refer to (E3).)

$$\dot{x} = u \cos \psi \cos \theta + v(\cos \psi \sin \theta \sin \phi - \sin \psi \cos \phi)$$
$$+ w(\cos \psi \sin \theta \cos \phi + \sin \psi \sin \phi)$$
$$\dot{y} = u \sin \psi \cos \theta + v(\sin \psi \sin \theta \sin \phi + \cos \psi \cos \phi)$$
$$+ w(\sin \psi \sin \theta \cos \phi - \cos \psi \sin \phi)$$
$$\dot{z} = -u \sin \theta + v \cos \theta \sin \phi + w \cos \theta \cos \phi$$
$$\dot{u} = -g \sin \theta + rv - qw + \frac{X(\xi)}{m} + \frac{J\rho}{m}$$
$$\dot{v} = g \cos \theta \sin \phi + pw - ru + \frac{Y(\xi)}{m}$$
$$\dot{w} = g \cos \theta \cos \phi + qu - pv + \frac{Z(\xi)}{m}$$
$$\dot{\phi} = p + \tan \theta (q \sin \phi + r \cos \phi)$$
$$\dot{\theta} = q \cos \phi - r \sin \phi$$
$$\dot{\psi} = \frac{q \sin \phi + r \cos \phi}{\cos \theta}$$

where
(x, y, z) the center of mass in an absolute frame
(u, v, w) the velocity components in a relative frame
(ϕ, θ, ψ) roll, pitch, and yaw angles
(p, q, r) the components of the kinetic moment in the relative frame
ρ the thrust
g gravitational acceleration
$(X(\xi) + J\rho, Y(\xi), Z(\xi))$ the components of the force vector excepting gravity.

With the state variables

$$\xi \triangleq \begin{bmatrix} x\ y\ z\ u\ v\ w\ \phi\ \theta\ \psi \end{bmatrix}^{\mathsf{T}}$$

we obtain the following state equation.

$$\dot{\xi} = f(\xi) + g_1(\xi)p + g_2(\xi)q + g_3(\xi)r + g_4(\xi)\rho \tag{4.88}$$

$$f(\xi) = \begin{bmatrix} f_1(\xi) \\ f_2(\xi) \\ -\xi_4 \sin \xi_8 + \xi_5 \cos \xi_8 \sin \xi_7 + \xi_6 \cos \xi_8 \cos \xi_7 \\ -g \sin \xi_8 \\ g \cos \xi_8 \sin \xi_7 \\ g \cos \xi_8 \cos \xi_7 \\ 0 \\ 0 \\ 0 \end{bmatrix} + \begin{bmatrix} 0 \\ 0 \\ 0 \\ \frac{X(\xi)}{m} \\ \frac{Y(\xi)}{m} \\ \frac{Z(\xi)}{m} \\ 0 \\ 0 \\ 0 \end{bmatrix}$$

$$\triangleq f^0(\xi) + \tilde{f}(\xi)$$

$$\begin{bmatrix} g_1(\xi)\ g_2(\xi)\ g_3(\xi)\ g_4(\xi) \end{bmatrix} = \begin{bmatrix} 0 & 0 & 0 & 0 \\ 0 & 0 & 0 & 0 \\ 0 & 0 & 0 & 0 \\ 0 & -\xi_6 & \xi_5 & \frac{J}{m} \\ \xi_6 & 0 & -\xi_4 & 0 \\ -\xi_5 & \xi_4 & 0 & 0 \\ 1 & \tan \xi_8 \sin \xi_7 & \tan \xi_8 \cos \xi_7 & 0 \\ 0 & \cos \xi_7 & -\sin \xi_7 & 0 \\ 0 & \frac{\sin \xi_7}{\cos \xi_8} & \frac{\cos \xi_7}{\cos \xi_8} & 0 \end{bmatrix}$$

where

$$f_1(\xi) = \xi_4 \cos \xi_9 \cos \xi_8 + \xi_5(\cos \xi_9 \sin \xi_8 \sin \xi_7 - \sin \xi_9 \cos \xi_7)$$
$$+ \xi_6(\cos \xi_9 \sin \xi_8 \cos \xi_7 + \sin \xi_9 \sin \xi_7)$$

and

$$f_2(\xi) = \xi_4 \sin \xi_9 \cos \xi_8 + \xi_5(\sin \xi_9 \sin \xi_8 \sin \xi_7 + \cos \xi_9 \cos \xi_7)$$
$$+ \xi_6(\sin \xi_9 \sin \xi_8 \cos \xi_7 - \cos \xi_9 \sin \xi_7).$$

Since

$$[g_2(\xi), g_4(\xi)] = \begin{bmatrix} 0 \\ 0 \\ 0 \\ 0 \\ 0 \\ \frac{J}{m} \\ 0 \\ 0 \\ 0 \end{bmatrix} \notin \Delta_0(x) = \text{span}\{g_1(x), g_2(x), g_3(x), g_4(x)\}$$

it is clear that $\Delta_0(x)$ is not an involutive distribution and thus condition (ii) of Theorem 4.3 is not satisfied. Therefore, by Theorem 4.3, system (4.88) is not feedback linearizable. Consider system (4.88) with $\rho = 0$. It is easy to see that

$$\begin{bmatrix} \text{ad}_f g_1(\xi) & \text{ad}_f g_2(\xi) & \text{ad}_f g_3(\xi) \end{bmatrix}$$

$$= \begin{bmatrix} 0 & 0 & 0 \\ 0 & 0 & 0 \\ 0 & 0 & 0 \\ -\frac{L_{g_1} X(\xi)}{m} & -\frac{Z(\xi)}{m} - \frac{L_{g_2} X(\xi)}{m} & \frac{Y(\xi)}{m} - \frac{L_{g_3} X(\xi)}{m} \\ \frac{Z(\xi)}{m} - \frac{L_{g_1} Y(\xi)}{m} & -\frac{L_{g_2} Y(\xi)}{m} & -\frac{X(\xi)}{m} - \frac{L_{g_3} Y(\xi)}{m} \\ -\frac{Y(\xi)}{m} - \frac{L_{g_1} Z(\xi)}{m} & \frac{X(\xi)}{m} - \frac{L_{g_2} Z(\xi)}{m} & -\frac{L_{g_3} Z(\xi)}{m} \\ 0 & 0 & 0 \\ 0 & 0 & 0 \\ 0 & 0 & 0 \end{bmatrix}$$

and

$$\left[\mathrm{ad}_f^2 g_1(\xi)\ \mathrm{ad}_f^2 g_2(\xi)\ \mathrm{ad}_f^2 g_3(\xi)\right]\Big|_{\xi=0}$$

$$= \begin{bmatrix} \begin{bmatrix} \dfrac{L_{g_1}X(\xi)}{m} & \dfrac{Z(\xi)}{m}+\dfrac{L_{g_2}X(\xi)}{m} & -\dfrac{Y(\xi)}{m}+\dfrac{L_{g_3}X(\xi)}{m} \\[2mm] -\dfrac{Z(\xi)}{m}+\dfrac{L_{g_1}Y(\xi)}{m} & \dfrac{L_{g_2}Y(\xi)}{m} & \dfrac{X(\xi)}{m}+\dfrac{L_{g_3}Y(\xi)}{m} \\[2mm] \dfrac{Y(\xi)}{m}+\dfrac{L_{g_1}Z(\xi)}{m} & -\dfrac{X(\xi)}{m}+\dfrac{L_{g_2}Z(\xi)}{m} & \dfrac{L_{g_3}Z(\xi)}{m} \\[2mm] 0 & 0 & 0 \\ 0 & 0 & 0 \\ 0 & 0 & 0 \\ 0 & 0 & 0 \\ 0 & 0 & 0 \\ 0 & 0 & 0 \end{bmatrix} \end{bmatrix}_{\xi=0}$$

mod span $\left\{g_1(\xi),\, g_2(\xi),\, g_3(\xi),\, \mathrm{ad}_f g_1(\xi),\, \mathrm{ad}_f g_2(\xi),\, \mathrm{ad}_f g_3(\xi)\right\}$

which imply that $(\kappa_1, \kappa_2, \kappa_3) = (3, 3, 3)$ and thus condition (i) of Theorem 4.3 is satisfied. Also, it is easy to see that $\Delta_0(x) \left(= \mathrm{span}\left\{g_1(\xi), g_2(\xi), g_3(\xi)\right\}\right)$ and $\Delta_1(x) \left(= \mathrm{span}\left\{g_1(\xi), g_2(\xi), g_3(\xi), \mathrm{ad}_f g_1(\xi), \mathrm{ad}_f g_2(\xi), \mathrm{ad}_f g_3(\xi)\right\}\right)$ are involutive distributions and thus condition (ii) of Theorem 4.3 is satisfied. Therefore, by Theorem 4.3, system (4.88) is feedback linearizable with

$$z = S(\xi) = m \begin{bmatrix} \xi_1 \\ f_1(\xi) \\ L_f f_1(\xi) \\ \xi_2 \\ f_2(\xi) \\ L_f f_2(\xi) \\ \xi_3 \\ f_3(\xi) \\ L_f f_3(\xi) \end{bmatrix} = m \begin{bmatrix} \xi_1 \\ f_1(\xi) \\ L_{\tilde{f}} f_1(\xi) \\ \xi_2 \\ f_2(\xi) \\ L_{\tilde{f}} f_2(\xi) \\ \xi_3 \\ f_3(\xi) \\ g + L_{\tilde{f}} f_3(\xi) \end{bmatrix}$$

and

$$\begin{bmatrix} p \\ q \\ r \end{bmatrix} = \begin{bmatrix} L_{g_1}L_{\tilde{f}}f_1 & L_{g_2}L_{\tilde{f}}f_1 & L_{g_3}L_{\tilde{f}}f_1 \\ L_{g_1}L_{\tilde{f}}f_2 & L_{g_2}L_{\tilde{f}}f_2 & L_{g_3}L_{\tilde{f}}f_2 \\ L_{g_1}L_{\tilde{f}}f_3 & L_{g_2}L_{\tilde{f}}f_3 & L_{g_3}L_{\tilde{f}}f_3 \end{bmatrix}^{-1} \left(-\begin{bmatrix} L_f L_{\tilde{f}} f_1(\xi) \\ L_f L_{\tilde{f}} f_2(\xi) \\ L_f L_{\tilde{f}} f_3(\xi) \end{bmatrix} + \frac{1}{m}\begin{bmatrix} v_1 \\ v_2 \\ v_3 \end{bmatrix} \right)$$

$$= \alpha(\xi) + \beta(\xi)v$$

where

$$L_{f^0} f_1(\xi) = 0; \quad L_{f^0} f_2(\xi) = 0; \quad L_{f^0} f_3(\xi) = g.$$

Refer to the MATLAB program in Sect. 4.5. Then it is easy to see that

$$
\begin{bmatrix} \dot{z}_1 \\ \dot{z}_2 \\ \dot{z}_3 \\ \dot{z}_4 \\ \dot{z}_5 \\ \dot{z}_6 \\ \dot{z}_7 \\ \dot{z}_8 \\ \dot{z}_9 \end{bmatrix} = \begin{bmatrix} 0\,1\,0\,0\,0\,0\,0\,0\,0 \\ 0\,0\,1\,0\,0\,0\,0\,0\,0 \\ 0\,0\,0\,0\,0\,0\,0\,0\,0 \\ 0\,0\,0\,0\,1\,0\,0\,0\,0 \\ 0\,0\,0\,0\,0\,1\,0\,0\,0 \\ 0\,0\,0\,0\,0\,0\,0\,0\,0 \\ 0\,0\,0\,0\,0\,0\,0\,1\,0 \\ 0\,0\,0\,0\,0\,0\,0\,0\,1 \\ 0\,0\,0\,0\,0\,0\,0\,0\,0 \end{bmatrix} \begin{bmatrix} z_1 \\ z_2 \\ z_3 \\ z_4 \\ z_5 \\ z_6 \\ z_7 \\ z_8 \\ z_9 \end{bmatrix} + \begin{bmatrix} 0\,0\,0 \\ 0\,0\,0 \\ 1\,0\,0 \\ 0\,0\,0 \\ 0\,0\,0 \\ 0\,1\,0 \\ 0\,0\,0 \\ 0\,0\,0 \\ 0\,0\,1 \end{bmatrix} \begin{bmatrix} v_1 \\ v_2 \\ v_3 \end{bmatrix}
$$

$$= Az + Bv.$$

Suppose that the system is not feedback linearizable. Then we can consider the approximate linearization(Problem 4-12), the partial linearization(Problem 4-13), and the dynamic feedback linearization(Problem 4-14). In fact, system (4.88) with $\rho \neq 0$ is dynamic feedback linearizable (Refer to Problem 4-15). The dynamic feedback linearization is considered in more detail in Chap. 6.

4.5 MATLAB Programs

In this section, the following subfunctions in Appendix C are needed:
adfg, adfgk, adfgM, adfgkM, ChExact, ChInvolutive, ChZero, Codi, CXexact, Delta, Kindex0, Kindex, Lfh, Lfhk, S1, S1M

MATLAB program for Theorem 4.1:

```
clear all syms x1 x2 x3 x4 x5 x6 x7 x8 x9 real

f=[x2; x1^2]; g=[x1-x1; 1]; %Ex:4.2.2

% f=[-x1-2*x2^3; -x2^2]; g=[1+2*x2; 1]; %Ex:4.2.3

% f=[x2; x1; x2+x1*x3]; g=[1; x1-x1; 0]; %Ex:4.2.4

% f=[x2-x1^2; x3+2*x1*x2-x1^3; x1^2-3*x1^2*x2+3*x1^4];
% g=[0; 0; 1+x1]; %Ex:4.2.5

% f=[x2; x3+x2^2; 0]; g=[x1-x1; 1; 1]; %Ex:4.2.7

% f=[x2+2*x3*x1^2; x3; x1^2];
% g=[2*x3*(1+x2^2); 0; 1+x2^2]; %P:4-1

% f=[x2; 0]; g=[x1; 1]; %P:4-3(a)

% f=[x2+x3^2; x3; 0]; g=[x1-x1; 0; 1]; %P:4-3(b)

% f=[x2+x3^2; x3; x4; 0]; g=[x1-x1; 0; 0; 1]; %P:4-3(c)
```

```
% f=[x2; x3^2; x4; 0]; g=[x1-x1; 0; 0; 1]; %P:4-4

f=simplify(f)
g=simplify(g)

[n,m]=size(g);
x=sym('x',[n,1]);

T(:,1)=g;
for k=2:n
  T(:,k)=adfg(f,T(:,k-1),x);
end
T=simplify(T)
T0=simplify(subs(T,x,x-x));
TD=T(:,1:n-1);

if rank(T0) < n
  display('condition (i) of Thm 4.1 is not satisfied.')
  display('System is NOT feedback linearizable.')
  return
end

if ChInvolutive(TD,x) == 0
  display('condition (ii) of Thm 4.1 is not satisfied.')
  display('System is NOT feedback linearizable.')
  return
end

display('System is feedback linearizable.')

[flag,S1]=S1(f,g,x)

if flag==0
  display('Find out z=S(x) without MATLAB.')
  return
end

S=x-x;
for k=1:n
  S(k)=Lfhk(f,S1,x,k-1);
end
S=simplify(S)
beta=simplify(inv(Lfh(g,S(n),x)))
alpha=simplify(-beta*Lfh(f,S(n),x))

hg=simplify(g*beta)
hf=simplify(f+g*alpha)

dS=simplify(jacobian(S,x));
idS=simplify(inv(dS));
AS=simplify(dS*hf);
dAS=simplify(jacobian(AS,x));
A=simplify(dAS*idS)
```

```
B=simplify(dS*hg)

return
```

MATLAB program for Theorem 4.3:

```
clear all
syms x1 x2 x3 x4 x5 x6 x7 x8 x9 real

f=[x2; -x1+x2^2; x3^2]; g=[0  0; 1+x1^2  0; 0  1]; %Ex:4.3.4

%  f=[x2+2*x4*x5; x3; 0; x5; 0];
%  g=[0 0; 0 0; 1 x1; 0 0; 0 1]; %Ex:4.3.5

%  f=[x2-x4*(x3+x4); 0; x4; 0];
%  g=[0 x1; 0 1; -1 -x1; 1 x1]; %Ex:4.3.6

%  f=[x2; x3; x4; 0]; g=[0 x1; 0 0; 1 0; 0 1]; %Ex:4.3.7

%  f=[x2; 0; 0]; g=[0 x2^2; 1 0; 0 1]; %Ex:4.3.8

%  f=[x2; x4; x4+3*x2^2*x4; 0];
%  g=[2*x1*x4  2*x4; 1  0; 3*x2^2  0; x1  1]; %P:4-5

%  f=[-x1+x2^2; -2*x2+sin(x2)]; g=[1 x1-x1; 0 1]; %P:4-6

%  g=[1 0 1; 2*(x1-x5) 0 0; 0 1 x3; -2*(x1-x5) 0 0; 0 0 1];
%  f=[x1^2; x3; 0; x5; x1^2]; %P:4-7

%  g=[0 x3; 0 0; 1 0; 0 1]; f=[0; x3; 0; 0]; %P:4-8(a)

%  g=[0 x3; 0 0; 1 0; 0 0]; f=[x2; x3; 0; x1]; %P:4-8(b)

%  g=[0 x3; 0 0; 1 0; 0 1]; f=[x2; x3; 0; 0]; %P:4-8(c)

%  f=[x2; x3; 0; 0]; g=[2*x1 0; 2*x2 0; 1 0; 0 1]; %P:4-9

%  f=[x2; 0; x4; 0]; g=[0 0; 1+x2 0; 0 0; 0 1]; %P:4-10

%  f=[x2; 0; x4; 0]; g=[0 x2^3; 1+x2 0; 0 0; 0 1]; %P:4-11

%  fo=[x2; 0; x4; 0]; go=[0 x2^3; 1+x2 0; 0 0; 0 1];
%  f=[fo+x5*go(:,2); 0]; g(:,1)=[go(:,1); 0]
%  g(:,2)=[go(:,2)-go(:,2); 1]; %P:4-14  (x5=eta)

f=simplify(f)
g=simplify(g)

[n,m]=size(g);
x=sym('x',[n,1]);

[kappa,D]=Kindex0(f,g,x)
```

```
if sum(kappa) < n
  display('condition (i) of Thm 4.3 is not satisfied.')
  display('System is NOT feedback linearizable.')
  return
end

for k=1:max(kappa)-1
  TD=D(:,1:k*m);
  if rank(TD) ~= rank(subs(TD,x,x-x))
    display('NOT (constant dimensional) distribution.')
    display('condition (ii) of Thm 4.3 is not satisfied.')
    display('System is NOT feedback linearizable.')
    return
  end
  if ChInvolutive(TD,x) == 0
    display('condition (ii) of Thm 4.3 is not satisfied.')
    display('System is NOT feedback linearizable.')
    return
  end
end

display('System is feedback linearizable.')

[flag,S1]=S1M(f,g,x,kappa);

if flag==0
  display('Find out z=S(x) without MATLAB.')
  return
end

S=x1-x1;
for k1=1:m
  for k=1:kappa(k1)
    t1=Lfhk(f,S1(k1),x,k-1);
    S=[S; t1];
  end
end
S=simplify(S(2:n+1))
t2=S1-S1;
for k1=1:m
  t2(k1)=Lfhk(f,S1(k1),x,kappa(k1)-1);
end
t2=simplify(t2);

ibeta=simplify(Lfh(g,t2,x));
beta=simplify(inv(ibeta))
t3=simplify(Lfh(f,t2,x));
alpha=simplify(-beta*t3)

hg=simplify(g*beta)
hf=simplify(f+g*alpha)

dS=simplify(jacobian(S,x));
```

```
idS=simplify(inv(dS));
AS=simplify(dS*hf);
dAS=simplify(jacobian(AS,x));
A=simplify(dAS*idS)
B=simplify(dS*hg)

return
```

MATLAB program for Example 4.4.1:

```
clear all
syms x1 x2 x3 x4 x5 real
syms R L M g y0 real

bg=[x1-x1; 0; 1/L];
bf=[x2; g-(1/M)*((x3+sqrt(M*g*y0))^2)/(x1+y0); -(R/L)*x3];

[n,m]=size(bg);
x=sym('x',[n,1]);

bf=simplify(bf)
bg=simplify(bg)

T(:,1)=bg;
for k=2:n
  T(:,k)=adfg(bf,T(:,k-1),x);
end
T=simplify(T)
T0=simplify(subs(T,x,x-x));
TD=T(:,1:n-1);

if rank(T0) < n
  display('condition (i) of Thm 4.1 is not satisfied.')
  display('System is NOT feedback linearizable.')
  return
end

if ChInvolutive(TD,x) == 0
  display('condition (ii) of Thm 4.1 is not satisfied.')
  display('System is NOT feedback linearizable.')
  return
end

display('System is feedback linearizable.')

[flag,S1]=S1(bf,bg,x)

if flag==0
  display('Find out z=S(x) without MATLAB.')
  return
end
S1=-(2*(M*g*y0)^(1/2))/(L*M*y0)*S1
```

```
S=x-x;
for k=1:n
  S(k)=Lfhk(bf,S1,x,k-1);
end
S=simplify(S)
beta=simplify(inv(Lfh(bg,S(n),x)))
alpha=simplify(-beta*Lfh(bf,S(n),x))

hg=simplify(bg*beta)
hf=simplify(bf+bg*alpha)

dS=simplify(jacobian(S,x));
idS=simplify(inv(dS));
AS=simplify(dS*hf);
dAS=simplify(jacobian(AS,x));
A=simplify(dAS*idS)
B=simplify(dS*hg)

return
```

MATLAB program for Example 4.4.3:

```
clear all
syms x1 x2 x3 x4 x5 real
syms u1 u2 u3 real
syms a1 a2 a3 a4 a5 real
syms p c J D Te TL KT real

g=[c 0 x2; 0 c -x1; 0 0 x4; 0 0 -x3; 0 0 0];
[n,m]=size(g);
x=sym('x',[n,1]);
f=x-x; f(5)=(-D*x5+KT*(x2*x3-x1*x4)-TL)/J;
f(1:2)=[-a1*x1+a2*x3+p*a3*x4*x5; -a1*x2-p*a3*x3*x5+a2*x4];
f(3:4)=[a5*x1-a4*x3-p*x4*x5; a5*x2+p*x3*x5-a4*x4];

f=simplify(f)
g=simplify(g)

[ka,D]=Kindex(f,g,x)

if sum(ka) < n
  display('condition (i) of Thm 4.3 is not satisfied.')
  return
end

for k=1:max(ka)-1
  if ChInvolutive(D(:,1:k*m),x) == 0
    display('condition (ii) of Thm 4.3 is not satisfied.')
    return
  end
end

display('System is feedback linearizable.')
```

```
S1=[x3^2+x4^2; x5; x1]

S=x1-x1;
for k1=1:m
  for k=1:ka(k1)
    t1=Lfhk(f,S1(k1),x,k-1);
    S=[S; t1];
  end
end
S=simplify(S(2:n+1))

t2=S1-S1;
for k1=1:m
  t2(k1)=Lfhk(f,S1(k1),x,ka(k1)-1);
end
t2=simplify(t2);

ibeta=simplify(Lfh(g,t2,x))
beta=simplify(inv(ibeta))
t3=simplify(Lfh(f,t2,x))
alpha=simplify(-beta*t3)

hg=simplify(g*beta)
hf=simplify(f+g*alpha)

dS=simplify(jacobian(S,x));
idS=simplify(inv(dS));
AS=simplify(dS*hf);
dAS=simplify(jacobian(AS,x));
A=simplify(dAS*idS)
B=simplify(dS*hg)

return
```

MATLAB program for Example 4.4.4 and Problem 4-15:

```
clear all
syms E1 E2 E3 E4 E5 E6 E7 E8 E9 E10 real
syms U1 U2 U3 real
syms a1 a2 a3 a4 a5 real
syms g J m X Y Z real
syms X11 X12 X13 Y11 Y12 Y13 Z11 Z12 Z13 real

G1=[0 0 0 0; 0 0 0 0; 0 0 0 0];
G2=[0 -E6 E5 J/m; E6 0 -E4 0; -E5 E4 0 0];
G3=[1 tan(E8)*sin(E7) tan(E8)*cos(E7) 0; 0 cos(E7) -sin(E7) 0;
    0 sin(E7)/cos(E8) cos(E7)/cos(E8) 0];
TG=[G1; G2; G3];

f0=TG(:,1)-TG(:,1); f0t1=sin(E8)*sin(E7);
f0(1)=E4*cos(E9)*cos(E8)+E5*(cos(E9)*f0t1-sin(E9)*cos(E7));
f0(1)=f0(1)+E6*(cos(E9)*sin(E8)*cos(E7)+sin(E9)*sin(E7)) ;
```

```
f0(2)=E4*sin(E9)*cos(E8)+E5*(sin(E9)*f0t1+cos(E9)*cos(E7));
f0(2)=f0(2)+E6*(sin(E9)*sin(E8)*cos(E7)-cos(E9)*sin(E7));
f0(3)=-E4*sin(E8)+E5*cos(E8)*sin(E7)+E6*cos(E8)*cos(E7);
f0(4)= -g*sin(E8);
f0(5)= g*cos(E8)*sin(E7);
f0(6)= g*cos(E8)*cos(E7);

G=TG; %Ex:4.4.4 (bm=4)
[n,bm]=size(G);
E=sym('E',[n,1]);
EYE=jacobian(E,E)

t1=adfg(G(:,2),G(:,4),E)

if rank([t1 G]) > rank(G)
  display('condition (ii) of Thm 4.3.1 is not satisfied.')
  display('The system with rho is NOT FB linearizable.')
end

G=TG(:,1:3); %Ex:4.4.4 (bm=3, rho=0)
[n,bm]=size(G);

tf=(X/m)*EYE(:,4)+(Y/m)*EYE(:,5)+(Z/m)*EYE(:,6)
f=f0+tf;
f=simplify(f)
G=simplify(G)

D=G
temp1=adfgM(f,G,E)
t11=(X11/m)*EYE(:,4)+(Y11/m)*EYE(:,5)+(Z11/m)*EYE(:,6);
t12=(X12/m)*EYE(:,4)+(Y12/m)*EYE(:,5)+(Z12/m)*EYE(:,6);
t13=(X13/m)*EYE(:,4)+(Y13/m)*EYE(:,5)+(Z13/m)*EYE(:,6);
temp2=[t11 t12 t13]
D=[D temp1-temp2]

temp3=adfgM(f,D(:,4:6),E)
temp30=subs(temp3,E,E-E)

Flag1=ChInvolutive(D(:,1:3),E)
Flag2=ChInvolutive(D(:,1:6),E)

display('System with rho=0 is feedback linearizable.')

ANS1=simplify(Lfh(f0,f(1),E))
ANS2=simplify(Lfh(f0,f(2),E))
ANS3=simplify(Lfh(f0,f(3),E))

display('Problem 4-15')

G=TG;
[n,bm]=size(G);
E=sym('E',[n+1,1]);
```

```
FE=[f+E(10)*G(:,4); 0]
GE=[G(:,1:3) G(:,4)-G(:,4)];
GE=[GE; [0 0 0 1]]

Flag1E=ChInvolutive(D(:,1:4),E)
Flag2E=ChInvolutive(D(:,1:8),E)

display('Extended system (4.6.4) is feedback linearizable.')

return
```

4.6 Problems

4-1. Show that the following nonlinear control system is feedback linearizable.
Also, find a linearizing state transformation and feedback.

$$\dot{x} = \begin{bmatrix} x_2 + 2x_3x_1^2 \\ x_3 \\ x_1^2 \end{bmatrix} + \begin{bmatrix} 2x_3(1 + x_2^2) \\ 0 \\ 1 + x_2^2 \end{bmatrix} u$$

4-2. Find a nonlinear feedback $u = \alpha(x)$ that causes $\lim_{t \to \infty} x(t) = 0$ for the nonlinear control system of Problem 4-1.

4-3. Find out whether the following nonlinear control system is feedback linearizable. If it is feedback linearizable, find a linearizing state transformation and feedback.

(a)

$$\dot{x} = \begin{bmatrix} x_2 \\ 0 \end{bmatrix} + \begin{bmatrix} x_1 \\ 1 \end{bmatrix} u$$

(b)

$$\dot{x} = \begin{bmatrix} x_2 + x_3^2 \\ x_3 \\ 0 \end{bmatrix} + \begin{bmatrix} 0 \\ 0 \\ 1 \end{bmatrix} u$$

(c)

$$\dot{x} = \begin{bmatrix} x_2 + x_3^2 \\ x_3 \\ x_4 \\ 0 \end{bmatrix} + \begin{bmatrix} 0 \\ 0 \\ 0 \\ 1 \end{bmatrix} u$$

4-4. Show that the following nonlinear control system is not locally feedback linearizable on a neighborhood of the origin:

$$\dot{x} = \begin{bmatrix} x_2 \\ x_3^2 \\ x_4 \\ 0 \end{bmatrix} + \begin{bmatrix} 0 \\ 0 \\ 0 \\ 1 \end{bmatrix} u$$

4-5. Linearize the following nonlinear control system by state transformation and feedback.

$$\dot{x} = \begin{bmatrix} x_2 \\ x_4 \\ x_4 + 3x_2^2 x_4 \\ 0 \end{bmatrix} + \begin{bmatrix} 2x_1 x_4 & 2x_4 \\ 1 & 0 \\ 3x_2^2 & 0 \\ x_1 & 1 \end{bmatrix} u$$

4-6. Show that the following nonlinear control system is feedback linearizable:

$$\dot{x} = \begin{bmatrix} -x_1 + x_2^2 \\ -2x_2 + \sin x_2 \end{bmatrix} + \begin{bmatrix} 1 \\ 0 \end{bmatrix} u_1 + \begin{bmatrix} 0 \\ 1 \end{bmatrix} u_2$$

4-7. Find out whether the following nonlinear control system is feedback linearizable. If it is feedback linearizable, find a linearizing state transformation and feedback.

$$\dot{x} = \begin{bmatrix} x_1^2 \\ x_3 \\ 0 \\ x_5 \\ x_1^2 \end{bmatrix} + u_1 \begin{bmatrix} 1 \\ 2(x_1 - x_5) \\ 0 \\ -2(x_1 - x_5) \\ 0 \end{bmatrix} + u_2 \begin{bmatrix} 0 \\ 0 \\ 1 \\ 0 \\ 0 \end{bmatrix} + u_3 \begin{bmatrix} 1 \\ 0 \\ x_3 \\ 0 \\ 1 \end{bmatrix}$$

4-8. Show that the following nonlinear control systems are not locally feedback linearizable on a neighborhood of the origin:

(a)

$$\dot{x} = \begin{bmatrix} 0 \\ x_3 \\ 0 \\ 0 \end{bmatrix} + \begin{bmatrix} 0 \\ 0 \\ 1 \\ 0 \end{bmatrix} u_1 + \begin{bmatrix} x_3 \\ 0 \\ 0 \\ 1 \end{bmatrix} u_2 = f(x) + g_1(x)u_1 + g_2(x)u_2$$

(b)

$$
\dot{x} = \begin{bmatrix} x_2 \\ x_3 \\ 0 \\ x_1 \end{bmatrix} + \begin{bmatrix} 0 \\ 0 \\ 1 \\ 0 \end{bmatrix} u_1 + \begin{bmatrix} x_3 \\ 0 \\ 0 \\ 0 \end{bmatrix} u_2 = f(x) + g_1(x)u_1 + g_2(x)u_2
$$

(c)

$$
\dot{x} = \begin{bmatrix} x_2 \\ x_3 \\ 0 \\ 0 \end{bmatrix} + \begin{bmatrix} 0 \\ 0 \\ 1 \\ 0 \end{bmatrix} u_1 + \begin{bmatrix} x_3 \\ 0 \\ 0 \\ 1 \end{bmatrix} u_2 = f(x) + g_1(x)u_1 + g_2(x)u_2
$$

4-9. Show that the following nonlinear control system is feedback linearizable. Also, find a linearizing state transformation and feedback.

$$
\dot{x} = \begin{bmatrix} x_2 \\ x_3 \\ 0 \\ 0 \end{bmatrix} + \begin{bmatrix} 2x_1 \\ 2x_2 \\ 1 \\ 0 \end{bmatrix} u_1 + \begin{bmatrix} 0 \\ 0 \\ 0 \\ 1 \end{bmatrix} u_2 = f(x) + g_1(x)u_1 + g_2(x)u_2
$$

4-10. Show that the following nonlinear control system is feedback linearizable. Also, find a linearizing state transformation and feedback.

$$
\dot{x} = \begin{bmatrix} x_2 \\ 0 \\ x_4 \\ 0 \end{bmatrix} + \begin{bmatrix} 0 \\ 1 + x_2 \\ 0 \\ 0 \end{bmatrix} u_1 + \begin{bmatrix} 0 \\ 0 \\ 0 \\ 1 \end{bmatrix} u_2 = f(x) + g_1(x)u_1 + g_2(x)u_2 \quad (4.89)
$$

4-11. Show that the following nonlinear control system is not feedback linearizable:

$$
\dot{x} = \begin{bmatrix} x_2 \\ 0 \\ x_4 \\ 0 \end{bmatrix} + \begin{bmatrix} 0 \\ 1 + x_2 \\ 0 \\ 0 \end{bmatrix} u_1 + \begin{bmatrix} x_2^3 \\ 0 \\ 0 \\ 1 \end{bmatrix} u_2 = f(x) + g_1(x)u_1 + g_2(x)u_2 \quad (4.90)
$$

4-12. Big O notation can be used to describe the error term in an approximation to a mathematical function. The most significant terms are written explicitly, and then the least-significant terms are summarized in a single big O term. For example, $e^x = 1 + x + \frac{1}{2}x^2 + \frac{1}{3!}x^3 + \frac{1}{4!}x^4 + \cdots = 1 + x + \frac{1}{2}x^2 + O(x^3) = 1 + x + O(x^2)$. Show that if the state transformation and nonsingular feedback of Problem 4-10 are used, then system (4.90) satisfies, in z-coordinates

$$\dot{z}(t) = Az(t) + Bv(t) + O\left((z, v)^4\right)$$

In other words, system (4.90) is not feedback linearizable. But, it can be approximated to a linear system (up to the third-order) more accurately than the classical first-order approximation technique, by using feedback and state transformation. It is called the approximate linearization.

4-13. For system (4.90), find a state transformation $z = S(x)$ and nonsingular feedback $u = \alpha(x) + \beta(x)v$ such that

$$\dot{z}^1 = Az^1 + Bv$$
$$\dot{z}^2 = \phi(z^1, z^2) + \psi(z^1, z^2)v$$

where $z = \begin{bmatrix} z^1 \\ z^2 \end{bmatrix}$, $z^1 \in \mathbb{R}^d$, and $z^2 \in \mathbb{R}^{n-d}$. Find the maximum of d. It is called the partial linearization.

4-14. Consider system (4.90). With the dynamic feedback

$$\begin{bmatrix} u_1 \\ u_2 \end{bmatrix} = \begin{bmatrix} w_1 \\ \eta_1 \end{bmatrix} \; ; \; \dot{\eta}_1 = w_2$$

we have the extended system

$$\dot{x}_E = \begin{bmatrix} \dot{x} \\ \dot{\eta}_1 \end{bmatrix} = \begin{bmatrix} f(x) + \eta_1 g_2(x) \\ 0 \end{bmatrix} + \begin{bmatrix} g_1(x) & 0 \\ 0 & 1 \end{bmatrix} w \qquad (4.91)$$
$$= f_E(x_E) + g_E(x_E)w$$

where $x_E = \begin{bmatrix} x \\ \eta_1 \end{bmatrix}$. Show that the extended system (4.91) is feedback linearizable. Also, find the extended state transformation $z_E = S_E(x_E)$ and nonsingular feedback $w = \alpha(x_E) + \beta(x_E)v$. In other words, system (4.90) is linearizable by the extended state transformation $z_E = S_E(x_E)$ and the dynamic feedback

$$\begin{bmatrix} u_1 \\ u_2 \end{bmatrix} = \begin{bmatrix} 0 \\ \eta_1 \end{bmatrix} + \begin{bmatrix} 1 & 0 \\ 0 & 0 \end{bmatrix} \alpha(x_E) + \begin{bmatrix} 1 & 0 \\ 0 & 0 \end{bmatrix} \beta(x_E)v$$
$$\dot{\eta}_1 = \begin{bmatrix} 0 & 1 \end{bmatrix} \alpha(x_E) + \begin{bmatrix} 0 & 1 \end{bmatrix} \beta(x_E)v.$$

It is called the dynamic feedback linearization.

4-15. Consider system (4.88) in Example 4.4.4. With the dynamic feedback

$$p = w_1; \quad q = w_2; \quad r = w_3; \quad \rho = \eta$$
$$\dot{\eta} = w_4$$

we have the extended system

$$\dot{\xi}_E = \begin{bmatrix} \dot{\xi} \\ \dot{\eta} \end{bmatrix} = \begin{bmatrix} f(\xi) + \eta g_4(\xi) \\ 0 \end{bmatrix} + \begin{bmatrix} g_1(\xi) & g_2(\xi) & g_3(\xi) & 0 \\ 0 & 0 & 0 & 1 \end{bmatrix} w$$

$$= f_E(\xi_E) + g_E(\xi_E) w \tag{4.92}$$

where $\xi_E = \begin{bmatrix} \xi \\ \eta \end{bmatrix}$. Show that the extended system (4.92) is feedback linearizable.

Chapter 5
Linearization with Output Equation

In Chaps. 3 and 4, only the state equation is considered for linearizing the system, and the output equation is still a nonlinear function of the new state variable. Here, we can also find the more restrictive problem that the output equation should also be a linear function for the new state variable. In this chapter, we discuss the linearization of a nonlinear control system with output. Linearization with output equation requires that both input-state and input-output relationships are linear.

5.1 Introduction

Consider the following smooth nonlinear control system with output:

$$\dot{x}(t) = f(x(t)) + g(x(t))u(t)$$
$$y(t) = h(x(t)) \tag{5.1}$$

where $x \in \mathbb{R}^n$, $u \in \mathbb{R}^m$, and $y \in \mathbb{R}^p$.

Example 5.1.1 Consider system (1.2) once again.

$$\begin{bmatrix} \dot{x}_1 \\ \dot{x}_2 \end{bmatrix} = \begin{bmatrix} 2x_1 x_2 - 2x_2^3 + (1 + 2x_2)u \\ x_1 - x_2^2 + u \end{bmatrix} \tag{5.2}$$
$$y = x_1 - x_2^2 + x_2.$$

It is shown, in Example 3.2.3, that system (5.2) is state equivalent, with state transformation $\begin{bmatrix} z_1 \\ z_2 \end{bmatrix} = S(x) = \begin{bmatrix} x_1 - x_2^2 \\ x_2 \end{bmatrix}$, to the following linear system:

© The Author(s), under exclusive license to Springer Nature Singapore Pte Ltd. 2022
H.-G. Lee, *Linearization of Nonlinear Control Systems*,
https://doi.org/10.1007/978-981-19-3643-2_5

$$\begin{bmatrix} \dot{z}_1 \\ \dot{z}_2 \end{bmatrix} = \begin{bmatrix} 0 & 0 \\ 1 & 0 \end{bmatrix} \begin{bmatrix} z_1 \\ z_2 \end{bmatrix} + \begin{bmatrix} 1 \\ 1 \end{bmatrix} u$$

Since $h \circ S^{-1}(z) = z_1 + z_2$, we have the following linear output equation:

$$y = \begin{bmatrix} 1 & 1 \end{bmatrix} \begin{bmatrix} z_1 \\ z_2 \end{bmatrix}.$$

But, if output equation is $y = x_1$ instead of $y = x_1 - x_2^2 + x_2$, then output equation is $y = z_1 + z_2^2$ which is nonlinear in z-coordinates.

As in the above Example, if not only state equation but also output equation is linearized by state transformation, then both the states and the output can be controlled very easily.

Definition 5.1 (*state equivalence to a LS with output*)
System (5.1) is said to be state equivalent to a linear system (LS) with output, if there exists a state transformation $z = S(x)$ such that system (5.1) satisfies, in z−coordinates, the following controllable linear system:

$$\begin{aligned} \dot{z} &= Az + Bu \\ y &= Cz. \end{aligned} \tag{5.3}$$

In other words,

$$\begin{aligned} S_*(f(x)) &= Az, \quad \begin{bmatrix} S_*(g_1(x)) & \cdots & S_*(g_m(x)) \end{bmatrix} = B \\ h \circ S^{-1}(z) &= Cz. \end{aligned} \tag{5.4}$$

That is, not only is the state equation of the system linear in the new coordinate system, but the output equation must also be linear in the new coordinate system. By using feedback in addition to state transformation, we can linearize the larger class of nonlinear control systems with output.

Definition 5.2 (*Feedback linearization with output*)
System (5.1) is said to be feedback linearizable with output, if there exist a feedback $u = \alpha(x) + \beta(x)v$ and a state transformation $z = S(x)$ such that the closed-loop system satisfies, in z−coordinates, the following controllable linear system:

$$\begin{aligned} \dot{z} &= Az + Bv \\ y &= Cz. \end{aligned} \tag{5.5}$$

$$\dot{z} = Az + Bu$$
$$y = Cz$$

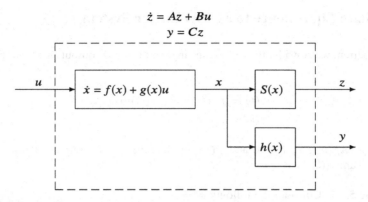

Fig. 5.1 State equivalence to a linear system with output

$$\dot{z} = Az + Bv$$
$$y = Cz$$

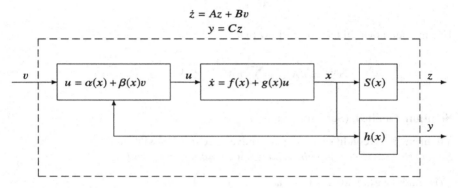

Fig. 5.2 Feedback linearization with output

In other words,

$$S_*(f(x) + g(x)\alpha(x)) = Az \; ; \quad h \circ S^{-1}(z) = Cz$$
$$\left[S_*(g(x)\beta_1(x)) \; \cdots \; S_*(g(x)\beta_m(x))\right] = B$$

(5.6)

where $\beta(x) = \left[\beta_1(x) \; \cdots \; \beta_m(x)\right]$.

Figures 5.1 and 5.2 give the block diagrams of the two linearization problems with output in Definitions 5.1 and 5.2. It is obvious that the conditions for the linearization problems with output would be more restricted than those for the linearization problems without output. In the next sections, the conditions for the linearization problems with output are considered.

5.2 State Equivalence to a SISO Linear System

In this section, we consider the following single input single output (SISO) nonlinear system:

$$\dot{x}(t) = f(x(t)) + g(x(t))u(t)$$
$$y(t) = h(x(t)) \tag{5.7}$$

where $x \in \mathbb{R}^n$, $u \in \mathbb{R}$, $y \in \mathbb{R}$, and $f(x)$, $g(x)$, and $h(x)$ are smooth functions with $f(0) = 0$ and $h(0) = 0$.

Example 5.2.1 Consider the following linear system:

$$\dot{z} = Az + bu ; \quad y = cz.$$

Use mathematical induction to show that

$$y^{(\ell)} = cA^{\ell}z + \sum_{j=0}^{\ell-1} cA^{\ell-1-j}bu^{(j)}, \quad \ell \geq 1.$$

Solution Omitted. (See Problem 5–1.) $\qquad\square$

Theorem 5.1 (conditions for state equivalence to a LS with output)
System (5.7) is state equivalent to a LS with output, if and only if

(i) $\text{rank} \left(\left[g(x) \ \text{ad}_f g(x) \ \cdots \ \text{ad}_f^{n-1} g(x) \right] \Big|_{x=0} \right) = n.$

(ii) $\left[\text{ad}_f^{i-1} g(x), \text{ad}_f^{j-1} g(x) \right] = 0, \quad 1 \leq i \leq n+1, \ 1 \leq j \leq n+1.$

(iii) $L_{\text{ad}_f^{k-1} g} h(x) = \text{const}, \ 1 \leq k \leq n.$

Furthermore, a state transformation $z = S(x)$ can be obtained by

$$\frac{\partial S(x)}{\partial x} = \left[g(x) \ \text{ad}_f g(x) \ \cdots \ \text{ad}_f^{n-1} g(x) \right]^{-1}. \tag{5.8}$$

Proof Necessity. Suppose that system (5.7) is state equivalent to a linear system with output. Then there exists a state transformation $z = S(x)$ such that

$$\tilde{f}(z) \triangleq S_*(f(x)) = Az ; \quad \tilde{g}(z) \triangleq S_*(g(x)) = b$$
$$\tilde{h}(z) \triangleq h \circ S^{-1}(z) = cz. \tag{5.9}$$

It is clear, by Theorem 3.1, that condition (i) and (ii) of Theorem 5.1 are satisfied. It is easy to see, by Example 2.4.14, that for $k \geq 0$,

$$S_*(\text{ad}_f^k g(x)) = \text{ad}_{\tilde{f}}^k \tilde{g}(z) = (-1)^k A^k b.$$

Thus, we have, by Theorem 2.5, that for $1 \leq k \leq n$,

$$L_{\mathrm{ad}_f^{k-1}g}h(x) = L_{\mathrm{ad}_f^{k-1}\tilde{g}}\tilde{h}(z)\Big|_{z=S(x)} = (-1)^{k-1}cA^{k-1}b.$$

Therefore, condition (iii) is satisfied.

Sufficiency. Suppose that condition (i)–(iii) are satisfied. Then, by Theorem 2.7, there exists a state transformation $z = S(x)$ such that for $1 \leq i \leq n$,

$$S_* \left(\mathrm{ad}_f^{i-1} g(x) \right) = \frac{\partial}{\partial z_i} \tag{5.10}$$

or

$$\frac{\partial S(x)}{\partial x} \left[g(x) \ \mathrm{ad}_f g(x) \cdots \mathrm{ad}_f^{n-1} g(x) \right] = I.$$

It is easy to see, by (3.11)–(3.12), that $S_*(g(x)) = \frac{\partial}{\partial z_1} = \begin{bmatrix} 1 & 0 & \cdots & 0 \end{bmatrix}^T \triangleq b$ and $S_*(f(x)) = Az$ for some constant matrix A. (For this, refer to the sufficiency proof of Theorem 3.1.) Let $\tilde{h}(z) = h \circ S^{-1}(z)$. Then we have, by (2.30), (5.10) and condition (iii), that for $1 \leq k \leq n$,

$$\frac{\partial \tilde{h}(z)}{\partial z_k} = L_{S_*(\mathrm{ad}_f^{k-1}g)}\tilde{h}(z) = L_{\mathrm{ad}_f^{k-1}g}h(x)\Big|_{x=S^{-1}(z)} = \mathrm{const} \triangleq c_k.$$

Since $\tilde{h}(0) = 0$, it is clear that $\tilde{h}(z) \triangleq h \circ S^{-1}(z) = \begin{bmatrix} c_1 & \cdots & c_n \end{bmatrix} z$. Therefore, system (5.7) is state equivalent to a linear system via $z = S(x)$ in (5.8). $\qquad\square$

Example 5.2.2 Show that the following nonlinear system is state equivalent to a LS with output:

$$\begin{bmatrix} \dot{x}_1 \\ \dot{x}_2 \end{bmatrix} = \begin{bmatrix} 0 \\ x_1 \cos^2 x_2 \end{bmatrix} + \begin{bmatrix} 1 \\ 0 \end{bmatrix} u = f(x) + g(x)u \tag{5.11}$$

$$y = 2x_1 + \tan x_2 = h(x).$$

Also, find a state transformation $z = S(x)$ in (5.8).

Solution It is easy to see that

$$\mathrm{ad}_f g(x) = \begin{bmatrix} 0 \\ -\cos^2 x_2 \end{bmatrix}, \quad \mathrm{ad}_f^2 g(x) = \begin{bmatrix} 0 \\ 0 \end{bmatrix}$$

and

$$L_g h(x) = 2, \quad L_{\mathrm{ad}_f g}h(x) = -1.$$

(Refer to Example 3.2.5.) It is easy to see that condition (i)–(iii) of Theorem 5.1 are satisfied. Therefore, by Theorem 5.1, system (5.11) is state equivalent to a LS with output. It is clear, by (5.8), that

$$\frac{\partial S(x)}{\partial x} = \begin{bmatrix} 1 & 0 \\ 0 & -\cos^2 x_2 \end{bmatrix}^{-1} = \begin{bmatrix} 1 & 0 \\ 0 & -\sec^2 x_2 \end{bmatrix}$$

and

$$\begin{bmatrix} z_1 \\ z_2 \end{bmatrix} = S(x) = \begin{bmatrix} x_1 \\ -\tan x_2 \end{bmatrix}.$$

Then we have that

$$\begin{bmatrix} \dot{z}_1 \\ \dot{z}_2 \end{bmatrix} = S_*(f(x)) + S_*(g(x))u = \begin{bmatrix} 0 & 0 \\ -1 & 0 \end{bmatrix}\begin{bmatrix} z_1 \\ z_2 \end{bmatrix} + \begin{bmatrix} 1 \\ 0 \end{bmatrix} u$$

$$y = h \circ S^{-1}(z) = \begin{bmatrix} 2 & -1 \end{bmatrix}\begin{bmatrix} z_1 \\ z_2 \end{bmatrix}. \qquad \Box$$

Example 5.2.3 Show that the following nonlinear system is not state equivalent to a LS with output:

$$\begin{bmatrix} \dot{x}_1 \\ \dot{x}_2 \end{bmatrix} = \begin{bmatrix} 0 \\ x_1 \cos^2 x_2 \end{bmatrix} + \begin{bmatrix} 1 \\ 0 \end{bmatrix} u = f(x) + g(x)u \qquad (5.12)$$

$$y = x_2 = h(x).$$

Solution It is easy to see that

$$\text{ad}_f g(x) = \begin{bmatrix} 0 \\ -\cos^2 x_2 \end{bmatrix}, \quad \text{ad}_f^2 g(x) = \begin{bmatrix} 0 \\ 0 \end{bmatrix}$$

and

$$L_g h(x) = 0, \quad L_{\text{ad}_f g} h(x) = -\cos^2 x_2.$$

Since condition (iii) of Theorem 5.1 is not satisfied, system (5.12) is not state equivalent to a LS with output. $\qquad \Box$

Example 5.2.4 Show that the following nonlinear system is not state equivalent to a LS with output:

$$\begin{bmatrix} \dot{x}_1 \\ \dot{x}_2 \end{bmatrix} = \begin{bmatrix} x_2 \\ x_1^2 \end{bmatrix} + \begin{bmatrix} 0 \\ 1 \end{bmatrix} u = f(x) + g(x)u \tag{5.13}$$

$$y = x_1 = h(x).$$

Solution It is easy to see that

$$\text{ad}_f g(x) = \begin{bmatrix} -1 \\ 0 \end{bmatrix}, \quad \text{ad}_f^2 g(x) = \begin{bmatrix} 0 \\ 2x_1 \end{bmatrix}$$

and

$$\left[\text{ad}_f g(x), \ \text{ad}_f^2 g(x) \right] = \begin{bmatrix} 0 \\ -2 \end{bmatrix} \neq 0$$

which implies that condition (ii) of Theorem 5.1 is not satisfied. Therefore, by Theorem 5.1, system (5.13) is not state equivalent to a LS with output. $\qquad \square$

It is clear that system (5.13) becomes a linear system with nonsingular feedback $u = -x_1^2 + v$. In other words, the larger class of input output systems can be linearized by using nonsingular feedback. It will be discussed in Sect. 5.4.

5.3 State Equivalence to a MIMO Linear System

In this section, we consider the following multi-input multi-output (MIMO) nonlinear system:

$$\dot{x}(t) = f(x(t)) + g(x(t))u(t)$$
$$y(t) = h(x(t)) \tag{5.14}$$

where $x \in \mathbb{R}^n$, $u \in \mathbb{R}^m$, $y \in \mathbb{R}^q$, and $f(x)$, $g(x)$, and $h(x)$ are smooth functions with $f(0) = 0$ and $h(0) = 0$. Suppose that $(\kappa_1, \ldots, \kappa_m)$ is the Kronecker indices of system (5.14).

Theorem 5.2 (conditions for state equivalence to a LS with output)
System (5.14) is state equivalent to a LS with output via state transformation $z = S(x)$, if and only if

(i) $\sum_{i=1}^m \kappa_i = n$
(ii) *for $1 \leq i \leq m$, $1 \leq j \leq m$, $1 \leq \ell_i \leq \kappa_i + 1$, and $1 \leq \ell_j \leq \kappa_j + 1$,*

$$[\text{ad}_f^{\ell_i - 1} g_i(x), \text{ad}_f^{\ell_j - 1} g_j(x)] = 0 \tag{5.15}$$

(iii) *for* $1 \leq i \leq m$, $1 \leq j \leq q$, *and* $1 \leq k \leq \kappa_i$,

$$L_{\mathrm{ad}_f^{k-1} g_i} h_j(x) = \text{constant}. \tag{5.16}$$

Furthermore, a state transformation $z = S(x)$ *can be obtained by*

$$\frac{\partial S(x)}{\partial x} = \left[g_1 \, \mathrm{ad}_f g_1 \, \cdots \, \mathrm{ad}_f^{\kappa_1 - 1} g_1 \, \cdots \, g_m \, \cdots \, \mathrm{ad}_f^{\kappa_m - 1} g_m \right]^{-1}. \tag{5.17}$$

Proof Necessity. Suppose that system (5.14) is state equivalent to a linear system. Then there exists a state transformation $z = S(x)$ such that for $1 \leq i \leq m$ and $1 \leq j \leq q$,

$$\tilde{f}(z) \triangleq S_*(f(x)) = Az \; ; \; \tilde{g}_i(z) \triangleq S_*(g_i(x)) = b_i$$
$$\tilde{h}_j(z) \triangleq h_j \circ S^{-1}(z) = c^j z \tag{5.18}$$

where A, $B \triangleq [b_1 \, \cdots \, b_m]$, and $C \triangleq \begin{bmatrix} c^1 \\ \vdots \\ c^q \end{bmatrix}$ are constant matrices. It is clear, by Theorem 3.2, that condition (i) and (ii) of Theorem 5.2 are satisfied. It is easy to see, by Example 2.4.14, that for $1 \leq i \leq m$ and $k \geq 0$,

$$\mathrm{ad}_{\tilde{f}}^k \tilde{g}_i(z) = S_*(\mathrm{ad}_f^k g_i(x)) = (-1)^k A^k b_i. \tag{5.19}$$

Thus, we have, by Theorem 2.5, that for $1 \leq i \leq m$, $1 \leq j \leq q$, and $1 \leq k \leq \kappa_i$,

$$L_{\mathrm{ad}_f^{k-1} g_i} h_j(x) = L_{\mathrm{ad}_{\tilde{f}}^{k-1} \tilde{g}_i} \tilde{h}_j(z) \Big|_{z=S(x)} = c^j A^{k-1} b_i.$$

Therefore, condition (iii) is satisfied.

Sufficiency. Suppose that condition (i)–(iii) are satisfied. Then, by Theorem 2.7, there exists a state transformation $z = S(x)$ such that for $1 \leq i \leq m$ and $1 \leq \ell \leq \kappa_i$,

$$S_*\left(\mathrm{ad}_f^{\ell-1} g_i(x) \right) = \frac{\partial}{\partial z_\ell^i}$$

$$z = \begin{bmatrix} z^1 \\ \vdots \\ z^m \end{bmatrix}, \quad z^i = \begin{bmatrix} z_1^i \\ \vdots \\ z_{\kappa_i}^i \end{bmatrix} \tag{5.20}$$

or

$$\frac{\partial S(x)}{\partial x} \left[g_1 \, \mathrm{ad}_f g_1 \, \cdots \, \mathrm{ad}_f^{\kappa_1 - 1} g_1 \, \cdots \, g_m \, \cdots \, \mathrm{ad}_f^{\kappa_m - 1} g_m \right] = I.$$

It is easy to see, by (3.36), that $S_*(g_i(x)) = \frac{\partial}{\partial z_1^i} \triangleq b_i$ and $S_*(f(x)) = Az$ for some constant matrix A. (For this, refer to the sufficiency proof of Theorem 3.2.) Let $\tilde{h}_j(z) = h_j \circ S^{-1}(z)$ for $1 \leq j \leq q$. Then we have, by (2.30), (5.20) and condition (iii), that for $1 \leq i \leq m$, $1 \leq j \leq q$, and $1 \leq k \leq \kappa_i$,

$$\frac{\partial \tilde{h}_j(z)}{\partial z_k^i} = L_{S_*(\mathrm{ad}_f^{k-1} g_i)} \tilde{h}_j(z) = L_{\mathrm{ad}_f^{k-1} g_i} h_j(x) \Big|_{x = S^{-1}(z)} = \mathrm{const} \triangleq c_{ik}^j.$$

Since $\tilde{h}(0) = 0$, it is clear that $\tilde{h}(z) \triangleq h \circ S^{-1}(z) = Cz$, where $C \triangleq \begin{bmatrix} c^1 \\ \vdots \\ c^q \end{bmatrix}$ and $c^j \triangleq$

$\begin{bmatrix} c_{11}^j \cdots c_{1\kappa_1}^j \cdots c_{m1}^j \cdots c_{m\kappa_m}^j \end{bmatrix}$. Therefore, system (5.14) is state equivalent to a linear system via $z = S(x)$ in (5.17). $\qquad \square$

Example 5.3.1 Show that the following nonlinear system is state equivalent to a LS with output:

$$\begin{bmatrix} \dot{x}_1 \\ \dot{x}_2 \\ \dot{x}_3 \end{bmatrix} = \begin{bmatrix} -2x_2(x_2 + x_2 + x_2^2) \\ x_1 + x_2 + x_2^2 \\ -2x_2(x_1 + x_2 + x_2^2) \end{bmatrix} + \begin{bmatrix} 1 & 0 \\ 0 & 0 \\ 0 & 1 \end{bmatrix} \begin{bmatrix} u_1 \\ u_2 \end{bmatrix}$$

$$= f(x) + g_1(x)u_1 + g_2(x)u_2 \qquad (5.21)$$

$$\begin{bmatrix} y_1 \\ y_2 \end{bmatrix} = \begin{bmatrix} x_1 + x_2^2 \\ x_2 + x_2^2 + x_3 \end{bmatrix} = \begin{bmatrix} h_1(x) \\ h_2(x) \end{bmatrix} = h(x).$$

Also, find a state transformation $z = S(x)$ in (5.17).

Solution By simple calculation, we have that $(\kappa_1, \kappa_2) = (2, 1)$ and

$$\mathrm{ad}_f g_1(x) = \begin{bmatrix} 2x_2 \\ -1 \\ 2x_2 \end{bmatrix}, \quad \mathrm{ad}_f g_2(x) = \begin{bmatrix} 0 \\ 0 \\ 0 \end{bmatrix}, \quad \mathrm{ad}_f^2 g_1(x) = \begin{bmatrix} -2x_2 \\ 1 \\ -2x_2 \end{bmatrix}.$$

(Refer to Example 3.3.3.) Also, we have that

$$L_{g_1} h(x) = \begin{bmatrix} 1 \\ 0 \end{bmatrix}, \quad L_{g_2} h(x) = \begin{bmatrix} 0 \\ 1 \end{bmatrix}, \quad L_{\mathrm{ad}_f g_1} h(x) = \begin{bmatrix} 0 \\ -1 \end{bmatrix}.$$

It is easy to see that condition (i)–(iii) of Theorem 5.2 are satisfied. Therefore, by Theorem 5.2, system (5.21) is state equivalent to a LS with output. It is clear, by (5.17), that

$$\frac{\partial S(x)}{\partial x} = \begin{bmatrix} 1 & 2x_2 & 0 \\ 0 & -1 & 0 \\ 0 & 2x_2 & 1 \end{bmatrix}^{-1} = \begin{bmatrix} 1 & 2x_2 & 0 \\ 0 & -1 & 0 \\ 0 & 2x_2 & 1 \end{bmatrix}$$

and

$$\begin{bmatrix} z_1 \\ z_2 \\ z_3 \end{bmatrix} = S(x) = \begin{bmatrix} x_1 + x_2^2 \\ -x_2 \\ x_3 + x_2^2 \end{bmatrix}.$$

Then we have that

$$\begin{bmatrix} \dot{z}_1 \\ \dot{z}_2 \end{bmatrix} = S_*(f(x)) + S_*(g_1(x))u_1 + S_*(g_2(x))u_2$$

$$= \begin{bmatrix} 0 & 0 & 0 \\ -1 & 1 & 0 \\ 0 & 0 & 0 \end{bmatrix} \begin{bmatrix} z_1 \\ z_2 \\ z_3 \end{bmatrix} + \begin{bmatrix} 1 & 0 \\ 0 & 0 \\ 0 & 1 \end{bmatrix} \begin{bmatrix} u_1 \\ u_2 \end{bmatrix}$$

$$y = h \circ S^{-1}(z) = \begin{bmatrix} 1 & 0 & 0 \\ 0 & -1 & 1 \end{bmatrix} \begin{bmatrix} z_1 \\ z_2 \\ z_3 \end{bmatrix}.$$

$\square$

5.4 Feedback Linearization with Output of SISO Systems

In this section, we consider the following single input single output (SISO) nonlinear system:

$$\begin{aligned} \dot{x}(t) &= f(x(t)) + g(x(t))u(t) \\ y(t) &= h(x(t)) \end{aligned} \tag{5.22}$$

where $x \in \mathbb{R}^n$, $u \in \mathbb{R}$, $y \in \mathbb{R}$, and $f(x)$, $g(x)$, and $h(x)$ are smooth functions with $f(0) = 0$ and $h(0) = 0$.

Definition 5.3 (*feedback linearization with output*)
System (5.22) is said to be feedback linearizable with output, if there exist a feedback $u = \alpha(x) + \beta(x)v$ and a state transformation $z = S(x)$ such that the closed-loop satisfies, in $z-$coordinates, the following Brunovsky canonical form:

$$
\dot{z} = \begin{bmatrix} 0 & 1 & 0 & \cdots & 0 & 0 \\ 0 & 0 & 1 & \cdots & 0 & 0 \\ \vdots & \vdots & \vdots & & \vdots & \vdots \\ 0 & 0 & 0 & \cdots & 0 & 1 \\ 0 & 0 & 0 & \cdots & 0 & 0 \end{bmatrix} z + \begin{bmatrix} 0 \\ 0 \\ \vdots \\ 0 \\ 1 \end{bmatrix} v = Az + bv
$$

(5.23)

$$
y = cz.
$$

In other words,

$$
S_* \left(f(x) + g(x)\alpha(x) \right) = Az \; ; \quad S_* \left(g(x)\beta(x) \right) = b
$$
$$
h \circ S^{-1}(z) = cz.
$$

(5.24)

Definition 5.4 (*characteristic number*)
The characteristic number ρ of the output is defined as the smallest natural number such that $L_g L_f^{\rho-1} h(x) \neq 0$. In other words,

$$
L_g L_f^{k-1} h(x) = 0, \; 1 \le k \le \rho - 1 \; \text{ and } \; L_g L_f^{\rho-1} h(x) \neq 0
$$

(5.25)

or (by Example 2.4.16)

$$
L_{\mathrm{ad}_f^{k-1} g} h(x) = 0, \; 1 \le k \le \rho - 1 \; \text{ and } \; L_{\mathrm{ad}_f^{\rho-1} g} h(x) \neq 0.
$$

(5.26)

If $L_g L_f^k h(x) = 0$ for $k \ge 0$, then we let $\rho \triangleq \infty$.

It is easy to see, by mathematical induction, that

$$
L_{f+g\alpha}^k h(x) = L_f^k h(x), \quad 1 \le k \le \rho - 1
$$
$$
L_{f+g\alpha}^\rho h(x) = L_f^\rho h(x) + L_g L_f^{\rho-1} h(x)\, \alpha(x).
$$

(5.27)

Example 5.4.1 Suppose that ρ is the characteristic number of the system output. Let $\rho \le n$. Find the nonsingular feedback $u = \alpha(x) + \beta(x)v$ such that the transfer function of the closed-loop system is

$$
G_c(s) \triangleq \frac{Y(s)}{V(s)} = \frac{1}{s^\rho + a_{\rho-1} s^{\rho-1} + \cdots + a_1 s + a_0}.
$$

Solution It is easy to see, by (5.27), that

$$
y^{(k)}(t) \triangleq \frac{d^k y(t)}{dt^k} = L_f^k h(x), \quad 1 \le k \le \rho - 1
$$
$$
y^{(\rho)}(t) = L_{f+gu} L_f^{\rho-1} h(x) = L_f^\rho h(x) + L_g L_f^{\rho-1} h(x)\, u.
$$

We need to find the feedback such that

$$y^{(\rho)} = -a_{\rho-1}y^{\rho-1} - \cdots - a_1\dot{y} - a_0 y + v$$
$$= -a_{\rho-1}L_f^{\rho-1}h(x) - \cdots - a_1 L_f h(x) - a_0 h(x) + v.$$

Thus, we have

$$u = \frac{-L_f^{\rho}h(x) - a_{\rho-1}L_f^{\rho-1}h(x) - \cdots - a_1 L_f h(x) - a_0 h(x) + v}{L_g L_f^{\rho-1}h(x)}.$$

□

Example 5.4.2 Show that if $\rho = n$ and $L_g L_f^{\rho-1}h(x)\big|_{x=0} \neq 0$, then system (5.22) is feedback linearizable with output.

Solution Suppose that $\rho = n$ and $L_g L_f^{\rho-1}h(x)\big|_{x=0} \neq 0$. Then, we have, by (5.25), that

$$L_g L_f^{k-1}h(x) = 0, \ 1 \le k \le n - 1 \ \text{ and } \ L_g L_f^{n-1}h(x) \neq 0.$$

Thus, conditions of Lemma 4.1 are satisfied with $S_1(x) = h(x)$. Therefore, by Lemma 4.1, system (5.22) is feedback linearizable with state transformation

$$z = S(x) = \begin{bmatrix} h(x) \\ L_f h(x) \\ \vdots \\ L_f^{\rho-1}h(x) \end{bmatrix} \tag{5.28}$$

and feedback

$$u = -\frac{L_f^{\rho}h(x)}{L_g L_f^{\rho-1}h(x)} + \frac{1}{L_g L_f^{\rho-1}h(x)}v = \alpha(x) + \beta(x)v. \tag{5.29}$$

Since $\tilde{h} = h \circ S^{-1}(z) = z_1$, it is easy to see that (5.23) is satisfied with $c = \begin{bmatrix} 1 \ 0 \cdots 0 \end{bmatrix}$.   □

Suppose that $\rho < n$. Let $z = S(x) = \begin{bmatrix} z^1 \\ z^2 \end{bmatrix}$, where $z^2 \in \mathbb{R}^{n-\rho}$ and

$$z^1 = \begin{bmatrix} h(x) \ L_f h(x) \ \cdots \ L_f^{\rho-1}h(x) \end{bmatrix}^{\mathsf{T}}.$$

Then, the closed-loop system with state feedback (5.29) satisfies, in z-coordinates, the following system:

$$\begin{bmatrix} \dot{z}^1(t) \\ \dot{z}^2(t) \end{bmatrix} = \begin{bmatrix} A_\rho z^1(t) \\ \tilde{f}^2(z(t)) \end{bmatrix} + \begin{bmatrix} b_\rho \\ \tilde{g}^2(z(t)) \end{bmatrix} v(t)$$

$$y(t) = \begin{bmatrix} 1 & 0 & \cdots & 0 \end{bmatrix} z(t).$$

Theorem 5.3 (conditions for feedback linearization with output)
Let $\rho \leq n$. System (5.22) is feedback linearizable with output, if and only if

(i) $\operatorname{rank}\left(\begin{bmatrix} g(x) & \operatorname{ad}_f g(x) & \cdots & \operatorname{ad}_f^{n-1} g(x) \end{bmatrix} \big|_{x=0} \right) = n$

(ii) $L_{\operatorname{ad}_f^{\rho-1} g} h(x) \big|_{x=0} \neq 0$

(iii) *one form $\omega(x)$ is exact or* $\dfrac{\partial \omega(x)^{\mathsf{T}}}{\partial x} = \left(\dfrac{\partial \omega(x)^{\mathsf{T}}}{\partial x} \right)^{\mathsf{T}}$.

(iv) $L_{\hat{g}} L_{\hat{f}}^{k-1} h(x) = \operatorname{const}$ *for $\rho \leq k \leq n$,*

where

$$\omega(x) \triangleq \begin{bmatrix} 0 & \cdots & 0 & \dfrac{(-1)^{n-1}}{\beta(x)} \end{bmatrix} \begin{bmatrix} g(x) & \operatorname{ad}_f g(x) & \cdots & \operatorname{ad}_f^{n-1} g(x) \end{bmatrix}^{-1} \tag{5.30}$$

$$\frac{\partial S_1(x)}{\partial x} = \omega(x) \tag{5.31}$$

$$\hat{g}(x) \triangleq \frac{1}{L_g L_f^{\rho-1} h(x)} g(x) = g(x)\beta(x) \tag{5.32}$$

$$\hat{f}(x) \triangleq f(x) - \frac{L_f^n S_1(x)}{L_g L_f^{\rho-1} h(x)} g(x) = f(x) + g(x)\alpha(x). \tag{5.33}$$

Furthermore, a state transformation $z = S(x)$ is given by

$$S(x) = \begin{bmatrix} S_1(x) & L_f S_1(x) & \cdots & L_f^{n-1} S_1(x) \end{bmatrix}^{\mathsf{T}}. \tag{5.34}$$

Proof Necessity. Suppose that system (5.22) is feedback linearizable with output. Then there exist a state transformation $z = S(x)$ and a nonsingular feedback $u = \alpha(x) + \beta(x)v$ such that

$$\tilde{f}(z) \triangleq S_*(f(x) + g(x)\alpha(x)) = Az ; \quad \tilde{g}(z) \triangleq S_*(g(x)\beta(x)) = b$$

$$\tilde{h}(z) \triangleq h \circ S^{-1}(z) = cz = \begin{bmatrix} c_1 & c_2 & \cdots & c_n \end{bmatrix} z \tag{5.35}$$

where $\hat{f}(x) = f(x) + g(x)\alpha(x)$ and $\hat{g}(x) = g(x)\beta(x)$. It is clear, by Theorem 4.1, that condition (i) of Theorem 5.3 is satisfied. Since $S_*(f(x) + g(x)\alpha(x) + g(x)\beta(x)v) = Az + bv$ by (5.35), it is clear that

$$\frac{\partial S(x)}{\partial x}\{f(x) + g(x)\alpha(x) + g(x)\beta(x)v\} = AS(x) + bv = \begin{bmatrix} S_2(x) \\ \vdots \\ S_n(x) \\ v \end{bmatrix}.$$

Thus, we have that for $1 \leq i \leq n - 1$,

$$S_{i+1}(x) = \frac{\partial S_i(x)}{\partial x}\{f(x) + g(x)\alpha(x) + g(x)\beta(x)v\} = L_{f+g\alpha+g\beta v}S_i(x)$$
$$= L_f S_i(x) + L_g S_i(x)\{\alpha(x) + \beta(x)v\}$$

and

$$v = L_{f+g(\alpha+\beta v)}S_n(x) = L_f S_n(x) + L_g S_n(x)\{\alpha(x) + \beta(x)v\}.$$

Since $\beta(0) \neq 0$, it is easy to see that for $1 \leq i \leq n - 1$,

$$S_{i+1}(x) = L_f S_i(x) ; \quad L_g S_i(x) = 0$$

and

$$L_f S_n(x) + L_g S_n(x)\alpha(x) = 0 ; \quad L_g S_n(x)\beta(x) = 1$$

which imply that

$$L_g L_f^i S_1(x) = 0, \quad 0 \leq i \leq n - 2 ; \quad L_g L_f^{n-1} S_1(x)\beta(x) = 1$$
$$\alpha(x) = -\beta(x)L_f^n S_1(x).$$

Then, it is clear, by Example 2.4.16, that

$$L_{\text{ad}_f^i g}S_1(x) = 0, \ 0 \leq i \leq n - 2 ; \quad L_{\text{ad}_f^{n-1}g(x)}S_1(x) = \frac{(-1)^{n-1}}{\beta(x)}$$

or

$$\frac{\partial S_1(x)}{\partial x}\left[g(x) \ \text{ad}_f g(x) \ \cdots \ \text{ad}_f^{n-1}g(x)\right] = \left[0 \cdots 0 \ \frac{(-1)^{n-1}}{\beta(x)}\right]$$

which implies that (5.30) and (5.31) are satisfied. Since $\omega(x) = \frac{\partial S_1(x)}{\partial x}$, it is clear that condition (ii) is satisfied. Also, it is easy to see, by Example 2.4.14 and (5.35), that for $1 \le k \le n$,

$$L_{\hat{g}} L_{\hat{f}}^{k-1} h(x) = L_{\tilde{g}} L_{\tilde{f}}^{k-1} \tilde{h}(z)\Big|_{z=S(x)} = cA^{k-1}b = c_{n+1-k} \qquad (5.36)$$

which implies that condition (iv) is satisfied. Note, by (5.27), that $c_k = 0$, $n + 2 - \rho \le k \le n$. Finally, it is easy to see, by (2.16) and (5.36), that

$$c_{n+1-\rho} = L_{g\beta} L_f^{\rho-1} h(x) = L_g L_f^{\rho-1} h(x)\, \beta(x)$$

which implies that (5.32) is satisfied. (Without loss of generality, we can let $c_{n+1-\rho} = 1$.) Since $\beta(0) \ne 0$ and $\frac{1}{\beta(0)} \ne 0$, condition (ii) is satisfied.

Sufficiency. Suppose that condition (i)–(iv) are satisfied. Then, it is clear that

$$\frac{\partial S_1(x)}{\partial x} \big[g(x)\ \mathrm{ad}_f g(x)\ \cdots\ \mathrm{ad}_f^{n-1} g(x) \big] = \big[0 \cdots 0\ \tfrac{(-1)^{n-1}}{\beta(x)} \big] \qquad (5.37)$$

or

$$L_{\mathrm{ad}_f^i g} S_1(x) = 0, \quad 0 \le i \le n - 2 ; \quad L_{\mathrm{ad}_f^{n-1} g} S_1(x) = \frac{(-1)^{n-1}}{\beta(x)}. \qquad (5.38)$$

Thus, it is easy to see, by Example 2.4.16, that

$$L_g L_f^i S_1(x) = 0, \quad 0 \le i \le n - 2 ; \quad L_g L_f^{n-1} S_1(x)\beta(x) = 1. \qquad (5.39)$$

By mathematical induction, it is easy to see that

$$L_{\hat{f}}^i S_1(x) = L_f^i S_1(x), \quad 0 \le i \le n - 1 \qquad (5.40)$$

which implies that

$$\begin{aligned} L_{\hat{f}}^n S_1(x) = L_{\hat{f}} L_f^{n-1} S_1(x) &= L_f^n S_1(x) + L_g L_f^{n-1} S_1(x)\alpha(x) \\ &= L_f^n S_1(x) + \frac{1}{\beta(x)} \alpha(x) = 0. \end{aligned} \qquad (5.41)$$

Let $S_i(x) = L_{\hat{f}}^{i-1} S_1(x) = L_f^{i-1} S_1(x)$, $2 \le i \le n$. Then, we have, by (5.39), (5.40), and (5.41), that

$$S_* \left(\hat{f} + \hat{g}v \right) = \frac{\partial S(x)}{\partial x} \left(\hat{f} + \hat{g}v \right) \Bigg|_{x=S^{-1}(z)}$$

$$= \begin{bmatrix} \frac{\partial S_1(x)}{\partial x} \\ \frac{\partial L_{\hat{f}} S_1(x)}{\partial x} \\ \vdots \\ \frac{\partial L_{\hat{f}}^{n-1} S_1(x)}{\partial x} \end{bmatrix} \left(\hat{f} + \hat{g}v \right) \Bigg|_{x=S^{-1}(z)}$$

$$= \begin{bmatrix} L_{\hat{f}} S_1(x) + L_{\hat{g}} S_1(x)v \\ \vdots \\ L_{\hat{f}}^{n-1} S_1(x) + L_{\hat{g}} L_{\hat{f}}^{n-2} S_1(x)v \\ L_{\hat{f}}^{n} S_1(x) + L_{\hat{g}} L_{\hat{f}}^{n-1} S_1(x)v \end{bmatrix} \Bigg|_{x=S^{-1}(z)} \qquad (5.42)$$

$$= \begin{bmatrix} L_f S_1(x) \\ \vdots \\ L_f^{n-1} S_1(x) \\ v \end{bmatrix} \Bigg|_{x=S^{-1}(z)} = \begin{bmatrix} z_2 \\ \vdots \\ z_n \\ v \end{bmatrix} = Az + bv$$

which implies that

$$S_* \left(\hat{f}(x) \right) = Az \quad \text{and} \quad S_* \left(\hat{g}(x) \right) = b. \qquad (5.43)$$

It is easy to see, by Example 2.4.14, that for $1 \le k \le n$,

$$S_* \left(\text{ad}_{\hat{f}}^{k-1} \hat{g}(x) \right) = \text{ad}_{\hat{f}}^{k-1} \tilde{g}(z) = (-1)^{k-1} A^{k-1} b = (-1)^{k-1} \frac{\partial}{\partial z_{n+1-k}}. \qquad (5.44)$$

Let $\tilde{h}(z) = h \circ S^{-1}(z)$. Then we have, by (2.30), (5.26), (5.44), Example 2.4.16, and condition (iv), that for $1 \le k \le n$,

$$(-1)^{k-1} \frac{\partial \tilde{h}(z)}{\partial z_{n+1-k}} = L_{S_* \left(\text{ad}_{\hat{f}}^{k-1} \hat{g} \right)} \tilde{h}(z) = L_{\text{ad}_{\hat{f}}^{k-1} \hat{g}} h(x) \Bigg|_{x=S^{-1}(z)}$$

$$= \begin{cases} 0, & 1 \le k \le \rho - 1 \\ \text{const} \triangleq (-1)^{k-1} c_{n+1-k}, & \rho \le k \le n \end{cases}. \qquad (5.45)$$

Since $\tilde{h}(0) = 0$, it is clear that

$$\tilde{h}(z) \triangleq h \circ S^{-1}(z) = \begin{bmatrix} c_1 & \cdots & c_{n+1-\rho} & 0 & \cdots & 0 \end{bmatrix} z. \qquad (5.46)$$

Hence, system (5.22) is feedback linearizable with output.   □

Theorem 5.4 (conditions for feedback linearization with output)
Let $\rho \le n$. System (5.22) is feedback linearizable with output, if and only if

(i) $\text{rank} \left(\left[g(x) \ \text{ad}_f g(x) \ \cdots \ \text{ad}_f^{n-1} g(x) \right] \big|_{x=0} \right) = n$

(ii) $L_{\text{ad}_f^{\rho-1} g} h(x) \big|_{x=0} \ne 0$

(iii) for $1 \le i \le n+1$ and $1 \le j \le n+1$,

$$\left[\text{ad}_{\bar{f}}^{i-1} \hat{g}(x), \ \text{ad}_{\bar{f}}^{j-1} \hat{g}(x) \right] = 0 \tag{5.47}$$

where

$$\hat{g}(x) \triangleq \frac{1}{L_g L_f^{\rho-1} h(x)} g(x) = g(x)\beta(x) \tag{5.48}$$

and

$$\bar{f}(x) \triangleq f(x) - \frac{L_f^\rho h(x)}{L_g L_f^{\rho-1} h(x)} g(x). \tag{5.49}$$

Proof Necessity. Suppose that system (5.22) is feedback linearizable with output. Then, by Theorem 5.3, condition (i) and (ii) of Theorem 5.4 are satisfied. Also, there exists a scalar function $S_1(x)$ such that

$$\frac{\partial S_1(x)}{\partial x} \left[g \ \text{ad}_f g \ \cdots \ \text{ad}_f^{n-1} g \right] = \left[0 \ \cdots \ 0 \ (-1)^{n-1} L_g L_f^{\rho-1} h(x) \right]$$

and

$$L_{\text{ad}_f^{k-1} g} h(x) = \begin{cases} 0, & 1 \le k \le \rho - 1 \\ (-1)^{k-1} c_{n+1-k} = \text{const}, & \rho \le k \le n \end{cases} \tag{5.50}$$

where

$$\hat{g}(x) \triangleq \frac{1}{L_g L_f^{\rho-1} h(x)} g(x) = g(x)\beta(x) \tag{5.51}$$

and

$$\hat{f}(x) \triangleq f(x) - \frac{L_f^n S_1(x)}{L_g L_f^{\rho-1} h(x)} g(x) = f(x) - L_f^n S_1(x)\hat{g}(x). \tag{5.52}$$

Thus, it is easy to see, by (5.37)–(5.46), that

$$S_* \left(\hat{f}(x) \right) = Az ; \quad S_* \left(\hat{g}(x) \right) = b \tag{5.53}$$

and

$$\tilde{h}(z) \triangleq h \circ S^{-1}(z) = \begin{bmatrix} c_1 & \cdots & c_{n+1-\rho} & 0 & \cdots & 0 \end{bmatrix} z \tag{5.54}$$

where

$$z = S(x) = \begin{bmatrix} S_1(x) & L_f S_1(x) & \cdots & L_f^{n-1} S_1(x) \end{bmatrix}^\mathsf{T}.$$

It is clear, by Example 2.4.16, (5.26), and (5.50)–(5.52), that

$$c_{n+1-\rho} = (-1)^{\rho-1} L_{\mathrm{ad}_f^{\rho-1} \hat{g}} h(x) = L_{\hat{g}} L_{\hat{f}}^{\rho-1} h(x) = L_g L_f^{\rho-1} h(x) \beta(x) = 1.$$

Thus, we have, by (5.54), that

$$h(x) = L_f^{n-\rho} S_1(x) + \sum_{i=1}^{n-\rho} c_i L_f^{i-1} S_1(x)$$

which implies, together with (5.49), that

$$L_f^\rho h(x) = L_f^n S_1(x) + \sum_{i=1}^{n-\rho} c_i L_f^{i-1+\rho} S_1(x)$$

and

$$\bar{f}(x) \triangleq \hat{f}(x) - \left(\sum_{i=1}^{n-\rho} c_i L_f^{i-1+\rho} S_1(x) \right) \hat{g}(x). \tag{5.55}$$

Therefore, it is easy to see, by (2.49), (5.53), and (5.55), that

$$S_* \left(\bar{f}(x) \right) = S_* \left(\hat{f}(x) \right) - \left(\sum_{i=1}^{n-\rho} c_i L_f^{i-1+\rho} S_1(x) \right) \Bigg|_{x=S^{-1}(z)} S_* \left(\hat{g}(x) \right)$$

$$= Az - b \begin{bmatrix} 0 & \cdots & 0 & c_1 & \cdots & c_{n-\rho} \end{bmatrix} z$$

$$\triangleq (A - b\bar{c})z \triangleq \bar{A}z.$$

Hence, by Example 2.4.14, condition (iii) of Theorem 5.4 is satisfied.

Sufficiency. Suppose that condition (i)–(iii) of Theorem 5.4 are satisfied. Then, it is clear, by (4.51), condition (i), and condition (iii), that

$$\text{rank}\left(\left[\hat{g}(x)\ \text{ad}_{\hat{f}}\hat{g}(x)\ \cdots\ \text{ad}_{\hat{f}}^{n-1}\hat{g}(x)\right]\Big|_{x=0}\right) = n.$$

Also, it is easy to see, by (5.25) and (5.49), that for $1 \le k \le \rho$,

$$L_{\hat{f}}^{k-1}h(x) = L_f^{k-1}h(x)$$

and

$$L_{\hat{f}}^{\rho}h(x) = L_{f-\frac{L_f^{\rho}h(x)}{L_g L_f^{\rho-1}h}g}\,L_f^{\rho-1}h(x) = 0$$

which imply, together with (5.48), that for $k \ge 1$,

$$L_{\hat{g}}L_{\hat{f}}^{k-1}h(x) = \begin{cases} 1, & k = \rho \\ 0, & k \ne \rho \end{cases}$$

or

$$L_{\text{ad}_{\hat{f}}^{k-1}\hat{g}}h(x) = \text{constant}, \quad 1 \le k \le n. \tag{5.56}$$

Hence, by Theorem 5.1, the closed-loop system of system (5.22) with nonsingular feedback $u = -\frac{L_f^{\rho}h(x)}{L_g L_f^{\rho-1}h(x)} + \frac{1}{L_g L_f^{\rho-1}h(x)}w$ is state equivalent to a LS with output and thus system (5.22) is feedback linearizable with output. □

If $L_g L_f^{\rho-1}h(x)\Big|_{x=0} \ne 0$, then we have nonsingular feedback

$$u = -\frac{L_f^{\rho}h(x)}{L_g L_f^{\rho-1}h(x)} + \frac{1}{L_g L_f^{\rho-1}h(x)}w \tag{5.57}$$

such that the closed-loop system

$$\begin{aligned} \dot{x} &= \bar{f}(x) + \hat{g}(x)w \\ y &= h(x) \end{aligned} \tag{5.58}$$

has linear input-output relation $y^{(\rho)} = w$, where $\hat{g}(x)$ and $\bar{f}(x)$ are given in (5.48) and (5.49). (Refer to Example 5.4.1.) Theorem 5.4 shows that system (5.22) is feedback linearizable with output, if and only if the closed-loop system (5.58) is state equivalent to a linear system (without feedback).

Example 5.4.3 Use Theorem 5.3 to show that the following nonlinear system is feedback linearizable with output:

$$\begin{bmatrix} \dot{x}_1 \\ \dot{x}_2 \end{bmatrix} = \begin{bmatrix} 0 \\ x_1 \cos^2 x_2 \end{bmatrix} + \begin{bmatrix} 1 \\ 0 \end{bmatrix} u = f(x) + g(x)u$$

$$y = x_2 = h(x).$$

(5.59)

Solution It is easy to see that

$$\text{ad}_f g(x) = \begin{bmatrix} 0 \\ -\cos^2 x_2 \end{bmatrix}, \quad L_g h(x) = 0, \quad L_{\text{ad}_f g} h(x) = -\cos^2 x_2.$$

Thus, we have that $\rho = 2$, $L_g L_f h(x) = -L_{\text{ad}_f g} h(x) = \cos^2 x_2$, $\beta(x) = \frac{1}{L_g L_f h(x)} = \frac{1}{\cos^2 x_2}$, and

$$\omega(x) \triangleq \begin{bmatrix} 0 & \frac{-1}{\beta(x)} \end{bmatrix} \begin{bmatrix} g(x) & \text{ad}_f g(x) \end{bmatrix}^{-1} = \begin{bmatrix} 0 & -\cos^2 x_2 \end{bmatrix} \begin{bmatrix} 1 & 0 \\ 0 & -\cos^2 x_2 \end{bmatrix}^{-1}$$

$$= \begin{bmatrix} 0 & -\cos^2 x_2 \end{bmatrix} \begin{bmatrix} 1 & 0 \\ 0 & -\frac{1}{\cos^2 x_2} \end{bmatrix} = \begin{bmatrix} 0 & 1 \end{bmatrix}.$$

Since $\frac{\partial \omega(x)^{\text{T}}}{\partial x} = O_{2\times 2}$ is symmetric, condition (iii) of Theorem 5.3 is satisfied. Since $\frac{\partial S_1(x)}{\partial x} = \omega(x) = \begin{bmatrix} 0 & 1 \end{bmatrix}$, it is clear that $S_1(x) = x_2$. Also, it is clear that condition (i) and (ii) of Theorem 5.3 are satisfied. We have, by (5.32) and (5.33), that $\alpha(x) = -\frac{L_f^2 S_1(x)}{L_g L_f h(x)} = -\frac{-2x_1^2 \cos^3 x_2 \sin x_2}{\cos^2 x_2} = x_1^2 \sin 2x_2$ and

$$\hat{f}(x) = f(x) + g(x)\alpha(x) = \begin{bmatrix} x_1^2 \sin 2x_2 \\ x_1 \cos^2 x_2 \end{bmatrix}; \quad \hat{g}(x) = g(x)\beta(x) = \begin{bmatrix} \frac{1}{\cos^2 x_2} \\ 0 \end{bmatrix}.$$

By simple calculation, we have that

$$\text{ad}_{\hat{f}} \hat{g}(x) = \begin{bmatrix} -\frac{2x_1 \sin x_2}{\cos x_2} \\ -1 \end{bmatrix}, \quad L_{\hat{g}} h(x) = 0, \quad L_{\text{ad}_{\hat{f}} \hat{g}} h(x) = -1$$

which implies that (iv) of Theorem 5.3 is satisfied. Hence, by Theorem 5.3, system (5.59) is feedback linearizable with output via state transformation $z = S(x) = \begin{bmatrix} S_1(x) \\ L_f S_1(x) \end{bmatrix} = \begin{bmatrix} x_2 \\ x_1 \cos^2 x_2 \end{bmatrix}$ and nonsingular feedback $u = \alpha(x) + \beta(x)v = x_1^2 \sin 2x_2 + \frac{1}{\cos^2 x_2} v$. It is easy to see that

$$\begin{bmatrix} \dot{z}_1 \\ \dot{z}_2 \end{bmatrix} = S_*(\hat{f}(x)) + S_*(\hat{g}(x))v = \begin{bmatrix} 0 & 1 \\ 0 & 0 \end{bmatrix} \begin{bmatrix} z_1 \\ z_2 \end{bmatrix} + \begin{bmatrix} 0 \\ 1 \end{bmatrix} v$$

$$y = h \circ S^{-1}(z) = \begin{bmatrix} 1 & 0 \end{bmatrix} \begin{bmatrix} z_1 \\ z_2 \end{bmatrix}.$$

$\square$

Example 5.4.4 Consider the following nonlinear system:

$$\begin{bmatrix} \dot{x}_1 \\ \dot{x}_2 \\ \dot{x}_3 \end{bmatrix} = \begin{bmatrix} x_2 \\ x_3 \\ x_1^2 \end{bmatrix} + \begin{bmatrix} 0 \\ 0 \\ e^{x_1} \end{bmatrix} u = f(x) + g(x)u \tag{5.60}$$

$$y = 2x_1 + x_2 = h(x).$$

(a) Use Theorem 5.3 to show that the above system is feedback linearizable with output.
(b) Use Theorem 5.4 to show that the above system is feedback linearizable with output.

Solution (a) It is easy to see that

$$\text{ad}_f g(x) = \begin{bmatrix} 0 \\ -e^{x_1} \\ x_2 e^{x_1} \end{bmatrix}, \quad \text{ad}_f^2 g(x) = \begin{bmatrix} e^{x_1} \\ -2x_2 e^{x_1} \\ (x_2^2 + x_3)e^{x_1} \end{bmatrix} \tag{5.61}$$

and

$$L_g h(x) = 0, \quad L_{\text{ad}_f g} h(x) = -e^{x_1}.$$

Thus, condition (i) and (ii) of Theorem 5.3 are satisfied. Also, we have that $\rho = 2$, $L_g L_f h(x) = -L_{\text{ad}_f g} h(x) = e^{x_1}$, $\beta(x) = \frac{1}{L_g L_f h(x)} = e^{-x_1}$, and

$$\omega(x) \triangleq \begin{bmatrix} 0 & 0 & \frac{1}{\beta(x)} \end{bmatrix} \begin{bmatrix} g(x) & \text{ad}_f g(x) & \text{ad}_f^2 g(x) \end{bmatrix}^{-1}$$

$$= \begin{bmatrix} 0 & 0 & e^{x_1} \end{bmatrix} \begin{bmatrix} 0 & 0 & e^{x_1} \\ 0 & -e^{x_1} & -2x_2 e^{x_1} \\ e^{x_1} & x_2 e^{x_1} & (x_2^2 + x_3)e^{x_1} \end{bmatrix}^{-1}$$

$$= \begin{bmatrix} 0 & 0 & e^{x_1} \end{bmatrix} \begin{bmatrix} (x_2^2 - x_3)e^{-x_1} & x_2 e^{-x_1} & e^{-x_1} \\ -2x_2 e^{-x_1} & -e^{-x_1} & 0 \\ e^{-x_1} & 0 & 0 \end{bmatrix} = \begin{bmatrix} 1 & 0 & 0 \end{bmatrix}.$$

Since $\frac{\partial \omega(x)^{\mathsf{T}}}{\partial x} = O_{3\times 3}$ is symmetric, condition (iii) of Theorem 5.3 is satisfied. Since $\frac{\partial S_1(x)}{\partial x} = \omega(x) = \begin{bmatrix} 1 & 0 & 0 \end{bmatrix}$, it is clear that $S_1(x) = x_1$. We have, by (5.32) and (5.33), that $\alpha(x) = -\frac{L_f^3 S_1(x)}{L_g L_f h(x)} = -\frac{x_1^2}{e^{x_1}} = -x_1^2 e^{-x_1}$ and

$$\hat{f}(x) = f(x) + g(x)\alpha(x) = \begin{bmatrix} x_2 \\ x_3 \\ 0 \end{bmatrix} \; ; \; \hat{g}(x) = g(x)\beta(x) = \begin{bmatrix} 0 \\ 0 \\ 1 \end{bmatrix}.$$

By simple calculation, we have that

$$\mathrm{ad}_{\hat{f}}\hat{g}(x) = \begin{bmatrix} 0 \\ -1 \\ 0 \end{bmatrix}, \quad \mathrm{ad}_{\hat{f}}^2\hat{g}(x) = \begin{bmatrix} 1 \\ 0 \\ 0 \end{bmatrix}$$

and

$$L_{\hat{g}}h(x) = 0, \quad L_{\mathrm{ad}_{\hat{f}}\hat{g}}h(x) = -1, \quad L_{\mathrm{ad}_{\hat{f}}^2\hat{g}}h(x) = 2$$

which implies that (iv) of Theorem 5.3 is satisfied. Hence, by Theorem 5.3, system (5.60) is feedback linearizable with output via state transformation $z = S(x) = \begin{bmatrix} S_1(x) \\ L_f S_1(x) \\ L_f^2 S_1(x) \end{bmatrix} = \begin{bmatrix} x_1 \\ x_2 \\ x_3 \end{bmatrix}$ and nonsingular feedback $u = \alpha(x) + \beta(x)v = -x_1^2 e^{-x_1} + e^{-x_1}v$. It is easy to see that

$$\begin{bmatrix} \dot{z}_1 \\ \dot{z}_2 \\ \dot{z}_3 \end{bmatrix} = S_*(\hat{f}(x)) + S_*(\hat{g}(x))v = \begin{bmatrix} 0 & 1 & 0 \\ 0 & 0 & 1 \\ 0 & 0 & 0 \end{bmatrix} \begin{bmatrix} z_1 \\ z_2 \\ z_3 \end{bmatrix} + \begin{bmatrix} 0 \\ 0 \\ 1 \end{bmatrix} v$$

$$y = h \circ S^{-1}(z) = \begin{bmatrix} 2 & 1 & 0 \end{bmatrix} \begin{bmatrix} z_1 \\ z_2 \\ z_3 \end{bmatrix}.$$

(5.62)

(b) Since $\rho = 2$, $L_g L_f h(x) = e^{x_1}$, and $L_f^2 h(x) = x_1^2 + 2x_3$, we have, by (5.48) and (5.49), that

$$\hat{g}(x) \triangleq \frac{1}{L_g L_f^{\rho-1} h(x)} g(x) = \begin{bmatrix} 0 \\ 0 \\ 1 \end{bmatrix}$$

and

$$\bar{f}(x) \triangleq f(x) - \frac{L_f^\rho h(x)}{L_g L_f^{\rho-1} h(x)} g(x) = \begin{bmatrix} x_2 \\ x_3 \\ -2x_3 \end{bmatrix}.$$

By simple calculation, we have that

$$\left[\text{ad}_{\bar{f}} \hat{g}(x) \ \text{ad}_{\bar{f}}^2 \hat{g}(x) \ \text{ad}_{\bar{f}}^3 \hat{g}(x) \right] = \begin{bmatrix} 0 & 1 & 2 \\ -1 & -2 & -4 \\ 2 & 4 & 8 \end{bmatrix}$$

which implies that (iii) of Theorem 5.4 is satisfied. Hence, by Theorem 5.4, system (5.60) is feedback linearizable with output.

$\square$

Example 5.4.5 Use Theorem 5.3 to show that the following nonlinear system is not feedback linearizable with output:

$$\begin{bmatrix} \dot{x}_1 \\ \dot{x}_2 \\ \dot{x}_3 \end{bmatrix} = \begin{bmatrix} x_2 \\ x_3 \\ x_1^2 \end{bmatrix} + \begin{bmatrix} 0 \\ 0 \\ e^{x_1} \end{bmatrix} u = f(x) + g(x)u \tag{5.63}$$

$$y = 2x_1 + e^{x_2} - 1 = h(x).$$

Solution By simple calculation, we have

$$L_g h(x) = 0, \quad L_{\text{ad}_f g} h(x) = -e^{x_1 + x_2}$$

where $\text{ad}_f g(x)$ and $\text{ad}_f^2 g(x)$ are given by (5.61). Thus, condition (i) and (ii) of Theorem 5.3 are satisfied. Also, we have that $\rho = 2$, $L_g L_f h(x) = -L_{\text{ad}_f g} h(x) = e^{x_1 + x_2}$, $\beta(x) = \frac{1}{L_g L_f h(x)} = e^{-x_1 - x_2}$, and

$$\omega(x) \triangleq \begin{bmatrix} 0 & 0 & \frac{1}{\beta(x)} \end{bmatrix} \left[g(x) \ \text{ad}_f g(x) \ \text{ad}_f^2 g(x) \right]^{-1}$$

$$= \begin{bmatrix} 0 & 0 & e^{x_1 + x_2} \end{bmatrix} \begin{bmatrix} (x_2^2 - x_3)e^{-x_1} & x_2 e^{-x_1} & e^{-x_1} \\ -2x_2 e^{-x_1} & -e^{-x_1} & 0 \\ e^{-x_1} & 0 & 0 \end{bmatrix} = \begin{bmatrix} e^{x_2} & 0 & 0 \end{bmatrix}.$$

Since

$$\frac{\partial \omega(x)^T}{\partial x} = \begin{bmatrix} 0 & e^{x_2} & 0 \\ 0 & 0 & 0 \\ 0 & 0 & 0 \end{bmatrix} \neq \left(\frac{\partial \omega(x)^T}{\partial x} \right)^T$$

$\omega(x)$ is not exact and condition (iii) of Theorem 5.3 is not satisfied. Hence, by Theorem 5.3, system (5.63) is not feedback linearizable with output. $\square$

Example 5.4.6 Consider the following nonlinear system:

$$\begin{bmatrix} \dot{x}_1 \\ \dot{x}_2 \\ \dot{x}_3 \end{bmatrix} = \begin{bmatrix} x_2 \\ x_3 \\ x_1^2 \end{bmatrix} + \begin{bmatrix} 0 \\ 0 \\ e^{x_1} \end{bmatrix} u = f(x) + g(x)u$$

$$y = \sin x_1 + x_2 = h(x).$$

(5.64)

(a) Use Theorem 5.3 to show that the above system is not feedback linearizable with output.
(b) Use Theorem 5.4 to show that the above system is not feedback linearizable with output.

Solution (a) By simple calculation, we have

$$L_g h(x) = 0, \quad L_{ad_f g} h(x) = -e^{x_1}$$

where $ad_f g(x)$ and $ad_f^2 g(x)$ are given by (5.61). Thus, condition (i) and (ii) of Theorem 5.3 are satisfied. Also, we have that $\rho = 2$, $L_g L_f h(x) = -L_{ad_f g} h(x) = e^{x_1}$, $\beta(x) = \frac{1}{L_g L_f h(x)} = e^{-x_1}$, and

$$\omega(x) \triangleq \begin{bmatrix} 0 & 0 & \frac{1}{\beta(x)} \end{bmatrix} \begin{bmatrix} g(x) & ad_f g(x) & ad_f^2 g(x) \end{bmatrix}^{-1}$$

$$= \begin{bmatrix} 0 & 0 & e^{x_1} \end{bmatrix} \begin{bmatrix} (x_2^2 - x_3)e^{-x_1} & x_2 e^{-x_1} & e^{-x_1} \\ -2x_2 e^{-x_1} & -e^{-x_1} & 0 \\ e^{-x_1} & 0 & 0 \end{bmatrix} = \begin{bmatrix} 1 & 0 & 0 \end{bmatrix}.$$

Since $\frac{\partial \omega(x)^T}{\partial x} = O_{3 \times 3} = \left(\frac{\partial \omega(x)^T}{\partial x} \right)^T$, condition (iii) of Theorem 5.3 is satisfied.
Since $\frac{\partial S_1(x)}{\partial x} = \omega(x) = \begin{bmatrix} 1 & 0 & 0 \end{bmatrix}$, it is clear that $S_1(x) = x_1$. We have, by (5.32) and (5.33), that $\alpha(x) = -\frac{L_f^3 S_1(x)}{L_g L_f h(x)} = -\frac{x_1^2}{e^{x_1}} = -x_1^2 e^{-x_1}$ and

$$\hat{f}(x) = f(x) + g(x)\alpha(x) = \begin{bmatrix} x_2 \\ x_3 \\ 0 \end{bmatrix} ; \quad \hat{g}(x) = g(x)\beta(x) = \begin{bmatrix} 0 \\ 0 \\ 1 \end{bmatrix}.$$

By simple calculation, we have that

$$ad_{\hat{f}} \hat{g}(x) = \begin{bmatrix} 0 \\ -1 \\ 0 \end{bmatrix}, \quad ad_{\hat{f}}^2 \hat{g}(x) = \begin{bmatrix} 1 \\ 0 \\ 0 \end{bmatrix}$$

and

$$L_{\hat{g}}h(x) = 0, \quad L_{\text{ad}_{\bar{f}}\hat{g}}h(x) = -1, \quad L_{\text{ad}_{\bar{f}}^2\hat{g}}h(x) = \cos x_1$$

which implies that (iv) of Theorem 5.3 is not satisfied. Hence, by Theorem 5.3, system (5.64) is not feedback linearizable with output.

(b) Since $\rho = 2$, $L_g L_f h(x) = e^{x_1}$, and $L_f^2 h(x) = x_3 \cos x_1 - x_2^2 \sin x_1 + x_1^2$, we have, by (5.48) and (5.49), that

$$\hat{g}(x) \triangleq \frac{1}{L_g L_f^{\rho-1}h(x)} g(x) = \begin{bmatrix} 0 \\ 0 \\ 1 \end{bmatrix}$$

and

$$\bar{f}(x) \triangleq f(x) - \frac{L_f^\rho h(x)}{L_g L_f^{\rho-1}h(x)} g(x) = \begin{bmatrix} x_2 \\ x_3 \\ x_2^2 \sin x_1 - x_3 \cos x_1 \end{bmatrix}.$$

By simple calculation, we have that

$$\begin{bmatrix} \text{ad}_{\bar{f}}\hat{g}(x) & \text{ad}_{\bar{f}}^2\hat{g}(x) \end{bmatrix} = \begin{bmatrix} 0 & 1 \\ -1 & -\cos x_1 \\ \cos x_1 & x_2 \sin x_1 - \sin^2 x_1 + 1 \end{bmatrix}$$

$$\text{ad}_{\bar{f}}^3\hat{g}(x) = \begin{bmatrix} \cos x_1 \\ \sin^2 x_1 - 1 \\ \frac{x_2 \sin(2x_1)}{2} + \cos^3 x_1 \end{bmatrix}$$

and

$$\begin{bmatrix} \text{ad}_{\bar{f}}^2\hat{g}(x), \text{ad}_{\bar{f}}^3\hat{g}(x) \end{bmatrix} = \begin{bmatrix} -\sin x_1 \\ \frac{\sin(2x_1)}{2} \\ -\sin x_1 \left(x_2 \sin x_1 - \sin^2 x_1 + 1\right) \end{bmatrix} \neq 0$$

which implies that (iii) of Theorem 5.4 is not satisfied. Hence, by Theorem 5.4, system (5.64) is not feedback linearizable with output.

□

5.5   Input-Output Linearization of MIMO Systems

5.5.1   Introduction

The feedback linearization problem of the nonlinear system with output is to obtain feedback that makes both the relationship between the state variable and the input and the relationship between the output and the input linear. In the previous section, we considered feedback linearization with the output for the single input single output nonlinear system. For the single input system, the state and output expressions must be made linear with one input. For the multi-input system, if we use only a part of the inputs to linearize the input-output relation, we could use the remaining inputs to linearize the state equation. Therefore, it is a less restrictive problem than the single input problem. First, the input-output linearization problem of a multi-input multi-output system is explained. Consider the following nonlinear systems:

$$
\begin{aligned}
\dot{x}(t) &= f(x(t)) + g(x(t))u(t) \\
y(t) &= h(x(t))
\end{aligned}
\tag{5.65}
$$

where $x \in \mathbb{R}^n$, $u \in \mathbb{R}^m$, $y \in \mathbb{R}^q$, and $f(x)$, $g(x)$, and $h(x)$ are analytic functions with $f(0) = 0$ and $h(0) = 0$. The input-output relation of system (5.65) is expressed in the following Volterra series:

$$
\begin{aligned}
y(t) = {}& w^{(0)}(t, x) + \sum_{i_1=1}^{m} \int_0^t w_{i_1}^{(1)}(t, \tau_1, x)u_{i_1}(\tau_1)d\tau_1 \\
&+ \sum_{i_1,i_2=1}^{m} \int_0^t \int_0^{\tau_1} w_{i_1,i_2}^{(2)}(t, \tau_1, \tau_2, x)u_{i_1}(\tau_1)u_{i_2}(\tau_2)d\tau_2 d\tau_1 + \cdots
\end{aligned}
\tag{5.66}
$$

where j-th triangular Volterra kernel $w_{i_1,i_2,\cdots,i_j}^{(j)}$ satisfies the following Taylor series:

$$
w^{(0)}(t, x) = \sum_{k=0}^{\infty} L_f^k h(x)\frac{t^k}{k!}
$$

$$
w_i^{(1)}(t, \tau_1, x) = \sum_{k_1=0}^{\infty}\sum_{k_2=0}^{\infty} L_f^{k_2} L_{g_i} L_f^{k_1} h(x)\frac{(t-\tau_1)^{k_1}}{k_1!}\frac{\tau_1^{k_2}}{k_2!}
$$

$$
w_{i_1,i_2}^{(2)}(t, \tau_1, \tau_2, x) = \sum_{k_3,k_2,k_1=0}^{\infty} L_f^{k_3} L_{g_{i_2}} L_f^{k_2} L_{g_{i_1}} L_f^{k_1} h(x)\frac{(t-\tau_1)^{k_1}(\tau_1-\tau_2)^{k_2}\tau_2^{k_3}}{k_1!k_2!k_3!}
$$

$$
\vdots
$$

$$
\tag{5.67}
$$

The first term on the right-hand side of (5.66) is the zero input response, whereas the rest is the part of the output depending on the input. System (5.65) is said to have a linear input-output relation if the following is satisfied:

$$y(t) = w^{(0)}(t, x) + \sum_{i=1}^{m} \int_{0}^{t} w_i^{(1)}(t - \tau)u_i(\tau)d\tau. \tag{5.68}$$

Example 5.5.1 By using (5.66) and (5.67), show that linear time-invariant system

$$\begin{aligned} \dot{z}(t) &= Az(t) + Bv(t) \\ y(t) &= Cz(t) \end{aligned} \tag{5.69}$$

has the following input-output relation:

$$y(t) = Ce^{At}z(0) + \int_{0}^{t} Ce^{A(t-\tau)}Bv(\tau)d\tau. \tag{5.70}$$

Solution Omitted. (See Problem 5–6.) □

Example 5.5.2 Show the following:

(a) The first Volterra kernel $w^{(1)}(t, \tau, x)$ depends only on $t - \tau$, if and only if

$$L_g L_f^k h(x) = \text{const}, \quad k \geq 0. \tag{5.71}$$

(b) If $w^{(1)}(t, \tau, x)$ depends only on $t - \tau$, then $w^{(i)} = 0$ for $i \geq 2$.

Solution (a) Suppose that $w^{(1)}(t, \tau, x)$ depends only on $t - \tau$. Then it is clear that $w^{(1)}(t, 0, x)$ depends only on t. Thus, we have, by (5.67), that

$$\begin{aligned} w_i^{(1)}(t, \tau_1, x) &= \sum_{k_1=0}^{\infty} L_{g_i} L_f^{k_1} h(x) \frac{(t - \tau_1)^{k_1}}{k_1!} \\ &+ \sum_{k_1=0}^{\infty} \sum_{k_2=1}^{\infty} L_f^{k_2} L_{g_i} L_f^{k_1} h(x) \frac{(t - \tau_1)^{k_1}}{k_1!} \frac{\tau_1^{k_2}}{k_2!} \end{aligned} \tag{5.72}$$

which implies that

$$w_i^{(1)}(t, 0, x) = \sum_{k_1=0}^{\infty} L_{g_i} L_f^{k_1} h(x) \frac{t^{k_1}}{k_1!}$$

and (5.71) is satisfied. Conversely, suppose that (5.71) is satisfied. Then

$$w_i^{(1)}(t, \tau_1, x) = \sum_{k_1=0}^{\infty} L_{g_i} L_f^{k_1} h(x) \frac{(t - \tau_1)^{k_1}}{k_1!}.$$

Since $L_{g_i} L_f^{k_1} h(x) = $ constant for $k_1 \geq 0$, it is clear that $w^{(1)}(t, \tau, x)$ depends only on $t - \tau$.

(b) Suppose that $w^{(1)}(t, \tau, x)$ depends only on $t - \tau$. Then, it is clear, by (a), that $L_{g_i} L_f^{k_1} h(x) = $ constant for $k_1 \geq 0$. Thus, it is easy to see, by (5.67), that for $k_1 \geq 0$ and $k_2 \geq 0$,

$$L_{g_{i_2}} L_f^{k_2} L_{g_i} L_f^{k_1} h(x) = 0$$

which implies that $w^{(i)} = 0$ for $i \geq 2$.

$\square$

The input-output linearization problem of a nonlinear system is to find nonsingular state feedback $u = \alpha(x) + \beta(x)v$ such that the closed-loop system

$$\dot{x}(t) = f(x) + g(x)\alpha(x) + g(x)\beta(x)v(t) = \hat{f}(x(t)) + \hat{g}(x(t))v(t)$$
$$y(t) = h(x(t))$$
(5.73)

satisfies

$$y(t) = \hat{w}^{(0)}(t, x) + \sum_{i=1}^{m} \int_0^t \hat{w}_i^{(1)}(t - \tau)v_i(\tau)d\tau \tag{5.74}$$

where

$$\hat{w}^{(0)}(t, x) = \sum_{k=0}^{\infty} L_{\hat{f}}^k h(x) \frac{t^k}{k!}$$
$$\hat{w}_i^{(1)}(t - \tau_1) = \sum_{k_1=0}^{\infty} L_{\hat{g}_i} L_{\hat{f}}^{k_1} h(x) \frac{(t - \tau_1)^{k_1}}{k_1!}. \tag{5.75}$$

Definition 5.5 (*input-output linearization*)
System (5.65) is said to be locally input-output linearizable, if there exists a nonsingular state feedback $u = \alpha(x) + \beta(x)v$ on a neighborhood U of $0 \in \mathbb{R}^n$ such that input-output relation of the closed-loop system of system (5.65) satisfies (5.74). In other words,

$$L_{\hat{g}} L_{\hat{f}}^k h(x) = \text{const}, \quad k \geq 0 \tag{5.76}$$

where $\hat{f}(x) = f(x) + g(x)\alpha(x)$ and $\hat{g}(x) = g(x)\beta(x)$.

We use the nonsingular feedback (i.e., nonsingular $\beta(x)$) for input-output linearization. If we use singular feedback, we are giving up some of the input. For example, the feedback $u = \alpha(x) + \beta(x)v = \alpha(x) + Ov$ obtains the linear relation between the output and the new input v. In the previous section, the characteristic number of the output has been defined for the single output system. For multi-output systems, each output can have a different characteristic number.

Definition 5.6 (*relative degree of output y_i*)
The relative degree ρ_i of the output y_i is defined as the smallest integer such that $L_g L_f^{\rho_i-1} h_i(x) \neq 0$, where

$$L_g h_i(x) \triangleq \frac{\partial h_i(x)}{\partial x} g(x) = \left[L_{g_1} h_i(x) \cdots L_{g_m} h_i(x) \right].$$

In other words, ρ_i is the characteristic number of y_i such that

$$L_g L_f^{\ell-1} h_i(x) = 0, \quad \ell \leq \rho_i - 1 ; \quad L_g L_f^{\rho_i-1} h_i(x) \neq 0 \tag{5.77}$$

or

$$L_{f+gu}^{\ell-1} h_i(x) = L_f^{\ell-1} h_i(x), \quad \ell \leq \rho_i ; \quad \frac{\partial}{\partial u} \left(L_{f+gu}^{\rho_i} h_i(x) \right) \neq 0. \tag{5.78}$$

$\rho \triangleq \min(\rho_1, \ldots, \rho_q)$ is said to be the characteristic number of the output. (Refer to Definition 5.4.) Suppose that ρ_i is the relative degree of the output y_i. Then, it is clear that

$$\begin{bmatrix} y_1^{(\rho_1)} \\ \vdots \\ y_q^{(\rho_q)} \end{bmatrix} = \begin{bmatrix} L_f^{\rho_1} h_1(x) \\ \vdots \\ L_f^{\rho_m} h_q(x) \end{bmatrix} + \begin{bmatrix} L_g L_f^{\rho_1-1} h_1(x) \\ \vdots \\ L_g L_f^{\rho_m-1} h_q(x) \end{bmatrix} u \tag{5.79}$$

$$\triangleq E(x) + D(x)u.$$

If $q = m$ and $m \times m$ matrix $D(0)$ is invertible, then the closed-loop system has decoupled input-output relationship

$$\begin{bmatrix} y_1^{(\rho_1)} \\ \vdots \\ y_m^{(\rho_m)} \end{bmatrix} = \begin{bmatrix} v_1 \\ \vdots \\ v_m \end{bmatrix} \tag{5.80}$$

with static feedback

$$u = -D(x)^{-1} E(x) + D(x)^{-1} v. \tag{5.81}$$

Therefore, $D(x)$ is called the decoupling matrix of system (5.65). (Refer to Chap. 9.)

Lemma 5.1 *Suppose that V is a $q \times q$ nonsingular constant matrix. If system (5.65) is locally input-output linearizable, then*

$$\dot{x}(t) = f(x(t)) + g(x(t))u(t)$$
$$\tilde{y}(t) = Vh(x(t)) = \tilde{h}(x(t))$$

(5.82)

is also locally input-output linearizable with the same feedback, and vice versa.

Proof Omitted. (See Problem 5–8.)   □

Example 5.5.3 Find out a state feedback for input-output linearization of the following nonlinear system.

$$\begin{bmatrix} \dot{x}_1 \\ \dot{x}_2 \\ \dot{x}_3 \end{bmatrix} = \begin{bmatrix} x_2^2 \\ x_3 \\ 0 \end{bmatrix} + \begin{bmatrix} 1+x_1 & 0 \\ 0 & 0 \\ 0 & 1 \end{bmatrix} u = f(x) + g(x)u$$

$$y = \begin{bmatrix} x_1 \\ x_2 \end{bmatrix} = h(x).$$

(5.83)

Solution By simple calculation, we have $(\rho_1, \rho_2) = (1, 2)$ and

$$\begin{bmatrix} \dot{y}_1 \\ \ddot{y}_2 \end{bmatrix} = \begin{bmatrix} x_2^2 \\ 0 \end{bmatrix} + \begin{bmatrix} 1+x_1 & 0 \\ 0 & 1 \end{bmatrix} \begin{bmatrix} u_1 \\ u_2 \end{bmatrix}.$$

(5.84)

Therefore, it is easy to see that system (5.83) is input-output linearizable by nonsingular state feedback

$$\begin{bmatrix} u_1 \\ u_2 \end{bmatrix} = \begin{bmatrix} \frac{-x_2^2}{1+x_1} \\ 0 \end{bmatrix} + \begin{bmatrix} \frac{1}{1+x_1} & 0 \\ 0 & 1 \end{bmatrix} \begin{bmatrix} v_1 \\ v_2 \end{bmatrix}.$$

(5.85)

□

Example 5.5.4 Find out a state feedback for input-output linearization of the following nonlinear system.

$$\begin{bmatrix} \dot{x}_1 \\ \dot{x}_2 \\ \dot{x}_3 \end{bmatrix} = \begin{bmatrix} x_2^2 \\ x_3 \\ 0 \end{bmatrix} + \begin{bmatrix} 1+x_1 & 0 \\ 0 & 0 \\ 0 & 1 \end{bmatrix} u = f(x) + g(x)u$$

$$y = \begin{bmatrix} x_1 \\ x_1 + x_2 \end{bmatrix} = h(x).$$

(5.86)

Solution By simple calculation, we have $(\rho_1, \rho_2) = (1, 1)$ and

$$\begin{bmatrix} \dot{y}_1 \\ \dot{y}_2 \end{bmatrix} = \begin{bmatrix} x_2^2 \\ x_2^2 + x_3 \end{bmatrix} + \begin{bmatrix} 1+x_1 & 0 \\ 1+x_1 & 0 \end{bmatrix} \begin{bmatrix} u_1 \\ u_2 \end{bmatrix} \triangleq E(x) + D(x)u.$$

(5.87)

Since $D(0)$ is not invertible, there is no nonsingular feedback such that $\dot{y}_1 = v_1$ and $\dot{y}_2 = v_2$ as in Example 5.5.3. By premultiplying equation (5.87) by constant nonsingular matrix $\begin{bmatrix} 1 & 0 \\ -1 & 1 \end{bmatrix}$, we have

$$\begin{bmatrix} \dot{\tilde{y}}_1 \\ \dot{\tilde{y}}_2 \end{bmatrix} = \begin{bmatrix} x_2^2 \\ x_3 \end{bmatrix} + \begin{bmatrix} 1 + x_1 & 0 \\ 0 & 0 \end{bmatrix} \begin{bmatrix} u_1 \\ u_2 \end{bmatrix} \tag{5.88}$$

where

$$\begin{bmatrix} \tilde{y}_1 \\ \tilde{y}_2 \end{bmatrix} = \begin{bmatrix} 1 & 0 \\ -1 & 1 \end{bmatrix} \begin{bmatrix} y_1 \\ y_2 \end{bmatrix}. \tag{5.89}$$

Since the relative degree of $\tilde{y}_2$ is not 1 but 2, we have

$$\begin{bmatrix} \dot{\tilde{y}}_1 \\ \ddot{\tilde{y}}_2 \end{bmatrix} = \begin{bmatrix} x_2^2 \\ 0 \end{bmatrix} + \begin{bmatrix} 1 + x_1 & 0 \\ 0 & 1 \end{bmatrix} \begin{bmatrix} u_1 \\ u_2 \end{bmatrix} \triangleq \tilde{E}(x) + \tilde{D}(x)u. \tag{5.90}$$

Since $\tilde{D}(0)$ is invertible, the new output $\tilde{y}$ and the new input v has linear input-output relation with nonsingular feedback (5.85). Therefore, by Lemma 5.1, it is clear that system (5.86) is also input-output linearizable by nonsingular feedback (5.85). $\square$

As in (5.89) of Example 5.5.4, we can find the linear transformation $\tilde{y}$ of the output y (and its derivatives) such that the decoupling matrix $\tilde{D}(x)$ of $\tilde{y}$ has the maximal rank. We call the above procedure the structure algorithm and introduce it in the next section.

5.5.2 Structure Algorithm

We define the structure algorithm for system (5.65).

5.5.2.1 Structure Algorithm of the Nonlinear System

Step 1: Let $\rho \triangleq \min(\rho_1, \ldots, \rho_q)$ and $\text{rank} \left(L_g L_f^{\rho-1} h(x) \big|_{x=0} \right) = \sigma_1$. If $\sigma_1 = q$, then the algorithm terminates with $P_1 = I_q$ (or $V_1 = I$) and $\gamma_1(x) = L_f^{\rho-1} h(x)$. Otherwise, we can find, by elementary row operations, a nonsingular constant matrix $V_1 = \begin{bmatrix} P_1 \\ K_1^1 \end{bmatrix}$ such that

$$V_1 L_g L_f^{\rho-1} h(x) \big|_{x=0} = \begin{bmatrix} E_1(0) \\ O_{(q-\sigma_1) \times m} \end{bmatrix}$$

and

$$V_1 L_g L_f^{\rho-1} h(x) = \begin{bmatrix} E_1(x) \\ \hat{E}_1(x) \end{bmatrix} \triangleq \begin{bmatrix} \bar{E}_1(x) \\ \hat{E}_1(x) \end{bmatrix}$$

where P_1 and K_1^1 are $\sigma_1 \times q$ matrix and $(q - \sigma_1) \times q$ matrix, respectively. Let

$$\gamma_1(x) = P_1 L_f^{\rho-1} h(x) \text{ and } \bar{\gamma}_1(x) = K_1^1 L_f^{\rho-1} h(x). \tag{5.91}$$

In other words, we have that $\hat{E}_1(0) = O(q - \sigma_1) \times 1$,

$$\begin{bmatrix} L_g \gamma_1(x) \\ L_g \bar{\gamma}_1(x) \end{bmatrix} = \begin{bmatrix} E_1(x) \\ \hat{E}_1(x) \end{bmatrix} \text{ and rank } (E_1(x)) = \text{rank } (E_1(0)) = \sigma_1.$$

If $\hat{E}_1(x) \neq O$, then the algorithm terminates. (System (5.65) is not, by Theorem 5.5, locally input-output linearizable.)

Step i: Suppose that

$$\text{rank} \left(\begin{bmatrix} \bar{E}_{i-1}(x) \\ L_g L_f \bar{\gamma}_{i-1}(x) \end{bmatrix} \Big|_{x=0} \right) = \sum_{j=1}^{i} \sigma_j \triangleq \bar{\sigma}_i.$$

If $\bar{\sigma}_i = q$, then the algorithm terminates with $P_i = I_{\sigma_i}$ (or $V_i = I$) and $\gamma_i(x) = P_i L_f \bar{\gamma}_{i-1}(x) = L_f \bar{\gamma}_{i-1}(x)$. In this case, $\bar{\gamma}_i(x)$ is not defined. Otherwise, we can find, by elementary row operations, a nonsingular constant matrix

$$V_i = \begin{bmatrix} I_{\sigma_1} & 0 & \cdots & 0 & 0 \\ 0 & I_{\sigma_2} & & 0 & 0 \\ \vdots & & \ddots & & \vdots \\ 0 & 0 & & I_{\sigma_{i-1}} & 0 \\ 0 & 0 & \cdots & 0 & P_i \\ K_1^i & K_2^i & \cdots & K_{i-1}^i & K_i^i \end{bmatrix} \triangleq \begin{bmatrix} I_{\bar{\sigma}_{i-1}} & O_{\bar{\sigma}_{i-1} \times (q - \bar{\sigma}_{i-1})} \\ O_{\sigma_i \times \bar{\sigma}_{i-1}} & P_i \\ \bar{K}_i & K_i^i \end{bmatrix}$$

such that

$$V_i \begin{bmatrix} \bar{E}_{i-1}(x) \\ L_g L_f \bar{\gamma}_{i-1}(x) \end{bmatrix} \Big|_{x=0} = \begin{bmatrix} \bar{E}_{i-1}(0) \\ E_i(0) \\ O_{(q-\bar{\sigma}_i) \times m} \end{bmatrix}$$

and

$$V_i \begin{bmatrix} \bar{E}_{i-1}(x) \\ L_g L_f \bar{\gamma}_{i-1}(x) \end{bmatrix} = \begin{bmatrix} \bar{E}_{i-1}(x) \\ E_i(x) \\ \hat{E}_i(x) \end{bmatrix} \triangleq \begin{bmatrix} \bar{E}_i(x) \\ \hat{E}_i(x) \end{bmatrix}$$

where P_i and K_i^i are $\sigma_i \times (q - \bar{\sigma}_{i-1})$ matrix and $(q - \bar{\sigma}_i) \times (q - \bar{\sigma}_{i-1})$ matrix, respectively. Let

$$
\begin{aligned}
\gamma_i(x) &= P_i L_f \bar{\gamma}_{i-1}(x) \\
\bar{\gamma}_i(x) &= K_1^i \gamma_1(x) + \cdots + K_{i-1}^i \gamma_{i-1}(x) + K_i^i L_f \bar{\gamma}_{i-1}(x) \\
&= \bar{K}_i \Gamma_{i-1}(x) + K_i^i L_f \bar{\gamma}_{i-1}(x)
\end{aligned}
\tag{5.92}
$$

where $\bar{K}_i \triangleq \begin{bmatrix} K_1^i & K_2^i & \cdots & K_{i-1}^i \end{bmatrix}$ and

$$
\Gamma_i(x) \triangleq \begin{bmatrix} \gamma_1(x) \\ \gamma_2(x) \\ \vdots \\ \gamma_i(x) \end{bmatrix} \quad (\bar{\sigma}_i \times 1).
$$

In other words, we have that

$$
\begin{aligned}
L_g \Gamma_i(x) &= \bar{E}_i(x) \, ; \quad \mathrm{rank}\left(\bar{E}_i(x) \right) = \mathrm{rank}\left(\bar{E}_i(0) \right) = \bar{\sigma}_i \\
L_g \bar{\gamma}_i(x) &= \hat{E}_i(x) \, ; \quad \hat{E}_i(0) = O_{(q - \bar{\sigma}_i) \times m}.
\end{aligned}
\tag{5.93}
$$

If $\sigma_i = 0$, the step is said to be degenerated with $\bar{E}_i(x) = \bar{E}_{i-1}(x)$ and $0 \times (q - \bar{\sigma}_{i-1})$ matrix P_i. If $\hat{E}_i(x) \neq O$, then the algorithm terminates. (System (5.65) is not, by Theorem 5.5, locally input-output linearizable.)

If the algorithm terminates at finite step $\bar{k}$, then we obtain $\Gamma_{\bar{k}}(x)$ such that $\mathrm{rank}\left(L_g \Gamma_{\bar{k}}(x) \right) = \mathrm{rank}\left(\bar{E}_{\bar{k}}(x) \right) = \bar{\sigma}_{\bar{k}} = q$. If the algorithm does not end in a finite step, we define k as the last nondegenerate step $\bar{k}$, and we obtain $\Gamma_{\bar{k}}(x)$ and $\bar{\gamma}_{\bar{k}}(x)$. In other words, there exist $(q - \bar{\sigma}_{\bar{k}}) \times \bar{\sigma}_{\bar{k}}$ matrices $\bar{K}_{\bar{k}+i}$ such that for $i \geq 1$,

$$
\begin{aligned}
\bar{\gamma}_{\bar{k}+i}(x) &= \bar{K}_{\bar{k}+i} \Gamma_{\bar{k}}(x) + L_f \bar{\gamma}_{\bar{k}+i-1}(x) \\
L_g \bar{\gamma}_{\bar{k}+i}(x) &= O_{(q - \bar{\sigma}_{\bar{k}}) \times m}.
\end{aligned}
\tag{5.94}
$$

Structure algorithm can be found in (A3) and (G13). Structure algorithm for the discrete time nonlinear systems can also be found in (G14).

Example 5.5.5 For system (5.86) in Example 5.5.4, use the structure algorithm to find out $\Gamma_{\bar{k}}(x)$ and $\bar{\gamma}_{\bar{k}}(x)$.

Solution It is easy to see that $(\rho_1, \rho_2) = (1, 1)$ and

$$
L_g h(x) = \begin{bmatrix} 1 + x_1 & 0 \\ 1 + x_1 & 0 \end{bmatrix}.
\tag{5.95}
$$

Since $\mathrm{rank}\left(L_g h(x) \big|_{x=0} \right) = 1 = \sigma_1 < 2$, we obtain, by elementary row operations, constant matrix V_1 such that

$$V_1\,L_g h(x)\big|_{x=0} = \begin{bmatrix} 1 & 0 \\ -1 & 1 \end{bmatrix}\begin{bmatrix} 1 & 0 \\ 1 & 0 \end{bmatrix} = \begin{bmatrix} 1 & 0 \\ 0 & 0 \end{bmatrix} = \begin{bmatrix} \bar{E}_1(0) \\ O_{1\times 2} \end{bmatrix}$$

which implies that $P_1 = \begin{bmatrix} 1 & 0 \end{bmatrix}$ and $K_1^1 = \begin{bmatrix} -1 & 1 \end{bmatrix}$. Let

$$\begin{bmatrix} \gamma_1(x) \\ \bar{\gamma}_1(x) \end{bmatrix} = V_1\begin{bmatrix} h_1(x) \\ h_2(x) \end{bmatrix} = \begin{bmatrix} P_1 \\ K_1^1 \end{bmatrix}\begin{bmatrix} x_1 \\ x_1 + x_2 \end{bmatrix} = \begin{bmatrix} x_1 \\ x_2 \end{bmatrix}.$$

Since $\hat{E}_1(x) \triangleq L_g\bar{\gamma}_1(x) = O_{1\times 2}$, we go to step 2. Note that

$$\text{rank}\left(\begin{bmatrix} \bar{E}_1(x) \\ L_g L_f \bar{\gamma}_1(x) \end{bmatrix}\bigg|_{x=0}\right) = \text{rank}\left(\begin{bmatrix} 1 & 0 \\ 0 & 1 \end{bmatrix}\right) = 2 = \bar{\sigma}_2.$$

Since $\bar{\sigma}_2 = q = 2$, the algorithm terminates with $P_2 = I_1$ (or $V_2 = I$) and $\gamma_2(x) = P_2 L_f \bar{\gamma}_1(x) = x_3$. Since $\bar{\sigma}_2 = q$, $\bar{\gamma}_2(x)$ is not defined. In other words, we have $\Gamma_2(x) = \begin{bmatrix} \gamma_1(x) \\ \gamma_2(x) \end{bmatrix} = \begin{bmatrix} x_1 \\ x_3 \end{bmatrix}$ such that $\text{rank}\left(L_g\Gamma_2(x)\big|_{x=0}\right) = 2 = q$, where

$$\frac{d}{dt}\Gamma_2(x) = L_f\Gamma_2(x) + L_g\Gamma_2(x)u.$$

□

Example 5.5.6 Use the structure algorithm to find out $\Gamma_{\bar{k}}(x)$ and $\bar{\gamma}_{\bar{k}}(x)$ for the following system.

$$\begin{bmatrix} \dot{x}_1 \\ \dot{x}_2 \\ \dot{x}_3 \end{bmatrix} = \begin{bmatrix} x_1^2 \\ x_1 \\ x_3^2 \end{bmatrix} + \begin{bmatrix} 1+x_1 & 1 \\ 0 & 0 \\ 0 & 1 \end{bmatrix} u = f(x) + g(x)u$$

$$y = \begin{bmatrix} x_1 \\ x_1 + x_2 \end{bmatrix} = h(x)$$

(5.96)

Solution It is easy to see that $(\rho_1, \rho_2) = (1, 1)$ and

$$L_g h(x) = \begin{bmatrix} 1+x_1 & 1 \\ 1+x_1 & 1 \end{bmatrix}.$$

Since $\text{rank}\left(L_g h(x)\big|_{x=0}\right) = 1 = \sigma_1 < 2$, we obtain, by elementary row operations, constant matrix V_1 such that

$$V_1\,L_g h(x)\big|_{x=0} = \begin{bmatrix} 1 & 0 \\ -1 & 1 \end{bmatrix}\begin{bmatrix} 1 & 1 \\ 1 & 1 \end{bmatrix} = \begin{bmatrix} 1 & 1 \\ 0 & 0 \end{bmatrix} = \begin{bmatrix} \bar{E}_1(0) \\ O_{1\times 2} \end{bmatrix}$$

which implies that $P_1 = \begin{bmatrix} 1 & 0 \end{bmatrix}$, $K_1^1 = \begin{bmatrix} -1 & 1 \end{bmatrix}$, and

$$\begin{bmatrix} \gamma_1(x) \\ \bar{\gamma}_1(x) \end{bmatrix} = V_1 \begin{bmatrix} h_1(x) \\ h_2(x) \end{bmatrix} = \begin{bmatrix} P_1 \\ K_1^1 \end{bmatrix} \begin{bmatrix} x_1 \\ x_1 + x_2 \end{bmatrix} = \begin{bmatrix} x_1 \\ x_2 \end{bmatrix}. \tag{5.97}$$

Since $\hat{E}_1(x) \triangleq L_g \bar{\gamma}_1(x) = O_{1 \times 2}$, we go to step 2. Note that

$$\text{rank} \left(\begin{bmatrix} \bar{E}_1(x) \\ L_g L_f \bar{\gamma}_1(x) \end{bmatrix} \bigg|_{x=0} \right) = \text{rank} \left(\begin{bmatrix} 1 & 1 \\ 1 & 1 \end{bmatrix} \right) = 1 = \bar{\sigma}_2.$$

Since $\bar{\sigma}_2 < q = 2$, we obtain, by elementary row operations, constant matrix V_2 such that

$$V_2 \begin{bmatrix} \bar{E}_1(x) \\ L_g L_f \bar{\gamma}_1(x) \end{bmatrix} \bigg|_{x=0} = \begin{bmatrix} 1 & 0 \\ -1 & 1 \end{bmatrix} \begin{bmatrix} 1 & 1 \\ 1 & 1 \end{bmatrix} = \begin{bmatrix} 1 & 1 \\ 0 & 0 \end{bmatrix} = \begin{bmatrix} \bar{E}_2(0) \\ O_{1 \times 2} \end{bmatrix}.$$

Since $\sigma_2 = 0$, step 2 is degenerated with $\bar{E}_2(x) = \bar{E}_1(x)$, 0×1 matrix P_2, and

$$\bar{\gamma}_2(x) = \begin{bmatrix} -1 & 1 \end{bmatrix} \begin{bmatrix} \gamma_1(x) \\ L_f \bar{\gamma}_1(x) \end{bmatrix} = \begin{bmatrix} -1 & 1 \end{bmatrix} \begin{bmatrix} x_1 \\ x_1 \end{bmatrix} = 0.$$

Since $\hat{E}_2(x) \triangleq L_g \bar{\gamma}_2(x) = O_{1 \times 2}$, we go to step 3. In this manner, it is easy to see that step i is also degenerated for $i \geq 3$. Therefore, $\bar{k} = 1$ is the last nondegenerated step with $\gamma_1(x)$ and $\bar{\gamma}_1(x)$ in (5.97). □

Remark 5.1 The structure algorithm is state transformation invariant. Suppose that $z = S(x)$ is a state transformation for system (5.65). Then we have the following system:

$$\begin{aligned} \dot{z}(t) &= \tilde{f}(z(t)) + \tilde{g}(z(t))u(t) \\ y(t) &= \tilde{h}(z(t)) \end{aligned} \tag{5.98}$$

where $\tilde{h}(z) = h \circ S^{-1}(z)$, $\tilde{f}(z) = S_*(f(x))$, and

$$\tilde{g}(z) = \begin{bmatrix} S_*(g_1(x)) & \cdots & S_*(g_m(x)) \end{bmatrix} \triangleq S_*(g(x)).$$

Thus, it is easy to see, by Example 2.4.14, that for $k \geq 0$,

$$L_{\tilde{g}} L_{\tilde{f}}^k \tilde{h}(z) = L_{S_*(g)} L_{S_*(f)}^k \tilde{h}(z) = L_g L_f^k h(x) \big|_{x=S^{-1}(z)}.$$

Therefore, if we obtain V_i, $1 \leq i \leq \bar{k}$, $\Gamma_{\bar{k}}(x)$, and $\bar{\gamma}_{\bar{k}}(x)$ by the structure algorithm for system (5.65), then we also have V_i, $1 \leq i \leq \bar{k}$, $\Gamma_{\bar{k}} \circ S^{-1}(z)$, and $\bar{\gamma}_{\bar{k}} \circ S^{-1}(z)$ by the structure algorithm for system (5.98).

Remark 5.2 The structure algorithm is feedback invariant. Suppose that $u = \alpha(x) + \beta(x)v$ is a nonsingular feedback for system (5.65). Then we have the following closed-loop system:

$$\dot{x}(t) = \hat{f}(x(t)) + \hat{g}(x(t))v(t)$$
$$y(t) = h(x(t)) \tag{5.99}$$

where $\hat{f}(x) = f(x) + g(x)\alpha(x)$ and $\hat{g}(x) = g(x)\beta(x)$. Note, by Definition 5.4, that $L_{f+g\alpha}^{\rho-1}h(x) = L_f^{\rho-1}h(x)$. Since $L_{\hat{g}}L_{\hat{f}}^{\rho-1}h(x) = L_g L_f^{\rho-1}h(x)\beta(x)$, it is clear, by (5.91), that

$$\hat{\gamma}_1(x) \triangleq P_1 L_{\hat{f}}^{\rho-1}h(x) = P_1 L_f^{\rho-1}h(x) = \gamma_1(x)$$

and

$$\bar{\hat{\gamma}}_1(x) \triangleq K_1^1 L_{\hat{f}}^{\rho-1}h(x) = K_1^1 L_f^{\rho-1}h(x) = \bar{\gamma}_1(x).$$

Since $L_g \bar{\gamma}_{i-1}(x) = O$ for $2 \le i \le \bar{k}$, we have, by (5.92), that for $2 \le i \le \bar{k}$,

$$\hat{\gamma}_i(x) = P_i L_{\hat{f}}\bar{\gamma}_{i-1}(x) = P_i L_f \bar{\gamma}_{i-1}(x) = \gamma_i(x)$$

$$\bar{\hat{\gamma}}_i(x) = \bar{K}_i \Gamma_{i-1}(x) + K_i^i L_{\hat{f}}\bar{\gamma}_{i-1}(x) = \bar{K}_i \Gamma_{i-1}(x) + K_i^i L_f \bar{\gamma}_{i-1}(x) = \bar{\gamma}_i(x)$$

and

$$\text{rank}\left(\begin{bmatrix} L_{\hat{g}}\hat{\Gamma}_{i-1}(x) \\ L_{\hat{g}}L_{\hat{f}}\bar{\hat{\gamma}}_{i-1}(x) \end{bmatrix}\right) = \text{rank}\left(\begin{bmatrix} L_g \Gamma_{i-1}(x) \\ L_g L_f \bar{\gamma}_{i-1}(x) \end{bmatrix}\beta(x)\right)$$
$$= \text{rank}\left(\begin{bmatrix} L_g \Gamma_{i-1}(x) \\ L_g L_f \bar{\gamma}_{i-1}(x) \end{bmatrix}\right).$$

Therefore, if we obtain V_i, $1 \le i \le \bar{k}$, $\Gamma_{\bar{k}}(x)$, and $\bar{\gamma}_{\bar{k}}(x)$ by the structure algorithm for system (5.65), then we also have V_i, $1 \le i \le \bar{k}$, $\Gamma_{\bar{k}}(x)$, and $\bar{\gamma}_{\bar{k}}(x)$ by the structure algorithm for system (5.99).

5.5.3 Conditions for Input-Output Linearization

Suppose that the structure algorithm for system (5.65) satisfies

$$\hat{E}_i(x) \triangleq L_g \bar{\gamma}_i(x) = O_{(q-\bar{\sigma}_i)\times m}, \ 1 \le i \le \bar{k}.$$

Then, by structure algorithm, we have

$$\Gamma_{\bar{k}}(x) = \begin{bmatrix} \gamma_1(x) \\ \gamma_2(x) \\ \vdots \\ \gamma_{\bar{k}}(x) \end{bmatrix}$$

such that

$$\text{rank}\left(L_g \Gamma_{\bar{k}}(x)\big|_{x=0} \right) = \text{rank}\left(\bar{E}_{\bar{k}}(0) \right) = \sum_{i=1}^{\bar{k}} \sigma_i = \bar{\sigma}_{\bar{k}}.$$

By column operation, we can obtain a $m \times m$ permutation matrix R_1 such that

$$L_g \Gamma_{\bar{k}}(x) R_1 = \bar{E}_{\bar{k}}(x) R_1 = \left[\bar{E}_{\bar{k}}^1(x) \; \bar{E}_{\bar{k}}^2(x) \right]$$

where $\bar{\sigma}_{\bar{k}} \times \bar{\sigma}_{\bar{k}}$ matrix $\bar{E}_{\bar{k}}^1(0)$ is invertible. Let

$$\bar{\beta}(x) = R_1 \begin{bmatrix} \left(\bar{E}_{\bar{k}}^1(x) \right)^{-1} & -\left(\bar{E}_{\bar{k}}^1(x) \right)^{-1} \bar{E}_{\bar{k}}^2(x) \\ O_{(m-\bar{\sigma}_{\bar{k}}) \times \bar{\sigma}_{\bar{k}}} & I_{m-\bar{\sigma}_{\bar{k}}} \end{bmatrix} \triangleq R_1 R_2(x)$$

$$\bar{\alpha}(x) = -\bar{\beta}(x) \begin{bmatrix} L_f \Gamma_{\bar{k}}(x) \\ O_{(m-\bar{\sigma}_{\bar{k}}) \times 1} \end{bmatrix}.$$

$$(5.100)$$

Then, it is easy to see that

$$L_{\bar{g}} \Gamma_{\bar{k}}(x) = L_g \Gamma_{\bar{k}}(x) \bar{\beta}(x) = \left[I_{\bar{\sigma}_{\bar{k}}} \; O_{\bar{\sigma}_{\bar{k}} \times (m-\bar{\sigma}_{\bar{k}})} \right]$$
$$L_{\bar{f}} \Gamma_{\bar{k}}(x) = L_f \Gamma_{\bar{k}}(x) + L_g \Gamma_{\bar{k}}(x) \bar{\alpha}(x) = O_{\bar{\sigma}_{\bar{k}} \times 1}$$

$$(5.101)$$

where $\bar{f}(x) = f(x) + g(x)\bar{\alpha}(x)$ and $\bar{g}(x) = g(x)\bar{\beta}(x)$. In other words, we have that for $1 \leq i \leq \bar{k}$,

$$L_{\bar{f}} \gamma_i(x) = L_{f+g\bar{\alpha}} \gamma_i(x) = O_{\sigma_i \times 1}. \tag{5.102}$$

Example 5.5.7 Show that for $1 \leq i \leq \bar{k}$,

$$V_i V_{i-1} \cdots V_1 L_{f+g\alpha}^{\rho-2+i} h(x) = \begin{bmatrix} L_{f+g\alpha}^{i-1} \gamma_1(x) \\ \vdots \\ L_{f+g\alpha} \gamma_{i-1}(x) \\ \gamma_i(x) \\ \bar{\gamma}_i(x) + Q_{i-1}(x) \end{bmatrix} \triangleq \begin{bmatrix} \tilde{\Gamma}_i(x) \\ \bar{\gamma}_i(x) + Q_{i-1}(x) \end{bmatrix} \tag{5.103}$$

$$V_i V_{i-1} \cdots V_1 L_{f+g\tilde{\alpha}}^{\rho-2+i} h(x) = \begin{bmatrix} O_{\tilde{\sigma}_{i-1} \times 1} \\ \gamma_i(x) \\ \bar{\gamma}_i(x) - \bar{K}_i \Gamma_{i-1} \end{bmatrix} \tag{5.104}$$

and for $i \geq 1$,

$$L_{f+g\tilde{\alpha}}^i \bar{\gamma}_{\bar{k}}(x) = \bar{\gamma}_{\bar{k}+i}(x) - \bar{K}_{\bar{k}+i} \Gamma_{\bar{k}}(x) \tag{5.105}$$

where $\bar{\sigma}_0 = 0$, $Q_0(x) = O_{(q-\tilde{\sigma}_1) \times 1}$, and for $1 \leq j \leq \bar{k} - 1$,

$$Q_j(x) = \bar{K}_{j+1} L_{f+g\alpha} \tilde{\Gamma}_j(x) - \bar{K}_{j+1} \Gamma_j(x) + K_{j+1}^{j+1} L_{f+g\alpha} Q_{j-1}(x). \tag{5.106}$$

Solution It is clear, by (5.27) and (5.91), that

$$V_1 L_{f+g\tilde{\alpha}}^{\rho-1} h(x) = V_1 L_f^{\rho-1} h(x) = \begin{bmatrix} \gamma_1(x) \\ \bar{\gamma}_1(x) \end{bmatrix}.$$

Thus, (5.103) is satisfied when $i = 1$. Assume that (5.103) is satisfied when $i = j$ and $1 \leq j \leq \bar{k} - 1$. Then it is easy to see, by (5.92), (5.93), and (5.106), that

$$V_{j+1} V_j \cdots V_1 L_{f+g\alpha}^{\rho-1+j} h(x) = V_{j+1} L_{f+g\alpha} \left(V_j \cdots V_1 L_{f+g\alpha}^{\rho-2+j} h(x) \right)$$

$$= V_{j+1} L_{f+g\alpha} \left(\begin{bmatrix} \tilde{\Gamma}_j(x) \\ \bar{\gamma}_j(x) + Q_{j-1}(x) \end{bmatrix} \right) = V_{j+1} \begin{bmatrix} L_{f+g\alpha} \tilde{\Gamma}_j(x) \\ L_f \bar{\gamma}_j(x) + L_{f+g\alpha} Q_{j-1}(x) \end{bmatrix}$$

$$= \begin{bmatrix} L_{f+g\alpha} \tilde{\Gamma}_j(x) \\ \gamma_{j+1}(x) \\ \bar{K}_{j+1} L_{f+g\alpha} \tilde{\Gamma}_j(x) + K_{j+1}^{j+1} L_f \bar{\gamma}_j(x) + K_{j+1}^{j+1} L_{f+g\alpha} Q_{j-1}(x) \end{bmatrix}$$

$$= \begin{bmatrix} \tilde{\Gamma}_{j+1}(x) \\ \bar{K}_{j+1} L_{f+g\alpha} \tilde{\Gamma}_j(x) + \bar{\gamma}_{j+1}(x) - \bar{K}_{j+1} \Gamma_j(x) + K_{j+1}^{j+1} L_{f+g\alpha} Q_{j-1}(x) \end{bmatrix}$$

$$= \begin{bmatrix} \tilde{\Gamma}_{j+1}(x) \\ \bar{\gamma}_{j+1}(x) + Q_j(x) \end{bmatrix}$$

which implies that (5.103) is satisfied when $i = j + 1$. Therefore, by mathematical induction, (5.103) is satisfied for $1 \leq i \leq \bar{k}$. It is clear, by (5.102), (5.103), and (5.106), that for $1 \leq i \leq \bar{k}$,

$$V_i V_{i-1} \cdots V_1 L_{f+g\tilde{\alpha}}^{\rho-2+i} h(x) = \begin{bmatrix} L_{f+g\tilde{\alpha}}^{i-1} \gamma_1(x) \\ \vdots \\ L_{f+g\tilde{\alpha}} \gamma_{i-1}(x) \\ \gamma_i(x) \\ \bar{\gamma}_i(x) + Q_{i-1}(x) \end{bmatrix} = \begin{bmatrix} O_{\tilde{\sigma}_{i-1} \times 1} \\ \gamma_i(x) \\ \bar{\gamma}_i(x) + Q_{i-1}(x) \end{bmatrix}$$

and for $2 \leq i \leq \bar{k}$,

$$Q_{i-1}(x) = \bar{K}_i L_{f+g\bar{\alpha}} \bar{\Gamma}_{i-1}(x) - \bar{K}_i \Gamma_{i-1}(x) + K_i^i L_{f+g\bar{\alpha}} Q_{i-2}(x)$$
$$= -\bar{K}_i \Gamma_{i-1}(x)$$

which imply that (5.104) is satisfied for $1 \le i \le \bar{k}$. Since $L_g \bar{\gamma}_{\bar{k}}(x) = 0$, it is clear, by (5.94), that

$$L_{f+g\bar{\alpha}} \bar{\gamma}_{\bar{k}}(x) = L_f \bar{\gamma}_{\bar{k}}(x) = \bar{\gamma}_{\bar{k}+1}(x) - \bar{K}_{\bar{k}+1} \Gamma_{\bar{k}}(x)$$

which implies that (5.105) is satisfied when $i = 1$. Assume that (5.105) is satisfied when $i = j$ and $j \ge 1$. Then it is easy to see, by (5.94) and (5.101), that

$$L_{f+g\bar{\alpha}}^{j+1} \bar{\gamma}_{\bar{k}}(x) = L_{f+g\bar{\alpha}} \left(\bar{\gamma}_{\bar{k}+j}(x) - \bar{K}_{\bar{k}+j} \Gamma_{\bar{k}}(x) \right)$$
$$= L_{f+g\bar{\alpha}} \bar{\gamma}_{\bar{k}+j}(x) - \bar{K}_{\bar{k}+j} L_{f+g\bar{\alpha}} \Gamma_{\bar{k}}(x) = L_f \bar{\gamma}_{\bar{k}+j}(x)$$
$$= \bar{\gamma}_{\bar{k}+j+1}(x) - \bar{K}_{\bar{k}+j+1} \Gamma_{\bar{k}}(x)$$

which implies that (5.105) is satisfied when $i = j + 1$. Therefore, by mathematical induction, (5.105) is satisfied for $i \ge 1$. $\qquad \square$

Theorem 5.5 (conditions for input-output linearization)
System (5.65) is locally input-output linearizable, if and only if for $1 \le i \le \bar{k}$,

$$\hat{E}_i(x) \triangleq L_g \bar{\gamma}_i(x) = O_{(q-\bar{\sigma}_i) \times m}. \tag{5.107}$$

Proof Necessity. Suppose that system (5.65) is input-output linearizable on a neighborhood U of $0 \in \mathbb{R}^n$. Then, by Definition 5.5, there exists a nonsingular feedback $u = \alpha(x) + \beta(x)v$ on a neighborhood U of $0 \in \mathbb{R}^n$ such that for $k \ge 0$,

$$L_{g\beta} L_{f+g\alpha}^k h(x) = L_g L_{f+g\alpha}^k h(x) \beta(x) = \text{const}$$

on $x \in U$. Thus, it is clear that for $k \ge 0$,

$$L_{g\beta} L_{f+g\alpha}^k \gamma_1(x) = P_1 L_{g\beta} L_{f+g\alpha}^k L_f^{\rho-1} h(x)$$
$$= P_1 L_{g\beta} L_{f+g\alpha}^k L_{f+g\alpha}^{\rho-1} h(x) = \text{const}$$

and

$$L_{g\beta} L_{f+g\alpha}^k \bar{\gamma}_1(x) = K_1^1 L_{g\beta} L_{f+g\alpha}^k L_f^{\rho-1} h(x)$$
$$= K_1^1 L_{g\beta} L_{f+g\alpha}^{k+\rho-1} h(x) = \text{const} \tag{5.108}$$

on $x \in U$. Thus, we have, by (5.108) with $k = 0$, that on $x \in U$,

$$\hat{E}_1(x)\beta(x) = L_{g\beta} \bar{\gamma}_1(x) \triangleq \tilde{E}_1 = \text{const}.$$

Since $\hat{E}_1(0) = O_{(q-\bar{\sigma}_1) \times m}$ and $\beta(x)$ is nonsingular on $x \in U$, it is clear that $\tilde{E}_1 = \hat{E}_1(0)\beta(0) = O_{(q-\bar{\sigma}_1) \times m}$ and

$$\hat{E}_1(x) = \tilde{E}_1\beta(x)^{-1} = O_{(q-\bar{\sigma}_1) \times m}$$

which implies that (5.107) is satisfied when $i = 1$. We will show, by induction, that for $1 \le i \le \bar{k}$ and $k \ge 0$,

$$L_{g\beta}L_{f+g\alpha}^k \gamma_i(x) = \text{const} \quad \text{and} \quad L_{g\beta}L_{f+g\alpha}^k \bar{\gamma}_i(x) = \text{const} \tag{5.109}$$

on $x \in U$. Assume that (5.107) and (5.109) are satisfied for $1 \le i \le \ell - 1$ and $2 \le \ell \le \bar{k}$. Then, it is clear, by (5.92) and (5.109), that for $k \ge 0$,

$$\begin{aligned} L_{g\beta}L_{f+g\alpha}^k \gamma_\ell(x) &= P_\ell L_{g\beta}L_{f+g\alpha}^k L_f \bar{\gamma}_{\ell-1}(x) \\ &= P_\ell L_{g\beta}L_{f+g\alpha}^k L_{f+g\alpha} \bar{\gamma}_{\ell-1}(x) = \text{const} \end{aligned}$$

and

$$\begin{aligned} L_{g\beta}L_{f+g\alpha}^k \bar{\gamma}_\ell(x) &= \sum_{j=1}^{\ell-1} K_j^\ell L_{g\beta}L_{f+g\alpha}^k \gamma_j(x) + K_\ell^\ell L_{g\beta}L_{f+g\alpha}^k L_f \bar{\gamma}_{\ell-1}(x) \\ &= \sum_{j=1}^{\ell-1} K_j^\ell L_{g\beta}L_{f+g\alpha}^k \gamma_j(x) + K_\ell^\ell L_{g\beta}L_{f+g\alpha}^k L_{f+g\alpha} \bar{\gamma}_{\ell-1}(x) = \text{const} \end{aligned} \tag{5.110}$$

which imply that (5.109) is satisfied for $i = \ell$. Thus, we have, by (5.110) with $k = 0$, that on $x \in U$,

$$\hat{E}_\ell(x)\beta(x) = L_{g\beta}\bar{\gamma}_\ell(x) \triangleq \tilde{E}_\ell = \text{const}.$$

Since $\hat{E}_\ell(0) = O_{(q-\bar{\sigma}_\ell) \times m}$ and $\beta(x)$ is nonsingular on $x \in U$, it is clear that $\tilde{E}_\ell = \hat{E}_\ell(0)\beta(0) = O_{(q-\bar{\sigma}_\ell) \times m}$ and

$$\hat{E}_\ell(x) = \tilde{E}_\ell\beta(x)^{-1} = O_{(q-\bar{\sigma}_\ell) \times m}$$

which implies that (5.107) is also satisfied when $i = \ell$. Therefore, (5.107) is, by mathematical induction, satisfied for $1 \le i \le \bar{k}$.

Sufficiency. Let $u = \bar{\alpha}(x) + \bar{\beta}(x)v$, where $\bar{\alpha}(x)$ and $\bar{\beta}(x)$ are defined by (5.100). Also, let $\bar{f}(x) = f(x) + g(x)\bar{\alpha}(x)$ and $\bar{g}(x) = g(x)\bar{\beta}(x)$. It will be shown that for $i \ge 0$,

$$L_{\bar{g}}L_{\bar{f}}^i h(x) = L_g L_{f+g\bar{\alpha}}^i h(x) \, \bar{\beta}(x) = \text{const}. \tag{5.111}$$

It is clear, by (5.25) and (5.27), that (5.111) is satisfied for $0 \leq i \leq \rho - 2$. Also, it is easy to see, by (5.93), (5.101), and (5.104), that for $1 \leq i \leq \bar{k}$,

$$V_i V_{i-1} \cdots V_1 L_{g\bar{\beta}} L_{f+g\bar{\alpha}}^{\rho-2+i} h(x) = L_{g\bar{\beta}} \left(V_i V_{i-1} \cdots V_1 L_{f+g\bar{\alpha}}^{\rho-2+i} h(x) \right)$$

$$= \begin{bmatrix} O_{\bar{\sigma}_{i-1} \times m} \\ L_{g\bar{\beta}} \gamma_i(x) \\ L_{g\bar{\beta}} \bar{\gamma}_i(x) \end{bmatrix} = \begin{bmatrix} O_{\bar{\sigma}_{i-1} \times \bar{\sigma}_{i-1}} & O_{\bar{\sigma}_{i-1} \times \sigma_i} & O_{\bar{\sigma}_{i-1} \times (m-\bar{\sigma}_i)} \\ O_{\sigma_i \times \bar{\sigma}_{i-1}} & I_{\sigma_i} & O_{\sigma_i \times (m-\bar{\sigma}_i)} \\ -\bar{K}_i & O_{(q-\bar{\sigma}_i) \times \sigma_i} & O_{(q-\bar{\sigma}_i) \times (m-\bar{\sigma}_i)} \end{bmatrix}$$

which implies that (5.111) is satisfied for $\rho - 1 \leq i \leq \rho - 2 + \bar{k}$. Finally, we have, by (5.94), (5.101), (5.104), and (5.105), that for $i \geq 1$,

$$V_{\bar{k}} V_{\bar{k}-1} \cdots V_1 L_{g\bar{\beta}} L_{f+g\bar{\alpha}}^{\rho-2+\bar{k}+i} h(x) = L_{g\bar{\beta}} L_{f+g\bar{\alpha}}^i \left(V_{\bar{k}} V_{\bar{k}-1} \cdots V_1 L_{f+g\bar{\alpha}}^{\rho-2+\bar{k}} h(x) \right)$$

$$= \begin{bmatrix} O_{\bar{\sigma}_{i-1} \times m} \\ L_{g\bar{\beta}} L_{f+g\bar{\alpha}}^i \gamma_{\bar{k}}(x) \\ L_{g\bar{\beta}} L_{f+g\bar{\alpha}}^i \bar{\gamma}_{\bar{k}}(x) \end{bmatrix} = \begin{bmatrix} O_{\bar{\sigma}_{i-1} \times m} \\ O_{\sigma_i \times m} \\ L_{g\bar{\beta}} \bar{\gamma}_{\bar{k}+i}(x) - L_{g\bar{\beta}} \bar{K}_{\bar{k}+i} \Gamma_{\bar{k}}(x) \end{bmatrix}$$

$$= \begin{bmatrix} O_{\bar{\sigma}_{\bar{k}} \times \bar{\sigma}_{\bar{k}}} & O_{\bar{\sigma}_{\bar{k}} \times (m-\bar{\sigma}_{\bar{k}})} \\ -\bar{K}_{\bar{k}+i} & O_{(q-\bar{\sigma}_{\bar{k}}) \times (m-\bar{\sigma}_{\bar{k}})} \end{bmatrix}$$

which implies that (5.111) is satisfied for $i \geq \rho - 1 + \bar{k}$. Since (5.76) is satisfied, system (5.65) is input-output linearizable. $\qquad \square$

Example 5.5.8 Find the nonsingular feedback $u = \alpha(x) + \beta(x)v$ for the input-output linearization of system (5.96) in Example 5.5.6.

$$\begin{bmatrix} \dot{x}_1 \\ \dot{x}_2 \\ \dot{x}_3 \end{bmatrix} = \begin{bmatrix} x_1^2 \\ x_1 \\ x_3^2 \end{bmatrix} + \begin{bmatrix} 1+x_1 & 1 \\ 0 & 0 \\ 0 & 1 \end{bmatrix} u = f(x) + g(x)u$$

$$y = \begin{bmatrix} x_1 \\ x_1 + x_2 \end{bmatrix} = h(x). \tag{5.112}$$

Solution In Example 5.5.6, we have, by structure algorithm, that $\bar{k} = 1$ and

$$\begin{bmatrix} \Gamma_1(x) \\ \bar{\gamma}_1(x) \end{bmatrix} = \begin{bmatrix} \gamma_1(x) \\ \bar{\gamma}_1(x) \end{bmatrix} = \begin{bmatrix} x_1 \\ x_2 \end{bmatrix}.$$

Since

$$L_g \Gamma_1(x) = \begin{bmatrix} 1+x_1 & 1 \end{bmatrix}$$

we have, by (5.100), that $R_1 = I$ and

$$\beta(x) = \begin{bmatrix} \frac{1}{1+x_1} & -\frac{1}{1+x_1} \\ 0 & 1 \end{bmatrix}$$

$$\alpha(x) = -\beta(x) \begin{bmatrix} x_1^2 \\ 0 \end{bmatrix} = \begin{bmatrix} -\frac{x_1^2}{1+x_1} \\ 0 \end{bmatrix}.$$

Then, it is easy to see that

$$\begin{bmatrix} \dot{x}_1 \\ \dot{x}_2 \\ \dot{x}_3 \end{bmatrix} = \begin{bmatrix} 0 \\ x_1 \\ x_3^2 \end{bmatrix} + \begin{bmatrix} 1 & 0 \\ 0 & 0 \\ 0 & 1 \end{bmatrix} v = \hat{f}(x) + \hat{g}(x)v$$

$$y = \begin{bmatrix} x_1 \\ x_1 + x_2 \end{bmatrix} = h(x)$$

and for $i \geq 0$,

$$L_{\hat{g}} L_{\hat{f}}^i h(x) = \text{constant}.$$

$\square$

Example 5.5.9 Show that the following system is not locally input-output linearizable:

$$\begin{bmatrix} \dot{x}_1 \\ \dot{x}_2 \\ \dot{x}_3 \end{bmatrix} = \begin{bmatrix} x_1^2 \\ x_3 \\ x_2^2 \end{bmatrix} + \begin{bmatrix} 1 & 0 \\ x_1 & 0 \\ 0 & 1 \end{bmatrix} u = f(x) + g(x)u$$

$$y = \begin{bmatrix} x_1 \\ x_1 + x_2 \end{bmatrix} = h(x). \tag{5.113}$$

Solution It is easy to see that $(\rho_1, \rho_2) = (1, 1)$ and

$$L_g h(x) = \begin{bmatrix} 1 & 0 \\ 1 + x_1 & 0 \end{bmatrix}.$$

Since rank $\left(L_g h(x)\big|_{x=0} \right) = 1 = \sigma_1 < 2$, we obtain, by elementary row operations, constant matrix V_1 such that

$$V_1 L_g h(x)\big|_{x=0} = \begin{bmatrix} 1 & 0 \\ -1 & 1 \end{bmatrix} \begin{bmatrix} 1 & 0 \\ 1 & 0 \end{bmatrix} = \begin{bmatrix} 1 & 0 \\ 0 & 0 \end{bmatrix} = \begin{bmatrix} \bar{E}_1(0) \\ O_{1 \times 2} \end{bmatrix}$$

which implies that $P_1 = \begin{bmatrix} 1 & 0 \end{bmatrix}$, $K_1^1 = \begin{bmatrix} -1 & 1 \end{bmatrix}$, and

$$\begin{bmatrix} \gamma_1(x) \\ \bar{\gamma}_1(x) \end{bmatrix} = V_1 \begin{bmatrix} h_1(x) \\ h_2(x) \end{bmatrix} = \begin{bmatrix} P_1 \\ K_1^1 \end{bmatrix} \begin{bmatrix} x_1 \\ x_1 + x_2 \end{bmatrix} = \begin{bmatrix} x_1 \\ x_2 \end{bmatrix}.$$

Since $\hat{E}_1(x) \triangleq L_g\bar{\gamma}_1(x) = \begin{bmatrix} x_1 & 0 \end{bmatrix} \neq O_{1\times2}$, system (5.113) is, by Theorem 5.5, not input-output linearizable.   □

5.6 Feedback Linearization with Multi Output

In this section, we deal with the multi output version of Sect. 5.4. Consider the following nonlinear systems.

$$
\begin{aligned}
\dot{x}(t) &= f(x(t)) + g(x(t))u(t) \\
y(t) &= h(x(t))
\end{aligned}
\tag{5.114}
$$

where $x \in \mathbb{R}^n$, $u \in \mathbb{R}^m$, $y \in \mathbb{R}^q$, and $f(x)$, $g(x)$, and $h(x)$ are analytic functions with $f(0) = 0$ and $h(0) = 0$. Suppose that $(\kappa_1, \ldots, \kappa_m)$ is the Kronecker indices of system (5.114).

Definition 5.7 (*feedback linearization with output*)
System (5.114) is said to be feedback linearizable with output, if there exist a feedback $u = \alpha(x) + \beta(x)v$ and a state transformation $z = S(x)$ such that the closed-loop system satisfies, in $z-$coordinates, the following Brunovsky canonical form:

$$
\begin{aligned}
\dot{z} &= \begin{bmatrix} \hat{A}_{11} & O & \cdots & O \\ O & \hat{A}_{22} & \cdots & O \\ \vdots & \vdots & \ddots & \vdots \\ O & O & \cdots & \hat{A}_{mm} \end{bmatrix} z + \begin{bmatrix} \hat{B}_{11} & O & \cdots & O \\ O & \hat{B}_{22} & \cdots & O \\ \vdots & \vdots & \ddots & \vdots \\ O & O & \cdots & \hat{B}_{mm} \end{bmatrix} v \\
&= Az + Bv \\
y &= Cx
\end{aligned}
\tag{5.115}
$$

or

$$
\begin{aligned}
S_*\left(f(x) + g(x)\alpha(x)\right) + S_*\left(g\beta(x)v\right) &= Az + Bv \\
h \circ S^{-1}(z) &= Cz
\end{aligned}
\tag{5.116}
$$

where $\sum_{i=1}^{m} \kappa_i = n$ and

$$
\hat{A}_{ii} = \begin{bmatrix} 0 & 1 & 0 & \cdots & 0 & 0 \\ 0 & 0 & 1 & \cdots & 0 & 0 \\ \vdots & \vdots & \vdots & & \vdots & \vdots \\ 0 & 0 & 0 & \cdots & 0 & 1 \\ 0 & 0 & 0 & \cdots & 0 & 0 \end{bmatrix} (\kappa_i \times \kappa_i) ; \quad \hat{B}_{ii} = \begin{bmatrix} 0 \\ 0 \\ \vdots \\ 0 \\ 1 \end{bmatrix} (\kappa_i \times 1).
$$

Theorem 5.6 *Suppose that $q = m$ and $\displaystyle\sum_{i=1}^{n} \rho_i = n$. If*

$$
\text{rank}\left(\left.\begin{bmatrix} L_g L_f^{\rho_1 - 1} h_1(x) \\ \vdots \\ L_g L_f^{\rho_q - 1} h_q(x) \end{bmatrix}\right|_{x=0}\right) = m \tag{5.117}
$$

then system (5.114) is feedback linearizable with output.

Proof Suppose that $q = m$, $\displaystyle\sum_{i=1}^{n} \rho_i = n$, and (5.117) is satisfied. Then, we have, by (5.77), that for $1 \le i \le q$ and $1 \le k \le \rho_i - 1$,

$$
L_g L_f^{k-1} h_i(x) = 0 \text{ and rank}\left(\left.\begin{bmatrix} L_g L_f^{\rho_1 - 1} h_1(x) \\ \vdots \\ L_g L_f^{\rho_q - 1} h_q(x) \end{bmatrix}\right|_{x=0}\right) = m.
$$

Thus, conditions of Lemma 4.3 are satisfied with $S_{i1}(x) = h_i(x)$ and $\kappa_i = \rho_i$ for $1 \le i \le m = q$. Therefore, by Lemma 4.3, system (5.114) is feedback linearizable with state transformation

$$
z = S(x) = \begin{bmatrix} h_1(x) \\ \vdots \\ L_f^{\rho_1 - 1} h_1(x) \\ \vdots \\ h_q(x) \\ \vdots \\ L_f^{\rho_q - 1} h_q(x) \end{bmatrix} = \begin{bmatrix} z_{11} \\ \vdots \\ z_{1\rho_1} \\ \vdots \\ z_{q1} \\ \vdots \\ z_{q\rho_q} \end{bmatrix}
$$

and feedback

$$
u = \begin{bmatrix} L_g L_f^{\rho_1 - 1} h_1(x) \\ \vdots \\ L_g L_f^{\rho_q - 1} h_q(x) \end{bmatrix}^{-1} \left(-\begin{bmatrix} L_f^{\rho_1} h_1(x) \\ \vdots \\ L_f^{\rho_q} h_q(x) \end{bmatrix} + v \right)
$$

$$
= \alpha(x) + \beta(x)v.
$$

Since $\tilde{h}_i = h_i \circ S^{-1}(z) = z_{i1}$, $1 \leq i \leq q$, it is easy to see that (5.116) is satisfied

with $C = \begin{bmatrix} 1\,0\cdots 0\cdots 0\,0\cdots 0 \\ \vdots\,\vdots \quad \vdots \quad \vdots\,\vdots \quad \vdots \\ 0\,0\cdots 0\cdots 1\,0\cdots 0 \end{bmatrix}$. □

Example 5.6.1 Use Theorem 5.6 to show that the following nonlinear system is feedback linearizable with output:

$$\begin{bmatrix} \dot{x}_1 \\ \dot{x}_2 \\ \dot{x}_3 \end{bmatrix} = \begin{bmatrix} x_2^2 \\ x_3 + x_1^2 \\ 0 \end{bmatrix} + \begin{bmatrix} 1 + x_1 & 0 \\ 0 & 0 \\ 0 & 1 \end{bmatrix} u = f(x) + g(x)u$$

$$y = \begin{bmatrix} x_1 \\ x_2 \end{bmatrix} = h(x).$$

(5.118)

Solution It is easy to see that $L_{f+gu}h_2(x) = x_3 + x_1^2$ and

$$\begin{bmatrix} y_1^{(1)} \\ y_2^{(2)} \end{bmatrix} = \begin{bmatrix} L_f h_1(x) \\ L_f^2 h_2(x) \end{bmatrix} + \begin{bmatrix} L_g h_1(x) \\ L_g L_f h_2(x) \end{bmatrix} \begin{bmatrix} u_1 \\ u_2 \end{bmatrix}$$

$$= \begin{bmatrix} x_2^2 \\ 2x_1 x_2^2 \end{bmatrix} + \begin{bmatrix} 1 + x_1 & 0 \\ 2x_1(1 + x_1) & 1 \end{bmatrix} \begin{bmatrix} u_1 \\ u_2 \end{bmatrix}$$

which implies that $(\rho_1, \rho_2) = (1, 2)$, $\rho_1 + \rho_2 = 3 = n$, and (5.117) is satisfied. Hence, by Theorem 5.6, system (5.118) is feedback linearizable with output. Let

$$z = S(x) = \begin{bmatrix} h_1(x) \\ h_2(x) \\ L_f h_2(x) \end{bmatrix} = \begin{bmatrix} x_1 \\ x_2 \\ x_3 + x_1^2 \end{bmatrix}$$

and

$$u = \begin{bmatrix} 1 + x_1 & 0 \\ 2x_1(1 + x_1) & 1 \end{bmatrix}^{-1} \left(-\begin{bmatrix} x_2^2 \\ 2x_1 x_2^2 \end{bmatrix} + \begin{bmatrix} v_1 \\ v_2 \end{bmatrix} \right)$$

$$= \begin{bmatrix} -\frac{x_2^2}{1+x_1} \\ 0 \end{bmatrix} + \begin{bmatrix} \frac{1}{1+x_1} & 0 \\ -2x_1 & 1 \end{bmatrix} \begin{bmatrix} v_1 \\ v_2 \end{bmatrix} = \alpha(x) + \beta(x)v.$$

Then we have that

$$\begin{bmatrix} \dot{z}_1 \\ \dot{z}_2 \\ \dot{z}_3 \end{bmatrix} = \begin{bmatrix} 0\,0\,0 \\ 0\,0\,1 \\ 0\,0\,0 \end{bmatrix} z + \begin{bmatrix} 1\,0 \\ 0\,0 \\ 0\,1 \end{bmatrix} \begin{bmatrix} v_1 \\ v_2 \end{bmatrix}$$

$$y = \begin{bmatrix} 1\,0\,0 \\ 0\,1\,0 \end{bmatrix} z.$$

□

Lemma 5.2 *System (5.114) is feedback linearizable with output via state transformation $z = S(x)$ and nonsingular feedback $u = \alpha(x) + \beta(x)v$, if and only if*

(i) system (5.114) is input-output linearizable.
(ii)

$$
\dot{x} = f(x) + g(x)u
$$
$$
\bar{y} = \Gamma_{\bar{k}}(x) \triangleq \bar{h}(x)
$$
(5.119)

is feedback linearizable with output via state transformation $z = S(x)$ and nonsingular feedback $u = \alpha(x) + \beta(x)v$.

Proof Necessity. Suppose that system (5.114) is feedback linearizable with output via state transformation $z = S(x)$ and feedback $u = \alpha(x) + \beta(x)v$. In other words, we have that

$$
\tilde{f}(z) + \tilde{g}(z)v \triangleq S_* \left(\hat{f}(x) + \hat{g}(x)v \right) = Az + Bv
$$
$$
\tilde{h}(z) \triangleq h \circ S^{-1}(z) = Cz
$$

where $\hat{f}(x) = f(x) + g(x)\alpha(x)$ and $\hat{g}(x) = g(x)\beta(x)$. Then system (5.114) is input-output linearizable. It is also clear, by Remark 5.1, that

$$
\tilde{\bar{h}}(z) \triangleq \bar{h} \circ S^{-1}(z) = \Gamma_{\bar{k}} \circ S^{-1}(z) = \bar{C}z
$$

where $\bar{C}z$ is the function $\Gamma_{\bar{k}}(z)$ that is obtained by the structure algorithm for linear system

$$
\dot{z} = Az + Bv ; \quad y = Cz.
$$

Hence, system (5.119) is feedback linearizable with output via state transformation $z = S(x)$ and feedback $u = \alpha(x) + \beta(x)v$.

Sufficiency. Suppose that system (5.114) is input-output linearizable and system (5.119) is feedback linearizable with output via state transformation $z = S(x)$ and feedback $u = \alpha(x) + \beta(x)v$. In other words, we have that

$$
\tilde{f}(z) + \tilde{g}(z)v \triangleq S_* \left(\hat{f}(x) + \hat{g}(x)v \right) = Az + Bv
$$
$$
\tilde{\bar{h}}(z) \triangleq \bar{h} \circ S^{-1}(z) = \Gamma_{\bar{k}} \circ S^{-1}(z) = \bar{C}z
$$
(5.120)

where $\hat{f}(x) = f(x) + g(x)\alpha(x)$ and $\hat{g}(x) = g(x)\beta(x)$. Thus, it is clear, by (2.30), that for $1 \leq i \leq \bar{k}$ and $j \geq 0$,

$$
L_{\hat{g}} L_{\hat{f}}^j \bar{h}(x) = L_{\tilde{g}} L_{\tilde{f}}^j \tilde{\bar{h}}(z) \Big|_{z = S(x)} = \bar{C} A^j B = \text{const}
$$

and

$$L_{\hat{g}}L_{\hat{f}}^{j}\gamma_i(x) = \text{const.} \tag{5.121}$$

We need to show that

$$\tilde{h}(z) \triangleq h \circ S^{-1}(z) = Cz. \tag{5.122}$$

It will be shown that for $i \geq 0$,

$$L_{\hat{g}}L_{\hat{f}}^{i}h(x) = \text{const.} \tag{5.123}$$

It is clear, by (5.25) and (5.27), that (5.123) is satisfied for $0 \leq i \leq \rho - 2$. Also, it is easy to see, by (5.93), (5.103), (5.106), and (5.121), that for $2 \leq i \leq \bar{k}$,

$$L_{\hat{g}}Q_{i-1}(x) = \bar{K}_iL_{\hat{g}}L_{\hat{f}}\tilde{\Gamma}_{i-1}(x) - \bar{K}_iL_{\hat{g}}\Gamma_{i-1}(x) + K_i^iL_{\hat{g}}L_{\hat{f}}Q_{i-2}(x)$$
$$= \text{const}$$

and for $1 \leq i \leq \bar{k}$,

$$V_iV_{i-1}\cdots V_1L_{\hat{g}}L_{\hat{f}}^{\rho-2+i}h(x) = L_{\hat{g}}\left(V_iV_{i-1}\cdots V_1L_{\hat{f}}^{\rho-2+i}h(x)\right)$$
$$= \left[\begin{array}{c}L_{\hat{g}}\tilde{\Gamma}_i(x)\\L_{\hat{g}}\bar{\gamma}_i(x) + L_{\hat{g}}Q_{i-1}(x)\end{array}\right] = \left[\begin{array}{c}L_{\hat{g}}\tilde{\Gamma}_i(x)\\L_{\hat{g}}Q_{i-1}(x)\end{array}\right] = \text{const}$$

which implies that (5.123) is satisfied for $\rho - 1 \leq i \leq \rho - 2 + \bar{k}$. It is easy to see, by (5.94) and mathematical induction, that for $i \geq 1$,

$$L_{\hat{f}}^{i}\bar{\gamma}_{\bar{k}}(x) = \bar{\gamma}_{\bar{k}+i}(x) - \sum_{j=1}^{i}\bar{K}_{\bar{k}+j}L_{\hat{f}}^{i-j}\Gamma_{\bar{k}}(x)$$

which implies, together with (5.94) and (5.121), that for $i \geq 1$,

$$L_{\hat{g}}L_{\hat{f}}^{i}\bar{\gamma}_{\bar{k}}(x) = L_{g}\bar{\gamma}_{\bar{k}+i}(x)\beta(x) - \sum_{j=1}^{i}\bar{K}_{\bar{k}+j}L_{\hat{g}}L_{\hat{f}}^{i-j}\Gamma_{\bar{k}}(x) = \text{const.} \tag{5.124}$$

Finally, we have, by (5.103), (5.106), (5.121), and (5.124), that for $i \geq 1$,

$$L_{\hat{g}}L_{\hat{f}}^{i}Q_{\bar{k}-1}(x) = \bar{K}_iL_{\hat{g}}L_{\hat{f}}^{i+1}\tilde{\Gamma}_{\bar{k}-1}(x) - \bar{K}_iL_{\hat{g}}L_{\hat{f}}^{i}\Gamma_{\bar{k}-1}(x) + K_i^iL_{\hat{g}}L_{\hat{f}}^{i+1}Q_{\bar{k}-2}(x)$$
$$= \text{const}$$

and

$$V_{\bar{k}}V_{\bar{k}-1}\cdots V_1 L_{\hat{g}}L_{\hat{f}}^{\rho-2+\bar{k}+i}h(x) = L_{\hat{g}}L_{\hat{f}}^i\left(V_{\bar{k}}V_{\bar{k}-1}\cdots V_1 L_{\hat{f}+g\bar{\alpha}}^{\rho-2+\bar{k}}h(x)\right)$$

$$= \begin{bmatrix} L_{\hat{g}}L_{\hat{f}}^i\tilde{\Gamma}_{\bar{k}}(x) \\ L_{\hat{g}}L_{\hat{f}}^i\bar{\gamma}_{\bar{k}}(x) + L_{\hat{g}}L_{\hat{f}}^i Q_{\bar{k}-1}(x) \end{bmatrix} = \text{const}$$

which implies that (5.123) is satisfied for $i \geq \rho - 1 + \bar{k}$. Since (5.123) is satisfied, it is clear, by Example 2.4.16, that for $1 \leq j \leq m$ and $i \geq 0$,

$$L_{\text{ad}_f^i \hat{g}_j} h(x) = \text{const}$$

which implies, together with (2.30), that for $1 \leq j \leq m$ and $i \geq 0$,

$$L_{S_*(\text{ad}_f^i \hat{g}_j)}\tilde{h}(z) = L_{\text{ad}_f^i \hat{g}_j}h(x)\Big|_{x=S^{-1}(z)} = \text{const.} \tag{5.125}$$

Note, by (2.38), (5.115), and (5.120), that

$$\begin{bmatrix} (-1)^{\kappa_1-1}S_*(\text{ad}_f^{\kappa_1-1}\hat{g}_1(x)) & \cdots & S_*(\hat{g}_1(x)) & \cdots \\ & (-1)^{\kappa_m-1}S_*(\text{ad}_f^{\kappa_m-1}\hat{g}_m(x)) & \cdots & S_*(\hat{g}_m(x)) \end{bmatrix}$$

$$= \begin{bmatrix} A^{\kappa_1-1}b_1 & \cdots & b_1 & \cdots & A^{\kappa_m-1}b_m & \cdots & b_m \end{bmatrix} = I_n$$

which implies, together with (5.125), that

$$\frac{\partial \tilde{h}(z)}{\partial z} = \frac{\partial \tilde{h}(z)}{\partial z}\Big[(-1)^{\kappa_1-1}S_*(\text{ad}_f^{\kappa_1-1}\hat{g}_1(x)) \cdots S_*(\hat{g}_1(x)) \cdots$$

$$(-1)^{\kappa_m-1}S_*(\text{ad}_f^{\kappa_m-1}\hat{g}_m(x)) \cdots S_*(\hat{g}_m(x))\Big]$$

$$= \Big[(-1)^{\kappa_1-1}L_{S_*(\text{ad}_f^{\kappa_1-1}\hat{g}_1)}\tilde{h}(z) \cdots L_{S_*(\hat{g}_1)}\tilde{h}(z) \cdots$$

$$(-1)^{\kappa_m-1}L_{S_*(\text{ad}_f^{\kappa_m-1}\hat{g}_m)}\tilde{h}(z) \cdots L_{S_*(\hat{g}_m)}\tilde{h}(z)\Big] = \text{const} \triangleq C.$$

Therefore, (5.122) is satisfied. Hence, system (5.114) is feedback linearizable with output via state transformation $z = S(x)$ and feedback $u = \alpha(x) + \beta(x)v$. $\square$

Suppose that system (5.114) is feedback linearizable with output. Then, it is clear that system (5.114) is input-output linearizable. Thus, by structure algorithm, we can obtain

$$\Gamma_{\bar{k}}(x) = \begin{bmatrix} \gamma_1(x) \\ \gamma_2(x) \\ \vdots \\ \gamma_{\bar{k}}(x) \end{bmatrix}$$

such that

$$\text{rank}\left(\left.L_g\Gamma_{\bar{k}}(x)\right|_{x=0}\right) = \text{rank}\left(\bar{E}_{\bar{k}}(0)\right) = \sum_{i=1}^{\bar{k}} \sigma_i = \bar{\sigma}_{\bar{k}}. \tag{5.126}$$

By column operation, we can obtain a nonsingular constant $m \times m$ matrix R_1 such that

$$L_g\Gamma_{\bar{k}}(x)R_1 = \bar{E}_{\bar{k}}(x)R_1 = \left[\bar{E}_{\bar{k}}^1(x) \ \bar{E}_{\bar{k}}^2(x)\right] \tag{5.127}$$

where $\bar{\sigma}_{\bar{k}} \times \bar{\sigma}_{\bar{k}}$ matrix $\bar{E}_{\bar{k}}^1(0)$ is invertible. Let

$$u = \bar{\alpha}(x) + \bar{\beta}(x)w \tag{5.128}$$

where

$$\bar{\beta}(x) = R_1 \begin{bmatrix} \left(\bar{E}_{\bar{k}}^1(x)\right)^{-1} & -\left(\bar{E}_{\bar{k}}^1(x)\right)^{-1}\bar{E}_{\bar{k}}^2(x) \\ O_{(m-\bar{\sigma}_{\bar{k}})\times\bar{\sigma}_{\bar{k}}} & I_{m-\bar{\sigma}_{\bar{k}}} \end{bmatrix} \triangleq R_1 R_2(x) \tag{5.129}$$

and

$$\bar{\alpha}(x) = -\bar{\beta}(x)\begin{bmatrix} L_f\Gamma_{\bar{k}}(x) \\ O_{(m-\bar{\sigma}_{\bar{k}})\times\bar{\sigma}_{\bar{k}}} \end{bmatrix}. \tag{5.130}$$

Then, it is easy to see that

$$L_{\bar{g}}\Gamma_{\bar{k}}(x) = L_g\Gamma_{\bar{k}}(x)\bar{\beta}(x) = \left[I_{\bar{\sigma}_{\bar{k}}} \ O_{\bar{\sigma}_{\bar{k}}\times(m-\bar{\sigma}_{\bar{k}})}\right] \tag{5.131}$$

and

$$L_{\bar{f}}\Gamma_{\bar{k}}(x) = L_f\Gamma_{\bar{k}}(x) + L_g\Gamma_{\bar{k}}(x)\bar{\alpha}(x) = O \tag{5.132}$$

where

$$\bar{f}(x) = f(x) + g(x)\bar{\alpha}(x) \tag{5.133}$$

and

$$\bar{g}(x) = g(x)\bar{\beta}(x) \triangleq \left[\bar{g}^1(x) \ \bar{g}^2(x)\right]. \tag{5.134}$$

In other words, we have that

$$\frac{d}{dt}\Gamma_{\bar{k}}(x(t)) = L_{\bar{f}}\Gamma_{\bar{k}}(x) + L_{\bar{g}}\Gamma_{\bar{k}}(x)w$$

$$= \begin{bmatrix} I_{\bar{\sigma}_{\bar{k}}} & O_{\bar{\sigma}_{\bar{k}} \times (m - \bar{\sigma}_{\bar{k}})} \end{bmatrix} \begin{bmatrix} w^1 \\ w^2 \end{bmatrix} = w^1.$$

Consider the following closed-loop system with output $\bar{y}$ and nonsingular feedback $u = \bar{\alpha}(x) + \bar{\beta}(x)w$:

$$\dot{x} = \bar{f}(x) + \bar{g}(x(t))w$$
$$\bar{y} = \Gamma_{\bar{k}}(x) \triangleq \bar{h}(x). \tag{5.135}$$

Suppose that $(\bar{k}_1, \ldots, \bar{k}_m)$ is the Kronecker indices of system (5.135). The following Corollary is a direct consequence of Lemma 5.2.

Corollary 5.1 *System (5.114) is feedback linearizable with output via state transformation $z = S(x)$ and nonsingular feedback $u = \alpha(x) + \beta(x)v$, if and only if*

(i) *system (5.114) is input-output linearizable.*
(ii) *system (5.135) is feedback linearizable with output via state transformation $z = S(x)$ and nonsingular feedback*

$$w = \bar{\beta}(x)^{-1}(\alpha(x) - \bar{\alpha}(x)) + \bar{\beta}(x)^{-1}\beta(x)v$$
$$\triangleq \hat{\alpha}(x) + \hat{\beta}(x)v. \tag{5.136}$$

Theorem 5.7 (conditions for feedback linearization with output)
Let $\bar{\sigma}_{\bar{k}} = m$. System(5.114) is feedback linearizable with output, if and only if

(i) *system (5.114) is input-output linearizable or for $1 \le i \le \bar{k}$,*

$$\hat{E}_i(x) \triangleq L_g\bar{\gamma}_i(x) = O_{(q - \bar{\sigma}_i) \times m}$$

(ii) $\sum_{j=1}^{m} \bar{k}_j = n$
(iii) *for $1 \le i \le m$, $1 \le j \le m$, $1 \le \ell_i \le \bar{k}_i + 1$, and $1 \le \ell_j \le \bar{k}_j + 1$,*

$$\left[\text{ad}_{\bar{f}}^{\ell_i - 1}\bar{g}_i(x), \ \text{ad}_{\bar{f}}^{\ell_j - 1}\bar{g}_j(x) \right] = 0 \tag{5.137}$$

where $\bar{f}(x)$ and $\bar{g}(x)$ are given in (5.133) and (5.134).

Proof Necessity. Suppose that system (5.114) is feedback linearizable with output via state transformation $z = S(x)$ and nonsingular feedback $u = \alpha(x) + \beta(x)v$. Then, by Corollary 5.1, condition (i) is satisfied and system (5.135) is feedback linearizable with output via state transformation $z = S(x)$ and nonsingular feedback (5.136). Thus, it is clear that condition (ii) is satisfied and

$$\tilde{f}(z) \triangleq S_*(\hat{f}(x)) = Az ; \quad \tilde{g}(z) \triangleq \left[S_*(\hat{g}_1(x)) \cdots S_*(\hat{g}_m(x)) \right] = B$$

$$\tilde{h}(z) \triangleq \bar{h} \circ S^{-1}(z) = \bar{C}z \tag{5.138}$$

where

$$\hat{f}(x) = \bar{f}(x) + \bar{g}(x)\hat{\alpha}(x) \text{ and } \hat{g}(x) = \bar{g}(x)\hat{\beta}(x). \tag{5.139}$$

Let

$$z = S(x) = \left[S_{11}(x) \cdots S_{1\kappa_1}(x) \cdots S_{m1}(x) \cdots S_{m\kappa_m}(x) \right]^{\mathsf{T}}.$$

Since $S_*(\bar{f}(x) + \bar{g}(x)\hat{\alpha}(x) + \bar{g}(x)\hat{\beta}(x)v) = Az + Bv$ by (5.138), it is clear that for $1 \le i \le m$,

$$\begin{bmatrix} \frac{\partial S_{i1}(x)}{\partial x} \\ \vdots \\ \frac{\partial S_{i(\kappa_i-1)}(x)}{\partial x} \\ \frac{\partial S_{i\kappa_i}(x)}{\partial x} \end{bmatrix} \left\{ \bar{f}(x) + \bar{g}(x)\hat{\alpha}(x) + \bar{g}(x)\hat{\beta}(x)v \right\} = \begin{bmatrix} S_{i2}(x) \\ \vdots \\ S_{i\kappa_i}(x) \\ v_i \end{bmatrix}.$$

Thus, it is easy to see that for $1 \le i \le m$ and $2 \le k \le \kappa_i$,

$$S_{ik}(x) = L_{\bar{f}}^{k-1} S_{i1}(x); \quad L_{\bar{g}} L_{\bar{f}}^{k-2} S_{i1}(x) = 0 \tag{5.140}$$

and

$$\hat{\beta}(x) = \begin{bmatrix} L_{\bar{g}} S_{1\kappa_1}(x) \\ \vdots \\ L_{\bar{g}} S_{m\kappa_m}(x) \end{bmatrix}^{-1} ; \quad \hat{\alpha}(x) = -\hat{\beta}(x) \begin{bmatrix} L_{\bar{f}} S_{1\kappa_1}(x) \\ \vdots \\ L_{\bar{f}} S_{m\kappa_m}(x) \end{bmatrix}. \tag{5.141}$$

(Refer to Lemma 4.3.) Also, it is easy to see, by Example 2.4.14, (5.131), (5.138), and (5.139), that

$$\hat{\beta}(x) = L_{\bar{g}} \bar{h}(x) \hat{\beta}(x) = L_{\hat{g}} \bar{h}(x) = L_{\tilde{g}} \tilde{h}(z) \Big|_{z=S(x)}$$

$$= \bar{C}B = \begin{bmatrix} \bar{c}_{1\kappa_1}^1 \cdots \bar{c}_{m\kappa_m}^1 \\ \vdots \quad\quad \vdots \\ \bar{c}_{1\kappa_1}^m \cdots \bar{c}_{m\kappa_m}^m \end{bmatrix} \tag{5.142}$$

and

$$\bar{g}(x) = \hat{g}(x) \left(\bar{C}B \right)^{-1} \tag{5.143}$$

where

$$\bar{C} = \begin{bmatrix} \bar{c}_{11}^1 & \cdots & \bar{c}_{1\kappa_1}^1 & \cdots & \bar{c}_{m1}^1 & \cdots & \bar{c}_{m\kappa_m}^1 \\ \vdots & & \vdots & & & & \\ \bar{c}_{11}^m & \cdots & \bar{c}_{1\kappa_1}^m & \cdots & \bar{c}_{m1}^m & \cdots & \bar{c}_{m\kappa_m}^m \end{bmatrix}. \tag{5.144}$$

Thus, we have, by (5.138), that

$$\bar{h}(x) = \bar{C}S(x) = \begin{bmatrix} \displaystyle\sum_{i=1}^m \sum_{j=1}^{\kappa_i} \bar{c}_{ij}^1 S_{ij}(x) \\ \vdots \\ \displaystyle\sum_{i=1}^m \sum_{j=1}^{\kappa_i} \bar{c}_{ij}^m S_{ij}(x) \end{bmatrix}$$

$$= \begin{bmatrix} \bar{c}_{1\kappa_1}^1 & \cdots & \bar{c}_{m\kappa_m}^1 \\ \vdots & & \vdots \\ \bar{c}_{1\kappa_1}^m & \cdots & \bar{c}_{m\kappa_m}^m \end{bmatrix} \begin{bmatrix} S_{1\kappa_1}(x) \\ \vdots \\ S_{m\kappa_m}(x) \end{bmatrix} + \begin{bmatrix} \displaystyle\sum_{i=1}^m \sum_{j=1}^{\kappa_i-1} \bar{c}_{ij}^1 S_{ij}(x) \\ \vdots \\ \displaystyle\sum_{i=1}^m \sum_{j=1}^{\kappa_i-1} \bar{c}_{ij}^m S_{ij}(x) \end{bmatrix}$$

which implies, together with (5.139)–(5.144), that

$$0 = L_{\bar{f}}\bar{h}(x) = \bar{C}B \begin{bmatrix} L_{\bar{f}}S_{1\kappa_1}(x) \\ \vdots \\ L_{\bar{f}}S_{m\kappa_m}(x) \end{bmatrix} + \begin{bmatrix} \displaystyle\sum_{i=1}^m \sum_{j=1}^{\kappa_i-1} \bar{c}_{ij}^1 S_{i(j+1)}(x) \\ \vdots \\ \displaystyle\sum_{i=1}^m \sum_{j=1}^{\kappa_i-1} \bar{c}_{ij}^m S_{i(j+1)}(x) \end{bmatrix}$$

$$= -\hat{\alpha}(x) + \bar{C}AS(x)$$

and

$$\bar{f}(x) = \hat{f}(x) - \bar{g}(x)\hat{\alpha}(x)$$
$$= \hat{f}(x) - \hat{g}(x)\left(\bar{C}B\right)^{-1}\bar{C}AS(x). \tag{5.145}$$

Therefore, it is easy to see, by (2.49), (5.143), and (5.145), that

$$\left[S_*(\bar{g}_1(x)) \cdots S_*(\bar{g}_m(x))\right] = \left[S_*(\hat{g}_1(x)) \cdots S_*(\hat{g}_m(x))\right] \left(\bar{C}B\right)^{-1}$$
$$= B\left(\bar{C}B\right)^{-1} \triangleq \bar{B} \tag{5.146}$$

and

$$S_*\left(\bar{f}(x)\right) = S_*\left(\hat{f}(x)\right)$$
$$- \left[S_*(\hat{g}_1(x)) \cdots S_*(\hat{g}_m(x))\right] \left(\bar{C}B\right)^{-1} \bar{C}AS(x)\Big|_{x=S^{-1}(z)} \tag{5.147}$$
$$= Az - B\left(\bar{C}B\right)^{-1}\bar{C}Az = \left\{A - B\left(\bar{C}B\right)^{-1}\bar{C}A\right\}z \triangleq \bar{A}z.$$

Hence, by Example 2.4.14, (5.146), and (5.147), condition (iii) is satisfied.

Sufficiency. Suppose that condition (i)–(iii) of Theorem 5.7 are satisfied. Then, by Theorem 2.7, there exists a state transformation $z = S(x)$ such that for $1 \le i \le m$ and $1 \le j \le \kappa_i$,

$$S_*\left(\mathrm{ad}_{\bar{f}}^{j-1} \bar{g}_i(x)\right) = \frac{\partial}{\partial z_{ij}} \tag{5.148}$$

or

$$\frac{\partial S(x)}{\partial x}\left[\bar{g}_1 \; \mathrm{ad}_{\bar{f}}\bar{g}_1 \cdots \mathrm{ad}_{\bar{f}}^{\kappa_1-1}\bar{g}_1 \cdots \bar{g}_m \cdots \mathrm{ad}_{\bar{f}}^{\kappa_m-1}\bar{g}_m\right] = I$$

where

$$z = \left[z_{11} \cdots z_{1\kappa_1} \cdots z_{m1} \cdots z_{m\kappa_m}\right]^{\mathsf{T}}.$$

Thus, it is clear that

$$S_*(\bar{g}_i(x)) = \frac{\partial}{\partial z_{i1}} \triangleq \bar{b}_i. \tag{5.149}$$

It is also easy to see that

$$S_*(\bar{f}(x)) = \bar{A}z \tag{5.150}$$

for some constant matrix $\bar{A}$. (Refer to the sufficiency part of Theorem 3.2.) Also, it is easy to see, by Example 2.4.16, (5.131), and (5.132), that for $1 \le k \le m$, $1 \le i \le m$ and $1 \le j \le \kappa_i$,

$$L_{\bar{g}_i}L_{\bar{f}}^{j-1}\bar{h}_k(x) = \begin{cases} 1, & \text{if } i = k \text{ and } j = 1 \\ 0, & \text{otherwise} \end{cases}$$

and

$$L_{ad_{\bar{f}}^{j-1}\bar{g}_i}\bar{h}_k(x) = \begin{cases} 1, & \text{if } i = k \text{ and } j = 1 \\ 0, & \text{otherwise} \end{cases}$$

which implies, together with Example 2.4.14 and (5.148), that for $1 \leq k \leq m$, $1 \leq i \leq m$ and $1 \leq j \leq \kappa_i$,

$$\frac{\partial}{\partial z_{ij}}\tilde{\bar{h}}_k(z) = L_{S_*\left(ad_{\bar{f}}^{j-1}\bar{g}_i\right)}\bar{h}_k \circ S^{-1}(z) = L_{ad_{\bar{f}}^{j-1}\bar{g}_i}\bar{h}_k(x)\bigg|_{x=S^{-1}(z)}$$

$$= \begin{cases} 1, & \text{if } i = k \text{ and } j = 1 \\ 0, & \text{otherwise} \end{cases}$$

and

$$\tilde{\bar{h}}_k(z) = \bar{h}_k \circ S^{-1}(z) = \begin{bmatrix} z_{11} \\ \vdots \\ z_{m1} \end{bmatrix} \triangleq \hat{C}z \qquad (5.151)$$

where $\tilde{\bar{h}}(z) \triangleq \bar{h} \circ S^{-1}(z)$. Therefore, it is clear, by (5.149), (5.150), and (5.151), that system (5.135) is state equivalent to a controllable linear MIMO system via state transformation $z = S(x)$. It is well-known that there exist a nonsingular matrices P, G, and an $m \times n$ matrix F such that

$$P^{-1}(\bar{A} + \bar{B}F)P = A \text{ and } P^{-1}\bar{B}G = B.$$

(Refer to Problem 5–13.) In other words, system (5.135) is feedback linearizable with output via state transformation $z = P^{-1}S(x)$ and nonsingular feedback $w = FS(x) + Gv$. Hence, by Corollary 5.1, system (5.114) is feedback linearizable with output via state transformation $z = P^{-1}S(x)$ and nonsingular feedback

$$u = \alpha(x) + \beta(x)v = \bar{\alpha}(x) + \bar{\beta}(x)\left(FS(x) + Gv\right).$$

□

Example 5.6.2 Use Theorem 5.7 to show that the following nonlinear system is feedback linearizable with output:

$$\begin{bmatrix} \dot{x}_1 \\ \dot{x}_2 \\ \dot{x}_3 \end{bmatrix} = \begin{bmatrix} x_2 \\ x_2^2 \\ x_1 \end{bmatrix} + \begin{bmatrix} 0 \\ 1 \\ 0 \end{bmatrix} u = f(x) + g(x)u$$

$$y = \begin{bmatrix} x_1 \\ x_2 \end{bmatrix} = h(x). \qquad (5.152)$$

Solution Let us consider the structure algorithm for system (5.152). Since

$$L_g h(x) = \begin{bmatrix} 0 \\ 1 \end{bmatrix}$$

we have that $\rho \triangleq \min(\rho_1, \rho_2) = 1$ and rank $\left(L_g h(x)\big|_{x=0} \right) = 1 = \sigma_1 < 2$. Thus, we obtain, by elementary row operations, constant matrix V_1 such that

$$V_1 \, L_g h(x)\big|_{x=0} = \begin{bmatrix} 0 & 1 \\ 1 & 0 \end{bmatrix} \begin{bmatrix} 0 \\ 1 \end{bmatrix} = \begin{bmatrix} 1 \\ 0 \end{bmatrix} = \begin{bmatrix} \bar{E}_1(0) \\ 0 \end{bmatrix}$$

which implies that $P_1 = \begin{bmatrix} 0 & 1 \end{bmatrix}$, $K_1^1 = \begin{bmatrix} 1 & 0 \end{bmatrix}$, and

$$\begin{bmatrix} \gamma_1(x) \\ \bar{\gamma}_1(x) \end{bmatrix} = V_1 \begin{bmatrix} h_1(x) \\ h_2(x) \end{bmatrix} = \begin{bmatrix} P_1 \\ K_1^1 \end{bmatrix} \begin{bmatrix} x_1 \\ x_2 \end{bmatrix} = \begin{bmatrix} x_2 \\ x_1 \end{bmatrix}.$$

Since

$$\hat{E}_1(x) \triangleq L_g \bar{\gamma}_1(x) = 0 \tag{5.153}$$

we go to step 2. Note that $L_f \bar{\gamma}_1(x) = x_2$ and

$$\text{rank}\left(\begin{bmatrix} \bar{E}_1(x) \\ L_g L_f \bar{\gamma}_1(x) \end{bmatrix}\bigg|_{x=0} \right) = \text{rank}\left(\begin{bmatrix} 1 \\ 1 \end{bmatrix} \right) = 1 = \bar{\sigma}_2.$$

Since $\bar{\sigma}_2 = 1 < q$, we obtain, by elementary row operations, constant matrix V_2 such that

$$V_2 \begin{bmatrix} \bar{E}_1(x) \\ L_g L_f \bar{\gamma}_1(x) \end{bmatrix}\bigg|_{x=0} = \begin{bmatrix} 1 & 0 \\ -1 & 1 \end{bmatrix} \begin{bmatrix} 1 \\ 1 \end{bmatrix} = \begin{bmatrix} 1 \\ 0 \end{bmatrix} = \begin{bmatrix} \bar{E}_1(0) \\ 0 \end{bmatrix}$$

which implies that $\begin{bmatrix} K_1^2 & K_2^2 \end{bmatrix} = \begin{bmatrix} -1 & 1 \end{bmatrix}$, and

$$\bar{\gamma}_2(x) = \begin{bmatrix} K_1^2 & K_2^2 \end{bmatrix} \begin{bmatrix} \gamma_1(x) \\ L_f \bar{\gamma}_1(x) \end{bmatrix} = \begin{bmatrix} -1 & 1 \end{bmatrix} \begin{bmatrix} x_2 \\ x_2 \end{bmatrix} = 0.$$

Since $\bar{\sigma}_2 = \bar{\sigma}_3 = \cdots$, the algorithm does not end in a finite step. Thus, we have that the final step $\bar{k} = 1$ and

$$\begin{bmatrix} \Gamma_1(x) \\ \bar{\gamma}_1(x) \end{bmatrix} = \begin{bmatrix} x_2 \\ 0 \end{bmatrix}.$$

It is clear, by Theorem 5.5 and (5.153), that system (5.152) is input-output linearizable. Therefore, we have, by (5.129) and (5.130), that

$$\bar{\beta}(x) = \left(L_g\Gamma_1(x)\right)^{-1} = 1$$

and

$$\bar{\alpha}(x) = -\bar{\beta}(x)L_f\Gamma_1(x) = -x_2^2$$

which imply that

$$\bar{f}(x) = f(x) + g(x)\bar{\alpha}(x) = \begin{bmatrix} x_2 \\ 0 \\ x_1 \end{bmatrix} \quad \text{and} \quad \bar{g}(x) = g(x)\bar{\beta}(x) = \begin{bmatrix} 0 \\ 1 \\ 0 \end{bmatrix}.$$

It is easy to see that

$$\left[\mathrm{ad}_{\bar{f}}\bar{g}(x) \ \mathrm{ad}_{\bar{f}}^2\bar{g}(x) \ \mathrm{ad}_{\bar{f}}^3\bar{g}(x) \right] = \begin{bmatrix} -1 & 0 & 0 \\ 0 & 0 & 0 \\ 0 & 1 & 0 \end{bmatrix}$$

which implies that $\bar{\kappa}_1 = 3 = n$ and condition (ii) of Theorem 5.7 is satisfied. It is also easy to see that condition (iii) of Theorem 5.7 is satisfied. Hence, by Theorem 5.7, system (5.152) is feedback linearizable with output.   □

Example 5.6.3 Use Theorem 5.7 to show that the following nonlinear system is not feedback linearizable with output:

$$\begin{bmatrix} \dot{x}_1 \\ \dot{x}_2 \\ \dot{x}_3 \end{bmatrix} = \begin{bmatrix} x_2 \\ x_2^2 \\ x_1 + x_2^2 \end{bmatrix} + \begin{bmatrix} 0 \\ 1 \\ x_1 \end{bmatrix} u = f(x) + g(x)u \tag{5.154}$$

$$y = \begin{bmatrix} x_1 \\ x_2 \end{bmatrix} = h(x).$$

Solution By the structure algorithm, we have that $\bar{k} = 1$ and

$$\begin{bmatrix} \Gamma_1(x) \\ \bar{\gamma}_1(x) \end{bmatrix} = \begin{bmatrix} x_2 \\ 0 \end{bmatrix}.$$

Therefore, we have, by (5.129) and (5.130), that

$$\bar{\beta}(x) = \left(L_g\Gamma_1(x)\right)^{-1} = 1; \quad \bar{\alpha}(x) = -\bar{\beta}(x)L_f\Gamma_1(x) = -x_2^2$$

which imply that

$$\bar{f}(x) = f(x) + g(x)\bar{\alpha}(x) = \begin{bmatrix} x_2 \\ 0 \\ x_1 + x_2^2(1 - x_1) \end{bmatrix}; \quad \bar{g}(x) = g(x)\bar{\beta}(x) = \begin{bmatrix} 0 \\ 1 \\ x_1 \end{bmatrix}.$$

It is easy to see that

$$\left[\text{ad}_{\bar{f}}\bar{g}(x) \ \text{ad}_{\bar{f}}^2\bar{g}(x) \ \text{ad}_{\bar{f}}^3\bar{g}(x) \right] = \begin{bmatrix} -1 & 0 & 0 \\ 0 & 0 & 0 \\ x_2(2x_1-1) & 1+x_2^2 & 0 \end{bmatrix}$$

which implies that $\bar{\kappa}_1 = 3 = n$ and condition (ii) of Theorem 5.7 is satisfied. However, we have that

$$\left[\bar{g}(x), \text{ad}_{\bar{f}}\bar{g}(x) \right] = \begin{bmatrix} 0 \\ 0 \\ 2x_1 \end{bmatrix} \neq \begin{bmatrix} 0 \\ 0 \\ 0 \end{bmatrix}$$

which implies that condition (iii) of Theorem 5.7 is not satisfied. Hence, by Theorem 5.7, system (5.152) is not feedback linearizable with output. □

5.7 MATLAB Programs

In this section, the following subfunctions in Appendix C are needed:
adfg, adfgk, adfgM, adfgkM, CharacterNum, ChExact, ChZero, ChCommute, ChConst, Codi, Delta, Kindex0, Lfh, Lfhk, RowReorder, TauFG

MATLAB program for Theorem 5.1:

```
clear all
syms x1 x2 x3 x4 x5 x6 x7 x8 x9 real

f=[0; x1*cos(x2)^2]; g=[1; x1-x1]; h=2*x1+tan(x2); %Ex:5.2.2

% f=[0; x1*cos(x2)^2]; g=[1; x1-x1]; h=x2; %Ex:5.2.3

% f=[x2; x1^2]; g=[x1-x1; 1]; h=x1; %Ex:5.2.4

% f=[x2+x3^2; x3; 0]; g=[2*x3; -2*x3; 1]; h=x1; %P:5.2(a)

f=simplify(f)
g=simplify(g)
h=simplify(h)
[n,m]=size(g); x=sym('x',[n,1]);

T(:,1)=g;
for k=2:n+1
    T(:,k)=adfg(f,T(:,k-1),x);
end
T=simplify(T)
BD=T(:,1:n)
```

```
if rank(BD) < n
  display('condition (i) of Thm 5.1 is not satisfied.')
  return
end

if ChCommute(T,x) == 0
  display('condition (ii) of Thm 5.1 is not satisfied.')
  return
end

tC=simplify(Lfh(BD,h,x))
if ChConst(tC,x) == 0
  display('condition (iii) of Thm 5.1 is not satisfied.')
  return
end

dS=simplify(inv(BD));

display('By Thm 5.1, state equivalent to a LS with')

S=Codi(dS,x)
AS=simplify(dS*f);
dAS=simplify(jacobian(AS,x));
A=simplify(dAS*BD)
B=simplify(dS*g)
C=tC

return
```

MATLAB program for Theorem 5.2:

```
clear all
syms x1 x2 x3 x4 x5 x6 x7 x8 x9 real

f=[-2*x2*(x1+x2+x2^2); x1+x2+x2^2; -2*x2*(x1+x2+x2^2)];
g=[1 x1-x1; 0 0; 0 1]; h=[x1+x2^2; x2+x2^2+x3]; %Ex:5.3.1

% f=[x2+x3^2; x3; 0];
% g=[2*x3; -2*x3; 1]; h=[x1-x3^2; x3]; %P:5.2(b)

% f=[x2; x4; x4+3*x2^2*x4; 0];
% g=[0 2*x4; 1 0; 3*x2^2 0; 0 1]; h=x2; %P:5.2(c)

% g=[0 2*x4; 1 0; 3*x2^2 0; 0 1];
% f=[x2; x4; x4+3*x2^2*x4; 0]; h=[x1-x4^2; x3-x2^3]; %P:5.2(d)

f=simplify(f)
g=simplify(g)
h=simplify(h)
[n,m]=size(g); x=sym('x',[n,1]);
```

```
[ka,D]=Kindex0(f,g,x)
if sum(ka) < n
  display('condition (i) of Thm 5.2 is not satisfied.')
  return
end

BDD=TauFG(f,g,x,ka+1)
if ChCommute(BDD,x) == 0
  display('condition (ii) of Thm 5.2 is not satisfied.')
  return
end

BD=TauFG(f,g,x,ka)
tC=simplify(Lfh(BD,h,x))
if ChConst(tC,x) == 0
  display('condition (iii) of Thm 5.2 is not satisfied.')
  return
end

display('By Thm 5.2, state equivalent to a LS with')

dS=simplify(inv(BD));
S=Codi(dS,x)
AS=simplify(dS*f);
dAS=simplify(jacobian(AS,x));
A=simplify(dAS*BD)
B=simplify(dS*g)
C=tC

return
```

MATLAB program for Theorem 5.3:

```
clear all
syms x1 x2 x3 x4 x5 x6 x7 x8 x9 real

f=[0; x1*cos(x2)^2]; g=[1; x1-x1]; h=x2; %Ex:5.4.3

% f=[x2; x3; x1^2]; g=[0; 0; exp(x1)]; h=2*x1+x2; %Ex:5.4.4

% f=[x2; x3; x1^2]; g=[0; 0; exp(x1)];
% h=2*x1+exp(x2)-1; %Ex:5.4.5

% f=[x2; x3; x1^2]; g=[0; 0; exp(x1)]; h=sin(x1)+x2; %Ex:5.4.6

% f=[x2; x1^2]; g=[x1-x1; 1]; h=x1; %Ex:P:5.3

% f=[x2; x1^2]; g=[x1-x1; 1]; h=x2+x1^2; %Ex:P:5.4

% f=[x2+x3^2; x3; 0]; g=[2*x3; -2*x3; 1]; h=x1; %P:5.5(a)

% f=[x2; x3; x1^2]; g=[x1-x1; 0; 1]; h=x2; %P:5.5(b)
```

```
% f=[x2; x3; x1^2]; g=[x1-x1; 0; 1]; h=x2+x1^2; %P:5.5(c)

f=simplify(f)
g=simplify(g)
h=simplify(h)
[n,m]=size(g); x=sym('x',[n,1]);

T(:,1)=g;
for k=2:n
  T(:,k)=adfg(f,T(:,k-1),x);
end
T=simplify(T)

BD=T(:,1:n)
if rank(BD) < n
  display('condition (i) of Thm 5.3 is not satisfied.')
  return
end

rho=CharacterNum(f,g,h,x)
CON2=Lfh(T(:,rho),h,x)
CON20=simplify(subs(CON2,x,x-x))
if ChZero(CON20) == 1
  display('condition (ii) of Thm 5.3 is not satisfied.')
  return
end

beta=(-1)^(rho-1)/CON2
iBD=simplify(inv(BD))
omega=(-1)^(n-1)/beta*iBD(n,:)
if ChExact(omega,x) == 0
  display('condition (iii) of Thm 5.3 is not satisfied.')
  return
end

S1=Codi(omega,x)
S=x-x; S(1)=S1;
for k=2:n
  S(k)=Lfh(f,S(k-1),x);
end
S=simplify(S)

tt2=Lfhk(f,S1,x,n);
alpha=-tt2*beta
hf=simplify(f+alpha*g)
hg=simplify(beta*g)

hT(:,1)=hg;
for k=2:n
  hT(:,k)=adfg(hf,hT(:,k-1),x);
end
hT=simplify(hT)
```

```
CON4=simplify(Lfh(hT,h,x))
if ChConst(CON4,x) == 0
  display('condition (iv) of Thm 5.3 is not satisfied.')
  return
end

display('By Thm 5.3, feedback linearizable with output.')

dS=simplify(jacobian(S,x));
idS=simplify(inv(dS));
AS=simplify(dS*hf);
dAS=simplify(jacobian(AS,x));
A=simplify(dAS*idS)
B=simplify(dS*hg)
dh=simplify(jacobian(h,x));
C=simplify(dh*idS)

return
```

MATLAB program for Theorem 5.4:

```
clear all
syms x1 x2 x3 x4 x5 x6 x7 x8 x9 real

f=[0; x1*cos(x2)^2]; g=[1; x1-x1]; h=x2; %Ex:5.4.3

% f=[x2; x3; x1^2]; g=[0; 0; exp(x1)]; h=2*x1+x2; %Ex:5.4.4

% f=[x2; x3; x1^2]; g=[0; 0; exp(x1)];
% h=2*x1+exp(x2)-1; %Ex:5.4.5

% f=[x2; x3; x1^2]; g=[0; 0; exp(x1)]; h=sin(x1)+x2; %Ex:5.4.6

% f=[x2; x1^2]; g=[x1-x1; 1]; h=x1; %Ex:P:5.3

% f=[x2; x1^2]; g=[x1-x1; 1]; h=x2+x1^2; %Ex:P:5.4

% f=[x2+x3^2; x3; 0]; g=[2*x3; -2*x3; 1]; h=x1; %P:5.5(a)

% f=[x2; x3; x1^2]; g=[x1-x1; 0; 1]; h=x2; %P:5.5(b)

% f=[x2; x3; x1^2]; g=[x1-x1; 0; 1]; h=x2+x1^2; %P:5.5(c)

f=simplify(f)
g=simplify(g)
h=simplify(h)
[n,m]=size(g); x=sym('x',[n,1]);

T(:,1)=g;
for k=2:n
  T(:,k)=adfg(f,T(:,k-1),x);
end
T
T=simplify(T)
```

```
T0=subs(T,x,x-x)
if rank(T0) < n
  display('condition (i) of Thm 5.4 is not satisfied.')
  return
end

rho=CharacterNum(f,g,h,x)
CON2=Lfh(T(:,rho),h,x)
CON20=simplify(subs(CON2,x,x-x))
if ChZero(CON20) == 1
  display('condition (ii) of Thm 5.4 is not satisfied.')
  return
end

beta=(-1)^(rho-1)/CON2
talpha=Lfhk(f,h,x,rho);
alpha=-beta*talpha
bf=simplify(f+alpha*g)
hg=simplify(beta*g)

hT(:,1)=hg;
for k=2:n+1
  hT(:,k)=adfg(bf,hT(:,k-1),x);
end
hT=simplify(hT)

if ChCommute(hT,x)==0
  display('condition (iii) of Thm 5.4 is not satisfied.')
  return
end

display('By Thm 5.4, feedback linearizable with output.')

idS=hT(:,1:n);
dS=inv(idS);
S=Codi(dS,x)

AS=simplify(dS*bf);
dAS=simplify(jacobian(AS,x));
A=simplify(dAS*idS)
B=simplify(dS*hg)
dh=simplify(jacobian(h,x));
C=simplify(dh*idS)

return
```

The following is a MATLAB subfunction program for Theorem 5.5.

```
function [r,V]=RowOperation(D)

q=size(D,1); r=rank(D); Iq=eye(q);
if r==q
  V=Iq;
  return
end
R0=RowReorder(D);
D1=R0*D;
t1=D1(1:r,:);
t2=D1((r+1):q,:);
K=t2*t1'*inv(t1*t1');
R1=Iq;
R1((r+1):q,1:r)=-K;
V=R1*R0;
```

The following is a MATLAB subfunction program for Theorem 5.5. (Refer to Structure Algorithm.)

```
function [flag,kf,GAMMA,bargammak]=StructureA(f,g,h,x)

flag=1; bargammak=x(1)-x(1);
n=size(g,1); q=length(h);
rho=CharacterNum(f,g,h,x);
T1=Lfhk(f,h,x,rho-1);
E=Lfh(g,T1,x);
E0=subs(E,x,x-x);
[s1,V]=RowOperation(E0);
VT1=V*T1;
VE=V*E;
GAMMA=VT1(1:s1);
kf=1;
if s1==q
  return
end
bargamma(1:q-s1)=VT1((s1+1):q);
hatSi=VE((s1+1):q,:)
if ChZero(hatSi) == 0
  flag=0;
  return
end
if s1>0
  oldbargammak=bargamma(1:q-s1);
end
s=s1;
for k1=2:n
  T1=[GAMMA; Lfh(f,bargamma(1:q-s1),x)];
  E=Lfh(g,T1,x);
  E0=subs(E,x,x-x);
  [s1,V]=RowOperation(E0);
```

```
  VT1=V*T1;
  VE=V*E;
  GAMMA=VT1(1:s1);
  kf=k1;
  olds=max(s);
  s=[s; s1];
  if s1==q
    return
  end
  bargamma(1:q-s1)=VT1((s1+1):q);
  hatSi=VE((s1+1):q,:)
  if ChZero(hatSi) == 0
    flag=0;
    return
  end
  if s1>max(s)
    oldbargammak=bargamma(1:q-s1);
  end
  s=[s; s1];
end
for k2=1:n
  if s(k2)==s(n)
    kf=k2;
    bargammak=oldbargammak;
    return
  end
end
```

MATLAB program for Theorem 5.5:

```
clear all
syms x1 x2 x3 x4 x5 x6 x7 x8 x9 real

f=[x2^2; x3; 0];
g=[1+x1 0; 0 0; 0 1]; h=[x1; x1+x2]; %Ex:5.5.6

% f=[x1^2; x1; x3^2];
% g=[1+x1 1; 0 0; 0 1]; h=[x1; x1+x2]; %Ex:5.5.8

% f=[x1^2; x3; x2^2];
% g=[1 0; x1 0; 0 1]; h=[x1; x1+x2]; %Ex:5.5.9

% f=[x1^2; x3; x3^2];
% g=[1+x1 1; 0 0; x1 x2]; h=[x1; x1+x2]; %Ex:P5-10

% f=[x2; x1^2; x4; x5; x3^2];
% g=[x1-x1 0; 1 0; 0 0; 0 0; 0 1]; h=[x1; x4]; %Ex:P5-11a

% f=[x2; x1^2; x4; x5; x3^2];
% g=[x1-x1 0; 1 0; 0 0; 0 0; 0 1]; h=[x1; x3]; %Ex:P5-11b

% f=[x2; x1^2; x4; x5; x3^2];
```

```
% g=[x1-x1 0; 1 0; 0 0; 0 0; 0 1]; h=[x1; 2*x1+x3]; %Ex:P5-11c

% f=[x2; x1^2; x4; x5; x3^2];
% g=[x1-x1 0; 1 0; 0 0; 0 0; 0 1]; h=[x1; x1^2+x3]; %Ex:P5-11d

% f=[x2; x1^2; x4; x5; x3^2];
% g=[x1-x1 0; 1 0; 0 0; 0 0; 0 1]; h=[x1+x5; x3]; %Ex:P5-11e

% f=[x1^2+x4; x3+x1*x4; x2^2; 0];
% g=[x1-x1 0; 0 0; 0 1; 1 0]; h=[x1; x1+x2]; %Ex:P5-12

f=simplify(f)
g=simplify(g)
h=simplify(h)
[n,m]=size(g); x=sym('x',[n,1]); u=sym('u',[m,1]);

[flag,kf,GAMMA,bargammak]=StructureA(f,g,h,x)

if flag==0
  display('By Thm 5.5, NOT locally i-o linearizable.')
  return
end

display('By Thm 5.5, system is locally i-o linearizable.')

bsk=length(GAMMA);
D=Lfh(g,GAMMA,x);
D0=subs(D,x,x-x);
L1=RowReorder(D0');
R1=L1';
bD=D*R1;
bS1=bD(:,1:bsk);
bS2=bD(:,bsk+1:m);
R2=jacobian(u,u);
R2(1:bsk,:)=[inv(bS1) -inv(bS1)*bS2];
beta=R1*R2
t3=Lfh(f,GAMMA,x);
alpha=beta(:,1)-beta(:,1);
alpha(1:bsk)=t3;
alpha=-beta*alpha

fc=simplify(f+g*alpha)
gc=simplify(g*beta)

Tc=h;
for k=2:n
  Tc=[Tc; Lfhk(fc,h,x,k-1)];
end
Tc=simplify(Tc)
cc=Lfh(gc,Tc,x)

return
```

5.8 Problems

5–1 Solve Example 5.2.1.

5–2 Find out whether the following nonlinear control systems are state equivalence to a linear system with output or not. If it is state equivalence to a linear system with output, find a linearizing state transformation.

(a)

$$
\begin{bmatrix} \dot{x}_1 \\ \dot{x}_2 \\ \dot{x}_3 \end{bmatrix} = \begin{bmatrix} x_2 + x_3^2 \\ x_3 \\ 0 \end{bmatrix} + \begin{bmatrix} 2x_3 \\ -2x_3 \\ 1 \end{bmatrix} u ; \quad y = x_1
$$

(b)

$$
\begin{bmatrix} \dot{x}_1 \\ \dot{x}_2 \\ \dot{x}_3 \end{bmatrix} = \begin{bmatrix} x_2 + x_3^2 \\ x_3 \\ 0 \end{bmatrix} + \begin{bmatrix} 2x_3 \\ -2x_3 \\ 1 \end{bmatrix} u ; \quad \begin{bmatrix} y_1 \\ y_2 \end{bmatrix} = \begin{bmatrix} x_1 - x_3^2 \\ x_3 \end{bmatrix}
$$

(c)

$$
\begin{bmatrix} \dot{x}_1 \\ \dot{x}_2 \\ \dot{x}_3 \\ \dot{x}_4 \end{bmatrix} = \begin{bmatrix} x_2 \\ x_4 \\ x_4 + 3x_2^2 x_4 \\ 0 \end{bmatrix} + \begin{bmatrix} 0 & 2x_4 \\ 1 & 0 \\ 3x_2^2 & 0 \\ 0 & 1 \end{bmatrix} \begin{bmatrix} u_1 \\ u_2 \end{bmatrix} ; \quad y = x_2
$$

(d)

$$
\begin{bmatrix} \dot{x}_1 \\ \dot{x}_2 \\ \dot{x}_3 \\ \dot{x}_4 \end{bmatrix} = \begin{bmatrix} x_2 \\ x_4 \\ x_4 + 3x_2^2 x_4 \\ 0 \end{bmatrix} + \begin{bmatrix} 0 & 2x_4 \\ 1 & 0 \\ 3x_2^2 & 0 \\ 0 & 1 \end{bmatrix} \begin{bmatrix} u_1 \\ u_2 \end{bmatrix} ; \quad \begin{bmatrix} y_1 \\ y_2 \end{bmatrix} = \begin{bmatrix} x_1 - x_4^2 \\ x_3 - x_2^3 \end{bmatrix}
$$

5–3 Use Theorem 5.3 or Theorem 5.4 to show that (5.13) is feedback linearizable with output.

5–4 Use Theorem 5.3 or Theorem 5.4 to show that (5.13) with output equation $y = h(x) = x_2 + x_1^2$ is not feedback linearizable with output.

5–5 Find out whether the following nonlinear control systems are feedback linearizable with output or not. If it is feedback linearizable with output, find a linearizing state transformation and feedback.

(a)

$$\begin{bmatrix} \dot{x}_1 \\ \dot{x}_2 \\ \dot{x}_3 \end{bmatrix} = \begin{bmatrix} x_2 + x_3^2 \\ x_3 \\ 0 \end{bmatrix} + \begin{bmatrix} 2x_3 \\ -2x_3 \\ 1 \end{bmatrix} u \; ; \quad y = x_1$$

(b)

$$\begin{bmatrix} \dot{x}_1 \\ \dot{x}_2 \\ \dot{x}_3 \end{bmatrix} = \begin{bmatrix} x_2 \\ x_3 \\ x_1^2 \end{bmatrix} + \begin{bmatrix} 0 \\ 0 \\ 1 \end{bmatrix} u \; ; \quad y = x_2$$

(c)

$$\begin{bmatrix} \dot{x}_1 \\ \dot{x}_2 \\ \dot{x}_3 \end{bmatrix} = \begin{bmatrix} x_2 \\ x_3 \\ x_1^2 \end{bmatrix} + \begin{bmatrix} 0 \\ 0 \\ 1 \end{bmatrix} u \; ; \quad y = x_2 + x_1^2$$

5–6 Solve Example 5.5.1.

5–7 Show that the relative degree of system (5.65) is invariant with nonsingular feedback.

5–8 Prove Lemma 5.1.

5–9 Show that the following system is input-output linearizable. Also, find the nonsingular feedback $u = \alpha(x) + \beta(x)v$ for the input-output linearization.

$$\begin{bmatrix} \dot{x}_1 \\ \dot{x}_2 \\ \dot{x}_3 \end{bmatrix} = \begin{bmatrix} x_2 \\ x_3 \\ x_1^2 \end{bmatrix} + \begin{bmatrix} 0 \\ 0 \\ 1 \end{bmatrix} u \; ; \quad y = x_2 + x_1^2$$

5–10 Show that the following system is not locally input-output linearizable.

$$\begin{bmatrix} \dot{x}_1 \\ \dot{x}_2 \\ \dot{x}_3 \end{bmatrix} = \begin{bmatrix} x_1^2 \\ x_3 \\ x_3^2 \end{bmatrix} + \begin{bmatrix} 1 + x_1 & 1 \\ 0 & 0 \\ x_1 & x_2 \end{bmatrix} u = f(x) + g(x)u$$

$$y = \begin{bmatrix} x_1 \\ x_1 + x_2 \end{bmatrix} = h(x)$$

5–11 Use Theorem 5.5 to determine whether the following system is input-output linearizable. If it is input-output linearizable, find the nonsingular feedback $u = \alpha(x) + \beta(x)v$ for the input-output linearization.

$$
\begin{bmatrix} \dot{x}_1 \\ \dot{x}_2 \\ \dot{x}_3 \\ \dot{x}_4 \\ \dot{x}_5 \end{bmatrix} = \begin{bmatrix} x_2 \\ x_1^2 \\ x_4 \\ x_5 \\ x_3^2 \end{bmatrix} + \begin{bmatrix} 0 & 0 \\ 1 & 0 \\ 0 & 0 \\ 0 & 0 \\ 0 & 1 \end{bmatrix} u \qquad (5.155)
$$

(a)

$$
\begin{bmatrix} y_1 \\ y_2 \end{bmatrix} = \begin{bmatrix} x_1 \\ x_4 \end{bmatrix}
$$

(b)

$$
\begin{bmatrix} y_1 \\ y_2 \end{bmatrix} = \begin{bmatrix} x_1 \\ x_3 \end{bmatrix}
$$

(c)

$$
\begin{bmatrix} y_1 \\ y_2 \end{bmatrix} = \begin{bmatrix} x_1 \\ 2x_1 + x_3 \end{bmatrix}
$$

(d)

$$
\begin{bmatrix} y_1 \\ y_2 \end{bmatrix} = \begin{bmatrix} x_1 \\ x_1^2 + x_3 \end{bmatrix}
$$

(e)

$$
\begin{bmatrix} y_1 \\ y_2 \end{bmatrix} = \begin{bmatrix} x_1 + x_5 \\ x_3 \end{bmatrix}
$$

5–12 For the system (5.113), consider the following dynamic feedback:

$$
u_1 = \eta ; \quad u_2 = w_2;
$$
$$
\dot{\eta} = w_1
$$

Then we have the following extended system:

$$
\begin{bmatrix} \dot{x}_1 \\ \dot{x}_2 \\ \dot{x}_3 \\ \dot{\eta} \end{bmatrix} = \begin{bmatrix} x_1^2 + \eta \\ x_3 + x_1\eta \\ x_2^2 \\ 0 \end{bmatrix} + \begin{bmatrix} 0 & 0 \\ 0 & 0 \\ 0 & 1 \\ 1 & 0 \end{bmatrix} w ; \quad \begin{bmatrix} y_1 \\ y_2 \end{bmatrix} = \begin{bmatrix} x_1 \\ x_1 + x_2 \end{bmatrix} . \qquad (5.156)
$$

Show that the extended system (5.156) is locally input-output linearizable. Find the nonsingular feedback $w = \alpha(x, \eta) + \beta(x, \eta)v$ for the input-output linearization. In other words, system (5.113) of Example 5.5.9 is locally input-output linearizable not by static feedback but by dynamic feedback. It is called the dynamic input-output linearization.

5–13 Suppose that $(\bar{A}, \bar{B})$ is a controllable pair. By using the controllable canonical form to show that there exist a nonsingular matrices P, G, and an $m \times n$ matrix F such that

$$P^{-1}(\bar{A} + \bar{B}F)P = A \quad \text{and} \quad P^{-1}\bar{B}G = B$$

where (A, B) is a Brunovsky canonical form in (5.115).

Chapter 6
Dynamic Feedback Linearization

6.1 Introduction

In Chap. 4, we have studied feedback linearization of the following affine nonlinear system:

$$\dot{x} = f(x) + \sum_{i=1}^{m} u_i g_i(x), \quad x \in \mathbb{R}^n. \qquad (6.1)$$

Some of the systems that cannot be linearized only by coordinate transformations can be linearized using feedback in addition to coordinate transformations. This chapter shows that more nonlinear systems can be linearized using the more general dynamic feedback than the static feedback used in Chap. 4. For example, consider system (4.80), which is not feedback linearizable, in Example 4.3.8.

$$\dot{x} = \begin{bmatrix} x_2 \\ 0 \\ 0 \end{bmatrix} + \begin{bmatrix} 0 \\ 1 \\ 0 \end{bmatrix} u_1 + \begin{bmatrix} x_2^2 \\ 0 \\ 1 \end{bmatrix} u_2 \\ = f(x) + g_1(x)u_1 + g_2(x)u_2. \qquad (6.2)$$

Since $\text{span}\{g_1(x), g_2(x)\}$ is not involutive, system (6.2) is not (static) feedback linearizable. Consider the following linear dynamic compensation:

$$\dot{z} = w_2 = A_{\mathrm{d}}z + B_{\mathrm{d}}w \\ \begin{bmatrix} u_1 \\ u_2 \end{bmatrix} = \begin{bmatrix} w_1 \\ z \end{bmatrix} = C_{\mathrm{d}}z + D_{\mathrm{d}}w \qquad (6.3)$$

where z and w are the state and new input of the dynamic compensation, respectively. Then the extended system of system (6.2) with dynamic compensation (6.3) is as follows.

© The Author(s), under exclusive license to Springer Nature Singapore Pte Ltd. 2022
H.-G. Lee, *Linearization of Nonlinear Control Systems*,
https://doi.org/10.1007/978-981-19-3643-2_6

$$\begin{bmatrix} \dot{x}_1 \\ \dot{x}_2 \\ \dot{x}_3 \\ \dot{z} \end{bmatrix} = \begin{bmatrix} x_2 + x_2^2 z \\ 0 \\ z \\ 0 \end{bmatrix} + \begin{bmatrix} 0 \\ 1 \\ 0 \\ 0 \end{bmatrix} w_1 + \begin{bmatrix} 0 \\ 0 \\ 0 \\ 1 \end{bmatrix} w_2 \tag{6.4}$$
$$= f_E(x, z) + g_{E1}(x, z) w_1 + g_{E2}(x, z) w_2.$$

Since span$\{g_{E1}(x_E), g_{E2}(x_E)\}$ and span$\{g_{E1}(x_E), g_{E2}(x_E), \mathrm{ad}_{f_E} g_{E1}(x_E),$
$\mathrm{ad}_{f_E} g_{E2}(x_E)\}$ are involutive distributions, it is easy to see, by Theorem 4.3, that
extended system (6.4) is feedback linearizable with state transformation
$\xi = S_E(x, z) = [x_1 \ \ x_2 + x_2^2 z \ \ x_3 \ \ z]^T$ and feedback

$$\begin{bmatrix} w_1 \\ w_2 \end{bmatrix} = \begin{bmatrix} \frac{1}{1+2x_2 z} & \frac{-x_2^2}{1+2x_2 z} \\ 0 & 1 \end{bmatrix} \begin{bmatrix} v_1 \\ v_2 \end{bmatrix}. \tag{6.5}$$

Extended system (6.4) satisfies, in the extended new states ξ, the following linear
system:

$$\begin{bmatrix} \dot{\xi}_1 \\ \dot{\xi}_2 \\ \dot{\xi}_3 \\ \dot{\xi}_4 \end{bmatrix} = \begin{bmatrix} 0 & 1 & 0 & 0 \\ 0 & 0 & 0 & 0 \\ 0 & 0 & 0 & 1 \\ 0 & 0 & 0 & 0 \end{bmatrix} \begin{bmatrix} \xi_1 \\ \xi_2 \\ \xi_3 \\ \xi_4 \end{bmatrix} + \begin{bmatrix} 0 \\ 1 \\ 0 \\ 0 \end{bmatrix} v_1 + \begin{bmatrix} 0 \\ 0 \\ 0 \\ 1 \end{bmatrix} v_2 \tag{6.6}$$
$$= A_E \xi + b_{E1} v_1 + b_{E2} v_2.$$

In other words, system (6.2) can be linearized by extended state transformation
$\xi = S_E(x, z) = [x_1 \ \ x_2 + x_2^2 z \ \ x_3 \ \ z]^T$ and dynamic feedback

$$\dot{z} = v_2 \tag{6.7a}$$

$$\begin{bmatrix} u_1 \\ u_2 \end{bmatrix} = \begin{bmatrix} 0 \\ z \end{bmatrix} + \begin{bmatrix} \frac{1}{1+2x_2 z} & \frac{-x_2^2}{1+2x_2 z} \\ 0 & 0 \end{bmatrix} \begin{bmatrix} v_1 \\ v_2 \end{bmatrix}. \tag{6.7b}$$

If we consider dynamic feedback

$$\dot{z} = a(x, z) + b(x, z)v, \quad z \in \mathbb{R}^d \tag{6.8a}$$
$$u = c(x, z) + d(x, z)v, \quad v \in \mathbb{R}^m \tag{6.8b}$$

then the extended system of system (6.1) can be obtained as follows.

$$\dot{x}_E = \begin{bmatrix} \dot{x} \\ \dot{z} \end{bmatrix} = \begin{bmatrix} f(x) + g(x)c(x, z) \\ a(x, z) \end{bmatrix} + \begin{bmatrix} g(x)d(x, z) \\ b(x, z) \end{bmatrix} v \tag{6.9}$$
$$= f_E(x_E) + g_E(x_E)v.$$

$$\dot{\xi} = A_E \xi + B_E v$$

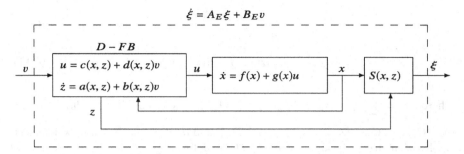

Fig. 6.1 Dynamic feedback linearization

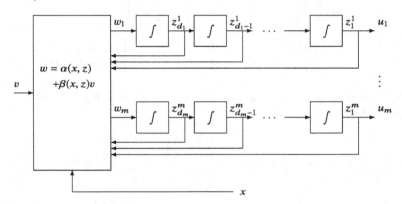

Fig. 6.2 Restricted dynamic feedback

Definition 6.1 (*dynamic feedback linearization*) System (6.1) is said to be locally dynamic feedback linearizable, if there exist $z^0 \in \mathbb{R}^d$, a neighborhood U_z of z^0, a neighborhood U_x of $0 \in \mathbb{R}^n$, a regular dynamic feedback (6.8), and an extended state transformation $\xi = S_E(x, z) = S_E(x_E) : U_x \times U_z \to \mathbb{R}^{n+d}$ such that the extended system (6.9) satisfies, in the new extended state ξ,

$$\dot{\xi} = A_E \xi + B_E v, \quad \xi \in \mathbb{R}^{n+d}$$

where A_E and B_E are Brunovsky canonical form.

Block diagram for dynamic feedback linearization is given in Fig. 6.1. However, the conditions for dynamic feedback linearization problem are very complicated and necessary and sufficient conditions have not to be known. Some results on restricted dynamic feedback linearization have been reported and will be introduced in this chapter.

Definition 6.2 (*restricted dynamic feedback*) For system (6.1), restricted dynamic feedback with indices $\mathbf{d} = (d_1, \ldots, d_m)$ is defined by the following dynamic system:

$$\dot{z} = A_{\mathbf{d}}z + B_{\mathbf{d}}w$$
$$u = C_{\mathbf{d}}z + D_{\mathbf{d}}w \tag{6.10}$$

$$w = \alpha(x, z) + \beta(x, z)v \tag{6.11}$$

where $\quad d \triangleq \sum_{i=1}^{m} d_i, \quad A_{\mathbf{d}} = \text{diag}\{A_1, \ldots, A_m\}, \quad B_{\mathbf{d}} = \text{diag}\{B_1, \ldots, B_m\}, \quad C_{\mathbf{d}} =$
$\text{diag}\{C_1, \ldots, C_m\}, D_{\mathbf{d}} = \text{diag}\{D_1, \ldots, D_m\}$, and $d_i \times d_i$ matrix A_i, $d_i \times 1$ matrix B_i, $1 \times d_i$ matrix C_i, and 1×1 matrix D_i are defined by

$$A_i = \begin{bmatrix} 0 & I_{(d_i-1)\times(d_i-1)} \\ 0 & O_{1\times(d_i-1)} \end{bmatrix}; \quad B_i = \begin{bmatrix} O_{(d_i-1)\times1} \\ 1 \end{bmatrix}$$

$$C_i = \begin{cases} \begin{bmatrix} 1 & O_{1\times(d_i-1)} \end{bmatrix}, & \text{if } d_i \geq 1 \\ O_{1\times0}, & \text{if } d_i = 0 \end{cases}; \quad D_i = \begin{cases} 0, & \text{if } d_i \geq 1 \\ 1, & \text{if } d_i = 0. \end{cases}$$

Block diagram for restricted dynamic feedback is given in Fig. 6.2. For example, if $\mathbf{d} = (1, 0, 3)$, then

$$\dot{z} = A_{\mathbf{d}}z + B_{\mathbf{d}}w = \begin{bmatrix} 0 & 0 & 0 & 0 \\ 0 & 0 & 1 & 0 \\ 0 & 0 & 0 & 1 \\ 0 & 0 & 0 & 0 \end{bmatrix} z + \begin{bmatrix} 1 & 0 & 0 \\ 0 & 0 & 0 \\ 0 & 0 & 0 \\ 0 & 0 & 1 \end{bmatrix} w$$

$$u = C_{\mathbf{d}}z + D_{\mathbf{d}}w = \begin{bmatrix} 1 & 0 & 0 & 0 \\ 0 & 0 & 0 & 0 \\ 0 & 1 & 0 & 0 \end{bmatrix} z + \begin{bmatrix} 0 & 0 & 0 \\ 0 & 1 & 0 \\ 0 & 0 & 0 \end{bmatrix} w$$

where $z \triangleq [z_1^1 \ z_1^3 \ z_2^3 \ z_3^3]^{\mathsf{T}}$, $z^1 \triangleq z_1^1$, $z^3 \triangleq [z_1^3 \ z_2^3 \ z_3^3]^{\mathsf{T}}$, and $\dot{z}^i = A_i z^i + B_i w$, for $i = 1, 3$. Also, it is easy to see, by (6.10), that for $1 \leq i \leq m$,

$$u_i^{(d_i)} = w_i. \tag{6.12}$$

Definition 6.3 (*restricted dynamic feedback linearization*) System (6.1) is said to be locally restricted dynamic feedback linearizable with indices $\mathbf{d} = (d_1, \ldots, d_m)$, if there exist $z^0 \in \mathbb{R}^d$, a neighborhood U_z of z^0, a neighborhood U_x of $0 \in \mathbb{R}^n$, a restricted dynamic feedback with indices $\mathbf{d} = (d_1, \ldots, d_m)$, and an extended state transformation $\xi = S_E(x, z) = S_E(x_E) : U_x \times U_z \to \mathbb{R}^{n+d}$ such that the extended system satisfies, in the new extended state ξ,

$$\dot{\xi} = A_E\xi + B_E v, \quad \xi \in \mathbb{R}^{n+d} \tag{6.13}$$

where A_E and B_E are Brunovsky canonical form.

In other words, restricted dynamic feedback linearization is to find a linear dynamic compensator (6.10) such that the extended system

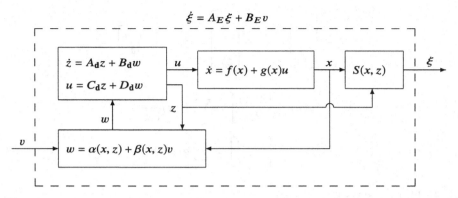

Fig. 6.3 Restricted dynamic feedback linearization

$$\dot{x}_E = \begin{bmatrix} \dot{x} \\ \dot{z} \end{bmatrix} = \begin{bmatrix} f(x) + g(x)C_\mathbf{d}z \\ A_\mathbf{d}z \end{bmatrix} + \begin{bmatrix} g(x)D_\mathbf{d} \\ B_\mathbf{d} \end{bmatrix} w$$
$$= F(x_E) + G(x_E)w$$

is locally (static) feedback linearizable on a neighborhood U_E ($= U_x \times U_z$) of $(x, z) = (0, z^0) \in \mathbb{R}^{n+d}$.

Block diagram for restricted dynamic feedback linearization is given in Fig. 6.3. Restricted dynamic feedback is composed of two parts, linear dynamic compensator (6.10) and extended state feedback (6.11). For this reason, restricted dynamic feedback linearization is sometimes called linearization by pure integrators followed by static feedback.

6.2 Preliminary

Suppose that system (6.1) is said to be restricted dynamic feedback linearizable with indices $\mathbf{d} = (d_1, \dots, d_m)$. Then extended system of system (6.1) with linear dynamic compensator (6.10)

$$\dot{x}_E = \begin{bmatrix} \dot{x} \\ \dot{z} \end{bmatrix} = \begin{bmatrix} f(x) + g(x)C_\mathbf{d}z \\ A_\mathbf{d}z \end{bmatrix} + \begin{bmatrix} g(x)D_\mathbf{d} \\ B_\mathbf{d} \end{bmatrix} w \qquad (6.14)$$
$$= F(x_E) + G(x_E)w$$

is (static) feedback linearizable on a neighborhood U_E of $(x, z) = (0, z^0)$. If we let $d \triangleq \sum_{j=1}^{m} d_j$ and for $1 \leq j \leq m$,

$$
z = \begin{bmatrix} z^1 \\ \vdots \\ z^m \end{bmatrix}, \quad z^j = \begin{bmatrix} z_1^j \\ \vdots \\ z_{d_j}^j \end{bmatrix}, \quad \bar{A}_\mathbf{d} \triangleq \begin{bmatrix} O_{n \times d} \\ A_\mathbf{d} \end{bmatrix}
$$

$$
\bar{f}(x_E) \triangleq \begin{bmatrix} f(x) \\ O_{d \times 1} \end{bmatrix} = \sum_{k=1}^{n} f_k(x) \frac{\partial}{\partial x_k}
$$

$$
\bar{g}_j(x_E) \triangleq \begin{bmatrix} g_j(x) \\ O_{d \times 1} \end{bmatrix} = \sum_{k=1}^{n} g_{j,k}(x) \frac{\partial}{\partial x_k}
$$

then it is easy to see that

$$
F(x_E) = \bar{f}(x, z) + \sum_{\substack{j=1 \\ d_j \geq 1}}^{m} z_1^j \bar{g}_j(x, z) + \bar{A}_\mathbf{d} z
$$

$$
G_j(x_E) = \begin{cases} \bar{g}_j(x_E), & \text{if } d_j = 0 \\ \frac{\partial}{\partial z_{d_j}^j}, & \text{if } d_j \geq 1. \end{cases}
$$

For example, if $\mathbf{d} = (0, 2)$, $z = \begin{bmatrix} z_1^2 \\ z_2^2 \end{bmatrix}$, $G_1(x_E) = \bar{g}_1(x, z)$, $G_2(x_E) = \frac{\partial}{\partial z_2^2}$, and

$$
F(x_E) = \bar{f}(x, z) + z_1^2 \bar{g}_2(x, z) + z_2^2 \frac{\partial}{\partial z_1^2} = \begin{bmatrix} f(x) + z_1^2 g_2(x) \\ z_2^2 \\ 0 \end{bmatrix}. \text{ For extended system}
$$

(6.14), let for $i \geq 0$,

$$
D_i(x_E) \triangleq \text{span} \left\{ \text{ad}_{F(x_E)}^k G_j(x_E) \mid 1 \leq j \leq m, \ 0 \leq k \leq i \right\}. \tag{6.15}
$$

Then, by Theorem 4.3, the following two conditions are satisfied:

(i) $\dim \left(D_{n+d-1}(0, z^0) \right) = n + d$
(ii) $D_i(x_E), i \geq 0$ are involutive distributions on U_E.

Also, define distributions $Q_i(x_E), \ i \geq 0$ by

$$
Q_0(x_E) \triangleq \text{span} \left\{ \bar{g}_j(x_E) \mid d_j = 0 \right\}
$$

$$
Q_i(x_E) \triangleq Q_{i-1}(x_E) + \text{ad}_F Q_{i-1}(x_E) + \text{span} \left\{ \bar{g}_j(x_E) \mid 1 \leq j \leq m, \ d_j = i \right\}
$$
$$
= \text{span} \left\{ \text{ad}_F^k \bar{g}_j(x_E) \mid 1 \leq j \leq m, \ 0 \leq k \leq i - d_j \right\}.
$$

$$\tag{6.16}$$

Example 6.2.1 Show the followings:

(a) for $1 \leq j \leq m$ and $k \geq 0$,

$$\mathrm{ad}_F^k G_j(x_E) = \begin{cases} (-1)^k \frac{\partial}{\partial z_{d_j-k}^j}, & \text{if } 0 \leq k < d_j \\ (-1)^{d_j} \mathrm{ad}_F^{k-d_j} \bar{g}_j(x_E), & \text{if } k \geq d_j. \end{cases} \quad (6.17)$$

(b) for $i \geq 0$,

$$D_i(x_E) = Q_i(x_E) + \text{span} \left\{ \frac{\partial}{\partial z_k^j} \,\Big|\, 1 \leq j \leq m, \ d_j \geq 1, \ d_j - i \leq k \leq d_j \right\}. \quad (6.18)$$

(c)

$$Q_i(x_E) \subset \text{span} \left\{ \frac{\partial}{\partial x_k} \,\Big|\, 1 \leq k \leq n \right\}. \quad (6.19)$$

Solution Omitted. (See Problem 6-1.) ☐

Example 6.2.2 Show the followings:

(a) For $1 \leq i \leq m$ and $p \geq 1$,

$$\mathrm{ad}_F^p \bar{g}_i(x_E) = X_p^i(x, z_{p-1}) + \sum_{\substack{j=1 \\ d_j \geq p}}^{m} z_p^j \mathrm{ad}_{\bar{g}_j} \bar{g}_i(x_E) \quad (6.20)$$

where $X_1^i(x) \triangleq \mathrm{ad}_{\bar{f}} \bar{g}_i(x_E)$,

$$z_p \triangleq \left\{ z_k^j \,\big|\, 1 \leq j \leq m, \ 1 \leq k \leq \min(p, d_j) \right\}$$

and for $p \geq 1$,

$$X_{p+1}^i(x, z_p) \triangleq \mathrm{ad}_F X_p^i(x, z_{p-1}) + \sum_{d_j \geq p} z_p^j \mathrm{ad}_F \mathrm{ad}_{\bar{g}_j} \bar{g}_i(x_E).$$

(b) For $1 \leq i \leq m$, $1 \leq j \leq m$, $p \geq 1$, and $p + 1 \leq k \leq d_j$,

$$\left[\frac{\partial}{\partial z_k^j}, \ \mathrm{ad}_F^p \bar{g}_i(x_E) \right] = 0. \quad (6.21)$$

(c) For $1 \le j \le m$, $i \ge 0$, and $k \ge i+1$,

$$\left[\frac{\partial}{\partial z_k^j}, Q_i(x_E) \right] \subset Q_i(x_E). \tag{6.22}$$

Solution It is clear that (6.20) is satisfied when $p = 1$. Since

$$L_F z_p^j = \begin{cases} z_{p+1}^j, & \text{if } p \le d_j - 1 \\ 0, & \text{if } p = d_j \end{cases}$$

it is easy to see that for $p \ge 1$,

$$\begin{aligned}
\mathrm{ad}_F^{p+1} \bar{g}_i &= \mathrm{ad}_F X_p^i + \sum_{d_j \ge p} \mathrm{ad}_F \left(z_p^j \mathrm{ad}_{\bar{g}_j} \bar{g}_i \right) \\
&= \mathrm{ad}_F X_p^i + \sum_{d_j \ge p} z_p^j \mathrm{ad}_F \mathrm{ad}_{\bar{g}_j} \bar{g}_i + \sum_{d_j \ge p} L_F(z_p^j) \mathrm{ad}_{\bar{g}_j} \bar{g}_i \\
&= X_{p+1}^i(x, \mathbf{z}_p) + \sum_{d_j \ge p+1} z_{p+1}^j \mathrm{ad}_{\bar{g}_j} \bar{g}_i
\end{aligned}$$

which implies that (6.20) is satisfied. Since $\mathrm{ad}_F^p \bar{g}_i$ is a function of x and $\mathbf{z}_p \left(= \left\{ z_k^j \mid 1 \le j \le m, \; k \le \min(p, d_j) \right\} \right)$ only by (6.20), it is easy to see, by (6.16), that (6.21) and (6.22) are satisfied. $\square$

Example 6.2.3 Use (6.18) and (6.19) to show that if $D_i(x_E)$, $i \ge 0$ are involutive distributions on a neighborhood U_E of $(x, z) = (0, z^0)$, then the following two conditions are satisfied:

(a) $Q_i(x_E)$, $i \ge 0$ are involutive distributions on U_E.
(b) For $1 \le j \le m$, $d_j \ge 1$, and $d_j - i \le k \le d_j$,

$$\left[\frac{\partial}{\partial z_k^j}, Q_i(x_E) \right] \subset Q_i(x_E). \tag{6.23}$$

Also, show that the converse is true.

Solution Obvious. (Refer to Problem 6-2.) $\square$

Lemma 6.1 *Suppose that* $\sigma \ge 0$, $\bar{g}_j(x_E) \in Q_\sigma(x_E)$, *and*

$$Q_\sigma(x_E) = Q_{\sigma+1}(x_E). \tag{6.24}$$

Then the followings are satisfied:

(i) for $k \geq 1$,

$$\mathrm{ad}_F^k \bar{g}_j(x_E) \in Q_\sigma(x_E). \tag{6.25}$$

(ii) for $1 \leq j \leq m$, $d_j \geq \sigma$ *and* $1 \leq k \leq d_j$,

$$\left[\frac{\partial}{\partial z_k^j}, Q_\sigma(x_E) \right] \subset Q_\sigma(x_E). \tag{6.26}$$

Proof Assume that that (6.25) is satisfied for $k = s$ and $s \geq 0$. Then we have that

$$\mathrm{ad}_F^{s+1} \bar{g}_j(x_E) \in [F(x_E), Q_\sigma(x_E)] \subset Q_{\sigma+1}(x_E) = Q_\sigma(x_E)$$

which implies that (6.25) is also satisfied for $k = s + 1$. Hence, by mathematical induction, (6.25) is satisfied. It is clear, by (6.22) and (6.24), that (6.26) is satisfied for $\sigma \leq k \leq d_j$. Suppose that $2 \leq s \leq \sigma$ and (6.26) is satisfied for $s \leq k \leq d_j$. In other words,

$$\left[\frac{\partial}{\partial z_s^j}, Q_\sigma(x_E) \right] \subset Q_\sigma(x_E). \tag{6.27}$$

Since $[F(x_E), Q_\sigma(x_E)] \subset Q_{\sigma+1}(x_E) = Q_\sigma(x_E)$ by (6.16), we have, by Jacobi identity (2.18) and (6.27), that for any vector field $\tau(x_E) \in Q_\sigma(x_E)$,

$$\left[\frac{\partial}{\partial z_{s-1}^j}, \tau \right] = -\left[\left[F, \frac{\partial}{\partial z_s^j} \right], \tau \right] = \left[\left[\frac{\partial}{\partial z_s^j}, \tau \right], F \right] + \left[\tau, F \right], \frac{\partial}{\partial z_s^j} \right]$$

$$= -\left[F, \left[\frac{\partial}{\partial z_s^j}, \tau \right] \right] + \left[\frac{\partial}{\partial z_s^j}, [F, \tau] \right]$$

$$\in [F, Q_\sigma] + \left[\frac{\partial}{\partial z_s^j}, Q_\sigma \right] \subset Q_\sigma(x_E)$$

which implies that (6.26) is also satisfied for $k = s - 1$. Hence, by mathematical induction, (6.26) is satisfied. $\qquad\square$

6.3 Restricted Dynamic Feedback Linearization

In this section, we derive the conditions of the restricted dynamic feedback linearization using the basic relations in the previous section and obtain the necessary and sufficient conditions that can be easily checked by using them.

Let $\mathbf{I} \subset \{1, 2, \ldots, m\}, \bar{\mathbf{I}} = \{1, 2, \ldots, m\} - \mathbf{I}$, and

$$\tilde{d}_i = \begin{cases} d_i, & i \in \mathbf{I} \\ d_i - 1, & i \in \bar{\mathbf{I}}. \end{cases}$$

In other words,

$$\mathbf{I} \triangleq \left\{ i \mid 1 \le i \le m, \ d_i = \tilde{d}_i \right\} \quad \text{and} \quad \bar{\mathbf{I}} \triangleq \left\{ i \mid 1 \le i \le m, \ d_i > \tilde{d}_i \right\}.$$

Let us denote the extended system of system (6.1) with linear dynamic compensator (6.10) with indices $\tilde{\mathbf{d}} = (\tilde{d}_1, \ldots, \tilde{d}_m)$ by

$$\dot{x}_E = \begin{bmatrix} \dot{x} \\ \dot{\tilde{z}} \end{bmatrix} = \begin{bmatrix} f(x) + g(x)C_{\tilde{\mathbf{d}}}\tilde{z} \\ A_{\tilde{\mathbf{d}}}\tilde{z} \end{bmatrix} + \begin{bmatrix} g(x)D_{\tilde{\mathbf{d}}} \\ B_{\tilde{\mathbf{d}}} \end{bmatrix} w$$
$$= \tilde{F}(\tilde{x}_E) + \tilde{G}(\tilde{x}_E)w \tag{6.28}$$

where for $1 \le j \le m$,

$$\tilde{z} = \begin{bmatrix} \tilde{z}^1 \\ \vdots \\ \tilde{z}^m \end{bmatrix} \quad \text{and} \quad \tilde{z}^j = \begin{bmatrix} \tilde{z}^j_1 \\ \vdots \\ \tilde{z}^j_{\tilde{d}_j} \end{bmatrix}.$$

If we let $\tilde{d} \triangleq \sum_{j=1}^{m} \tilde{d}_j$ and for $1 \le j \le m$,

$$\tilde{f}(\tilde{x}_E) \triangleq \begin{bmatrix} f(x) \\ O_{\tilde{d} \times 1} \end{bmatrix} = \sum_{k=1}^{n} f_k(x)\frac{\partial}{\partial x_k}$$
$$\tilde{g}_j(\tilde{x}_E) \triangleq \begin{bmatrix} g_j(x) \\ O_{\tilde{d} \times 1} \end{bmatrix} ; \quad \bar{A}_{\tilde{\mathbf{d}}} \triangleq \begin{bmatrix} O_{n \times \tilde{d}} \\ A_{\tilde{\mathbf{d}}} \end{bmatrix}$$

then it is easy to see that

$$\tilde{F}(\tilde{x}_E) = \tilde{f}(x, \tilde{z}) + \sum_{d_j \ge 1} \tilde{z}^j_1 \tilde{g}_j(x, \tilde{z}) + \bar{A}_{\tilde{\mathbf{d}}}\tilde{z}$$

$$\tilde{G}_j(\tilde{x}_E) = \begin{cases} \tilde{g}_j(\tilde{x}_E), & \text{if } \tilde{d}_j = 0 \\ \frac{\partial}{\partial \tilde{z}^j_{\tilde{d}_j}}, & \text{if } \tilde{d}_j \ge 1 \end{cases}$$

and for $i \geq 0$,

$$\tilde{D}_i = \text{span} \left\{ \text{ad}_{\tilde{F}}^{\ell} \tilde{G}_j(\tilde{x}_E) \mid 1 \leq j \leq m, \ 0 \leq \ell \leq i \right\}. \tag{6.29}$$

Let $\begin{bmatrix} x \\ \tilde{z} \end{bmatrix} = \pi(x, z)$ and $\begin{bmatrix} 0 \\ \tilde{z}^0 \end{bmatrix} = \pi(0, z^0)$, where canonical projection map $\pi : \mathbb{R}^{n+d} \to \mathbb{R}^{n+\tilde{d}}$ is defined by

$$\pi(x, z_1^1, \ldots, z_{d_1}^1, \ldots, z_1^m, \ldots, z_{d_m}^m) = (x, z_1^1, \ldots, z_{\tilde{d}_1}^1, \ldots, z_1^m, \ldots, z_{\tilde{d}_m}^m).$$

In other words,

$$\pi_* \left(\frac{\partial}{\partial x_k} \right) = \frac{\partial}{\partial x_k}, \ 1 \leq k \leq n$$

$$\pi_* \left(\frac{\partial}{\partial z_\ell^j} \right) = \frac{\partial}{\partial \tilde{z}_\ell^j}, \ 1 \leq j \leq m, \ 1 \leq \ell \leq \tilde{d}_j$$

$$\pi_* \left(\frac{\partial}{\partial z_{d_j}^j} \right) = 0, \ j \in \bar{\mathbf{I}}$$

and

$$\ker(\pi_*) = \text{span} \left\{ \frac{\partial}{\partial z_{d_j}^j} \ \middle| \ j \in \bar{\mathbf{I}} \right\}. \tag{6.30}$$

Let $\bar{z} \triangleq \{ z_{d_\ell}^\ell \mid \ell \in \bar{\mathbf{I}} \}$.

Lemma 6.2 *Suppose that extended system* (6.14) *is (static) feedback linearizable on a neighborhood U_E of $(x, z) = (0, z^0)$. Then the followings are satisfied:*

(i) $\pi_* \left(\text{ad}_F^k G_j(x_E) \big|_{\bar{z}=0} \right)$, $1 \leq j \leq m$, $k \geq 0$ *are well-defined vector fields on a neighborhood $\tilde{U}_E (= \pi(U_E))$ of $(x, \tilde{z}) = (0, \tilde{z}^0)(= \pi(0, z^0))$. In other words, for $j \in \mathbf{I}$ and $k \geq 0$,*

$$\pi_* \left(\text{ad}_F^k G_j(x_E) \big|_{\bar{z}=0} \right) = \text{ad}_{\tilde{F}}^k \tilde{G}_j(\tilde{x}_E) \tag{6.31}$$

and for $j \in \bar{\mathbf{I}}$,

$$\pi_* \left(\text{ad}_F^k G_j(x_E) \big|_{\bar{z}=0} \right) = \begin{cases} 0, & \text{if } k = 0 \\ -\text{ad}_{\tilde{F}}^{k-1} \tilde{G}_j(\tilde{x}_E), & \text{if } k \geq 1. \end{cases} \tag{6.32}$$

(ii) $\pi_* (D_i(x_E))$, $i \geq 0$ *are well-defined involutive distributions on a neighborhood $\tilde{U}_E (= \pi(U_E))$ of $(x, \tilde{z}) = (0, \tilde{z}^0)(= \pi(0, z^0))$. In other words, for $i \geq 0$,*

$$\pi_* \left(D_i(x_E) \right) = \text{span} \left\{ \text{ad}_{\tilde{F}}^k \tilde{G}_j(\tilde{x}_E) \,\middle|\, j \in \mathbf{I},\ 0 \leq k \leq i \right\}$$

$$+ \text{span} \left\{ \text{ad}_{\tilde{F}}^k \tilde{G}_j(\tilde{x}_E) \,\middle|\, j \in \bar{\mathbf{I}},\ 0 \leq k \leq i - 1 \right\}$$

$$= \pi_* \left(Q_i(x_E) \right) + \text{span} \left\{ \frac{\partial}{\partial \tilde{z}_\ell^j} \,\middle|\, j \in \mathbf{I},\ \tilde{d}_j - i \leq \ell \leq \tilde{d}_j \right\} \tag{6.33}$$

$$+ \text{span} \left\{ \frac{\partial}{\partial \tilde{z}_\ell^j} \,\middle|\, j \in \bar{\mathbf{I}},\ \tilde{d}_j + 1 - i \leq \ell \leq \tilde{d}_j \right\}$$

and

$$dim(\pi_* \left(D_{n+d-1}(0, z^0) \right)) = n + \tilde{d}. \tag{6.34}$$

(iii) $\pi_* \left(\text{ad}_F^k \bar{g}_j(x_E) \big|_{\bar{z}=0} \right)$, $1 \leq j \leq m$, $k \geq 0$ *are well-defined vector fields on* $\tilde{U}_E$ *and for* $1 \leq j \leq m$ *and* $k \geq 0$,

$$\pi_* \left(\text{ad}_F^k \bar{g}_j(x_E) \big|_{\bar{z}=0} \right) = \text{ad}_{\tilde{F}}^k \bar{g}_j(\tilde{x}_E). \tag{6.35}$$

(iv) $\pi_* \left(Q_i(x_E) \right)$, $i \geq 0$ *are well-defined involutive distributions on a neighborhood* $\hat{U}_E (= \pi(U_E))$ *and for* $i \geq 0$,

$$\pi_* \left(Q_i(x_E) \right) = \text{span} \left\{ \text{ad}_{\tilde{F}}^k \bar{g}_j(\tilde{x}_E) \,\middle|\, j \in \mathbf{I},\ 0 \leq k \leq i - \tilde{d}_j \right\}$$

$$+ \text{span} \left\{ \text{ad}_{\tilde{F}}^k \bar{g}_j(\tilde{x}_E) \,\middle|\, j \in \bar{\mathbf{I}},\ 0 \leq k \leq i - 1 - \tilde{d}_j \right\}. \tag{6.36}$$

Proof Suppose that extended system (6.14) is (static) feedback linearizable on a neighborhood U_E of $(x, z) = (0, z^0)$. Thus, by Theorem 4.3, we have that

(a) $\dim \left(D_{n+d-1}(0, z^0) \right) = n + d$
(b) $D_i(x_E)$, $i \geq 0$ are involutive distributions on U_E

where for $i \geq 0$,

$$D_i(x_E) = \text{span} \left\{ \text{ad}_F^k G_j(x_E) \,\middle|\, 1 \leq j \leq m,\ 0 \leq k \leq i \right\}$$

$$= Q_i(x_E) + \text{span} \left\{ \frac{\partial}{\partial z_k^j} \,\middle|\, 1 \leq j \leq m,\ d_j \geq 1,\ d_j - i \leq k \leq d_j \right\} \tag{6.37}$$

and

$$Q_i(x_E) = \text{span} \left\{ \text{ad}_F^k \bar{g}_j(x_E) \mid 1 \leq j \leq m,\ 0 \leq k \leq i - d_j \right\}. \tag{6.38}$$

Since $\ker(\pi_*) \subset D_0(x_E) \subset D_i(x_E)$ by (6.17), (6.30) and (6.37) and distributions $D_i(x_E), i \geq 0$ are involutive on U_E, it is clear that distributions $\ker(\pi_*) + D_i(x_E) (= D_i(x_E)), i \geq 0$ are involutive on U_E. Therefore, $\pi_*(D_i(x_E)), i \geq 0$ are, by Theorem 2.10, well-defined involutive distributions on $\tilde{U}_E(= \pi_*(U_E))$. In other words, we have, by Definition 2.19 and (6.37), that for $i \geq 0$,

$$
\begin{aligned}
\pi_* (D_i(x_E)) &= \pi_* (D_i(x_E|_{\bar{z}=0})) \\
&= \operatorname{span} \left\{ \pi_* \left(\operatorname{ad}_F^k G_j(x_E)\big|_{\bar{z}=0} \right) \,\Big|\, 1 \leq j \leq m,\, 0 \leq k \leq i \right\}.
\end{aligned}
\tag{6.39}
$$

Since for $1 \leq j \leq m, k \geq 0$, and $i \in \bar{\mathbf{I}}$,

$$
\left[\frac{\partial}{\partial z_{d_i}^i}, \ \operatorname{ad}_F^k G_j(x_E)\big|_{\bar{z}=0} \right] = 0
$$

it is clear, by Theorem 2.6, that $\pi_* \left(\operatorname{ad}_F^k G_j(x_E)\big|_{\bar{z}=0} \right)$, $1 \leq j \leq m$, $k \geq 0$ are well-defined vector fields on $\tilde{U}_E$. It is easy to see, by (6.17) and mathematical induction, that for $j \in \mathbf{I}$ and $0 \leq k \leq d_j$,

$$
\pi_* \left(\operatorname{ad}_F^k G_j(x_E)\big|_{\bar{z}=0} \right) =
\begin{cases}
(-1)^k \pi_* \left(\frac{\partial}{\partial z_{d_j-k}^j} \right), & \text{if } 0 \leq k < d_j \\
(-1)^{d_j} \pi_* \left(\bar{g}_j(x_E) \right), & \text{if } k = d_j
\end{cases}
$$

$$
= \operatorname{ad}_{\tilde{F}}^k \tilde{G}_j(\tilde{x}_E)
$$

and for $j \in \mathbf{I}$ and $k \geq d_j$,

$$
\begin{aligned}
&\pi_* \left(\operatorname{ad}_F^{k+1} G_j(x_E)\big|_{\bar{z}=0} \right) \\
&= (-1)^{d_j} \frac{\partial \pi(x_E)}{\partial x_E} \left(\frac{\partial \operatorname{ad}_F^{k-d_j} \bar{g}_j(x_E)}{\partial x_E} F(x_E) - \frac{\partial F(x_E)}{\partial x_E} \operatorname{ad}_F^{k-d_j} \bar{g}_j(x_E) \right)\Bigg|_{\bar{z}=0} \\
&= (-1)^{d_j} \frac{\partial \pi(x_E)}{\partial x_E} \frac{\partial \operatorname{ad}_F^{k-d_j} \bar{g}_j(x_E)}{\partial \tilde{x}_E}\Bigg|_{\bar{z}=0} \tilde{F}(\tilde{x}_E) \\
&\quad - (-1)^{d_j} \frac{\partial \pi(x_E)}{\partial x_E} \frac{\partial F(x_E|_{\bar{z}=0})}{\partial x} \frac{\partial x}{\partial x_E} \operatorname{ad}_F^{k-d_j} \bar{g}_j(x_E)\Bigg|_{\bar{z}=0} \\
&= \frac{\partial \pi_* \left(\operatorname{ad}_F^{k-d_j} \bar{g}_j(x_E)\big|_{\bar{z}=0} \right)}{\partial \tilde{x}_E} \tilde{F}(\tilde{x}_E) - \frac{\partial \tilde{F}(x_E)}{\partial \tilde{x}_E} \pi_* \left(\operatorname{ad}_F^{k-d_j} \bar{g}_j(x_E)\big|_{\bar{z}=0} \right) \\
&= \frac{\partial \operatorname{ad}_{\tilde{F}}^k \tilde{G}_j(\tilde{x}_E)}{\partial \tilde{x}_E} \tilde{F}(\tilde{x}_E) - \frac{\partial \tilde{F}(x_E)}{\partial \tilde{x}_E} \operatorname{ad}_{\tilde{F}}^k \tilde{G}_j(\tilde{x}_E) = \operatorname{ad}_{\tilde{F}}^{k+1} \tilde{G}_j(\tilde{x}_E)
\end{aligned}
$$

which imply that (6.31) is satisfied. Similarly, it can be shown, by (6.17) and mathematical induction, that for $j \in \bar{\mathbf{I}}$ and $0 \le k \le d_j$,

$$
\pi_* \left(\mathrm{ad}_F^k G_j(x_E) \big|_{\bar{z}=0} \right) =
\begin{cases}
\pi_* \left(\dfrac{\partial}{\partial z_{d_j}^j} \right), & \text{if } k = 0 \\[2mm]
(-1)^k \pi_* \left(\dfrac{\partial}{\partial z_{d_j-k}^j} \right), & \text{if } 1 \le k < d_j \\[2mm]
(-1)^{d_j} \pi_* \left(\bar{g}_j(x_E) \right), & \text{if } k = d_j
\end{cases}
$$

$$
=
\begin{cases}
0, & \text{if } k = 0 \\[2mm]
-\mathrm{ad}_{\tilde{F}}^{k-1} \tilde{G}_j(\tilde{x}_E), & \text{if } 1 \le k \le d_j
\end{cases}
$$

and for $j \in \bar{\mathbf{I}}$ and $k \ge d_j$,

$$
\pi_* \left(\mathrm{ad}_F^{k+1} G_j(x_E) \big|_{\bar{z}=0} \right) = -\mathrm{ad}_{\tilde{F}}^k \tilde{G}_j(\tilde{x}_E)
$$

which imply that (6.32) is satisfied. Therefore, we have, by (6.31), (6.32), and (6.39), that for $i \ge 0$,

$$
\begin{aligned}
\pi_* \left(D_i(x_E) \right) &= \mathrm{span} \left\{ \mathrm{ad}_{\tilde{F}}^k \tilde{G}_j(\tilde{x}_E) \ \big| \ j \in \mathbf{I}, \ 0 \le k \le i \right\} \\
&\quad + \mathrm{span} \left\{ \mathrm{ad}_{\tilde{F}}^k \tilde{G}_j(\tilde{x}_E) \ \big| \ j \in \bar{\mathbf{I}}, \ 0 \le k \le i - 1 \right\} \\
&= \pi_* \left(Q_i(x_E) \right) + \mathrm{span} \left\{ \dfrac{\partial}{\partial \tilde{z}_\ell^j} \ \big| \ j \in \mathbf{I}, \ \tilde{d}_j - i \le \ell \le \tilde{d}_j \right\} \\
&\quad + \mathrm{span} \left\{ \dfrac{\partial}{\partial \tilde{z}_\ell^j} \ \big| \ j \in \bar{\mathbf{I}}, \ \tilde{d}_j + 1 - i \le \ell \le \tilde{d}_j \right\}.
\end{aligned}
$$

Also, it is clear that

$$
\dim(\pi_* \left(D_{n+d-1}(0, z^0) \right)) = n + d - \dim(\ker(\pi_*)) = n + \tilde{d}.
$$

Since for $1 \le j \le m$, $k \ge 0$, and $i \in \bar{\mathbf{I}}$,

$$
\left[\frac{\partial}{\partial z_{d_i}^i}, \ \mathrm{ad}_F^k \bar{g}_j(x_E) \big|_{\bar{z}=0} \right] = 0
$$

it is clear, by Theorem 2.6, that $\pi_* \left(\mathrm{ad}_F^k \bar{g}_j(x_E) \big|_{\bar{z}=0} \right)$, $1 \le j \le m$, $k \ge 0$ are well-defined vector fields on $\tilde{U}_E$. It is easy to see, by (6.17), (6.31), and (6.32), that for $1 \le j \le m$ and $k \ge 0$,

$$\pi_* \left(\mathrm{ad}_F^k \bar{g}_j(x_E) \big|_{\bar{z}=0} \right) = (-1)^{d_j} \pi_* \left(\mathrm{ad}_F^{k+d_j} G_j(x_E) \big|_{\bar{z}=0} \right)$$

$$= \begin{cases} (-1)^{d_j} \mathrm{ad}_{\tilde{F}}^{k+d_j} \tilde{G}_j(\tilde{x}_E), & j \in \mathbf{I} \\ (-1)^{d_j-1} \mathrm{ad}_{\tilde{F}}^{k+\tilde{d}_j} G_j(\tilde{x}_E), & j \in \bar{\mathbf{I}} \end{cases} = \begin{cases} \mathrm{ad}_{\tilde{F}}^k \bar{g}_j(\tilde{x}_E), & j \in \mathbf{I} \\ \mathrm{ad}_{\tilde{F}}^k \bar{g}_j(\tilde{x}_E), & j \in \bar{\mathbf{I}} \end{cases}$$

$$= \mathrm{ad}_{\tilde{F}}^k \bar{g}_j(\tilde{x}_E)$$

which imply that (6.35) is satisfied. Since $\ker(\pi_*)$ and $D_i(x_E)$, $i \geq 0$ are involutive distributions on U_E, it is clear, by Example 6.2.3, that distributions $\ker(\pi_*) + Q_i(x_E)$, $i \geq 0$ are involutive on U_E. Therefore, $\pi_*(Q_i(x_E))$, $i \geq 0$ are, by Theorem 2.10, well-defined involutive distributions on $\tilde{U}_E (= \pi_*(U_E))$. In other words, we have, by Definition 2.19, (6.35), and (6.38), that for $i \geq 0$,

$$\pi_* (Q_i(x_E)) = \pi_* (Q_i(x_E|_{\bar{z}=0}))$$

$$= \mathrm{span} \left\{ \pi_* \left(\mathrm{ad}_F^k \bar{g}_j(x_E) \big|_{\bar{z}=0} \right) \,\Big|\, 1 \leq j \leq m, \, 0 \leq k \leq i - d_j \right\}$$

$$= \mathrm{span} \left\{ \mathrm{ad}_{\tilde{F}}^k \bar{g}_j(\tilde{x}_E) \,\Big|\, 1 \leq j \leq m, \, 0 \leq k \leq i - d_j \right\}$$

which implies that (6.36) is satisfied.   □

Theorem 6.1 *If system* (6.1) *is restricted dynamic feedback linearizable with indices* $\mathbf{d} = (d_1, \ldots, d_m)$ *and* $d_i \geq 1$ *for* $1 \leq i \leq m$, *then system* (6.1) *is also restricted dynamic feedback linearizable with indices* $\tilde{\mathbf{d}} = (\tilde{d}_1, \ldots, \tilde{d}_m)$, *where* $\tilde{d}_i = d_i - 1$ *for* $1 \leq i \leq m$.

Proof Suppose that system (6.1) is restricted dynamic feedback linearizable with indices $\mathbf{d} = (d_1, \ldots, d_m)$ and $d_i \geq 1$ for $1 \leq i \leq m$. Then extended system (6.14) is (static) feedback linearizable on a neighborhood U_E of $(x, z) = (0, z^0)$. We need to show that extended system (6.28) with indices $\tilde{\mathbf{d}} = (\tilde{d}_1, \ldots, \tilde{d}_m)$ is (static) feedback linearizable on a neighborhood $\tilde{U}_E$ of $(x, \tilde{z}) = (0, \tilde{z}^0)$. In other words, we need to show, by Theorem 4.3, that

(a) $\dim \left(\tilde{D}_{n+d-2}(0, \tilde{z}^0) \right) = n + \tilde{d}$

(b) $\tilde{D}_i(\tilde{x}_E)$, $i \geq 0$ are involutive distributions on $\tilde{U}_E$

where for $i \geq 0$,

$$\tilde{D}_i(\tilde{x}_E) = \mathrm{span} \left\{ \mathrm{ad}_{\tilde{F}}^k \tilde{G}_j(\tilde{x}_E) \,\Big|\, 1 \leq j \leq m, \, 0 \leq k \leq i \right\}. \qquad (6.40)$$

Let $\begin{bmatrix} x \\ \tilde{z} \end{bmatrix} = \pi(x, z)$ and $\begin{bmatrix} 0 \\ \tilde{z}^0 \end{bmatrix} = \pi(0, z^0)$, where canonical projection map $\pi : \mathbb{R}^{n+d} \to \mathbb{R}^{n+\tilde{d}}$ is defined by

$$\pi(x, z_1^1, \ldots, z_{d_1}^1, \ldots, z_1^m, \ldots, z_{d_m}^m) = (x, z_1^1, \ldots, z_{d_1-1}^1, \ldots, z_1^m, \ldots, z_{d_m-1}^m).$$

In other words, let

$$\mathbf{I} = \phi \quad \text{and} \quad \bar{\mathbf{I}} = \{1, 2, \ldots, m\}.$$

Then, it is clear, by (6.33), that for $i \geq 0$,

$$\tilde{D}_i(\tilde{x}_E) = \pi_* \left(D_{i+1}(x_E) \right). \tag{6.41}$$

Therefore, by Lemma 6.2, $\tilde{D}_i(\tilde{x}_E)$, $i \geq 0$ are involutive distributions on $\tilde{U}_E (= \pi_*(U_E))$ and

$$\dim \left(\tilde{D}_{n+d-2} \left(0, \tilde{z}^0 \right) \right) = \dim \left(\pi_* \left(D_{n+d-1} \left(0, z^0 \right) \right) \right) = n + \tilde{d}.$$

Hence, system (6.1) is also restricted dynamic feedback linearizable with indices $\tilde{\mathbf{d}} = (\tilde{d}_1, \ldots, \tilde{d}_m)$. $\qquad\qquad\qquad\qquad\qquad\qquad\qquad\qquad\square$

It can be easily shown that the converse of Theorem 6.1 also holds. (See Problem 6-3.) The following results can be obtained by repeated use of Theorem 6.1.

Corollary 6.1 *If the system* (6.1) *is restricted dynamic feedback linearizable with indices* $\mathbf{d} = (d_1, \ldots, d_m)$, *then the system* (6.1) *is also restricted dynamic feedback linearizable with indices* $\tilde{\mathbf{d}} = (\tilde{d}_1, \ldots, \tilde{d}_m)$, *where*

$$\tilde{d}_i = d_i - d_{\min} \quad \text{and} \quad d_{\min} = \min \{d_1, \ldots, d_m\}. \tag{6.42}$$

If system (6.1) is restricted dynamic feedback linearizable, then it is linearizable without a pure integrator for at least one of the input channels (i.e., $d_{\min} = 0$). Thus, we have the following interesting results.

Corollary 6.2 *If single input system* (6.1) *(with* $m = 1$*) is restricted dynamic feedback linearizable, then system* (6.1) *is static feedback linearizable. It is obvious that the converse also holds.*

In other words, dynamic feedback linearization is meaningful only for multi-input nonlinear systems. Without loss of generality, we can assume that

$$0 = d_1 \leq d_2 \leq \cdots \leq d_m$$

and $\{g_1(0), g_2(0), \ldots, g_m(0)\}$ are linearly independent. Now we will find the upper limit of indices d_i, $i \geq 2$. In other words, it will be shown that, if system (6.1) is restricted dynamic feedback linearizable, then system (6.1) is restricted dynamic feedback linearizable with indices $\mathbf{d} = (d_1, \ldots, d_m)$, $d_{\min} = 0$, and $d_i \leq 2n - 3$, $1 \leq i \leq m$. Define the smallest positive integer σ_1 by

$$Q_0(x_E) \neq \cdots \neq Q_{\sigma_1 - 1}(x_E) = Q_{\sigma_1}(x_E) \tag{6.43}$$

on a neighborhood U_E of $(x, z) = (0, z^0)$. For example, if $Q_0(x_E) = Q_1(x_E)$, then $\sigma_1 = 1$. It is clear that $\dim(Q_{\sigma_1 - 1}) = \dim(Q_{\sigma_1}) \geq \sigma_1$.

Lemma 6.3 *Suppose that system (6.1) is restricted dynamic feedback linearizable with indices* $\mathbf{d} = (d_1, \ldots, d_m)$ *and* $d_1 = 0$. *Also assume that*

$$Q_{\sigma_1 - 1}(x_E) = Q_{\sigma_1}(x_E) = \cdots = Q_{2(\sigma_1 - 1)}(x_E) \qquad (6.44)$$

on a neighborhood U_E *of* $(x, z) = (0, z^0)$. *Then system (6.1) is also restricted dynamic feedback linearizable with indices* $\tilde{\mathbf{d}} = (\tilde{d}_1, \ldots, \tilde{d}_m)$, *where*

$$\tilde{d}_i = \begin{cases} d_i - 1, & \text{if } d_i \geq max(2\sigma_1 - 2, 1) \triangleq \bar{d} \\ d_i, & \text{otherwise.} \end{cases} \qquad (6.45)$$

Proof Suppose that system (6.1) is restricted dynamic feedback linearizable with indices $\mathbf{d} = (d_1, \ldots, d_m)$ and $d_1 = 0$. Then extended system (6.14) is (static) feedback linearizable on a neighborhood U_E of $(x, z) = (0, z^0)$. We need to show that extended system (6.28) with indices $\tilde{\mathbf{d}} = (\tilde{d}_1, \ldots, \tilde{d}_m)$ is (static) feedback linearizable on a neighborhood $\tilde{U}_E$ of $(x, \tilde{z}) = (0, \tilde{z}^0)$. In other words, we need to show, by Theorem 4.3, that

(a) $\dim\left(\tilde{D}_{n+d-2}(0, \tilde{z}^0)\right) = n + \tilde{d}$

(b) $\tilde{D}_i(\tilde{x}_E), i \geq 0$ are involutive distributions on $\tilde{U}_E$

where for $i \geq 0$,

$$\begin{aligned}
\tilde{D}_i(\tilde{x}_E) &= \text{span}\left\{ad_{\tilde{F}}^k \tilde{G}_j(\tilde{x}_E) \,\middle|\, 1 \leq j \leq m, \, 0 \leq k \leq i\right\} \\
&= \tilde{Q}_i(\tilde{x}_E) + \text{span}\left\{\frac{\partial}{\partial \tilde{z}_k^j} \,\middle|\, j \in \mathbf{I}, \, \tilde{d}_j - i \leq k \leq \tilde{d}_j\right\} \\
&\quad + \text{span}\left\{\frac{\partial}{\partial \tilde{z}_k^j} \,\middle|\, j \in \bar{\mathbf{I}}, \, \tilde{d}_j - i \leq k \leq \tilde{d}_j\right\}
\end{aligned} \qquad (6.46)$$

and

$$\tilde{Q}_i(\tilde{x}_E) = \text{span}\left\{ad_{\tilde{F}}^k \tilde{g}_j(\tilde{x}_E) \,\middle|\, 1 \leq j \leq m, \, 0 \leq k \leq i - \tilde{d}_j\right\}. \qquad (6.47)$$

Let

$$\mathbf{I} \triangleq \left\{i \mid 1 \leq i \leq m, \, d_i < \bar{d}\right\} \quad \text{and} \quad \bar{\mathbf{I}} \triangleq \left\{i \mid 1 \leq i \leq m, \, d_i \geq \bar{d}\right\}.$$

Since $Q_{\bar{d} - 1}(x_E) = Q_{\bar{d}}(x_E)$ and $\bar{g}_j(x_E) \in Q_{\bar{d} - 1}(x_E)$, $j \in \mathbf{I}$, it is easy to see, by Lemma 6.1 and (6.35), that for $i \geq 0$,

$$\text{ad}_{\bar{F}}^{i}\bar{g}_j(x_E) \in Q_{\bar{d}-1}(x_E)$$
$$= \text{span}\left\{\text{ad}_{\bar{F}}^{k}\bar{g}_j(x_E) \,\Big|\, j \in \mathbf{I},\ 0 \le k \le \bar{d} - 1 - d_j\right\}$$

and for $i \ge \bar{d}$,

$$\text{span}\left\{\text{ad}_{\bar{F}}^{k}\bar{g}_j(\tilde{x}_E) \,\Big|\, j \in \mathbf{I},\ 0 \le k \le i - \tilde{d}_j\right\}$$
$$= \text{span}\left\{\text{ad}_{\bar{F}}^{k}\bar{g}_j(\tilde{x}_E) \,\Big|\, j \in \mathbf{I},\ 0 \le k \le \bar{d} - 1 - \tilde{d}_j\right\} \qquad (6.48)$$
$$= \text{span}\left\{\text{ad}_{\bar{F}}^{k}\bar{g}_j(\tilde{x}_E) \,\Big|\, j \in \mathbf{I},\ 0 \le k \le i - 1 - \tilde{d}_j\right\}.$$

Therefore, it is easy to see, by (6.36), (6.47), and (6.48), that for $0 \le i \le \bar{d} - 2$,

$$\pi_*\left(Q_i(x_E)\right) = \text{span}\left\{\text{ad}_{\bar{F}}^{k}\bar{g}_j(\tilde{x}_E) \,\Big|\, j \in \mathbf{I},\ 0 \le k \le i - \tilde{d}_j\right\} = \tilde{Q}_i(\tilde{x}_E)$$

$$\pi_*\left(Q_{\bar{d}}(x_E)\right) = \text{span}\left\{\text{ad}_{\bar{F}}^{k}\bar{g}_j(\tilde{x}_E) \,\Big|\, j \in \mathbf{I},\ 0 \le k \le \bar{d} - \tilde{d}_j\right\}$$
$$+ \text{span}\left\{\text{ad}_{\bar{F}}^{k}\bar{g}_j(\tilde{x}_E) \,\Big|\, j \in \bar{\mathbf{I}},\ 0 \le k \le \bar{d} - 1 - \tilde{d}_j\right\}$$
$$= \text{span}\left\{\text{ad}_{\bar{F}}^{k}\bar{g}_j(\tilde{x}_E) \,\Big|\, j \in \mathbf{I},\ 0 \le k \le \bar{d} - 1 - \tilde{d}_j\right\}$$
$$+ \text{span}\left\{\text{ad}_{\bar{F}}^{k}\bar{g}_j(\tilde{x}_E) \,\Big|\, j \in \bar{\mathbf{I}},\ 0 \le k \le \bar{d} - 1 - \tilde{d}_j\right\} = \tilde{Q}_{\bar{d}-1}(\tilde{x}_E)$$

and for $i \ge \bar{d}$,

$$\pi_*\left(Q_{i+1}(x_E)\right) = \text{span}\left\{\text{ad}_{\bar{F}}^{k}\bar{g}_j(\tilde{x}_E) \,\Big|\, j \in \mathbf{I},\ 0 \le k \le i + 1 - \tilde{d}_j\right\}$$
$$+ \text{span}\left\{\text{ad}_{\bar{F}}^{k}\bar{g}_j(\tilde{x}_E) \,\Big|\, j \in \bar{\mathbf{I}},\ 0 \le k \le i - \tilde{d}_j\right\}$$
$$= \text{span}\left\{\text{ad}_{\bar{F}}^{k}\bar{g}_j(\tilde{x}_E) \,\Big|\, j \in \mathbf{I},\ 0 \le k \le i - \tilde{d}_j\right\}$$
$$+ \text{span}\left\{\text{ad}_{\bar{F}}^{k}\bar{g}_j(\tilde{x}_E) \,\Big|\, j \in \bar{\mathbf{I}},\ 0 \le k \le i - \tilde{d}_j\right\} = \tilde{Q}_i(\tilde{x}_E)$$

which imply that

$$\tilde{Q}_i(\tilde{x}_E) = \begin{cases} \pi_*\left(Q_i(x_E)\right), & \text{if } 0 \le i \le \bar{d} - 1 \\ \pi_*\left(Q_{i+1}(x_E)\right), & \text{if } i \ge \bar{d}. \end{cases} \qquad (6.49)$$

Thus, by Lemma 6.2, $\tilde{Q}_i(\tilde{x}_E)$, $i \ge 0$ are involutive distributions. We also have, by (6.33) and (6.46), that for $0 \le i \le \bar{d} - 1$,

$$\tilde{D}_i(\tilde{x}_E) = \pi_* \left(D_i(x_E) \right) + \text{span} \left\{ \frac{\partial}{\partial \tilde{z}^j_{\tilde{d}_j - i}} \, \Big| \, j \in \bar{\mathbf{I}} \right\} \tag{6.50}$$

and for $i \geq \bar{d}$,

$$\tilde{D}_i(\tilde{x}_E) = \pi_* \left(Q_{i+1}(x_E) \right) + \text{span} \left\{ \frac{\partial}{\partial \tilde{z}^j_\ell} \, \Big| \, j \in \mathbf{I}, \, 1 \leq \ell \leq \tilde{d}_j \right\} \tag{6.51}$$

$$+ \text{span} \left\{ \frac{\partial}{\partial \tilde{z}^j_\ell} \, \Big| \, j \in \bar{\mathbf{I}}, \, \tilde{d}_j - i \leq \ell \leq \tilde{d}_j \right\} = \pi_* \left(D_{i+1}(x_E) \right).$$

It will be shown that for $0 \leq i \leq \bar{d} - 1$,

$$\left[\frac{\partial}{\partial z^j_{\tilde{d}_j - i - 1}}, Q_i(x_E) \right] \subset Q_i(x_E), \text{ for } j \in \bar{\mathbf{I}} \tag{6.52}$$

or

$$\left[\frac{\partial}{\partial \tilde{z}^j_{\tilde{d}_j - i}}, \tilde{Q}_i(\tilde{x}_E) \right] \subset \tilde{Q}_i(\tilde{x}_E), \text{ for } j \in \bar{\mathbf{I}}. \tag{6.53}$$

If $\bar{d} = 1$ or $\sigma_1 = 1$, (6.52) is obviously satisfied, since $Q_0(x_E) = \text{span}\{\bar{g}_j(x_E) \, | \, d_j = 0\}$. Thus, let $\bar{d} = 2\sigma_1 - 2$ or $\sigma_1 \geq 2$. If $0 \leq i \leq \sigma_1 - 2$ and $d_j \geq 2\sigma_1 - 2$, then $d_j - i - 1 \geq \sigma_1 - 1 > i$. Therefore, it is easy to see, by (6.22), that (6.52) holds for $0 \leq i \leq \sigma_1 - 2$. Also, it is clear, by Lemma 6.1 and (6.44), that (6.52) holds for $\sigma_1 - 1 \leq i \leq 2\sigma_1 - 3(= \bar{d} - 1)$. In other words, (6.53) is satisfied for $0 \leq i \leq \bar{d} - 1$. Thus, by (6.50), distributions $\tilde{D}_i(\tilde{x}_E)$, $0 \leq i \leq \bar{d} - 1$ are involutive. It is clear, by Lemma 6.2 and (6.51), that $\tilde{Q}_i(\tilde{x}_E)$, $i \geq \bar{d}$ are also involutive distributions on $\tilde{U}_E(= \pi_*(U_E))$ and

$$\dim \left(\tilde{D}_{n+d-2} \left(0, \tilde{z}^0 \right) \right) = \dim \left(\pi_* \left(D_{n+d-1} \left(0, z^0 \right) \right) \right) = n + \tilde{d}.$$

Hence, system (6.1) is also restricted dynamic feedback linearizable with indices $\tilde{\mathbf{d}} = (\tilde{d}_1, \ldots, \tilde{d}_m)$. $\square$

Remark 6.1 Suppose that system (6.1) is restricted dynamic feedback linearizable with indices $\mathbf{d} = (d_1, \ldots, d_m)$, where $0 = d_1 \leq d_2 \leq \cdots \leq d_m$, and

$$Q_0(x_E) \neq \cdots \neq Q_{\sigma_1 - 1}(x_E) = Q_{\sigma_1}(x_E). \tag{6.54}$$

If $\dim(Q_{\sigma_1 - 1}(x_E)) = n$, then $\sigma_1 \leq n$ and (6.44) is satisfied. Thus, it can be assumed, by repeated use of Lemma 6.3, that $d_i \leq 2(\sigma_1 - 1) - 1 \leq 2n - 3$, for $2 \leq i \leq m$.

If $\dim(Q_{\sigma_1-1}(x_E)) < n$, define index p_2 $(d_{p_2} > \sigma_1)$ and the smallest positive integer σ_2 by

$$p_2 \triangleq \min\{j \mid \bar{g}_j(x_E) \notin Q_{\sigma_1}(x_E)\} \tag{6.55}$$

and

$$\begin{aligned}
Q_0 \neq \cdots &\neq Q_{\sigma_1-1} = Q_{\sigma_1} = \cdots = Q_{d_{p_2}-1} \\
&\neq Q_{d_{p_2}} \neq \cdots \neq Q_{d_{p_2}+\sigma_2-1} = Q_{d_{p_2}+\sigma_2}.
\end{aligned} \tag{6.56}$$

If $d_{p_2} > 2(\sigma_1 - 1)$, then (6.44) is satisfied. Thus, it can be assumed, by repeated use of Lemma 6.3, that $\sigma_1 \geq 2$ and

$$d_{p_2} \leq 2\sigma_1 - 2.$$

Note that

$$\sigma_1 + \sigma_2 \leq \dim(Q_{d_{p_2}+\sigma_2}(x_E)) \leq n. \tag{6.57}$$

Let $p_1 = 1$. In this manner, if $\dim(Q_{d_{p_{k-1}}+\sigma_{k-1}}(x_E)) < n$, we can define, for $2 \leq k \leq r$, index $p_k(d_{p_k} > d_{p_{k-1}} + \sigma_{k-1})$ and the smallest positive integer σ_k by

$$p_k \triangleq \min\{j \mid \bar{g}_j(x_E) \notin Q_{d_{p_{k-1}}+\sigma_{k-1}}(x_E)\} \tag{6.58}$$

$$Q_{d_{p_k}-1} \neq Q_{d_{p_k}} \neq \cdots \neq Q_{d_{p_k}+\sigma_k-1} = Q_{d_{p_k}+\sigma_k} \tag{6.59}$$

and

$$\sigma_1 + \cdots + \sigma_r \leq \dim(Q_{d_{p_r}+\sigma_r}(x_E)) = n. \tag{6.60}$$

Lemma 6.4 *Suppose that system* (6.1) *is restricted dynamic feedback linearizable with indices* $\mathbf{d} = (d_1, \ldots, d_m)$ *and* $d_1 = 0$. *Also, assume that* $k \geq 2$, $\mu_k \triangleq \sum_{j=1}^{k} 2(\sigma_j - 1)$, $\mu_{k-2} + \sigma_{k-1} + 1 \leq d_{p_k} \leq \mu_{k-1}$, *and*

$$Q_{d_{p_k}+\sigma_k-1}(x_E) = Q_{d_{p_k}+\sigma_k}(x_E) = \cdots = Q_{\mu_k}(x_E). \tag{6.61}$$

Then system (6.1) *is also restricted dynamic feedback linearizable with indices* $\tilde{\mathbf{d}} = (\tilde{d}_1, \ldots, \tilde{d}_m)$, *where*

$$\tilde{d}_i = \begin{cases} d_i - 1, & \text{if } d_i \geq \mu_k \\ d_i, & \text{otherwise.} \end{cases} \tag{6.62}$$

Proof The proof of Lemma 6.4 is the same as that of Lemma 6.3 with $\bar{d} \triangleq \mu_k$. $\square$

Theorem 6.2 *Let $n > m \geq 2$. Suppose that system (6.1) is restricted dynamic feedback linearizable. Then system (6.1) is also restricted dynamic feedback linearizable with indices* $\mathbf{d} = (d_1, \dots, d_m)$, *where* $d_{\min} = 0$ *and for* $1 \leq i \leq m$,

$$d_i \leq 2n - 3. \tag{6.63}$$

Proof Suppose that system (6.1) is restricted dynamic feedback linearizable with indices $\mathbf{d} = (d_1, \dots, d_m)$. We assume, without loss of generality, that $d_1 \leq d_2 \leq \cdots \leq d_m$. Also, we can assume $d_1 = 0$ by Theorem 6.1. Define positive integer σ_1 by (6.54). If $\sigma_1 = n$, then $\dim(Q_{\sigma_1-1}(x_E)) = n$ and (6.61) is satisfied. Thus, it can be assumed, by repeated use of Lemma 6.3, that $d_i \leq 2(\sigma_1 - 1) - 1 = 2n - 3$, for $2 \leq i \leq m$. As explained in Remark 6.1, if $\dim(Q_{\sigma_1-1}(x_E)) < n$, define index $p_2(d_{p_2} > \sigma_1)$ and the smallest positive integer σ_2 such that (6.55) and (6.56) are satisfied. Note that (6.57) is satisfied. If $d_{p_2} > 2(\sigma_1 - 1)$, then (6.61) is satisfied. Thus, it can be assumed, by repeated use of Lemma 6.4, that $\sigma_1 \geq 2$ and

$$d_{p_2} \leq 2\sigma_1 - 2.$$

If $\dim(Q_{d_{p_2}+\sigma_2}(x_E)) = n$, then (6.61) holds for $k = 2$. Thus, it can be assumed, by repeated use of Lemma 6.4, that for $p_2 + 1 \leq i \leq m$,

$$d_i \leq 2(\sigma_1 + \sigma_2 - 2) - 1 \leq 2n - 5. \tag{6.64}$$

If $\dim(Q_{d_{p_2}+\sigma_2}(x_E)) < n$, define index $p_3(d_{p_2} > \sigma_1)$ and the smallest positive integer σ_3 such that (6.58) and (6.59) are satisfied. Note that $\sigma_1 + \sigma_2 + \sigma_3 \leq \dim(Q_{d_{p_3}+\sigma_3}(x_E)) \leq n$. If $d_{p_3} > \mu_2 = 2(\sigma_1 + \sigma_2 - 2)$, then (6.61) holds for $k = 3$. Thus, it can be assumed, by repeated use of Lemma 6.4, that

$$d_{p_3} \leq \mu_2 = 2(\sigma_1 + \sigma_2 - 2).$$

If $\dim(Q_{d_{p_3}+\sigma_3}(x_E)) = n$, then (6.61) holds for $k = 3$. Thus, it can be assumed, by repeated use of Lemma 6.4, that for $p_3 + 1 \leq i \leq m$,

$$d_i \leq 2(\sigma_1 + \sigma_2 + \sigma_3 - 3) - 1 \leq 2n - 7. \tag{6.65}$$

In this manner, if $\dim(Q_{d_{p_3}+\sigma_3}(x_E)) < n$, we can define, for $2 \leq k \leq r$, index $p_k(d_{p_k} > d_{p_{k-1}} + \sigma_{k-1})$ and the smallest positive integer σ_k by

$$p_k \triangleq \min\left\{ j \mid \bar{g}_j(x_E) \notin Q_{d_{p_{k-1}}+\sigma_{k-1}}(x_E) \right\}$$

$$Q_{d_{p_k}-1} \neq Q_{d_{p_k}} \neq \cdots \neq Q_{d_{p_k}+\sigma_k-1} = Q_{d_{p_k}+\sigma_k}$$

$$d_{p_k} \le \mu_{k-1} = \sum_{j=1}^{k-1} 2(\sigma_j - 1)$$

and

$$\sigma_1 + \cdots + \sigma_r \le \dim(Q_{d_{p_r}+\sigma_r}(x_E)) = n.$$

Finally, since $\dim(Q_{d_{p_{r-1}}+\sigma_r-1}(x_E)) = n$, (6.61) holds for $k = r$. Thus, it can be assumed, by repeated use of Lemma 6.4, that for $p_r + 1 \le i \le m$,

$$d_i \le \mu_r - 1 = 2(\sigma_1 + \cdots + \sigma_r - r) - 1 \le 2n - (2r + 1). \tag{6.66}$$

$\square$

Let $n > m \ge 2$. If system (6.1) is restricted dynamic feedback linearizable, then system (6.1) is also restricted dynamic feedback linearizable such that the number d of extended state z satisfies

$$d \le (m - 1)(2n - 3). \tag{6.67}$$

The upper limit d_i and d in Theorem 6.2 are sharp. When $\sigma_1 = n$, d is the maximum. It can be seen in Example 6.3.1.

Example 6.3.1 Let $n \ge m + 2$. Show that the following nonlinear system is not restricted dynamic feedback linearizable with indices $\mathbf{d} = (d_1, \ldots, d_m)$ and $d_i < 2n - 3$, $1 \le i \le m$. Also, show that the following nonlinear system is restricted dynamic feedback linearizable with indices $\mathbf{d} = (d_1, \ldots, d_m) = (0, 2n - 3, \ldots, 2n - 3)$.

$$\dot{x}_1 = x_2 + x_1 \sum_{i=2}^{m} u_i$$

$$\dot{x}_j = x_{j+1}, \ 2 \le j \le n - m \tag{6.68}$$

$$\dot{x}_{n-m+1} = u_1$$

$$\dot{x}_{n-m+i} = (1 + x_{n-m+1})u_i, \ 2 \le i \le m.$$

Solution For simplicity, we consider the case of $m = 2$. Let $d_1 = 0$ and $d_2 < 2n - 3$. It is easy to see that $\bar{g}_1(x_E) = \frac{\partial}{\partial x_{n-1}}$ and for $1 \le k \le n - 2$,

$$\mathrm{ad}_F^k \bar{g}_1(x_E) = (-1)^k \frac{\partial}{\partial x_{n-k-1}} - z_k^2 \frac{\partial}{\partial x_n}. \tag{6.69}$$

Let d_2 is even and $d_2 \le 2n - 4$. Then we have that $\frac{d_2}{2} \le n - 2$ and

$$Q_{\frac{d_2}{2}}(x_E) = \text{span}\left\{\text{ad}_F^k \bar{g}_1(x_E) \,\middle|\, 0 \le k \le \frac{d_2}{2}\right\}.$$

Since

$$\left[\frac{\partial}{\partial z_{\frac{d_2}{2}}^2}, \text{ad}_F^{\frac{d_2}{2}} \bar{g}_1(x_E)\right] = -\frac{\partial}{\partial x_n} \notin Q_{\frac{d_2}{2}}(x_E)$$

$D_{\frac{d_2}{2}}(x_E)$ is not involutive, where

$$D_{\frac{d_2}{2}}(x_E) = Q_{\frac{d_2}{2}}(x_E) + \text{span}\left\{\frac{\partial}{\partial z_k^2} \,\middle|\, \frac{d_2}{2} \le k \le d_2\right\}.$$

Let d_2 is odd and $d_2 \le 2n - 5$. Then we have that $\frac{d_2+1}{2} \le n - 2$ and

$$Q_{\frac{d_2+1}{2}}(x_E) = \text{span}\left\{\text{ad}_F^k \bar{g}_1(x_E) \,\middle|\, 0 \le k \le \frac{d_2+1}{2}\right\}.$$

Since

$$\left[\frac{\partial}{\partial z_{\frac{d_2+1}{2}}^2}, \text{ad}_F^{\frac{d_2+1}{2}} \bar{g}_1(x_E)\right] = -\frac{\partial}{\partial x_n} \notin Q_{\frac{d_2+1}{2}}(x_E)$$

$D_{\frac{d_2+1}{2}}(x_E)$ is not involutive, where

$$D_{\frac{d_2+1}{2}}(x_E) = Q_{\frac{d_2+1}{2}}(x_E) + \text{span}\left\{\frac{\partial}{\partial z_k^2} \,\middle|\, \frac{d_2-1}{2} \le k \le d_2\right\}.$$

Therefore, system (6.68) is not restricted dynamic feedback linearizable with indices $d_1 = 0$ and $d_2 < 2n - 3$. In the same manner, it can be shown that system (6.68) is not restricted dynamic feedback linearizable with indices $d_2 = 0$ and $d_1 < 2n - 3$. Now let $d_1 = 0$ and $d_2 = 2n - 3$. Then it is easy to see that

$$\text{ad}_F^{n-1} \bar{g}_1(x_E) = z_1^2 \frac{\partial}{\partial x_1} - z_{n-1}^2 \frac{\partial}{\partial x_n} \tag{6.70}$$

which implies, together with (6.70), that $Q_{n-1}(x_E)$ is an n-dimensional involutive distribution on $U_E(= \{(x, z) \mid z_1^2 = 0, \ z_{n-1}^2 \neq 0\})$, where

$$Q_{n-1}(x_E) = \text{span}\left\{\frac{\partial}{\partial x_k} \,\middle|\, 1 \le k \le n\right\}. \tag{6.71}$$

It is clear that $Q_i(x_E)$, $0 \le i \le n - 2$ is an $(i + 1)$-dimensional involutive distributions. For $0 \le i \le n - 2$ and $2n - 3 - i = d_2 - i \le j \le d_2$, we have, by (6.22), that $i < n - 1 \le d_2 - i \le j$ and

$$\left[\frac{\partial}{\partial z_j^2}, Q_i(x_E) \right] \subset Q_i(x_E)$$

which implies that $D_i(x_E)$, $0 \le i \le n - 2$ are involutive, where for $0 \le i \le n - 2$,

$$D_i(x_E) = Q_i(x_E) + \text{span} \left\{ \frac{\partial}{\partial z_j^2} \, \middle| \, d_2 - i \le j \le d_2 \right\}.$$

Finally, It is obvious, by (6.71), that $D_i(x_E)$, $0 \le i \le n - 2$ are involutive distributions on U_E and

$$\dim(D_{2n-4}(x_E)) = 3n - 3 = n + d_2 \quad \text{for} \quad x_E \in U_E.$$

Hence, system (6.68) is restricted dynamic feedback linearizable with indices $\mathbf{d} = (0, 2n - 3)$. □

By Corollary 6.1 and Theorem 6.2, only a finite set of indices need to be considered, in order to determine whether system (6.1) is restricted dynamic feedback linearizable or not. Therefore, the conditions of Theorem 6.2 is verifiable. For example, if $m = 2$, then check whether system (6.1) is restricted dynamic feedback linearizable with indices $\mathbf{d} = (0, 0), (0, 1), \ldots, (0, 2n - 3), (1, 0), (2, 0), \ldots, (2n - 3, 0)$ in sequence. If system (6.1) is not restricted dynamic feedback linearizable with the above indices, then system (6.1) is not restricted dynamic feedback linearizable.

6.4 Examples

Example 6.4.1 Find out whether the following nonlinear system is restricted dynamic feedback linearizable or not. If it is restricted dynamic feedback linearizable, find out the restricted dynamic feedback and state transformation.

$$\dot{x} = \begin{bmatrix} x_2 \\ x_3 \\ 0 \\ 0 \end{bmatrix} + \begin{bmatrix} 0 \\ 0 \\ 1 \\ 0 \end{bmatrix} u_1 + \begin{bmatrix} 0 \\ x_4 \\ 0 \\ 1 + x_3 \end{bmatrix} u_2 = f(x) + g_1(x)u_1 + g_2(x)u_2. \quad (6.72)$$

Solution Since $[g_1(x), g_2(x)] = \begin{bmatrix} 0 & 0 & 0 & 1 \end{bmatrix}^\mathsf{T} \notin \text{span}\{g_1(x), g_2(x)\}$, system (6.72) is not feedback linearizable. Consider the following dynamic compensator with index $(d_1, d_2) = (0, 1)$:

$$u_1 = w_1 \; ; \quad u_2 = z$$
$$\dot{z} = w_2.$$

Then we have the extended system

$$
\begin{bmatrix} \dot{x}_1 \\ \dot{x}_2 \\ \dot{x}_3 \\ \dot{x}_4 \\ \dot{z} \end{bmatrix} = \begin{bmatrix} x_2 \\ x_3 + x_4 z \\ 0 \\ (1 + x_3)z \\ 0 \end{bmatrix} + \begin{bmatrix} 0 \\ 0 \\ 1 \\ 0 \\ 0 \end{bmatrix} w_1 + \begin{bmatrix} 0 \\ 0 \\ 0 \\ 0 \\ 1 \end{bmatrix} w_2 \tag{6.73}
$$
$$= F(x, z) + G_1(x, z)w_1 + G_2(x, z)w_2.$$

Since

$$
\begin{bmatrix} \mathrm{ad}_F G_1(x, z) & \mathrm{ad}_F G_2(x, z) \end{bmatrix} = \begin{bmatrix} 0 & 0 \\ -1 & -x_4 \\ 0 & 0 \\ -z & -1 - x_3 \\ 0 & 0 \end{bmatrix} \quad \text{and} \quad \mathrm{ad}_F^2 G_1(x, z) = \begin{bmatrix} 1 \\ z^2 \\ 0 \\ 0 \\ 0 \end{bmatrix}
$$

it is easy to see that $(\kappa_1, \kappa_2) = (3, 2)$, $\dim(D_2(x, z)) = 5 = n + d$, and distributions $D_0 = \mathrm{span}\{G_1, G_2\}$ and $D_1 = \mathrm{span}\{G_1, G_2, \mathrm{ad}_F G_1, \mathrm{ad}_F G_2\}$ are involutive. Hence, by Theorem 4.3, system (6.73) is (static) feedback linearizable. Functions $S_{11}(x, z) = x_1$ and $S_{21}(x, z) = x_4$ satisfying the conditions of Lemma 4.3 can be easily found. Thus, extended state transformation $\xi = S_E(x, z)$ and extended static feedback $w = \alpha(x, z) + \beta(x, z)v$ can be obtained by (4.56) and (4.57), respectively.

$$
\xi = S_E(x, z) = \begin{bmatrix} S_{11}(x, z) \\ L_F S_{11}(x, z) \\ L_F^2 S_{11}(x, z) \\ S_{21}(x, z) \\ L_F S_{21}(x, z) \end{bmatrix} = \begin{bmatrix} x_1 \\ x_2 \\ x_3 + x_4 z \\ x_4 \\ (1 + x_3)z \end{bmatrix} \tag{6.74}
$$

and

$$
\begin{bmatrix} w_1 \\ w_2 \end{bmatrix} = \begin{bmatrix} 1 & x_4 \\ z & 1 + x_3 \end{bmatrix}^{-1} \left(- \begin{bmatrix} (1 + x_3)z^2 \\ 0 \end{bmatrix} + \begin{bmatrix} v_1 \\ v_2 \end{bmatrix} \right)
$$
$$
= \begin{bmatrix} \frac{-(1+x_3)^2 z^2}{1+x_3-x_4 z} \\ \frac{(1+x_3)z^3}{1+x_3-x_4 z} \end{bmatrix} + \begin{bmatrix} \frac{1+x_3}{1+x_3-x_4 z} & \frac{-x_4}{1+x_3-x_4 z} \\ \frac{-z}{1+x_3-x_4 z} & \frac{1}{1+x_3-x_4 z} \end{bmatrix} \begin{bmatrix} v_1 \\ v_2 \end{bmatrix}.
$$

In other words, the restricted dynamic feedback for system (6.72) is

$$\begin{bmatrix} u_1 \\ u_2 \end{bmatrix} = \begin{bmatrix} \frac{-(1+x_3)^2 z^2}{1+x_3-x_4 z} \\ z \end{bmatrix} + \begin{bmatrix} \frac{1+x_3}{1+x_3-x_4 z} & \frac{-x_4}{1+x_3-x_4 z} \\ 0 & 0 \end{bmatrix} \begin{bmatrix} v_1 \\ v_2 \end{bmatrix}$$

$$\dot{z} = \frac{(1+x_3)z^3}{1+x_3-x_4 z} + \begin{bmatrix} \frac{-z}{1+x_3-x_4 z} & \frac{1}{1+x_3-x_4 z} \end{bmatrix} \begin{bmatrix} v_1 \\ v_2 \end{bmatrix}. \tag{6.75}$$

$\square$

Example 6.4.2 Show that system (6.76) is not restricted dynamic feedback linearizable.

$$\dot{x} = \begin{bmatrix} x_2 \\ x_3 \\ 0 \\ 0 \end{bmatrix} + \begin{bmatrix} 0 \\ x_4 \\ 1 \\ 1+x_3 \end{bmatrix} u_1 + \begin{bmatrix} 0 \\ x_4 \\ 0 \\ 1+x_3 \end{bmatrix} u_2 \tag{6.76}$$

$$= f(x) + g_1(x)u_1 + g_2(x)u_2.$$

But, system (6.76) is dynamic feedback linearizable.

Solution By simple calculation or MATLAB program in Sect. 6.5, it is easy to see that system (6.76) is not restricted dynamic feedback linearizable with indices $\mathbf{d} = (0, 0), (0, 1), \ldots, (0, 5), (1, 0), (2, 0), \ldots, (5, 0)$. Hence, system (6.76) is not restricted dynamic feedback linearizable by Theorem 6.2. If we let

$$\begin{bmatrix} u_1 \\ u_2 \end{bmatrix} = \begin{bmatrix} 1 & 0 \\ -1 & 1 \end{bmatrix} \begin{bmatrix} \bar{u}_1 \\ \bar{u}_2 \end{bmatrix}$$

then we have

$$\dot{x} = \begin{bmatrix} x_2 \\ x_3 \\ 0 \\ 0 \end{bmatrix} + \begin{bmatrix} 0 \\ 0 \\ 1 \\ 0 \end{bmatrix} \bar{u}_1 + \begin{bmatrix} 0 \\ x_4 \\ 0 \\ 1+x_3 \end{bmatrix} \bar{u}_2 \tag{6.77}$$

$$= f(x) + g'_1(x)\bar{u}_1 + g'_2(x)\bar{u}_2.$$

In Example 6.4.1, it is shown that system (6.77) is restricted dynamic feedback linearizable. In other words, system (6.76) is linearizable by the extended state transformation (6.74) and the dynamic feedback

$$\begin{bmatrix} u_1 \\ u_2 \end{bmatrix} = \begin{bmatrix} \frac{-(1+x_3)^2 z^2}{1+x_3-x_4 z} \\ z + \frac{(1+x_3)^2 z^2}{1+x_3-x_4 z} \end{bmatrix} + \begin{bmatrix} \frac{1+x_3}{1+x_3-x_4 z} & \frac{-x_4}{1+x_3-x_4 z} \\ \frac{-(1+x_3)}{1+x_3-x_4 z} & \frac{x_4}{1+x_3-x_4 z} \end{bmatrix} \begin{bmatrix} v_1 \\ v_2 \end{bmatrix}$$

$$\dot{z} = \frac{(1+x_3)z^3}{1+x_3-x_4 z} + \begin{bmatrix} \frac{-z}{1+x_3-x_4 z} & \frac{1}{1+x_3-x_4 z} \end{bmatrix} \begin{bmatrix} v_1 \\ v_2 \end{bmatrix}. \tag{6.78}$$

However, dynamic feedback (6.78) is not a restricted dynamic feedback. $\square$

Example 6.4.3 Consider system (6.79) that is not reachable on a neighborhood of $0 \in \mathbb{R}^n$.

$$\dot{x} = \begin{bmatrix} 0 \\ x_3 \\ 0 \\ 0 \end{bmatrix} + \begin{bmatrix} 0 \\ 0 \\ 1 \\ 0 \end{bmatrix} u_1 + \begin{bmatrix} x_2 \\ 0 \\ 0 \\ 1 \end{bmatrix} u_2 = f(x) + g_1(x)u_1 + g_2(x)u_2. \tag{6.79}$$

(a) Show that system (6.79) is restricted dynamic feedback linearizable with $\mathbf{d} = (0, 2)$. Also, find out the restricted dynamic feedback and extended state transformation.

(b) Let $x(0) = \begin{bmatrix} 1 & 1 & 1 & 1 \end{bmatrix}^\mathsf{T}$. Find an input $u(t)$, $0 \le t \le t_f$ such that $t_f = 2$ and $x(t_f) = \begin{bmatrix} 0 & 0 & 0 & 0 \end{bmatrix}^\mathsf{T}$.

Solution (a) Since $(\kappa_1, \kappa_2) = (2, 1)$ and $\kappa_1 + \kappa_2 < 4 = n$, system (6.79) is not feedback linearizable. It is also easy to see that system (6.79) is not restricted dynamic feedback linearizable with $\mathbf{d} = (0, 1)$. Consider the following dynamic compensator with index $(d_1, d_2) = (0, 2)$:

$$\begin{bmatrix} u_1 \\ u_2 \end{bmatrix} = \begin{bmatrix} w_1 \\ z_1 \end{bmatrix} ; \quad \begin{bmatrix} \dot{z}_1 \\ \dot{z}_2 \end{bmatrix} = \begin{bmatrix} z_2 \\ w_2 \end{bmatrix}.$$

Then we have the extended system

$$\begin{bmatrix} \dot{x}_1 \\ \dot{x}_2 \\ \dot{x}_3 \\ \dot{x}_4 \\ \dot{z}_1 \\ \dot{z}_2 \end{bmatrix} = \begin{bmatrix} x_2 z_1 \\ x_3 \\ 0 \\ z_1 \\ z_2 \\ 0 \end{bmatrix} + \begin{bmatrix} 0 \\ 0 \\ 1 \\ 0 \\ 0 \\ 0 \end{bmatrix} w_1 + \begin{bmatrix} 0 \\ 0 \\ 0 \\ 0 \\ 0 \\ 1 \end{bmatrix} w_2 \tag{6.80}$$

$$= F(x, z) + G_1(x, z)w_1 + G_2(x, z)w_2.$$

Since

$$\begin{bmatrix} \mathrm{ad}_F G_1 & \mathrm{ad}_F G_2 & \mathrm{ad}_F^2 G_1 & \mathrm{ad}_F^2 G_2 \end{bmatrix} = \begin{bmatrix} 0 & 0 & z_1 & x_2 \\ -1 & 0 & 0 & 0 \\ 0 & 0 & 0 & 0 \\ 0 & 0 & 0 & 1 \\ 0 & -1 & 0 & 0 \\ 0 & 0 & 0 & 0 \end{bmatrix}$$

it is easy to see that $(\kappa_1, \kappa_2) = (3, 3)$ on $U_E \left(= \{(x, z) \in \mathbb{R}^6 \mid z_1 \neq 0\}\right)$, $\kappa_1 + \kappa_2 = 6 = n + d$, and distributions $D_0(x, z)$ and $D_1(x, z)$ are involutive distributions on U_E, where

$$D_0(x, z) = \text{span}\, \{G_1(x, z), G_2(x, z)\}$$

and

$$D_1(x, z) = \text{span}\, \{G_1(x, z), G_2(x, z), \text{ad}_F G_1(x, z), \text{ad}_F G_2(x, z)\}\,.$$

Hence, by Theorem 4.3, system (6.80) is (static) feedback linearizable on U_E. Scalar functions $S_{11}(x, z) = x_1$ and $S_{21}(x, z) = x_4$ satisfying the conditions of Lemma 4.3 can be easily found. Thus, extended state transformation $\xi = S_E(x, z)$ and extended static feedback $w = \alpha(x, z) + \beta(x, z)v$ can be obtained by (4.56) and (4.57), respectively.

$$\xi = S_E(x, z) \triangleq \begin{bmatrix} S_E^1(x, z) \\ S_E^2(x, z) \end{bmatrix} = \begin{bmatrix} S_{11}(x, z) \\ L_F S_{11}(x, z) \\ L_F^2 S_{11}(x, z) \\ S_{21}(x, z) \\ L_F S_{21}(x, z) \\ L_F^2 S_{21}(x, z) \end{bmatrix} = \begin{bmatrix} x_1 \\ x_2 z_1 \\ x_3 z_1 + x_2 z_2 \\ x_4 \\ z_1 \\ z_2 \end{bmatrix} \qquad (6.81)$$

and

$$\begin{bmatrix} w_1 \\ w_2 \end{bmatrix} = \begin{bmatrix} z_1 & x_2 \\ 0 & 1 \end{bmatrix}^{-1} \left(- \begin{bmatrix} 2x_3 z_2 \\ 0 \end{bmatrix} + \begin{bmatrix} v_1 \\ v_2 \end{bmatrix} \right)$$

$$= \begin{bmatrix} -\frac{2x_3 z_2}{z_1} \\ 0 \end{bmatrix} + \begin{bmatrix} \frac{1}{z_1} & \frac{-x_2}{z_1} \\ 0 & 1 \end{bmatrix} \begin{bmatrix} v_1 \\ v_2 \end{bmatrix}.$$

In other words, the restricted dynamic feedback for system (6.79) is

$$\begin{bmatrix} u_1 \\ u_2 \end{bmatrix} = \begin{bmatrix} -\frac{2x_3 z_2}{z_1} \\ z_1 \end{bmatrix} + \begin{bmatrix} \frac{1}{z_1} & \frac{-x_2}{z_1} \\ 0 & 0 \end{bmatrix} \begin{bmatrix} v_1 \\ v_2 \end{bmatrix}$$

$$\begin{bmatrix} \dot{z}_1 \\ \dot{z}_2 \end{bmatrix} = \begin{bmatrix} z_2 \\ 0 \end{bmatrix} + \begin{bmatrix} 0 & 0 \\ 0 & 1 \end{bmatrix} \begin{bmatrix} v_1 \\ v_2 \end{bmatrix} \qquad (6.82)$$

and the extended system of system (6.79) with the dynamic feedback (6.82) satisfies, in ξ-coordinates, the following controllable linear system:

$$\begin{bmatrix} \dot{\xi_1} \\ \dot{\xi_2} \\ \dot{\xi_3} \\ \dot{\xi_4} \\ \dot{\xi_5} \\ \dot{\xi_6} \end{bmatrix} = \begin{bmatrix} 0\,1\,0\,0\,0\,0 \\ 0\,0\,1\,0\,0\,0 \\ 0\,0\,0\,0\,0\,0 \\ 0\,0\,0\,0\,1\,0 \\ 0\,0\,0\,0\,0\,1 \\ 0\,0\,0\,0\,0\,0 \end{bmatrix} \begin{bmatrix} \xi_1 \\ \xi_2 \\ \xi_3 \\ \xi_4 \\ \xi_5 \\ \xi_6 \end{bmatrix} + \begin{bmatrix} 0\,0 \\ 0\,0 \\ 1\,0 \\ 0\,0 \\ 0\,0 \\ 0\,1 \end{bmatrix} \begin{bmatrix} v_1 \\ v_2 \end{bmatrix}$$

$$= \begin{bmatrix} A_{11} & O \\ O & A_{22} \end{bmatrix} \xi + \begin{bmatrix} B_{11} & O \\ O & B_{22} \end{bmatrix} v. \tag{6.83}$$

(b) Since $z_1(t) \neq 0$ for $0 \leq t \leq t_f$, we let $x_4(t) = 1 - \frac{1}{2}t$, $z_1(t) = \dot{x}_4(t) = -\frac{1}{2}$, $z_2(t) = \dot{z}_1(t) = 0$, and $v_2(t) = \dot{z}_2(t) = 0$, for $0 \leq t \leq t_f$, where $z(0) = \begin{bmatrix} -\frac{1}{2} \\ 0 \end{bmatrix}$.
In order to control $\{x_1(t), x_2(t), x_3(t)\}$ or $\{\xi_1(t), \xi_2(t), \xi_3(t)\}$, consider the following controllability Gramian of linear subsystem in (6.83):

$$W_{11}(0, t) \triangleq \int_0^t e^{-A_{11}\tau} B_{11} B_{11}^T (e^{-A_{11}\tau})^T d\tau = \begin{bmatrix} \frac{t^5}{20} & -\frac{t^4}{8} & \frac{t^3}{6} \\ -\frac{t^4}{8} & \frac{t^3}{3} & -\frac{t^2}{2} \\ \frac{t^3}{6} & -\frac{t^2}{2} & t \end{bmatrix}.$$

Since $z(0) = \begin{bmatrix} -\frac{1}{2} \\ 0 \end{bmatrix}$ and $z(t_f) = \begin{bmatrix} -\frac{1}{2} \\ 0 \end{bmatrix}$, it is clear that

$$\xi^1(0) = S_E^1(x(0), z(0)) = \begin{bmatrix} 1 \\ -\frac{1}{2} \\ -\frac{1}{2} \end{bmatrix} \text{ and } \xi^1(t_f) = S_E^1(x(t_f), z(t_f)) = \begin{bmatrix} 0 \\ 0 \\ 0 \end{bmatrix}$$

where $\xi^1 = S_E^1(x, z)$ is given in (6.81). Thus, it is easy to see that

$$v_1(t) = B_{11}^T (e^{-A_{11}t})^T W_{11}(0, t_f)^{-1} \left[e^{-A_{11}t_f} \xi^1(t_f) - \xi^1(0) \right]$$

$$= -\frac{3}{4} + 6t - \frac{15}{4}t^2, \quad 0 \leq t \leq 2$$

is an input such that $\xi(2) = \begin{bmatrix} 0\,0\,0\,0 -\frac{1}{2}\,0 \end{bmatrix}^T$ and $\begin{bmatrix} x(2) \\ z(2) \end{bmatrix} = \begin{bmatrix} 0\,0\,0\,0 -\frac{1}{2}\,0 \end{bmatrix}^T.$

□

6.5   MATLAB Programs

In this section, the following subfunctions in Appendix C are needed:
   **adfg, adfgk, adfgM, ChExact, ChZero, ChInvolutive, Codi,
   CXexact, Delta, Kindex0, Lfh, Lfhk, S1M**

The following is a MATLAB subfunction program for Theorem 6.2.

```
function d=dec2N(a,N,m)

d=zeros(1,m);
for k=1:m-1
  d(k)=fix(a/power(N,m-k));
  a=rem(a,power(N,m-k));
end
d(m)=a;
```

The following is a MATLAB subfunction program for Theorem 6.2.

```
function [kappa,D]=KindexE0z(fe,ge,xe,x)

[N,m]=size(ge);
D1=Delta(fe,ge,xe);
D0=subs(D1,x,x-x);
kappa=zeros(m,1); DD=xe-xe;
for k1=1:N
  for k2=1:m
    t1=[DD D0(:,m*(k1-1)+k2)];
    if rank(t1)>rank(DD)
      kappa(k2)=kappa(k2)+1;
      DD=t1;
      if rank(DD)==rank(D0)
        D=D1(:,1:m*max(kappa));
        return
      end
    end
  end
end
end
```

The following is a MATLAB subfunction program for Theorem 6.2.

```
function [out,dd,F,G,xe]=dRDFL(d,fx,g,x)

out=0; dd=d;
[n,m]=size(g);
sumd=sum(d);
for k=1:m
  s(k)=sum(d(1:k));
end
z0=sym('z',[2*n-3,m]);
```

```
zz=x(1)-x(1);
for k=1:m
  zz=[zz; z0(1:d(k),k)];
end
z=zz(2:sumd+1);
xe=[x; z];
F=xe-xe;
bz=xe-xe;
for k=1:m
  G(:,k)=xe-xe;
end
bf=[fx; z-z];
for k=1:m
  bg(:,k)= [g(:,k); z-z];
end
bz=x-x;
for k=1:m
  if d(k) >= 1
    bz=[bz; z0(2:d(k),k); x(1)-x(1)];
  end
end
bz=simplify(bz);
F=bf;
for k=1:m
  if d(k) ~= 0
    F=F+z0(1,k)*bg(:,k);
    G(:,k)=xe-xe;
    G(n+s(k),k)=1;
  else
    F=F;
    G(:,k)=bg(:,k);
  end
end
F=simplify(F+bz);
G=simplify(G);
[kappaE,De]=KindexE0z(F,G,xe,x);
if sum(kappaE) < n+sumd
  return
end
for k=1:max(kappaE)-1
  TD=De(:,1:k*m);
  if rank(TD) ~= rank(subs(TD,x,x-x))
    return
  end
  if ChInvolutive(TD,xe)==0
    return
  end
end
out=1;
```

MATLAB program for Theorem 6.2:

```
clear all
syms x1 x2 x3 x4 x5 x6 x7 x8 x9 x10 x11 x12 real

fx=[x2; x3; x4; 0; 0];
g=[ 0 x1; 0 0; 0 0; 1 0; 0 (1+x4)]; %Ex:6.3.1, n=5

% fx=[x2; x3; 0; 0];
% g=[ 0 x1; 0 0; 1 0; 0 (1+x3)]; %Ex:6.3.1, n=4

% g=[ 0 x1 x1; 0 0 0; 1 0 0; 0 (1+x3) 0; 0 0 (1+x3)];
% fx=[x2; x3; 0; 0; 0]; %Ex:6.3.1, m=3

% fx=[x2; x3; 0; 0]; g=[0 0; 0 x4; 1 0; 0 1+x3]; %Ex:6.4.1

% fx=[x2; x3; 0; 0]; g=[0 0; x4 x4; 1 0; 1+x3 1+x3]; %Ex:6.4.2

% fx=[0; x3; 0; 0]; g=[0 x2; 0 0; 1 0; 0 1]; %Ex:6.4.3

% fx=[x2; x3; 0; 0]; g=[0 x1^3; 0 0; 1 0; 0 1+x1]; %P:6-6(a)

% fx=[x2; x3; 0; 0]; g=[0 x2^3; 0 0; 1 0; 0 1+x1]; %P:6-6(b)

% fx=[x2; x3; 0; 0]; g=[0 x3; 0 0; 1 0; 0 1+x3]; %P:6-6(c)

% fx=[x2; x3; 0; 0]; g=[0 x3^2; 0 0; 1 0; 0 1+x3]; %P:6-6(d)

% fx=[x2; x3+x2^2; x1^2]; g=[x1-x1; 1; 1]; %P:6-6(e)

% fx=[x2; x3; 0; 0; 0];
% g=[0 0 x2^2; 0 x3 0; 1 0 0; 0 1 0; 0 0 1]; %P:6-6(f)

% fx=[x1-x1; 0; 0]; g=[0 x2; 1 0; 0 1]; %P:6-7(a)

% fx=[0; x3; 0; 0]; g=[0 x3; 0 0; 1 0; 0 1]; %P:6-7(b)

fx=simplify(fx)
g=simplify(g)
[n,m]=size(g); x=sym('x',[n,1]);

% d=[0 1 2]
% [out,dd,F,G,xe]=dRDFL(d,fx,g,x)
% [kappaE,D]=KindexE0z(F,G,xe,x)
% return

N=2*n-2;
for k=0:N^m-1
  d=dec2N(k,N,m)
  if min(d)==0
    [out,dd,F,G,xe]=dRDFL(d,fx,g,x);
```

```
      if out==1
        break
      end
    end
end

if out ==0
  display('System is not restricted dynamic FB linearizable.')
  return
end

display('System is restricted dynamic FB linearizable with')
d=dd

[FLAG,dd,F,G,xe]=dRDFL(d,fx,g,x)
[kappaE,D]=KindexE0z(F,G,xe,x)

[kappa,D]=Kindex0(F,G,xe);
if sum(kappa)<length(xe)
  display('Find out xi=Se(xe) without MATLAB.')
  return
end

[flag,Se1]=S1M(F,G,xe,kappa)
if flag==0
  display('Find out xi=Se(xe) without MATLAB.')
  return
end

Se=xe(1)-xe(1);
for k1=1:m
  for k=1:kappa(k1)
    t1=Lfhk(F,Se1(k1),xe,k-1);
    Se=[Se; t1];
  end
end
Se=simplify(Se(2:length(xe)+1))
t2=Se1-Se1;
for k1=1:m
  t2(k1)=Lfhk(F,Se1(k1),xe,kappa(k1)-1);
end
t2=simplify(t2);

ibeta=simplify(Lfh(G,t2,xe));
beta=simplify(inv(ibeta))
t3=simplify(Lfh(F,t2,xe));
alpha=simplify(-beta*t3)

hG=simplify(G*beta)
hF=simplify(F+G*alpha)

dSe=simplify(jacobian(Se,xe));
idSe=simplify(inv(dSe));
```

```
AS=simplify(dSe*hF);
dAS=simplify(jacobian(AS,xe));
A=simplify(dAS*idSe)
B=simplify(dSe*hG)

return
```

6.6 Problems

6-1. Solve Example 6.2.1.

6-2. Solve Example 6.2.3.

6-3. Show that the converse of Theorem 6.1 also holds.

6-4. Prove Corollary 6.2.

6-5. Solve Example 6.3.1 for $m \geq 3$.

6-6. Find out whether or not the following nonlinear system is restricted dynamic feedback linearizable. If it is restricted dynamic feedback linearizable, find out the restricted dynamic feedback and state transformation.

(a) $\dot{x} = \begin{bmatrix} x_2 \\ x_3 \\ 0 \\ 0 \end{bmatrix} + \begin{bmatrix} 0 \\ 0 \\ 1 \\ 0 \end{bmatrix} u_1 + \begin{bmatrix} x_1^3 \\ 0 \\ 0 \\ 1+x_1 \end{bmatrix} u_2$

(b) $\dot{x} = \begin{bmatrix} x_2 \\ x_3 \\ 0 \\ 0 \end{bmatrix} + \begin{bmatrix} 0 \\ 0 \\ 1 \\ 0 \end{bmatrix} u_1 + \begin{bmatrix} x_2^3 \\ 0 \\ 0 \\ 1+x_1 \end{bmatrix} u_2$

(c) $\dot{x} = \begin{bmatrix} x_2 \\ x_3 \\ 0 \\ 0 \end{bmatrix} + \begin{bmatrix} 0 \\ 0 \\ 1 \\ 0 \end{bmatrix} u_1 + \begin{bmatrix} x_3 \\ 0 \\ 0 \\ 1+x_3 \end{bmatrix} u_2$

(d) $\dot{x} = \begin{bmatrix} x_2 \\ x_3 \\ 0 \\ 0 \end{bmatrix} + \begin{bmatrix} 0 \\ 0 \\ 1 \\ 0 \end{bmatrix} u_1 + \begin{bmatrix} x_3^2 \\ 0 \\ 0 \\ 1+x_3 \end{bmatrix} u_2$

(e) $\dot{x} = \begin{bmatrix} x_2 \\ x_3 + x_2^2 \\ x_1^2 \end{bmatrix} + \begin{bmatrix} 0 \\ 1 \\ 1 \end{bmatrix} u$

$$\text{(f)} \quad \dot{x} = \begin{bmatrix} x_2 \\ x_3 \\ 0 \\ 0 \\ 0 \end{bmatrix} + \begin{bmatrix} 0 \\ 0 \\ 1 \\ 0 \\ 0 \end{bmatrix} u_1 + \begin{bmatrix} 0 \\ x_3 \\ 0 \\ 1 \\ 0 \end{bmatrix} u_2 + \begin{bmatrix} x_2^2 \\ 0 \\ 0 \\ 0 \\ 1 \end{bmatrix} u_3$$

6-7. Consider the following systems that are not reachable on a neighborhood of $0 \in \mathbb{R}^n$. Show that they are restricted dynamic feedback linearizable. Also, find out the restricted dynamic feedback and extended state transformation.

(a)

$$\dot{x} = \begin{bmatrix} 0 \\ 0 \\ 0 \end{bmatrix} + \begin{bmatrix} 0 \\ 1 \\ 0 \end{bmatrix} u_1 + \begin{bmatrix} x_2 \\ 0 \\ 1 \end{bmatrix} u_2$$

(b)

$$\dot{x} = \begin{bmatrix} 0 \\ x_3 \\ 0 \\ 0 \end{bmatrix} + \begin{bmatrix} 0 \\ 0 \\ 1 \\ 0 \end{bmatrix} u_1 + \begin{bmatrix} x_3 \\ 0 \\ 0 \\ 1 \end{bmatrix} u_2$$

6-8. Define the dynamic feedback linearization problem of the discrete time control systems. In other words, obtain the discrete version of Definition 6.1.

6-9. Define the restricted dynamic feedback linearization problem of the discrete time control systems. In other words, obtain the discrete version of Definition 6.2.

6-10. Find out the necessary and sufficient conditions for the discrete time control systems to be restricted dynamic feedback linearizable. In other words, obtain the discrete version of Theorem 6.2.

Chapter 7
Linearization of Discrete Time Control Systems

7.1 Introduction

In Chaps. 3–6, we have discussed the linearization of continuous nonlinear control systems. The following four different linearization problems can be defined depending on the use of feedback and the consideration of output.

- State equivalence to a linear system without the output.
- Feedback linearization without the output.
- State equivalence to a linear system with the output.
- Feedback linearization with the output.

In this chapter, it will be shown that the idea of linearization can be applied to the discrete time nonlinear control systems. The discrete version of Chap. 6 can also be found in (E12).

Example 7.1.1 Consider the following discrete linear control system:

$$\begin{bmatrix} z_1(t+1) \\ z_2(t+1) \end{bmatrix} = \begin{bmatrix} 0 & 1 \\ 1 & 0 \end{bmatrix} \begin{bmatrix} z_1(t) \\ z_2(t) \end{bmatrix} + \begin{bmatrix} 0 \\ 1 \end{bmatrix} u(t) = \tilde{f}(z(t), u(t)). \tag{7.1}$$

Find the state equation in x-coordinates with state transformation

$$\begin{bmatrix} x_1 \\ x_2 \end{bmatrix} = \begin{bmatrix} z_1 + z_2^2 \\ z_2 \end{bmatrix} \triangleq S(z). \tag{7.2}$$

Solution It is clear that

$$\begin{bmatrix} z_1 \\ z_2 \end{bmatrix} = \begin{bmatrix} x_1 - x_2^2 \\ x_2 \end{bmatrix} = S^{-1}(x). \tag{7.3}$$

© The Author(s), under exclusive license to Springer Nature Singapore Pte Ltd. 2022
H.-G. Lee, *Linearization of Nonlinear Control Systems*,
https://doi.org/10.1007/978-981-19-3643-2_7

Thus, we have

$$x_1(t+1) = z_1(t+1) + z_2(t+1)^2 = z_2(t) + (z_1(t) + u(t))^2$$
$$= x_2(t) + \{x_1(t) - x_2(t)^2 + u(t)\}^2$$
$$x_2(t+1) = z_2(t+1) = z_1(t) + u(t) = x_1(t) - x_2(t)^2 + u(t)$$

or

$$\begin{bmatrix} x_1(t+1) \\ x_2(t+1) \end{bmatrix} = S \circ \tilde{f}\left(S^{-1}(x(t)), u(t)\right)$$
$$= \begin{bmatrix} x_2(t) + \{x_1(t) - x_2(t)^2 + u(t)\}^2 \\ x_1(t) - x_2(t)^2 + u(t) \end{bmatrix}. \tag{7.4}$$

$\square$

In the above example, we have obtained the nonlinear system (7.4) from a linear system (7.1) with a state transformation (7.2). Conversely, we can obtain, by using state transformation (7.3), the linear system (7.1) from a nonlinear system (7.4). System (7.4) is not an affine system, whereas the system (7.1) is affine. Unlike the continuous case, the state equivalent system to a discrete linear system may not be affine. Therefore, for the linearization problems of the discrete systems, the general nonlinear systems should be considered. For the feedback linearization problems, the general feedback $u = \gamma(x, v)$ should also be used rather than the affine form $u = \alpha(x) + \beta(x)v$. Consider the following discrete nonlinear control system:

$$x(t+1) = F(x(t), u(t)) \triangleq F_u(x)$$
$$y(t) = h(x(t)) \tag{7.5}$$

where $x \in \mathbb{R}^n$, $u \in \mathbb{R}^m$, $y \in \mathbb{R}^q$, and $F(x, u) : \mathbb{R}^{n+m} \to \mathbb{R}^n$ and $h(x)$ are smooth functions with $f(0, 0) = 0$ and $h(0) = 0$.

Definition 7.1 (*state equivalence to a linear system*)
System (7.5) is said to be state equivalent to a linear system if there exists a state transformation $z = S(x)$ such that

$$z(t+1) = Az(t) + Bu(t)$$
$$y(t) = h \circ S^{-1}(z(t)) \tag{7.6}$$

or

$$\tilde{F}_u(z) \triangleq S \circ F_u \circ S^{-1}(z) = Az + Bu \tag{7.7}$$

where

$$\mathrm{rank}\left([B \; AB \; \cdots \; A^{n-1}B]\right) = n.$$

Definition 7.2 (*state equivalence to a linear system with output*)
System (7.5) is said to be state equivalent to a linear system with output if there exists
a state transformation $z = S(x)$ such that

$$z(t+1) = Az(t) + Bu(t)$$
$$y(t) = Cz(t)$$

or

$$\tilde{F}_u(z) \triangleq S \circ F_u \circ S^{-1}(z) = Az + Bu \; ; \;\; \tilde{h}(z) \triangleq h \circ S^{-1}(z) = Cz$$

where

$$\text{rank}\left([B \; AB \; \cdots \; A^{n-1}B]\right) = n.$$

Definition 7.3 (*feedback linearization*)
System (7.5) is said to be feedback linearizable if there exist a nonsingular feedback
$u = \gamma(x, v) \left(\det\left\{\frac{\partial \gamma(x,v)}{\partial v}\right\} \neq 0\right)$ and a state transformation $z = S(x)$ such that the
closed-loop system satisfies, in z–coordinates, the following Brunovsky canonical
form:

$$z(t+1) = \begin{bmatrix} A_{11} & O & \cdots & O \\ O & A_{22} & \cdots & O \\ \vdots & \vdots & \ddots & \vdots \\ O & O & \cdots & A_{mm} \end{bmatrix} z(t) + \begin{bmatrix} B_{11} & O & \cdots & O \\ O & B_{22} & \cdots & O \\ \vdots & \vdots & \ddots & \vdots \\ O & O & \cdots & B_{mm} \end{bmatrix} v(t)$$
$$= Az(t) + Bv(t)$$

or

$$\tilde{F}_v(z) \triangleq S \circ F(S^{-1}(z), \gamma(S^{-1}(z), v)) = Az + Bv$$

where $\sum_{i=1}^{m} \kappa_i = n$, $z = \begin{bmatrix} z^1 \\ \vdots \\ z^m \end{bmatrix}$, $z^i = \begin{bmatrix} z_1^i \\ \vdots \\ z_{\kappa_i}^i \end{bmatrix}$, and

$$A_{ii} = \begin{bmatrix} 0 & 1 & 0 & \cdots & 0 & 0 \\ 0 & 0 & 1 & \cdots & 0 & 0 \\ \vdots & \vdots & \vdots & & \vdots & \vdots \\ 0 & 0 & 0 & \cdots & 0 & 1 \\ 0 & 0 & 0 & \cdots & 0 & 0 \end{bmatrix} (\kappa_i \times \kappa_i), \quad B_{ii} = \begin{bmatrix} 0 \\ 0 \\ \vdots \\ 0 \\ 1 \end{bmatrix} (\kappa_i \times 1).$$

$$z(t + 1) = Az(t) + Bu(t)$$

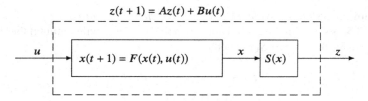

Fig. 7.1 Linearization of discrete time system by state transformation

$$z(t + 1) = Az(t) + Bu(t)$$
$$y = Cz(t)$$

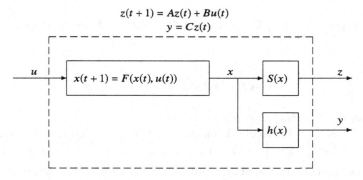

Fig. 7.2 Discrete state equivalence to a linear system with output

Definition 7.4 (*feedback linearization with output*)
System (7.5) is said to be feedback linearizable with output if there exist a nonsingular
feedback $u = \gamma(x, v)$ $\left(\det \left\{ \frac{\partial \gamma(x, v)}{\partial v} \right\} \neq 0 \right)$ and a state transformation $z = S(x)$ such
that the closed-loop system satisfies, in z−coordinates, the following Brunovsky
canonical form:

$$z(t + 1) = Az(t) + Bv(t)$$
$$y(t) = Cz(t) \tag{7.8}$$

or

$$\tilde{F}_v(z) \triangleq S \circ F(S^{-1}(z), \gamma(S^{-1}(z), v)) = Az + Bv$$
$$\tilde{h}(z) \triangleq h \circ S^{-1}(z) = Cz$$

where A and B are defined in Definition 7.3 and C is a $q \times n$ constant matrix.

Block diagrams of the linearization problems defined above are shown in Figs. 7.1,
7.2, 7.3, and 7.4. The state equation of the continuous system is a differential equation,
whereas that of the discrete system is a difference equation. Therefore, the conditions
for discrete linearization problems are very different from those for the continuous
case. Suppose that $z = S(x)$ is a state transformation. Then, it is easy to see that
system (7.5) satisfies, in z-coordinates,

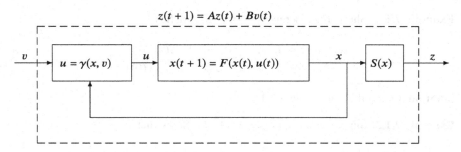

Fig. 7.3 Feedback linearization of discrete time system

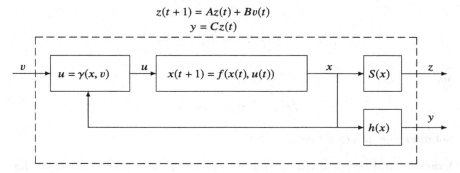

Fig. 7.4 Feedback linearization of discrete time system with output

$$z(t+1) = S(x(t+1)) = S \circ F_{u(t)}(x(t))$$
$$= S \circ F_{u(t)} \circ S^{-1}(z(t)) \triangleq \tilde{F}_{u(t)}(z(t)). \tag{7.9}$$

For the continuous system, the vector fields $f(x)$ and $g_j(x)$ are $S_*(f(x))$ and $S_*(g_j(x))$ in z-coordinates when $z = S(x)$. (Refer to (3.4) and (3.27).) And we use that $S_* \left(\mathrm{ad}_f^k g_j(x) \right)$, $1 \le j \le m$, $k \ge 0$, are constant vector fields when linearizable. In other words, if $S_*(f(x)) = Az$ and $S_*(g_j(x)) = b_j$, then we have that for $1 \le j \le m$ and $k \ge 0$,

$$S_* \left(\mathrm{ad}_f^k g_j(x) \right) = (-1)^k A^k b_j.$$

For discrete system (7.5), the right-hand side of the state equation is, in z-coordinates, the composite function

$$\tilde{F}_u(z) \triangleq S \circ F_u \circ S^{-1}(z).$$

Let $F_0^0(x) = x$, $\hat{F}_u^0(x) = x$, and for $k \ge 1$,

$$F_0^k = F_0^{k-1} \circ F_0(x) \quad \text{and} \quad \hat{F}_u^k(x) = F_0^{k-1} \circ F_u(x). \tag{7.10}$$

Example 7.1.2 Show that discrete system (7.5) satisfies

$$x(t+k) = F_{u(t+k-1)} \circ \cdots \circ F_{u(t)}(x(t))$$
$$y(k) = h \circ F_{u(k-1)} \circ \cdots \circ F_{u(0)}(x(0)).$$

Solution Omitted. (See Problem 7.1.)   □

Example 7.1.3 Suppose that $F_u(x) = Ax + Bu$. Show that

$$F_{u^1} \circ \cdots \circ F_{u^k}(x) = A^k x + \sum_{\ell=1}^{k} A^{\ell-1} B u^{\ell}$$

$$\left(F_{u^1} \circ \cdots \circ F_{u^k}(0) = \sum_{\ell=1}^{k} A^{\ell-1} B u^{\ell} \right)$$

and

$$\hat{F}_u^k(x) = A^k x + A^{k-1} B u.$$

Solution Omitted. (See Problem 7.2.)   □

Example 7.1.4 Suppose that $\tilde{F}_u(z) = S \circ F_u \circ S^{-1}(z)$ and $S(0) = 0$. Show that for $i \geq 1$,

$$F_{u^1} \circ \cdots \circ F_{u^i}(x) = S^{-1} \circ \tilde{F}_{u^1} \circ \cdots \circ \tilde{F}_{u^i} \circ S(x)$$
$$F_{u^1} \circ \cdots \circ F_{u^i}(0) = S^{-1} \circ \tilde{F}_{u^1} \circ \cdots \circ \tilde{F}_{u^i}(0)$$

and

$$\hat{F}_u^i(x) = S^{-1} \circ \hat{\tilde{F}}_u^i \circ S(x).$$

Solution Omitted. (See Problem 7.3.)   □

7.2   Single Input Discrete Time Systems

In this section, we consider the following single input discrete nonlinear system:

$$x(t+1) = F(x(t), u(t)) \triangleq F_{u(t)}(x(t)) \tag{7.11}$$

where $x \in \mathbb{R}^n, u \in \mathbb{R}$, and $F(x, u) : \mathbb{R}^{n+1} \to \mathbb{R}^n$ is a smooth function with $F(0, 0) = 0$. Let us define composite functions $\Psi : \mathbb{R}^n \to \mathbb{R}^n$ and $\mathcal{F} : \mathbb{R}^{n+1} \to \mathbb{R}^n$ as follows:

$$\Psi(U) = \Psi(u^1, \ldots, u^n) \triangleq F_{u^1} \circ F_{u^2} \circ \cdots \circ F_{u^n}(0) \tag{7.12}$$

$$\begin{aligned} \mathcal{F}(\tilde{U}) = \mathcal{F}(u^1, \ldots, u^{n+1}) &\triangleq F_{u^1} \circ \cdots \circ F_{u^n} \circ F_{u^{n+1}}(0) \\ &= F_{u^1} \circ \Psi(u^2, \ldots, u^{n+1}) \end{aligned} \tag{7.13}$$

where

$$U \triangleq \begin{bmatrix} u^1 \cdots u^n \end{bmatrix}^{\mathrm{T}} \text{ and } \tilde{U} \triangleq \begin{bmatrix} u^1 \cdots u^n \ u^{n+1} \end{bmatrix}^{\mathrm{T}}.$$

It is clear that

$$\Psi\left(u^1, \ldots, u^n\right) = \mathcal{F}\left(u^1, \ldots, u^n, 0\right). \tag{7.14}$$

If $\left.\frac{\partial \Psi(U)}{\partial U}\right|_{U=0}$ is nonsingular, then ker $\Psi_* = \text{span}\{0\}$. Thus, it is clear, by Theorem 2.6, that $\Psi_*\left(\frac{\partial}{\partial u^i}\right)$, $1 \le i \le n$ are well-defined vector fields and

$$\Psi_*\left(\frac{\partial}{\partial u^i}\right) = \left.\frac{\partial \Psi(U)}{\partial u^i}\right|_{U=\Psi^{-1}(z)}.$$

However, $\mathcal{F}(\tilde{U})$ is not 1-1 and ker $\mathcal{F}_* \ne \text{span}\{0\}$. Therefore, if

$$\left[\frac{\partial}{\partial u^i}, \text{ ker } \mathcal{F}_*\right] \not\subset \text{ker } \mathcal{F}_*,$$

then $\mathcal{F}_*\left(\frac{\partial}{\partial u^i}\right)$ is not a well-defined vector field. Suppose that $\mathcal{F}_*\left(\frac{\partial}{\partial u^i}\right)$ is a well-defined vector field. Then, it is clear, by Definition 2.11, that

$$\frac{\partial \mathcal{F}(U, u^{n+1})}{\partial u^i} = \left.\frac{\partial \mathcal{F}(U, u^{n+1})}{\partial u^i}\right|_{\tilde{U}=\begin{bmatrix} U \\ 0 \end{bmatrix}}$$

where $z = \mathcal{F}(U, u^{n+1}) = \mathcal{F}(U, 0) = \Psi(U)$. Therefore, we have, by Definition 2.12, that

$$\mathcal{F}_*\left(\frac{\partial}{\partial u^i}\right) = \left.\frac{\partial \mathcal{F}(U, u^{n+1})}{\partial u^i}\right|_{\begin{bmatrix} U \\ u^{n+1} \end{bmatrix}=\begin{bmatrix} \Psi^{-1}(z) \\ 0 \end{bmatrix}}. \tag{7.15}$$

Example 7.2.1 Let $F_u(x) = \begin{bmatrix} x_2 \\ x_1 x_2 + u \end{bmatrix}$. Find out $\Psi_*\left(\frac{\partial}{\partial u^1}\right)$, $\Psi_*\left(\frac{\partial}{\partial u^2}\right)$, $\mathcal{F}_*\left(\frac{\partial}{\partial u^1}\right)$, and $\mathcal{F}_*\left(\frac{\partial}{\partial u^3}\right)$. Also, show that $\mathcal{F}_*\left(\frac{\partial}{\partial u^2}\right)$ is not a well-defined vector field.

Solution It is clear that

$$\mathcal{F}(u^1, u^2, u^3) \triangleq F_{u^1} \circ F_{u^2} \circ F_{u^3}(0) = F_{u^1} \circ F_{u^2}\left(\begin{bmatrix} 0 \\ u^3 \end{bmatrix}\right)$$

$$= F_{u^1}\left(\begin{bmatrix} u^3 \\ u^2 \end{bmatrix}\right) = \begin{bmatrix} u^2 \\ u^2 u^3 + u^1 \end{bmatrix}$$

and

$$\Psi(u^1, u^2) \triangleq F_{u^1} \circ F_{u^2}(0) = \mathcal{F}(u^1, u^2, 0) = \begin{bmatrix} u^2 \\ u^1 \end{bmatrix}.$$

Since $\frac{\partial \mathcal{F}(u^1, u^2, u^3)}{\partial U} = \begin{bmatrix} 0 & 1 & 0 \\ 1 & u^3 & u^2 \end{bmatrix}$, we have that

$$\ker \mathcal{F}_* = \mathrm{span}\left\{-u^2 \frac{\partial}{\partial u^1} + \frac{\partial}{\partial u^3}\right\}$$

and

$$\left[\frac{\partial}{\partial u^i}, -u^2\frac{\partial}{\partial u^1} + \frac{\partial}{\partial u^3}\right] = 0 \in \ker \mathcal{F}_*, \text{ for } i = 1, 3$$

$$\left[\frac{\partial}{\partial u^2}, -u^2\frac{\partial}{\partial u^1} + \frac{\partial}{\partial u^3}\right] = -\frac{\partial}{\partial u^1} \notin \ker \mathcal{F}_*.$$

Thus, it is clear, by Theorem 2.6, that $\mathcal{F}_*\left(\frac{\partial}{\partial u^2}\right)$ is not a well-defined vector field. Finally, it is easy to see that

$$\Psi_*\left(\frac{\partial}{\partial u^1}\right) = \frac{\partial \Psi(u^1, u^2)}{\partial u^1}\bigg|_{\begin{bmatrix} u^1 \\ u^2 \end{bmatrix}=\begin{bmatrix} z_2 \\ z_1 \end{bmatrix}} = \begin{bmatrix} 0 \\ 1 \end{bmatrix} = \frac{\partial}{\partial z_2}$$

$$\Psi_*\left(\frac{\partial}{\partial u^2}\right) = \frac{\partial \Psi(u^1, u^2)}{\partial u^2}\bigg|_{\begin{bmatrix} u^1 \\ u^2 \end{bmatrix}=\begin{bmatrix} z_2 \\ z_1 \end{bmatrix}} = \begin{bmatrix} 1 \\ 0 \end{bmatrix} = \frac{\partial}{\partial z_1}$$

$$\mathcal{F}_*\left(\frac{\partial}{\partial u^1}\right) = \frac{\partial \mathcal{F}(u^1, u^2, u^3)}{\partial u^1}\bigg|_{\begin{bmatrix} u^1 \\ u^2 \\ u^3 \end{bmatrix}=\begin{bmatrix} z_2 \\ z_1 \\ 0 \end{bmatrix}} = \begin{bmatrix} 0 \\ 1 \end{bmatrix} = \frac{\partial}{\partial z_2}$$

$$\mathcal{F}_*\left(\frac{\partial}{\partial u^3}\right) = \frac{\partial \mathcal{F}(u^1, u^2, u^3)}{\partial u^3}\bigg|_{\begin{bmatrix} u^1 \\ u^2 \\ u^3 \end{bmatrix}=\begin{bmatrix} z_2 \\ z_1 \\ 0 \end{bmatrix}} = \begin{bmatrix} 0 \\ z_1 \end{bmatrix} = z_1\frac{\partial}{\partial z_2}.$$

$\square$

Example 7.2.2 Let $\frac{\partial \Psi(U)}{\partial U}\bigg|_{U=0}$ be nonsingular. Suppose that $\mathcal{F}_*\left(\frac{\partial}{\partial u^i}\right)$ is a well-defined vector field for $1 \leq i \leq n$. Show that for $1 \leq i \leq n$,

$$\mathcal{F}_* \left(\frac{\partial}{\partial u^i} \right) = \Psi_* \left(\frac{\partial}{\partial u^i} \right). \tag{7.16}$$

Solution Suppose that $\mathcal{F}_* \left(\frac{\partial}{\partial u^i} \right)$ is a well-defined vector field for $1 \leq i \leq n$. Then, it is easy to see, by (7.15), that for $1 \leq i \leq n$,

$$
\begin{aligned}
\mathcal{F}_* \left(\frac{\partial}{\partial u^i} \right) &= \frac{\partial \mathcal{F}(U, u^{n+1})}{\partial u^i} \Bigg|_{\left[{U \atop u^{n+1}} \right] = \left[{\Psi^{-1}(z) \atop 0} \right]} \\
&= \frac{\partial F_{u^1} \circ \cdots \circ F_{u^n} \circ F_{u^{n+1}}(0)}{\partial u^i} \Bigg|_{\left[{U \atop u^{n+1}} \right] = \left[{\Psi^{-1}(z) \atop 0} \right]} \\
&= \frac{\partial F_{u^1} \circ \cdots \circ F_{u^n}(0)}{\partial u^i} \Bigg|_{U = \Psi^{-1}(z)} = \Psi_* \left(\frac{\partial}{\partial u^i} \right).
\end{aligned}
$$

$\square$

Theorem 7.1 (conditions for state equivalence to a linear system)
System (7.11) is state equivalent to a linear system, if and only if

(i) $\frac{\partial \Psi(U)}{\partial U} \Big|_{U=0}$ *is nonsingular.*
(ii) $\mathcal{F}_* \left(\frac{\partial}{\partial u^i} \right)$, $1 \leq i \leq n+1$, *are well-defined vector fields or*

$$\left[\frac{\partial}{\partial u^i}, \ker \mathcal{F}_* \right] \subset \ker \mathcal{F}_*, \ 1 \leq i \leq n+1. \tag{7.17}$$

Furthermore, $z = S(x) = \Psi^{-1}(x)$ is a linearizing state transformation.

Proof Necessity. Suppose that system (7.11) is state equivalent to a linear system with state transformation $z = S(x)$. Then we have

$$
\begin{aligned}
z(t+1) &= S \circ F_{u(t)} \circ S^{-1}(z(t)) \\
&\triangleq \tilde{F}_{u(t)}(z(t)) = Az(t) + bu(t)
\end{aligned} \tag{7.18}
$$

where

$$\text{rank} \left([b \ Ab \ \cdots \ A^{n-1}b] \right) = n. \tag{7.19}$$

Since $F_u(x) = S^{-1} \circ \tilde{F}_u \circ S(x)$, $\tilde{F}_u(z) = Az + bu$, and $S(0) = 0$, it is easy to see, by Examples 7.1.3 and 7.1.4, that

$$
\begin{aligned}
\Psi(u^1, \ldots, u^n) &= S^{-1}(A^{n-1}bu^n + \cdots + bu^1) \\
\mathcal{F}(u^1, \ldots, u^{n+1}) &= S^{-1}(A^n bu^{n+1} + \cdots + bu^1)
\end{aligned} \tag{7.20}
$$

which implies that

$$\left.\frac{\partial \Psi(u^1, \ldots, u^n)}{\partial U}\right|_{U=0} = \left.\frac{\partial S^{-1}(z)}{\partial z}\right|_{z=0} \begin{bmatrix} b & Ab & \cdots & A^{n-1}b \end{bmatrix}$$

where $U = [u^1 \ u^2 \ \cdots \ u^n]^{\mathsf{T}}$. Since $\left.\frac{\partial S^{-1}(z)}{\partial z}\right|_{z=0}$ is nonsingular, it is clear, by (7.19), that $\left.\frac{\partial \Psi(U)}{\partial U}\right|_{U=0}$ is nonsingular. Also, it is clear, by (7.20), that

$$\mathcal{F}_*\left(\frac{\partial}{\partial u^i}\right) = (S^{-1})_*(A^{i-1}b), \ 1 \le i \le n+1 \tag{7.21}$$

which implies that $\mathcal{F}_*\left(\frac{\partial}{\partial u^i}\right)$, $1 \le i \le n+1$, are well-defined vector fields and (7.17) is, by Theorem 2.6, satisfied.

Sufficiency. Suppose that system (7.11) satisfies conditions (i) and (ii). By condition (i), it is clear that $z = S(x) = \Psi^{-1}(x)$ is a state transformation on a neighborhood of the origin. We will show that system (7.11) satisfies, in z-coordinates, a linear system. In other words, we will show that

$$\tilde{F}_u(z) \triangleq \Psi^{-1} \circ F_u \circ \Psi(z) = Az + bu \tag{7.22}$$

for some constant matrices A and b. If we let

$$Y^i = \mathcal{F}_*\left(\frac{\partial}{\partial u^i}\right), \ 1 \le i \le n+1, \tag{7.23}$$

then we have, by Theorem 2.4 and (7.16), that for $1 \le i \le n+1$ and $1 \le j \le n+1$,

$$[Y^i, Y^j] = \mathcal{F}_*\left(\left[\frac{\partial}{\partial u^i}, \frac{\partial}{\partial u^j}\right]\right) = \mathcal{F}_*(0) = 0 \tag{7.24}$$

and for $1 \le i \le n$,

$$\Psi_*\left(\frac{\partial}{\partial u^i}\right) = Y^i. \tag{7.25}$$

Thus, it is clear, by condition (i), that $\{Y^1, Y^2, \ldots, Y^n\}$ is a set of linearly independent vector fields on a neighborhood of the origin. Also, we have, by (7.24) and Example 2.4.20, that

$$Y^{n+1} = \sum_{i=1}^{n} a_i Y^i \tag{7.26}$$

for some constant $a_i \in \mathbb{R}$. It is easy to see, by (7.13), that $F_u \circ \Psi(z) = \mathcal{F}(u, z)$ and

$$\tilde{F}(z, u) \triangleq \tilde{F}_u(z) = \Psi^{-1} \circ F_u \circ \Psi(z) = \Psi^{-1} \circ \mathcal{F}(u, z). \tag{7.27}$$

Thus, we have, by (7.23), (7.25), and (7.27), that

$$\tilde{F}(z, u)_* (\frac{\partial}{\partial u}) = (\Psi^{-1} \circ \mathcal{F}(u, z))_* (\frac{\partial}{\partial u}) = (\Psi^{-1})_* (\mathcal{F}(u, z))_* (\frac{\partial}{\partial u})$$

$$= (\Psi^{-1})_* (Y^1) = \frac{\partial}{\partial z_1}. \tag{7.28}$$

Similarly, it is easy to see, by (7.23), (7.25), (7.26), and (7.27), that for $1 \leq i \leq n-1$,

$$\tilde{F}(z, u)_* \left(\frac{\partial}{\partial z_i} \right) = (\Psi^{-1} \circ \mathcal{F}(u, z))_* \left(\frac{\partial}{\partial z_i} \right)$$

$$= (\Psi^{-1})_* (Y^{i+1}) = \frac{\partial}{\partial z_{i+1}} \tag{7.29}$$

and

$$\tilde{F}(z, u)_* \left(\frac{\partial}{\partial z_n} \right) = (\Psi^{-1} \circ \mathcal{F}(u, z))_* \left(\frac{\partial}{\partial z_n} \right) = (\Psi^{-1})_* (Y^{n+1})$$

$$= (\Psi^{-1})_* \left(\sum_{i=1}^{n} a_i Y^i \right) = \sum_{i=1}^{n} a_i \frac{\partial}{\partial z_i}. \tag{7.30}$$

Therefore, it is clear, by (7.28)–(7.30), that

$$\tilde{F}(z, u) = \begin{bmatrix} 0 & 0 & \cdots & 0 & a_1 \\ 1 & 0 & \cdots & 0 & a_2 \\ 0 & 1 & \cdots & 0 & a_3 \\ \vdots & \vdots & & \vdots & \vdots \\ 0 & 0 & \cdots & 1 & a_n \end{bmatrix} z + \begin{bmatrix} 1 \\ 0 \\ 0 \\ \vdots \\ 0 \end{bmatrix} u. \tag{7.31}$$

It is easy to see, by (7.14), that $\frac{\partial \mathcal{F}(\tilde{U})}{\partial U} \big|_{\tilde{U}=0} = \frac{\partial \Psi(U)}{\partial U} \big|_{U=0}$. Thus, if condition (i) of Theorem 7.1 is satisfied, then it is clear that $\frac{\partial \mathcal{F}(\tilde{U})}{\partial U} = \left[\frac{\partial \mathcal{F}(\tilde{U})}{\partial U} \ \frac{\partial \mathcal{F}(\tilde{U})}{\partial u^{n+1}} \right]$ and

$$\ker \mathcal{F}_* = \operatorname{span} \left\{ \begin{bmatrix} -\left(\frac{\partial \mathcal{F}(\tilde{U})}{\partial U} \right)^{-1} \frac{\partial \mathcal{F}(\tilde{U})}{\partial u^{n+1}} \\ 1 \end{bmatrix} \right\}. \tag{7.32}$$

(Refer to MATLAB subfunction **ker-sF**.)

Example 7.2.3 Show that the following discrete time system is state equivalent to a linear system:

$$\begin{bmatrix} x_1(t+1) \\ x_2(t+1) \end{bmatrix} = \begin{bmatrix} x_2(t) - u(t)^2 \\ u(t) \end{bmatrix} = F(x(t), u(t)) = F_{u(t)}(x(t)). \qquad (7.33)$$

Solution It is easy to see, by (7.13) and (7.14), that

$$\mathcal{F}(u^1, u^2, u^3) \triangleq F_{u^1} \circ F_{u^2} \circ F_{u^3}(0) = F_{u^1} \circ F_{u^2} \left(\begin{bmatrix} -(u^3)^2 \\ u^3 \end{bmatrix} \right)$$

$$= F_{u^1} \left(\begin{bmatrix} u^3 - (u^2)^2 \\ u^2 \end{bmatrix} \right) = \begin{bmatrix} u^2 - (u^1)^2 \\ u^1 \end{bmatrix}$$

and

$$\Psi(u^1, u^2) \triangleq F_{u^1} \circ F_{u^2}(0) = \mathcal{F}(u^1, u^2, 0) = \begin{bmatrix} u^2 - (u^1)^2 \\ u^1 \end{bmatrix}.$$

Since $\det \left(\frac{\partial \Psi(U)}{\partial U} \right) = \det \left(\begin{bmatrix} -2u^1 & 1 \\ 1 & 0 \end{bmatrix} \right) = -1 \neq 0$, condition (i) of Theorem 7.1 is satisfied. Since $\frac{\partial \mathcal{F}(u^1, u^2, u^3)}{\partial U} = \begin{bmatrix} -2u^1 & 1 & 0 \\ 1 & 0 & 0 \end{bmatrix}$, we have that

$$\ker \mathcal{F}_* = \operatorname{span} \left\{ \frac{\partial}{\partial u^3} \right\}$$

and for $1 \leq i \leq 3$,

$$\left[\frac{\partial}{\partial u^i}, \frac{\partial}{\partial u^3} \right] = 0 \in \ker \mathcal{F}_*.$$

Thus, it is clear, by Theorem 2.6, that $\mathcal{F}_* \left(\frac{\partial}{\partial u^i} \right)$, $1 \leq i \leq 3$, are well-defined vector fields and condition (ii) of Theorem 7.1 is satisfied. Hence, by Theorem 7.1, system (7.33) is state equivalent to a linear system. Let

$$\begin{bmatrix} z_1 \\ z_2 \end{bmatrix} = S(x) = \Psi^{-1}(x) = \begin{bmatrix} x_2 \\ x_1 + x_2^2 \end{bmatrix}.$$

Then it is easy to see that

$$\tilde{F}_u(z) = S \circ F_u \circ S^{-1}(z) = S \left(\begin{bmatrix} z_1 - u^2 \\ u \end{bmatrix} \right) = \begin{bmatrix} 0 & 0 \\ 1 & 0 \end{bmatrix} \begin{bmatrix} z_1 \\ z_2 \end{bmatrix} + \begin{bmatrix} 1 \\ 0 \end{bmatrix} u.$$

$\square$

Theorem 7.1 is the discrete version of Theorem 3.1. Even though the reachability condition (i) is the same, the condition (ii) is quite different. For a discrete system, $\mathcal{F}_*\left(\frac{\partial}{\partial u^i}\right)$, $1 \leq i \leq n+1$, may not be well-defined vector fields. If they are well-defined vector fields, then they commute. Another difference is that the partial differential equation must be solved to obtain a linearizing state transformation for a continuous system. But in the case of a discrete system, it can be obtained directly as $z = S(x) = \Psi^{-1}(x)$.

Example 7.2.4 Show that the following discrete time system is not state equivalent to a linear system:

$$\begin{bmatrix} x_1(t+1) \\ x_2(t+1) \end{bmatrix} = \begin{bmatrix} x_2(t) \\ x_1(t)^2 + u(t) \end{bmatrix} = F_{u(t)}(x(t)). \tag{7.34}$$

Solution It is easy to see, by (7.13) and (7.14), that

$$\mathcal{F}(u^1, u^2, u^3) \triangleq F_{u^1} \circ F_{u^2} \circ F_{u^3}(0) = F_{u^1} \circ F_{u^2}\left(\begin{bmatrix} 0 \\ u^3 \end{bmatrix}\right)$$

$$= F_{u^1}\left(\begin{bmatrix} u^3 \\ u^2 \end{bmatrix}\right) = \begin{bmatrix} u^2 \\ u^1 + (u^3)^2 \end{bmatrix}$$

and

$$\Psi(u^1, u^2) \triangleq F_{u^1} \circ F_{u^2}(0) = \mathcal{F}(u^1, u^2, 0) = \begin{bmatrix} u^2 \\ u^1 \end{bmatrix}.$$

Since $\det\left(\frac{\partial \Psi(U)}{\partial U}\right) = \det\left(\begin{bmatrix} -2u^1 & 1 \\ 1 & 0 \end{bmatrix}\right) = -1 \neq 0$, condition (i) of Theorem 7.1 is satisfied. Since $\frac{\partial \mathcal{F}(u^1, u^2, u^3)}{\partial \bar{U}} = \begin{bmatrix} 0 & 1 & 0 \\ 1 & 0 & 2u^3 \end{bmatrix}$, we have, by (7.32), that

$$\ker \mathcal{F}_* = \text{span}\left\{-2u^3 \frac{\partial}{\partial u^1} + \frac{\partial}{\partial u^3}\right\}$$

and

$$\left[\frac{\partial}{\partial u^3}, \ -2u^3 \frac{\partial}{\partial u^1} + \frac{\partial}{\partial u^3}\right] = -2\frac{\partial}{\partial u^1} \notin \ker \mathcal{F}_*.$$

Thus, it is clear, by Theorem 2.6, that $\mathcal{F}_*\left(\frac{\partial}{\partial u^3}\right)$ is not a well-defined vector field and condition (ii) of Theorem 7.1 is not satisfied. Hence, by Theorem 7.1, system (7.34) is not state equivalent to a linear system.   □

It is clear that system (7.34) is linearizable by using feedback $u(t) = -x_1(t)^2 + v(t)$. In the following, the discrete version of Theorem 4.1 will be obtained.

Example 7.2.5 Suppose that for $0 \leq i \leq n-1$,

$$\frac{\partial}{\partial u}\left(S_1 \circ \hat{F}_u^i(x)\right) = 0 ; \quad \frac{\partial}{\partial u}\left(S_1 \circ \hat{F}_u^n(x)\right)\bigg|_{(0,0)} \neq 0. \tag{7.35}$$

Show that

$$\text{rank}\left(\begin{bmatrix} \frac{\partial S_1(x)}{\partial x} \\ \frac{\partial(S_1 \circ F_0(x))}{\partial x} \\ \vdots \\ \frac{\partial\left(S_1 \circ F_0^{n-1}(x)\right)}{\partial x} \end{bmatrix}\right) = n \tag{7.36}$$

and

$$\text{rank}\left(\left[\frac{\partial F_u(x)}{\partial u} \frac{\partial F_0(x)}{\partial x} \frac{\partial F_u(x)}{\partial u} \cdots \left(\frac{\partial F_0(x)}{\partial x}\right)^{n-1} \frac{\partial F_u(x)}{\partial u}\right]\bigg|_{(0,0)}\right)$$

$$= \text{rank}\left(\frac{\partial}{\partial U}\Psi(U)\bigg|_{U=0}\right) = n. \tag{7.37}$$

Solution It is easy to see, by (7.12) and chain rule, that for $1 \leq i \leq n$,

$$\frac{\partial}{\partial u^i}\Psi(U)\bigg|_{U=0} = \frac{\partial}{\partial u^i}\left(F_{u^1} \circ F_{u^2} \circ \cdots F_{u^n}(0)\right)\bigg|_{U=0}$$

$$= \frac{\partial F_{u^1}(x)}{\partial x}\bigg|_{(0,0)} \cdots \frac{\partial F_{u^{i-1}}(x)}{\partial x}\bigg|_{(0,0)} \frac{\partial F_{u^i}(x)}{\partial u^i}\bigg|_{(0,0)}$$

$$= \left(\frac{\partial F_0(x)}{\partial x}\bigg|_{(0,0)}\right)^{n-1} \frac{\partial F_u(x)}{\partial u}\bigg|_{(0,0)}$$

which implies that

$$\frac{\partial}{\partial U}\Psi(U)\bigg|_{U=0} = \left[\frac{\partial F_u(x)}{\partial u} \frac{\partial F_0(x)}{\partial x} \frac{\partial F_u(x)}{\partial u} \cdots \left(\frac{\partial F_0(x)}{\partial x}\right)^{n-1} \frac{\partial F_u(x)}{\partial u}\right]\bigg|_{(0,0)}.$$

Also, it is easy to see, by (7.35) and chain rule, that

$$
\left.\begin{bmatrix} \dfrac{\partial S_1(x)}{\partial x} \\ \dfrac{\partial S_1 \circ F_0(x)}{\partial x} \\ \vdots \\ \dfrac{\partial S_1 \circ F_0^{n-1}(x)}{\partial x} \end{bmatrix} \begin{bmatrix} \dfrac{\partial F_u(x)}{\partial u} & \dfrac{\partial F_0(x)}{\partial x}\dfrac{\partial F_u(x)}{\partial u} & \cdots & \left(\dfrac{\partial F_0(x)}{\partial x}\right)^{n-1}\dfrac{\partial F_u(x)}{\partial u} \end{bmatrix}\right|_{(0,0)}
$$

$$
= \left.\begin{bmatrix} \dfrac{\partial\left(S_1 \circ \hat{F}_u(x)\right)}{\partial u} & \dfrac{\partial\left(S_1 \circ \hat{F}_u^2(x)\right)}{\partial u} & \cdots & \dfrac{\partial\left(S_1 \circ \hat{F}_u^n(x)\right)}{\partial u} \\ \vdots & \vdots & & \vdots \\ \dfrac{\partial\left(S_1 \circ \hat{F}_u^{n-1}(x)\right)}{\partial u} & \dfrac{\partial\left(S_1 \circ \hat{F}_u^n(x)\right)}{\partial u} & \cdots & \dfrac{\partial\left(S_1 \circ \hat{F}_u^{2n-2}(x)\right)}{\partial u} \\ \dfrac{\partial\left(S_1 \circ \hat{F}_u^n(x)\right)}{\partial u} & \dfrac{\partial\left(S_1 \circ \hat{F}_u^{n+1}(x)\right)}{\partial u} & \cdots & \dfrac{\partial\left(S_1 \circ \hat{F}_u^{2n-1}(x)\right)}{\partial u} \end{bmatrix}\right|_{(0,0)}
$$

$$
= \begin{bmatrix} 0 & 0 & \cdots & \left.\dfrac{\partial\left(S_1 \circ \hat{F}_u^n(x)\right)}{\partial u}\right|_{(0,0)} \\ \vdots & \vdots & & * \\ 0 & \left.\dfrac{\partial\left(S_1 \circ \hat{F}_u^n(x)\right)}{\partial u}\right|_{(0,0)} & \cdots & * \\ \left.\dfrac{\partial\left(S_1 \circ \hat{F}_u^n(x)\right)}{\partial u}\right|_{(0,0)} & * & \cdots & * \end{bmatrix}.
$$

Since the matrix of the right-hand side has rank n, it is clear that (7.36) and (7.37) are satisfied. $\qquad\square$

Lemma 7.1 *System* (7.11) *is feedback linearizable, if and only if there exists a scalar smooth function* $S_1(x)$ *such that*

(i) $\frac{\partial}{\partial u}\left(S_1 \circ \hat{F}_u^i(x)\right) = 0,\ 1 \le i \le n-1.$

(ii) $\left.\frac{\partial}{\partial u}\left(S_1 \circ \hat{F}_u^n(x)\right)\right|_{(0,0)} \ne 0.$

Furthermore, state transformation $z = S(x)$ *and feedback* $u = \gamma(x, v)$ *satisfy*

$$
z = S(x) = \begin{bmatrix} S_1(x) & S_1 \circ F_0(x) & \cdots & S_1 \circ F_0^{n-1}(x) \end{bmatrix}^{\mathsf{T}} \tag{7.38}
$$

and

$$
v = S_1 \circ F_0^{n-1} \circ F_{\gamma(x,v)}(x) = S_1 \circ \hat{F}_{\gamma(x,v)}^n(x). \tag{7.39}
$$

Proof Necessity. Suppose that system (7.11) is feedback linearizable. Then, there exist a state transformation $z = S(x)$ and a nonsingular feedback $u = \gamma(x, v)$ $\left(\frac{\partial \gamma(x,v)}{\partial v} \ne 0\right)$ such that

$$
\tilde{F}_v(z) \triangleq S \circ F_{\gamma(x,v)} \circ S^{-1}(z) = Az + bv.
$$

Thus, we have that

$$
\begin{bmatrix} S_1 \circ F_{\gamma(x,v)}(x) \\ \vdots \\ S_{n-1} \circ F_{\gamma(x,v)}(x) \\ S_n \circ F_{\gamma(x,v)}(x) \end{bmatrix} = AS(x) + bv = \begin{bmatrix} S_2(x) \\ \vdots \\ S_n(x) \\ v \end{bmatrix}.
$$

In other words, for $1 \le i \le n-1$,

$$
S_{i+1}(x) = S_i \circ F_{\gamma(x,v)}(x)
$$

and

$$
v = S_n \circ F_{\gamma(x,v)}(x)
$$

which imply that for $1 \le i \le n-1$,

$$
0 = \frac{\partial \left(S_i \circ F_{\gamma(x,v)}(x) \right)}{\partial v} = \frac{\partial \left(S_i \circ F_u(x) \right)}{\partial u} \bigg|_{u=\gamma(x,v)} \frac{\partial \gamma(x, v)}{\partial v}
$$

and

$$
1 = \frac{\partial \left(S_n \circ F_{\gamma(x,v)}(x) \right)}{\partial v} = \frac{\partial \left(S_n \circ F_u(x) \right)}{\partial u} \bigg|_{u=\gamma(x,v)} \frac{\partial \gamma(x, v)}{\partial v}.
$$

Since $\frac{\partial \gamma(x,v)}{\partial v} \ne 0$, it is easy to see that for $1 \le i \le n-1$,

$$
S_{i+1}(x) = S_i \circ F_0(x) ; \quad \frac{\partial \left(S_i \circ F_u(x) \right)}{\partial u} = 0
$$

and

$$
v = S_n \circ F_{\gamma(x,v)}(x) ; \quad \frac{\partial \left(S_n \circ F_u(x) \right)}{\partial u} \ne 0.
$$

In other words, for $1 \le i \le n-1$,

$$
S_{i+1}(x) = S_1 \circ F_0^i(x) = S_1 \circ \hat{F}_u^i(x) ; \quad \frac{\partial \left(S_1 \circ \hat{F}_u^i(x) \right)}{\partial u} = 0
$$

and

$$
v = S_1 \circ \hat{F}_{\gamma(x,v)}^n(x) ; \quad \frac{\partial \left(S_1 \circ \hat{F}_u^n(x) \right)}{\partial u} \ne 0
$$

which imply that conditions (i), (ii), (7.38), and (7.39) are satisfied.

Sufficiency. Suppose that there exists a scalar function $S_1(x)$ such that conditions (i) and (ii) are satisfied. Let us define $z = S(x) = [S_1(x) \cdots S_n(x)]^{\mathsf{T}}$ and feedback $u = \gamma(x, v)$ as (7.38) and (7.39), respectively. Then it is clear, by Example 7.2.5, that $z = S(x)$ is a state transformation. Also, it is easy to see, by conditions (i), (7.38), and (7.39), that

$$
\begin{aligned}
\tilde{F}_v(z) &\triangleq S \circ F_{\gamma(x,v)} \circ S^{-1}(z) \\
&= \begin{bmatrix} S_1 \circ F_{\gamma(x,v)} \circ S^{-1}(z) \\ S_1 \circ F_0 \circ F_{\gamma(x,v)} \circ S^{-1}(z) \\ \vdots \\ S_1 \circ F_0^{n-1} \circ F_{\gamma(x,v)} \circ S^{-1}(z) \end{bmatrix} \\
&= \begin{bmatrix} S_1 \circ \hat{F}_{\gamma(x,v)} \circ S^{-1}(z) \\ S_1 \circ \hat{F}_{\gamma(x,v)}^2 \circ S^{-1}(z) \\ \vdots \\ S_1 \circ \hat{F}_{\gamma(x,v)}^n \circ S^{-1}(z) \end{bmatrix} = \begin{bmatrix} S_1 \circ \hat{F}_0 \circ S^{-1}(z) \\ \vdots \\ S_1 \circ \hat{F}_0^{n-1} \circ S^{-1}(z) \\ S_1 \circ \hat{F}_{\gamma(x,v)}^n(x)\Big|_{x=S^{-1}(z)} \end{bmatrix} \\
&= \begin{bmatrix} S_1 \circ F_0 \circ S^{-1}(z) \\ \vdots \\ S_1 \circ F_0^{n-1} \circ S^{-1}(z) \\ S_1 \circ \hat{F}_{\gamma(x,v)}^n(x)\Big|_{x=S^{-1}(z)} \end{bmatrix} = \begin{bmatrix} z_2 \\ \vdots \\ z_n \\ v \end{bmatrix}.
\end{aligned}
$$

By Lemma 7.1, the necessary and sufficient conditions for feedback linearization can be obtained as follows.

Theorem 7.2 (conditions for feedback linearization)
System (7.11) is feedback linearizable, if and only if

(i) $\dfrac{\partial \Psi(U)}{\partial U}\Big|_{U=0}$ *is nonsingular.*
(ii) $\mathcal{F}_*(\Delta_i)$, $1 \le i \le n-1$, *are well-defined involutive distributions or*

$$
\left[\frac{\partial}{\partial u^i},\ \ker \mathcal{F}_* \right] \subset \ker \mathcal{F}_* + \Delta_i(\tilde{U}), \quad 1 \le i \le n-1 \tag{7.40}
$$

where for $1 \le i \le n-1$

$$
\Delta_i(\tilde{U}) = \mathrm{span}\left\{ \frac{\partial}{\partial u^1}, \ldots, \frac{\partial}{\partial u^i} \right\}. \tag{7.41}
$$

Proof Necessity. Suppose that system (7.11) is feedback linearizable. Then, by Lemma 7.1, there exists a smooth function $S_1(x)$ such that

$$\frac{\partial}{\partial u}\left(S_1 \circ \hat{F}_u^i(x)\right) = 0 ; \quad \frac{\partial}{\partial u}\left(S_1 \circ \hat{F}_u^n(x)\right)\Big|_{(0,0)} \neq 0. \qquad (7.42)$$

Thus, by Example 7.2.5, condition (i) is satisfied. Let $\tilde{F}_u(z) \triangleq S \circ F_u \circ S^{-1}(z)$, where

$$z = S(x) = \left[S_1(x) \; S_1 \circ F_0(x) \; \cdots \; S_1 \circ F_0^{n-1}(x)\right]^{\mathsf{T}}.$$

Then it is easy to see, by (7.42), that $\frac{\partial \alpha_u(z)}{\partial u}\Big|_{(0,0)} \neq 0$ and for $2 \leq i \leq n+1$,

$$\tilde{F}_u(z) = \begin{bmatrix} z_2 \\ \vdots \\ z_n \\ \alpha_u(z) \end{bmatrix}, \quad \tilde{F}_{u^i} \circ \cdots \circ \tilde{F}_{u^{n+1}}(0) = \begin{bmatrix} O_{(i-2)\times 1} \\ \alpha_{u^n+1}(0) \\ \alpha_{u^n} \circ \tilde{F}_{u^{n+1}}(0) \\ \vdots \\ \alpha_{u^{i+1}} \circ \tilde{F}_{u^{i+2}} \circ \cdots \circ \tilde{F}_{u^{n+1}}(0) \\ \alpha_{u^i} \circ \tilde{F}_{u^{i+1}} \circ \cdots \circ \tilde{F}_{u^{n+1}}(0) \end{bmatrix}$$

$$\tilde{\mathcal{F}}(u^1, \ldots, u^{n+1}) \triangleq \tilde{F}_{u^1} \circ \cdots \circ \tilde{F}_{u^{n+1}}(0)$$

$$= \begin{bmatrix} \alpha_{u^n} \circ \tilde{F}_{u^{n+1}}(0) \\ \vdots \\ \alpha_{u^2} \circ \tilde{F}_{u^3} \circ \cdots \circ \tilde{F}_{u^{n+1}}(0) \\ \alpha_{u^1} \circ \tilde{F}_{u^2} \circ \cdots \circ \tilde{F}_{u^{n+1}}(0) \end{bmatrix}$$

where $\alpha_u(z) \triangleq S_1 \circ F_0^{n-1} \circ F_u \circ S^{-1}(z) = S_1 \circ \hat{F}_u^n \circ S^{-1}(z)$. Thus, it is easy to see that for $1 \leq i \leq n$,

$$\tilde{\mathcal{F}}_*(\Delta_i) = \tilde{\mathcal{F}}_*\left(\text{span}\left\{\frac{\partial}{\partial u^1}, \ldots, \frac{\partial}{\partial u^i}\right\}\right) = \text{span}\left\{\frac{\partial}{\partial z_{n+1-i}}, \ldots, \frac{\partial}{\partial z_n}\right\}$$

which implies that $\tilde{\mathcal{F}}_*(\Delta_i)$, $1 \leq i \leq n$, are well-defined involutive distributions. It is clear, by Example 7.1.4, that

$$\mathcal{F}(u^1, \ldots, u^{n+1}) = S^{-1} \circ \tilde{\mathcal{F}}(u^1, \ldots, u^{n+1})$$

and for $1 \leq i \leq n$,

$$\mathcal{F}_*(\Delta_i) = \left(S^{-1} \circ \tilde{\mathcal{F}}\right)_*(\Delta_i) = S_*^{-1}\left(\text{span}\left\{\frac{\partial}{\partial z_{n+1-i}}, \ldots, \frac{\partial}{\partial z_n}\right\}\right).$$

Since $x = S^{-1}(z)$ is invertible, $\mathcal{F}_*(\Delta_i)$, $1 \leq i \leq n$, are also well-defined involutive distributions and (7.40) is, by Theorem 2.10, satisfied. Therefore, condition (ii) is satisfied.

Sufficiency. Suppose that conditions (i) and (ii) are satisfied. Then, $D_i = \mathcal{F}_*(\Delta_i)$ is a i-dimensional well-defined involutive distribution for $1 \le i \le n$ and $D_1 \subset D_2 \subset \cdots \subset D_n$. Thus, there exists, by the Frobenius Theorem (or Theorem 2.8), a state transformation $\xi = \tilde{S}(x)$ such that for $1 \le i \le n$,

$$\tilde{D}_i \triangleq \tilde{S}_* \left(\mathcal{F}_* \left(\text{span} \left\{ \frac{\partial}{\partial u^1}, \ldots, \frac{\partial}{\partial u^i} \right\} \right) \right) = \text{span} \left\{ \frac{\partial}{\partial \xi_1}, \ldots, \frac{\partial}{\partial \xi_i} \right\}. \tag{7.43}$$

Let $\tilde{F}_u(\xi) \triangleq \tilde{S} \circ F_u \circ \tilde{S}^{-1}(\xi)$. Then, we have, by Example 7.1.4, that

$$\tilde{\mathcal{F}}(u^1, \ldots, u^{n+1}) \triangleq \tilde{F}_{u^1} \circ \cdots \circ \tilde{F}_{u^{n+1}}(0) = \tilde{S} \circ \mathcal{F}(u^1, \ldots, u^{n+1}).$$

Therefore, it is easy to see, by (7.43), that for $1 \le i \le n$,

$$\frac{\partial}{\partial u^\ell} \tilde{\mathcal{F}}_i(u^1, \ldots, u^{n+1}) = 0, \ 1 \le \ell \le i - 1$$

$$\frac{\partial}{\partial u^i} \tilde{\mathcal{F}}_i(u^1, \ldots, u^{n+1}) \ne 0.$$

In other words, we have that

$$\tilde{\mathcal{F}}(\tilde{U}) = \begin{bmatrix} \alpha_1(u^1, \ldots, u^{n+1}) \\ \alpha_2(u^2, \ldots, u^{n+1}) \\ \vdots \\ \alpha_{n-1}(u^{n-1}, u^n, u^{n+1}) \\ \alpha_n(u^n, u^{n+1}) \end{bmatrix}; \ \tilde{\Psi}(U) = \begin{bmatrix} \hat{\alpha}_1(u^1, \ldots, u^n) \\ \hat{\alpha}_2(u^2, \ldots, u^n) \\ \vdots \\ \hat{\alpha}_{n-1}(u^{n-1}, u^n) \\ \hat{\alpha}_n(u^n) \end{bmatrix} \tag{7.44}$$

where $\alpha_i(u^i, \ldots, u^n, u^{n+1}) \triangleq \tilde{\mathcal{F}}_i(0, \ldots, 0, u^i, \ldots, u^{n+1})$ and for $1 \le i \le n$,

$$\frac{\partial}{\partial u^i} \alpha_i(u^i, \ldots, u^n, u^{n+1}) \ne 0 \tag{7.45}$$

$$\hat{\alpha}_i(u^i, \ldots, u^n) \triangleq \alpha_i(u^i, \ldots, u^n, 0) ; \ \frac{\partial}{\partial u^i} \hat{\alpha}_i(u^i, \ldots, u^n) \ne 0.$$

Thus, it is clear that there exist smooth functions $\bar{\alpha}_i(\xi_i, \ldots, \xi_n) : \mathbb{R}^{n+1-i} \to \mathbb{R}, 1 \le i \le n$ such that for $1 \le i \le n$,

$$\hat{\alpha}_i(\bar{\alpha}_i(\xi_i, \ldots, \xi_n), \ldots, \bar{\alpha}_n(\xi_n)) = \xi_i$$

or

$$U = \tilde{\Psi}^{-1}(\xi) = \begin{bmatrix} \bar{\alpha}_1(\xi_1, \ldots, \xi_n) \\ \bar{\alpha}_2(\xi_2, \ldots, \xi_n) \\ \vdots \\ \bar{\alpha}_{n-1}(\xi_{n-1}, \xi_n) \\ \bar{\alpha}_n(\xi_n) \end{bmatrix}. \tag{7.46}$$

Since $\tilde{F}_u \circ \tilde{\Psi}(U) = \tilde{\mathcal{F}}(u, U)$ by (7.13), we have

$$\tilde{F}_u(\xi) = \tilde{\mathcal{F}}(u, \Psi^{-1}(\xi)) = \begin{bmatrix} \tilde{\alpha}_1(u, \xi_1, \ldots, \xi_n) \\ \tilde{\alpha}_2(\xi_1, \ldots, \xi_n) \\ \vdots \\ \tilde{\alpha}_{n-1}(\xi_{n-2}, \xi_{n-1}, \xi_n) \\ \tilde{\alpha}_n(\xi_{n-1}, \xi_n) \end{bmatrix} \tag{7.47}$$

where for $2 \leq i \leq n$,

$$\tilde{\alpha}_1(u, \xi_1, \ldots, \xi_n) \triangleq \alpha_1(u, \bar{\alpha}_1(\xi_1, \ldots, \xi_n), \ldots, \bar{\alpha}_n(\xi_n))$$
$$\tilde{\alpha}_i(\xi_{i-1}, \ldots, \xi_n) \triangleq \alpha_i(\bar{\alpha}_{i-1}(\xi_{i-1}, \ldots, \xi_n), \ldots, \bar{\alpha}_n(\xi_n)).$$

Let $\tilde{h}(\xi) = \xi_n$. Then, we have that $\tilde{h} \circ \hat{\tilde{F}}_u(\xi) = \tilde{\alpha}_n(\xi_{n-1}, \xi_n) \triangleq H_1(\xi_{n-1}, \xi_n) = \tilde{h}(\xi)$ $\circ \hat{\tilde{F}}_0(\xi)$ and

$$\tilde{h} \circ \hat{\tilde{F}}_u^2(\xi) = H_1(\tilde{\alpha}_{n-1}(\xi_{n-2}, \xi_{n-1}, \xi_n), \tilde{\alpha}_n(\xi_{n-1}, \xi_n))$$
$$\triangleq H_2(\xi_{n-2}, \xi_{n-1}, \xi_n) = \tilde{h}(\xi) \circ \hat{\tilde{F}}_0^2(\xi).$$

In this manner, it is easy to show, by (7.45) and (7.47), that for $1 \leq i \leq n - 1$,

$$\tilde{h} \circ \hat{\tilde{F}}_u^i(\xi) = \tilde{h} \circ \hat{\tilde{F}}_0^i(\xi) ; \quad \frac{\partial}{\partial u}\left(\tilde{h} \circ \hat{\tilde{F}}_u^n(\xi)\right) \neq 0$$

or

$$\tilde{h} \circ \tilde{S} \circ \hat{F}_u^i(x) = \tilde{h} \circ \tilde{S} \circ \hat{F}_0^i(x) ; \quad \frac{\partial}{\partial u}\left(\tilde{h} \circ \tilde{S} \circ \hat{F}_u^n(x)\right) \neq 0.$$

Therefore, $S_1(x) \triangleq \tilde{h} \circ \tilde{S}(x) = \tilde{S}_n(x)$ satisfies conditions (i) and (ii) of Lemma 7.1. Hence, by Lemma 7.1, system (7.11) is feedback linearizable.

If condition (ii) of Theorem 7.1 is satisfied, then condition (ii) of Theorem 7.2 is satisfied. In other words, if the system is state equivalent to a linear system, then it is also feedback linearizable. The following example shows that system (7.34) of Example 7.2.4 is not state equivalent to a linear system but feedback linearizable.

Example 7.2.6 Show that system (7.34) of Example 7.2.4 is feedback linearizable.

$$\begin{bmatrix} x_1(t+1) \\ x_2(t+1) \end{bmatrix} = \begin{bmatrix} x_2(t) \\ x_1(t)^2 + u(t) \end{bmatrix} = F_{u(t)}(x(t)).$$

Solution In Example 7.2.4, it is shown that condition (i) of Theorem 7.2 is satisfied. Since

$$\left[\frac{\partial}{\partial u^1}, -2u^3 \frac{\partial}{\partial u^1} + \frac{\partial}{\partial u^3} \right] = 0 \in \ker \mathcal{F}_* + \Delta_1,$$

it is clear, by Theorem 2.10, that $\mathcal{F}_*(\Delta_1) = \mathcal{F}_* \left(\text{span} \left\{ \frac{\partial}{\partial u^1} \right\} \right)$ is a well-defined involutive distribution and

$$\mathcal{F}_*(\Delta_1) = \text{span} \left\{ \frac{\partial}{\partial x_2} \right\}.$$

Thus, condition (ii) of Theorem 7.2 is satisfied. Hence, by Theorem 7.2, system (7.34) is feedback linearizable. Since $dx_1 \in (\mathcal{F}_*(\Delta_1))^{\perp}$, scalar function $S_1(x) = x_1$ satisfies conditions of Lemma 7.1. Thus, it is easy to see that

$$\begin{bmatrix} z_1 \\ z_2 \end{bmatrix} = \begin{bmatrix} S_1(x) \\ S_1 \circ F_0(x) \end{bmatrix} = \begin{bmatrix} x_1 \\ x_2 \end{bmatrix}$$

and

$$v = S_1 \circ \hat{F}_u^2(x) = x_1^2 + u \text{ or } u = -x_1^2 + v = \gamma(x, v).$$

Then it is clear that $\tilde{F}_v(z) \triangleq S \circ F_{\gamma(x,v)} \circ S^{-1}(z) = \begin{bmatrix} z_2 \\ v \end{bmatrix}$ and

$$\begin{bmatrix} z_1(t+1) \\ z_2(t+1) \end{bmatrix} = \begin{bmatrix} 0 & 1 \\ 0 & 0 \end{bmatrix} \begin{bmatrix} z_1(t) \\ z_2(t) \end{bmatrix} + \begin{bmatrix} 0 \\ 1 \end{bmatrix} v(t).$$

$\square$

Example 7.2.7 Show that the following discrete system is feedback linearizable:

$$\begin{bmatrix} x_1(t+1) \\ x_2(t+1) \\ x_3(t+1) \end{bmatrix} = \begin{bmatrix} x_2(t) + (1 + x_2(t))^2 u(t)^2 \\ x_3(t) \\ (1 + x_2(t))u(t) \end{bmatrix} = F_{u(t)}(x(t)). \tag{7.48}$$

Solution By simple calculations, we have that

$$
\mathcal{F}(u^1, u^2, u^3, u^4) \triangleq F_{u^1} \circ F_{u^2} \circ F_{u^3} \circ F_{u^4}(0) = F_{u^1} \circ F_{u^2} \circ F_{u^3}\left(\begin{bmatrix} (u^4)^2 \\ 0 \\ u^4 \end{bmatrix}\right)
$$

$$
= F_{u^1} \circ F_{u^2}\left(\begin{bmatrix} (u^3)^2 \\ u^4 \\ u^3 \end{bmatrix}\right) = F_{u^1}\left(\begin{bmatrix} u^4 + (u^2)^2(1 + u^4)^2 \\ u^3 \\ u^2(1 + u^4) \end{bmatrix}\right)
$$

$$
= \begin{bmatrix} u^3 + (u^1)^2(1 + u^3)^2 \\ u^2(1 + u^4) \\ u^1(1 + u^3) \end{bmatrix}
$$

and

$$
\Psi(u^1, u^2, u^3) \triangleq F_{u^1} \circ F_{u^2} \circ F_{u^3}(0) = \mathcal{F}(u^1, u^2, u^3, 0) = \begin{bmatrix} u^3 + (u^1)^2(1 + u^3)^2 \\ u^2 \\ u^1(1 + u^3) \end{bmatrix}.
$$

Since $\det\left(\left.\frac{\partial \Psi(U)}{\partial U}\right|_{U=0}\right) = \det\left(\begin{bmatrix} 0 & 0 & 1 \\ 0 & 1 & 0 \\ 1 & 0 & 0 \end{bmatrix}\right) = -1 \neq 0$, condition (i) of Theorem 7.2 is satisfied. Since

$$
\frac{\partial \mathcal{F}(\tilde{U})}{\partial \tilde{U}} = \begin{bmatrix} 2u^1(1 + u^3)^2 & 0 & 1 + 2(u^1)^2(1 + u^3) & 0 \\ 0 & 1 + u^4 & 0 & u^2 \\ 1 + u^3 & 0 & u^1 & 0 \end{bmatrix},
$$

we have that

$$
\ker \mathcal{F}_* = \mathrm{span}\left\{ -\frac{u^2}{1 + u^4}\frac{\partial}{\partial u^2} + \frac{\partial}{\partial u^4} \right\}
$$

and

$$
\left[\frac{\partial}{\partial u^1}, \quad -\frac{u^2}{1 + u^4}\frac{\partial}{\partial u^2} + \frac{\partial}{\partial u^4} \right] = 0 \in \ker \mathcal{F}_* + \Delta_1
$$

$$
\left[\frac{\partial}{\partial u^2}, \quad -\frac{u^2}{1 + u^4}\frac{\partial}{\partial u^2} + \frac{\partial}{\partial u^4} \right] = -\frac{1}{1 + u^4}\frac{\partial}{\partial u^2} \in \ker \mathcal{F}_* + \Delta_2.
$$

Therefore, condition (ii) of Theorem 7.2 is satisfied and $\mathcal{F}_*(\Delta_i)$, $i = 1, 2$, are well-defined involutive distributions with

$$\mathcal{F}_*(\Delta_1) = \text{span}\left\{2x_3\frac{\partial}{\partial x_1} + \frac{\partial}{\partial x_3}\right\} = \text{span}\left\{\begin{bmatrix} 2x_3 \\ 0 \\ 1 \end{bmatrix}\right\}$$

$$\mathcal{F}_*(\Delta_2) = \text{span}\left\{\begin{bmatrix} 2x_3 \\ 0 \\ 1 \end{bmatrix}, \begin{bmatrix} 0 \\ 1 \\ 0 \end{bmatrix}\right\}.$$

Hence, by Theorem 7.2, system (7.48) is feedback linearizable. Since $d(x_1 - x_3^2) \in (\mathcal{F}_*(\Delta_1))^\perp$, we have that $S_1(x) = x_1 - x_3^2$,

$$\begin{bmatrix} z_1 \\ z_2 \\ z_3 \end{bmatrix} = \begin{bmatrix} S_1(x) \\ S_1 \circ F_0(x) \\ S_1 \circ \hat{F}_0^2(x) \end{bmatrix} = \begin{bmatrix} x_1 - x_3^2 \\ x_2 \\ x_3 \end{bmatrix},$$

and

$$v = S_1 \circ \hat{F}_u^3(x) = (1 + x_2)u \text{ or } u = \frac{v}{1 + x_2} = \gamma(x, v).$$

Then it is clear that $\tilde{F}_v(z) \triangleq S \circ F_{\gamma(x,v)} \circ S^{-1}(z) = \begin{bmatrix} z_2 \\ z_3 \\ v \end{bmatrix}$ and

$$\begin{bmatrix} z_1(t+1) \\ z_2(t+1) \\ z_3(t+1) \end{bmatrix} = \begin{bmatrix} 0 & 1 & 0 \\ 0 & 0 & 1 \\ 0 & 0 & 0 \end{bmatrix} \begin{bmatrix} z_1(t) \\ z_2(t) \\ z_3(t) \end{bmatrix} + \begin{bmatrix} 0 \\ 0 \\ 1 \end{bmatrix} v(t).$$

$\square$

Example 7.2.8 Show that the following discrete system is not feedback linearizable:

$$\begin{bmatrix} x_1(t+1) \\ x_2(t+1) \end{bmatrix} = \begin{bmatrix} x_2(t) + u(t)^2 \\ x_1(t) + u(t) \end{bmatrix} = F_{u(t)}(x(t)). \tag{7.49}$$

Solution By simple calculations, we have that

$$\mathcal{F}(u^1, u^2, u^3) \triangleq F_{u^1} \circ F_{u^2} \circ F_{u^3}(0) = F_{u^1} \circ F_{u^2}\left(\begin{bmatrix} (u^3)^2 \\ u^3 \end{bmatrix}\right)$$

$$= F_{u^1}\left(\begin{bmatrix} (u^2)^2 + u^3 \\ u^2 + (u^3)^2 \end{bmatrix}\right) = \begin{bmatrix} (u^1)^2 + u^2 + (u^3)^2 \\ u^1 + (u^2)^2 + u^3 \end{bmatrix}$$

and

$$\Psi(u^1, u^2) \triangleq F_{u^1} \circ F_{u^2}(0) = \mathcal{F}(u^1, u^2, 0) = \begin{bmatrix} (u^1)^2 + u^2 \\ u^1 + (u^2)^2 \end{bmatrix}.$$

Since $\det\left(\left.\frac{\partial\Psi(U)}{\partial U}\right|_{U=0}\right) = \det\left(\begin{bmatrix} 0 & 1 \\ 1 & 0 \end{bmatrix}\right) \neq 0$, condition (i) of Theorem 7.2 is satisfied. Since

$$\frac{\partial\mathcal{F}(\tilde{U})}{\partial\tilde{U}} = \begin{bmatrix} 2u^1 & 1 & 2u^3 \\ 1 & 2u^2 & 1 \end{bmatrix},$$

we have that

$$\ker\mathcal{F}_* = \text{span}\left\{\frac{1 - 4u^2u^3}{4u^1u^2 - 1}\frac{\partial}{\partial u^1} + \frac{2(u^3 - u^1)}{4u^1u^2 - 1}\frac{\partial}{\partial u^2} + \frac{\partial}{\partial u^3}\right\}$$

and

$$\left[\frac{\partial}{\partial u^1}, -\frac{u^2}{1 + u^4}\frac{\partial}{\partial u^2} + \frac{\partial}{\partial u^4}\right] = \frac{4u^2(4u^2u^3 - 1)}{(4u^1u^2 - 1)^2}\frac{\partial}{\partial u^1} + \frac{2(1 - 4u^2u^3)}{(4u^1u^2 - 1)^2}\frac{\partial}{\partial u^2}$$
$$\notin \ker\mathcal{F}_* + \Delta_1.$$

Therefore, (7.40) is not satisfied and $\mathcal{F}_*(\Delta_1)$ is not a well-defined involutive distribution. Since condition (ii) of Theorem 7.2 is not satisfied, system (7.49) is not feedback linearizable.   □

7.3 Multi-input Discrete Time Systems

In this section, we consider the following multi-input discrete nonlinear system:

$$\begin{aligned} x(t + 1) &= F(x(t), u(t)) \triangleq F_u(x) \\ y(t) &= h(x(t)) \end{aligned}$$ (7.50)

where $x \in \mathbb{R}^n$, $u \in \mathbb{R}^m$, $y \in \mathbb{R}^q$, and $F(x, u) : \mathbb{R}^{n+m} \to \mathbb{R}^n$ and $h(x)$ are smooth functions with $F(0, 0) = 0$ and $h(0) = 0$.

Definition 7.5 (*Kronecker indices*)
For the list of mn vectors of the form

$$\left(\left.\frac{\partial F_u}{\partial u_1}\right|_{(0,0)}, \dots, \left.\frac{\partial F_u}{\partial u_m}\right|_{(0,0)}, \left.\frac{\partial\hat{F}_u^2}{\partial u_1}\right|_{(0,0)}, \dots, \left.\frac{\partial\hat{F}_u^2}{\partial u_m}\right|_{(0,0)}, \dots, \left.\frac{\partial\hat{F}_u^n}{\partial u_1}\right|_{(0,0)}, \dots, \right.$$
$$\left.\left.\frac{\partial\hat{F}_u^n}{\partial u_m}\right|_{(0,0)}\right),$$

delete all vector fields that are linearly dependent on the set of preceding vector fields and obtain the unique set of linearly independent vectors

$$\left\{ \left.\frac{\partial F_u}{\partial u_1}\right|_{(0,0)}, \left.\frac{\partial \hat{F}_u^2}{\partial u_1}\right|_{(0,0)}, \ldots, \left.\frac{\partial \hat{F}_u^{\kappa_1}}{\partial u_1}\right|_{(0,0)}, \ldots, \left.\frac{\partial F_u}{\partial u_m}\right|_{(0,0)}, \ldots, \left.\frac{\partial \hat{F}_u^{\kappa_m}}{\partial u_m}\right|_{(0,0)} \right\}$$

or

$$\left\{ \bar{b}_1, \bar{A}\bar{b}_1, \ldots, \bar{A}^{\kappa_1-1}\bar{b}_1, \ldots, \bar{b}_m, \ldots, \bar{A}^{\kappa_m-1}\bar{b}_m \right\}$$

where $\bar{A} \triangleq \left.\frac{\partial F_u(x)}{\partial x}\right|_{(0,0)}$ and $\bar{b}_j \triangleq \left.\frac{\partial F_u(x)}{\partial u_j}\right|_{(0,0)}$. Then, $(\kappa_1, \ldots, \kappa_m)$ are said to be the Kronecker indices of system (7.50).

In other words, κ_i is the smallest nonnegative integer such that for $1 \leq i \leq m$,

$$\left.\frac{\partial \hat{F}_u^{\kappa_i+1}}{\partial u_i}\right|_{(0,0)} \in \text{span}\left\{ \left.\frac{\partial \hat{F}_u^{\ell}}{\partial u_j}\right|_{(0,0)} \middle| 1 \leq j \leq m, \quad 1 \leq \ell \leq \kappa_i \right\}$$
$$+ \text{span}\left\{ \left.\frac{\partial \hat{F}_u^{\kappa_i+1}}{\partial u_j}\right|_{(0,0)} \middle| 1 \leq j \leq i-1 \right\}.$$

If $\sum_{i=1}^{n} \kappa_i = n$, then system (7.5) is reachable on a neighborhood of the origin. Let $\kappa_{\max} \triangleq \max\{\kappa_i, \ 1 \leq i \leq m\}$ and for $1 \leq i \leq \kappa_{\max} + 1$,

$$\bar{u}^i \triangleq \left\{ u_j^i \mid \kappa_j \geq i \right\} \text{ and } \tilde{u}^i \triangleq \left\{ u_j^i \mid \kappa_j + 1 \geq i \right\}.$$

For example, if $(\kappa_1, \kappa_2, \kappa_3) = (3, 1, 2)$, then we have that $\bar{u}^1 = [u_1^1 \ u_2^1 \ u_3^1]^\mathsf{T} = \tilde{u}^1$, $\bar{u}^2 = [u_1^2 \ u_3^2]^\mathsf{T}, \tilde{u}^2 = [u_1^2 \ u_2^2 \ u_3^2]^\mathsf{T}, \bar{u}^3 = u_1^3, \tilde{u}^3 = [u_1^3 \ u_3^3]^\mathsf{T}$, and $\tilde{u}^4 = u_1^4$. Let us define composite functions $\Psi : \mathbb{R}^n \to \mathbb{R}^n$ and $\mathcal{F} : \mathbb{R}^{m+\sum_{i=1}^{m}\kappa_i} \to \mathbb{R}^n$ as follows:

$$\begin{aligned}
\Psi(U) &\triangleq F_{u^1} \circ \cdots \circ F_{u^n}(0)|_{u_j^i=0, \ i \geq \kappa_j+1} \\
&= F_{u^1} \circ \cdots \circ F_{u^{\kappa_{\max}}}(0)|_{u_j^i=0, \ i \geq \kappa_j+1}
\end{aligned} \tag{7.51}$$

$$\begin{aligned}
\mathcal{F}(\tilde{U}) &\triangleq F_{u^1} \circ \cdots \circ F_{u^n} \circ F_{u^{n+1}}(0)|_{u_j^i=0, \ i \geq \kappa_j+2} \\
&= F_{u^1} \circ \cdots \circ F_{u^{\kappa_{\max}}} \circ F_{u^{\kappa_{\max}+1}}(0)|_{u_j^i=0, \ i \geq \kappa_j+2}
\end{aligned} \tag{7.52}$$

where $u^i = [u_1^i \cdots u_m^i]^\mathsf{T}, \ i \geq 1$, and

$$U \triangleq \left[(\bar{u}^1)^\mathsf{T} \ \cdots \ (\bar{u}^{\kappa_{\max}})^\mathsf{T} \right]^\mathsf{T}$$
$$\tilde{U} \triangleq \left[(\tilde{u}^1)^\mathsf{T} \ \cdots \ (\tilde{u}^{\kappa_{\max}})^\mathsf{T} \ (\tilde{u}^{\kappa_{\max}+1})^\mathsf{T} \right]^\mathsf{T}.$$

Then, it is clear that

$$\Psi(U) = \mathcal{F}(\tilde{U})\Big|_{u_i^{\kappa_i+1}=0} \tag{7.53}$$

and

$$\mathcal{F}(\tilde{U}) = F_{\tilde{u}^1} \circ \Psi(\tilde{u}^2, \ldots, \tilde{u}^{\kappa_{\max}+1}). \tag{7.54}$$

Also, it is easy to see that for $1 \leq j \leq m$ and $1 \leq i \leq \kappa_j$,

$$\frac{\partial \Psi(U)}{\partial u_j^i}\Big|_{U=0} = \left(\frac{\partial F_u(x)}{\partial x}\Big|_{(0,0)}\right)^{i-1} \frac{\partial F_u(x)}{\partial u_j}\Big|_{(0,0)}.$$

Thus, it is clear that $\frac{\partial \Psi(U)}{\partial U}\Big|_{U=0}$ is nonsingular, if and only if $\sum\limits_{j=1}^{m} \kappa_j = n$.

Example 7.3.1 Let $\frac{\partial \Psi(U)}{\partial U}\Big|_{U=0}$ be nonsingular. Suppose that $\mathcal{F}_* \left(\frac{\partial}{\partial u_j^i}\right)$ is a well-defined vector field for $1 \leq j \leq m$ and $1 \leq i \leq \kappa_j$. Show that for $1 \leq j \leq m$ and $1 \leq i \leq \kappa_j$,

$$\mathcal{F}_* \left(\frac{\partial}{\partial u_j^i}\right) = \Psi_* \left(\frac{\partial}{\partial u_j^i}\right). \tag{7.55}$$

Solution Omitted. (See Problem 7.4.)   □

Example 7.3.2 The Kronecker indices of the nonlinear discrete time system are invariant under state transformation. In other words, the Kronecker indices of system (7.5) and those of system (7.9) are the same.

Solution Omitted. (See Problem 7.5.)   □

Suppose that $(\kappa_1, \ldots, \kappa_m)$ are the Kronecker indices of system (7.50). Then, the multi-input version of Theorem 7.1 can be obtained as follows.

Theorem 7.3 (conditions for linearization by state transformation)
System (7.50) is state equivalent to a linear system, if and only if

(i) $\frac{\partial \Psi(U)}{\partial U}\Big|_{U=0}$ *is nonsingular or* $\sum\limits_{j=1}^{m} \kappa_j = n$.

(ii) $\mathcal{F}_* \left(\frac{\partial}{\partial u_j^i}\right)$, $1 \leq j \leq m$, $1 \leq i \leq \kappa_j + 1$, *are well-defined vector fields or*

$$\left[\frac{\partial}{\partial u_j^i}, \ker(\mathcal{F}_*)\right] \subset \ker(\mathcal{F}_*), \quad 1 \leq j \leq m, \ 1 \leq i \leq \kappa_j + 1. \tag{7.56}$$

Furthermore, $z = S(x) = \Psi^{-1}(x)$ is a linearizing state transformation.

Proof Necessity. Suppose that system (7.50) is state equivalent to a linear system with state transformation $z = S(x)$. Then we have

$$
\begin{aligned}
z(t+1) &= S \circ F_{u(t)} \circ S^{-1}(z(t)) \\
&\triangleq \tilde{F}_{u(t)}(z(t)) = Az(t) + Bu(t)
\end{aligned}
\tag{7.57}
$$

where

$$
\text{rank}\left([b_1 \ Ab_1 \ \cdots \ A^{\kappa_1-1}b_1 \ \cdots \ b_m \ \cdots \ A^{\kappa_m-1}b_m]\right) = n.
\tag{7.58}
$$

Since $F_u(x) = S^{-1} \circ \tilde{F}_u \circ S(x)$, $\tilde{F}_u(z) = Az + Bu$, and $S(0) = 0$, it is easy to see, by Examples 7.1.3 and 7.1.4, that

$$
\Psi(U) = S^{-1}\left(\sum_{i=1}^{\kappa_{max}} \sum_{\substack{j=1 \\ \kappa_j \geq i}}^{m} A^{i-1}b_j u_j^i\right)
$$

$$
\mathcal{F}(\tilde{U}) = S^{-1}\left(\sum_{i=1}^{\kappa_{max}+1} \sum_{\substack{j=1 \\ \kappa_j+1 \geq i}}^{m} A^{i-1}b_j u_j^i\right)
\tag{7.59}
$$

which implies that

$$
\left.\frac{\partial \Psi(U)}{\partial \hat{U}}\right|_{U=0} = \left.\frac{\partial S^{-1}(z)}{\partial z}\right|_{z=0} [b_1 \ \cdots \ A^{\kappa_1-1}b_1 \ \cdots \ b_m \ \cdots \ A^{\kappa_m-1}b_m]
\tag{7.60}
$$

where $\hat{U} = [u_1^1 \ \cdots \ u_1^{\kappa_1} \ \cdots \ u_m^1 \ \cdots \ u_m^{\kappa_m}]^{\mathsf{T}}$. Since $\left.\frac{\partial S^{-1}(z)}{\partial z}\right|_{z=0}$ is nonsingular, it is clear, by (7.58), that $\left.\frac{\partial \Psi(U)}{\partial \hat{U}}\right|_{U=0}$ (or $\left.\frac{\partial \Psi(U)}{\partial U}\right|_{U=0}$) is nonsingular. Also, it is clear, by (7.59), that for $1 \leq j \leq m$ and $1 \leq i \leq \kappa_j + 1$,

$$
\mathcal{F}_*\left(\frac{\partial}{\partial u_j^i}\right) = (S^{-1})_*(A^{i-1}b_j)
\tag{7.61}
$$

which implies that $\mathcal{F}_*\left(\frac{\partial}{\partial u_j^i}\right)$, $1 \leq j \leq m$, $1 \leq i \leq \kappa_j + 1$, are well-defined vector fields and (7.56) is, by Theorem 2.6, satisfied..

Sufficiency. Suppose that system (7.50) satisfies conditions (i) and (ii). By condition (i), it is clear that $z = S(x) = \Psi^{-1}(x)$ is a state transformation on a neighborhood of the origin. We will show that system (7.50) satisfies, in z-coordinates, a linear system. In other words, we will show that

$$\tilde{F}_u(z) \triangleq \Psi^{-1} \circ F_u \circ \Psi(z) = Az + Bu \tag{7.62}$$

for some constant matrices A and B. If we let

$$Y_j^i = \mathcal{F}_* \left(\frac{\partial}{\partial u_j^i} \right), \quad 1 \le j \le m, \ 1 \le i \le \kappa_j + 1, \tag{7.63}$$

then we have, by Theorem 2.4 and (7.55), that for $1 \le j \le m, 1 \le \bar{j} \le m, 1 \le i \le \kappa_j + 1$, and $1 \le \bar{i} \le \kappa_{\bar{j}} + 1$,

$$\left[Y_j^i, \ Y_{\bar{j}}^{\bar{i}} \right] = \mathcal{F}_* \left(\left[\frac{\partial}{\partial u_j^i}, \frac{\partial}{\partial u_{\bar{j}}^{\bar{i}}} \right] \right) = \mathcal{F}_*(0) = 0 \tag{7.64}$$

and for $1 \le j \le m$ and $1 \le i \le \kappa_j$,

$$\Psi_* \left(\frac{\partial}{\partial u_j^i} \right) = Y_j^i. \tag{7.65}$$

Thus, it is clear, by condition (i), that $\{Y_j^i \mid 1 \le j \le m, \ 1 \le i \le \kappa_j\}$ is a set of linearly independent vector fields on a neighborhood of the origin. Also, we have, by Example 2.4.20, that for $1 \le j \le m$,

$$Y_j^{\kappa_j + 1} = \sum_{k=1}^{m} \sum_{i=1}^{\kappa_k} a_{k,i}^j Y_k^i \tag{7.66}$$

for some constant $a_{k,i}^j \in \mathbb{R}$. It is easy to see, by (7.13), that $F_u \circ \Psi(z) = \mathcal{F}(u, z)$ and

$$\tilde{F}(z, u) \triangleq \tilde{F}_u(z) = \Psi^{-1} \circ F_u \circ \Psi(z) = \Psi^{-1} \circ \mathcal{F}(u, z). \tag{7.67}$$

Thus, we have, by (7.63), (7.65), and (7.67), that for $1 \le j \le m$,

$$\tilde{F}(z, u)_* \left(\frac{\partial}{\partial u_j} \right) = \left(\Psi^{-1} \circ \mathcal{F}(u, z) \right)_* \left(\frac{\partial}{\partial u_j} \right) = (\Psi^{-1})_* (\mathcal{F}(u, z))_* \left(\frac{\partial}{\partial u_j} \right)$$

$$= (\Psi^{-1})_* \left(Y_j^1 \right) = \frac{\partial}{\partial z_j^1}. \tag{7.68}$$

Similarly, it is easy to see, by (7.63), (7.65), (7.66), and (7.67), that for $1 \le j \le m$ and $1 \le i \le \kappa_j - 1$,

$$\tilde{F}(z,u)_* \left(\frac{\partial}{\partial z_j^i} \right) = (\Psi^{-1} \circ \mathcal{F}(u,z))_* \left(\frac{\partial}{\partial z_j^i} \right)$$

$$= (\Psi^{-1})_* (Y_j^{i+1}) = \frac{\partial}{\partial z_j^{i+1}} \tag{7.69}$$

and

$$\tilde{F}(z,u)_* \left(\frac{\partial}{\partial z_j^{\kappa_j}} \right) = (\Psi^{-1} \circ \mathcal{F}(u,z))_* \left(\frac{\partial}{\partial z_j^{\kappa_j}} \right) = (\Psi^{-1})_* \left(Y_j^{\kappa_j+1} \right)$$

$$= (\Psi^{-1})_* \left(\sum_{k=1}^{m} \sum_{i=1}^{\kappa_k} a_{k,i}^j Y_k^i \right) = \sum_{k=1}^{m} \sum_{i=1}^{\kappa_k} a_{k,i}^j \frac{\partial}{\partial z_k^i}. \tag{7.70}$$

Therefore, it is clear, by (7.68)–(7.70), that

$$\tilde{F}(z,u) = Az + Bu$$

for some constant matrices A and B.

Let $\hat{u} = \left[u_1^{\kappa_1+1} \cdots u_m^{\kappa_m+1} \right]^\mathsf{T}$, $\bar{U} = \left[U^\mathsf{T} \, \hat{u}^\mathsf{T} \right]^\mathsf{T}$, and $\frac{\partial \bar{U}}{\partial \tilde{U}} = P$. It is easy to see that $\frac{\partial \mathcal{F}(\bar{U})}{\partial U} \Big|_{\tilde{U}=0} = \frac{\partial \Psi(U)}{\partial U} \Big|_{U=0}$. Thus, if condition (i) of Theorem 7.3 is satisfied, then it is clear that

$$\frac{\partial \mathcal{F}(\tilde{U})}{\partial \tilde{U}} = \frac{\partial \mathcal{F}(\tilde{U})}{\partial \bar{U}} \frac{\partial \bar{U}}{\partial \tilde{U}} = \left[\begin{array}{cc} \frac{\partial \mathcal{F}(\tilde{U})}{\partial U} & \frac{\partial \mathcal{F}(\tilde{U})}{\partial \hat{u}} \end{array} \right] P$$

and

$$\ker \mathcal{F}_* = \mathrm{span} \left\{ Y_1(\tilde{U}), \ldots, Y_m(\tilde{U}) \right\} \tag{7.71}$$

where

$$Y(\tilde{U}) = \left[Y_1(\tilde{U}) \cdots Y_m(\tilde{U}) \right] \triangleq P^{-1} \left[\begin{array}{c} -\left(\frac{\partial \mathcal{F}(\tilde{U})}{\partial U} \right)^{-1} \frac{\partial \mathcal{F}(\tilde{U})}{\partial \hat{u}} \\ I_m \end{array} \right].$$

(Refer to MATLAB subfunction **ker-sF-M**.)

Example 7.3.3 Show that the following discrete time system is state equivalent to a linear system:

$$\begin{bmatrix} x_1(t+1) \\ x_2(t+1) \\ x_3(t+1) \end{bmatrix} = \begin{bmatrix} x_2(t) - u_1(t)^2 \\ u_1(t) \\ u_2(t) - u_1(t)^2 \end{bmatrix} = F_{u(t)}(x(t)). \tag{7.72}$$

Solution Since $\frac{\partial F_u(x)}{\partial x}\Big|_{(0,0)} = \begin{bmatrix} 0 & 1 & 0 \\ 0 & 0 & 0 \\ 0 & 0 & 0 \end{bmatrix}$ and $\frac{\partial F_u(x)}{\partial u}\Big|_{(0,0)} = \begin{bmatrix} 0 & 0 \\ 1 & 0 \\ 0 & 1 \end{bmatrix}$, we have, by simple calculation, that $(\kappa_1, \kappa_2) = (2, 1)$. Since $\kappa_1 + \kappa_2 = 3$, condition (i) of Theorem 7.3 is satisfied. Also, it is easy to see, by (7.51) and (7.52), that

$$
\mathcal{F}(u_1^1, u_2^1, u_1^2, u_2^2, u_1^3) \triangleq F_{u^1} \circ F_{u^2} \circ F_{u^3}(0)|_{u_2^3 = 0} = F_{u^1} \circ F_{u^2}\left(\begin{bmatrix} -(u_1^3)^2 \\ u_1^3 \\ -(u_1^3)^2 \end{bmatrix}\right)
$$

$$
= F_{u^1}\left(\begin{bmatrix} u_1^3 - (u_1^2)^2 \\ u_1^2 \\ u_2^2 - (u_1^2)^2 \end{bmatrix}\right) = \begin{bmatrix} u_1^2 - (u_1^1)^2 \\ u_1^1 \\ u_2^1 - (u_1^1)^2 \end{bmatrix}
$$

and

$$
\Psi(u_1^1, u_2^1, u_1^2) \triangleq F_{u^1} \circ F_{u^2}(0)|_{u_2^2 = 0} = \mathcal{F}(u_1^1, u_2^1, u_1^2, 0, 0) = \begin{bmatrix} u_1^2 - (u_1^1)^2 \\ u_1^1 \\ u_2^1 - (u_1^1)^2 \end{bmatrix}.
$$

Since $\frac{\partial \mathcal{F}(u_1^1, u_2^1, u_1^2, u_2^2, u_1^3)}{\partial \bar{U}} = \begin{bmatrix} -2u_1^1 & 0 & 1 & 0 & 0 \\ 1 & 0 & 0 & 0 & 0 \\ -2u_1^1 & 1 & 0 & 0 & 0 \end{bmatrix}$, we have that

$$
\ker \mathcal{F}_* = \text{span}\left\{ \frac{\partial}{\partial u_2^2}, \frac{\partial}{\partial u_1^3} \right\}
$$

and for $1 \leq j \leq 2$ and $1 \leq i \leq \kappa_j + 1$,

$$
\left[\frac{\partial}{\partial u_j^i}, \frac{\partial}{\partial u_2^2} \right] = 0 \in \ker \mathcal{F}_* ; \quad \left[\frac{\partial}{\partial u_j^i}, \frac{\partial}{\partial u_1^3} \right] = 0 \in \ker \mathcal{F}_*.
$$

Thus, it is clear, by Theorem 2.6, that $\mathcal{F}_*\left(\frac{\partial}{\partial u_j^i}\right)$, $1 \leq j \leq 2$, $1 \leq i \leq \kappa_j + 1$, are well-defined vector fields and condition (ii) of Theorem 7.3 is satisfied. Hence, by Theorem 7.3, system (7.72) is state equivalent to a linear system. Let

$$
\begin{bmatrix} x_1 \\ x_2 \\ x_3 \end{bmatrix} = S^{-1}(z) = \Psi(z) = \begin{bmatrix} z_3 - z_1^2 \\ z_1 \\ z_2 - z_1^2 \end{bmatrix}
$$

or

$$
\begin{bmatrix} z_1 \\ z_2 \\ z_3 \end{bmatrix} = S(x) = \Psi^{-1}(x) = \begin{bmatrix} x_2 \\ x_3 + x_2^2 \\ x_1 + x_2^2 \end{bmatrix}.
$$

Then it is easy to see that

$$
\tilde{F}_u(z) = S \circ F_u \circ S^{-1}(z) = S\left(\begin{bmatrix} z_1 - u_1^2 \\ u_1 \\ u_2 - u_1^2 \end{bmatrix}\right) = \begin{bmatrix} 0 & 0 & 0 \\ 0 & 0 & 0 \\ 1 & 0 & 0 \end{bmatrix}\begin{bmatrix} z_1 \\ z_2 \\ z_3 \end{bmatrix} + \begin{bmatrix} 1 & 0 \\ 0 & 1 \\ 0 & 0 \end{bmatrix} u. \qquad \square
$$

Example 7.3.4 Show that the following discrete time system is not state equivalent to a linear system:

$$
\begin{bmatrix} x_1(t+1) \\ x_2(t+1) \\ x_3(t+1) \end{bmatrix} = \begin{bmatrix} x_2(t) \\ x_1(t)^2 + u_1(t) \\ x_1(t)u_1(t) + u_2(t) \end{bmatrix} = F_{u(t)}(x(t)). \tag{7.73}
$$

Solution Since $\left.\frac{\partial F_u(x)}{\partial x}\right|_{(0,0)} = \begin{bmatrix} 0 & 1 & 0 \\ 0 & 0 & 0 \\ 0 & 0 & 0 \end{bmatrix}$ and $\left.\frac{\partial F_u(x)}{\partial u}\right|_{(0,0)} = \begin{bmatrix} 0 & 0 \\ 1 & 0 \\ 0 & 1 \end{bmatrix}$, we have, by simple calculation, that $(\kappa_1, \kappa_2) = (2, 1)$. Since $\kappa_1 + \kappa_2 = 3$, condition (i) of Theorem 7.3 is satisfied. Also, it is easy to see, by (7.52), that

$$
\mathcal{F}(u_1^1, u_2^1, u_1^2, u_2^2, u_1^3) \triangleq F_{u^1} \circ F_{u^2} \circ F_{u^3}(0)|_{u_2^3=0} = F_{u^1} \circ F_{u^2}\left(\begin{bmatrix} 0 \\ u_1^3 \\ 0 \end{bmatrix}\right)
$$

$$
= F_{u^1}\left(\begin{bmatrix} u_1^3 \\ u_1^2 \\ u_2^2 \end{bmatrix}\right) = \begin{bmatrix} u_1^2 \\ (u_1^3)^2 + u_1^1 \\ u_1^3 u_1^1 + u_2^1 \end{bmatrix}.
$$

Since $\frac{\partial \mathcal{F}(u_1^1, u_2^1, u_1^2, u_2^2, u_1^3)}{\partial \tilde{U}} = \begin{bmatrix} 0 & 0 & 1 & 0 & 0 \\ 1 & 0 & 0 & 0 & 2u_1^3 \\ u_1^3 & 1 & 0 & 0 & u_1^1 \end{bmatrix}$, we have that

$$
\ker \mathcal{F}_* = \mathrm{span}\left\{\frac{\partial}{\partial u_2^2}, -2u_1^3\frac{\partial}{\partial u_1^1} + (2(u_1^3)^2 - u_1^1)\frac{\partial}{\partial u_2^1} + \frac{\partial}{\partial u_1^3}\right\}
$$

$$
= \mathrm{span}\left\{\begin{bmatrix} 0 \\ 0 \\ 0 \\ 1 \\ 0 \end{bmatrix}, \begin{bmatrix} -2u_1^3 \\ 2(u_1^3)^2 - u_1^1 \\ 0 \\ 0 \\ 1 \end{bmatrix}\right\}
$$

and

$$
\left[\frac{\partial}{\partial u_1^1}, -2u_1^3\frac{\partial}{\partial u_1^1} + (2(u_1^3)^2 - u_1^1)\frac{\partial}{\partial u_2^1} + \frac{\partial}{\partial u_1^3}\right] = -\frac{\partial}{\partial u_2^1} \notin \ker \mathcal{F}_*.
$$

Thus, it is clear, by Theorem 2.6, that $\mathcal{F}_*\left(\frac{\partial}{\partial u_i^1}\right)$ is not a well-defined vector field and condition (ii) of Theorem 7.3 is not satisfied. Hence, by Theorem 7.3, system (7.73) is not state equivalent to a linear system.   □

It is clear that system (7.73) is linearizable by using feedback $\begin{bmatrix} u_1(t) \\ u_2(t) \end{bmatrix} = \gamma(x(t))$,

$v(t)) = \begin{bmatrix} -x_1(t)^2 + v_1(t) \\ x_1(t)^3 - x_1(t)v_1(t) + v_2(t) \end{bmatrix}$. In the following, we consider the feedback linearization of multi-input discrete time systems. In other words, the discrete version of Lemma 4.3 and Theorem 4.3 will be obtained.

Example 7.3.5 Suppose that for $1 \le i \le m$ and $1 \le \ell \le \kappa_i - 1$,

$$\frac{\partial}{\partial u}\left(S_{i1} \circ \hat{F}_u^\ell(x)\right) = 0 ; \quad \det\left(\begin{bmatrix} \left.\frac{\partial\left(S_{11} \circ \hat{F}_u^{\kappa_1}(x)\right)}{\partial u}\right|_{(0,0)} \\ \vdots \\ \left.\frac{\partial\left(S_{m1} \circ \hat{F}_u^{\kappa_m}(x)\right)}{\partial u}\right|_{(0,0)} \end{bmatrix}\right) \neq 0. \tag{7.74}$$

Show that

$$\mathrm{rank}\left(\begin{bmatrix} \frac{\partial S_{11}(x)}{\partial x} \\ \vdots \\ \frac{\partial\left(S_{11} \circ F_0^{\kappa_1-1}(x)\right)}{\partial x} \\ \vdots \\ \frac{\partial S_{m1}(x)}{\partial x} \\ \vdots \\ \frac{\partial\left(S_{m1} \circ F_0^{\kappa_m-1}(x)\right)}{\partial x} \end{bmatrix}\right) = n \tag{7.75}$$

and

$$\mathrm{rank}\left(\begin{bmatrix} \bar{b}_1 \ \bar{A}\bar{b}_1, \ldots, \bar{A}^{\kappa_1-1}\bar{b}_1 \cdots \bar{b}_m \cdots \bar{A}^{\kappa_m-1}\bar{b}_m \end{bmatrix}\right) = n \tag{7.76}$$

where $\bar{A} \triangleq \left.\frac{\partial F_u(x)}{\partial x}\right|_{(0,0)}$ and $\bar{b}_j \triangleq \left.\frac{\partial F_u(x)}{\partial u_j}\right|_{(0,0)}$.

Solution Omitted. (See Problem 7.6.)   □

Lemma 7.2 *System (7.50) is feedback linearizable, if and only if there exist smooth functions $S_{i1}(x) : \mathbb{R}^n \to \mathbb{R}$, $1 \le i \le m$, such that for $1 \le i \le m$,*

(i) $\frac{\partial}{\partial u}\left(S_{i1} \circ \hat{F}_u^\ell(x)\right) = 0$, $1 \le \ell \le \kappa_i - 1$.

$$(ii) \quad \det \left(\begin{bmatrix} \left. \dfrac{\partial \left(S_{11} \circ \hat{F}_u^{\kappa_1}(x) \right)}{\partial u} \right|_{(0,0)} \\ \vdots \\ \left. \dfrac{\partial \left(S_{m1} \circ \hat{F}_u^{\kappa_m}(x) \right)}{\partial u} \right|_{(0,0)} \end{bmatrix} \right) \neq 0.$$

Furthermore, state transformation $z = S(x)$ *and feedback* $u = \gamma(x, v)$ *satisfy*

$$S(x) = \begin{bmatrix} S_{11}(x) & \cdots & S_{11} \circ F_0^{\kappa_1 - 1}(x) & \cdots & S_{m1}(x) & \cdots & S_{m1} \circ F_0^{\kappa_m - 1}(x) \end{bmatrix}^{\mathsf{T}} \quad (7.77)$$

and

$$\begin{bmatrix} v_1 \\ \vdots \\ v_m \end{bmatrix} = \begin{bmatrix} S_{11} \circ \hat{F}_{\gamma(x,v)}^{\kappa_1}(x) \\ \vdots \\ S_{m1} \circ \hat{F}_{\gamma(x,v)}^{\kappa_m}(x) \end{bmatrix}. \quad (7.78)$$

Proof Necessity. Suppose that system (7.50) is feedback linearizable. Then, there exist a state transformation $z = S(x)$ and a nonsingular feedback $u = \gamma(x, v)$ $\left(\det \left(\left. \dfrac{\partial \gamma(x,v)}{\partial v} \right|_{(0,0)} \right) \neq 0 \right)$ such that

$$\tilde{F}_v(z) \triangleq S \circ F_{\gamma(x,v)} \circ S^{-1}(z) = Az + Bv.$$

Thus, we have that for $1 \leq i \leq m$,

$$\begin{bmatrix} S_{i1} \circ F_{\gamma(x,v)}(x) \\ \vdots \\ S_{i(\kappa_i - 1)} \circ F_{\gamma(x,v)}(x) \\ S_{i\kappa_i} \circ F_{\gamma(x,v)}(x) \end{bmatrix} = AS(x) + Bv = \begin{bmatrix} S_{i2}(x) \\ \vdots \\ S_{i\kappa_i}(x) \\ v_i \end{bmatrix}.$$

In other words, for $1 \leq i \leq m$ and $1 \leq \ell \leq \kappa_i - 1$,

$$S_{i(\ell+1)}(x) = S_{i\ell} \circ F_{\gamma(x,v)}(x)$$

and

$$v_i = S_{i\kappa_i} \circ F_{\gamma(x,v)}(x)$$

which imply that for $1 \leq i \leq m$ and $1 \leq \ell \leq \kappa_i - 1$,

$$0 = \frac{\partial \left(S_{i\ell} \circ F_{\gamma(x,v)}(x) \right)}{\partial v} = \left. \frac{\partial \left(S_{i\ell} \circ F_u(x) \right)}{\partial u} \right|_{u=\gamma(x,v)} \frac{\partial \gamma(x, v)}{\partial v}$$

and

$$
I_m = \begin{bmatrix} \frac{\partial\left(S_{1\kappa_1} \circ F_{\gamma(x,v)}(x)\right)}{\partial v} \\ \vdots \\ \frac{\partial\left(S_{m\kappa_m} \circ F_{\gamma(x,v)}(x)\right)}{\partial v} \end{bmatrix} = \begin{bmatrix} \frac{\partial\left(S_{1\kappa_1} \circ F_u(x)\right)}{\partial u} \\ \vdots \\ \frac{\partial\left(S_{m\kappa_m} \circ F_u(x)\right)}{\partial u} \end{bmatrix}\Bigg|_{u=\gamma(x,v)} \frac{\partial\gamma(x,v)}{\partial v}.
$$

Since $\frac{\partial\gamma(x,v)}{\partial v} \neq 0$, it is easy to see that for $1 \le i \le m$ and $1 \le \ell \le \kappa_i - 1$,

$$
S_{i(\ell+1)}(x) = S_{i\ell} \circ F_0(x) \; ; \quad \frac{\partial\left(S_{i\ell} \circ F_u(x)\right)}{\partial u} = 0
$$

and

$$
\begin{bmatrix} v_1 \\ \vdots \\ v_m \end{bmatrix} = \begin{bmatrix} S_{1\kappa_1} \circ F_{\gamma(x,v)}(x) \\ \vdots \\ S_{m\kappa_m} \circ F_{\gamma(x,v)}(x) \end{bmatrix} ; \quad \det\left(\begin{bmatrix} \frac{\partial\left(S_{1\kappa_1} \circ F_u(x)\right)}{\partial u}\Big|_{(0,0)} \\ \vdots \\ \frac{\partial\left(S_{m\kappa_m} \circ F_u(x)\right)}{\partial u}\Big|_{(0,0)} \end{bmatrix} \right) \neq 0.
$$

In other words, for $1 \le i \le m$ and $1 \le \ell \le \kappa_i - 1$,

$$
S_{i(\ell+1)}(x) = S_{i1} \circ F_0^\ell(x) = S_{i1} \circ \hat{F}_u^\ell(x) \; ; \quad \frac{\partial\left(S_{i1} \circ \hat{F}_u^\ell(x)\right)}{\partial u} = 0
$$

and

$$
\begin{bmatrix} v_1 \\ \vdots \\ v_m \end{bmatrix} = \begin{bmatrix} S_{11} \circ \hat{F}_{\gamma(x,v)}^{\kappa_1}(x) \\ \vdots \\ S_{m1} \circ \hat{F}_{\gamma(x,v)}^{\kappa_m}(x) \end{bmatrix} ; \quad \det\left(\begin{bmatrix} \frac{\partial\left(S_{11} \circ \hat{F}_u^{\kappa_1}(x)\right)}{\partial u}\Big|_{(0,0)} \\ \vdots \\ \frac{\partial\left(S_{m1} \circ \hat{F}_u^{\kappa_m}(x)\right)}{\partial u}\Big|_{(0,0)} \end{bmatrix} \right) \neq 0
$$

which imply that conditions (i), (ii), (7.77), and (7.78) are satisfied.

Sufficiency. Suppose that there exist smooth functions $S_{i1}(x)$, $1 \le i \le m$, such that conditions (i) and (ii) are satisfied. Let us define state transformation

$$
z = \begin{bmatrix} z_{11} & \cdots & z_{1\kappa_1} & \cdots & z_{m1} & \cdots & z_{m\kappa_m} \end{bmatrix}^\mathsf{T} = S(x)
$$

and feedback $u = \gamma(x, v)$ as (7.77) and (7.78), respectively. Then it is clear, by Example 7.3.5, that $z = S(x)$ is a state transformation. Also, it is easy to see, by conditions (i), (7.77), and (7.78), that

$$\tilde{F}_v(z) \triangleq S \circ F_{\gamma(x,v)} \circ S^{-1}(z)$$

$$= \begin{bmatrix} S_{11} \circ \hat{F}_{\gamma(x,v)} \circ S^{-1}(z) \\ S_{11} \circ \hat{F}^2_{\gamma(x,v)} \circ S^{-1}(z) \\ \vdots \\ S_{11} \circ \hat{F}^{\kappa_1}_{\gamma(x,v)} \circ S^{-1}(z) \\ \vdots \\ S_{m1} \circ \hat{F}_{\gamma(x,v)} \circ S^{-1}(z) \\ S_{m1} \circ \hat{F}^2_{\gamma(x,v)} \circ S^{-1}(z) \\ \vdots \\ S_{m1} \circ \hat{F}^{\kappa_m}_{\gamma(x,v)} \circ S^{-1}(z) \end{bmatrix} = \begin{bmatrix} S_{11} \circ \hat{F}_0 \circ S^{-1}(z) \\ \vdots \\ S_{11} \circ \hat{F}_0^{\kappa_1 - 1} \circ S^{-1}(z) \\ S_{11} \circ \hat{F}^{\kappa_1}_{\gamma(x,v)}(x)\Big|_{x = S^{-1}(z)} \\ \vdots \\ S_{m1} \circ \hat{F}_0 \circ S^{-1}(z) \\ \vdots \\ S_{m1} \circ \hat{F}_0^{\kappa_m - 1} \circ S^{-1}(z) \\ S_{m1} \circ \hat{F}^{\kappa_m}_{\gamma(x,v)}(x)\Big|_{x = S^{-1}(z)} \end{bmatrix} = \begin{bmatrix} z_{12} \\ \vdots \\ z_{1\kappa_1} \\ v_1 \\ \vdots \\ z_{12} \\ \vdots \\ z_{1\kappa_1} \\ v_1 \end{bmatrix}.$$

By Lemma 7.2, the necessary and sufficient conditions for feedback linearization can be obtained as follows.

Theorem 7.4 (conditions for feedback linearization)
System (7.50) is feedback linearizable, if and only if

(i) $\frac{\partial \Psi(U)}{\partial U}\Big|_{U=0}$ *is nonsingular or* $\sum_{j=1}^{m} \kappa_j = n$.

(ii) $\mathcal{F}_*(\Delta_i)$, $1 \le i \le \kappa_{\max} - 1$, *are well-defined involutive distributions or*

$$\left[\frac{\partial}{\partial u_\ell^i}, \ker \mathcal{F}_*\right] \subset \ker \mathcal{F}_* + \Delta_i, \quad 1 \le i \le \kappa_{\max} - 1, \ \kappa_\ell \ge i - 1 \qquad (7.79)$$

where for $1 \le i \le \kappa_{\max} - 1$,

$$\Delta_i \triangleq \mathrm{span}\left\{\frac{\partial}{\partial u_j^\ell} \ \Bigg|\ 1 \le j \le m, \ 1 \le \ell \le \min(i, \kappa_j + 1)\right\}. \qquad (7.80)$$

Proof Necessity. Suppose that system (7.50) is feedback linearizable. Then, by Lemma 7.2, there exist smooth functions $S_{i1}(x)$, $1 \le i \le m$, such that for $1 \le i \le m$ and $1 \le \ell \le \kappa_i - 1$,

$$\frac{\partial}{\partial u}\left(S_{i1} \circ \hat{F}_u^\ell(x)\right) = 0; \quad \det\left(\begin{bmatrix} \frac{\partial\left(S_{11} \circ \hat{F}_u^{\kappa_1}(x)\right)}{\partial u}\Big|_{(0,0)} \\ \vdots \\ \frac{\partial\left(S_{m1} \circ \hat{F}_u^{\kappa_m}(x)\right)}{\partial u}\Big|_{(0,0)} \end{bmatrix}\right) \ne 0. \qquad (7.81)$$

Thus, by Example 7.3.5, condition (i) is satisfied. Let $\tilde{F}_u(z) \triangleq S \circ F_u \circ S^{-1}(z)$, where

$$
\begin{aligned}
z &\triangleq \begin{bmatrix} z_{11} & z_{12} & \cdots & z_{1\kappa_1} & \cdots & z_{m1} & \cdots & z_{m\kappa_m} \end{bmatrix}^\mathsf{T} = S(x) \\
&= \begin{bmatrix} S_{11}(x) & \cdots & S_{11} \circ F_0^{\kappa_1 - 1}(x) & \cdots & S_{m1}(x) & \cdots & S_{m1} \circ F_0^{\kappa_m - 1}(x) \end{bmatrix}^\mathsf{T}.
\end{aligned}
$$

Then it is easy to see, by (7.81), that for $1 \le i \le m$,

$$
\det \left(\begin{bmatrix} \left. \dfrac{\partial \alpha_{1,u}(z)}{\partial u} \right|_{(0,0)} \\ \vdots \\ \left. \dfrac{\partial \alpha_{m,u}(z)}{\partial u} \right|_{(0,0)} \end{bmatrix} \right) \neq 0 \tag{7.82}
$$

and

$$
\tilde{F}_u(z) \triangleq \begin{bmatrix} \tilde{F}_{1,u}(z) \\ \vdots \\ \tilde{F}_{m,u}(z) \end{bmatrix}, \quad \tilde{F}_{i,u}(z) = \begin{bmatrix} z_{i2} \\ \vdots \\ z_{i\kappa_i} \\ \alpha_{i,u}(z) \end{bmatrix}, \quad \tilde{\mathcal{F}}(\tilde{U}) \triangleq \begin{bmatrix} \tilde{\mathcal{F}}_1(\tilde{U}) \\ \vdots \\ \tilde{\mathcal{F}}_m(\tilde{U}) \end{bmatrix}
$$

$$
\tilde{\mathcal{F}}_i(\tilde{U}) \triangleq \tilde{F}_{i,u^1} \circ \cdots \circ \tilde{F}_{u^{n+1}}(0) \tag{7.83}
$$

$$
= \begin{bmatrix} \alpha_{i,u^{\kappa_i}} \circ \tilde{F}_{u^{\kappa_i+1}} \circ \cdots \circ \tilde{F}_{u^{n+1}}(0) \\ \vdots \\ \alpha_{i,u^2} \circ \tilde{F}_{u^3} \circ \cdots \circ \tilde{F}_{u^{n+1}}(0) \\ \alpha_{i,u^1} \circ \tilde{F}_{u^2} \circ \cdots \circ \tilde{F}_{u^{n+1}}(0) \end{bmatrix} \Bigg|_{u_j^i = 0, \ i \ge \kappa_j + 2}
$$

where $\alpha_{i,u}(z) \triangleq S_{i1} \circ F_0^{\kappa_i - 1} \circ F_u \circ S^{-1}(z) = S_{i1} \circ \hat{F}_u^{\kappa_i} \circ S^{-1}(z)$. Thus, it is clear, by (7.82) and (7.83), that

$$
\tilde{\mathcal{F}}_*(\Delta_1) = \tilde{\mathcal{F}}_* \left(\operatorname{span} \left\{ \frac{\partial}{\partial u_1^1}, \cdots, \frac{\partial}{\partial u_m^1} \right\} \right) = \operatorname{span} \left\{ \frac{\partial}{\partial z_{1\kappa_1}}, \ldots, \frac{\partial}{\partial z_{m\kappa_m}} \right\}.
$$

In this manner, we can show, by (7.82) and (7.83), that for $1 \le i \le \kappa_{\max}$,

$$
\tilde{\mathcal{F}}_*(\Delta_i) = \tilde{\mathcal{F}}_* \left(\operatorname{span} \left\{ \frac{\partial}{\partial u_j^\ell} \ \bigg| \ 1 \le j \le m, \ 1 \le \ell \le \min(i, \kappa_j + 1) \right\} \right)
$$

$$
= \operatorname{span} \left\{ \frac{\partial}{\partial z_{j(\kappa_j + 1 - \ell)}} \ \bigg| \ 1 \le j \le m, \ 1 \le \ell \le \min(i, \kappa_j) \right\}
$$

which implies that $\tilde{\mathcal{F}}_*(\Delta_i)$, $1 \le i \le \kappa_{\max}$, are well-defined involutive distributions. It is clear, by Example 7.1.4, that

$$\mathcal{F}(\tilde{U}) = S^{-1} \circ \tilde{\mathcal{F}}(\tilde{U})$$

and for $1 \le i \le \kappa_{\max}$,

$$
\begin{aligned}
\mathcal{F}_*(\Delta_i) &= \left(S^{-1} \circ \tilde{\mathcal{F}} \right)_* (\Delta_i) \\
&= S_*^{-1} \left(\text{span} \left\{ \left. \frac{\partial}{\partial z_{j(\kappa_j + 1 - \ell)}} \, \right| \, 1 \le j \le m, \ 1 \le \ell \le \min(i, \kappa_j) \right\} \right).
\end{aligned}
$$

Since $x = S^{-1}(z)$ is invertible, $\mathcal{F}_*(\Delta_i)$, $1 \le i \le \kappa_{\max}$, are also well-defined involutive distributions and (7.79) is, by Theorem 2.10, satisfied. Therefore, condition (ii) is satisfied.

Sufficiency. Suppose that conditions (i) and (ii) are satisfied. Without loss of generality, we can assume that $\kappa_{\max} = \kappa_1$ and

$$\kappa_1 \ge \kappa_2 \ge \cdots \kappa_m.$$

Let $\mu_i \triangleq \text{card}\{j \mid 1 \le j \le m \text{ and } \kappa_j \ge i\}$ for $0 \le i \le \kappa_1$. For example, if $(\kappa_1, \kappa_2, \kappa_3) = (4, 2, 1)$, then it is clear that $(\mu_0, \mu_1, \mu_2, \mu_3, \mu_4, \mu_5) = (3, 3, 2, 1, 1, 0)$. Then, it is easy to see that $\mu_0 = m$, $\sum_{\ell=1}^{\kappa_1} \mu_\ell = \sum_{\ell=1}^{m} \kappa_\ell = n$, and for $1 \le i \le \kappa_1$,

$$\Delta_i \triangleq \text{span} \left\{ \left. \frac{\partial}{\partial u_j^\ell} \, \right| \, 1 \le j \le \mu_{\ell-1}, \ 1 \le \ell \le i \right\}.$$

By condition (ii), $D_i = \mathcal{F}_*(\Delta_i)$ is a $\left(\sum_{\ell=1}^{i} \mu_\ell \right)$-dimensional well-defined involutive distribution for $1 \le i \le \kappa_{\max} = \kappa_1$ and $D_1 \subset D_2 \subset \cdots \subset D_{\kappa_1}$. Thus, there exists, by the Frobenius Theorem (or Theorem 2.8), a state transformation $\xi = [\xi_{11} \cdots \xi_{1\kappa_1} \cdots \xi_{m1} \cdots \xi_{m\kappa_m}]^T = \tilde{S}(x)$ such that for $1 \le i \le \kappa_1$,

$$
\begin{aligned}
\tilde{D}_i &\triangleq \tilde{S}_* \left(\mathcal{F}_* \left(\text{span} \left\{ \left. \frac{\partial}{\partial u_j^\ell} \, \right| \, 1 \le j \le \mu_{\ell-1}, \ 1 \le \ell \le i \right\} \right) \right) \\
&= \text{span} \left\{ \left. \frac{\partial}{\partial \xi_j^{(\kappa_j + 1 - \ell)}} \, \right| \, 1 \le j \le m, \ 1 \le \ell \le \min(i, \kappa_j) \right\}.
\end{aligned}
\tag{7.84}
$$

Let $\tilde{F}_u(\xi) \triangleq \tilde{S} \circ F_u \circ \tilde{S}^{-1}(\xi)$. Then, we have, by Example 7.1.4, that

$$\tilde{\mathcal{F}}(\tilde{U}) \triangleq F_{\tilde{u}^1} \circ \cdots \circ F_{\tilde{u}^{\kappa_1 + 1}}(0) = \tilde{S} \circ \mathcal{F}(\tilde{U})$$

where for $1 \le i \le m$, $1 \le \ell \le \kappa_1 + 1$,

$$\tilde{u}^\ell \triangleq \begin{bmatrix} u_1^\ell \\ \vdots \\ u_{\mu_{\ell-1}}^\ell \\ O_{(m-\mu_\ell)\times 1} \end{bmatrix} \triangleq \begin{bmatrix} \tilde{U}^\ell \\ O_{(m-\mu_\ell)\times 1} \end{bmatrix} ; \quad \tilde{U} \triangleq \begin{bmatrix} \tilde{U}^1 \\ \vdots \\ \tilde{U}^{\kappa_1+1} \end{bmatrix}$$

$$U^\ell \triangleq \begin{bmatrix} u_1^\ell \\ \vdots \\ u_{\mu_\ell}^\ell \end{bmatrix} ; \quad U \triangleq \begin{bmatrix} U^1 \\ \vdots \\ U^{\kappa_1} \end{bmatrix} ; \quad \bar{U}^\ell \triangleq \begin{bmatrix} U^\ell \\ O_{(\mu_{\ell-1}-\mu_\ell)\times 1} \end{bmatrix}$$

$$\tilde{\mathcal{F}}(\tilde{U}) \triangleq \begin{bmatrix} \tilde{\mathcal{F}}_1(\tilde{U}) \\ \vdots \\ \tilde{\mathcal{F}}_m(\tilde{U}) \end{bmatrix} ; \quad \tilde{\mathcal{F}}_i(\tilde{U}) \triangleq \begin{bmatrix} \tilde{\mathcal{F}}_{i,1}(\tilde{U}) \\ \vdots \\ \tilde{\mathcal{F}}_{i,\kappa_i}(\tilde{U}) \end{bmatrix} .$$

Therefore, it is easy to see, by (7.84), that for $1 \le i \le m$, $1 \le j \le \kappa_i$,

$$\frac{\partial}{\partial \tilde{U}^\ell} \tilde{\mathcal{F}}_{ij}(\tilde{U}) = 0, \ 1 \le \ell \le j-1 ; \quad \det\left(\begin{bmatrix} \frac{\partial}{\partial U^j} \tilde{\mathcal{F}}_{1j}(\tilde{U}) \\ \vdots \\ \frac{\partial}{\partial U^j} \tilde{\mathcal{F}}_{\mu_j j}(\tilde{U}) \end{bmatrix} \right) \ne 0.$$

In other words, we have that for $1 \le i \le m$,

$$\tilde{\mathcal{F}}_i(\tilde{U}) = \begin{bmatrix} \alpha_{i1}(\tilde{U}^1, \dots, \tilde{U}^{\kappa_i+1}) \\ \alpha_{i2}(\tilde{U}^2, \dots, \tilde{U}^{\kappa_i+1}) \\ \vdots \\ \alpha_{i(\kappa_i-1)}(\tilde{U}^{\kappa_i-1}, \tilde{U}^{\kappa_i}, \tilde{U}^{\kappa_i+1}) \\ \alpha_{i\kappa_i}(\tilde{U}^{\kappa_i}, \tilde{U}^{\kappa_i+1}) \end{bmatrix}$$

$$\tilde{\Psi}_i(U) = \tilde{\mathcal{F}}_i(\bar{U}^1, \dots, \bar{U}^{\kappa_1}, O) = \begin{bmatrix} \hat{\alpha}_{i1}(U^1, \dots, U^{\kappa_i}) \\ \hat{\alpha}_{i2}(U^2, \dots, U^{\kappa_i}) \\ \vdots \\ \hat{\alpha}_{i(\kappa_i-1)}(U^{\kappa_i-1}, U^{\kappa_i}) \\ \hat{\alpha}_{i\kappa_i}(U^{\kappa_i}) \end{bmatrix}$$

(7.85)

where $\alpha_{ij}(\tilde{U}^j, \dots, \tilde{U}^{\kappa_1}, \tilde{U}^{\kappa_1+1}) \triangleq \tilde{\mathcal{F}}_{ij}(O, \dots, O, \tilde{U}^j, \dots, \tilde{U}^{\kappa_1+1})$ and for $1 \le i \le m$, $1 \le j \le \kappa_i$,

$$\hat{\alpha}_{ij}(U^j, \dots, U^{\kappa_i}) \triangleq \alpha_{ij}(\bar{U}^j, \dots, \bar{U}^{\kappa_i}, O)$$

$$\det\left(\begin{bmatrix} \frac{\partial \alpha_{1j}(U^j, \dots, U^{\kappa_1+1})}{\partial U^j} \\ \vdots \\ \frac{\partial \alpha_{\mu_j j}(U^j, \dots, U^{\kappa_{\mu_j}+1})}{\partial U^j} \end{bmatrix} \right) \ne 0; \quad \det\left(\begin{bmatrix} \frac{\partial \hat{\alpha}_{1j}(U^j, \dots, U^{\kappa_1})}{\partial U^j} \\ \vdots \\ \frac{\partial \hat{\alpha}_{\mu_j j}(U^j, \dots, U^{\kappa_{\mu_j}})}{\partial U^j} \end{bmatrix} \right) \ne 0. \quad (7.86)$$

$$\hat{\alpha}_{ij}(U^j, \ldots, U^{\kappa_i}) \triangleq \alpha_{ij}(\bar{U}^j, \ldots, \bar{U}^{\kappa_i}, O)$$

$$\det\left(\begin{bmatrix} \frac{\partial}{\partial U^j}\hat{\alpha}_{1j}(U^j, \ldots, U^{\kappa_1}) \\ \vdots \\ \frac{\partial}{\partial U^j}\hat{\alpha}_{\mu_j j}(U^j, \ldots, U^{\kappa_{\mu_j}}) \end{bmatrix}\right) \neq 0.$$

Thus, it is clear that there exist smooth functions $\bar{\alpha}_i(\xi^i, \ldots, \xi^{\kappa_1}) : \mathbb{R}^{\sum_{\ell=i}^{\kappa_1}\mu_\ell} \to \mathbb{R}^{\mu_i}$, $1 \leq i \leq n$, such that for $1 \leq i \leq \kappa_1$,

$$\begin{bmatrix} \hat{\alpha}_{1i}(\bar{\alpha}_i(\xi^i, \ldots, \xi^{\kappa_1}), \ldots, \bar{\alpha}_{\kappa_1}(\xi^{\kappa_1})) \\ \vdots \\ \hat{\alpha}_{\mu_i i}(\bar{\alpha}_i(\xi^i, \ldots, \xi^{\kappa_1}), \ldots, \bar{\alpha}_{\kappa_{\mu_i}}(\xi^{\kappa_{\mu_i}}, \ldots, \xi^{\kappa_{\mu_1}})) \end{bmatrix} = \begin{bmatrix} \xi_1^i \\ \vdots \\ \xi_{\mu_i}^i \end{bmatrix} \triangleq \xi^i$$

or

$$\begin{bmatrix} U^1 \\ \vdots \\ U^{\kappa_1} \end{bmatrix} = \tilde{\Psi}^{-1}(\xi) = \begin{bmatrix} \bar{\alpha}_1(\xi^1, \ldots, \xi^{\kappa_1}) \\ \bar{\alpha}_2(\xi^2, \ldots, \xi^{\kappa_1}) \\ \vdots \\ \bar{\alpha}_{\kappa_1-1}(\xi^{\kappa_1-1}, \xi^{\kappa_1}) \\ \bar{\alpha}_{\kappa_1}(\xi^{\kappa_1}) \end{bmatrix}. \tag{7.87}$$

Since $\tilde{F}_u \circ \tilde{\Psi}(U) = \tilde{\mathcal{F}}(u, U)$ by (7.54), we have that for $1 \leq i \leq m$,

$$\tilde{F}_{i,u}(\xi) = \tilde{\mathcal{F}}_i(u, \Psi^{-1}(\xi)) = \begin{bmatrix} \tilde{\alpha}_{i1}(u, \xi^1, \ldots, \xi^{\kappa_i}) \\ \tilde{\alpha}_{i2}(\xi^1, \ldots, \xi^{\kappa_i}) \\ \vdots \\ \tilde{\alpha}_{i(\kappa_i-1)}(\xi^{\kappa_i-2}, \xi^{\kappa_i-1}, \xi^{\kappa_i}) \\ \tilde{\alpha}_{i\kappa_i}(\xi^{\kappa_i-1}, \xi^{\kappa_i}) \end{bmatrix} \tag{7.88}$$

where for $1 \leq i \leq m$ and $2 \leq \ell \leq \kappa_i$,

$$\tilde{\alpha}_{i1}(u, \xi^1, \ldots, \xi^{\kappa_i}) \triangleq \alpha_{i1}(u, \bar{\alpha}_1(\xi^1, \ldots, \xi^{\kappa_i}), \ldots, \bar{\alpha}_{\kappa_i}(\xi^{\kappa_i}))$$

$$\tilde{\alpha}_{i\ell}(\xi^{i-1}, \ldots, \xi^{\kappa_i}) \triangleq \alpha_{i\ell}(\bar{\alpha}_{i-1}(\xi^{i-1}, \ldots, \xi^{\kappa_i}), \ldots, \bar{\alpha}_{\kappa_i}(\xi^{\kappa_i})).$$

Let $\tilde{h}_i(\xi) = \xi_i^{\kappa_i}$ for $1 \leq i \leq m$. Then, we have that $\tilde{h}_i \circ \hat{\tilde{F}}_u(\xi) = \tilde{\alpha}_{i\kappa_i}(\xi^{\kappa_i-1}, \xi^{\kappa_i}) \triangleq H_{i1}(\xi^{\kappa_i-1}, \xi^{\kappa_i}) = \tilde{h}_i(\xi) \circ \hat{\tilde{F}}_0(\xi)$ and

$$\tilde{h}_i \circ \hat{\tilde{F}}_u^2(\xi) = H_{i1}\left(\begin{bmatrix} \tilde{\alpha}_{1(\kappa_i-1)}(\xi^{\kappa_i-2}, \xi^{\kappa_i-1}, \xi^{\kappa_i}) \\ \vdots \\ \tilde{\alpha}_{\mu_{\kappa_i}-1(\kappa_i-1)}(\xi^{\kappa_i-2}, \xi^{\kappa_i-1}, \xi^{\kappa_i}) \end{bmatrix}, \begin{bmatrix} \tilde{\alpha}_{1\kappa_i}(\xi^{\kappa_i-1}, \xi^{\kappa_i}) \\ \vdots \\ \tilde{\alpha}_{\mu_{\kappa_i}\kappa_i}(\xi^{\kappa_i-1}, \xi^{\kappa_i}) \end{bmatrix}\right)$$

$$\triangleq H_{i2}(\xi^{\kappa_i-2}, \xi^{\kappa_i-1}, \xi^{\kappa_i}) = \tilde{h}_i \circ \hat{\tilde{F}}_0^2(\xi).$$

In this manner, it is easy to show, by (7.86) and (7.88), that for $1 \le i \le m$ and $2 \le \ell \le \kappa_i - 1$,

$$\tilde{h}_i \circ \hat{\tilde{F}}_u^\ell(\xi) = \tilde{h}_i \circ \hat{\tilde{F}}_0^\ell(\xi); \quad \det\left(\begin{bmatrix} \dfrac{\partial\left(\tilde{h}_1 \circ \hat{\tilde{F}}_u^{\kappa_1}(\xi)\right)}{\partial u}\Bigg|_{(0,0)} \\ \vdots \\ \dfrac{\partial\left(\tilde{h}_m \circ \hat{\tilde{F}}_u^{\kappa_m}(\xi)\right)}{\partial u}\Bigg|_{(0,0)} \end{bmatrix}\right) \neq 0$$

or

$$\tilde{h}_i \circ \tilde{S} \circ \hat{F}_u^\ell(x) = \tilde{h}_i \circ \tilde{S} \circ \hat{F}_0^\ell(x); \quad \det\left(\begin{bmatrix} \dfrac{\partial\left(\tilde{h}_1 \circ \tilde{S} \circ \hat{F}_u^{\kappa_i}(x)\right)}{\partial u}\Bigg|_{(0,0)} \\ \vdots \\ \dfrac{\partial\left(\tilde{h}_m \circ \tilde{S} \circ \hat{F}_u^{\kappa_i}(x)\right)}{\partial u}\Bigg|_{(0,0)} \end{bmatrix}\right) \neq 0.$$

Therefore, $S_{i1}(x) \triangleq \tilde{h}_i \circ \tilde{S}(x) = \tilde{S}_{i\kappa_i}(x)$, $1 \le i \le m$, satisfy conditions (i) and (ii) of Lemma 7.2. Hence, by Lemma 7.2, system (7.50) is feedback linearizable.

It is clear that if condition (ii) of Theorem 7.3 is satisfied, then condition (ii) of Theorem 7.4 is satisfied. In other words, if a system is state equivalent to a linear system, then it is also feedback linearizable.

Example 7.3.6 Show that system (7.73) is feedback linearizable.

$$\begin{bmatrix} x_1(t+1) \\ x_2(t+1) \\ x_3(t+1) \end{bmatrix} = \begin{bmatrix} x_2(t) \\ x_1(t)^2 + u_1(t) \\ x_1(t)u_1(t) + u_2(t) \end{bmatrix} = F_{u(t)}(x(t)).$$

Solution In Example 7.3.4, we have that $(\kappa_1, \kappa_2) = (2, 1)$,

$$\mathcal{F}(u_1^1, u_2^1, u_1^2, u_2^2, u_1^3) = \begin{bmatrix} u_1^2 \\ (u_1^3)^2 + u_1^1 \\ u_1^3 u_1^1 + u_2^1 \end{bmatrix}$$

and

$$\ker \mathcal{F}_* = \text{span} \left\{ \frac{\partial}{\partial u_2^1}, -2u_1^3 \frac{\partial}{\partial u_1^1} + (2(u_1^3)^2 - u_1^1) \frac{\partial}{\partial u_2^1} + \frac{\partial}{\partial u_1^3} \right\}.$$

Since $\Delta_1 = \text{span} \left\{ \frac{\partial}{\partial u_1^1}, \frac{\partial}{\partial u_2^1} \right\}$, it is easy to see that

$$\left[\frac{\partial}{\partial u_1^1}, \frac{\partial}{\partial u_2^2} \right] = 0 \in \ker \mathcal{F}_* + \Delta_1 \; ; \quad \left[\frac{\partial}{\partial u_2^1}, \frac{\partial}{\partial u_2^2} \right] = 0 \in \ker \mathcal{F}_* + \Delta_1$$

$$\left[\frac{\partial}{\partial u_1^1}, -2u_1^3 \frac{\partial}{\partial u_1^1} + (2(u_1^3)^2 - u_1^1) \frac{\partial}{\partial u_2^1} + \frac{\partial}{\partial u_1^3} \right] = -\frac{\partial}{\partial u_2^1} \in \ker \mathcal{F}_* + \Delta_1$$

$$\left[\frac{\partial}{\partial u_2^1}, -2u_1^3 \frac{\partial}{\partial u_1^1} + (2(u_1^3)^2 - u_1^1) \frac{\partial}{\partial u_2^1} + \frac{\partial}{\partial u_1^3} \right] = 0 \in \ker \mathcal{F}_* + \Delta_1$$

which imply that condition (ii) of Theorem 7.4 is satisfied. Hence, system (7.73) is feedback linearizable. Since

$$\mathcal{F}_*(\Delta_1) = \text{span} \left\{ \frac{\partial}{\partial x_2}, \frac{\partial}{\partial x_3} \right\}, \tag{7.89}$$

we have that $\text{span}\{dx_1\} = \mathcal{F}_*(\Delta_1)^\perp$ and $\text{span}\{dx_1, d(x_1 \circ F_0(x)), dx_3\} = \mathcal{F}_*(\Delta_0)^\perp$, where $\Delta_0 = \text{span}\{0\}$. Thus, $S_{11}(x) = x_1$ and $S_{21}(x) = x_3$ satisfy the conditions of Lemma 7.2. Let

$$\begin{bmatrix} z_1 \\ z_2 \\ z_3 \end{bmatrix} = S(x) = \begin{bmatrix} S_{11}(x) \\ S_{11} \circ F_0(x) \\ S_{21}(x) \end{bmatrix} = \begin{bmatrix} x_1 \\ x_2 \\ x_3 \end{bmatrix}$$

and

$$\begin{bmatrix} S_{11} \circ \hat{F}_u^2(x) \\ S_{11} \circ \hat{F}_u(x) \end{bmatrix} = \begin{bmatrix} x_1^2 + u_1 \\ x_1 u_1 + u_2 \end{bmatrix} = \begin{bmatrix} v_1 \\ v_2 \end{bmatrix} \text{ or } \begin{bmatrix} u_1 \\ u_2 \end{bmatrix} = \begin{bmatrix} -x_1^2 + v_1 \\ x_1^3 - x_1 v_1 + v_2 \end{bmatrix} = \gamma(x, v).$$

Then it is clear that

$$\tilde{F}_u(z) = S \circ F_{\gamma(x,v)} \circ S^{-1}(z) = \begin{bmatrix} 0 & 1 & 0 \\ 0 & 0 & 0 \\ 0 & 0 & 0 \end{bmatrix} \begin{bmatrix} z_1 \\ z_2 \\ z_3 \end{bmatrix} + \begin{bmatrix} 0 & 0 \\ 1 & 0 \\ 0 & 1 \end{bmatrix} u.$$

$\square$

Example 7.3.7 Show that the following discrete system is not feedback linearizable:

$$x(t+1) = \begin{bmatrix} x_2(t) + x_1(t)u_2(t) \\ u_1(t) \\ u_2(t) \end{bmatrix} = F_{u(t)}(x(t)). \tag{7.90}$$

Solution Since $\left.\frac{\partial F_u(x)}{\partial x}\right|_{(0,0)} = \begin{bmatrix} 0 & 1 & 0 \\ 0 & 0 & 0 \\ 0 & 0 & 0 \end{bmatrix}$ and $\left.\frac{\partial F_u(x)}{\partial u}\right|_{(0,0)} = \begin{bmatrix} 0 & 0 \\ 1 & 0 \\ 0 & 1 \end{bmatrix}$, we have, by sim-

ple calculation, that $(\kappa_1, \kappa_2) = (2, 1)$. Since $\kappa_1 + \kappa_2 = 3$, condition (i) of Theorem 7.4 is satisfied. Also, it is easy to see, by (7.52), that

$$\mathcal{F}(u_1^1, u_2^1, u_1^2, u_2^2, u_1^3) \triangleq F_{u^1} \circ F_{u^2} \circ F_{u^3}(0)|_{u_2^3=0} = F_{u^1} \circ F_{u^2}\left(\begin{bmatrix} 0 \\ u_1^3 \\ 0 \end{bmatrix}\right)$$

$$= F_{u^1}\left(\begin{bmatrix} u_1^3 \\ u_1^2 \\ u_2^2 \end{bmatrix}\right) = \begin{bmatrix} u_1^2 + u_1^3 u_2^1 \\ u_1^1 \\ u_2^1 \end{bmatrix}.$$

Since $\frac{\partial \mathcal{F}(u_1^1, u_2^1, u_1^2, u_2^2, u_1^3)}{\partial \bar{U}} = \begin{bmatrix} 0 & u_1^3 & 1 & 0 & u_2^1 \\ 1 & 0 & 0 & 0 & 0 \\ 0 & 1 & 0 & 0 & 0 \end{bmatrix}$, we have that

$$\ker \mathcal{F}_* = \text{span}\left\{ \frac{\partial}{\partial u_2^2}, -u_2^1 \frac{\partial}{\partial u_1^2} + \frac{\partial}{\partial u_1^3} \right\}$$

and

$$\left[\frac{\partial}{\partial u_2^1}, -u_2^1 \frac{\partial}{\partial u_1^2} + \frac{\partial}{\partial u_1^3} \right] = -\frac{\partial}{\partial u_1^2} \notin \ker \mathcal{F}_* + \text{span}\left\{ \frac{\partial}{\partial u_1^1}, \frac{\partial}{\partial u_2^1} \right\}$$

which imply that $\mathcal{F}_*(\Delta_1)$ is not a well-defined vector field. Hence, condition (ii) of Theorem 7.4 is not satisfied and system (7.90) is not feedback linearizable. □

In Example 7.3.7, it is shown that system (7.90) is not feedback linearizable. If we consider the dynamic feedback

$$\begin{bmatrix} u_1(t) \\ u_2(t) \end{bmatrix} = \begin{bmatrix} w_1(t) \\ \eta(t) \end{bmatrix} \tag{7.91}$$

$$\eta(t+1) = w_2(t), \tag{7.92}$$

then we have the following extended system:

$$\begin{bmatrix} x_1(t+1) \\ x_2(t+1) \\ x_3(t+1) \\ \eta(t+1) \end{bmatrix} = \begin{bmatrix} x_2(t) + x_1(t)\eta(t) \\ w_1(t) \\ \eta(t) \\ w_2(t) \end{bmatrix} = \bar{F}_{w(t)}(x(t), \eta(t)). \qquad (7.93)$$

It is easy to see that extended system (7.93) is feedback linearizable with

$$z_E = S(x, \eta) = \begin{bmatrix} x_1 \\ x_2 + x_1\eta \\ x_3 \\ \eta \end{bmatrix} \quad \text{and} \quad \begin{bmatrix} w_1 \\ w_2 \end{bmatrix} = \begin{bmatrix} v_1 - (x_2 + x_1\eta)v_2 \\ v_2 \end{bmatrix}. \qquad (7.94)$$

In other words, system (7.90) is not linearizable by static feedback. However, system (7.90) is linearizable by dynamic feedback

$$\begin{bmatrix} u_1(t) \\ u_2(t) \end{bmatrix} = \begin{bmatrix} v_1(t) - (x_2(t) + x_1(t)\eta(t))v_2(t) \\ \eta(t) \end{bmatrix}$$
$$\eta(t+1) = v_2(t).$$

It is called the dynamic feedback linearization of the discrete time systems.

7.4   Linearization of Discrete Time Systems with Single Output

In this section, we consider the following single input single output discrete nonlinear system:

$$x(t+1) = F(x(t), u(t)) \triangleq F_{u(t)}(x(t))$$
$$y(t) = h(x(t)) \qquad (7.95)$$

where $x \in \mathbb{R}^n$, $u \in \mathbb{R}$, $y \in \mathbb{R}$, and $F(x, u) : \mathbb{R}^{n+1} \to \mathbb{R}^n$ and $h(x) : \mathbb{R}^n \to \mathbb{R}$ are smooth functions with $F(0, 0) = 0$ and $h(0) = 0$.

Theorem 7.5 (conditions for state equivalence to a LS with output)
System (7.95) is state equivalent to a LS with output via state transformation $z = S(x)$, if and only if

(i) $\left.\frac{\partial \Psi(U)}{\partial U}\right|_{U=0}$ *is nonsingular.*
(ii) $\mathcal{F}_* \left(\frac{\partial}{\partial u^i}\right)$, $1 \le i \le n+1$, *are well-defined vector fields.*
(iii) $\frac{\partial(h \circ \Psi(U))}{\partial U} = \bar{c} = \text{const}.$

Furthermore, $z = S(x) = \Psi^{-1}(x)$ is a linearizing state transformation.

Proof Necessity. Suppose that system (7.95) is state equivalent to a linear system with output. Then there exists a state transformation $z = S(x)$ such that

$$\tilde{F}_u(z) \triangleq S \circ F_u \circ S^{-1}(z) = Az + bu$$
$$\tilde{h}(z) \triangleq h \circ S^{-1}(z) = cz$$

where

$$\text{rank}\left([b \ Ab \ \cdots \ A^{n-1}b]\right) = n.$$

It is clear, by Theorem 7.1, that conditions (i) and (ii) of Theorem 7.5 are satisfied. Since $\tilde{F}_u(z) = S \circ F_u \circ S^{-1}(z)$, $\tilde{F}_u(z) = Az + bu$, and $S(0) = 0$, it is easy to see, by Examples 7.1.3 and 7.1.4, that $\tilde{\Psi}(U) = S \circ \Psi(U)$ and

$$\tilde{\Psi}(U) \triangleq \tilde{F}_{u^1} \circ \cdots \circ \tilde{F}_{u^n}(0) = S \circ \Psi(U)$$
$$\tilde{h} \circ \tilde{\Psi}(U) = c\left(A^{n-1}bu^n + \cdots + bu^1\right)$$

which imply that

$$\frac{\partial\,(h \circ \Psi(U))}{\partial U} = \frac{\partial\left(h \circ S^{-1} \circ S \circ \Psi(U)\right)}{\partial U} = \frac{\partial\left(\tilde{h} \circ \tilde{\Psi}(U)\right)}{\partial U}$$
$$= c\left[b \ Ab \ \cdots \ A^{n-1}b\right] \triangleq \bar{c}$$

where $U = [u^1 \ u^2 \ \cdots \ u^n]^{\mathsf{T}}$. Therefore, condition (iii) is satisfied.

Sufficiency. Suppose that conditions (i)–(iii) are satisfied. Then, by Theorem 7.1, we have that

$$\tilde{F}_u(z) \triangleq S \circ F_u \circ S^{-1}(z) = Az + bu \qquad (7.96)$$

where $z = S(x) = \Psi^{-1}(x)$ and

$$\text{rank}\left([b \ Ab \ \cdots \ A^{n-1}b]\right) = n.$$

Also, it is easy to see, by condition (iii), that

$$\tilde{h}(z) \triangleq h \circ S^{-1}(z) = h \circ \Psi(z) = \bar{c}z. \qquad (7.97)$$

Therefore, by (7.96) and (7.97), system (7.95) is state equivalent to a linear system with output via $z = S(x) = \Psi^{-1}(x)$.

Example 7.4.1 Show that the following discrete time system is state equivalent to a linear system with output:

$$\begin{bmatrix} x_1(t+1) \\ x_2(t+1) \end{bmatrix} = \begin{bmatrix} x_2(t) - u(t)^2 \\ u(t) \end{bmatrix} = F_{u(t)}(x(t))$$

$$y(t) = x_1(t) + x_2(t) + x_2(t)^2 = h(x(t)).$$
(7.98)

Solution In Example 7.2.3, it has been shown that condition (i) and condition (ii) of Theorem 7.5 are satisfied with

$$\mathcal{F}(u^1, u^2, u^3) \triangleq F_{u^1} \circ F_{u^2} \circ F_{u^3}(0) = \begin{bmatrix} u^2 - (u^1)^2 \\ u^1 \end{bmatrix}$$

and

$$\Psi(u^1, u^2) \triangleq F_{u^1} \circ F_{u^2}(0) = \begin{bmatrix} u^2 - (u^1)^2 \\ u^1 \end{bmatrix}.$$

Since

$$\frac{\partial (h \circ \Psi(U))}{\partial U} = \frac{\partial (u^2 + u^1)}{\partial U} = \begin{bmatrix} 1 & 1 \end{bmatrix} = \bar{c},$$

condition (iii) of Theorem 7.5 is also satisfied. Hence, by Theorem 7.5, system (7.98) is state equivalent to a linear system with output. Let

$$\begin{bmatrix} z_1 \\ z_2 \end{bmatrix} = S(x) = \Psi^{-1}(x) = \begin{bmatrix} x_2 \\ x_1 + x_2^2 \end{bmatrix}.$$

Then it is easy to see that

$$\tilde{F}_u(z) \triangleq S \circ F_u \circ S^{-1}(z) = \begin{bmatrix} 0 & 0 \\ 1 & 0 \end{bmatrix} \begin{bmatrix} z_1 \\ z_2 \end{bmatrix} + \begin{bmatrix} 1 \\ 0 \end{bmatrix} u$$

and

$$\tilde{h}(z) \triangleq h \circ S^{-1}(z) = \begin{bmatrix} 1 & 1 \end{bmatrix} \begin{bmatrix} z_1 \\ z_2 \end{bmatrix}.$$

□

Example 7.4.2 Show that the following discrete time system is not state equivalent to a linear system with output:

$$\begin{bmatrix} x_1(t+1) \\ x_2(t+1) \end{bmatrix} = \begin{bmatrix} x_2(t) + (1 + x_1(t))u(t) \\ (1 + x_1(t))u(t) \end{bmatrix} = F_{u(t)}(x(t))$$

$$y(t) = 2x_1(t) - x_2(t) = h(x(t)). \tag{7.99}$$

Solution It is easy to see, by (7.13) and (7.14), that

$$\mathcal{F}(u^1, u^2, u^3) \triangleq F_{u^1} \circ F_{u^2} \circ F_{u^3}(0) = F_{u^1} \circ F_{u^2} \left(\begin{bmatrix} u^3 \\ u^3 \end{bmatrix} \right)$$

$$= F_{u^1} \left(\begin{bmatrix} u^3 + u^2 + u^2 u^3 \\ u^2 + u^2 u^3 \end{bmatrix} \right) = \begin{bmatrix} (u^1 + u^2 + u^1 u^2)(1 + u^3) \\ u^1(1 + u^2)(1 + u^3) \end{bmatrix}$$

and

$$\Psi(u^1, u^2) \triangleq F_{u^1} \circ F_{u^2}(0) = \mathcal{F}(u^1, u^2, 0) = \begin{bmatrix} u^1 + u^2 + u^1 u^2 \\ u^1 + u^1 u^2 \end{bmatrix}.$$

Since $\det \left(\frac{\partial \Psi(U)}{\partial U} \Big|_{U=0} \right) = \det \left(\begin{bmatrix} 1 & 1 \\ 1 & 0 \end{bmatrix} \right) = -1 \neq 0$, condition (i) of Theorem 7.5 is satisfied. Since

$$\frac{\partial \mathcal{F}(u^1, u^2, u^3)}{\partial \tilde{U}} = \begin{bmatrix} (1+u^2)(1+u^3) & (1+u^1)(1+u^3) & u^1 + u^2 + u^1 u^2 \\ (1+u^2)(1+u^3) & u^1(1+u^3) & u^1(1+u^2) \end{bmatrix},$$

we have that

$$\ker \mathcal{F}_* = \mathrm{span} \left\{ -\frac{u^1}{(1+u^2)(1+u^3)} \frac{\partial}{\partial u^1} - \frac{u^2}{1+u^3} \frac{\partial}{\partial u^2} + \frac{\partial}{\partial u^3} \right\}$$

and

$$\left[\frac{\partial}{\partial u^1}, \frac{\partial}{\partial u^3} \right] = -\frac{1}{(1+u^2)(1+u^3)} \frac{\partial}{\partial u^1} \notin \ker \mathcal{F}_*.$$

Thus, it is clear, by Theorem 2.6, that $\mathcal{F}_* \left(\frac{\partial}{\partial u^\top} \right)$ is not a well-defined vector field and condition (ii) of Theorem 7.5 is not satisfied. Hence, by Theorem 7.5, system (7.99) is not state equivalent to a linear system with output.   □

Definition 7.6 (*feedback linearization with output*)
System (7.95) is said to be feedback linearizable with output, if there exist a nonsingular feedback $u = \gamma(x, v) \left(\frac{\partial \gamma(x,v)}{\partial v} \Big|_{(0,0)} \neq 0 \text{ and } \gamma(0, 0) = 0 \right)$ and a state transformation $z = S(x)$ such that the closed-loop satisfies, in z-coordinates, the following Brunovsky canonical form:

$$z(t+1) = \begin{bmatrix} 0 & 1 & 0 & \cdots & 0 & 0 \\ 0 & 0 & 1 & \cdots & 0 & 0 \\ \vdots & \vdots & \vdots & & \vdots & \vdots \\ 0 & 0 & 0 & \cdots & 0 & 1 \\ 0 & 0 & 0 & \cdots & 0 & 0 \end{bmatrix} z(t) + \begin{bmatrix} 0 \\ 0 \\ \vdots \\ 0 \\ 1 \end{bmatrix} v(t) = Az(t) + bv(t)$$

$$y(t) = cz(t).$$

In other words,

$$\tilde{F}_v(z) \triangleq S \circ F_{\gamma(x,v)} \circ S^{-1}(z) = Az + bv$$
$$\tilde{h}(z) \triangleq h \circ S^{-1}(z) = cz. \tag{7.100}$$

For the continuous system, the characteristic number of the output is defined as the nonnegative integer ρ if ρth derivative of the output is a function of the input for the first time. (See Definition 5.4.) Similarly, the characteristic number of the output can also be defined for the discrete systems.

Definition 7.7 (*characteristic number*)
The characteristic number ρ of the output is defined as the smallest natural number such that $\frac{\partial}{\partial u}\left(h \circ \hat{F}_u^\rho(x)\right) \neq 0$. In other words,

$$\frac{\partial}{\partial u}\left(h \circ \hat{F}_u^k(x)\right) = 0, \ 1 \le k \le \rho - 1 \ \text{and} \ \frac{\partial}{\partial u}\left(h \circ \hat{F}_u^\rho(x)\right) \neq 0. \tag{7.101}$$

If $\frac{\partial}{\partial u}\left(h \circ \hat{F}_u^k(x)\right) = 0$ for $k \ge 1$, then we let $\rho \triangleq \infty$.

It is easy to see, by mathematical induction, that

$$y(k) = h \circ F_{u(k-1)} \circ \cdots \circ F_{u(0)}(x(0)) = h \circ F_0^k(x(0)), \ 1 \le k \le \rho - 1$$
$$y(\rho) = h \circ F_{u(\rho-1)} \circ \cdots \circ F_{u(0)}(x(0)) = h \circ \hat{F}_{u(0)}^\rho(x(0)).$$

In other words, the output $y(\rho)$ first becomes a function of the input $u(0)$.

Example 7.4.3 Suppose that ρ is the characteristic number of the system (7.95) and $\frac{\partial\left(h \circ \hat{F}_u^\rho(x)\right)}{\partial u}\bigg|_{(0,0)} \neq 0$. Find the nonsingular feedback $u = \gamma(x, v)$ such that the transfer function of the closed-loop system is $G_c(z) \triangleq \frac{Y(z)}{V(z)} = \frac{1}{z^\rho + a_{\rho-1}z^{\rho-1} + \cdots + a_1 z + a_0}$.

Solution It is easy to see, by (7.101), that

$$y(t+k) = h \circ F_0^k(x(t)), \ 1 \le k \le \rho - 1$$
$$y(t+\rho) = h \circ F_0^{\rho-1} \circ F_{u(t)}(x(t)) = h \circ \hat{F}_{u(t)}^\rho(x(t)).$$

We need to find the feedback such that

$$y(t + \rho) = -a_{\rho-1} y(t + \rho - 1) - \cdots - a_1 y(t + 1) - a_0 y(t) + v(t)$$

or

$$h \circ \hat{F}_{u(t)}^{\rho}(x(t)) + a_{\rho-1} h \circ F_0^{\rho-1}(x(t)) + \cdots + a_1 h \circ F_0(x(t)) + a_0 h(x(t)) = v(t).$$

By inverse function Theorem (or Theorem 2.2), there exists a nonsingular feedback $u = \gamma(x, v)$ such that

$$h \circ \hat{F}_{\gamma(x,v)}^{\rho}(x) + a_{\rho-1} h \circ F_0^{\rho-1}(x) + \cdots + a_1 h \circ F_0(x) + a_0 h(x) = v.$$

$\square$

Example 7.4.4 Show that if $\rho = n$ and $\left. \dfrac{\partial\left(h \circ \hat{F}_u^{\rho}(x)\right)}{\partial u} \right|_{(0,0)} \neq 0$, then system (7.95) is feedback linearizable with output.

Solution Suppose that $\rho = n$ and $\left. \dfrac{\partial\left(h \circ \hat{F}_u^{\rho}(x)\right)}{\partial u} \right|_{(0,0)} \neq 0$. Then, we have, by (7.101), that

$$\frac{\partial}{\partial u}\left(h \circ \hat{F}_u^i(x)\right) = 0, \ 1 \leq i \leq n - 1 \quad \text{and} \quad \left. \frac{\partial}{\partial u}\left(h \circ \hat{F}_u^n(x)\right) \right|_{(0,0)} \neq 0.$$

Thus, conditions of Lemma 7.1 are satisfied with $S_1(x) = h(x)$. Therefore, by Lemma 7.1, system (7.95) is feedback linearizable with state transformation

$$z = S(x) = \left[h(x) \ h \circ F_0(x) \ \cdots \ h \circ F_0^{n-1}(x) \right]^{\mathsf{T}}$$

and feedback $u = \gamma(x, v)$ such that

$$v = S_1 \circ \hat{F}_{\gamma(x,v)}^n(x).$$

Since $\tilde{h} = h \circ S^{-1}(z) = z_1$, it is easy to see that (7.100) is satisfied with $c = [1 \ 0 \cdots 0]$. $\square$

Theorem 7.6 (conditions for feedback linearization with output)
Let $\rho \leq n$. System (7.95) is feedback linearizable with output, if and only if

(i) $\left. \dfrac{\partial\left(h \circ \hat{F}_u^{\rho}(x)\right)}{\partial u} \right|_{(0,0)} \neq 0.$

(ii) $\left. \dfrac{\partial \tilde{\Psi}(V)}{\partial V} \right|_{V=0}$ is nonsingular.

(iii) $\bar{\mathcal{F}}_* \left(\frac{\partial}{\partial v^i} \right)$, $1 \leq i \leq n+1$, *are well-defined vector fields or*

$$\left[\frac{\partial}{\partial v^i}, \ \ker \bar{\mathcal{F}}_* \right] \subset \ker \bar{\mathcal{F}}_*, \ \ 1 \leq i \leq n+1$$

where $V \triangleq [v^1 \ \cdots \ v^n]^{\mathsf{T}}$,

$$h \circ \hat{F}^{\rho}_{\bar{\gamma}(x,v)}(x) \triangleq v \tag{7.102}$$

$$\bar{F}_v(x) \triangleq F_{\bar{\gamma}(x,v)}(x) \tag{7.103}$$

$$\bar{\mathcal{F}}(v^1, \ldots, v^n, v^{n+1}) \triangleq \bar{F}_{v^1} \circ \cdots \circ \bar{F}_{v^n} \circ \bar{F}_{v^{n+1}}(0) \tag{7.104}$$

$$\bar{\Psi}(V) = \bar{\Psi}(v^1, \ldots, v^n) \triangleq \bar{\mathcal{F}}(v^1, \ldots, v^n, 0) = \bar{F}_{v^1} \circ \cdots \circ \bar{F}_{v^n}(0). \tag{7.105}$$

Furthermore, state transformation $z = S(x)$ *and nonsingular feedback* $u = \gamma(x, v)$
are given by

$$S(x) = P^{-1} \bar{\Psi}^{-1}(x) \ \text{and} \ \gamma(x, v) = \bar{\gamma}(x, v - \bar{a}S(x)) \tag{7.106}$$

where $\bar{a} \triangleq [a_1 \ a_2 \ \cdots \ a_{n-1} \ a_n]$ *and*

$$\bar{\mathcal{F}}_* \left(\frac{\partial}{\partial v^{n+1}} \right) = \sum_{i=1}^{n} a_i \bar{\mathcal{F}}_* \left(\frac{\partial}{\partial v^i} \right) = \sum_{i=1}^{n} a_i \bar{\Psi}_* \left(\frac{\partial}{\partial v^i} \right) \tag{7.107}$$

$$P = \begin{bmatrix} -a_2 & -a_3 & \cdots & -a_n & 1 \\ -a_3 & -a_4 & \cdots & 1 & 0 \\ \vdots & \vdots & & \vdots & \vdots \\ -a_{n-1} & -a_n & \cdots & 0 & 0 \\ -a_n & 1 & \cdots & 0 & 0 \\ 1 & 0 & \cdots & 0 & 0 \end{bmatrix}. \tag{7.108}$$

Proof Necessity. Suppose that system (7.95) is feedback linearizable with output. Then there exist a state transformation $z = S(x)$ and a nonsingular feedback $u = \gamma(x, v)$ such that

$$\tilde{F}_v(z) \triangleq S \circ F_{\gamma(x,v)} \circ S^{-1}(z) = Az + bv$$
$$\tilde{h}(z) \triangleq h \circ S^{-1}(z) = cz = \begin{bmatrix} c_1 & c_2 & \cdots & c_n \end{bmatrix} z \tag{7.109}$$

where $\bar{F}_v(x) \triangleq F_{\gamma(x,v)}(x)$. It is easy to see, by Example 7.1.3 and Example 7.1.4, that

$$h \circ \hat{F}_{\gamma(x,v)}^\rho(x) = \tilde{h} \circ \hat{\bar{F}}_v^\rho \circ S(x) = cA^\rho S(x) + c_{n+1-\rho}v$$
$$\triangleq \alpha(x) + c_{n+1-\rho}v \tag{7.110}$$

which implies that

$$\frac{\partial \left(h \circ \hat{F}_{\gamma(x,v)}^\rho(x) \right)}{\partial v} = \frac{\partial \left(h \circ \hat{F}_u^\rho(x) \right)}{\partial u} \Bigg|_{u=\gamma(x,v)} \frac{\partial \gamma(x,v)}{\partial v} = c_{n+1-\rho} \neq 0$$

and

$$\frac{\partial \left(h \circ \hat{F}_u^\rho(x) \right)}{\partial u} \Bigg|_{(0,0)} \frac{\partial \gamma(x,v)}{\partial v} \Bigg|_{(0,0)} = c_{n+1-\rho} \neq 0.$$

Therefore, it is clear that condition (i) of Theorem 7.6 is satisfied. Without loss of generality, we can let $c_{n+1-\rho} = 1$. If we let

$$\bar{\gamma}(x,v) \triangleq \gamma(x, -\alpha(x) + v) \text{ or } \gamma(x,v) \triangleq \bar{\gamma}(x, \alpha(x) + v) \tag{7.111}$$

then it is clear, by (7.110), that (7.102) is satisfied. Also, we have, by (7.103), (7.109), and (7.111), that

$$\tilde{F}_v(z) \triangleq S \circ F_{\bar{\gamma}(x,\alpha(x)+v)} \circ S^{-1}(z) = S \circ \bar{F}_{\alpha(x)+v} \circ S^{-1}(z) = Az + bv$$

or

$$\bar{F}_v = S^{-1} \circ \tilde{F}_{-\bar{\alpha}(z)+v}(z) \circ S(x) \triangleq S^{-1} \circ F_v'(z) \circ S(x)$$
$$F_v'(z) \triangleq \tilde{F}_{-\bar{\alpha}(z)+v}(z) = Az + b(-cA^\rho z + v) \triangleq \bar{A}z + bv \tag{7.112}$$

where $\bar{\alpha}(z) \triangleq \alpha \circ S^{-1}(z) = cA^\rho z$,

$$\bar{a} = \begin{bmatrix} a_1 & a_2 & \cdots & a_{n-1} & a_n \end{bmatrix} \triangleq -cA^\rho = -[0 \cdots 0 \, c_{\rho+1} \cdots c_{n-\rho}],$$

and

$$\bar{A} \triangleq A - bcA^{\rho} = \begin{bmatrix} 0 & 1 & 0 & \cdots & 0 & 0 \\ 0 & 0 & 1 & \cdots & 0 & 0 \\ \vdots & \vdots & \vdots & & \vdots & \vdots \\ 0 & 0 & 0 & \cdots & 0 & 1 \\ a_1 & a_2 & a_3 & \cdots & a_{n-1} & a_n \end{bmatrix}.$$

Therefore, it is easy to see, by Example 7.1.3 and Example 7.1.4, that

$$\bar{\mathcal{F}}(\tilde{V}) \triangleq \bar{F}_{v^1} \circ \cdots \circ \bar{F}_{v^n} \circ \bar{F}_{v^{n+1}}(0) = S^{-1} \circ F'_{v^1} \circ \cdots \circ F'_{v^n} \circ F'_{v^{n+1}}(0)$$

$$= S^{-1} \left(\sum_{k=1}^{n+1} \bar{A}^{k-1} b v^k \right)$$

and

$$\bar{\Psi}(V) = \bar{\mathcal{F}}(V, 0) = S^{-1} \left(\sum_{k=1}^{n} \bar{A}^{k-1} b v^k \right).$$

Since $\left. \frac{\partial \bar{\Psi}(V)}{\partial V} \right|_{V=0} = \left. \frac{\partial S^{-1}(z)}{\partial z} \right|_{z=0} \begin{bmatrix} b & \bar{A}b & \cdots & \bar{A}^{n-1}b \end{bmatrix}$, it is clear that condition (ii) of Theorem 7.6 is satisfied. Also, it is easy to see that $\bar{\mathcal{F}}_* \left(\frac{\partial}{\partial v^i} \right) = S_*^{-1} \left(\bar{A}^{i-1}b \right)$, $1 \le i \le n+1$, are well-defined vector fields and condition (iii) of Theorem 7.6 is satisfied.

Sufficiency. Suppose that conditions (i)–(iii) are satisfied. By condition (ii), it is clear that $\bar{z} = \bar{S}(x) = \bar{\Psi}^{-1}(x)$ is a state transformation on a neighborhood of the origin. By Theorem 7.1 or (7.31), we have that

$$\bar{\Psi}^{-1} \circ \bar{F}_{\bar{v}} \circ \bar{\Psi}(\bar{z}) = \bar{A}\bar{z} + \bar{b}\bar{v}$$

$$= \begin{bmatrix} 0 & 0 & \cdots & 0 & a_1 \\ 1 & 0 & \cdots & 0 & a_2 \\ 0 & 1 & \cdots & 0 & a_3 \\ \vdots & \vdots & & \vdots & \vdots \\ 0 & 0 & \cdots & 1 & a_n \end{bmatrix} \bar{z} + \begin{bmatrix} 1 \\ 0 \\ 0 \\ \vdots \\ 0 \end{bmatrix} \bar{v}$$

where

$$\bar{\mathcal{F}}_* \left(\frac{\partial}{\partial v^{n+1}} \right) = \sum_{i=1}^{n} a_i \bar{\mathcal{F}}_* \left(\frac{\partial}{\partial v^i} \right) = \sum_{i=1}^{n} a_i \bar{\Psi}_* \left(\frac{\partial}{\partial v^i} \right).$$

Let $z = S(x) \triangleq P^{-1}\bar{z} = P^{-1}\bar{S}(x) = P^{-1}\bar{\Psi}^{-1}(x)$, where P is defined by (7.107). Then, it is easy to see that $P^{-1}AP = A'$, $P^{-1}\bar{b} = b$, and

$$S \circ \bar{F}_v \circ S^{-1}(z) = P^{-1}\bar{\Psi}^{-1} \circ \bar{F}_{\bar{v}} \circ \bar{\Psi}(Pz) = A'z + b\bar{v}$$

$$= \begin{bmatrix} 0 & 1 & 0 & \cdots & 0 & 0 \\ 0 & 0 & 1 & \cdots & 0 & 0 \\ \vdots & \vdots & \vdots & & \vdots & \vdots \\ 0 & 0 & 0 & \cdots & 0 & 1 \\ a_1 & a_2 & a_3 & \cdots & a_{n-1} & a_n \end{bmatrix} z + \begin{bmatrix} 0 \\ 0 \\ \vdots \\ 0 \\ 1 \end{bmatrix} \bar{v}.$$

Therefore, if we let $u = \gamma(x, v) = \bar{\gamma}(x, v - \bar{a}S(x))$, then we have that

$$\tilde{F}_v(\bar{z}) \triangleq S \circ F_{\gamma(x,v)} \circ S^{-1}(z) = S \circ \bar{F}_{v-\bar{a}S(x)} \circ S^{-1}(z) = Az + bv$$

$$= \begin{bmatrix} 0 & 1 & 0 & \cdots & 0 & 0 \\ 0 & 0 & 1 & \cdots & 0 & 0 \\ \vdots & \vdots & \vdots & & \vdots & \vdots \\ 0 & 0 & 0 & \cdots & 0 & 1 \\ 0 & 0 & 0 & \cdots & 0 & 0 \end{bmatrix} z + \begin{bmatrix} 0 \\ 0 \\ \vdots \\ 0 \\ 1 \end{bmatrix} v \tag{7.113}$$

where $\bar{a} \triangleq \begin{bmatrix} a_1 & a_2 & \cdots & a_{n-1} & a_n \end{bmatrix}$. Finally, it is easy to see, by (7.101) and (7.102), that

$$h \circ \bar{\Psi}(V) = h \circ \bar{F}_{v^1} \circ \cdots \circ \bar{F}_{v^n}(0) = h \circ \bar{F}_0^{\rho-1} \circ \bar{F}_{v^\rho} \circ \cdots \circ \bar{F}_{v^n}(0)$$
$$= v^\rho \triangleq \bar{c}V$$

and

$$\tilde{h} = h \circ S^{-1}(z) = h \circ \bar{\Psi}(Pz) = \bar{c}Pz \triangleq cz. \tag{7.114}$$

Hence, by (7.113) and (7.114), system (7.95) is feedback linearizable with output via $z = S(x) = P^{-1}\bar{\Psi}^{-1}(x)$ and $u = \gamma(x, v) = \bar{\gamma}(x, v - \bar{a}S(x))$.

Example 7.4.5 Show that system (7.99) is feedback linearizable with output.

$$\begin{bmatrix} x_1(t+1) \\ x_2(t+1) \end{bmatrix} = \begin{bmatrix} x_2(t) + (1 + x_1(t))u(t) \\ (1 + x_1(t))u(t) \end{bmatrix} = F_{u(t)}(x(t))$$
$$y(t) = 2x_1(t) - x_2(t) = h(x(t)).$$

Solution Since $h \circ F_u(x) = 2x_2 + (1 + x_1)u$, it is clear that $\rho = 1$ and

$$\left. \frac{\partial \left(h \circ \hat{F}_u^\rho(x) \right)}{\partial u} \right|_{(0,0)} = 1 \neq 0$$

and

$$h \circ F_{\bar{\gamma}(x,v)}(x) = 2x_2 + (1 + x_1)\bar{\gamma}(x, v) = v \text{ or } \bar{\gamma}(x, v) = \frac{v - 2x_2}{1 + x_1}.$$

Thus, condition (i) of Theorem 7.6 is satisfied and

$$\bar{F}_v(x) \triangleq F_{\bar{\gamma}(x,v)}(x) = \begin{bmatrix} -x_2 + v \\ -2x_2 + v \end{bmatrix}.$$

Thus, we have that

$$\tilde{\mathcal{F}}(v^1, v^2, v^3) \triangleq \bar{F}_{v^1} \circ \bar{F}_{v^2} \circ \bar{F}_{v^3}(0) = \bar{F}_{v^1} \circ \bar{F}_{v^2} \left(\begin{bmatrix} v^3 \\ v^3 \end{bmatrix} \right)$$

$$= \bar{F}_{v^1} \left(\begin{bmatrix} v^2 - v^3 \\ v^2 - 2v^3 \end{bmatrix} \right) = \begin{bmatrix} v^1 - v^2 + 2v^3 \\ v^1 - 2v^2 + 4v^3 \end{bmatrix}$$

and

$$\bar{\Psi}(v^1, v^2) \triangleq \bar{F}_{v^1} \circ \bar{F}_{v^2}(0) = \tilde{\mathcal{F}}(v^1, v^2, 0) = \begin{bmatrix} v^1 - v^2 \\ v^1 - 2v^2 \end{bmatrix}.$$

Since $\det\left(\frac{\partial \bar{\Psi}(V)}{\partial V} \Big|_{V=0} \right) = \det\left(\begin{bmatrix} 1 & -1 \\ 1 & -2 \end{bmatrix} \right) = -1 \neq 0$, condition (ii) of Theorem 7.6 is satisfied. Since

$$\frac{\partial \tilde{\mathcal{F}}(\tilde{V})}{\partial \tilde{V}} = \begin{bmatrix} 1 & -1 & 2 \\ 1 & -2 & 4 \end{bmatrix},$$

we have that

$$\ker \tilde{\mathcal{F}}_* = \mathrm{span}\left\{ 2\frac{\partial}{\partial v^2} + \frac{\partial}{\partial v^3} \right\}$$

and for $1 \leq i \leq 3$,

$$\left[\frac{\partial}{\partial v^i}, 2\frac{\partial}{\partial v^2} + \frac{\partial}{\partial v^3} \right] = 0 \in \ker \tilde{\mathcal{F}}_*.$$

Thus, it is clear, by Theorem 2.6, that $\tilde{\mathcal{F}}_*\left(\frac{\partial}{\partial v^i} \right)$, $1 \leq i \leq 3$, are well-defined vector fields and condition (iii) of Theorem 7.6 is satisfied. Hence, by Theorem 7.6, system (7.99) is feedback linearizable with output. Note that

$$\tilde{\mathcal{F}}_*\left(\frac{\partial}{\partial v^3} \right) = \begin{bmatrix} 2 \\ 4 \end{bmatrix}, \ \bar{\Psi}_*\left(\frac{\partial}{\partial v^1} \right) = \begin{bmatrix} 1 \\ 1 \end{bmatrix}, \ \bar{\Psi}_*\left(\frac{\partial}{\partial v^2} \right) = \begin{bmatrix} -1 \\ -2 \end{bmatrix}$$

which imply, together with (7.107) and (7.108), that $\bar{a} \triangleq \begin{bmatrix} a_1 & a_2 \end{bmatrix} = \begin{bmatrix} 0 & -2 \end{bmatrix}$ and

$$P = \begin{bmatrix} 2 & 1 \\ 1 & 0 \end{bmatrix}.$$

Thus, we have, by (7.106), that

$$z = S(x) = P^{-1}\bar{\Psi}^{-1}(x) = \begin{bmatrix} 0 & 1 \\ 1 & -2 \end{bmatrix} \begin{bmatrix} 2x_1 - x_2 \\ x_1 - x_2 \end{bmatrix} = \begin{bmatrix} x_1 - x_2 \\ x_2 \end{bmatrix}$$

and

$$\gamma(x, v) = \bar{\gamma}(x, v - \bar{a}S(x)) = \bar{\gamma}(x, v + 2x_2)$$
$$= \frac{(v + 2x_2) - 2x_2}{1 + x_1} = \frac{v}{1 + x_1}.$$

Then it is easy to see that

$$\tilde{F}_v(z) \triangleq S \circ F_{\gamma(x,v)}(x) \circ S^{-1}(z) = S \circ \begin{bmatrix} x_2 + v \\ v \end{bmatrix} \circ S^{-1}(z)$$
$$= S\left(\begin{bmatrix} z_2 + v \\ v \end{bmatrix}\right) = \begin{bmatrix} z_2 \\ v \end{bmatrix} = Az + bv$$

and

$$\tilde{h}(z) \triangleq h \circ S^{-1}(z) = \begin{bmatrix} 2 & 1 \end{bmatrix} z = cz.$$

□

Example 7.4.6 Show that the following system is not feedback linearizable with output:

$$\begin{bmatrix} x_1(t+1) \\ x_2(t+1) \\ x_3(t+1) \end{bmatrix} = \begin{bmatrix} x_2(t) \\ x_3(t) + u(t)^2 \\ u(t) \end{bmatrix} = F_{u(t)}(x(t)) \qquad (7.115)$$
$$y(t) = -x_1(t) + x_3(t) = h(x(t)).$$

Solution Since $h \circ F_u(x) = -x_2 + u$, it is clear that $\rho = 1$ and

$$\frac{\partial \left(h \circ \hat{F}_u^\rho(x)\right)}{\partial u}\Bigg|_{(0,0)} = 1 \neq 0$$

and

$$h \circ \hat{F}_{\bar{\gamma}(x,v)}(x) = -x_2 + \bar{\gamma}(x, v) = v \text{ or } \bar{\gamma}(x, v) = x_2 + v.$$

Thus, condition (i) of Theorem 7.6 is satisfied and

$$\bar{F}_v(x) \triangleq F_{\bar{\gamma}(x,v)}(x) = \begin{bmatrix} x_2 \\ x_3 + (x_2 + v)^2 \\ x_2 + v \end{bmatrix}.$$

Thus, we have that

$$\bar{\mathcal{F}}(\tilde{V}) \triangleq \bar{F}_{v_1} \circ \bar{F}_{v_2} \circ \bar{F}_{v_3} \circ \bar{F}_{v_4}(0) = \bar{F}_{v_1} \circ \bar{F}_{v_2} \circ \bar{F}_{v_3} \left(\begin{bmatrix} 0 \\ v_4^2 \\ v_4 \end{bmatrix} \right)$$

$$= \bar{F}_{v_1} \circ \bar{F}_{v_2} \left(\begin{bmatrix} v_4^2 \\ v_4 + (v_4^2 + v_3)^2 \\ v_4^2 + v_3 \end{bmatrix} \right)$$

$$= \bar{F}_{v_1} \left(\begin{bmatrix} v_4 + (v_4^2 + v_3)^2 \\ v_4^2 + v_3 + (v_4 + (v_4^2 + v_3)^2 + v_2)^2 \\ v_4 + (v_4^2 + v_3)^2 + v_2 \end{bmatrix} \right)$$

$$= \begin{bmatrix} v_4^2 + v_3 + (v_4 + (v_4^2 + v_3)^2 + v_2)^2 \\ v_4 + (v_4^2 + v_3)^2 + v_2 + (v_4^2 + v_3 + (v_4 + (v_4^2 + v_3)^2 + v_2)^2 + v_1)^2 \\ v_4^2 + v_3 + (v_4 + (v_4^2 + v_3)^2 + v_2)^2 + v_1 \end{bmatrix}$$

and

$$\bar{\Psi}(V) \triangleq \bar{F}_{v_1} \circ \bar{F}_{v_2} \circ \bar{F}_{v_3}(0) = \bar{\mathcal{F}}(v_1, v_2, v_3, 0)$$

$$= \begin{bmatrix} v_3 + (v_3^2 + v_2)^2 \\ v_3^2 + v_2 + (v_3 + (v_3^2 + v_2)^2 + v_1)^2 \\ v_3 + (v_3^2 + v_2)^2 + v_1 \end{bmatrix}$$

where $\tilde{V} = \begin{bmatrix} v_1 & v_2 & v_3 & v_4 \end{bmatrix}^{\mathsf{T}}$ and $V = \begin{bmatrix} v_1 & v_2 & v_3 \end{bmatrix}^{\mathsf{T}}$. Since

$$\det \left(\left. \frac{\partial \bar{\Psi}(V)}{\partial V} \right|_{V=0} \right) = \det \left(\begin{bmatrix} 0 & 0 & 1 \\ 0 & 1 & 0 \\ 1 & 0 & 1 \end{bmatrix} \right) = -1 \neq 0,$$

condition (ii) of Theorem 7.6 is satisfied. By complicated calculations or MATLAB program **ker-sF**$(\bar{\mathcal{F}}, \tilde{V}, n)$, we have that

$$\ker \bar{\mathcal{F}}_* = \text{span} \left\{ -\frac{\partial}{\partial v_2} - 2v_4 \frac{\partial}{\partial v_3} + \frac{\partial}{\partial v_4} \right\}$$

and

$$\left[\frac{\partial}{\partial v_4}, \ -\frac{\partial}{\partial v_2} - 2v_4\frac{\partial}{\partial v_3} + \frac{\partial}{\partial v_4}\right] = -2\frac{\partial}{\partial v_3} \notin \ker \bar{\mathcal{F}}_*.$$

Thus, it is clear, by Theorem 2.6, that $\bar{\mathcal{F}}_* \left(\frac{\partial}{\partial v_4}\right)$ is not a well-defined vector field and condition (iii) of Theorem 7.6 is not satisfied. Hence, by Theorem 7.6, system (7.115) is not feedback linearizable with output.   □

7.5 MATLAB Programs

In this section, the following subfunctions in Appendix C are needed:
adfg, adfgM, ChConst, ChZero, DeltaDT, HatF, ker-sF,
ker-sF-M, KindexDT0, Psi-sF, Psi-sF-M, transp

MATLAB program for Theorem 7.1.

```
clear all
syms x1 x2 x3 x4 x5 x6 x7 x8 x9 real
syms u real
syms w1 w2 w3 w4 w5 w6 w7 w8 w9 real

Fu=[x2-u^2; u]; %Ex:7.2.3

%Fu=[x2; x1^2+u]; %Ex:7.2.4 or Ex:7.2.6

%Fu=[x2+(1+x2)^2*u^2; x3; (1+x2)*u]; %Ex:7.2.7

%Fu=[x2+u^2; x1+u]; %Ex:7.2.8

%Fu=[x2-(x1+u)^2; x1+u]; %P:7-8(a)

%Fu=[x2-(x1+x2^2+u)^2; x1+x2^2+u]; %P:7-8(b)

%Fu=[x2*exp(u); x3; u]; %P:7-8(c)

%Fu=[x2*exp(x1+u); x3; x1+u]; %P:7-8(d)

%Fu=[x2+x3^2; x1+x3; u]; %P:7-8(e)

Fu=simplify(Fu)
n=length(Fu);
x=sym('x',[n,1]);
w=sym('w',[n+1,1]);
W=w(1:n);

[Psi,sF]=Psi_sF(Fu,x,u,w,n)
```

```
dPsi=jacobian(Psi,W);
dPsi0=simplify(subs(dPsi,W,W-W))

if rank(dPsi0)<n
  disp('cond (i) of Thm 7.1 is not satisfied.')
  disp('System is not state equivalent to a LS.')
  return
end

kersF=ker_sF(sF,w,n)

U=jacobian(w,w);
for k=1:n+1
  cc=adfg(U(:,k),kersF,w);
  cc1=[ kersF cc];
  if rank(cc1)>rank(kersF)
    disp('cond (ii) of Thm 7.1 is not satisfied.')
    disp('System is not state equivalent to a LS.')
    return
  end
end

disp('System is, by Thm 7.1, state equivalent to a LS.')

return
```

MATLAB program for Theorem 7.2.

```
clear all
syms x1 x2 x3 x4 x5 x6 x7 x8 x9 real
syms u real
syms w1 w2 w3 w4 w5 w6 w7 w8 w9 real

Fu=[x2-u^2; u]; %Ex:7.2.3

%Fu=[x2; x1^2+u]; %Ex:7.2.4 or Ex:7.2.6

%Fu=[x2+(1+x2)^2*u^2; x3; (1+x2)*u]; %Ex:7.2.7

%Fu=[x2+u^2; x1+u]; %Ex:7.2.8

%Fu=[x2-(x1+u)^2; x1+u]; %P:7-8(a)

%Fu=[x2-(x1+x2^2+u)^2; x1+x2^2+u]; %P:7-8(b)

%Fu=[x2*exp(u); x3; u]; %P:7-8(c)

%Fu=[x2*exp(x1+u); x3; x1+u]; %P:7-8(d)

%Fu=[x2+x3^2; x1+x3; u]; %P:7-8(e)

Fu=simplify(Fu)
n=length(Fu);
```

```
x=sym('x',[n,1]);
w=sym('w',[n+1,1]);
W=w(1:n);

[Psi,sF]=Psi_sF(Fu,x,u,w,n)

dPsi=jacobian(Psi,W);
dPsi0=simplify(subs(dPsi,W,W-W))

if rank(dPsi0)<n
  disp('cond (i) of Thm 7.2 is not satisfied.')
  disp('System is not feedback linearizable.')
  return
end

kersF=ker_sF(sF,w,n)

U=jacobian(w,w);
for k=1:n-1
  Deltak=U(:,1:k);
  ccc=adfg(U(:,k),kersF,w);
  ccc0=[kersF Deltak];
  ccc1=[ ccc0 ccc];
  if rank(ccc1)>rank(ccc0)
    disp('cond (ii) of Thm 7.2 is not satisfied.')
    disp('System is not feedback linearizable.')
    return
  end
end

disp('System is, by Thm 7.2, feedback linearizable.')

return
```

MATLAB program for Theorem 7.3.

```
clear all
syms x1 x2 x3 x4 x5 x6 x7 x8 x9 real
syms u1 u2 u3 u4 u5 u6 u7 u8 u9 real

Fu=[x2-u1^2; u1; u2-u1^2]; m=2; %Ex:7.3.3

%Fu=[x2; x1^2 + u1; x1*u1+u2]; m=2; %Ex:7.3.4 & Ex:7.3.6

%Fu=[x2+x1*u2; u1; u2]; m=2; %Ex:7.3.7

%Fu=[u1; x3+u1^2; u2]; m=2; %P:7-10(a)

%Fu=[x1+u1; x3+(x1+u1)^2; x1^2+u2]; m=2; %P:7-10(b)

%Fu=[x2+x1*u2; x3; u1; u2]; m=2; %P:7-10(c)

n=length(Fu);
```

```
x=sym('x',[n,1]);
u=sym('u',[m,1]);

Fu=simplify(Fu)
ka=KindexDT0(Fu,x,u)

w=sym('w',[m,n+1]);

[Psi,sF,W,tW,bU]=Psi_sF_M(Fu,x,u,w,m,ka)

if sum(ka)<n
  disp('cond (i) of Thm 7.3 is not satisfied.')
  disp('System is not state equivalent to a LS.')
  return
end

kersF=ker_sF_M(sF,w,tW,ka,n,m)

U=jacobian(tW,tW);
for k=1:length(tW)
  cc=adfgM(U(:,k),kersF,tW);
  cc1=[ kersF cc];
  if rank(cc1)>rank(kersF)
    disp('cond (ii) of Thm 7.3 is not satisfied.')
    disp('System is not state equivalent to a LS.')
    return
  end
end

disp('System is, by Thm 7.3, state equivalent to a LS.')

return
```

MATLAB program for Theorem 7.4.

```
clear all
syms x1 x2 x3 x4 x5 x6 x7 x8 x9 real
syms u1 u2 u3 u4 u5 u6 u7 u8 u9 real

Fu=[x2-u1^2; u1; u2-u1^2]; m=2; %Ex:7.3.3

%Fu=[x2; x1^2 + u1; x1*u1+u2]; m=2; %Ex:7.3.4 & Ex:7.3.6

%Fu=[x2+x1*u2; u1; u2]; m=2; %Ex:7.3.7

%Fu=[u1; x3+u1^2; u2]; m=2; %P:7-10(a)

%Fu=[x1+u1; x3+(x1+u1)^2; x1^2+u2]; m=2; %P:7-10(b)

%Fu=[x2+x1*u2; x3; u1; u2]; m=2; %P:7-10(c)

n=length(Fu);
x=sym('x',[n,1]);
```

```
u=sym('u',[m,1]);
w=sym('w',[m,n+1]);

Fu=simplify(Fu)
ka=KindexDT0(Fu,x,u)
[Psi,sF,W,tW,U]=Psi_sF_M(Fu,x,u,w,m,ka)

if sum(ka)<n
  disp('cond (i) of Thm 7.4 is not satisfied.')
  disp('System is not feedback linearizable.')
  return
end

kersF=ker_sF_M(sF,w,tW,ka,n,m)

Delta=tW-tW;
for k2=1:max(ka)-1
  bU=transp(jacobian(U(:,k2),tW));
  Delta=[Delta bU];
  ccc0=[kersF Delta];
  for k1=1:m
    ccc=adfgM(bU(:,k1),kersF,tW);
    if rank([ccc0 ccc])>rank(ccc0)
      disp('cond (ii) of Thm 7.4 is not satisfied.')
      disp('System is not feedback linearizable.')
      return
    end
  end
end

disp('System is, by Thm 7.4, feedback linearizable.')

return
```

MATLAB program for Theorem 7.5.

```
clear all
syms x1 x2 x3 x4 x5 x6 x7 x8 x9 real
syms u real
syms w1 w2 w3 w4 w5 w6 w7 w8 w9 real

Fu=[x2-u^2; u]; h=x1+x2+x2^2; %Ex:7.4.1

%Fu=[x2+(1+x1)*u; (1+x1)*u]; h=2*x1-x2; %Ex:7.4.2

%Fu=[x2-u^2; u]; h=x1+x2; %P:7-13(a)

%Fu=[x2*exp(u); x3; u]; h=x2+x3; %P:7-13(b)

%Fu=[x2*exp(u); x3; u]; h=x1+x2; %P:7-13(c)

%Fu=[x2*exp(u); x3; u]; h=x2^2+x3; %P:7-13(d)
```

```
Fu=simplify(Fu)
h=simplify(h)

[n,m]=size(Fu);
x=sym('x',[n,1]);

w=sym('w',[n+1,1]);
W=w(1:n);

[Psi,sF]=Psi_sF(Fu,x,u,w,n)

dPsi=jacobian(Psi,W);
dPsi0=simplify(subs(dPsi,W,W-W))

if rank(dPsi0)<n
  disp('cond (i) of Thm 7.5 is not satisfied.')
  disp('System is NOT state equivalent to a LS with output.')
  return
end

kersF=ker_sF(sF,w,n)

U=jacobian(w,w);
for k=1:n+1
  cc=adfg(U(:,k),kersF,w);
  cc1=[ kersF cc];
  if rank(cc1)>rank(kersF)
    disp('cond (ii) of Thm 7.5 is not satisfied.')
    disp('System is NOT state equivalent to a LS with output.')
    return
  end
end

hiS=simplify(subs(h,x,Psi));
hC=jacobian(hiS,W)
if ChConst(hC,W)==0
  disp('cond (iii) of Thm 7.5 is not satisfied.')
  disp('System is NOT state equivalent to a LS with output.')
  return
end

disp('System is state equivalent to a LS with output.')

return
```

MATLAB program for Theorem 7.6.

```
clear all
syms x1 x2 x3 x4 x5 x6 x7 x8 x9 real
syms u v real
syms w1 w2 w3 w4 w5 w6 w7 w8 w9 real
syms z1 z2 z3 z4 z5 z6 z7 z8 z9 real

Fu=[x2+(1+x1)*u; (1+x1)*u]; h=2*x1-x2;
Bgamma=(v-2*x2)/(1+x1); %Ex:7.4.5

%Fu=[x2; x3+u^2 ; u]; h=-x1+x3; Bgamma=v+x2; %Ex:7.4.6

%Fu=[x2-u^2; u]; h=x1+x2;
%Bgamma=(1-sqrt(1-4*(v-x2)))/2; %P:7-13(a)

%Fu=[x2*exp(u); x3; u]; h=x2+x3; Bgamma=-x3+v; %P:7-13(b)

%Fu=[x2*exp(u); x3; u]; h=x1+x2; Bgamma=-x3+v; %P:7-13(c)

%Fu=[x2*exp(u); x3; u]; h=x2^2+x3; Bgamma=-x3^2+v; %P:7-13(d)

Fu=simplify(Fu)
h=simplify(h)

[n,m]=size(Fu);
x=sym('x',[n,1]);
xu=[x; u];

rho=CharacDT(Fu,h,x,u)
cc1=simplify(subs(h,x,HatF(Fu,x,u,rho)))
dcc1=jacobian(cc1,u)
dcc10=subs(dcc1,xu,xu-xu)
if ChZero(dcc10)==1
  disp('cond (i) of Thm 7.6 is not satisfied.')
  disp('System is NOT feedback linearizable with output.')
  return
end

Bgam=Bgamma
bFv=simplify(subs(Fu,xu,[x;Bgam]))
ccgam=simplify(subs(h,x,HatF(bFv,x,v,rho)))
if ChZero(ccgam-v)==0
  disp('Bgamma(x,v) is not correct.')
  return
end

w=sym('w',[n+1,1]);
W=w(1:n);

[bPsi,bsF]=Psi_sF(bFv,x,v,w,n)

dPsi=jacobian(bPsi,W);
dPsi0=simplify(subs(dPsi,W,W-W))
```

```
if rank(dPsi0) < n
  disp('cond (ii) of Thm 7.6 is not satisfied.')
  disp('System is NOT feedback linearizable with output.')
  return
end

kersF=ker_sF(bsF,w,n)

U=jacobian(w,w);
for k=1:n+1
  cc=adfg(U(:,k),kersF,w);
  cc3=[ kersF cc];
  if rank(cc3)>rank(kersF)
    disp('cond (iii) of Thm 7.6 is not satisfied.')
    disp('System is NOT feedback linearizable with output.')
    return
  end
end

disp('System is, by Thm 7.6, FB linearizable with output.')

ca1=jacobian(bsF,w)
ba=simplify(inv(ca1(:,1:n))*ca1(:,n+1))

P=jacobian(flip(W),W);
for k1=1:n-1
  P(1:n-k1,k1)=-ba(k1+1:n);
end
P=simplify(P)

z=sym('z',[n,1]);
iS=simplify(subs(bPsi,W,P*z))

hiS=simplify(subs(h,x,iS))
hC=jacobian(hiS,z)

return
```

7.6 Problems

7-1. Solve Example 7.1.2.
7-2. Solve Example 7.1.3.
7-3. Solve Example 7.1.4.
7-4. Solve Example 7.3.1.
7-5. Solve Example 7.3.2.
7-6. Solve Example 7.3.5.

7-7. Consider the following smooth functions $z = S(x) : \mathbb{R}^3 \to \mathbb{R}^2$ and distributions $D(x)$ on $\mathbb{R}^3$. Use Theorem 2.10 to determine whether $S_*(D(x))$ is a well-defined distribution on a neighborhood of $0 \in \mathbb{R}^2$ or not. If it is a well-defined distribution, then find $S_*(D(x))$.

(a) $S(x) = \begin{bmatrix} x_2 - x_1^2 \\ x_1 \end{bmatrix}$, $D(x) = \text{span}\{\frac{\partial}{\partial x_1}\}$.

(b) $S(x) = \begin{bmatrix} x_2 - x_1^2 \\ x_1 \end{bmatrix}$, $D(x) = \text{span}\{\frac{\partial}{\partial x_3}\}$.

(c) $S(x) = \begin{bmatrix} x_2 - x_1^2 \\ x_3 + x_1 x_3 \end{bmatrix}$, $D(x) = \text{span}\{\frac{\partial}{\partial x_3}\}$.

(d) $S(x) = \begin{bmatrix} x_2 - x_1(x_2^2 + x_3) \\ x_2^2 + x_3 \end{bmatrix}$, $D(x) = \text{span}\{\frac{\partial}{\partial x_2}\}$.

(e) $S(x) = \begin{bmatrix} x_2 - x_1(x_2^2 + x_3) \\ x_2^2 + x_3 \end{bmatrix}$, $D(x) = \text{span}\{\frac{\partial}{\partial x_2}, \frac{\partial}{\partial x_3}\}$.

7-8. Find out whether the following single input discrete time systems are state equivalent to a linear system or not. If not, find out whether it is feedback linearizable or not.

(a) $\begin{bmatrix} x_1(t+1) \\ x_2(t+1) \end{bmatrix} = \begin{bmatrix} x_2(t) - (x_1(t) + u(t))^2 \\ x_1(t) + u(t) \end{bmatrix}$.

(b) $\begin{bmatrix} x_1(t+1) \\ x_2(t+1) \end{bmatrix} = \begin{bmatrix} x_2(t) - (x_1(t) + x_2(t)^2 + u(t))^2 \\ x_1(t) + x_2(t)^2 + u(t) \end{bmatrix}$.

(c) $\begin{bmatrix} x_1(t+1) \\ x_2(t+1) \\ x_3(t+1) \end{bmatrix} = \begin{bmatrix} x_2(t)e^{u(t)} \\ x_3(t) \\ u(t) \end{bmatrix}$.

(d) $\begin{bmatrix} x_1(t+1) \\ x_2(t+1) \\ x_3(t+1) \end{bmatrix} = \begin{bmatrix} x_2(t)e^{x_1(t)+u(t)} \\ x_3(t) \\ x_1(t) + u(t) \end{bmatrix}$.

(e) $\begin{bmatrix} x_1(t+1) \\ x_2(t+1) \\ x_3(t+1) \end{bmatrix} = \begin{bmatrix} x_2(t) + x_3(t)^2 \\ x_1(t) + x_3(t) \\ u(t) \end{bmatrix}$.

7-9. Show that system (7.93) is feedback linearizable by extended state transformation $z_E = S(x, \eta)$ and feedback $w = \gamma(x, v)$ in (7.94).

7-10. Find out whether the following multi-input discrete time systems are state equivalent to a linear system or not. If not, find out whether it is feedback linearizable or not.

(a) $\begin{bmatrix} x_1(t+1) \\ x_2(t+1) \\ x_3(t+1) \end{bmatrix} = \begin{bmatrix} u_1(t) \\ x_3(t) + u_1(t)^2 \\ u_2(t) \end{bmatrix}$.

(b) $\begin{bmatrix} x_1(t+1) \\ x_2(t+1) \\ x_3(t+1) \end{bmatrix} = \begin{bmatrix} x_1(t) + u_1(t) \\ x_3(t) + (x_1(t) + u_1(t))^2 \\ x_1(t)^2 + u_2(t) \end{bmatrix}$.

(c) $\begin{bmatrix} x_1(t+1) \\ x_2(t+1) \\ x_3(t+1) \\ x_4(t+1) \end{bmatrix} = \begin{bmatrix} x_2(t) + x_1(t)u_2(t) \\ x_3(t) \\ u_1(t) \\ u_2(t) \end{bmatrix}.$

7-11. Consider the system in Problem 7.10c. With the dynamic feedback

$$\begin{bmatrix} u_1(t) \\ u_2(t) \end{bmatrix} = \begin{bmatrix} w_1(t) \\ \eta_1(t) \end{bmatrix} ; \quad \begin{bmatrix} \eta_1(t+1) \\ \eta_2(t+1) \end{bmatrix} = \begin{bmatrix} \eta_2(t) \\ w_2(t) \end{bmatrix}, \tag{7.116}$$

we have the following extended system:

$$\begin{bmatrix} x_1(t+1) \\ x_2(t+1) \\ x_3(t+1) \\ x_4(t+1) \\ \eta_1(t+1) \\ \eta_2(t+1) \end{bmatrix} = \begin{bmatrix} x_2(t) + x_1(t)\eta_1(t) \\ x_3(t) \\ w_1(t) \\ \eta_1(t) \\ \eta_2(t) \\ w_2(t) \end{bmatrix} = \bar{F}_{w(t)}(x_E(t)) \tag{7.117}$$

where $x_E = \begin{bmatrix} x \\ \eta \end{bmatrix}$. Show that the extended system (7.117) is feedback linearizable. In other words, the system in Problem 7.10c is restricted dynamic
feedback linearizable with indices $(d_1, d_2) = (0, 2)$.

7-12. Suppose that $\left. \frac{\partial h \circ \hat{F}_u^\rho(x)}{\partial u} \right|_{(0,0)} \neq 0$, where ρ is the characteristic number of system
(7.95). Show that

$$\left\{ \left. \frac{\partial h}{\partial x} \right|_{x=0}, \left. \frac{\partial h \circ F_0(x)}{\partial x} \right|_{x=0}, \ldots, \left. \frac{\partial h \circ F_0^{\rho-1}(x)}{\partial x} \right|_{x=0} \right\}$$

is a set of linearly independent one form.

7-13. Find out whether the following SISO discrete time systems are state equivalent
to a linear system with output or not. If not, find out whether it is feedback
linearizable with output or not.

(a) $\begin{bmatrix} x_1(t+1) \\ x_2(t+1) \end{bmatrix} = \begin{bmatrix} x_2(t) - u(t)^2 \\ u(t) \end{bmatrix} ; \quad y(t) = x_1(t) + x_2(t).$

(b) $\begin{bmatrix} x_1(t+1) \\ x_2(t+1) \\ x_3(t+1) \end{bmatrix} = \begin{bmatrix} x_2(t)e^{u(t)} \\ x_3(t) \\ u(t) \end{bmatrix} ; \quad y(t) = x_2(t) + x_3(t).$

(c) $\begin{bmatrix} x_1(t+1) \\ x_2(t+1) \\ x_3(t+1) \end{bmatrix} = \begin{bmatrix} x_2(t)e^{u(t)} \\ x_3(t) \\ u(t) \end{bmatrix} ; \quad y(t) = x_1(t) + x_2(t).$

(d) $\begin{bmatrix} x_1(t+1) \\ x_2(t+1) \\ x_3(t+1) \end{bmatrix} = \begin{bmatrix} x_2(t)e^{u(t)} \\ x_3(t) \\ u(t) \end{bmatrix} ; \quad y(t) = x_2(t)^2 + x_3(t).$

Chapter 8
Observer Error Linearization

8.1 Introduction

An observer is a dynamic system which estimates the state of the system from
the output and the input of the system. For the observable linear control systems,
one of the famous linear observers is Luenberger observer. Consider the following
observable linear system:

$$\dot{z} = Az + Bu \; ; \quad \bar{y} = Cz. \tag{8.1}$$

A Luenberger observer for system (8.1) is the following dynamic system whose
output is $\bar{z}(t)$:

$$\dot{\bar{z}} = (A - LC)\bar{z} + Bu + L\bar{y} \tag{8.2}$$

where L is a matrix such that $(A - LC)$ is an asymptotically stable matrix (or all the
eigenvalues of $(A - LC)$ are in the open left half plane of the complex plane). If we
let $\varepsilon(t) = \bar{z}(t) - z(t)$, then we have

$$\begin{aligned}
\dot{\varepsilon} = \dot{\bar{z}} - \dot{z} &= (A - LC)\bar{z} + Bu + LCz - Az - Bu \\
&= (A - LC)(\bar{z} - z) = (A - LC)\varepsilon
\end{aligned} \tag{8.3}$$

which implies that $\lim_{t \to \infty} \varepsilon(t) = \lim_{t \to \infty} e^{(A-LC)t}\varepsilon(0) = 0$. Block diagram for Luenberger
observer of a linear system can be found in Fig. 8.1. Note that (8.3) or Luenberger
observer is still valid even if we use vector function $\gamma(u)$ or $\gamma(\bar{y}, u)$ instead of Bu,
in (8.1) and (8.2). In other words, Luenberger observer can also be designed for the
following two observable systems:

$$\dot{z} = Az + \gamma(u) \; ; \quad \bar{y} = Cz \tag{8.4}$$

Fig. 8.1 Luenberger
observer of a linear system

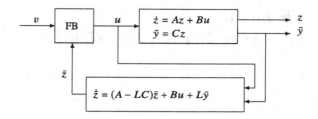

and

$$\dot{z} = Az + \gamma(\bar{y}, u) \; ; \quad \bar{y} = Cz \tag{8.5}$$

where the pair (C, A) is observable or

$$\mathrm{rank} \left(\begin{bmatrix} C \\ CA \\ \vdots \\ CA^{n-1} \end{bmatrix} \right) = n.$$

Thus, system (8.4) and system (8.5) are called a linear observer canonical form (LOCF) and a nonlinear observer canonical form (NOCF), respectively.

Consider the nonlinear system

$$\begin{aligned} \dot{x} &= F(x, u) \triangleq F_u(x) \\ y &= H(x) \end{aligned} \tag{8.6}$$

where $x \in \mathbb{R}^n$, $u \in \mathbb{R}^m$, $y \in \mathbb{R}^q$, and $F(x, u)$ and $H(x)$ are smooth functions with $F(0, 0) = 0$ and $H(0) = 0$. If we use state transformation and output transformation (OT), Luenberger-like observers are also feasible for some nonlinear systems. Suppose that system (8.6) is equivalent to NOCF (8.5) with state transformation $z = S(x)$ and output transformation $\bar{y} = \varphi(y)$. In other words, $S_*(F_u(x)) = Az + \gamma(Cz, u)$ and $\varphi \circ H \circ S^{-1}(z) = Cz$. Then Luenberger-like observers of system (8.6) is given by

$$\begin{aligned} \dot{\bar{z}} &= (A - LC)\bar{z} + \gamma(\bar{y}, u) + L\bar{y} \\ \bar{x}(t) &= S^{-1}(\bar{z}(t)). \end{aligned}$$

Similar arguments can be applied to LOCF. Block diagram for Luenberger-like observers of a nonlinear system can be found in Fig. 8.2. However, not all nonlinear systems are state equivalent to a NOCF with OT. In the following sections, the conditions for a nonlinear system to be equivalent to a NOCF or LOCF will be found.

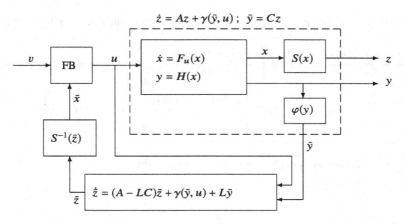

Fig. 8.2 Luenberger-like observers of a nonlinear system

Definition 8.1 (*observability indices*)
For the list of qn one forms of the form

$$\frac{\partial h_1(x)}{\partial x}\bigg|_{x=0}, \cdots, \frac{\partial h_q(x)}{\partial x}\bigg|_{x=0}, \frac{\partial \left(L_{F_0} h_1(x)\right)}{\partial x}\bigg|_{x=0}, \cdots, \frac{\partial \left(L_{F_0} h_q(x)\right)}{\partial x}\bigg|_{x=0},$$

$$\cdots, \frac{\partial \left(L_{F_0}^{n-1} h_1(x)\right)}{\partial x}\bigg|_{x=0}, \cdots, \frac{\partial \left(L_{F_0}^{n-1} h_q(x)\right)}{\partial x}\bigg|_{x=0},$$

delete all one forms that are linearly dependent on the set of preceding one forms and obtain the unique set of linearly independent one forms

$$\left\{\frac{\partial h_1(x)}{\partial x}, \cdots, \frac{\partial \left(L_{F_0}^{v_1-1} h_1(x)\right)}{\partial x}, \cdots, \frac{\partial h_q(x)}{\partial x}, \cdots, \frac{\partial \left(L_{F_0}^{v_q-1} h_q(x)\right)}{\partial x}\right\}\bigg|_{x=0}$$

or

$$\left\{\bar{c}_1, \bar{c}_1 \bar{A}, \cdots, \bar{c}_1 \bar{A}^{v_1-1}, \cdots, \bar{c}_q, \cdots, \bar{c}_q \bar{A}^{v_q-1}\right\}$$

where $\bar{c}_j \triangleq \frac{\partial h_j(x)}{\partial x}\bigg|_{x=0}$ and $\bar{A} \triangleq \frac{\partial F_0(x)}{\partial x}\bigg|_{x=0}$. Then, $(v_1, \cdots, v_q)$ are said to be the observability indices of system (8.6).

In other words, ν_i is the smallest nonnegative integer such that for $1 \le i \le q$,

$$\left. \frac{\partial \left(h_1 \circ F_0^{\nu_i}(x)\right)}{\partial x} \right|_{x=0} \in \text{span} \left\{ \left. \frac{\partial \left(h_j \circ F_0^{\ell-1}(x)\right)}{\partial x} \right|_{x=0} \middle| 1 \le j \le m, \quad 1 \le \ell \le \nu_i \right\}$$

$$+ \text{span} \left\{ \left. \frac{\partial \left(h_j \circ F_0^{\nu_i}(x)\right)}{\partial x} \right|_{x=0} \middle| 1 \le j \le i-1 \right\}.$$

If $\sum\limits_{i=1}^{q} \nu_i = n$, then system (8.6) is said to be observable.

8.2 Single Output Observer Error Linearization

Consider a single output control system of the form

$$\begin{aligned} \dot{x} &= F(x, u) \triangleq F_u(x) \\ y &= H(x) \end{aligned} \tag{8.7}$$

with $F_0(0) = 0$, $H(0) = 0$, state $x \in \mathbb{R}^n$, input $u \in \mathbb{R}^m$, and output $y \in \mathbb{R}$. By letting $u = 0$ in system (8.7), we obtain the following autonomous system:

$$\dot{x} = F_0(x); \quad y = H(x). \tag{8.8}$$

Definition 8.2 (*state equivalence to a LOCF*)
System (8.7) is said to be state equivalent to a LOCF, if there exist a diffeomorphism $z = S(x) : V_0 \to \mathbb{R}^n$, defined on some neighborhood V_0 of $0 \in \mathbb{R}^n$, such that

$$\begin{aligned} \dot{z} &= Az + \gamma(u) \triangleq S_*(F_u(x)) \\ y &= Cz \triangleq H \circ S^{-1}(z) \end{aligned}$$

where the pair (C, A) is observable and $\gamma(u) : \mathbb{R}^m \to \mathbb{R}^n$ is a smooth vector function with $\gamma(0) = 0$.

Definition 8.3 (*state equivalence to a NOCF*)
System (8.7) is said to be state equivalent to a NOCF, if there exist a diffeomorphism $z = S(x) : V_0 \to \mathbb{R}^n$, defined on some neighborhood V_0 of $0 \in \mathbb{R}^n$, such that

$$\begin{aligned} \dot{z} &= Az + \gamma(y, u) \triangleq \bar{f}_u(z) \\ y &= Cz \triangleq \bar{h}(z) \end{aligned}$$

where the pair (C, A) is observable and $\gamma(y, u) : \mathbb{R} \times \mathbb{R}^m \to \mathbb{R}^n$ is a smooth vector function with $\gamma(0, 0) = 0$.

For single output case, if the pair (C, A) is observable, there exists a linear state transform $z = P^{-1}x$ such that $(\hat{C}, \hat{A})(\triangleq (CP, \; P^{-1}AP))$ is an observable canonical form. In other words,

$$CP = \hat{C} = \begin{bmatrix} 1 \; 0 \; 0 \cdots 0 \end{bmatrix} = C_o$$

$$P^{-1}AP = \hat{A} = \begin{bmatrix} \hat{a}_{11} & 1 \; 0 \cdots 0 \\ \hat{a}_{21} & 0 \; 1 \cdots 0 \\ \vdots & \vdots \; \vdots & \vdots \\ \hat{a}_{(n-1)1} & 0 \; 0 \cdots 1 \\ \hat{a}_{n1} & 0 \; 0 \cdots 0 \end{bmatrix} = A_o + \begin{bmatrix} \hat{a}_{11} \\ \hat{a}_{21} \\ \vdots \\ \hat{a}_{(n-1)1} \\ \hat{a}_{n1} \end{bmatrix} \hat{C}$$

where

$$C_o = \begin{bmatrix} 1 \; 0 \; 0 \cdots 0 \end{bmatrix} \quad \text{and} \quad A_o = \begin{bmatrix} 0 \; 1 \; 0 \cdots 0 \\ 0 \; 0 \; 1 \cdots 0 \\ \vdots \; \vdots \; \vdots & \vdots \\ 0 \; 0 \; 0 \cdots 1 \\ 0 \; 0 \; 0 \cdots 0 \end{bmatrix}.$$

Let us call (C_o, A_o) a dual Brunovsky canonical form, even though the order of the states are reversed compared to Brunovsky canonical form (4.9). Since $[\hat{a}_{11} \; \cdots \; \hat{a}_{n1}]^{\mathsf{T}} \hat{C} z = z_1 [\hat{a}_{11} \; \cdots \; \hat{a}_{n1}]^{\mathsf{T}}$, it is clear that single output system (8.7) is state equivalent to a NOCF, if and only if single output system (8.7) is state equivalent to a dual Brunovsky NOCF which is defined by

$$\dot{z} = A_o z + \gamma(z_1, u) \triangleq \bar{f}_u(z)$$
$$y = C_o z \triangleq \bar{h}(z).$$

Definition 8.4 (*state equivalence to a dual Brunovsky NOCF with OT*)
System (8.7) is said to be state equivalent to a dual Brunovsky NOCF with output transformation (OT), if there exist a smooth function $\varphi(y)$ $\left(\frac{\partial \varphi(y)}{\partial y} \Big|_{y=0} = 1 \text{ and } \varphi(0) = 0 \right)$ and a state transformation $z = S(x)$ such that

$$\dot{z} = A_o z + \bar{\gamma}^u(\bar{y}) \triangleq \bar{f}_u(z)$$
$$\bar{y} = \varphi(y) = C_o z \triangleq \bar{h}(z)$$

where $\bar{\gamma}^u(\bar{y}) : \mathbb{R} \times \mathbb{R}^m \to \mathbb{R}^n$ is a smooth vector function with $\bar{\gamma}^0(0) = 0$. In other words,

$$\bar{h}(z) \triangleq \varphi \circ H \circ S^{-1}(z) = C_o z = z_1 \tag{8.9}$$

and

$$\begin{aligned} \bar{f}_u(z) &\triangleq S_*(F_u(x)) = A_o z + \bar{\gamma}^u(C_o z) \\ &= A_o z + \bar{\gamma}^u \circ \varphi(y) \triangleq A_o z + \gamma^u(y). \end{aligned} \tag{8.10}$$

State equivalence to a dual Brunovsky NOCF for autonomous system (8.8) can be similarly defined with $u = 0$. If $\bar{f}_u(z) \triangleq S_*(F_u(x)) = A_o z + \gamma(z_1, u)$, then it is clear that $\bar{f}_0(z) \triangleq S_*(F_0(x)) = A_o z + \gamma(z_1, 0)$. Thus, we have the following remark.

Remark 8.1 If system (8.7) is state equivalent to a dual Brunovsky NOCF with OT $\bar{y} = \varphi(y)$ and state transformation $z = S(x)$, then system (8.8) is also state equivalent to a dual Brunovsky NOCF with OT $\bar{y} = \varphi(y)$ and state transformation $z = S(x)$. But the converse is not true.

Since observability is invariant under state transformation, we assume the observability rank condition on the neighborhood of the origin. In other words,

$$\dim \text{span}\{dH(x), d(L_{F_0} H(x)), \cdots, d(L_{F_0}^{n-1} H(x))\} = n$$

or

$$\text{rank}\left(\frac{\partial T(x)}{\partial x} \bigg|_{x=0} \right) = n$$

where

$$\xi = T(x) \triangleq \begin{bmatrix} H(x) \\ L_{F_0} H(x) \\ \vdots \\ L_{F_0}^{n-1} H(x) \end{bmatrix}.$$

Definition 8.5 (*Canonical System*)
The canonical system of system (8.7) is defined by

$$\begin{bmatrix} \dot{\xi}_1 \\ \vdots \\ \dot{\xi}_{n-1} \\ \dot{\xi}_n \end{bmatrix} = \begin{bmatrix} \xi_2 + \alpha_1^u(\xi) \\ \vdots \\ \xi_n + \alpha_{n-1}^u(\xi) \\ \alpha_n^u(\xi) \end{bmatrix} \triangleq f_u(\xi) ; \quad y = \xi_1 \triangleq h(\xi) \tag{8.11}$$

where $\xi = T(x)$, $f_u(\xi) \triangleq T_*(F_u(x))$, $h(\xi) \triangleq H \circ T^{-1}(\xi)$,

$$\alpha_n^u(\xi) \triangleq L_{F_u} L_{F_0}^{n-1} H(x)\big|_{x=T^{-1}(\xi)}$$

and for $1 \le i \le n - 1$,

$$\alpha_i^u(\xi) \triangleq L_{F_u} L_{F_0}^{i-1} H(x)\big|_{x=T^{-1}(\xi)} - L_{F_0}^i H(x)\big|_{x=T^{-1}(\xi)}.$$

It is clear that $\alpha_i^0(\xi) = 0$ for $1 \le i \le n - 1$ and

$$f_0(\xi) \triangleq T_*(F_0(x)) = \begin{bmatrix} \xi_2 \\ \vdots \\ \xi_n \\ \alpha_n^0(\xi) \end{bmatrix}. \tag{8.12}$$

Remark 8.2 System (8.7) is state equivalent to a dual Brunovsky NOCF with OT (or without OT) via $z = S(x)$, if and only if canonical system (8.11) is state equivalent to a dual Brunovsky NOCF with OT (or without OT) via $z = \tilde{S}(\xi)$ ($\triangleq S \circ T^{-1}(\xi)$). Canonical system (8.11) is more convenient to solve the observer problems than system (8.7). Since geometric conditions are coordinate free, any geometric condition in $\xi-$ coordinates (for system (8.11)) can be expressed in $x-$ coordinates (for system (8.7)).

For system (8.7), we define vector fields $\{\mathbf{g}_i^0(x), \ i \ge 1\}$ and $\{\mathbf{g}_i^u(x), \ i \ge 1\}$ as follows.

$$L_{\mathbf{g}_1^0(x)} L_{F_0}^{k-1} H(x) = \delta_{k,n}, \ 1 \le k \le n$$

$$\left(\text{or } \mathbf{g}_1^0(x) \triangleq \left(\frac{\partial T(x)}{\partial x} \right)^{-1} [0 \ \cdots \ 0 \ 1]^T = T_*^{-1} \left(\frac{\partial}{\partial \xi_n} \right) \right) \tag{8.13}$$

and for $i \ge 2$,

$$\mathbf{g}_i^0(x) \triangleq \text{ad}_{F_0}^{i-1} \mathbf{g}_1^0(x)$$
$$\mathbf{g}_1^u(x) \triangleq \mathbf{g}_1^0(x) \ ; \ \mathbf{g}_i^u(x) \triangleq \text{ad}_{F_u}^{i-1} \mathbf{g}_1^u(x). \tag{8.14}$$

Then it is easy to see, by Example 2.4.16, that for $1 \le i \le n$ and $0 \le k \le n - 1$,

$$L_{\mathbf{g}_i^0(x)} L_{F_0}^k H(x) = \begin{cases} 0, & i + k < n \\ (-1)^{i+1}, & i + k = n. \end{cases} \tag{8.15}$$

Theorem 8.1 *System (8.7) is state equivalent to a LOCF, if and only if*

(i)

$$\mathbf{g}_i^u(x) = \mathbf{g}_i^0(x), \quad 2 \le i \le n+1$$

(ii)

$$[\mathbf{g}_i^0(x), \ \mathbf{g}_k^0(x)] = 0, \quad 1 \le i \le n, \ 1 \le k \le n+1.$$

Furthermore, a state transformation $z = S(x)$ can be obtained by

$$\frac{\partial S(x)}{\partial x} = \left[\, (-1)^{n-1}\mathbf{g}_n^0(x) \quad \cdots \quad -\mathbf{g}_2^0(x) \quad \mathbf{g}_1^0(x) \right]^{-1}.$$

Proof Proof is omitted. (If $u = 0$, this is dual of the linearization of control system by state coordinated change that is considered in Chap. 3.)   □

Example 8.2.1 Consider the following control system:

$$\dot{x} = \begin{bmatrix} x_2 + 2x_2(x_1 - x_2^2) + 2x_2u_1 + u_2^2 \\ x_1 - x_2^2 + u_1 \end{bmatrix} = F_u(x)$$

$$y = x_1 - x_2^2 = H(x).$$

(8.16)

Show that the above system is state equivalent to a LOCF without OT and find a state transformation $z = S(x)$ and the LOCF that the new state z satisfies.

Solution Since $T(x) \triangleq [H(x) \ \ L_{F_0}H(x)]^\mathsf{T} = [x_1 - x_2^2 \ \ x_2]^\mathsf{T}$, it is clear, by (8.13) and (8.14), that

$$\mathbf{g}_1^u(x) \triangleq \mathbf{g}_1^0(x) \triangleq \left(\frac{\partial T(x)}{\partial x}\right)^{-1} \begin{bmatrix} 0 \\ 1 \end{bmatrix} = \begin{bmatrix} 1 & -2x_2 \\ 0 & 1 \end{bmatrix}^{-1} \begin{bmatrix} 0 \\ 1 \end{bmatrix} = \begin{bmatrix} 2x_2 \\ 1 \end{bmatrix}$$

$$\mathbf{g}_2^u(x) \triangleq \mathrm{ad}_{F_u}\mathbf{g}_1^u(x) = \begin{bmatrix} -1 \\ 0 \end{bmatrix}$$

$$\mathbf{g}_3^u(x) \triangleq \mathrm{ad}_{F_u}^2\mathbf{g}_1^u(x) = \begin{bmatrix} 2x_2 \\ 1 \end{bmatrix}$$

which imply that condition (i) and condition (ii) of Theorem 8.1 are satisfied. Hence, system (8.16) is state equivalent to a LOCF with state transformation $z = S(x) = [x_1 - x_2^2 \ \ x_2]^\mathsf{T}$ and $\gamma(u) = [u_2^2 \ \ u_1]^\mathsf{T}$, where

$$\frac{\partial S(x)}{\partial x} = \left[\, -\mathbf{g}_2^0(x) \quad \mathbf{g}_1^0(x) \right]^{-1} = \begin{bmatrix} 1 & -2x_2 \\ 0 & 1 \end{bmatrix}$$

and

$$\dot{z} = S_*(F_u(x)) = \begin{bmatrix} 0 & 1 \\ 1 & 0 \end{bmatrix} z + \begin{bmatrix} u_2^2 \\ u_1 \end{bmatrix} ; \quad y = H \circ S^{-1}(z) = \begin{bmatrix} 1 & 0 \end{bmatrix} z.$$

$\square$

Theorem 8.2 *System (8.7) is state equivalent to a dual Brunovsky NOCF with state transformation* $z = S(x)$, *if and only if*

(i)

$$\mathbf{g}_i^u(x) = \mathbf{g}_i^0(x), \quad 2 \leq i \leq n \tag{8.17}$$

(ii)

$$[\mathbf{g}_i^0(x), \mathbf{g}_k^0(x)] = 0, \quad 1 \leq i \leq n, \ 1 \leq k \leq n \tag{8.18}$$

(iii)

$$\frac{\partial S(x)}{\partial x} = \begin{bmatrix} (-1)^{n-1}\mathbf{g}_n^0(x) & \cdots & -\mathbf{g}_2^0(x) & \mathbf{g}_1^0(x) \end{bmatrix}^{-1}. \tag{8.19}$$

Proof Proof is omitted. (Special case of Lemma 8.2 with $\varphi(y) = y$.) $\square$

Example 8.2.2 Consider the following control system:

$$\dot{x} = \begin{bmatrix} x_2 + 2x_2u + (x_1 - x_2^2)^2u^2 \\ u \end{bmatrix} = F_u(x) \tag{8.20}$$

$$y = x_1 - x_2^2 = H(x).$$

Show that system (8.20) is state equivalent to a dual Brunovsky NOCF without OT and find a state transformation $z = S(x)$ and the dual Brunovsky NOCF that new state z satisfies.

Solution Since $T(x) \triangleq [H(x) \ L_{F_0}H(x)]^\mathsf{T} = [x_1 - x_2^2 \ x_2]^\mathsf{T}$, it is clear, by (8.13) and (8.14), that

$$\mathbf{g}_1^u(x) \triangleq \mathbf{g}_1^0(x) \triangleq \left(\frac{\partial T(x)}{\partial x}\right)^{-1} \begin{bmatrix} 0 \\ 1 \end{bmatrix} = \begin{bmatrix} 2x_2 \\ 1 \end{bmatrix}$$

$$\mathbf{g}_2^u(x) \triangleq \mathrm{ad}_{F_u}\mathbf{g}_1^u(x) = \begin{bmatrix} -1 \\ 0 \end{bmatrix}$$

$$\mathbf{g}_3^u(x) \triangleq \mathrm{ad}_{F_u}^2\mathbf{g}_1^u(x) = \begin{bmatrix} 2(x_1 - x_2^2)u^2 \\ 0 \end{bmatrix}$$

which imply that $\mathbf{g}_3^u(x) \neq \mathbf{g}_3^0(x)$ and condition (i) of Theorem 8.1 is not satisfied. Therefore, by Theorem 8.1, system (8.20) is not state equivalent to a LOCF. However, since condition (i) and condition (ii) of Theorem 8.2 are satisfied, system (8.20) is state equivalent to a dual Brunovsky NOCF with state transformation $z = S(x) = [x_1 - x_2^2 \ \ x_2]^T$ and $\gamma(y, u) = [y^2 u^2 \ \ u]^T$, where

$$\frac{\partial S(x)}{\partial x} = \left[\ -\mathbf{g}_2^0(x) \ \ \mathbf{g}_1^0(x) \right]^{-1} = \begin{bmatrix} 1 & -2x_2 \\ 0 & 1 \end{bmatrix}$$

and

$$\dot{z} = S_*(F_u(x)) = \begin{bmatrix} z_2 \\ 0 \end{bmatrix} + \begin{bmatrix} z_1^2 u^2 \\ u \end{bmatrix} ; \quad y = H \circ S^{-1}(z) = z_1.$$

$\square$

Lemma 8.1 *System (8.7) is state equivalent to a dual Brunovsky NOCF with OT $\bar{y} = \varphi(y)$ and state transformation $z = S(x)$, if and only if there exist a diffeomorphism $\bar{y} = \varphi(y)$ and smooth functions $\gamma_k^u(y) : \mathbb{R}^{1+m} \to \mathbb{R}$, $1 \leq k \leq n$ such that for $1 \leq i \leq n$,*

$$S_i(x) = L_{F_0}^{i-1} (\varphi \circ H(x)) - \sum_{k=1}^{i-1} L_{F_0}^{i-1-k} \left(\gamma_k^0 \circ H(x) \right) \tag{8.21}$$

$$L_{F_u} L_{F_0}^{n-1} (\varphi \circ H(x)) = \sum_{k=1}^{n-1} L_{F_u} L_{F_0}^{n-1-k} \left(\gamma_k^0 \circ H(x) \right) + \gamma_n^u \circ H(x), \tag{8.22}$$

and

$$L_{F_u} S_i(x) - L_{F_0} S_i(x) = \varepsilon_i^u \circ H(x) \tag{8.23}$$

where for $1 \leq i \leq n$,

$$\gamma_i^u(y) \triangleq \gamma_i^0(y) + \varepsilon_i^u(y). \tag{8.24}$$

Proof Necessity. Suppose that system (8.7) is state equivalent to a dual Brunovsky NOCF with OT $\bar{y} = \varphi(y)$ and state transformation $z = S(x)$. Then, it is clear, by (8.9) and (8.10), that

$$\bar{h}(z) \triangleq \varphi \circ H \circ S^{-1}(z) = C_o z = z_1 \tag{8.25}$$

and

$$\bar{f}_u(z) \triangleq S_*(F_u(x)) = A_o z + \bar{\gamma}^u(z_1) = \begin{bmatrix} z_2 + \bar{\gamma}_1^u(z_1) \\ \vdots \\ z_n + \bar{\gamma}_{n-1}^u(z_1) \\ \bar{\gamma}_n^u(z_1) \end{bmatrix} \tag{8.26}$$

which imply that for $1 \le k \le n - 1$,

$$\begin{aligned} S_{k+1}(x) &= L_{F_u} S_k(x) - \bar{\gamma}_k^u(\varphi \circ H(x)) = L_{F_u} S_k(x) - \gamma_k^u \circ H(x) \\ &= L_{F_0} S_k(x) - \gamma_k^0 \circ H(x) \end{aligned} \tag{8.27}$$

and

$$L_{F_u} S_n(x) = \bar{\gamma}_n^u(\varphi \circ H(x)) = \gamma_n^u \circ H(x) \tag{8.28}$$

where $\bar{\gamma}_k^u \circ \varphi(y) \triangleq \gamma_k^u(y)$ for $1 \le k \le n$. Thus, it is clear, by (8.25), that (8.21) is satisfied when $i = 1$. Assume that (8.21) is satisfied when $1 \le i \le \ell \le n - 1$. Then we have, by (8.27), that

$$\begin{aligned} S_{\ell+1}(x) &= L_{F_0} S_\ell(x) - \gamma_\ell^0 \circ H(x) \\ &= L_{F_0}^\ell (\varphi \circ H(x)) - \sum_{k=1}^{\ell-1} L_{F_0}^{\ell-k} \left(\gamma_k^0 \circ H(x) \right) - \gamma_\ell^0 \circ H(x) \\ &= L_{F_0}^\ell (\varphi \circ H(x)) - \sum_{k=1}^{\ell} L_{F_0}^{\ell-k} \left(\gamma_k^0 \circ H(x) \right) \end{aligned} \tag{8.29}$$

which implies that (8.21) is satisfied when $i = \ell + 1 \le n$. Therefore, by mathematical induction, (8.21) is satisfied for $1 \le i \le n$. Since

$$S_n(x) = L_{F_0}^{n-1} (\varphi \circ H(x)) - \sum_{k=1}^{n-1} L_{F_0}^{n-1-k} \left(\gamma_k^0 \circ H(x) \right),$$

it is clear, by (8.28), that

$$L_{F_u} L_{F_0}^{n-1} (\varphi \circ H(x)) - \sum_{k=1}^{n-1} L_{F_u} L_{F_0}^{n-1-k} \left(\gamma_k^0 \circ H(x) \right) = \gamma_n^u \circ H(x) \tag{8.30}$$

which implies that (8.22) is satisfied. Finally, it is easy to see, by (8.27) and (8.28), that for $1 \le k \le n$,

$$L_{F_u} S_k(x) - L_{F_0} S_k(x) = \gamma_k^u \circ H(x) - \gamma_k^0 \circ H(x) \triangleq \varepsilon_k^u \circ H(x)$$

which implies that (8.23) is satisfied.

Sufficiency. Suppose that there exist a diffeomorphism $\bar{y} = \varphi(y)$ and smooth functions $\gamma_k^u(y)$, $1 \le k \le n$ such that (8.21)–(8.24) are satisfied. Let $z = S(x)$. Since $S_1(x) = \varphi \circ H(x)$, it is clear that

$$\bar{h}(z) \triangleq \varphi \circ H \circ S^{-1}(z) = z_1 = C_o z \tag{8.31}$$

and (8.9) is satisfied. Also, it is easy to see, by (8.21)–(8.24), that for $1 \le i \le n - 1$,

$$L_{F_u} S_i(x) = L_{F_0} S_i(x) + \varepsilon_i^u \circ H(x)$$

$$= L_{F_0}^i \left(\varphi \circ H(x) \right) - \sum_{k=1}^{i-1} L_{F_0}^{i-k} \left(\gamma_k^0 \circ H(x) \right) + \varepsilon_i^u \circ H(x)$$

$$= S_{i+1}(x) + \gamma_i^0 \circ H(x) + \varepsilon_i^u \circ H(x)$$

$$= S_{i+1}(x) + \gamma_i^u \circ H(x)$$

and

$$L_{F_u} S_n(x) = L_{F_u} L_{F_0}^{n-1} \left(\varphi \circ H(x) \right) - \sum_{k=1}^{n-1} L_{F_u} L_{F_0}^{n-1-k} \left(\gamma_k^0 \circ H(x) \right)$$

$$= \gamma_n^u \circ H(x)$$

which imply, together with (8.31), that

$$\bar{f}_u(z) \triangleq S_*(F_u(x)) = \begin{bmatrix} L_{F_u} S_1(x) \big|_{x=S^{-1}(z)} \\ \vdots \\ L_{F_u} S_{n-1}(x) \big|_{x=S^{-1}(z)} \\ L_{F_u} S_n(x) \big|_{x=S^{-1}(z)} \end{bmatrix} = \begin{bmatrix} z_2 + \bar{\gamma}_1(z_1) \\ \vdots \\ z_n + \bar{\gamma}_{n-1}(z_1) \\ \bar{\gamma}_n(z_1) \end{bmatrix}$$

$$= A_o z + \bar{\gamma}^u(z_1)$$

where $\bar{\gamma}_k^u \circ \varphi(y) \triangleq \gamma_k^u(y)$ for $1 \le k \le n$. Therefore, (8.10) is satisfied. In other words, system (8.7) is state equivalent to a dual Brunovsky NOCF with OT $\bar{y} = \varphi(y)$ and state transformation $z = S(x)$.   □

Corollary 8.1 *System (8.8) is state equivalent to a dual Brunovsky NOCF with OT* $\bar{y} = \varphi(y)$ *and state transformation* $z = S(x)$, *if and only if there exist a diffeomorphism* $\bar{y} = \varphi(y)$ *and smooth functions* $\gamma_k^0(y) : \mathbb{R}^{1+m} \to \mathbb{R}$, $1 \le k \le n$ *such that for* $1 \le i \le n$,

$$S_i(x) = L_{F_0}^{i-1} \left(\varphi \circ H(x) \right) - \sum_{k=1}^{i-1} L_{F_0}^{i-1-k} \left(\gamma_k^0 \circ H(x) \right) \tag{8.32}$$

and

$$L_{F_0}^n \left(\varphi \circ H(x) \right) = \sum_{k=1}^n L_{F_0}^{n-k} \left(\gamma_k^0 \circ H(x) \right). \tag{8.33}$$

Corollary 8.2 *System (8.8) is state equivalent to a dual Brunovsky NOCF with state transformation $z = S(x)$, if and only if there exist smooth functions $\gamma_k^0(y) : \mathbb{R}^{1+m} \to \mathbb{R}$, $1 \le k \le n$ such that for $1 \le i \le n$,*

$$S_i(x) = L_{F_0}^{i-1} H(x) - \sum_{k=1}^{i-1} L_{F_0}^{i-1-k} \left(\gamma_k^0 \circ H(x) \right) \tag{8.34}$$

and

$$L_{F_0}^n H(x) = \sum_{k=1}^n L_{F_0}^{n-k} \left(\gamma_k^0 \circ H(x) \right). \tag{8.35}$$

Lemma 8.2 *System (8.7) is state equivalent to a dual Brunovsky NOCF with OT $\bar{y} = \varphi(y)$ and state transformation $z = S(x)$, if and only if there exists a smooth function $\ell(y)$ $(\ne 0)$ such that*

(i)

$$\bar{\mathbf{g}}_i^u(x) = \bar{\mathbf{g}}_i^0(x), \ \ 2 \le i \le n \tag{8.36}$$

(ii)

$$[\bar{\mathbf{g}}_i^0(x), \ \bar{\mathbf{g}}_k^0(x)] = 0, \ \ 1 \le i \le n, \ 1 \le k \le n \tag{8.37}$$

where

$$\bar{\mathbf{g}}_1^u(x) = \bar{\mathbf{g}}_1^0(x) \triangleq \ell(H(x))\mathbf{g}_1^0(x) \tag{8.38}$$

$$\bar{\mathbf{g}}_i^u(x) \triangleq \mathrm{ad}_{F_u}^{i-1}\bar{\mathbf{g}}_1^0(x), \ \ i \ge 2 \tag{8.39}$$

$$\varphi(y) = \int_0^y \frac{1}{\ell(\bar{y})}d\bar{y} \tag{8.40}$$

$$\frac{\partial S(x)}{\partial x} = \left[\ (-1)^{n-1}\bar{\mathbf{g}}_n^0(x) \ \ \cdots \ \ -\bar{\mathbf{g}}_2^0(x) \ \ \bar{\mathbf{g}}_1^0(x) \right]^{-1}. \tag{8.41}$$

Proof Necessity. Suppose that system (8.7) is state equivalent to a dual Brunovsky NOCF with OT $\bar{y} = \varphi(y)$ and state transformation $z = S(x)$. Then, by Lemma 8.1, there exist a smooth function $\varphi(y)$ and smooth functions $\gamma_k^u(y)$, $1 \le k \le n$ such that for $1 \le i \le n$,

$$z_i = S_i(x) = L_{F_0}^{i-1}\left(\varphi \circ H(x)\right) - \sum_{k=1}^{i-1} L_{F_0}^{i-1-k}\left(\gamma_k^0 \circ H(x)\right)$$

or

$$\tilde{S}_i(\xi) \triangleq S_i \circ T^{-1}(\xi) = L_{f_0(\xi)}^{i-1}\varphi(\xi_1) - \sum_{k=1}^{i-1} L_{f_0(\xi)}^{i-1-k}\left(\bar{\gamma}_k \circ \varphi(\xi_1)\right) \tag{8.42}$$

where $\xi = T(x) = \begin{bmatrix} H(x) \\ L_{F_0}H(x) \\ \vdots \\ L_{F_0}^{n-1}H(x) \end{bmatrix}$, $f_0(\xi) \triangleq T_*(F_0(x))$, and $\tilde{S}(\xi) \triangleq S \circ T^{-1}(\xi)$.

Also, we have, by (8.9) and (8.10), that

$$\bar{h}(z) \triangleq \varphi \circ H \circ S^{-1}(z) = z_1$$

and

$$\bar{f}_u(z) \triangleq S_*\left(F_u(x)\right) = A_o z + \bar{\gamma}^u(z_1) = \begin{bmatrix} z_2 + \bar{\gamma}_1^u(z_1) \\ \vdots \\ z_n + \bar{\gamma}_{n-1}^u(z_1) \\ \bar{\gamma}_n^u(z_1) \end{bmatrix} \tag{8.43}$$

where $\bar{\gamma}_k^u \circ \varphi(y) \triangleq \gamma_k^u(y)$ for $1 \le k \le n$. Define vector fields $\{\bar{\psi}_1^u(z), \cdots, \bar{\psi}_n^u(z)\}$ by

$$\bar{\psi}_1^u(z) \triangleq \frac{\partial}{\partial z_n} \; ; \quad \bar{\psi}_i^u(z) \triangleq \mathrm{ad}_{\bar{f}_u}^{i-1}\bar{\psi}_1^u(z), \; i \ge 2. \tag{8.44}$$

Then, by (8.43) and (8.44), it is clear that for $1 \le i \le n$,

$$\bar{\psi}_i^u(z) = (-1)^{i-1}\frac{\partial}{\partial z_{n+1-i}} = \bar{\psi}_i^0(z) \tag{8.45}$$

which implies that

$$[\bar{\psi}_i^u(z), \bar{\psi}_k^u(z)] = 0, \quad 1 \le i \le n, \; 1 \le k \le n. \tag{8.46}$$

Note, by (8.42), that

$$\frac{\partial \tilde{S}_i(\xi)}{\partial \xi_n} = \begin{cases} 0, & \text{if } 1 \le i \le n-1 \\ \frac{d\varphi(\xi_1)}{d\xi_1}, & \text{if } i = n \end{cases}$$

which implies, together with (8.44), that

$$\tilde{S}_*(\frac{\partial}{\partial \xi_n}) = \sum_{i=1}^{n} \frac{\partial \tilde{S}_i(\xi)}{\partial \xi_n}\bigg|_{\xi=\tilde{S}^{-1}(z)} \frac{\partial}{\partial z_i} = \frac{d\varphi(\xi_1)}{d\xi_1}\bigg|_{\xi=\tilde{S}^{-1}(z)} \frac{\partial}{\partial z_n}$$

$$= \frac{d\varphi(\xi_1)}{d\xi_1}\bigg|_{\xi=\tilde{S}^{-1}(z)} \bar{\psi}_1^u(z)$$

and

$$\bar{\psi}_1^u(z) = \ell(\xi_1)\bigg|_{\xi=\tilde{S}^{-1}(z)} \tilde{S}_*(\frac{\partial}{\partial \xi_n})$$

where

$$\frac{1}{\ell(\xi_1)} = \frac{d\varphi(\xi_1)}{d\xi_1} \left(\text{or } \varphi(y) = \int_0^y \frac{1}{\ell(\xi_1)} d\xi_1 \right).$$

Therefore, we have, by (2.49), that

$$\tilde{S}_*^{-1}(\bar{\psi}_1^u(z)) = \tilde{S}_*^{-1}\left(\ell(\xi_1)\bigg|_{\xi=\tilde{S}^{-1}(z)} \tilde{S}_*(\frac{\partial}{\partial \xi_n})\right) = \ell(\xi_1)\frac{\partial}{\partial \xi_n}. \tag{8.47}$$

Hence, if we let $\bar{g}_1^u(x) \triangleq S_*^{-1}(\bar{\psi}_1^u(z))$, we have, by (2.49), (8.13), and (8.47), that

$$\bar{g}_1^u(x) = S_*^{-1}(\bar{\psi}_1^u(z)) = T_*^{-1} \circ \tilde{S}_*^{-1}(\bar{\psi}_1^u(z)) = T_*^{-1}\left(\ell(\xi_1)\frac{\partial}{\partial \xi_n}\right)$$

$$= \ell(H(x))T_*^{-1}\left(\frac{\partial}{\partial \xi_n}\right) = \ell(H(x))g_1^0(x)$$

which implies that (8.38) is satisfied. Also, since $\bar{f}_u(z) = S_*(F_u(x))$ or $F_u(x) = S_*^{-1}(\bar{f}_u(z))$, it is clear, by (2.37), (8.39), and (8.44), that for $i \ge 2$,

$$\bar{g}_i^u(x) = \text{ad}_{F_u}^{i-1}\bar{g}_1^u(x) = S_*^{-1}\left\{\text{ad}_{S_*(F_u)}^{i-1}S_*(\bar{g}_1^u(x))\right\}$$

$$= S_*^{-1}\left\{\text{ad}_{\bar{f}_u}^{i-1}\bar{\psi}_1^u(z)\right\} = S_*^{-1}(\bar{\psi}_i^u(z)) \tag{8.48}$$

and thus condition (i) and condition (ii) are satisfied by (8.45) and (8.46). Finally, it is easy to see, by (8.45) and (8.48), that

$$I = \left[(-1)^{n-1} S_*\left(\bar{\mathbf{g}}_n^0(x)\right) \cdots -S_*\left(\bar{\mathbf{g}}_2^0(x)\right) S_*\left(\bar{\mathbf{g}}_1^0(x)\right)\right]$$

$$= \left(\frac{\partial S(x)}{\partial x}\left[(-1)^{n-1}\bar{\mathbf{g}}_n^0(x) \cdots -\bar{\mathbf{g}}_2^0(x)\,\bar{\mathbf{g}}_1^0(x)\right]\right)_{x=S^{-1}(z)}$$

or

$$I = \frac{\partial S(x)}{\partial x}\left[(-1)^{n-1}\bar{\mathbf{g}}_n^0(x) \cdots -\bar{\mathbf{g}}_2^0(x)\,\bar{\mathbf{g}}_1^0(x)\right]$$

which implies that (8.41) is satisfied.

Sufficiency. Suppose that there exists $\ell(y)$ such that (8.36)–(8.40) are satisfied. Since $\{\bar{\mathbf{g}}_1^0(x),\ \bar{\mathbf{g}}_2^0(x),\ \cdots,\ \bar{\mathbf{g}}_n^0(x)\}$ is a set of commuting vector fields, there exists, by Theorem 2.7, a state transformation $z = S(x)$ such that

$$S_*\left(\bar{\mathbf{g}}_i^0(x)\right) = (-1)^{i-1}\frac{\partial}{\partial z_{n+1-i}},\quad 1 \le i \le n. \tag{8.49}$$

In fact, $z = S(x)$ can be calculated by (8.41). Now it will be shown that

$$\bar{h}(z) \triangleq \varphi \circ H \circ S^{-1}(z) = z_1 \tag{8.50}$$

and

$$\bar{f}_u(z) \triangleq S_*(F_u(x)) = A_o z + \bar{\gamma}^u(z_1). \tag{8.51}$$

Note, by (8.15), that for $1 \le i \le n$,

$$L_{\bar{\mathbf{g}}_1^0} L_{F_0}^k H(x) = L_{\ell(H)\mathbf{g}_1^0} L_{F_0}^k H(x) = \ell(H(x)) L_{\mathbf{g}_1^0} L_{F_0}^k H(x)$$

$$= \begin{cases} 0, & 0 \le k \le n-2 \\ \ell(H(x)), & k = n-1 \end{cases}$$

which implies, together with (2.30), (2.45), (8.14), (8.39), and (8.49), that for $1 \le i \le n$,

$$L_{\bar{\mathbf{g}}_i^0} H(x) = L_{\mathrm{ad}_{F_0}^{i-1}\bar{\mathbf{g}}_1^0} H(x) = \sum_{k=0}^{i-1}(-1)^k\binom{i-1}{k} L_{F_0}^{i-1-k} L_{\bar{\mathbf{g}}_1^0} L_{F_0}^k H(x)$$

$$= \begin{cases} 0, & 1 \le i \le n-1 \\ (-1)^{n-1}\ell(H(x)), & i = n \end{cases}$$

and

$$\frac{\partial \bar{h}(z)}{\partial z_{n+1-i}} = (-1)^{i-1} L_{S_*(\bar{g}_i^0)} \left(\varphi \circ H \circ S^{-1}(z)\right)$$

$$= (-1)^{i-1} \left\{ L_{\bar{g}_i^0(x)} \left(\varphi \circ H(x)\right) \right\}\Big|_{x=S^{-1}(z)}$$

$$= (-1)^{i-1} \left\{ \frac{\partial \varphi(y)}{\partial y}\Big|_{y=H(x)} L_{\bar{g}_i^0(x)} H(x) \right\}\Big|_{x=S^{-1}(z)}$$

$$= \begin{cases} 0, & 1 \le i \le n-1 \\ \left\{ \frac{\partial \varphi(y)}{\partial y}\Big|_{y=H(x)} \ell(H(x)) \right\}\Big|_{x=S^{-1}(z)}, & i = n \end{cases}$$

$$= \begin{cases} 0, & 1 \le i \le n-1 \\ 1, & i = n. \end{cases}$$

Therefore, $\bar{h}(z) = z_1$ and (8.50) is satisfied. Let

$$\bar{f}_u(z) \triangleq \sum_{k=1}^{n} \bar{f}_{u,k}(z) \frac{\partial}{\partial z_k} = \begin{bmatrix} \bar{f}_{u,1}(z) \\ \vdots \\ \bar{f}_{u,n}(z) \end{bmatrix}. \tag{8.52}$$

Since $\bar{f}_u(z) = S_*(F_u(x))$, it is clear that for $1 \le i \le n-1$,

$$S_* \left(\bar{g}_{i+1}^u(x)\right) = S_* \left(\text{ad}_{F_u} \bar{g}_i^u(x)\right) = \left[S_*(F_u(x)), S_*(\bar{g}_i^u(x))\right]$$
$$= \left[\bar{f}_u(z), S_*(\bar{g}_i^u(x))\right]. \tag{8.53}$$

Thus, we have, by (8.49), (8.52), and (8.53), that for $1 \le i \le n-1$,

$$(-1)^i \frac{\partial}{\partial z_{n-i}} = \left[\bar{f}_u(z), (-1)^{i-1} \frac{\partial}{\partial z_{n+1-i}}\right]$$
$$= (-1)^i \sum_{k=1}^{n} \frac{\partial \bar{f}_{u,k}(z)}{\partial z_{n+1-i}} \frac{\partial}{\partial z_k}$$

which implies that for $1 \le k \le n$ and $1 \le i \le n-1$,

$$\frac{\partial \bar{f}_{u,k}(z)}{\partial z_{n+1-i}} = \begin{cases} 1, & k = n-i \\ 0, & \text{otherwise} \end{cases} \quad \text{or} \quad \frac{\partial \bar{f}_{u,k}(z)}{\partial z_{i+1}} = \begin{cases} 1, & i = k \\ 0, & \text{otherwise}. \end{cases}$$

Therefore, it is clear that $\bar{f}_{u,n}(z) = \bar{\gamma}_n^u(z_1)$ and $\bar{f}_{u,k}(z) = z_{k+1} + \bar{\gamma}_k^u(z_1)$, $1 \le k \le n-1$, for some functions $\bar{\gamma}_k^u(z_1)$, $1 \le k \le n$. In other words, (8.51) is satisfied. Hence, by (8.50) and (8.51), system (8.7) is state equivalent to a dual Brunovsky NOCF with OT $\bar{y} = \varphi(y)$ and state transformation $z = S(x)$. $\qquad \square$

Theorem 8.3 *System (8.7) is state equivalent to a dual Brunovsky NOCF with OT* $\bar{y} = \varphi(y)$ *and state transformation* $z = S(x)$, *if and only if there exists a smooth function* $\beta(y)$ *such that*

(i)

$$[\mathbf{g}_1^0(x), \mathbf{g}_i^0(x)] = 0, \ 2 \leq i \leq n-1 \tag{8.54}$$

(ii)

$$[\mathbf{g}_1^0(x), \mathbf{g}_n^0(x)] = -2\beta(H(x))\mathbf{g}_1^0(x), \quad \text{for even } n \tag{8.55}$$

$$[\mathbf{g}_2^0(x), \mathbf{g}_n^0(x)] = n\beta(H(x))\mathbf{g}_2^0(x) \ \text{mod span}\{\mathbf{g}_1^0(x)\}, \text{for odd } n \tag{8.56}$$

(iii)

$$\bar{\mathbf{g}}_i^u(x) = \bar{\mathbf{g}}_i^0(x), \ 2 \leq i \leq n \tag{8.57}$$

(iv)

$$[\bar{\mathbf{g}}_i^0(x), \bar{\mathbf{g}}_k^0(x)] = 0, \ 1 \leq i \leq n, \ 1 \leq k \leq n \tag{8.58}$$

where

$$\ell(y) \triangleq e^{\int_0^y \beta(\bar{y})d\bar{y}} \tag{8.59}$$

$$\bar{\mathbf{g}}_1^u(x) \triangleq \ell(H(x))\mathbf{g}_1^0(x) \tag{8.60}$$

$$\bar{\mathbf{g}}_i^u(x) \triangleq \text{ad}_{F_u}^{i-1}\bar{\mathbf{g}}_1^0(x), \ i \geq 2 \tag{8.61}$$

$$\varphi(y) = \int_0^y \frac{1}{\ell(\bar{y})}d\bar{y} \tag{8.62}$$

$$\frac{\partial S(x)}{\partial x} = \left[(-1)^{n-1}\bar{\mathbf{g}}_n^0(x) \ \cdots \ -\bar{\mathbf{g}}_2^0(x) \ \bar{\mathbf{g}}_1^0(x) \right]^{-1}. \tag{8.63}$$

Proof Necessity. Suppose that system (8.7) is state equivalent to a dual Brunovsky NOCF with OT $\bar{y} = \varphi(y)$ and state transformation $z = S(x)$. Then, by Lemma 8.2, there exist smooth functions $\ell(y)$ ($\neq 0$) such that (8.36)–(8.41) are satisfied. Note, by (2.44), (8.14), (8.38), and (8.39), that for $1 \leq i \leq n$,

$$\bar{\mathbf{g}}_i^0(x) = \mathrm{ad}_{F_0}^{i-1}\bar{\mathbf{g}}_1^0(x) = \mathrm{ad}_{F_0}^{i-1}\left(\ell(H(x))\mathbf{g}_1^0(x)\right)$$

$$= \sum_{k=0}^{i-1}\binom{i-1}{k}L_{F_0}^k\ell(H(x))\mathrm{ad}_{F_0}^{i-1-k}\mathbf{g}_1^0(x) \tag{8.64}$$

$$= \sum_{k=0}^{i-1}\binom{i-1}{k}L_{F_0}^k\ell(H(x))\mathbf{g}_{i-k}^0(x).$$

Also note, by (8.15), that for $1 \le i \le n$ and $1 \le k \le n$,

$$L_{\mathbf{g}_i^0(x)}L_{F_0}^k\ell(H(x)) = \left.\frac{d\ell(y)}{dy}\right|_{y=H(x)}L_{\mathbf{g}_i^0(x)}L_{F_0}^k H(x)$$

$$= \begin{cases} 0, & i+k < n \\ (-1)^{i+1}\left.\frac{d\ell(y)}{dy}\right|_{y=H(x)}, & i+k = n. \end{cases} \tag{8.65}$$

Thus, we have, by (2.43), (8.37), (8.64), and (8.65), that for $2 \le i \le n-1$,

$$0 = [\bar{\mathbf{g}}_1^0(x), \bar{\mathbf{g}}_i^0(x)] = \left[\ell(H)\mathbf{g}_1^0, \sum_{k=1}^{i-1}\binom{i-1}{k}L_{F_0}^k\ell(H)\mathbf{g}_{i-k}^0 + \ell(H)\mathbf{g}_i^0\right] \tag{8.66}$$

$$= \ell(H)\sum_{k=1}^{i-1}\binom{i-1}{k}L_{F_0}^k\ell(H)\,[\mathbf{g}_1^0, \mathbf{g}_{i-k}^0] + \ell(H)^2[\mathbf{g}_1^0, \mathbf{g}_i^0].$$

Since $[\mathbf{g}_1^0(x), \mathbf{g}_1^0(x)] = 0$, it is easy to show, by (8.66) and mathematical induction, that condition (i) is satisfied. Also, we have, by (2.43), (8.37), (8.54), (8.64), and (8.65), that

$$0 = [\bar{\mathbf{g}}_1^0(x), \bar{\mathbf{g}}_n^0(x)]$$

$$= [\ell(H)\mathbf{g}_1^0, L_{F_0}^{n-1}\ell(H)\mathbf{g}_1^0 + \sum_{k=1}^{n-2}\binom{n-1}{k}L_{F_0}^k\ell(H)\mathbf{g}_{n-k}^0 + \ell(H)\mathbf{g}_n^0]$$

$$= [\ell(H)\mathbf{g}_1^0, L_{F_0}^{n-1}\ell(H)\mathbf{g}_1^0] + [\ell(H)\mathbf{g}_1^0, \ell(H)\mathbf{g}_n^0]$$

$$= \ell(H)L_{\mathbf{g}_1^0}L_{F_0}^{n-1}\ell(H)\mathbf{g}_1^0 + \ell(H)^2[\mathbf{g}_1^0, \mathbf{g}_n^0] - \ell(H)L_{\mathbf{g}_n^0}\ell(H)\mathbf{g}_1^0$$

which implies, together with (8.15), that

$$[\mathbf{g}_1^0(x), \mathbf{g}_n^0(x)] = \frac{1}{\ell(H(x))}\left\{L_{\mathbf{g}_n^0}\ell(H(x)) - L_{\mathbf{g}_1^0}L_{F_0}^{n-1}\ell(H(x))\right\}\mathbf{g}_1^0(x)$$

$$= \begin{cases} -2\left\{\frac{1}{\ell(y)}\frac{d\ell(y)}{dy}\right\}\Big|_{y=H(x)}\mathbf{g}_1^0(x), & \text{for even } n \\ 0, & \text{for odd } n. \end{cases} \tag{8.67}$$

Therefore, (8.55) and (8.59) are satisfied with $\beta(y) = \frac{1}{\ell(y)}\frac{d\ell(y)}{dy} = \frac{d\ln\ell(y)}{dy}$ for even n. Similarly, for odd n, we have, by (2.43), (8.38), (8.54), (8.64), (8.65), and (8.67), that

$$
\begin{aligned}
0 &= [\bar{\mathbf{g}}_2^0(x), \bar{\mathbf{g}}_n^0(x)] \\
&= [L_{F_0}\ell(H)\mathbf{g}_1^0 + \ell(H)\mathbf{g}_2^0,\ L_{F_0}^{n-1}\ell(H)\mathbf{g}_1^0 + (n-1)L_{F_0}^{n-2}\ell(H)\mathbf{g}_2^0 + \cdots + \ell(H)\mathbf{g}_n^0] \\
&= [\ell(H)\mathbf{g}_2^0,\ (n-1)L_{F_0}^{n-2}\ell(H)\mathbf{g}_2^0] + [\ell(H)\mathbf{g}_2^0,\ \ell(H)\mathbf{g}_n^0]\ \ \text{mod span}\{\mathbf{g}_1^0\} \\
&= (n-1)\ell(H)L_{\mathbf{g}_2^0}L_{F_0}^{n-2}\ell(H)\mathbf{g}_2^0 + \ell(H)^2[\mathbf{g}_2^0,\ \mathbf{g}_n^0] \\
&\quad - \ell(H)L_{\mathbf{g}_n^0}\ell(H)\mathbf{g}_2^0\ \ \text{mod span}\{\mathbf{g}_1^0\}
\end{aligned}
$$

which implies, together with (8.65), that for odd n,

$$
\begin{aligned}
[\mathbf{g}_2^0(x), \mathbf{g}_n^0(x)] &= \frac{1}{\ell(H)}\left\{ L_{\mathbf{g}_n^0}\ell(H) + (n-1)L_{\mathbf{g}_1^0}L_{F_0}^{n-1}\ell(H)\right\}\mathbf{g}_2^0\ \ \text{mod span}\{\mathbf{g}_1^0\} \\
&= n\left\{\frac{1}{\ell(y)}\frac{d\ell(y)}{dy}\right\}\bigg|_{y=H(x)}\mathbf{g}_2^0(x)\ \ \ \ \text{mod span}\{\mathbf{g}_1^0(x)\}
\end{aligned}
$$

and (8.56) and (8.59) are satisfied with $\beta(y) = \frac{1}{\ell(y)}\frac{d\ell(y)}{dy} = \frac{d\ln\ell(y)}{dy}$. Condition (iii) and condition (iv) are obviously satisfied by (8.36) and (8.37).

Sufficiency. It is obvious by Lemma 8.2.   □

Example 8.2.3 Consider the following control system:

$$
\dot{x} = \begin{bmatrix} x_2 \\ -x_2^2 + x_1^2 e^{-x_1} + u \end{bmatrix} = F_u(x)
$$

$$
y = x_1 = H(x).
$$

(8.68)

Show that the above system is state equivalent to a dual Brunovsky NOCF with OT. Also find a OT $\bar{y} = \varphi(y)$, a state transformation $z = S(x)$, and the dual Brunovsky NOCF that new state z satisfies.

Solution Since $T(x) \triangleq [H(x)\ \ L_{F_0}H(x)]^{\mathsf{T}} = x$, it is clear, by (8.13) and (8.14), that

$$
\mathbf{g}_1^0(x) \triangleq \left(\frac{\partial T(x)}{\partial x}\right)^{-1}\begin{bmatrix} 0 \\ 1 \end{bmatrix} = \begin{bmatrix} 0 \\ 1 \end{bmatrix}
$$

$$
\mathbf{g}_2^u(x) \triangleq \operatorname{ad}_{F_u}\mathbf{g}_1^0(x) = \begin{bmatrix} -1 \\ 2x_2 \end{bmatrix}
$$

which imply that $[\mathbf{g}_1^0(x), \mathbf{g}_2^0(x)] = 2\mathbf{g}_1^0(x) \neq 0$ and condition (ii) of Theorem 8.2 is not satisfied. Therefore, by Theorem 8.2, system (8.68) is not state equivalent to a dual Brunovsky NOCF without OT. Note that condition (i) and condition (ii) of Theorem 8.3 are satisfied with $\beta(x_1) = -1$. Thus, we have, by (8.59)–(8.62), that $\ell(y) \triangleq e^{\int_0^y \beta(\bar{y})d\bar{y}} = e^{-y}$ and

$$\bar{g}_1^u(x) \triangleq \ell(H(x))g_1^0(x) = \begin{bmatrix} 0 \\ e^{-x_1} \end{bmatrix}$$

$$\bar{g}_2^u(x) \triangleq \mathrm{ad}_{F_u}g_1^0(x) = \begin{bmatrix} -e^{-x_1} \\ x_2e^{-x_1} \end{bmatrix}$$

$$\varphi(y) \triangleq \int_0^y \frac{1}{\ell(\bar{y})}d\bar{y} = e^y - 1$$

which imply that condition (iii) and condition (iv) of Theorem 8.3 are also satisfied. Hence, system (8.68) is state equivalent to a dual Brunovsky NOCF with OT $\bar{y} = \varphi(y) = e^y - 1$ and state transformation $z = S(x) = [e^{x_1} - 1 \ \ x_2e^{x_1}]^T$, $\gamma(\varphi(y), u) = [0 \ \ y^2 + e^y u]^T$, and $\gamma(\bar{y}, u) = [0 \ \ \{\ln(\bar{y} + 1)\}^2 + (\bar{y} + 1)u]^T$, where

$$\frac{\partial S(x)}{\partial x} = [\ -g_2^0(x) \ \ g_1^0(x)]^{-1} = \begin{bmatrix} e^{-x_1} & 0 \\ -x_2e^{-x_1} & e^{-x_1} \end{bmatrix}^{-1}$$

and

$$\dot{z} = S_*(F_u(x)) = \begin{bmatrix} z_2 \\ 0 \end{bmatrix} + \begin{bmatrix} 0 \\ \{\ln(z_1 + 1)\}^2 + (z_1 + 1)u \end{bmatrix}$$

$$\bar{y} = \varphi \circ H \circ S^{-1}(z) = z_1.$$

$\square$

If n is odd, (8.56) should be used instead of (8.55) in condition (ii) of Theorem 8.3.

Example 8.2.4 Consider the following control system:

$$\dot{x} = \begin{bmatrix} x_2 \\ x_3 \\ -4x_1x_3 - 3x_2^2 - 6x_1^2x_2 + u \end{bmatrix} = F_u(x) \tag{8.69}$$

$$y = x_1 = H(x).$$

Show that the above system is not state equivalent to a dual Brunovsky NOCF with OT.

Solution Since $T(x) \triangleq [H(x) \ \ L_{F_0}H(x) \ \ L_{F_0}^2H(x)]^T = x$, it is clear, by (8.13) and (8.14), that

$$\mathbf{g}_1^0(x) \triangleq \left(\frac{\partial T(x)}{\partial x}\right)^{-1} \begin{bmatrix} 0 \\ 0 \\ 1 \end{bmatrix} = \begin{bmatrix} 0 \\ 0 \\ 1 \end{bmatrix}$$

$$\mathbf{g}_2^u(x) \triangleq \mathrm{ad}_{F_u}\mathbf{g}_1^0(x) = \begin{bmatrix} 0 \\ -1 \\ 4x_1 \end{bmatrix} ; \quad \mathbf{g}_3^u(x) \triangleq \mathrm{ad}_{F_u}\mathbf{g}_2^u(x) = \begin{bmatrix} 1 \\ -4x_1 \\ 10x_1^2 - 2x_2 \end{bmatrix}$$

which imply that $[\mathbf{g}_2^0(x), \mathbf{g}_3^0(x)] = -2\mathbf{g}_1^0(x) \neq 0$ and condition (ii) of Theorem 8.2 is not satisfied. Therefore, by Theorem 8.2, system (8.69) is not state equivalent to a dual Brunovsky NOCF without OT. Note that condition (i) and condition (ii) of Theorem 8.3 are satisfied with $\beta(x_1) = 0$. Thus, we have, by (8.59)–(8.62), that $\ell(y) \triangleq e^{\int_0^y \beta(\bar{y})d\bar{y}} = 1, \varphi(y) \triangleq \int_0^y \frac{1}{\ell(\bar{y})}d\bar{y} = y$, and

$$\bar{\mathbf{g}}_i^u(x) = \mathbf{g}_i^u(x), \quad 1 \leq i \leq n$$

which imply that condition (iv) of Theorem 8.3 is not satisfied, even though condition (iii) is satisfied. Hence, by Theorem 8.3, system (8.69) is not state equivalent to a dual Brunovsky NOCF with OT.   □

8.3 Dynamic Observer Error Linearization

Consider the following single output control system and autonomous system:

$$\dot{x} = F_u(x) ; \quad y = H(x) \tag{8.70}$$

$$\dot{x} = F_0(x) ; \quad y = H(x) \tag{8.71}$$

with $F_0(0) = 0$, $H(0) = 0$, state $x \in \mathbb{R}^n$, input $u \in \mathbb{R}^m$, and output $y \in \mathbb{R}$. Define the restricted dynamic system with index d (called auxiliary dynamics) by

$$\begin{bmatrix} \dot{w}_1 \\ \vdots \\ \dot{w}_{d-1} \\ \dot{w}_d \end{bmatrix} = \begin{bmatrix} w_2 \\ \vdots \\ w_d \\ y \end{bmatrix} \triangleq p(w, y). \tag{8.72}$$

Define the extended system of system (8.70) with index d by

$$\dot{x}^e \triangleq \begin{bmatrix} \dot{w} \\ \dot{x} \end{bmatrix} = \begin{bmatrix} p(w, H(x)) \\ F_u(x) \end{bmatrix} \triangleq F_u^e(x^e) \tag{8.73}$$

$$y_a = w_1$$

where $x^e \triangleq [w^{\mathsf{T}} \ x^{\mathsf{T}}]^{\mathsf{T}} \in \mathbb{R}^{d+n}$.

Definition 8.6 (*RDOEL with index d*)
System (8.70) is said to be restricted dynamic observer error linearizable (RDOEL) with index d, if there exists a local extended state transformation $z^e = S^e(w, x) = [w^\top \ z^\top]^\top = [w^\top \ S(w, x)^\top]^\top$, which transforms (8.73), in the new states z^e, to a generalized nonlinear observer canonical form (GNOCF) with index d defined by

$$\dot{z}^e = A_e z^e + \gamma(w, y, u) \ ; \ y_a = C_e z^e = w_1$$

where $\gamma(w, y, u) : \mathbb{R}^{d+1} \times \mathbb{R}^m \to \mathbb{R}^{d+n}$ is a smooth vector function with $\gamma_i = 0$ for $1 \le i \le d - 1$, $C_e = \begin{bmatrix} 1 \ O_{1 \times (n+d-1)} \end{bmatrix}$, and $A_e = \begin{bmatrix} O_{(n+d-1) \times 1} & I_{(n+d-1)} \\ 0 & O_{1 \times (n+d-1)} \end{bmatrix}$.

System (8.70) is said to be RDOEL, if system (8.70) is RDOEL with some index d. If we use a general nonlinear dynamic system $\dot{w} = \bar{p}(w, y)$ in Definition 8.6 instead of restricted (or linear) dynamic system (8.72), system (8.70) is said to be dynamic observer error linearizable (DOEL) with index d.

Let $S^{-1}(w, z)$ be the vector function such that $S(w, S^{-1}(w, z)) = z$ for all $w \in \mathbb{R}^d$. In other words,

$$x^e = \begin{bmatrix} w \\ x \end{bmatrix} = (S^e)^{-1}(w, z) = \begin{bmatrix} w \\ S^{-1}(w, z) \end{bmatrix}.$$

If system (8.70) is RDOEL with index d, then we can design a state estimator

$$\dot{\bar{z}}^e(t) = \begin{bmatrix} \dot{w} \\ \dot{\bar{z}} \end{bmatrix} = (A_e - L_e C_e) \begin{bmatrix} w \\ \bar{z} \end{bmatrix} + \gamma(w, y, u) + L_e w_1$$

$$\bar{x} \triangleq S^{-1}(w, \bar{z})$$

that yields an asymptotically vanishing error, i.e., $\lim_{t \to \infty} \|z^e(t) - \bar{z}^e(t)\| = 0$ or $\lim_{t \to \infty} \|x(t) - \bar{x}(t)\| = 0$, where $(A_e - L_e C_e)$ is an asymptotically stable $(d + n) \times (d + n)$ matrix. Block diagram for dynamic nonlinear observer can be found in Fig. 8.3.

RDOEL for autonomous system (8.71) can also be similarly defined with $u = 0$. If $\bar{f}^e_u(z^e) \triangleq (S^e)_*(F^e_u(x^e)) = A_e z^e + \gamma(w, y, u)$, then it is clear that $\bar{f}^e_0(z) \triangleq (S^e)_*(F^e_0(x^e)) = A_e z^e + \gamma(w, y, 0)$. Thus, we have the following remark.

Remark 8.3 If system (8.70) is RDOEL with index d and state transformation $z^e = S^e(w, x)$, then system (8.71) is also RDOEL with index d and state transformation $z^e = S^e(w, x)$. But the converse is not true.

Lemma 8.3 *System (8.71) is RDOEL with index* $d(\ge 1)$ *and state transformation* $z^e = S^e(w, x) = [w^\top \ S(w, x)^\top]^\top$, *if and only if there exist smooth functions* $\bar{\gamma}_k(w, y)$, $d \le k \le d + n$ *such that*

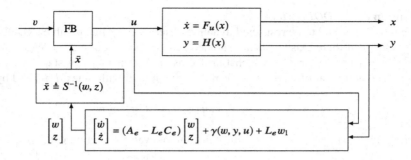

Fig. 8.3 Dynamic nonlinear observer

$$L_{F_0^e}^n H(x) = \sum_{k=0}^{n} L_{F_0^e}^{n-k} \bar{\gamma}_{d+k}(w, H(x)) \tag{8.74}$$

and for $1 \le i \le n$,

$$S_i(w, x) = L_{F_0^e}^{i-1} H(x) - \sum_{k=0}^{i-1} L_{F_0^e}^{i-1-k} \bar{\gamma}_{d+k}(w, H(x)). \tag{8.75}$$

Proof Proof is obvious.   □

For extended system (8.73), as in Definition 8.5, the canonical system can also be defined by

$$\dot{\xi}^e = f_0^e\left(\xi^e\right) \ ; \ \ y_a = \xi_1^e = w_1 \triangleq h_E(\xi^e) \tag{8.76}$$

where $\xi^e \triangleq \begin{bmatrix} w \\ \xi \end{bmatrix} = T_e(x^e) \triangleq \begin{bmatrix} w \\ T(x) \end{bmatrix}$, $f_u^e(\xi^e) \triangleq (T_e)_*(F_u^e(x^e))$,

$$\xi = T(x) \triangleq [H(x) \ L_{F_0} H(x) \ \cdots \ L_{F_0}^{n-1} H(x)]^\mathsf{T}$$
$$= [L_{F_0^e}^d w_1 \ L_{F_0^e}^{d+1} w_1 \ \cdots \ L_{F_0^e}^{d+n-1} w_1]^\mathsf{T}$$

$$\alpha_e(\xi^e) \triangleq L_{f_0^e}^{d+n} w_1 = L_{F_0^e}^{d+n} w_1 \Big|_{x^e = T_e^{-1}(\xi^e)} = L_{F_0}^n H(x) \Big|_{x=T^{-1}(\xi)} = \alpha_e(0, \xi),$$

and

$$f_0^e(\xi^e) = \begin{bmatrix} w_2 & \cdots & w_d & \xi_1 & \cdots & \xi_n & \alpha_e(\xi^e) \end{bmatrix}^\mathsf{T} = (T_e)_*(F_0^e(x^e)).$$

For extended system (8.73), we define vector fields $\{\mathbf{g}_1^0(x), \mathbf{g}_2^0(x), \cdots\}$ and $\{\mathbf{g}_1^u(x), \mathbf{g}_2^u(x), \cdots\}$ as follows:

$$L_{\mathbf{g}_1^0(x^e)} L_{F_0^e}^{k-1} w_1 = \delta_{k,d+n}, \ 1 \le k \le d+n$$

$$\left(\text{or } \mathbf{g}_1^0(x^e) \triangleq \left(\frac{\partial T_e(x^e)}{\partial x^e}\right)^{-1} [0 \ \cdots \ 0 \ 1]^{\mathsf{T}} = (T_e)_*^{-1}\left(\frac{\partial}{\partial \xi_n}\right)\right) \tag{8.77}$$

and for $i \ge 2$,

$$\mathbf{g}_i^0(x^e) \triangleq \mathrm{ad}_{F_0^e}^{i-1} \mathbf{g}_1^0(x^e)$$

$$\mathbf{g}_1^u(x^e) \triangleq \mathbf{g}_1^0(x^e); \quad \mathbf{g}_i^u(x^e) \triangleq \mathrm{ad}_{F_u^e}^{i-1} \mathbf{g}_1^u(x^e). \tag{8.78}$$

Then it is easy to see that for $1 \le i \le n+d$ and $0 \le k \le n+d-1$,

$$L_{\mathbf{g}_i^0(x^e)} L_{F_0^e}^k w_1 = \begin{cases} 0, & i+k < d+n \\ (-1)^{i+1}, & i+k = d+n. \end{cases} \tag{8.79}$$

Also, since $w_j = L_{F_0^e}^{j-1} w_1$, $1 \le j \le d$ and $H(x) = L_{F_0^e}^d w_1$, it is clear that for $1 \le k \le n$ and $1 \le j \le d$,

$$L_{\mathbf{g}_i^0(x^e)} L_{F_0^e}^k w_j = \begin{cases} 0, & i+k < d+n+1-j \\ (-1)^{i+1}, & i+k = d+n+1-j \end{cases}$$

$$L_{\mathbf{g}_i^0(x^e)} L_{F_0^e}^k H(x) = \begin{cases} 0, & i+k < n \\ (-1)^{i+1}, & i+k = n. \end{cases} \tag{8.80}$$

Example 8.3.1 Let

$$\tilde{\mathbf{g}}_1^s(x^e) = \begin{cases} \bar{\ell}_0(H(x))\mathbf{g}_1^0(x^e), & s = 1 \\ \bar{\ell}_{s-1}(w_{d+2-s}, \cdots, w_d, H(x))\mathbf{g}_1^0(x^e), & 2 \le s \le d \end{cases}$$

$$\tilde{\mathbf{g}}_i^s(x^e) \triangleq \mathrm{ad}_{F_e^e}^{i-1} \tilde{\mathbf{g}}_1^s(x^e), \ 1 \le i \le n$$

for some scalar function $\bar{\ell}_{s-1}(w_{d+2-s}, \cdots, w_d, y)$. Prove the following:

(a) for $1 \le i \le n$,

$$\mathbf{g}_i^0(w, x) = \mathbf{g}_i^0(0, x) \tag{8.81}$$

$$\tilde{\mathbf{g}}_i^s(w, x) = \tilde{\mathbf{g}}_i^s(w, x)\big|_{w_j=0, \ 1 \le j \le d+1-s} \tag{8.82}$$

(b) for $1 \leq i \leq n$ and $1 \leq k \leq n$,

$$\left[\mathbf{g}_i^0(w, x), \ \mathbf{g}_k^0(w, x) \right] = \left[\mathbf{g}_i^0(w, x), \ \mathbf{g}_k^0(w, x) \right] \big|_{w=0} \qquad (8.83)$$

$$\left[\tilde{\mathbf{g}}_i^s(w, x), \ \tilde{\mathbf{g}}_k^s(w, x) \right] = \left[\tilde{\mathbf{g}}_i^s(w, x), \ \tilde{\mathbf{g}}_k^s(w, x) \right] \big|_{w_j=0, \ 1 \leq j \leq d+1-s} . \qquad (8.84)$$

Solution For canonical system (8.76), let $\mathbf{r}_1(\xi^e) \triangleq \frac{\partial}{\partial \xi_{d+n}^e} = \frac{\partial}{\partial \xi_n}$ and $\mathbf{r}_i(\xi^e) \triangleq$ $\mathrm{ad}_{f_0^e}^{i-1} \mathbf{r}_1(\xi^e)$, $i \geq 2$. Since $f_0^e(\xi^e) \triangleq (T_e)_*(F_0^e(x^e))$ and $\mathbf{r}_1(\xi^e) = (T_e)_*(\mathbf{g}_1^0(x^e))$ by (8.77), it is clear, by (2.49) and (2.37), that for $1 \leq i \leq n$ and $1 \leq s \leq d$,

$$\mathbf{g}_i^0(w, x) = (T_e)_*^{-1}(\mathbf{r}_i(w, \xi))$$

$$\tilde{\mathbf{g}}_i^s(w, x) = (T_e)_*^{-1}(\tilde{\mathbf{r}}_i^s(w, \xi))$$

$$\tilde{\mathbf{r}}_1^s(w, \xi) = (T_e)_*(\tilde{\mathbf{g}}_1^s(w, x)) = \bar{\ell}_{s-1}(\xi_{d+2-s}^e, \cdots, \xi_{d+1}^e) \mathbf{r}_1(w, \xi)$$

where $\tilde{\mathbf{r}}_1^s(w, \xi) \triangleq (T_e)_*(\tilde{\mathbf{g}}_1^s(w, x))$ and $\tilde{\mathbf{r}}_i^s(\xi^e) \triangleq \mathrm{ad}_{f_0^e(\xi^e)}^{i-1} \tilde{\mathbf{r}}_1^s(\xi^e)$, $1 \leq i \leq n$. Note that $\mathbf{r}_1(w, \xi) = \mathbf{r}_1(0, \xi)$ and $\tilde{\mathbf{r}}_1^s(w, \xi) = \tilde{\mathbf{r}}_1^s(w, \xi) \big|_{w_j=0, \ 1 \leq j \leq d+1-s}$. Since $\mathbf{r}_1(w, \xi) = \mathbf{r}_1(0, \xi)$, $\alpha_e(w, \xi) = \alpha_e(0, \xi)$, $\frac{\partial f_0^e(\xi^e)}{\partial \xi^e} = \frac{\partial f_0^e(\xi^e)}{\partial \xi^e} \big|_{w=0}$, and $\frac{\partial \mathbf{r}_1(\xi^e)}{\partial \xi^e} f_0^e(\xi^e) = \left(\frac{\partial \mathbf{r}_1(\xi^e)}{\partial \xi^e} f_0^e(\xi^e) \right) \big|_{w=0}$, it is easy to see that

$$\begin{aligned} \mathbf{r}_2(w, \xi) = \mathrm{ad}_{f_0^e(\xi^e)} \mathbf{r}_1(\xi^e) &= \frac{\partial \mathbf{r}_1(\xi^e)}{\partial \xi^e} f_0^e(\xi^e) - \frac{\partial f_0^e(\xi^e)}{\partial \xi^e} \mathbf{r}_1(\xi^e) \\ &= \left(\frac{\partial \mathbf{r}_1(\xi^e)}{\partial \xi^e} f_0^e(\xi^e) \right) \Big|_{w=0} - \frac{\partial f_0^e(\xi^e)}{\partial \xi^e} \Big|_{w=0} \mathbf{r}_1(0, \xi) \\ &= \left(\frac{\partial \mathbf{r}_1(\xi^e)}{\partial \xi^e} f_0^e(\xi^e) - \frac{\partial f_0^e(\xi^e)}{\partial \xi^e} \mathbf{r}_1(\xi^e) \right) \Big|_{w=0} = \mathbf{r}_2(0, \xi). \end{aligned}$$

By mathematical induction, it can also be easily shown that $\mathbf{r}_i(w, \xi) = \mathbf{r}_i(0, \xi)$ for $i \geq 1$. Thus, we have, by $T_e^{-1}(\xi^e) = \begin{bmatrix} w \\ T^{-1}(\xi) \end{bmatrix}$ and $\frac{\partial T_e^{-1}(\xi^e)}{\partial \xi^e} = \frac{\partial T_e^{-1}(\xi^e)}{\partial \xi^e} \big|_{w=0}$, that for $1 \leq i \leq n$,

$$\begin{aligned} \mathbf{g}_i^0(w, x) = (T_e)_*^{-1} \mathbf{r}_i(w, \xi) &= \left(\frac{\partial T_e^{-1}(\xi^e)}{\partial \xi^e} \mathbf{r}_i(w, \xi) \right) \Big|_{\xi = T^{-1}(x)} \\ &= \left(\frac{\partial T_e^{-1}(\xi^e)}{\partial \xi^e} \mathbf{r}_i(w, \xi) \right) \Big|_{\xi = T^{-1}(x), w=0} = \mathbf{g}_i^0(0, x) \end{aligned}$$

which implies that (8.81) and (8.83) are satisfied. In the same manner, it can be proved that (8.82) and (8.84) are also satisfied.   □

Theorem 8.4 *System (8.70) is RDOEL with index $d(\geq 1)$ and state transformation $z^e = S^e(w, x) = [w^T \, S(w, x)^T]^T$, if and only if there exists a smooth function $\ell(w, y)$ ($\neq 0$) such that*

(i)

$$\bar{\mathbf{g}}_i^u(x^e) = \bar{\mathbf{g}}_i^0(x^e), \quad 2 \leq i \leq n \tag{8.85}$$

(ii)

$$[\bar{\mathbf{g}}_i^0(x^e), \, \bar{\mathbf{g}}_k^0(x^e)] = 0, \quad 1 \leq i \leq n, \, 1 \leq k \leq n \tag{8.86}$$

where

$$\bar{\mathbf{g}}_1^u(x^e) \triangleq \ell(w, H(x))\mathbf{g}_1^0(x^e) \tag{8.87}$$

$$\bar{\mathbf{g}}_i^u(x^e) \triangleq \mathrm{ad}_{F_u^e}^{i-1} \bar{\mathbf{g}}_1^0(x^e), \quad i \geq 2 \tag{8.88}$$

$$\frac{\partial S(w, x)}{\partial x} = D(x^e)^{-1} \tag{8.89}$$

$$\left[(-1)^{n-1}\bar{\mathbf{g}}_n^0(x^e) \quad \cdots \quad -\bar{\mathbf{g}}_2^0(x^e) \; \bar{\mathbf{g}}_1^0(x^e) \right] \triangleq \begin{bmatrix} O_{d \times n} \\ D(x^e) \end{bmatrix}. \tag{8.90}$$

Proof Necessity. Suppose that system (8.70) is RDOEL with index d and state transformation $z^e = [w^T \quad z^T]^T = S^e(w, x) = [w^T \quad S(w, x)^T]^T$. Then, by Remark 8.10, autonomous system (8.71) is RDOEL with index d and state transformation $z^e = S^e(w, x)$. Therefore, by Lemma 8.3, there exist smooth functions $\bar{\gamma}_k(w, y)$, $1 \leq k \leq n$ such that (8.74) is satisfied and for $1 \leq i \leq n$,

$$z_i = S_i(w, x) = L_{F_0^e}^{i-1} H(x) - \sum_{k=0}^{i-1} L_{F_0^e}^{i-1-k} \bar{\gamma}_{d+k}(w, H(x)) \tag{8.91}$$

or

$$z_i = \tilde{S}_i(w, \xi) \triangleq S_i\left(w, T^{-1}(\xi)\right) = \xi_i - \sum_{k=0}^{i-1} L_{f_0^e(\xi)}^{i-1-k} \bar{\gamma}_{d+k}(w, \xi_1) \tag{8.92}$$

where

$$\xi = T(x) \triangleq \left[H(x)\ L_{F_0}H(x)\ \cdots\ L_{F_0}^{n-1}H(x) \right]^{\mathsf{T}}$$

$$= \left[L_{F_e^e}^d w_1\ L_{F_0^e}^{d+1}w_1\ \cdots\ L_{F_e^e}^{d+n-1}w_1 \right]^{\mathsf{T}}$$

$\xi^e = \begin{bmatrix} w \\ \xi \end{bmatrix} = T_e(w, x) \triangleq \begin{bmatrix} w \\ T(x) \end{bmatrix}$, $\quad f_0^e(\xi^e) \triangleq (T_e)_*(F_0^e(x))$, $\quad$ and $\quad \tilde{S}_e(\xi^e) \triangleq S^e \circ T_e^{-1}(\xi^e)$. Note, by (8.91) and (8.92), that $z_1 = H(x) - \bar{\gamma}_d(w, H(x)) = y - \bar{\gamma}_d(w, y) = \xi_1 - \bar{\gamma}_d(w, \xi_1) = \tilde{S}_1(w, \xi) = \tilde{S}_1(w, \xi_1, 0, \cdots, 0)$ and $\frac{\partial \tilde{S}_1(w, \xi_1, 0, \cdots, 0)}{\partial \xi_1} \neq 0$. Thus, it is clear, by implicit function theorem, that there exists a function $y = q(w, z_1)$ such that

$$z_1 = q(w, z_1) + \bar{\gamma}_d(w, q(w, z_1)), \quad \text{for all } w \in \mathbb{R}^d. \tag{8.93}$$

In other words, we have

$$\begin{bmatrix} \dot{w}_1 \\ \vdots \\ \dot{w}_{d-1} \\ \dot{w}_d \\ \dot{z}_1 \\ \vdots \\ \dot{z}_{n-1} \\ \dot{z}_n \end{bmatrix} = \begin{bmatrix} w_2 \\ \vdots \\ w_d \\ z_1 + \gamma_d(w, z_1, u) \\ z_2 + \gamma_{d+1}(w, z_1, u) \\ \vdots \\ z_n + \gamma_{d+n-1}(w, z_1, u) \\ \gamma_{d+n}(w, z_1, u) \end{bmatrix} \triangleq \bar{f}_u^e(z^e) \tag{8.94}$$

$$y_a = w_1 = Cz^e \triangleq \bar{h}_E(z^e)$$

where $\quad \bar{f}_u^e(z^e) = (S^e)_* \left(F_u^e(x^e) \right) = (\tilde{S}_e)_* \left(f_u^e(\xi^e) \right) \quad$ and $\quad \bar{\gamma}_i(w, q(w, z_1)) = \gamma_i(w, z_1, 0)$ for $d \leq i \leq d + n$. For system (8.94), we define vector fields $\{ \bar{\psi}_1^u(z^e), \cdots, \bar{\psi}_n^u(z^e) \}$ by

$$\bar{\psi}_1^u(z^e) \triangleq \frac{\partial}{\partial z_n} = \frac{\partial}{\partial z_{d+n}^e}; \quad \bar{\psi}_i^u(z^e) \triangleq \mathrm{ad}_{\bar{f}_u^e}^{i-1} \bar{\psi}_1^u(z^e), \ i \geq 2. \tag{8.95}$$

Then, by (8.94), it is clear that

$$\bar{\psi}_i^u(z^e) = (-1)^{i-1} \frac{\partial}{\partial z_{n+1-i}} = \bar{\psi}_i^0(z^e), \ 1 \leq i \leq n \tag{8.96}$$

which implies that

$$[\bar{\psi}_i^u(z^e), \bar{\psi}_k^u(z^e)] = 0, \ 1 \leq i \leq n, \ 1 \leq k \leq n. \tag{8.97}$$

It is not difficult to show, by (8.92), that

$$\frac{\partial z_i^e}{\partial \xi_n} = \frac{\partial \tilde{S}_{e,i}(\xi^e)}{\partial \xi_n} = \begin{cases} 0, & \text{if } 1 \leq i \leq d+n-1 \\ 1 - \frac{d\bar{\gamma}_d(w,\xi_1)}{d\xi_1} \triangleq \frac{1}{\ell(w,\xi_1)}, & \text{if } i = d+n \end{cases} \tag{8.98}$$

which implies, together with (8.95), that

$$(\tilde{S}_e)_* \left(\frac{\partial}{\partial \xi_n} \right) = \sum_{i=1}^{n+d} \frac{\partial \tilde{S}_{e,i}(\xi^e)}{\partial \xi_n} \bigg|_{\xi^e = \tilde{S}_e^{-1}(z^e)} \frac{\partial}{\partial z_i^e}$$

$$= \frac{1}{\ell(w,\xi_1)} \bigg|_{\xi^e = \tilde{S}_e^{-1}(z^e)} \frac{\partial}{\partial z_n} = \frac{1}{\ell(w,\xi_1)} \bigg|_{\xi^e = \tilde{S}_e^{-1}(z^e)} \bar{\psi}_1^u(z^e).$$

Therefore

$$\bar{\psi}_1^u(z^e) = \ell(w,\xi_1) \bigg|_{\xi^e = \tilde{S}_e^{-1}(z^e)} (\tilde{S}_e)_* \left(\frac{\partial}{\partial \xi_n} \right)$$

and

$$(\tilde{S}_e)_*^{-1} \left(\bar{\psi}_1^u(z^e) \right) = (\tilde{S}_e)_*^{-1} \left(\ell(w,\xi_1) \bigg|_{\xi^e = \tilde{S}_e^{-1}(z^e)} (\tilde{S}_e)_* \left(\frac{\partial}{\partial \xi_n} \right) \right)$$

$$= \ell(w,\xi_1) \frac{\partial}{\partial \xi_n}. \tag{8.99}$$

Hence, if we let $\bar{\mathbf{g}}_1^u(x^e) \triangleq (S^e)_*^{-1}(\bar{\psi}_1^u(z^e))$, we have, by (2.49), (8.77), and (8.99), that

$$\bar{\mathbf{g}}_1^u(x^e) = (S^e)_*^{-1}(\bar{\psi}_1^u(z^e)) = (T_e)_*^{-1} \circ (\tilde{S}_e)_*^{-1}(\bar{\psi}_1^u(z^e))$$

$$= (T_e)_*^{-1} \left(\ell(w,\xi_1) \frac{\partial}{\partial \xi_n} \right)$$

$$= \ell(w, H(x))(T_e)_*^{-1} \left(\frac{\partial}{\partial \xi_n} \right) = \ell(w, H(x))\mathbf{g}_1^0(x^e)$$

which implies that (8.87) is satisfied. Also, since $\bar{f}_u^e(z) = (S^e)_*(F_u^e(x))$ or $F_u^e(x) = (S^e)_*^{-1}(\bar{f}_u^e(z))$, it is clear, by (2.37), (8.88), and (8.95), that for $i \geq 2$,

$$\bar{\mathbf{g}}_i^u(x^e) = \text{ad}_{F_u^e}^{i-1}\bar{\mathbf{g}}_1^u(x^e) = (S^e)_*^{-1} \left\{ \text{ad}_{(S^e)_*(F_u^e)}^{i-1}(S^e)_*(\bar{\mathbf{g}}_1^u(x^e)) \right\}$$

$$= (S^e)_*^{-1} \left\{ \text{ad}_{\bar{f}_u^e}^{i-1} \bar{\psi}_1^u(z^e) \right\} = (S^e)_*^{-1}(\bar{\psi}_i^u(z^e))$$

and thus condition (i) and condition (ii), and (8.89) are satisfied by (8.96) and (8.97).

Sufficiency. Suppose that condition (i) and condition (ii) are satisfied. Since $\{\bar{\mathbf{g}}_1^0(x^e),\ \bar{\mathbf{g}}_2^0(x^e),\ \cdots,\ \bar{\mathbf{g}}_n^0(x^e)\}$ is a set of commuting vector fields, there exists, by Corollary 2.1, a state transformation $z^e = S^e(x^e) = \begin{bmatrix} w \\ S(w,x) \end{bmatrix}$ such that

$$(S^e)_* \left((-1)^{i-1}\bar{\mathbf{g}}_i^0(x^e)\right) = \frac{\partial}{\partial z_{n+d+1-i}^e} = \frac{\partial}{\partial z_{n+1-i}}, \quad 1 \le i \le n. \tag{8.100}$$

In fact, $z^e = S^e(x^e)$ can be calculated by (8.89) and (8.90). Now we will show that $\tilde{f}_u^e(z^e) \triangleq (S^e)_* \left(F_u^e(x^e)\right) = A_e z^e + \gamma(z_1^e, \cdots, z_{d+1}^e, u(x^e))$. Note that for $1 \le i \le n-1$,

$$(S^e)_* \left(\bar{\mathbf{g}}_{i+1}^u(x^e)\right) = (S^e)_* \left(\mathrm{ad}_{F_u^e}\bar{\mathbf{g}}_i^u(x^e)\right) = [(S^e)_*(F_u^e(x^e)),\ (S^e)_* \left(\bar{\mathbf{g}}_i^u(x^e)\right)]. \tag{8.101}$$

Thus, if we let

$$\tilde{f}_u^e(z^e) = \sum_{k=1}^{n+d} c_k(z^e)\frac{\partial}{\partial z_k^e} = \begin{bmatrix} c_1(z^e) \\ \vdots \\ c_{n+d}(z^e) \end{bmatrix} \tag{8.102}$$

then we have, by (8.100) and (8.101), that for $1 \le i \le n-1$,

$$(-1)^i \frac{\partial}{\partial z_{n+d-i}^e} = \left[\tilde{f}_u^e(z^e),\ (-1)^{i-1}\frac{\partial}{\partial z_{n+d+1-i}^e} \right]$$
$$= \sum_{k=1}^{n+d}(-1)^i \frac{\partial c_k(z^e)}{\partial z_{n+d+1-i}^e}\frac{\partial}{\partial z_k^e}$$

which implies that for $1 \le i \le n-1$ and $1 \le k \le n+d$,

$$\frac{\partial c_k(z^e)}{\partial z_{n+d+1-i}^e} = \begin{cases} 1, & k = n+d-i \\ 0, & k \ne n+d-i \end{cases}$$

or, for $d+1 \le i \le n+d-1$ and $1 \le k \le n+d$,

$$\frac{\partial c_k(z^e)}{\partial z_{i+1}^e} = \begin{cases} 1, & i = k \\ 0, & i \ne k. \end{cases}$$

Thus, it is clear that

$$c_k(z^e) = \begin{cases} z_{k+1}^e + \tilde{\gamma}_k(z_1^e, \cdots, z_{d+1}^e), & 1 \le k \le n+d-1 \\ \tilde{\gamma}_{n+d}(z_1^e, \cdots, z_{d+1}^e), & k = n+d \end{cases} \tag{8.103}$$

for some functions $\tilde{\gamma}_k(z_1^e, \cdots, z_{d+1}^e)$, $1 \le k \le n + d$. Therefore, it is easy to see, by (8.102) and (8.103), that

$$\bar{f}_u^e(z^e) = A_e z^e + \tilde{\gamma}(z_1^e, \cdots, z_{d+1}^e).$$

Since $z_i^e = w_i$, $1 \le i \le d$, it is easy to show, by (8.73), that $\tilde{\gamma}_i(z_1^e, \cdots, z_{d+1}^e) = 0$, $1 \le i \le d - 1$, $y = z_{d+1}^e + \tilde{\gamma}_d(w, z_{d+1}^e)$, and

$$\bar{f}_u^e(z^e) = A_e z^e + \tilde{\gamma}(w, \tilde{q}(w, y)) \triangleq A_e z^e + \gamma(w, y)$$

where $\tilde{q}(w, y)$ is a function such that $y = \tilde{q}(w, y) + \tilde{\gamma}_d(w, \tilde{q}(w, y))$ for all $w \in \mathbb{R}^d$.

$\square$

Theorem 8.5 *System (8.70) is RDOEL with index $d(\ge 1)$ and state transformation $z^e = S^e(w, x) = [w^\mathsf{T} \ S(w, x)^\mathsf{T}]^\mathsf{T}$, if and only if there exist some constants β_i, $1 \le i \le d$, and smooth function $\beta_0(y) : \mathbb{R} \to \mathbb{R}$ such that*

(i)

$$[\mathbf{g}_1^0, \ \mathbf{g}_i^0] = 0, \ 2 \le i \le n - 1 \tag{8.104}$$

(ii)

$$[\mathbf{g}_1^0, \ \mathbf{g}_n^0] = -2\beta_0(H(x))\mathbf{g}_1^0, \ \text{for even } n \tag{8.105}$$

$$[\mathbf{g}_2^0, \ \mathbf{g}_n^0] = n\beta_0(H(x))\mathbf{g}_2^0 \ \text{mod} \ \{\mathbf{g}_1^0\}, \ \text{for odd } n \tag{8.106}$$

(iii) for $1 \le i \le \min(d, n - 2)$,

$$[\tilde{\mathbf{g}}_{i+1}^i, \ \tilde{\mathbf{g}}_n^i] = (-1)^{n-1} 2\bar{\ell}_{i-1}(x^e)\beta_i\tilde{\mathbf{g}}_1^i, \\ \text{for even } (n + i) \tag{8.107}$$

$$[\tilde{\mathbf{g}}_{i+2}^i, \ \tilde{\mathbf{g}}_n^i] = (-1)^{n-1}(n + i)\bar{\ell}_{i-1}(x^e)\beta_i\tilde{\mathbf{g}}_2^i \ \text{mod} \ \{\tilde{\mathbf{g}}_1^i\}, \\ \text{for odd } (n + i) \tag{8.108}$$

(iv)

$$\text{ad}_{F_u^e}^k \tilde{\mathbf{g}}_1^{d+1} = \text{ad}_{F_0^e}^k \tilde{\mathbf{g}}_1^{d+1}, \ 1 \le k \le n - 1 \tag{8.109}$$

(v)

$$[\tilde{\mathbf{g}}_i^{d+1}, \ \tilde{\mathbf{g}}_k^{d+1}] = 0, \ 1 \le i \le n, \ 1 \le k \le n \tag{8.110}$$

where $S(w, x)$ is defined by (8.89) and (8.90), with $\tilde{g}_i^{d+1}$, $1 \leq i \leq n$ instead of $\bar{g}_i^0$, $1 \leq i \leq n$, and

$$
\bar{\ell}_i(x^e) \triangleq
\begin{cases}
e^{\int_0^{H(x)} \beta_0(\bar{y}) d\bar{y}}, & i = 0 \\
e^{\int_0^{H(x)} \beta_0(y) dy} \displaystyle\prod_{k=1}^{i} e^{\beta_k w_{d+1-k}}, & 1 \leq i \leq d
\end{cases}
\tag{8.111}
$$

$$
\tilde{g}_1^{i+1}(x^e) \triangleq \bar{\ell}_i(x^e) g_1^0(x^e), \ 0 \leq i \leq d
\tag{8.112}
$$

$$
\tilde{g}_k^i(x^e) \triangleq \mathrm{ad}_{F_0^e}^{k-1} \tilde{g}_1^i(x^e), \ 1 \leq i \leq d+1 \text{ and } k \geq 2.
\tag{8.113}
$$

Proof Necessity. Suppose that system (8.70) is RDOEL with index d and state transformation $z^e = [w^\mathsf{T} \ z^\mathsf{T}]^\mathsf{T} = S^e(w, x) = [w^\mathsf{T} \ S(w, x)^\mathsf{T}]^\mathsf{T}$. Then, by Theorem 8.4, there exist smooth functions $\ell(w, y)$ ($\neq 0$) such that (8.85), (8.86), (8.87), and (8.88) are satisfied. It will be shown that

$$
\bar{g}_1^0(x^e) = \tilde{g}_1^{d+1}(x^e) \text{ or } \ell(w, H(x)) = \bar{\ell}_d(x^e).
\tag{8.114}
$$

Let $\ell_0(w) \triangleq 1$, $\ell_d(w) \triangleq \ell(w, 0)$, and for $1 \leq i \leq d$ and $1 \leq s \leq i$,

$$
\ell_i(w) \triangleq \ell_d(w_1, \cdots, w_i, O_{(d-i) \times 1})
$$

$$
\hat{\ell}_{i,s}(w) \triangleq \prod_{k=s}^{i} e^{\beta_k w_{d+1-k}} \text{ or } \tilde{g}_1^{i+1}(x^e) \triangleq \hat{\ell}_{i,s}(w) \tilde{g}_1^s(x^e).
\tag{8.115}
$$

Note, by (8.80) and Example 2.4.16, that for $1 \leq i \leq n$, $1 \leq k \leq n$, $1 \leq j \leq d$, and $s \leq q \leq d$,

$$
L_{g_i^0(x^e)} L_{F_0^e}^k \ell(w, H(x)) = \sum_{j=1}^{d} \frac{\partial \ell(w, H(x))}{\partial w_j} L_{g_i^0(x^e)} L_{F_0^e}^k w_j
$$

$$
+ \left. \frac{\partial \ell(w, y)}{\partial y} \right|_{y = H(x)} L_{g_i^0(x^e)} L_{F_0^e}^k H(x)
\tag{8.116}
$$

$$
= \begin{cases}
0, & i + k < n \\
(-1)^{i+1} \left. \frac{\partial \ell(w, y)}{\partial y} \right|_{y = H(x)}, & i + k = n
\end{cases}
$$

$$
L_{g_i^0(x^e)} L_{F_0^e}^k \ell_j(w) =
\begin{cases}
0, & i + k < n + d + 1 - j \\
(-1)^{i+1} \frac{\partial \ell_j(w)}{\partial w_j}, & i + k = n + d + 1 - j
\end{cases}
\tag{8.117}
$$

$$L_{\tilde{\mathbf{g}}_i^s} L_{F_0^e}^k \ell_{d+1-s}(w) = \begin{cases} 0, & i+k < n+s \\ (-1)^{i+1} \bar{\ell}_{s-1}(x^e) \frac{\partial \ell_{d+1-s}(w)}{\partial w_{d+1-s}}, & i+k = n+s, \end{cases} \quad (8.118)$$

and

$$L_{\tilde{\mathbf{g}}_i^s} L_{F_0^e}^k \hat{\ell}_{q,s}(w) = \begin{cases} 0, & i+k < n+s \\ (-1)^{i+1} \bar{\ell}_{s-1}(x^e) \beta_s \hat{\ell}_{q,s}(w), & i+k = n+s. \end{cases} \quad (8.119)$$

Also note, by (2.44), (8.78), (8.87), and (8.88), that for $1 \le i \le n$,

$$\begin{aligned}
\bar{\mathbf{g}}_i^0(x^e) &= \mathrm{ad}_{F_0^e}^{i-1} \bar{\mathbf{g}}_1^0(x^e) = \mathrm{ad}_{F_0^e}^{i-1} \{\ell(w, H(x)) \mathbf{g}_1^0(x^e)\} \\
&= \sum_{k=0}^{i-1} \binom{i-1}{k} L_{F_0^e}^k \ell(w, H(x)) \mathrm{ad}_{F_0^e}^{i-1-k} \mathbf{g}_1^0(x^e) \\
&= \sum_{k=0}^{i-1} \binom{i-1}{k} L_{F_0^e}^k \ell(w, H(x)) \mathbf{g}_{i-k}^0(x^e).
\end{aligned} \quad (8.120)$$

Thus, we have, by (2.43), (8.87), (8.116), and (8.120), that for $2 \le i \le n-1$,

$$\begin{aligned}
0 = [\bar{\mathbf{g}}_1^0, \bar{\mathbf{g}}_i^0] &= \left[\ell(w, H) \mathbf{g}_1^0, \ \sum_{k=1}^{i-1} \binom{i-1}{k} L_{F_0^e}^k \ell(w, H) \mathbf{g}_{i-k}^0 + \ell(w, H) \mathbf{g}_i^0 \right] \\
&= \ell(w, H) \sum_{k=1}^{i-1} \binom{i-1}{k} L_{F_0^e}^k \ell(w, H) [\mathbf{g}_1^0, \mathbf{g}_{i-k}^0] + \ell(w, H)^2 [\mathbf{g}_1^0, \mathbf{g}_i^0].
\end{aligned}$$
$$(8.121)$$

Since $[\mathbf{g}_1^0, \mathbf{g}_1^0] = 0$, it is easy to show, by (8.121) and mathematical induction, that condition (i) is satisfied. Thus, it is easy to see, by (2.43), (2.44), (8.104), (8.112), and, (8.116), that for $1 \le i \le d+1$ and $1 \le k \le n-1$,

$$\begin{aligned}
[\tilde{\mathbf{g}}_1^i, \ \tilde{\mathbf{g}}_k^i] &= \left[\bar{\ell}_{i-1}(x^e) \mathbf{g}_1^0, \ \mathrm{ad}_{F_0^e}^{k-1} \{\bar{\ell}_{i-1}(x^e) \mathbf{g}_1^0\} \right] \\
&= \left[\bar{\ell}_{i-1}(x^e) \mathbf{g}_1^0, \ \sum_{j=0}^{k-1} \binom{k-1}{j} L_{F_0^e}^j \bar{\ell}_{i-1}(x^e) \mathbf{g}_{k-j}^0 \right] = 0.
\end{aligned} \quad (8.122)$$

Now it will be shown, by mathematical induction, that for $1 \le i \le d+1$,

$$\bar{\mathbf{g}}_1^0(x^e) = \ell_{d+1-i}(w) \tilde{\mathbf{g}}_1^i(x^e) \quad (8.123)$$

and

$$[\tilde{\mathbf{g}}_k^q(x^e),\ \tilde{\mathbf{g}}_j^q(x^e)] = 0,\ i \le q \le d+1,\ 2 \le k+j \le n+i. \tag{8.124}$$

Let n be even. Note, by (2.43), (8.86), (8.87), (8.104), (8.116), and (8.120), that

$$0 = [\bar{\mathbf{g}}_1^0, \bar{\mathbf{g}}_n^0] = \left[\ell(w,H)\mathbf{g}_1^0,\ L_{F_0^e}^{n-1}\ell(w,H)\mathbf{g}_1^0 + \sum_{k=1}^{n-2} \binom{n-1}{k} L_{F_0^e}^k \ell(w,H)\mathbf{g}_{n-k}^0 \right.$$
$$\left. + \ell(w,H)\mathbf{g}_n^0 \right] = [\ell(w,H)\mathbf{g}_1^0,\ L_{F_0^e}^{n-1}\ell(w,H)\mathbf{g}_1^0] + [\ell(w,H)\mathbf{g}_1^0,\ \ell(w,H)\mathbf{g}_n^0]$$
$$= \ell(w,H)L_{\mathbf{g}_1^0}L_{F_0^e}^{n-1}\ell(w,H)\mathbf{g}_1^0 + \ell(w,H)^2[\mathbf{g}_1^0,\mathbf{g}_n^0] - \ell(w,H)L_{\mathbf{g}_n^0}\ell(w,H)\mathbf{g}_1^0$$

which implies, together with (8.83) and (8.116), that

$$[\mathbf{g}_1^0, \mathbf{g}_n^0] = \frac{1}{\ell(w,H(x))} \left\{ L_{\mathbf{g}_n^0}\ell(w,H(x)) - L_{\mathbf{g}_1^0}L_{F_0^e}^{n-1}\ell(w,H(x)) \right\} \mathbf{g}_1^0$$
$$= -2 \left\{ \frac{1}{\ell(w,y)} \frac{\partial \ell(w,y)}{\partial y} \right\} \Bigg|_{y=H(x)} \mathbf{g}_1^0$$
$$= -2 \left\{ \frac{1}{\ell(O_{d\times 1},y)} \frac{\partial \ell(O_{d\times 1},y)}{\partial y} \right\} \Bigg|_{y=H(x)} \mathbf{g}_1^0.$$

Thus, condition (ii) holds with $\beta_0(y) = \frac{1}{\ell(O,y)}\frac{\partial \ell(O,y)}{\partial y} = \frac{\partial \ln \ell(O,y)}{\partial y} = \frac{\partial \ln \ell(w,y)}{\partial y}$ or

$$\ell(w,y) = \ell(w,0)e^{\int_0^y \beta_0(\bar{y})d\bar{y}}.$$

Therefore, it is clear that $\bar{\mathbf{g}}_1^0 = \ell_d(w)\tilde{\mathbf{g}}_1^1$ and (8.123) holds for $i=1$. Also, we have, by (2.43), (2.44), (8.104), (8.105), (8.115), (8.116), (8.117), and (8.122), that

$$[\tilde{\mathbf{g}}_1^1, \tilde{\mathbf{g}}_n^1] = [\bar{\ell}_0 \mathbf{g}_1^0,\ \bar{\ell}_0 \mathbf{g}_n^0 + \cdots + L_{F_0^e}^{n-1}\bar{\ell}_0 \mathbf{g}_1^0]$$
$$= \bar{\ell}_0^2[\mathbf{g}_1^0, \mathbf{g}_n^0] + \bar{\ell}_0 \left\{ L_{\mathbf{g}_1^0}L_{F_0^e}^{n-1}\bar{\ell}_0 - L_{\mathbf{g}_n^0}\bar{\ell}_0 \right\} \mathbf{g}_1^0$$
$$= \bar{\ell}_0^2[\mathbf{g}_1^0, \mathbf{g}_n^0] + 2\bar{\ell}_0 \frac{\partial e^{\int_0^y \beta_0(\bar{y})d\bar{y}}}{\partial y} \Bigg|_{y=H(x)} \mathbf{g}_1^0 \tag{8.125}$$
$$= \bar{\ell}_0^2 \left\{ [\mathbf{g}_1^0, \mathbf{g}_n^0] + 2\beta_0(H(x))\mathbf{g}_1^0 \right\} = 0$$

and for $1 \le q \le d$,

$$\left[\tilde{\mathbf{g}}_1^{q+1}, \tilde{\mathbf{g}}_n^{q+1} \right] = \left[\hat{\ell}_{q,1}(w)\tilde{\mathbf{g}}_1^1,\ \sum_{k=0}^{n-1} \binom{n-1}{k} L_{F_0^e}^k \hat{\ell}_{q,1}(w)\,\tilde{\mathbf{g}}_{n-k}^1 \right] = 0. \tag{8.126}$$

Thus, it is easy to see, by (8.122), (8.125), (8.126), and Example 2.4.18, that

$$[\tilde{\mathbf{g}}_k^q(x^e), \, \tilde{\mathbf{g}}_j^q(x^e)] = 0, \; 1 \leq q \leq d+1, \; 2 \leq k+j \leq n+1.$$

Therefore, (8.124) also holds for $i = 1$. Assume that (8.123) and (8.124) hold for $i = s$ and s is odd with $1 \leq s \leq \min(d, n-2)$. (Even s will be considered later.) In other words,

$$\bar{\mathbf{g}}_1^0(x^e) = \ell_{d+1-s}(w)\tilde{\mathbf{g}}_1^s(x^e) \tag{8.127}$$

and

$$[\tilde{\mathbf{g}}_k^q(x^e), \, \tilde{\mathbf{g}}_j^q(x^e)] = 0, \; s \leq q \leq d+1, \; 2 \leq k+j \leq n+s. \tag{8.128}$$

Since $n + s$ is odd, then it is clear, by Example 2.4.19, that

$$[\tilde{\mathbf{g}}_k^q(x^e), \, \tilde{\mathbf{g}}_j^q(x^e)] = 0, \; s \leq q \leq d+1, \; 2 \leq k+j \leq n+s+1 \tag{8.129}$$

and (8.124) is satisfied when $i = s + 1$. Therefore, we have, by (2.43), (2.44), (8.113), (8.118), (8.127), and (8.129), that

$$
\begin{aligned}
0 = [\bar{\mathbf{g}}_{s+2}^0, \, \bar{\mathbf{g}}_n^0] &= \left[\mathrm{ad}_{F_0^e}^{s+1} \left(\ell_{d+1-s}(w)\tilde{\mathbf{g}}_1^s \right), \, \mathrm{ad}_{F_0^e}^{n-1} \left(\ell_{d+1-s}(w)\tilde{\mathbf{g}}_1^s \right) \right] \\
&= \left[\sum_{j=0}^{s+1} \binom{s+1}{j} L_{F_0^e}^j \ell_{d+1-s} \, \tilde{\mathbf{g}}_{s+2-j}^s, \; \sum_{j=0}^{n-1} \binom{n-1}{j} L_{F_0^e}^j \ell_{d+1-s} \, \tilde{\mathbf{g}}_{n-j}^s \right] \\
&= \ell_{d+1-s}^2 [\tilde{\mathbf{g}}_{s+2}^s, \, \tilde{\mathbf{g}}_n^s] + \ell_{d+1-s} \Big\{ (n-1) L_{\tilde{\mathbf{g}}_{s+2}^s} L_{F_0^e}^{n-2} \ell_{d+1-s} \\
&\quad - (s+1) L_{\tilde{\mathbf{g}}_n^s} L_{F_0^e}^s \ell_{d+1-s} \Big\} \tilde{\mathbf{g}}_2^s \; \mathrm{mod}\{\tilde{\mathbf{g}}_1^s\} \\
&= \ell_{d+1-s}^2 [\tilde{\mathbf{g}}_{s+2}^s, \, \tilde{\mathbf{g}}_n^s] + (n+s)\ell_{d+1-s}\bar{\ell}_{s-1}(x^e) \frac{\partial \ell_{d+1-s}(w)}{\partial w_{d+1-s}} \tilde{\mathbf{g}}_2^s \; \mathrm{mod}\{\tilde{\mathbf{g}}_1^s\}
\end{aligned}
$$

which implies that

$$[\tilde{\mathbf{g}}_{s+2}^s, \, \tilde{\mathbf{g}}_n^s] = -(n+s)\bar{\ell}_{s-1}(x^e) \frac{\partial \ln \ell_{d+1-s}(w)}{\partial w_{d+1-s}} \tilde{\mathbf{g}}_2^s \; \mathrm{mod}\{\tilde{\mathbf{g}}_1^s\}.$$

Since $[\tilde{\mathbf{g}}_{s+2}^s(x^e), \, \tilde{\mathbf{g}}_n^s(x^e)]$ does not depend, by (8.84), on $w_1, \cdots,$ and w_{d+1-s}, condition (iii) (or (8.108)) holds with constant $\beta_s = \frac{\partial \ln \ell_{d+1-s}(w)}{\partial w_{d+1-s}}$ when $i = s$. Thus, it is clear that $\ell_{d+1-s}(w) = \ell_{d-s}(w)e^{\beta_s w_{d+1-s}}$ and $\bar{\mathbf{g}}_1^0(x^e) = \ell_{d-s}(w)\tilde{\mathbf{g}}_1^{s+1}(x^e)$ from (8.115). Therefore, (8.123) is also satisfied when $i = s + 1$. Now assume that (8.123) and (8.124) hold for $i = s$ and s is even with $1 \leq s \leq \min(d, n-2)$. Then, we have, by (2.43), (2.44), (8.113), (8.118), (8.127), and (8.128), that

$$0 = [\bar{\mathbf{g}}_{s+1}, \ \bar{\mathbf{g}}_n] = \left[\mathrm{ad}_{F_0^e}^s \left(\ell_{d+1-s}(w)\tilde{\mathbf{g}}_1^s \right), \ \mathrm{ad}_{F_0^e}^{n-1} \left(\ell_{d+1-s}(w)\tilde{\mathbf{g}}_1^s \right) \right]$$

$$= \left[\sum_{j=0}^{s} \binom{s}{j} L_{F_0^e}^j \ell_{d+1-s} \ \tilde{\mathbf{g}}_{s+1-j}^s, \ \sum_{j=0}^{n-1} \binom{n-1}{j} L_{F_0^e}^j \ell_{d+1-s} \ \tilde{\mathbf{g}}_{n-j}^s \right]$$

$$= \ell_{d+1-s}^2 [\tilde{\mathbf{g}}_{s+1}^s, \ \tilde{\mathbf{g}}_n^s] + \ell_{d+1-s} \left\{ L_{\tilde{\mathbf{g}}_{s+1}^s} L_{F_0^e}^{n-1} \ell_{d+1-s} - L_{\tilde{\mathbf{g}}_n^s} L_{F_0^e}^s \ell_{d+1-s} \right\} \tilde{\mathbf{g}}_1^s$$

$$= \ell_{d+1-s}^2 [\tilde{\mathbf{g}}_{s+1}^s, \ \tilde{\mathbf{g}}_n^s] + 2\ell_{d+1-s}\bar{\ell}_{s-1}(x^e)\frac{\partial \ell_{d+1-s}(w)}{\partial w_{d+1-s}} \tilde{\mathbf{g}}_1^s$$

which implies that

$$[\tilde{\mathbf{g}}_{s+1}^s(x^e), \ \tilde{\mathbf{g}}_n^s(x^e)] = -2\bar{\ell}_{s-1}(x^e)\frac{\partial \ln \ell_{d+1-s}(w)}{\partial w_{d+1-s}} \tilde{\mathbf{g}}_1^s(x^e).$$

Since $[\tilde{\mathbf{g}}_{s+1}^s(x^e), \ \tilde{\mathbf{g}}_n^s(x^e)]$ does not depend, by (8.84), on $w_1, \cdots,$ and w_{d+1-s}, condition (iii) (or (8.107)) holds with constant $\beta_s = \frac{\partial \ln \ell_{d+1-s}(w)}{\partial w_{d+1-s}}$ when $i = s$. Thus, it is clear that $\ell_{d+1-s}(w) = \ell_{d-s}(w)e^{\beta_s w_{d+1-s}}$ and $\bar{\mathbf{g}}_1^0(x^e) = \ell_{d-s}(w)\tilde{\mathbf{g}}_1^{s+1}(x^e)$ from (8.115). Thus, (8.123) is satisfied when $i = s + 1$. Also, we have, by (2.43), (2.44), (8.107), (8.115), and (8.119), that for $s \le q \le d$,

$$[\tilde{\mathbf{g}}_{s+1}^{q+1}(x^e), \ \tilde{\mathbf{g}}_n^{q+1}(x^e)] = \left[\mathrm{ad}_{F_0^e}^s \left(\hat{\ell}_{q,s}(w)\tilde{\mathbf{g}}_1^s \right), \ \mathrm{ad}_{F_0^e}^{n-1} \left(\hat{\ell}_{q,s}(w)\tilde{\mathbf{g}}_1^s \right) \right]$$

$$= \left[\sum_{j=0}^{s} \binom{s}{j} L_{F_0^e}^j \hat{\ell}_{q,s} \tilde{\mathbf{g}}_{s+1-j}^s, \ \sum_{j=0}^{n-1} \binom{n-1}{j} L_{F_0^e}^j \hat{\ell}_{q,s} \tilde{\mathbf{g}}_{n-j}^s \right]$$

$$= \hat{\ell}_{q,s}^2 [\tilde{\mathbf{g}}_{s+1}^s, \ \tilde{\mathbf{g}}_n^s] + \hat{\ell}_{q,s} \left\{ L_{\tilde{\mathbf{g}}_{s+1}^s} L_{F_0^e}^{n-1} \hat{\ell}_{q,s} - L_{\tilde{\mathbf{g}}_n^s} L_{F_0^e}^s \hat{\ell}_{q,s} \right\} \tilde{\mathbf{g}}_1^s$$

$$= \hat{\ell}_{q,s}^2 \left\{ [\tilde{\mathbf{g}}_{s+1}^s, \ \tilde{\mathbf{g}}_n^s] + 2\bar{\ell}_{s-1}(x^e)\beta_s\tilde{\mathbf{g}}_1^s \right\} = 0$$

which implies, together with (8.128) and Example 2.4.18, that

$$[\tilde{\mathbf{g}}_k^q(x^e), \ \tilde{\mathbf{g}}_j^q(x^e)] = 0, \ s+1 \le q \le d+1, \ 2 \le k+j \le n+s+1$$

and (8.124) also holds for even s when $i = s + 1$. Hence, by mathematical induction, (8.123), (8.124), and condition (iii) are satisfied for even n. Similarly, it can be shown that (8.123), (8.124), and condition (iii) are satisfied for odd n. Finally, since $\bar{\mathbf{g}}_1^0(x^e) = \tilde{\mathbf{g}}_1^{d+1}(x^e)$, condition (iv) and condition (v) are satisfied by (8.85) and (8.86), respectively.

Sufficiency. Suppose that conditions of Theorem 8.5 are satisfied. Then it is clear that conditions of Theorem 8.4 are also satisfied with $\ell(w, H(x)) = \bar{\ell}_d(x^e)$ or $\bar{\mathbf{g}}_1^0(x^e) = \tilde{\mathbf{g}}_1^{d+1}(x^e)$. Hence, system (8.70) is, by Theorem 8.4, RDOEL with index d and state transformation $z^e = S^e(w, x) = [w^\top \ S(w, x)^\top]^\top$ where $S(w, x)$ is defined by (8.89) and (8.90), with $\tilde{\mathbf{g}}_i^{d+1}(x^e)$, $1 \le i \le n$ instead of $\bar{\mathbf{g}}_i^0(x^e)$, $1 \le i \le n$. $\square$

By letting $d = 0$ in Theorem 8.5, the necessary and sufficient conditions in Theorem 8.3 can be obtained, for the state equivalence to a dual Brunovsky NOCF with OT.

Remark 8.4 If system (8.70) is RDOEL with index d (β_i, $0 \le i \le d$), then it is RDOEL with index $d + 1$ ($\beta'_{d+1} = 0$ and $\beta'_i = \beta_i$, $0 \le i \le d$). But the converse does not hold.

If the system with $n = 2$ is not state equivalent to a dual Brunovsky NOCF with OT, then the system is not RDEOL either. In the next theorem, the bound of the index in the RDOEL problem will be given.

Theorem 8.6 If system (8.70) with $n \ge 3$ is not RDOEL with index $d \le n - 2$, then it is not RDOEL.

Proof If $d \ge n - 1$, then $\min(d, n - 2) = n - 2$ in condition (iii) of Theorem 8.5. Therefore, the necessary and sufficient conditions of Theorem 8.5 for $d \ge n - 1$ are the same as those for $d = n - 2$. □

Example 8.3.2 Consider system (8.69) in Example 8.2.4 again.

$$\dot{x} = \begin{bmatrix} x_2 \\ x_3 \\ -4x_1 x_3 - 3x_2^2 - 6x_1^2 x_2 + u \end{bmatrix} = F_u(x) \; ; \quad y = x_1 = H(x). \qquad (8.130)$$

Show that the above system is RDOEL.

Solution In Example 8.2.4, it has been shown, by Theorem 8.3, that system (8.130) is not state equivalent to a dual Brunovsky NOCF with OT. Thus, it will be investigated whether system (8.130) is RDOEL with index $d = 1$ or not. In other words, consider the following extended system with $\dot{w}_1 = y$:

$$\begin{bmatrix} \dot{w}_1 \\ \dot{x}_1 \\ \dot{x}_2 \\ \dot{x}_3 \end{bmatrix} = \begin{bmatrix} x_1 \\ x_2 \\ x_3 \\ -4x_1 x_3 - 3x_2^2 - 6x_1^2 x_2 + u \end{bmatrix} = F_u^e(x) \; ; \quad y_a = w_1.$$

Since $T_e(w, x) \triangleq [w \; T(x)^\mathsf{T}]^\mathsf{T} \triangleq [w \; H(x) \; L_{F_0} H(x) \; L_{F_0}^2 H(x)]^\mathsf{T} = [w \; x^\mathsf{T}]^\mathsf{T} = x^e$, it is clear, by (8.77) and (8.78), that

$$\mathbf{g}_1^0(x^e) \triangleq \left(\frac{\partial T_e(x^e)}{\partial x^e} \right)^{-1} \begin{bmatrix} 0 \\ 0 \\ 0 \\ 1 \end{bmatrix} = \begin{bmatrix} 0 \\ 0 \\ 0 \\ 1 \end{bmatrix}$$

$$\mathbf{g}_2^u(x^e) \triangleq \mathrm{ad}_{F_u^e} \mathbf{g}_1^0(x^e) = \begin{bmatrix} 0 \\ 0 \\ -1 \\ 4x_1 \end{bmatrix} \; ; \quad \mathbf{g}_3^u(x^e) = \begin{bmatrix} 0 \\ 1 \\ -4x_1 \\ 10x_1^2 - 2x_2 \end{bmatrix}$$

which imply that $[\mathbf{g}_1^0(x^e), \mathbf{g}_2^0(x^e)] = 0$ and $[\mathbf{g}_2^0(x^e), \mathbf{g}_3^0(x^e)] = -2\mathbf{g}_1^0(x^e) = 0$ mod $\{\mathbf{g}_1^0(x^e)\}$. Therefore, condition (i) and condition (ii) of Theorem 8.5 are satisfied with $\beta_0(y) = 0$ or $\bar{\ell}_0(x^e) = 1$, which implies, together with (8.111), (8.112), and (8.113), that $\tilde{\mathbf{g}}_i^1(x^e) = \mathbf{g}_i^0(x^e)$ for $1 \le i \le 3$. Since $[\tilde{\mathbf{g}}_2^1(x^e), \tilde{\mathbf{g}}_3^1(x^e)] = -2\tilde{\mathbf{g}}_1^1(x^e)$, condition (iii) of Theorem 8.5 is satisfied with $\beta_1 = -1$. Since $\bar{\ell}_1(x^e) \triangleq \bar{\ell}_0(x^e)e^{\beta_1 w_1} = e^{-w_1}$ by (8.111), it is clear, by (8.112), that

$$
\tilde{\mathbf{g}}_1^2(x^e) \triangleq \bar{\ell}_1(x^e)\mathbf{g}_1^0(x^e) = \begin{bmatrix} 0 \\ 0 \\ 0 \\ e^{-w_1} \end{bmatrix}
$$

$$
\mathrm{ad}_{F_u^e}\tilde{\mathbf{g}}_1^2(x^e) = \begin{bmatrix} 0 \\ 0 \\ -e^{-w_1} \\ 3x_1 e^{-w_1} \end{bmatrix} ; \quad \mathrm{ad}_{F_u^e}^2 \tilde{\mathbf{g}}_1^2(x^e) = \begin{bmatrix} 0 \\ e^{-w_1} \\ -2x_1 e^{-w_1} \\ -3(x_2 - x_1^2)e^{-w_1} \end{bmatrix}
$$

which imply that $\mathrm{ad}_{F_u^e}^{i-1}\tilde{\mathbf{g}}_2^2(x^e) = \mathrm{ad}_{F_0^e}^{i-1}\tilde{\mathbf{g}}_1^2(x^e) \triangleq \tilde{\mathbf{g}}_i^2(x^e)$, $2 \le i \le 3$ and thus condition (iv) of Theorem 8.5 is also satisfied. Finally, it is easy to see that condition (v) of Theorem 8.5 holds. Hence, by Theorem 8.5, system (8.130) is RDOEL with index $d = 1$. The extended state transformation $z^e = [w\ z^{\mathsf{T}}]^{\mathsf{T}} = S^e(w, x) = [w\ S(w, x)^{\mathsf{T}}]^{\mathsf{T}} = [w_1\ x_1 e^{w_1}\ (x_2 + x_1^2)e^{w_1}\ (x_3 + x_1^3 + 3x_1x_2)e^{w_1}]^{\mathsf{T}}$ can be obtained by (8.89) and (8.90), with $\tilde{\mathbf{g}}_i^2(x^e)$, $1 \le i \le 3$ instead of $\bar{\mathbf{g}}_i^0(x^e)$, $1 \le i \le 3$. In other words,

$$
\frac{\partial S(w, x)}{\partial x} = D(x^e)^{-1} = \begin{bmatrix} e^{-w_1} & 0 & 0 \\ -2x_1 e^{-w_1} & e^{-w_1} & 0 \\ -3(x_2 - x_1^2)e^{-w_1} & -3x_1 e^{-w_1} & e^{-w_1} \end{bmatrix}^{-1}
$$

$$
= \begin{bmatrix} e^{w_1} & 0 & 0 \\ 2x_1 e^{w_1} & e^{w_1} & 0 \\ 3(x_2 + x_1^2)e^{w_1} & 3x_1 e^{w_1} & e^{w_1} \end{bmatrix}.
$$

Finally, it is easy to see that

$$
\begin{bmatrix} \dot{w}_1 \\ \dot{z}_1 \\ \dot{z}_2 \\ \dot{z}_3 \end{bmatrix} = \begin{bmatrix} z_1 \\ z_2 \\ z_3 \\ 0 \end{bmatrix} + \begin{bmatrix} y(1 - e^{w_1}) \\ 0 \\ 0 \\ e^{w_1}(y^4 + u) \end{bmatrix} ; \quad y_a = w_1.
$$

$\square$

Example 8.3.3 Show that the following system is not RDOEL:

$$\dot{x} = \begin{bmatrix} x_2 \\ x_3 \\ x_2^3 + u \end{bmatrix} = F_u(x) \; ; \quad y = x_1 = H(x). \tag{8.131}$$

Solution It is easy to see that system (8.131) is not state equivalent to a dual Brunovsky NOCF with OT. (See Problem 8-2.) Thus, it will be investigated whether system (8.131) is RDOEL with index $d = 1$ or not. Consider the following extended system with $\dot{w}_1 = y$:

$$\begin{bmatrix} \dot{w}_1 \\ \dot{x}_1 \\ \dot{x}_2 \\ \dot{x}_3 \end{bmatrix} = \begin{bmatrix} x_1 \\ x_2 \\ x_3 \\ x_2^3 + u \end{bmatrix} = F_u^e(x) \; ; \quad y_a = w_1.$$

Since $T_e(w, x) \triangleq [w \ T(x)^\mathsf{T}]^\mathsf{T} \triangleq [w \ H(x) \ L_{F_0} H(x) \ L_{F_0}^2 H(x)]^\mathsf{T} = [w \ x^\mathsf{T}]^\mathsf{T} = x^e$, it is clear, by (8.77) and (8.78), that

$$\mathbf{g}_1^0(x^e) \triangleq \left(\frac{\partial T_e(x^e)}{\partial x^e} \right)^{-1} \begin{bmatrix} 0 \\ 0 \\ 0 \\ 1 \end{bmatrix} = \begin{bmatrix} 0 \\ 0 \\ 0 \\ 1 \end{bmatrix}$$

$$\mathbf{g}_2^u(x^e) \triangleq \mathrm{ad}_{F_u^e} \mathbf{g}_1^0(x^e) = \begin{bmatrix} 0 \\ 0 \\ -1 \\ 0 \end{bmatrix} \; ; \quad \mathbf{g}_3^u(x^e) = \begin{bmatrix} 0 \\ 1 \\ 0 \\ 3x_2^2 \end{bmatrix}$$

which imply that $[\mathbf{g}_1^0(x^e), \mathbf{g}_2^0(x^e)] = 0$ and $[\mathbf{g}_2^0(x^e), \mathbf{g}_3^0(x^e)] = -6x_2 \mathbf{g}_1^0(x^e) = 0 \bmod \{\mathbf{g}_1^0(x^e)\}$. Therefore, condition (i) and condition (ii) of Theorem 8.5 are satisfied with $\beta_0(y) = 0$ or $\bar{\ell}_0(x^e) = 1$, which implies, together with (8.111), (8.112), and (8.113), that $\tilde{\mathbf{g}}_i^1(x^e) = \mathbf{g}_i^0(x^e)$ for $1 \le i \le 3$. Since $[\tilde{\mathbf{g}}_2^1(x^e), \tilde{\mathbf{g}}_3^1(x^e)] = -6x_2 \tilde{\mathbf{g}}_1^1(x^e)$, there does not exist constant β_1 such that (8.107) is satisfied. Since condition (iii) of Theorem 8.5 is not satisfied, system (8.131) is not, by Theorem 8.5, RDOEL with index $d = 1$. Also, system (8.131) is not RDOEL by Theorem 8.6. $\qquad\square$

8.4 Multi Output Observer Error Linearization

Consider a multi output control system of the form

$$\dot{x} = F(x, u) \triangleq F_u(x)$$
$$y = H(x) \tag{8.132}$$

with $F_0(0) = 0, H(0) = 0$, state $x \in \mathbb{R}^n$, input $u \in \mathbb{R}^m$, and output $y \in \mathbb{R}^p$. By letting $u = 0$ in system (8.132), we obtain the following autonomous system:

$$\dot{x} = F_0(x) ; \quad y = H(x).\tag{8.133}$$

Definition 8.7 (*observability indices*)
For the list of pn one forms of the form

$$\frac{\partial H_1(x)}{\partial x}\bigg|_{x=0}, \cdots, \frac{\partial H_q(x)}{\partial x}\bigg|_{x=0}, \frac{\partial \left(L_{F_0} H_1(x)\right)}{\partial x}\bigg|_{x=0}, \cdots, \frac{\partial \left(L_{F_0} H_p(x)\right)}{\partial x}\bigg|_{x=0},$$

$$\cdots, \frac{\partial \left(L_{F_0}^{n-1} H_1(x)\right)}{\partial x}\bigg|_{x=0}, \cdots, \frac{\partial \left(L_{F_0}^{n-1} H_p(x)\right)}{\partial x}\bigg|_{x=0},$$

delete all one forms that are linearly dependent on the set of preceding one forms and obtain the unique set of linearly independent one forms

$$\left\{\frac{\partial H_1(x)}{\partial x}, \cdots, \frac{\partial \left(L_{F_0}^{v_1-1} H_1(x)\right)}{\partial x}, \cdots, \frac{\partial H_p(x)}{\partial x}, \cdots, \frac{\partial \left(L_{F_0}^{v_p-1} H_p(x)\right)}{\partial x}\right\}\bigg|_{x=0}$$

or

$$\left\{\bar{c}_1, \bar{c}_1 \bar{A}, \cdots, \bar{c}_1 \bar{A}^{v_1-1}, \cdots, \bar{c}_p, \cdots, \bar{c}_p \bar{A}^{v_p-1}\right\}$$

where $\bar{c}_j \triangleq \frac{\partial H_j(x)}{\partial x}\big|_{x=0}$ and $\bar{A} \triangleq \frac{\partial F_0(x)}{\partial x}\big|_{x=0}$. Then, $(v_1, \cdots, v_p)$ are said to be the observability indices of system (8.132).

In other words, v_i is the smallest nonnegative integer such that for $1 \le i \le p$,

$$dL_{F_0}^{v_i} H_i(x)\big|_{x=0} \in \text{span} \left\{ dL_{F_0}^{\ell-1} H_j(x)\big|_{x=0} \middle| 1 \le j \le m, \quad 1 \le \ell \le v_i \right\}$$
$$+ \text{span} \left\{ dL_{F_0}^{v_i} H_j(x)\big|_{x=0} \middle| 1 \le j \le i-1 \right\}.\tag{8.134}$$

If $\sum_{i=1}^{p} v_i = n$, then system (8.132) is said to be observable. Since observability is invariant under state transformation, we assume $\sum_{i=1}^{p} v_i = n$. Also, we assume, without loss of generality, that

$$v_1 \ge v_2 \ge \cdots \ge v_p \ge 1.\tag{8.135}$$

Definition 8.8 (*state equivalence to a NOCF*)
System (8.132) is said to be state equivalent to a NOCF, if there exist a state transformation $z = S(x)$ such that

$$\dot{z} = Az + \gamma(y, u) = Az + \gamma^u(y) \triangleq \bar{f}_u(z)$$
$$y = Cz \triangleq \bar{h}(z) \tag{8.136}$$

where the pair (C, A) is observable and $\gamma^u(y) : \mathbb{R}^p \times \mathbb{R}^m \to \mathbb{R}^n$ is a smooth vector function with $\gamma^0(0) = 0$. In other words,

$$\bar{h}(z) \triangleq H \circ S^{-1}(z) = Cz \tag{8.137}$$

and

$$\bar{f}_u(z) \triangleq S_*(F_u(x)) = Az + \gamma^u(Cz)$$
$$= Az + \gamma^u(y). \tag{8.138}$$

For single output case, if the pair (C, A) is observable, there exists a linear state transform $z = P^{-1}x$ such that $(\hat{C}, \hat{A})(\triangleq (CP, \ P^{-1}AP))$ is observable canonical form. In other words,

$$CP = \hat{C} = \begin{bmatrix} 1 \ 0 \ 0 \cdots 0 \end{bmatrix} = C_o$$

$$P^{-1}AP = \hat{A} = \begin{bmatrix} \hat{a}_{11} & 1 \ 0 \cdots 0 \\ \hat{a}_{21} & 0 \ 1 \cdots 0 \\ \vdots & \vdots \ \vdots \quad \vdots \\ \hat{a}_{(n-1)1} & 0 \ 0 \cdots 1 \\ \hat{a}_{n1} & 0 \ 0 \cdots 0 \end{bmatrix} = A_o + \begin{bmatrix} \hat{a}_{11} \\ \hat{a}_{21} \\ \vdots \\ \hat{a}_{(n-1)1} \\ \hat{a}_{n1} \end{bmatrix} \hat{C}$$

where

$$C_o = \begin{bmatrix} 1 \ 0 \ 0 \cdots 0 \end{bmatrix} \ \text{ and } \ A_o = \begin{bmatrix} 0 \ 1 \ 0 \cdots 0 \\ 0 \ 0 \ 1 \cdots 0 \\ \vdots \ \vdots \ \vdots \quad \vdots \\ 0 \ 0 \ 0 \cdots 1 \\ 0 \ 0 \ 0 \cdots 0 \end{bmatrix}.$$

Let us call (C_o, A_o) a dual Brunovsky canonical form, even though the order of the states are reversed compared to Brunovsky canonical form (4.9). Since

$$\begin{bmatrix} \hat{a}_{11} \\ \vdots \\ \hat{a}_{n1} \end{bmatrix} \hat{C}z = \begin{bmatrix} \hat{a}_{11}z_1 \\ \vdots \\ \hat{a}_{n1}z_1 \end{bmatrix},$$

it is clear that single output system (8.132) is state equivalent to a NOCF, if and only if single output system (8.132) is state equivalent to a dual Brunovsky NOCF which is defined by

$$\dot{z} = A_o z + \gamma^u(z_1) \triangleq \bar{f}_u(z)$$
$$y = C_o z \triangleq \bar{h}(z).$$

However, it is not true that if multi output system is state equivalent to a NOCF, then multi output system is also state equivalent to a dual Brunovsky NOCF. (For this, see Theorem 8.7.)

Definition 8.9 (*state equivalence to a dual Brunovsky NOCF with OT*)
System (8.132) is said to be state equivalent to a dual Brunovsky NOCF with output transformation (OT), if there exist a smooth function $\varphi(y)$ ($\varphi(0) = 0$ and rank $\left(\frac{\partial \varphi(y)}{\partial y} \Big|_{y=0} \right) = p$) and a state transformation $z = S(x)$ such that

$$\dot{z} = A_o z + \gamma^u(\bar{y}) \triangleq \bar{f}_u(z)$$
$$\bar{y} = \varphi(y) = C_o z \triangleq \bar{h}(z)$$

where $A_o = \text{blockdiag}\{A_o^1, \cdots, A_o^p\}$, $C_o = \text{blockdiag}\{C_o^1, \cdots, C_o^p\}$,

$$A_o^i = \begin{bmatrix} O_{(v_i-1)\times 1} & I_{(v_i-1)\times(v_i-1)} \\ 0 & O_{1\times(v_i-1)} \end{bmatrix} ; \quad C_o^i = \begin{bmatrix} 1 & O_{1\times(v_i-1)} \end{bmatrix}, \ 1 \le i \le p$$

and $\gamma^u(\bar{y}) : \mathbb{R}^p \times \mathbb{R}^m \to \mathbb{R}^n$ is a smooth vector function with $\gamma^0(0) = 0$. In other words,

$$\bar{h}(z) \triangleq \varphi \circ H \circ S^{-1}(z) = C_o z \tag{8.139}$$

and

$$\bar{f}_u(z) \triangleq S_*(F_u(x)) = A_o z + \bar{\gamma}^u(C_o z)$$
$$= A_o z + \bar{\gamma}^u \circ \varphi(y) \triangleq A_o z + \gamma^u(y) \tag{8.140}$$

where $\bar{\gamma}^u(\bar{y}) \triangleq \gamma^u \circ \varphi^{-1}(\bar{y})$.

If system (8.132) is state equivalent to a NOCF with OT $\varphi(y) = y$, then system (8.132) is state equivalent to a NOCF (without OT). State equivalence to a NOCF for autonomous system (8.133) can be similarly defined with $u = 0$. If $\bar{f}_u(z) \triangleq S_*(F_u(x)) = Az + \bar{\gamma}^u(Cz)$, then it is clear that $\bar{f}_0(z) \triangleq S_*(F_0(x)) = Az + \bar{\gamma}^0(Cz)$. Thus, we have the following remark.

Remark 8.5 If system (8.132) is state equivalent to a NOCF with OT $\bar{y} = \varphi(y)$ and state transformation $z = S(x)$, then system (8.133) is also state equivalent to a NOCF with OT $\bar{y} = \varphi(y)$ and state transformation $z = S(x)$. But the converse is not true.

Lemma 8.4 (observable canonical form of MO linear system)
Suppose that (A, C) is an observable pair and (8.135) is satisfied. Then there exist a nonsingular matrix Q, a lower triangular matrix R with 1's in the diagonal, and an $n \times p$ matrix $\tilde{A}$ such that

$$\hat{A} \triangleq QAQ^{-1} = A_o + \tilde{A}C_o \quad \text{and} \quad \hat{C} \triangleq CQ^{-1} = RC_o \tag{8.141}$$

where for $1 \leq i \leq p$ and $1 \leq i \leq v_i$,

$$c_i A^{v_i} = -\sum_{\ell=1}^{i-1} \bar{r}_{i\ell} c_\ell A^{v_i} \tag{8.142}$$

$$\text{mod span} \left\{ c_s A^{k-1}, \ 1 \leq s \leq p, \ 1 \leq k \leq \min(v_j, v_i) \right\}$$

$$R^{-1} = \begin{bmatrix} 1 & 0 & \cdots & 0 \\ \bar{r}_{21} & 1 & \cdots & 0 \\ \vdots & \vdots & \ddots & \vdots \\ \bar{r}_{p1} & \bar{r}_{p2} & \cdots & 1 \end{bmatrix} ; \quad \bar{C} = \begin{bmatrix} \bar{c}_1 \\ \vdots \\ \bar{c}_p \end{bmatrix} \triangleq R^{-1}C \tag{8.143}$$

$$\bar{c}_i A^{v_i} = \sum_{k=1}^{v_i} \sum_{\ell=1}^{p} \tilde{a}^{\ell}_{i(v_i-k+1)} \bar{c}_\ell A^{k-1} \triangleq -\sum_{k=1}^{v_i} \mathbf{m}_{i(v_i-k+1)} \bar{C} A^{k-1} \tag{8.144}$$

$$\tilde{A} \triangleq \begin{bmatrix} \tilde{A}_1 \\ \vdots \\ \tilde{A}_p \end{bmatrix} = \begin{bmatrix} \tilde{a}^1_{11} & \cdots & \tilde{a}^p_{11} \\ \vdots & & \vdots \\ \tilde{a}^1_{1v_1} & \cdots & \tilde{a}^p_{1v_1} \\ \vdots & & \vdots \\ \tilde{a}^1_{p1} & \cdots & \tilde{a}^p_{p1} \\ \vdots & & \vdots \\ \tilde{a}^1_{pv_p} & \cdots & \tilde{a}^p_{pv_p} \end{bmatrix} \triangleq -\begin{bmatrix} \mathbf{m}_{11} \\ \vdots \\ \mathbf{m}_{1v_1} \\ \vdots \\ \mathbf{m}_{p1} \\ \vdots \\ \mathbf{m}_{pv_p} \end{bmatrix} ; \quad \begin{bmatrix} \mathbf{m}_{10} \\ \vdots \\ \mathbf{m}_{p0} \end{bmatrix} = I_{p \times p} \tag{8.145}$$

$$M_{ij} = \begin{bmatrix} \mathbf{m}_{i(j-1)} & \cdots & \mathbf{m}_{i0} & O_{1 \times (v_1-j)p} \end{bmatrix}, \tag{8.146}$$

and

$$Q \triangleq \begin{bmatrix} Q_{11} \\ \vdots \\ Q_{1\nu_1} \\ \vdots \\ Q_{p1} \\ \vdots \\ Q_{p\nu_p} \end{bmatrix} = \begin{bmatrix} M_{11} \\ \vdots \\ M_{1\nu_1} \\ \vdots \\ M_{p1} \\ \vdots \\ M_{p\nu_p} \end{bmatrix} \begin{bmatrix} \bar{C} \\ \bar{C}A \\ \vdots \\ \bar{C}A^{\nu_1-1} \end{bmatrix} \triangleq M\bar{V}. \tag{8.147}$$

Proof It is easy to see, by (8.134), that there exist $\bar{r}_{i\ell}$, $2 \leq i \leq p$, $1 \leq \ell \leq i-1$ such that (8.142) is satisfied. It is clear, by (8.143) and (8.145)–(8.147), that

$$C_o Q = \begin{bmatrix} M_{11} \\ \vdots \\ M_{p1} \end{bmatrix} \begin{bmatrix} \bar{C} \\ \bar{C}A \\ \vdots \\ \bar{C}A^{\nu_1-1} \end{bmatrix} = \begin{bmatrix} \mathbf{m}_{i0} & O_{1\times(\nu_1-1)p} \\ \vdots & \vdots \\ \mathbf{m}_{p0} & O_{1\times(\nu_1-1)p} \end{bmatrix} \begin{bmatrix} \bar{C} \\ \bar{C}A \\ \vdots \\ \bar{C}A^{\nu_1-1} \end{bmatrix}$$

$$= \begin{bmatrix} I_{p\times p} & O_{p\times(\nu_1-1)p} \end{bmatrix} \begin{bmatrix} \bar{C} \\ \bar{C}A \\ \vdots \\ \bar{C}A^{\nu_1-1} \end{bmatrix} = \bar{C} = R^{-1}C \tag{8.148}$$

which implies that $\hat{C} \triangleq CQ^{-1} = RC_o$. It is easy to see, by (8.144)–(8.147), that for $1 \leq i \leq p$ and $1 \leq j \leq \nu_i$,

$$\sum_{k=1}^{\nu_i+1} \mathbf{m}_{i(\nu_i-k+1)} \bar{C}A^{k-1} = \sum_{k=1}^{\nu_i} \mathbf{m}_{i(\nu_i-k+1)} \bar{C}A^{k-1} + \mathbf{m}_{i0}\bar{C}A^{\nu_i}$$

$$= \sum_{k=1}^{\nu_i} \mathbf{m}_{i(\nu_i-k+1)} \bar{C}A^{k-1} + \bar{c}_i A^{\nu_i} = 0 \tag{8.149}$$

and

$$Q_{ij} = M_{ij}\bar{V} = \sum_{k=1}^{j} \mathbf{m}_{i(j-k)} \bar{C}A^{k-1}. \tag{8.150}$$

Thus, we have, by (8.149) and (8.150), that for $1 \leq i \leq p$ and $1 \leq j \leq \nu_i - 1$,

$$Q_{ij}A = M_{ij}\bar{V}A = \sum_{k=1}^{j} \mathbf{m}_{i(j-k)} \bar{C}A^k = \sum_{k=2}^{j+1} \mathbf{m}_{i(j+1-k)} \bar{C}A^{k-1}$$

$$= Q_{i(j+1)} - \mathbf{m}_{ij}\bar{C}$$

and

$$Q_{i\nu_i} A = M_{i\nu_i} \bar{V} A = \sum_{k=1}^{\nu_i} \mathbf{m}_{i(\nu_i-k)} \bar{C} A^k = \sum_{k=2}^{\nu_i+1} \mathbf{m}_{i(\nu_i-k+1)} \bar{C} A^{k-1}$$

$$= \sum_{k=1}^{\nu_i+1} \mathbf{m}_{i(\nu_i-k+1)} \bar{C} A^{k-1} - \mathbf{m}_{i\nu_i} \bar{C} = -\mathbf{m}_{i\nu_i} \bar{C}$$

which imply, together with (8.148), that for $1 \leq i \leq p$,

$$\begin{bmatrix} Q_{i1} \\ \vdots \\ Q_{i(\nu_i-1)} \\ Q_{i\nu_i} \end{bmatrix} A = \begin{bmatrix} Q_{i2} - \mathbf{m}_{i1}\bar{C} \\ \vdots \\ Q_{i\nu_i} - \mathbf{m}_{i(\nu_i-1)}\bar{C} \\ -\mathbf{m}_{i\nu_i}\bar{C} \end{bmatrix} = \begin{bmatrix} Q_{i2} \\ \vdots \\ Q_{i\nu_i} \\ 0 \end{bmatrix} - \begin{bmatrix} \mathbf{m}_{i1} \\ \vdots \\ \mathbf{m}_{i(\nu_i-1)} \\ \mathbf{m}_{i\nu_i} \end{bmatrix} C_o Q$$

$$= \left[O_{\nu_i \times \sum_{k=1}^{i-1} \nu_k} \quad A_o^i \quad O_{\nu_i \times \sum_{k=i+1}^{p} \nu_k} \right] Q + \tilde{A}_i C_o Q.$$

Hence, it is clear that

$$QA = A_o Q + \tilde{A} C_o Q \quad \text{or} \quad \hat{A} \triangleq QAQ^{-1} = A_o + \tilde{A} C_o.$$

$\square$

The dual version of Lemma 8.4 can be found in Section IV of [H25]. The following theorem, which can be easily proven by Lemma 8.4, shows the difference between the equivalence of the MO system to a NOCF and the equivalence of the MO systems to a dual Brunovsky NOCF.

Theorem 8.7 *System (8.132) is state equivalent to a NOCF, if and only if,*

(i) for $2 \leq i \leq p$ and some real constants $\bar{r}_{ij}$'s,

$$dL_{F_u}^{\nu_i} H_i(x) = - \sum_{j=1}^{i-1} \bar{r}_{ij} dL_{F_u}^{\nu_i} H_j(x) \tag{8.151}$$

$$\text{mod span}\{dL_{F_u}^{k-1} H_j(x), \ 1 \leq j \leq p, \ 1 \leq k \leq \min(\nu_j, \nu_i)\}.$$

(ii) System

$$\dot{x}(t) = F_u(x(t)) ; \quad \hat{y}(t) = \hat{H}(x(t)) \triangleq R^{-1} H(x(t)) \tag{8.152}$$

is state equivalent to a dual Brunovsky NOCF, where

$$R^{-1} = \begin{bmatrix} 1 & 0 & \cdots & 0 \\ \bar{r}_{21} & 1 & \cdots & 0 \\ \vdots & \vdots & \ddots & \vdots \\ \bar{r}_{p1} & \bar{r}_{p2} & \cdots & 1 \end{bmatrix}. \tag{8.153}$$

Proof Necessity. Suppose that system (8.132) is state equivalent to a NOCF with state transformation $z = S(x)$. Then, we have, by (8.137) and (8.138), that

$$\bar{h}(z) \triangleq H \circ S^{-1}(z) = Cz \tag{8.154}$$

and

$$\bar{f}_u(z) \triangleq S_*(F_u(x)) = Az + \gamma^u(Cz) \tag{8.155}$$

where (A, C) is an observable. Thus, by Lemma 8.4, there exists a nonsingular matrix Q such that

$$\hat{A} \triangleq QAQ^{-1} = A_o + \tilde{A}C_o \quad \text{and} \quad \hat{C} \triangleq CQ^{-1} = RC_o \tag{8.156}$$

where for $1 \leq i \leq p$,

$$c_i A^{\nu_i} = -\sum_{\ell=1}^{i-1} \bar{r}_{i\ell} c_\ell A^{\nu_i}$$

$$\text{mod span} \left\{ c_s A^{k-1}, \ 1 \leq s \leq p, \ 1 \leq k \leq \min(\nu_j, \nu_i) \right\}.$$

Let $\hat{z} \triangleq Qz = QS(x) \triangleq \hat{S}(x)$. Then it is easy to see, by (8.154)–(8.156), that

$$\hat{h}(\hat{z}) \triangleq \hat{H} \circ \hat{S}^{-1}(z) = R^{-1}\bar{h}(Q^{-1}\hat{z}) = R^{-1}CQ^{-1}\hat{z} = C_o\hat{z} \tag{8.157}$$

and

$$\begin{aligned}
\hat{f}_u(\hat{z}) &\triangleq \hat{S}_*(F_u(x)) = QAQ^{-1}\hat{z} + Q\gamma^u(CQ^{-1}\hat{z}) \\
&= A_o\hat{z} + \tilde{A}C_o\hat{z} + Q\gamma^u(RC_o\hat{z}) = A_o\hat{z} + \hat{\gamma}^u(C_o\hat{z})
\end{aligned} \tag{8.158}$$

where $\hat{\gamma}^u(y) \triangleq \tilde{A}y + Q\gamma^u(Ry)$. Therefore, system (8.152) is state equivalent to a dual Brunovsky NOCF and condition (ii) of Theorem 8.7 is satisfied. Since $\hat{h}(\hat{z}) = R^{-1}\bar{h}(Q^{-1}\hat{z})$, we have, by (2.30) and (8.153), that for $1 \leq i \leq p$ and $k \geq 1$,

$$\hat{h}_i(\hat{z}) = \bar{h}_i(Q^{-1}\hat{z}) + \sum_{j=1}^{i-1} \bar{r}_{ij}\bar{h}_j(Q^{-1}\hat{z})$$

and

$$L_{\hat{f}_u}^k \hat{h}_i(\hat{z}) = L_{\bar{f}_u}^k \bar{h}_i(z)\Big|_{z=Q^{-1}\hat{z}} + \sum_{j=1}^{i-1} \bar{r}_{ij} \, L_{\bar{f}_u}^k \bar{h}_j(z)\Big|_{z=Q^{-1}\hat{z}}$$

which imply that for $1 \le i \le p$ and $k \ge 0$,

$$
\begin{aligned}
dL^k_{\hat{f}_u} \hat{h}_i(\hat{z}) &= \frac{\partial}{\partial \hat{z}} \left(L^k_{\hat{f}_u} \hat{h}_i(\hat{z}) \right) \\
&= \left(dL^k_{\hat{f}_u} \bar{h}_i(z) + \sum_{j=1}^{i-1} \bar{r}_{ij} dL^k_{\hat{f}_u} \bar{h}_j(z) \right) \Bigg|_{z=Q^{-1}\hat{z}} Q^{-1}.
\end{aligned}
\tag{8.159}
$$

Let

$$
\hat{z} \triangleq \left[\hat{z}_{11} \cdots \hat{z}_{1\nu_1} \cdots \hat{z}_{p1} \cdots \hat{z}_{p\nu_p} \right]^{\mathsf{T}}
$$

$$
\tilde{\mathbf{z}}_k \triangleq \left\{ \hat{z}_{jk} \mid 1 \le j \le p \text{ and } \nu_j \ge k \right\}
$$

$$
\hat{f}^u(\hat{z}) \triangleq \left[\hat{f}^u_{11}(\hat{z}) \cdots \hat{f}^u_{1\nu_1}(\hat{z}) \cdots \hat{f}^u_{p1}(\hat{z}) \cdots \hat{f}^u_{p\nu_p}(\hat{z}) \right]^{\mathsf{T}}
$$

and for $1 \le i \le p$,

$$
\hat{f}^u_{ij}(\hat{z}) = \begin{cases} \hat{z}_{i(j+1)} + \hat{\gamma}^u_{ij}(\tilde{\mathbf{z}}_1), & 1 \le j \le \nu_i - 1 \\ \hat{\gamma}^u_{i\nu_i}(\tilde{\mathbf{z}}_1), & j = \nu_i. \end{cases}
\tag{8.160}
$$

Then it is easy to see, by (8.157), (8.158) and (8.160), that for $1 \le i \le p$ and $1 \le k \le \nu_i - 1$,

$$
\hat{h}_i(\hat{z}) = \hat{z}_{i1}
\tag{8.161}
$$

$$
L^k_{\hat{f}_u} \hat{h}_i(\hat{z}) = \hat{z}_{i(k+1)} + \phi^u_{i,k}(\tilde{\mathbf{z}}_1, \cdots, \tilde{\mathbf{z}}_k)
\tag{8.162}
$$

and

$$
L^{\nu_i}_{\hat{f}_u} \hat{h}_i(\hat{z}) = \phi^u_{i,\nu_i}(\tilde{\mathbf{z}}_1, \cdots, \tilde{\mathbf{z}}_{\nu_i})
\tag{8.163}
$$

where $\phi^u_{i,1}(\tilde{\mathbf{z}}_1) = \hat{\gamma}^u_{i1}(\tilde{\mathbf{z}}_1)$ and

$$
\phi^u_{i,k}(\tilde{\mathbf{z}}_1, \cdots, \tilde{\mathbf{z}}_k) = \hat{\gamma}^u_{ik}(\tilde{\mathbf{z}}_1) + \sum_{\ell=1}^{p} \sum_{j=1}^{\min(k-1,\nu_\ell)} \frac{\partial \phi^u_{i,k-1}(\tilde{\mathbf{z}}_1, \cdots, \tilde{\mathbf{z}}_{k-1})}{\partial \hat{z}_{\ell j}} \hat{f}^u_{\ell j}(\hat{z}).
$$

Therefore, it is clear, by (8.161)–(8.163), that for $1 \le i \le p$,

$$\text{span} \left\{ \frac{\partial}{\partial \hat{z}} \left(L_{\hat{f}_u}^{k-1} \hat{h}_j(\hat{z}) \right) \mid 1 \le j \le p, \ 1 \le k \le \min(\nu_j, \nu_i) \right\}$$

$$= \text{span} \left\{ \frac{\partial \hat{z}_{jk}}{\partial \hat{z}} \mid 1 \le j \le p, \ 1 \le k \le \min(\nu_j, \nu_i) \right\}$$

and

$$\frac{\partial}{\partial \hat{z}} \left(L_{\hat{f}_u}^{\nu_i} \hat{h}_i(\hat{z}) \right) \in \text{span} \left\{ \frac{\partial \hat{z}_{jk}}{\partial \hat{z}} \mid 1 \le j \le p, \ 1 \le k \le \min(\nu_j, \nu_i) \right\}$$

which implies, together with (8.159), that for $1 \le i \le p$,

$$dL_{\hat{f}_u}^{\nu_i} \hat{h}_i(\hat{z}) \in \text{span} \left\{ dL_{\hat{f}_u}^{k-1} \hat{h}_j(\hat{z}) \mid 1 \le j \le p, \ 1 \le k \le \min(\nu_j, \nu_i) \right\}$$

$$= \text{span} \left\{ dL_{\bar{f}_u}^{k-1} \bar{h}_j(\bar{z}) \Big|_{z=Q^{-1}\hat{z}} Q^{-1} \mid 1 \le j \le p, \ 1 \le k \le \min(\nu_j, \nu_i) \right\}$$

and

$$dL_{\bar{f}_u}^{\nu_i} \bar{h}_i(z) = - \sum_{j=1}^{i-1} \bar{r}_{ij} dL_{\bar{f}_u}^{\nu_i} \bar{h}_j(z) \tag{8.164}$$

$$\text{mod span} \left\{ dL_{\bar{f}_u}^{k-1} \bar{h}_j(\bar{z}) \mid 1 \le j \le p, \ 1 \le k \le \min(\nu_j, \nu_i) \right\}.$$

Since

$$L_{F_u}^k H(x) = L_{\bar{f}_u}^k \bar{h}(\bar{z}) \Big|_{\bar{z}=S(x)} \quad \text{or} \quad dL_{F_u}^k H(x) = dL_{\bar{f}_u}^k \bar{h}(\bar{z}) \Big|_{\bar{z}=S(x)} \frac{\partial S(x)}{\partial x},$$

it is easy to see, by (8.164), that condition (i) of Theorem 8.7 is satisfied.

Sufficiency. Suppose that system (8.152) is state equivalent to a dual Brunovsky NOCF with $z = S(x)$. Then, we have, by (8.139) and (8.140), that

$$\hat{h}(z) \triangleq R^{-1} H \circ S^{-1}(z) = C_o z$$

and

$$\hat{f}_u(z) \triangleq S_*(F_u(x)) = A_o z + \gamma^u(C_o z).$$

Therefore, it is clear that

$$\bar{h}(z) \triangleq H \circ S^{-1}(z) = R C_o z \triangleq C z$$

and

$$\bar{f}_u(z) \triangleq S_*(F_u(x)) = Az + \gamma^u(R^{-1}Cz)$$
$$= Az + \bar{\gamma}^u(y)$$

where $\bar{\gamma}^u(y) \triangleq \gamma^u(R^{-1}y)$. Hence, by Definition 8.8, system (8.132) is state equivalent to a NOCF with state transformation $z = S(x)$. $\square$

Define a state transformation $\xi = T(x)$ by

$$\xi = \begin{bmatrix} \xi_{11} \\ \xi_{12} \\ \vdots \\ \xi_{1\nu_1} \\ \vdots \\ \xi_{p1} \\ \vdots \\ \xi_{p\nu_p} \end{bmatrix} = T(x) \triangleq \begin{bmatrix} H_1(x) \\ L_{F_0}H_1(x) \\ \vdots \\ L_{F_0}^{\nu_1-1}H_1(x) \\ \vdots \\ H_p(x) \\ \vdots \\ L_{F_0}^{\nu_p-1}H_p(x) \end{bmatrix}. \tag{8.165}$$

Definition 8.10 (*canonical system*)
The canonical system of system (8.132) is defined by

$$\begin{bmatrix} \dot{\xi}_{11} \\ \vdots \\ \dot{\xi}_{1(\nu_1-1)} \\ \dot{\xi}_{1\nu_1} \\ \vdots \\ \dot{\xi}_{p1} \\ \vdots \\ \dot{\xi}_{p(\nu_p-1)} \\ \dot{\xi}_{p\nu_p} \end{bmatrix} = \begin{bmatrix} \xi_{12} + \alpha_{11}^u(\xi) \\ \vdots \\ \xi_{1\nu_1} + \alpha_{1(\nu_1-1)}^u(\xi) \\ \alpha_{1\nu_1}^u(\xi) \\ \vdots \\ \xi_{p2} + \alpha_{p1}^u(\xi) \\ \vdots \\ \xi_{p\nu_p} + \alpha_{p(\nu_p-1)}^u(\xi) \\ \alpha_{p\nu_p}^u(\xi) \end{bmatrix} \triangleq f_u(\xi) \, ; \quad y = \begin{bmatrix} \xi_{11} \\ \vdots \\ \xi_{p1} \end{bmatrix} \triangleq h(\xi) \tag{8.166}$$

where $\xi = T(x)$, $f_u(\xi) \triangleq T_*(F_u(x))$, $h(\xi) \triangleq H \circ T^{-1}(\xi)$,

$$\alpha_{i\nu_i}^u(\xi) \triangleq L_{F_u}L_{F_0}^{\nu_i-1}H_i(x)\Big|_{x=T^{-1}(\xi)}, \quad 1 \le i \le p$$

and for $1 \le i \le p$ and $1 \le k \le \nu_i - 1$,

$$\alpha_{ik}^u(\xi) \triangleq L_{F_u}L_{F_0}^{k-1}H_i(x)\Big|_{x=T^{-1}(\xi)} - L_{F_0}^k H_i(x)\Big|_{x=T^{-1}(\xi)}.$$

It is clear that $\alpha_{ik}^0(\xi) = 0$, $1 \le i \le p$, $1 \le k \le \nu_i - 1$ and

$$f_0(\xi) \triangleq T_*(F_0(x)) = \begin{bmatrix} \xi_{12} \\ \vdots \\ \xi_{1\nu_1} \\ \alpha^0_{1\nu_1}(\xi) \\ \vdots \\ \xi_{p2} \\ \vdots \\ \xi_{p\nu_p} \\ \alpha^0_{p\nu_p}(\xi) \end{bmatrix}. \tag{8.167}$$

System (8.132) is state equivalent to a NOCF with OT (or without OT) via $z = S(x)$, if and only if canonical system (8.166) is state equivalent to a NOCF with OT (or without OT) via $z = \tilde{S}(\xi)$ $(\triangleq S \circ T^{-1}(\xi))$. Canonical system (8.166) is more convenient to solve the observer problems than system (8.132). Since geometric conditions are coordinate free, any geometric condition in $\xi-$ coordinates (for system (8.166)) can be expressed in $x-$ coordinates (for system (8.132)).

For system (8.132), we define vector fields $\{\mathbf{g}^0_{ik}(x), \; 1 \leq i \leq p, \; 1 \leq k \leq \nu_i\}$ and $\{\mathbf{g}^u_{ik}(x), \; 1 \leq i \leq p, \; 1 \leq k \leq \nu_i\}$ as follows.

$$L_{\mathbf{g}^0_{i1}} L^{k-1}_{F_0} H_j(x) = \delta_{i,j} \, \delta_{k,\nu_i}, \; 1 \leq i \leq p, \; 1 \leq j \leq p, \; 1 \leq k \leq \nu_j$$
$$\left(\text{or} \; \mathbf{g}^0_{i1}(x) \triangleq T^{-1}_* \left(\frac{\partial}{\partial \xi_{i\nu_i}} \right), \; 1 \leq i \leq p \right) \tag{8.168}$$

and for $1 \leq i \leq p$ and $1 \leq k \leq \nu_i$,

$$\mathbf{g}^u_{i1}(x) \triangleq \mathbf{g}^0_{i1}(x); \; \mathbf{g}^u_{ik}(x) \triangleq \mathrm{ad}^{k-1}_{F_u} \mathbf{g}^u_{i1}(x)$$
$$\mathbf{g}^0_{ik}(x) \triangleq \mathrm{ad}^{k-1}_{F_0} \mathbf{g}^0_{i1}(x) = \mathbf{g}^u_{ik}(x) \big|_{u=0} . \tag{8.169}$$

Then it is easy to see, by Example 2.4.16 and (8.168), that for $1 \leq i \leq p, 1 \leq j \leq p$, $1 \leq k \leq \nu_i$, and $0 \leq \ell \leq \nu_j - 1$,

$$L_{\mathbf{g}^0_{ik}} L^{\ell}_{F_0} H_j(x) = \begin{cases} 0, & k + \ell < \nu_j \\ (-1)^{k+1} \delta_{i,j}, & k + \ell = \nu_j. \end{cases} \tag{8.170}$$

For system (8.166), we define vector fields $\{\tau^u_{ik}(\xi), \; 1 \leq i \leq p, \; 1 \leq k \leq \nu_i\}$ by

$$\tau^u_{ik}(\xi) \triangleq T_* \left(\mathbf{g}^u_{ik}(x) \right), \; 1 \leq i \leq p, \; 1 \leq k \leq \nu_i. \tag{8.171}$$

Then it is easy to see, by (2.37), (8.168), (8.169), and, (8.171), that for $1 \leq i \leq p$ and $1 \leq k \leq \nu_i$,

$$\tau^u_{i1}(\xi) = \tau^0_{i1}(\xi) = \frac{\partial}{\partial \xi_{i\nu_i}} \tag{8.172}$$

and

$$\tau_{ik}^u(\xi) = T_* \left(\mathrm{ad}_{F_u}^{k-1} \mathbf{g}_{i1}^u(x) \right) = \mathrm{ad}_{f_u}^{k-1} \tau_{i1}^u(\xi). \tag{8.173}$$

That is, $\tau_{ik}^u(\xi)$ is the ξ-coordinates expression of vector field $\mathbf{g}_{ik}^u(x)$. It is also easy to see, by (2.30), (8.170), and (8.171), that for $1 \le i \le p$, $1 \le j \le p$, $1 \le k \le \nu_i$, and $0 \le \ell \le \nu_j - 1$,

$$
\begin{aligned}
L_{\tau_{ik}^0} L_{f_0}^\ell h_j(\xi) &= L_{T_*(\mathbf{g}_{ik}^0)} L_{F_0}^\ell \left(H_j \circ T^{-1}(\xi) \right) = L_{\mathbf{g}_{ik}^0} L_{F_0}^\ell H_j(x) \Big|_{x = T^{-1}(\xi)} \\
&= \begin{cases} 0, & k + \ell < \nu_j \\ (-1)^{k+1} \delta_{i,j}, & k + \ell = \nu_j. \end{cases}
\end{aligned}
\tag{8.174}
$$

Lemma 8.5 *System (8.132) is state equivalent to a dual Brunovsky NOCF with OT $\bar{y} = \varphi(y)$ and state transformation $z = S(x)$, if and only if there exist a diffeomorphism $\varphi(y) : \mathbb{R}^p \to \mathbb{R}^p$ and smooth functions $\gamma_{ij}^u(y)$, $1 \le i \le p$, $1 \le j \le \nu_i$ such that for $1 \le i \le p$ and $1 \le k \le \nu_i$,*

$$z_{ik} = S_{ik}(x) = L_{F_0}^{k-1} \left(\varphi_i \circ H(x) \right) - \sum_{j=1}^{k-1} L_{F_0}^{k-1-j} \left(\gamma_{ij}^0 \circ H(x) \right) \tag{8.175}$$

$$L_{F_u} L_{F_0}^{\nu_i - 1} \left(\varphi_i \circ H(x) \right) = \sum_{j=1}^{\nu_i - 1} L_{F_u} L_{F_0}^{\nu_i - 1 - j} \left(\gamma_{ij}^0 \circ H(x) \right) + \gamma_{i\nu_i}^u \circ H(x) \tag{8.176}$$

and

$$L_{F_u} S_{ik}(x) - L_{F_0} S_{ik}(x) = \varepsilon_{ik}^u \circ H(x) \tag{8.177}$$

where

$$z \triangleq [z_{11} \ \cdots \ z_{1\nu_1} \ \cdots \ z_{p1} \ \cdots \ z_{p\nu_p}]^\mathsf{T}$$

and for $1 \le i \le p$ and $1 \le k \le \nu_i$,

$$\gamma_{ik}^u(y) \triangleq \gamma_{ik}^0(y) + \varepsilon_{ik}^u(y). \tag{8.178}$$

Proof Necessity. Suppose that system (8.132) is state equivalent to a dual Brunovsky NOCF with OT $\bar{y} = \varphi(y)$ and state transformation $z = S(x)$. Then, it is clear, by (8.139) and (8.140), that

$$\bar{h}(z) \triangleq \varphi \circ H \circ S^{-1}(z) = C_o z = [z_{11} \ \cdots \ z_{p1}]^\mathsf{T} \triangleq \tilde{\mathbf{z}}_1 \tag{8.179}$$

and

$$f_u(z) \triangleq S_*(F_u(x)) = A_o z + \bar{\gamma}^u(\tilde{z}_1) = \begin{bmatrix} z_{12} + \bar{\gamma}_{11}^u(\tilde{z}_1) \\ \vdots \\ z_{1v_1} + \bar{\gamma}_{1(v_1-1)}^u(\tilde{z}_1) \\ \bar{\gamma}_{1v_1}^u(\tilde{z}_1) \\ \vdots \\ z_{p2} + \bar{\gamma}_{p1}^u(\tilde{z}_1) \\ \vdots \\ z_{pv_p} + \bar{\gamma}_{p(v_p-1)}^u(\tilde{z}_1) \\ \bar{\gamma}_{pv_p}^u(\tilde{z}_1) \end{bmatrix} \tag{8.180}$$

which imply that for $1 \le i \le p$ and $1 \le k \le v_i - 1$,

$$S_{i(k+1)}(x) = L_{F_u} S_{ik}(x) - \bar{\gamma}_{ik}^u(\varphi \circ H(x)) = L_{F_u} S_{ik}(x) - \gamma_{ik}^u \circ H(x)$$
$$= L_{F_0} S_{ik}(x) - \gamma_{ik}^0 \circ H(x) \tag{8.181}$$

and

$$L_{F_u} S_{iv_i}(x) = \bar{\gamma}_{iv_i}^u(\varphi \circ H(x)) = \gamma_{iv_i}^u \circ H(x) \tag{8.182}$$

where $\bar{\gamma}_{ik}^u \circ \varphi(y) \triangleq \gamma_{ik}^u(y)$ for $1 \le i \le p$ and $1 \le k \le v_i$. Thus, it is clear, by (8.179), that (8.175) is satisfied when $1 \le i \le p$ and $k = 1$. Assume that (8.175) is satisfied when $1 \le i \le p$ and $1 \le k \le \ell \le v_i - 1$. Then we have, by (8.181), that for $1 \le i \le p$,

$$S_{i(\ell+1)}(x) = L_{F_0} S_{i\ell}(x) - \gamma_{i\ell}^0 \circ H(x)$$
$$= L_{F_0}^\ell (\varphi_i \circ H(x)) - \sum_{j=1}^{\ell-1} L_{F_0}^{\ell-j} (\gamma_{ij}^0 \circ H(x)) - \gamma_{i\ell}^0 \circ H(x)$$
$$= L_{F_0}^\ell (\varphi_i \circ H(x)) - \sum_{j=1}^{\ell} L_{F_0}^{\ell-j} (\gamma_{ij}^0 \circ H(x))$$

which implies that (8.175) is satisfied when $1 \le i \le p$ and $k = \ell + 1 \le v_i$. Therefore, by mathematical induction, (8.175) is satisfied for $1 \le i \le p$ and $1 \le k \le v_i$. Since

$$S_{iv_i}(x) = L_{F_0}^{v_i-1} (\varphi_i \circ H(x)) - \sum_{j=1}^{v_i-1} L_{F_0}^{v_i-1-j} (\gamma_{ij}^0 \circ H(x))$$

it is clear, by (8.182), that for $1 \le i \le p$,

$$L_{F_u} L_{F_0}^{v_i-1} (\varphi_i \circ H(x)) - \sum_{j=1}^{v_i-1} L_{F_u} L_{F_0}^{v_i-1-j} (\gamma_{ij}^0 \circ H(x)) = \gamma_{iv_i}^u \circ H(x)$$

which implies that (8.176) is satisfied. Finally, it is easy to see, by (8.181) and (8.182), that for $1 \le i \le p$ and $1 \le k \le v_i$,

$$L_{F_u} S_{ik}(x) - L_{F_0} S_{ik}(x) = \gamma_{ik}^u \circ H(x) - \gamma_{ik}^0 \circ H(x) \triangleq \varepsilon_{ik}^u \circ H(x)$$

which implies that (8.177) is satisfied.

Sufficiency. Suppose that there exist a diffeomorphism $\bar{y} = \varphi(y)$ and smooth functions $\gamma_{ij}^u(y)$, $1 \le i \le p$, $1 \le j \le v_i$ such that (8.175)–(8.178) are satisfied. Let $z = S(x)$. Since $S_1(x) = \varphi \circ H(x)$, it is clear that for $1 \le i \le p$,

$$\bar{h}_i(z) \triangleq \varphi_i \circ H \circ S^{-1}(z) = z_{i1} \tag{8.183}$$

and (8.139) is satisfied. Also, it is easy to see, by (8.175)–(8.178), that for $1 \le i \le p$ and $1 \le k \le v_i - 1$,

$$\begin{aligned}
L_{F_u} S_{ik}(x) &= L_{F_0} S_{ik}(x) + \varepsilon_{ik}^u \circ H(x) \\
&= L_{F_0}^k (\varphi_i \circ H(x)) - \sum_{j=1}^{k-1} L_{F_0}^{k-j} (\gamma_{ij}^0 \circ H(x)) + \varepsilon_{ik}^u \circ H(x) \\
&= S_{i(k+1)}(x) + \gamma_{ik}^0 \circ H(x) + \varepsilon_{ik}^u \circ H(x) \\
&= S_{i(k+1)}(x) + \gamma_{ik}^u \circ H(x)
\end{aligned}$$

and

$$\begin{aligned}
L_{F_u} S_{iv_i}(x) &= L_{F_u} L_{F_0}^{v_i-1} (\varphi_i \circ H(x)) - \sum_{j=1}^{v_i-1} L_{F_u} L_{F_0}^{n-1-j} (\gamma_{ij}^0 \circ H(x)) \\
&= \gamma_{iv_i}^u \circ H(x)
\end{aligned}$$

which imply, together with (8.183), that

$$\bar{f}_u(z) \triangleq S_*(F_u(x)) = \begin{bmatrix} L_{F_u}S_{11}(x)\big|_{x=S^{-1}(z)} \\ \vdots \\ L_{F_u}S_{1(\nu_1-1)}(x)\big|_{x=S^{-1}(z)} \\ L_{F_u}S_{1\nu_1}(x)\big|_{x=S^{-1}(z)} \\ \vdots \\ L_{F_u}S_{p1}(x)\big|_{x=S^{-1}(z)} \\ \vdots \\ L_{F_u}S_{p(\nu_p-1)}(x)\big|_{x=S^{-1}(z)} \\ L_{F_u}S_{p\nu_p}(x)\big|_{x=S^{-1}(z)} \end{bmatrix} = \begin{bmatrix} z_{12} + \bar{\gamma}_{11}(\tilde{z}_1) \\ \vdots \\ z_{1\nu_1} + \bar{\gamma}_{1(\nu_1-1)}(\tilde{z}_1) \\ \bar{\gamma}_{1\nu_1}(\tilde{z}_1) \\ \vdots \\ z_{p2} + \bar{\gamma}_{p1}(\tilde{z}_1) \\ \vdots \\ z_{p\nu_p} + \bar{\gamma}_{p(\nu_p-1)}(\tilde{z}_1) \\ \bar{\gamma}_{p\nu_p}(\tilde{z}_1) \end{bmatrix}$$

$$= A_o z + \bar{\gamma}^u(\tilde{z}_1)$$

where $\bar{\gamma}_{ik}^u \circ \varphi(y) \triangleq \gamma_{ik}^u(y)$ for $1 \le i \le p$ and $1 \le k \le \nu_i$. Therefore, (8.140) is satisfied. In other words, system (8.132) is state equivalent to a dual Brunovsky NOCF with OT $\bar{y} = \varphi(y)$ and state transformation $z = S(x)$.   □

Corollary 8.3 *System (8.132) is state equivalent to a dual Brunovsky NOCF with state transformation $z = S(x)$, if and only if there exist smooth functions $\gamma_{ij}^u(y)$, $1 \le i \le p$, $1 \le j \le \nu_i$ such that for $1 \le i \le p$ and $1 \le k \le \nu_i$,*

$$z_{ik} = S_{ik}(x) = L_{F_0}^{k-1}H_i(x) - \sum_{j=1}^{k-1} L_{F_0}^{k-1-j}\left(\gamma_{ij}^0 \circ H(x)\right) \tag{8.184}$$

$$L_{F_u}L_{F_0}^{\nu_i-1}H_i(x) = \sum_{j=1}^{\nu_i-1} L_{F_u}L_{F_0}^{\nu_i-1-j}\left(\gamma_{ij}^0 \circ H(x)\right) + \gamma_{i\nu_i}^u \circ H(x) \tag{8.185}$$

and

$$L_{F_u}S_{ik}(x) - L_{F_0}S_{ik}(x) = \varepsilon_{ik}^u \circ H(x) \tag{8.186}$$

where

$$z \triangleq [z_{11} \cdots z_{1\nu_1} \cdots z_{p1} \cdots z_{p\nu_p}]^\mathsf{T}$$

and for $1 \le i \le p$ and $1 \le k \le \nu_i$,

$$\gamma_{ik}^u(y) \triangleq \gamma_{ik}^0(y) + \varepsilon_{ik}^u(y). \tag{8.187}$$

Corollary 8.4 *System (8.133) is state equivalent to a dual Brunovsky NOCF with OT $\bar{y} = \varphi(y)$ and state transformation $z = S(x)$, if and only if there exist a diffeomorphism $\varphi(y) : \mathbb{R}^p \to \mathbb{R}^p$ and smooth functions $\gamma_{ij}^0(y)$, $1 \le i \le p$, $1 \le j \le \nu_i$ such that for $1 \le i \le p$ and $1 \le k \le \nu_i$,*

$$z_{ik} = S_{ik}(x) = L_{F_0}^{k-1} (\varphi_i \circ H(x)) - \sum_{j=1}^{k-1} L_{F_0}^{k-1-j} (\gamma_{ij}^0 \circ H(x)) \qquad (8.188)$$

and

$$L_{F_0}^{\nu_i} (\varphi_i \circ H(x)) = \sum_{j=1}^{\nu_i} L_{F_0}^{\nu_i - j} (\gamma_{ij}^0 \circ H(x)) \qquad (8.189)$$

where

$$z \triangleq [z_{11} \cdots z_{1\nu_1} \cdots z_{p1} \cdots z_{p\nu_p}]^{\mathsf{T}}.$$

Lemma 8.6 *Suppose that system (8.132) is state equivalent to a dual Brunovsky NOCF. Let $c(y)$ be a smooth function of $y(\in \mathbb{R}^p)$. Then*

(i) for $1 \le i \le p$ and $1 \le k \le \nu_i - 1$,

$$L_{\tau_{i1}^0} L_{f_0}^{k-1} c(h(\xi)) = 0 \qquad (8.190)$$

(ii) for $1 \le i \le p$,

$$dL_{f_0}^{\nu_i} h_i(\xi) \in \mathrm{span}\{dL_{f_0}^{k-1} h_j(\xi),\ 1 \le j \le p,\ 1 \le k \le \nu_i\} \qquad (8.191)$$

(iii) for $1 \le i \le p$ and $1 \le k \le \nu_i$,

$$\tau_{ik}^0 \equiv (-1)^{k-1} \frac{\partial}{\partial \xi_{i(\nu_i+1-k)}}$$
$$\mathrm{mod\ span}\{\frac{\partial}{\partial \xi_{j\ell}},\ 1 \le j \le p,\ \nu_i + 2 - k \le \ell \le \nu_j\} \qquad (8.192)$$

(iv)

$$L_{\tau_{ik}^0} L_{f_0}^{\ell} c(h(\xi)) = \begin{cases} 0, & k + \ell \le \nu_i - 1 \\ (-1)^{k-1} \frac{\partial c(h(\xi))}{\partial \xi_{i1}}, & k + \ell = \nu_i \end{cases} \qquad (8.193)$$

where $f_0(\xi) \triangleq T_*(F_0(x))$ in (8.167), $h(\xi) \triangleq H \circ T^{-1}(\xi)$, and

$$\xi \triangleq [\xi_{11} \cdots \xi_{1\nu_1} \cdots \xi_{p1} \cdots \xi_{p\nu_p}]^{\mathsf{T}}.$$

Proof Suppose that system (8.132) is state equivalent to a dual Brunovsky NOCF without OT. Thus, we have, by (2.30) and (8.185), that for $1 \le i \le p$ and $1 \le k \le \nu_i$,

$$\alpha_{i\nu_i}^0(\xi) \triangleq L_{F_0}^{\nu_i} H_i(x)\Big|_{x=T^{-1}(\xi)} = \sum_{j=1}^{\nu_i} L_{F_0}^{\nu_i-j}\left(\gamma_{ij}^0 \circ H(x)\right)\Big|_{x=T^{-1}(\xi)}$$

$$= \sum_{k=1}^{\nu_i} L_{f_0}^{\nu_i-k} \gamma_{ik}(h(\xi)). \tag{8.194}$$

Let for $1 \le k \le \nu_1$,

$$\tilde{\xi}_k \triangleq \{\xi_{jk} \mid 1 \le j \le p \text{ and } \nu_j \ge k\}$$

and for $1 \le i \le p$ and $1 \le j \le \nu_i$,

$$f_{ij}^0(\xi) \triangleq \begin{cases} \xi_{i(j+1)}, & 1 \le j \le \nu_i - 1 \\ \alpha_{i\nu_i}^0(\xi), & j = \nu_i. \end{cases}$$

Then it is easy to see, by (8.167) and (8.194), that for $1 \le i \le p$ and $1 \le k \le \nu_i$,

$$L_{f_0}^{k-1} c(h(\xi)) = \phi_k(\tilde{\xi}_1, \cdots, \tilde{\xi}_k) \tag{8.195}$$

and

$$\alpha_{i\nu_i}^0(\xi) = \sum_{k=1}^{\nu_i} L_{f_0}^{\nu_i-k} \gamma_{ik}(h(\xi)) = \bar{\alpha}_{i\nu_i}^0(\tilde{\xi}_1, \cdots, \tilde{\xi}_{\nu_i}) \tag{8.196}$$

where $\phi_1(\tilde{\xi}_1) = c(\tilde{\xi}_1)$ and for $2 \le k \le \nu_1$,

$$\phi_k(\tilde{\xi}_1, \cdots, \tilde{\xi}_k) = \sum_{\ell=1}^{p} \sum_{j=1}^{\min(k-1,\nu_\ell)} \frac{\partial \phi_{k-1}(\tilde{\xi}_1, \cdots, \tilde{\xi}_{k-1})}{\partial \xi_{\ell j}} f_{\ell j}^0(\xi).$$

Thus, it is clear, by (8.172) and (8.195), that for $1 \le i \le p$ and $1 \le k \le \nu_i - 1$,

$$L_{\tau_{i1}^0} L_{f_0}^{k-1} c(h(\xi)) = \frac{\partial}{\partial \xi_{i\nu_i}}\left(L_{f_0}^{k-1} c(h(\xi))\right) = \frac{\partial \phi_k(\tilde{\xi}_1, \cdots, \tilde{\xi}_k)}{\partial \xi_{i\nu_i}} = 0$$

which implies that condition (i) of Lemma 8.6 is satisfied. We have, by (8.165) and (8.196), that

$$dL_{f_0}^{\nu_i} h_i(\xi) = d\alpha_{i\nu_i}^0(\xi) \in \text{span}\{d\xi_{jk},\ 1 \le j \le p,\ 1 \le k \le \min(\nu_j, \nu_i)\}$$
$$\subset \text{span}\{dL_{f_0}^{k-1} h_j(\xi),\ 1 \le j \le p,\ 1 \le k \le \nu_i\}$$

which implies that condition (ii) of Lemma 8.6 is satisfied. It is also easy to see, by (8.165) and (8.174), that for $1 \le i \le p, 1 \le j \le p, 1 \le k \le \nu_i$, and $0 \le \ell \le \nu_j - 1$,

$$L_{\tau_{ik}^0} \xi_{j(\ell+1)} = L_{\tau_{ik}^0} L_{f_0}^{\ell} h_j(\xi) = \begin{cases} 0, & 0 \leq \ell < \nu_i - k \\ (-1)^{k-1} \delta_{i,j}, & \ell = \nu_i - k \end{cases}$$

which implies that condition (iii) of Lemma 8.6 is satisfied. Finally, we have, by (8.192) and (8.195), that for $1 \leq i \leq p$, $1 \leq k \leq \nu_i$, and $\ell \leq \nu_i - 1 - k$,

$$L_{\tau_{ik}^0} L_{f_0}^{\ell} c(h(\xi)) = 0. \qquad (8.197)$$

Therefore, it is easy to see, by Example 2.4.16, (8.192), and (8.197), that for $1 \leq i \leq p$ and $1 \leq k \leq \nu_i$,

$$\begin{aligned} L_{\tau_{ik}^0} L_{f_0}^{\nu_i - k} c(h(\xi)) &= (-1)^{\nu_i - k} L_{\mathrm{ad}_{f_0}^{\nu_i - k} \tau_{ik}^0(\xi)} c(h(\xi)) \\ &= (-1)^{\nu_i - k} L_{\tau_{i\nu_i}^0} c(h(\xi)) \\ &= (-1)^{k-1} \frac{\partial c(h(\xi))}{\partial \xi_{i1}} \end{aligned}$$

which implies, together with (8.197), that condition (iv) of Lemma 8.6 is satisfied. $\square$

Theorem 8.8 *System (8.132) is state equivalent to a dual Brunovsky NOCF, if and only if*

(i) for $1 \leq i \leq p$,

$$dL_{F_0}^{\nu_i} H_i(x) \in \mathrm{span}\{dL_{F_0}^{k-1} H_j(x), \ 1 \leq j \leq p, \ 1 \leq k \leq \nu_i\} \qquad (8.198)$$

(ii) there exist smooth vector fields $\bar{g}_{11}^u(x), \cdots, \bar{g}_{p1}^u(x)$ such that for $1 \leq i \leq p$, $1 \leq j \leq p$, $1 \leq k \leq \nu_i$, and $1 \leq \ell \leq \nu_j$,

$$L_{\bar{g}_{i1}^0} L_{F_0}^{k-1} H_j(x) = \delta_{i,j} \, \delta_{k,\nu_i} \qquad (8.199)$$

$$\bar{g}_{ik}^u(x) = \bar{g}_{ik}^u(x)\big|_{u=0} \triangleq \bar{g}_{ik}^0(x) \qquad (8.200)$$

and

$$[\bar{g}_{ik}^0(x), \bar{g}_{j\ell}^0(x)] = 0 \qquad (8.201)$$

where $\delta_{i,j}$ is the Kronecker delta function and for $1 \leq i \leq p$ and $1 \leq k \leq \nu_i$,

$$\bar{g}_{ik}^u(x) \triangleq \mathrm{ad}_{F_u}^{k-1} \bar{g}_{i1}^0(x). \qquad (8.202)$$

Furthermore, a state coordinates transformation $z = S(x)$ is given by

$$\frac{\partial S(x)}{\partial x} = \left[(-1)^{\nu_1-1}\bar{\mathbf{g}}^0_{1\nu_1}(x) \cdots - \bar{\mathbf{g}}^0_{12}(x)\,\bar{\mathbf{g}}^0_{11}(x)\right.$$
$$\left. \cdots (-1)^{\nu_p-1}\bar{\mathbf{g}}^0_{p\nu_p}(x) \cdots - \bar{\mathbf{g}}^0_{p2}(x)\,\bar{\mathbf{g}}^0_{p1}(x)\right]^{-1}. \tag{8.203}$$

Proof Necessity. Suppose that system (8.132) is state equivalent to a dual Brunovsky NOCF with state transformation $z = S(x)$. Then, we have, by (8.139) and (8.140), that

$$\bar{h}(z) \triangleq H \circ S^{-1}(z) = C_o z = [z_{11} \cdots z_{p1}]^{\mathsf{T}} \triangleq \tilde{\mathbf{z}}_1$$

and

$$\bar{f}_u(z) \triangleq S_*(F_u(x)) = A_o z + \gamma^u(C_o z) = A_o z + \gamma^u(\tilde{\mathbf{z}}_1) \tag{8.204}$$

where

$$z \triangleq \begin{bmatrix} z_{11} \\ z_{12} \\ \vdots \\ z_{1\nu_1} \\ \vdots \\ z_{p1} \\ z_{p2} \\ \vdots \\ z_{p\nu_p} \end{bmatrix}; \quad \bar{f}^u(z) \triangleq \begin{bmatrix} \bar{f}^u_{11}(z) \\ \vdots \\ \bar{f}^u_{1(\nu_1-1)}(z) \\ \bar{f}^u_{1\nu_1}(z) \\ \vdots \\ \bar{f}^u_{p1}(z) \\ \vdots \\ \bar{f}^u_{p(\nu_p-1)}(z) \\ \bar{f}^u_{p\nu_p}(z) \end{bmatrix} = \begin{bmatrix} z_{12} + \gamma^u_{11}(\tilde{\mathbf{z}}_1) \\ \vdots \\ z_{1\nu_1} + \gamma^u_{1(\nu_1-1)}(\tilde{\mathbf{z}}_1) \\ \gamma^u_{1\nu_1}(\tilde{\mathbf{z}}_1) \\ \vdots \\ z_{p2} + \gamma^u_{p1}(\tilde{\mathbf{z}}_1) \\ \vdots \\ z_{p\nu_p} + \gamma^u_{p(\nu_p-1)}(\tilde{\mathbf{z}}_1) \\ \gamma^u_{p\nu_p}(\tilde{\mathbf{z}}_1) \end{bmatrix}.$$

Since

$$L^k_{F_0} H_i(x) = L^k_{f_0} h_i(\xi)\big|_{\xi=T(x)} \quad \text{or} \quad dL^k_{F_0} H_i(x) = dL^k_{f_0} h_i(\xi)\big|_{\xi=T(x)} \frac{\partial T(x)}{\partial x}$$

for $1 \le j \le p$ and $k \ge 0$, it is easy to see, by (8.191), that condition (i) of Theorem 8.8 is satisfied. Define vector fields $\{\bar{\psi}^u_{ik}(z) \mid 1 \le i \le p, \ 1 \le k \le \nu_i\}$ by

$$\bar{\psi}^0_{i1}(z) \triangleq \frac{\partial}{\partial z_{i\nu_i}}; \quad \bar{\psi}^u_{ik}(z) \triangleq \mathrm{ad}^{k-1}_{\bar{f}_u}\bar{\psi}^0_{i1}(z). \tag{8.205}$$

Then, by (8.204) and (8.205), it is clear that for $1 \le i \le p$ and $1 \le k \le \nu_i$,

$$\bar{\psi}^u_{ik}(z) = (-1)^{k-1}\frac{\partial}{\partial z_{i(\nu_i+1-k)}} = \bar{\psi}^0_{ik}(z) \tag{8.206}$$

which implies that for $1 \leq i \leq p$, $1 \leq j \leq p$, and $1 \leq k \leq v_i$, and $1 \leq \ell \leq v_j$,

$$[\bar{\psi}_{ik}^0(z), \bar{\psi}_{j\ell}^0(z)] = 0. \tag{8.207}$$

Note that $\bar{f}_u(z) = S_*(F_u(x))$ or $F_u(x) = S_*^{-1}(\bar{f}_u(z))$. It is clear, by (8.205) and (8.206), that for $1 \leq i \leq p$, $1 \leq j \leq p$, and $1 \leq k \leq v_i$,

$$L_{\mathrm{ad}_{\bar{f}_0}^{k-1}\bar{\psi}_{i1}^0(z)}\bar{h}_j(z) = L_{\bar{\psi}_{ik}^0}\bar{h}_j(z) = (-1)^{k-1}\frac{\partial z_{j1}}{\partial z_{i(v_i+1-k)}}$$

$$= \begin{cases} 0, & k \leq v_i - 1 \\ (-1)^{k-1}\delta_{i,j}, & k = v_i \end{cases}$$

which implies, together with Example 2.4.16, that for $1 \leq i \leq p$, $1 \leq j \leq p$, and $1 \leq k \leq v_i$,

$$L_{\bar{\psi}_{i1}^0}L_{\bar{f}_0}^{k-1}\bar{h}_j(z) = \delta_{i,j}\,\delta_{k,v_i}. \tag{8.208}$$

Hence, if we let

$$\bar{g}_{i1}^u(x) = \bar{g}_{i1}^0(x) \triangleq S_*^{-1}(\bar{\psi}_{i1}^0(z)), \quad 1 \leq i \leq p$$

then we have, by (2.30) and (8.208), that for $1 \leq i \leq p$, $1 \leq j \leq p$, and $1 \leq k \leq v_i$,

$$L_{\bar{g}_{i1}^0}L_{F_0}^{k-1}H_j(x) = L_{\bar{\psi}_{i1}^0}L_{\bar{f}_u}^{k-1}\bar{h}_j(z)\Big|_{z=S(x)} = \delta_{i,j}\,\delta_{k,v_i}$$

which implies that (8.199) is satisfied. Also, it is clear, by (2.37), (8.202), and (8.205), that for $1 \leq i \leq p$ and $1 \leq k \leq v_i$,

$$\bar{g}_{ik}^u(x) = \mathrm{ad}_{F_u}^{k-1}\bar{g}_{i1}^0(x) = S_*^{-1}\left\{\mathrm{ad}_{S_*(F_u)}^{k-1}S_*(\bar{g}_{i1}^0(x))\right\}$$
$$= S_*^{-1}\left\{\mathrm{ad}_{\bar{f}_u}^{k-1}\bar{\psi}_{i1}^0(z)\right\} = S_*^{-1}(\bar{\psi}_{ik}^u(z)) \tag{8.209}$$

and thus (8.200) and (8.201) are satisfied by (8.206) and (8.207). Finally, it is easy to see, by (8.206) and (8.209), that

$$I = \left[(-1)^{v_1-1}S_*\left(\bar{g}_{1v_1}^0(x)\right) \cdots - S_*\left(\bar{g}_{12}^0(x)\right) S_*\left(\bar{g}_{11}^0(x)\right)\right.$$
$$\cdots (-1)^{v_p-1}S_*\left(\bar{g}_{pv_p}^0(x)\right) \cdots - S_*\left(\bar{g}_{p2}^0(x)\right) \left. S_*\left(\bar{g}_{p1}^0(x)\right)\right]$$
$$= \left(\frac{\partial S(x)}{\partial x}\left[(-1)^{v_1-1}\bar{g}_{1v_1}^0(x) \cdots - \bar{g}_{12}^0(x)\,\bar{g}_{11}^0(x)\right.\right.$$
$$\cdots (-1)^{v_p-1}\bar{g}_{pv_p}^0(x) \cdots - \bar{g}_{p2}^0(x)\,\bar{g}_{p1}^0(x)\bigg]\bigg)_{x=S^{-1}(z)}$$

or

$$I = \frac{\partial S(x)}{\partial x} \Big[(-1)^{\nu_1-1} \bar{\mathbf{g}}_{1\nu_1}^0(x) \cdots - \bar{\mathbf{g}}_{12}^0(x) \, \bar{\mathbf{g}}_{11}^0(x)$$

$$\cdots (-1)^{\nu_p-1} \bar{\mathbf{g}}_{p\nu_p}^0(x) \cdots - \bar{\mathbf{g}}_{p2}^0(x) \, \bar{\mathbf{g}}_{p1}^0(x) \Big]$$

which implies that (8.203) is satisfied.

Sufficiency. Suppose that condition (i) and condition (ii) of Theorem 8.8 are satisfied. Since $\{\bar{\mathbf{g}}_{ik}^0(x) \mid 1 \leq i \leq p, \; 1 \leq k \leq \nu_i\}$ is a set of commuting vector fields, there exists, by Theorem 2.7, a state transformation $z = S(x)$ such that for $1 \leq i \leq p$ and $1 \leq k \leq \nu_i$,

$$S_* \left(\bar{\mathbf{g}}_{ik}^0(x) \right) = (-1)^{k-1} \frac{\partial}{\partial z_{i(\nu_i+1-k)}} \tag{8.210}$$

where

$$z \triangleq \begin{bmatrix} z_{11} & \cdots & z_{1\nu_1} & \cdots & z_{p1} & \cdots & z_{p\nu_p} \end{bmatrix}^{\mathsf{T}}.$$

In fact, $z = S(x)$ can be calculated by (8.203). Now it will be shown that for $1 \leq j \leq p$

$$\bar{h}_j(z) \triangleq \varphi \circ H \circ S^{-1}(z) = z_{j1} \tag{8.211}$$

and

$$\bar{f}_u(z) \triangleq S_*(F_u(x)) = A_o z + \bar{\gamma}^u(z_{11}, \cdots, z_{p1}). \tag{8.212}$$

It is easy to see, by (2.30), (2.45), (8.199), (8.202), and (8.210), that for $1 \leq i \leq p$, $1 \leq j \leq p$, and $1 \leq k \leq \nu_i$,

$$L_{\bar{\mathbf{g}}_{ik}^0} H_j(x) = L_{\mathrm{ad}_{F_0}^{k-1} \bar{\mathbf{g}}_{i1}^0} H_j(x)$$

$$= \sum_{\ell=0}^{k-1} (-1)^k \binom{k-1}{\ell} L_{F_0}^{k-1-\ell} L_{\bar{\mathbf{g}}_{i1}^0} L_{F_0}^{\ell} H_j(x)$$

$$= L_{F_0}^{k-1} H_j(x) = \delta_{i,j} \, \delta_{k,\nu_i}$$

and

$$\frac{\partial \bar{h}_j(z)}{\partial z_{i(\nu_i+1-k)}} = \frac{\partial \left(H_j \circ S^{-1}(z) \right)}{\partial z_{i(\nu_i+1-k)}} = (-1)^{k-1} L_{S_*(\bar{\mathbf{g}}_{ik}^0)} \left(H_j \circ S^{-1}(z) \right)$$

$$= (-1)^{k-1} \left\{ L_{\bar{\mathbf{g}}_{ik}^0(x)} H_j(x) \right\} \Big|_{x=S^{-1}(z)} = \delta_{i,j} \, \delta_{k,\nu_i}$$

which implies that $\bar{h}_j(z) = z_{j1}$ for $1 \leq j \leq p$ and thus (8.211) is satisfied. Let

$$\bar{f}_u(z) \triangleq \sum_{j=1}^{p} \sum_{\ell=1}^{\nu_j} \bar{f}_{j\ell}^u(z) \frac{\partial}{\partial z_{j\ell}}$$

$$= \left[\bar{f}_{11}^u(z) \cdots \bar{f}_{1\nu_1}^u(z) \cdots \bar{f}_{p1}^u(z) \cdots \bar{f}_{p\nu_p}^u(z) \right]^{\mathsf{T}}.$$

(8.213)

Since $\bar{f}_u(z) = S_*(F_u(x))$, it is clear that for $1 \le i \le p$ and $1 \le k \le \nu_i - 1$,

$$S_* \left(\bar{\mathbf{g}}_{i(k+1)}^u(x) \right) = S_* \left(\mathrm{ad}_{F_u} \bar{\mathbf{g}}_{ik}^u(x) \right) = \left[S_*(F_u(x)), S_*(\bar{\mathbf{g}}_{ik}^u(x)) \right]$$

$$= \left[\bar{f}_u(z), S_*(\bar{\mathbf{g}}_{ik}^u(x)) \right].$$

(8.214)

Thus, we have, by (8.210), (8.213), and (8.214), that for $1 \le i \le p$ and $1 \le k \le \nu_i - 1$,

$$(-1)^k \frac{\partial}{\partial z_{i(\nu_i - k)}} = \left[\bar{f}_u(z), (-1)^{k-1} \frac{\partial}{\partial z_{i(\nu_i + 1 - k)}} \right]$$

$$= (-1)^k \sum_{j=1}^{p} \sum_{\ell=1}^{\nu_j} \frac{\partial \bar{f}_{j\ell}^u(z)}{\partial z_{i(\nu_i + 1 - k)}} \frac{\partial}{\partial z_{j\ell}}$$

which implies that for $1 \le i \le p$, $1 \le k \le \nu_i - 1$, $1 \le j \le p$, and $1 \le \ell \le \nu_j$,

$$\frac{\partial \bar{f}_{j\ell}^u(z)}{\partial z_{i(\nu_i + 1 - k)}} = \begin{cases} 1, & j = i, \ \ell = \nu_i - k \\ 0, & \text{otherwise} \end{cases}$$

or

$$\frac{\partial \bar{f}_{j\ell}^u(z)}{\partial z_{i(k+1)}} = \begin{cases} 1, & j = i, \ \ell = k \\ 0, & \text{otherwise}. \end{cases}$$

Therefore, it is clear that for $1 \le j \le p$ and $1 \le \ell \le \nu_j$,

$$\bar{f}_{j\ell}^u(z) = \begin{cases} z_{j(\ell+1)} + \bar{\gamma}_{j\ell}^u(z_{11}, \cdots, z_{p1}), & 1 \le \ell \le \nu_j - 1 \\ \bar{\gamma}_{j\nu_j}^u(z_{11}, \cdots, z_{p1}), & \ell = \nu_j \end{cases}$$

for some functions $\bar{\gamma}_{j\ell}^u(z_{11}, \cdots, z_{p1})$. In other words, (8.212) is satisfied. Hence, by (8.211) and (8.212), system (8.132) is state equivalent to a dual Brunovsky NOCF with state transformation $z = S(x)$. $\qquad\square$

Remark 8.6 Condition (i) of Theorem 8.8 is needed for the existence of the vector fields $\{\bar{\mathbf{g}}_{i1}^0(x), \ 1 \le i \le p\}$ which satisfy (8.199). For example, let $p = 2$ and $(\nu_1, \nu_2) = (2, 1)$. Then, by (8.199), 3×1 vector fields $\bar{\mathbf{g}}_{11}^0(x)$ and $\bar{\mathbf{g}}_{21}^0(x)$ satisfy the following equations:

$$\begin{bmatrix} dH_1(x) \\ dL_{F_0}H_1(x) \\ dH_2(x) \\ dL_{F_0}H_2(x) \end{bmatrix} \bar{\mathbf{g}}_{11}^0(x) = \begin{bmatrix} 0 \\ 1 \\ 0 \\ 0 \end{bmatrix} \qquad (8.215)$$

and

$$\begin{bmatrix} dH_1(x) \\ dH_2(x) \end{bmatrix} \bar{\mathbf{g}}_{21}^0(x) = \begin{bmatrix} 0 \\ 1 \end{bmatrix}. \qquad (8.216)$$

If $dL_{F_0}H_2(x) \notin \text{span}\{dH_1(x), dH_2(x)\}$ (or condition (i) of Theorem 8.8 is not satisfied), there does not exist $\bar{\mathbf{g}}_{11}^0(x)$ that satisfies equation (8.215).

A unique set of vector fields $\{\mathbf{g}_{i1}^0(x), 1 \le i \le p\}$ has been defined in (8.168) by using $1 \le k \le \nu_j$ instead of $1 \le k \le \nu_i$ in (8.199). For example, let $p = 2$ and $(\nu_1, \nu_2) = (2, 1)$. Then, we have, by (8.168), that

$$\left[\mathbf{g}_{11}^0(x) \ \mathbf{g}_{21}^0(x)\right] = \begin{bmatrix} dH_1(x) \\ dL_{F_0}H_1(x) \\ dH_2(x) \end{bmatrix}^{-1} \begin{bmatrix} 0 & 0 \\ 1 & 0 \\ 0 & 1 \end{bmatrix} \qquad (8.217)$$

whereas $\bar{\mathbf{g}}_{11}^0(x)$ and $\bar{\mathbf{g}}_{21}^0(x)$ satisfy

$$\begin{bmatrix} dH_1(x) \\ dL_{F_0}H_1(x) \\ dH_2(x) \end{bmatrix} \left[\bar{\mathbf{g}}_{11}^0(x) \ \bar{\mathbf{g}}_{21}^0(x)\right] = \begin{bmatrix} 0 & 0 \\ 1 & \tilde{\ell}(x) \\ 0 & 1 \end{bmatrix} = \begin{bmatrix} 0 & 0 \\ 1 & 0 \\ 0 & 1 \end{bmatrix} \begin{bmatrix} 1 & \tilde{\ell}(x) \\ 0 & 1 \end{bmatrix} \qquad (8.218)$$

if we let $L_{\bar{\mathbf{g}}_{21}^0} L_{F_0} H_1(x) = \tilde{\ell}(x)$ which is not defined in (8.216). Thus, we have, by (8.217) and (8.218), that

$$\begin{aligned} \left[\bar{\mathbf{g}}_{11}^0(x) \ \bar{\mathbf{g}}_{21}^0(x)\right] &= \left[\mathbf{g}_{11}^0(x) \ \mathbf{g}_{21}^0(x)\right] \begin{bmatrix} 1 & \tilde{\ell}(x) \\ 0 & 1 \end{bmatrix} \\ &= \left[\mathbf{g}_{11}^0(x) \ \mathbf{g}_{21}^0(x) + \tilde{\ell}(x)\mathbf{g}_{11}^0(x)\right]. \end{aligned}$$

Therefore, we need to find $\tilde{\ell}(x)$ such that (8.200) is satisfied. By this reason, the conditions in Theorem 8.8 are not verifiable necessary and sufficient conditions. In fact, we can restate condition (ii) of Theorem 8.8 as follows:

(ii)' there exist smooth functions $\tilde{\ell}_{i,j,k}(x)$, $2 \le i \le p$, $1 \le j \le i - 1$, $1 \le k \le \nu_j - \nu_i$ such that (8.200) and (8.201) are satisfied, where for $2 \le i \le p$,

$$\bar{\mathbf{g}}_{i1}^0(x) = \mathbf{g}_{i1}^0(x) + \sum_{j=1}^{i-1} \sum_{k=1}^{\nu_j - \nu_i} \tilde{\ell}_{i,j,k}(x)\bar{\mathbf{g}}_{jk}^0(x). \qquad (8.219)$$

Theorem 8.9 *System (8.132) is state equivalent to a dual Brunovsky NOCF, if and only if*

(i) *for* $1 \leq i \leq p$,

$$dL_{F_0}^{v_i} H_i(x) \in \text{span}\{dL_{F_0}^{k-1} H_j(x),\ 1 \leq j \leq p,\ 1 \leq k \leq v_i\} \qquad (8.220)$$

(ii) *there exist smooth functions* $\gamma_{jr}(y)$, $1 \leq j \leq p-1$, $1 \leq r \leq v_j - v_p$ *such that for* $1 \leq i \leq p$, $1 \leq j \leq p$, $1 \leq k \leq v_i$, *and* $1 \leq r \leq v_j$,

$$\bar{\mathbf{g}}_{ik}^u(x) = \bar{\mathbf{g}}_{ik}^u(x)\big|_{u=0} \triangleq \bar{\mathbf{g}}_{ik}^0(x) \qquad (8.221)$$

and

$$[\bar{\mathbf{g}}_{ik}^0(x), \bar{\mathbf{g}}_{jr}^0(x)] = 0 \qquad (8.222)$$

where for $1 \leq i \leq p$, $1 \leq j \leq p$, *and* $1 \leq k \leq v_i$,

$$\tilde{\ell}_{i,j,k}(x) = (-1)^{k-1} \sum_{r=1}^{v_j - v_i + 1 - k} L_{\mathbf{g}_{i1}^0} L_{F_0}^{v_j - k - r} \gamma_{jr}(H(x)) \qquad (8.223)$$

$$\bar{\mathbf{g}}_{i1}^0(x) = \mathbf{g}_{i1}^0(x) + \sum_{j=1}^{i-1} \sum_{k=1}^{v_j - v_i} \tilde{\ell}_{i,j,k}(x) \bar{\mathbf{g}}_{jk}^0(x) \qquad (8.224)$$

and

$$\bar{\mathbf{g}}_{ik}^u(x) \triangleq \text{ad}_{F_u}^{k-1} \bar{\mathbf{g}}_{i1}^0(x). \qquad (8.225)$$

Furthermore, a state coordinates transformation $z = S(x)$ *is given by*

$$\begin{aligned}
\frac{\partial S(x)}{\partial x} = \Big[& (-1)^{v_1 - 1} \bar{\mathbf{g}}_{1v_1}^0(x) \ \cdots\ - \bar{\mathbf{g}}_{12}^0(x)\ \bar{\mathbf{g}}_{11}^0(x) \\
& \cdots\ (-1)^{v_p - 1} \bar{\mathbf{g}}_{pv_p}^0(x) \ \cdots\ - \bar{\mathbf{g}}_{p2}^0(x)\ \bar{\mathbf{g}}_{p1}^0(x) \Big]^{-1}.
\end{aligned} \qquad (8.226)$$

Proof Suppose that system (8.132) is state equivalent to a dual Brunovsky NOCF with $z = S(x)$. Then, it is clear, by Theorem 8.8, that condition (i) of Theorem 8.9 is satisfied. Also, we have, by (8.139) and (8.140), that

$$\bar{h}(z) \triangleq \varphi \circ H \circ S^{-1}(z) = C_o z = [z_{11} \ \cdots \ z_{p1}]^{\mathsf{T}} \triangleq \tilde{\mathbf{z}}_1 \qquad (8.227)$$

and

$$
\bar{f}_u(z) \triangleq S_*(F_u(x)) = A_o z + \bar{\gamma}^u(\tilde{\mathbf{z}}_1) =
\begin{bmatrix}
z_{12} + \bar{\gamma}_{11}^u(\tilde{\mathbf{z}}_1) \\
\vdots \\
z_{1\nu_1} + \bar{\gamma}_{1(\nu_1-1)}^u(\tilde{\mathbf{z}}_1) \\
\bar{\gamma}_{1\nu_1}^u(\tilde{\mathbf{z}}_1) \\
\vdots \\
z_{p2} + \bar{\gamma}_{p1}^u(\tilde{\mathbf{z}}_1) \\
\vdots \\
z_{p\nu_p} + \bar{\gamma}_{p(\nu_p-1)}^u(\tilde{\mathbf{z}}_1) \\
\bar{\gamma}_{p\nu_p}^u(\tilde{\mathbf{z}}_1)
\end{bmatrix}
\tag{8.228}
$$

where

$$
z \triangleq [z_{11} \quad \cdots \quad z_{1\nu_1} \quad \cdots \quad z_{p1} \quad \cdots \quad z_{p\nu_p}]^\mathsf{T}.
$$

We define vector fields $\{\bar{\psi}_{ik}^u(z) \mid 1 \leq i \leq p, \ 1 \leq k \leq \nu_i\}$ by

$$
\bar{\psi}_{i1}^0(z) \triangleq \frac{\partial}{\partial z_{i\nu_i}} \ ; \quad \bar{\psi}_{ik}^u(z) \triangleq \mathrm{ad}_{\bar{f}_u}^{k-1} \bar{\psi}_{i1}^0(z).
\tag{8.229}
$$

Then, by (8.228) and (8.229), it is clear that for $1 \leq i \leq p$ and $1 \leq k \leq \nu_i$,

$$
\bar{\psi}_{ik}^u(z) = (-1)^{k-1} \frac{\partial}{\partial z_{i(\nu_i+1-k)}} = \bar{\psi}_{ik}^0(z)
\tag{8.230}
$$

which implies that for $1 \leq i \leq p$, $1 \leq j \leq p$, and $1 \leq k \leq \nu_i$, and $1 \leq r \leq \nu_j$,

$$
[\bar{\psi}_{ik}^0(z), \bar{\psi}_{jr}^0(z)] = 0.
\tag{8.231}
$$

Let $\xi = T(x)$ and $z = \bar{S}(\xi) \triangleq S \circ T^{-1}(\xi)$. Also, we let, for $1 \leq i \leq p$,

$$
\bar{\tau}_{i1}^0(\xi) \triangleq \bar{S}_*^{-1}(\bar{\psi}_{i1}^0(z))
\tag{8.232}
$$

and

$$
\bar{g}_{i1}^0(x) \triangleq S_*^{-1}(\bar{\psi}_{i1}^0(z)) = T_*^{-1}(\bar{\tau}_{i1}^0(\xi)).
\tag{8.233}
$$

Then, it is clear, by (2.37), (8.225), and (8.229), that for $1 \leq i \leq p$ and $1 \leq k \leq \nu_i$,

$$
\begin{aligned}
\bar{g}_{ik}^u(x) &= \mathrm{ad}_{F_u}^{k-1} \bar{g}_{i1}^0(x) = S_*^{-1} \left\{ \mathrm{ad}_{S_*(F_u)}^{k-1} S_*(\bar{g}_{i1}^0(x)) \right\} \\
&= S_*^{-1} \left\{ \mathrm{ad}_{\bar{f}_u}^{k-1} \bar{\psi}_{i1}^0(z) \right\} = S_*^{-1}(\bar{\psi}_{ik}^u(z))
\end{aligned}
\tag{8.234}
$$

and thus (8.221) and (8.222) are satisfied by (8.230) and (8.231). We need to show that (8.223) and (8.224) are satisfied. Note that we are assuming $\nu_1 \geq \nu_2 \geq \cdots \geq \nu_p \geq 1$. Thus, we have, by (2.30) and (8.184), that for $1 \leq j \leq p$ and $1 \leq k \leq \nu_j$,

$$
\begin{aligned}
z_{jk} \triangleq \bar{S}_{jk}(\xi) &= S_{jk} \circ T^{-1}(\xi) \\
&= L_{F_0}^{k-1} H_j(x)\big|_{x=T^{-1}(\xi)} - \sum_{r=1}^{k-1} L_{F_0}^{k-1-r} \left(\gamma_{jr}^0 \circ H(x)\right)\big|_{x=T^{-1}(\xi)} \\
&= L_{f_0}^{k-1} h_j(\xi) - \sum_{r=1}^{k-1} L_{f_0}^{k-1-r} \gamma_{jr}^0(h(\xi)) \\
&= \xi_{jk} - \sum_{r=1}^{k-1} L_{f_0}^{k-1-r} \gamma_{jr}^0(h(\xi))
\end{aligned}
\tag{8.235}
$$

where $f_0(\xi) \triangleq T_*(F_0(x))$ in (8.167). Note, by (8.190), that for $1 \leq i \leq p$, $1 \leq j \leq p$, $1 \leq r \leq k-1$, and $k \leq \nu_i$,

$$
\frac{\partial}{\partial \xi_{i\nu_i}} \left(L_{f_0}^{k-1-r} \gamma_{jr}^0(h(\xi)) \right) = L_{\tau_{i1}^0} L_{f_0}^{k-1-r} \gamma_{jr}^0(h(\xi)) = 0
$$

which implies, together with (8.235), that for $1 \leq j \leq p$, $1 \leq i \leq p$, and $1 \leq k \leq \nu_j$,

$$
\frac{\partial \bar{S}_{jk}(\xi)}{\partial \xi_{i\nu_i}} = \begin{cases} \delta_{i,j}\delta_{k,\nu_i}, & 1 \leq k \leq \nu_i \\ \epsilon_{i,j,k}, & \nu_i + 1 \leq k \leq \nu_j \ (\text{and } j < i) \end{cases}
\tag{8.236}
$$

where

$$
\begin{aligned}
\epsilon_{i,j,k}(\xi) &\triangleq -\sum_{r=1}^{k-1} \frac{\partial}{\partial \xi_{i\nu_i}} \left(L_{f_0}^{k-1-r} \gamma_{jr}(h(\xi)) \right) \\
&= -\sum_{r=1}^{k-\nu_i} L_{\tau_{i1}^0} L_{f_0}^{k-1-r} \gamma_{jr}(h(\xi)).
\end{aligned}
\tag{8.237}
$$

Now, we will express vector fields $\bar{S}_*(\tau_{i1}^0(\xi))$, $1 \leq i \leq p$ in terms of vector fields $\{\bar{\psi}_{jk}^0(z), \ 1 \leq j \leq p, \ 1 \leq k \leq \nu_j\}$. Note, by the definition of $\bar{S}_*(\tau_{i1}^0)$, that for $1 \leq i \leq p$,

$$
\bar{S}_*(\tau_{i1}^0) = \bar{S}_*\left(\frac{\partial}{\partial \xi_{i\nu_i}}\right) = \sum_{j=1}^{p} \sum_{k=1}^{\nu_j} \frac{\partial \bar{S}_{jk}(\xi)}{\partial \xi_{i\nu_i}}\bigg|_{\xi=S^{-1}(z)} \frac{\partial}{\partial z_{jk}}
$$

which implies, together with (8.230), (8.236), and (8.237), that for $1 \leq i \leq p$,

$$\bar{S}_*(\tau_{i1}^0(\xi)) = \frac{\partial}{\partial z_{i\nu_i}} + \sum_{j=1}^{i-1} \sum_{k=\nu_i+1}^{\nu_j} \epsilon_{i,j,k}(\xi)\Big|_{\xi=S^{-1}(z)} \frac{\partial}{\partial z_{jk}}$$

$$= \frac{\partial}{\partial z_{i\nu_i}} + \sum_{j=1}^{i-1} \sum_{k=1}^{\nu_j-\nu_i} \epsilon_{i,j,(\nu_j+1-k)}(\xi)\Big|_{\xi=S^{-1}(z)} \frac{\partial}{\partial z_{j(\nu_j+1-k)}} \qquad (8.238)$$

$$= \bar{\psi}_{i1}^0(z) - \sum_{j=1}^{i-1} \sum_{k=1}^{\nu_j-\nu_i} \ell_{i,j,k}(\xi)\Big|_{\xi=S^{-1}(z)} \bar{\psi}_{jk}^0(z)$$

where

$$\ell_{i,j,k}(\xi) \triangleq (-1)^k \epsilon_{i,j,(\nu_j+1-k)}(\xi)$$

$$= (-1)^{k-1} \sum_{r=1}^{\nu_j-\nu_i+1-k} L_{\tau_{i1}^0} L_{f_0}^{\nu_j-k-r} \gamma_{jr}(h(\xi)). \qquad (8.239)$$

Note, by (2.49), (8.232), (8.233), and (8.239), that

$$\bar{S}_*^{-1}\left(\ell_{i,j,k}(\xi)\Big|_{\xi=S^{-1}(z)} \bar{\psi}_{jk}^0(z)\right) = \ell_{i,j,k}(\xi)\bar{\tau}_{jk}^0(\xi)$$

and

$$T_*^{-1}\left(\ell_{i,j,k}(\xi)\bar{\tau}_{jk}^0(\xi)\right) = \ell_{i,j,k}(T(x))\,\bar{g}_{jk}^0(\xi) \triangleq \tilde{\ell}_{i,j,k}(x)\bar{g}_{jk}^0(\xi)$$

where

$$\tilde{\ell}_{i,j,k}(x) = (-1)^{k-1} \sum_{r=1}^{\nu_j-\nu_i+1-k} L_{\tau_{i1}^0} L_{f_0}^{\nu_j-k-r} \gamma_{jr}(h(\xi))\Big|_{\xi=T(x)}$$

$$= (-1)^{k-1} \sum_{r=1}^{\nu_j-\nu_i+1-k} L_{g_{i1}^0} L_{F_0}^{\nu_j-k-r} \gamma_{jr}(H(x)). \qquad (8.240)$$

Therefore, it is clear, by (8.240), that (8.223) is satisfied. Also, we have, by (8.232), (8.233), and (8.238), that for $1 \le i \le p$,

$$\bar{\tau}_{i1}^0(\xi) = \bar{S}_*^{-1}(\bar{\psi}_{i1}^0(z)) = \tau_{i1}^0(\xi) + \sum_{j=1}^{i-1} \sum_{k=1}^{\nu_j-\nu_i} \ell_{i,j,k}(\xi)\bar{\tau}_{jk}^0(\xi)$$

and

$$\bar{g}_{i1}^0(x) = T_*^{-1}(\bar{\tau}_{i1}^0(\xi)) = g_{i1}^0(x) + \sum_{j=1}^{i-1} \sum_{k=1}^{\nu_j-\nu_i} \tilde{\ell}_{i,j,k}(x)\bar{g}_{jk}^0(x)$$

which implies that (8.224) is satisfied.

Sufficiency. Suppose that condition (i) and condition (ii) of Theorem 8.9 are satisfied. If we show that for $1 \le i \le p, 1 \le j \le p$, and $1 \le k \le \nu_i$,

$$L_{\tilde{\mathbf{g}}_{i1}^0} L_{F_0}^{k-1} H_j(x) = \delta_{i,j}\, \delta_{k,\nu_i} \tag{8.241}$$

then system (8.132) is, by Theorem 8.8, state equivalent to a dual Brunovsky NOCF. Let $\xi = T(x)$, $h(\xi) \triangleq H \circ T^{-1}(\xi)$, $f_0(\xi) \triangleq T_*(F_0(x))$, and $\tau_{i1}^0(\xi) \triangleq T_*(\mathbf{g}_{i1}^0(x))$. First, it will be shown that for $1 \le i \le p, 1 \le j \le p$, and $1 \le k \le \nu_i$,

$$L_{\tau_{i1}^0} L_{f_0}^{k-1} h_j(\xi) = \delta_{i,j}\, \delta_{k,\nu_i} \tag{8.242}$$

or

$$L_{\mathbf{g}_{i1}^0} L_{F_0}^{k-1} H_j(x) = L_{\tau_{i1}^0} L_{f_0}^{k-1} h_j(\xi)\Big|_{\xi=T(x)} = \delta_{i,j}\, \delta_{k,\nu_i}. \tag{8.243}$$

Note, by (8.174), that for $1 \le i \le p, 1 \le j \le p$, and $1 \le k \le \nu_j$,

$$L_{\tau_{i1}^0} L_{f_0}^{k-1} h_j(\xi) = \delta_{i,j}\, \delta_{k,\nu_i}. \tag{8.244}$$

Therefore, (8.242) is satisfied when $\nu_i \le \nu_j$. Let for $1 \le k \le \nu_1$,

$$\tilde{\xi}_k \triangleq \{\xi_{rk} \mid 1 \le r \le p \text{ and } \nu_r \ge k\}$$

and for $1 \le i \le p$ and $1 \le j \le \nu_i$,

$$f_{ij}^0(\xi) \triangleq \begin{cases} \xi_{i(j+1)}, & 1 \le j \le \nu_i - 1 \\ \alpha_{i\nu_i}^0(\xi), & j = \nu_i. \end{cases}$$

It is easy to see, by (8.167) and (8.220), that for $1 \le j \le p$,

$$dL_{f_0}^{\nu_j} h_j(\xi) \in \text{span}\{dL_{f_0}^{k-1} h_r(\xi),\ 1 \le r \le p,\ 1 \le k \le \nu_j\}$$
$$\in \text{span}\{d\xi_{rk},\ 1 \le r \le p,\ 1 \le k \le \min(\nu_r, \nu_j)\}$$

and

$$\alpha_{j\nu_j}^0(\xi) = L_{f_0}^{\nu_j} h_j(x) \triangleq \bar{\alpha}_{j\nu_j}^0(\tilde{\xi}_1, \cdots, \tilde{\xi}_{\nu_j}). \tag{8.245}$$

Thus, we have, by (8.167) and (8.245), that for $1 \le j \le p$ and $1 \le k \le \nu_1 - \nu_j$,

$$L_{f_0}^{\nu_j+k-1} h_j(\xi) = \phi_{j,k}(\tilde{\xi}_1, \cdots, \tilde{\xi}_{\nu_j+k-1}) \tag{8.246}$$

where $\phi_{j,1}(\tilde{\xi}_1, \cdots, \tilde{\xi}_{\nu_j}) = \bar{\alpha}_{j\nu_j}^0(\tilde{\xi}_1, \cdots, \tilde{\xi}_{\nu_j})$ and for $2 \le k \le \nu_1 - \nu_j$,

$$\phi_{j,k}(\tilde{\xi}_1, \cdots, \tilde{\xi}_{v_j+k-1}) = \sum_{r=1}^{p} \sum_{j=1}^{\min(k-1,v_r)} \frac{\partial \phi_{j,(k-1)}(\tilde{\xi}_1, \cdots, \tilde{\xi}_{v_j+k-2})}{\partial \xi_{rj}} f_{rj}^0(\xi).$$

Thus, it is clear, by (8.172) and (8.246), that for $1 \leq i \leq p$, $1 \leq j \leq p$, and $v_j + 1 \leq k \leq v_i$,

$$L_{\tau_{i1}^0} L_{f_0}^{k-1} h_j(\xi) = \frac{\partial \phi_{j,k}(\tilde{\xi}_1, \cdots, \tilde{\xi}_{k-1})}{\partial \xi_{iv_i}} = 0$$

which implies that (8.242) is also satisfied when $v_i > v_j$. Therefore, (8.242) and (8.243) are satisfied, if condition (i) of Theorem 8.9 is satisfied. Finally, it will be shown, by mathematical induction, that (8.241) is satisfied. Since $\bar{\mathbf{g}}_{11}(x) = \mathbf{g}_{11}(x)$, it is clear, by (8.243), that (8.241) is satisfied for $i = 1$. Assume that (8.241) holds when $1 \leq i \leq q - 1$ and $2 \leq q \leq p$. Then, we have, by Example 2.4.16, that for $1 \leq i \leq q - 1$, $1 \leq j \leq p$, and $1 \leq k \leq v_i$,

$$L_{\bar{\mathbf{g}}_{ir}^0} L_{F_0}^{k-1} H_j(x) = 0, \quad r + k - 1 \leq v_i - 1. \tag{8.247}$$

Thus, it is clear, by (8.224), (8.243), and (8.247), that for $1 \leq j \leq p$ and $1 \leq k \leq v_q$,

$$L_{\bar{\mathbf{g}}_{q1}^0} L_{F_0}^{k-1} H_j(x) = L_{\mathbf{g}_{q1}^0} L_{F_0}^{k-1} H_j(x) + \sum_{s=1}^{q-1} \sum_{r=1}^{v_s - v_q} \tilde{\ell}_{q,s,r}(x) L_{\bar{\mathbf{g}}_{sr}^0} L_{F_0}^{k-1} H_j(x)$$

$$= L_{\mathbf{g}_{q1}^0} L_{F_0}^{k-1} H_j(x) = \delta_{q,j}\, \delta_{k,v_q}$$

which implies that (8.241) is satisfied for $i = q$. Hence, by mathematical induction, (8.241) is satisfied. $\square$

Corollary 8.5 *Suppose that $v_1 = \cdots = v_p$. System (8.132) is state equivalent to a dual Brunovsky NOCF, if and only if*

(i) for $1 \leq i \leq p$, and $1 \leq k \leq v_i$

$$\mathbf{g}_{ik}^u(x) = \mathbf{g}_{ik}^u(x)\big|_{u=0} \triangleq \mathbf{g}_{ik}^0(x)$$

(ii) for $1 \leq i \leq p$, $1 \leq j \leq p$, $1 \leq k \leq v_i$, and $1 \leq \ell \leq v_j$,

$$[\mathbf{g}_{ik}^0(x), \mathbf{g}_{j\ell}^0(x)] = 0.$$

Furthermore, a state coordinates transformation $z = S(x)$ is given by

$$\frac{\partial S(x)}{\partial x} = \left[(-1)^{v_1-1} \mathbf{g}_{1v_1}^0(x) \cdots - \mathbf{g}_{12}^0(x)\, \mathbf{g}_{11}^0(x) \right.$$

$$\left. \cdots (-1)^{v_p-1} \mathbf{g}_{pv_p}^0(x) \cdots - \mathbf{g}_{p2}^0(x)\, \mathbf{g}_{p1}^0(x) \right]^{-1}.$$

Suppose that system (8.132) is equivalent to a dual Brunovsky NOCF. Then, by Theorem 8.9, there exist smooth functions $\gamma_{jr}(y)$, $1 \le j \le p - 1$, $1 \le r \le \nu_j - \nu_p$ such that for $2 \le i \le p$ and $1 \le s \le \nu_1 - \nu_i$ (by changing the order of summations),

$$
\begin{aligned}
\bar{\mathbf{g}}_{i1}^0(x) &= \mathbf{g}_{i1}^0(x) + \sum_{j=1}^{i-1} \sum_{k=1}^{\nu_j-\nu_i} \sum_{r=1}^{\nu_j-\nu_i+1-k} (-1)^{k-1} L_{\mathbf{g}_{i1}^0} L_{F_0}^{\nu_j-k-r} \gamma_{jr}(H(x)) \bar{\mathbf{g}}_{jk}^0(x) \\
&= \mathbf{g}_{i1}^0(x) + \sum_{j=1}^{i-1} \sum_{r=1}^{\nu_j-\nu_i} \sum_{k=1}^{\nu_j-\nu_i+1-r} (-1)^{k-1} L_{\mathbf{g}_{i1}^0} L_{F_0}^{\nu_j-k-r} \gamma_{jr}(H(x)) \bar{\mathbf{g}}_{jk}^0(x) \\
&= \tilde{\mathbf{g}}_{i1}^{s-1}(x) + \sum_{j=1}^{i-1} \sum_{r=s}^{\nu_j-\nu_i} \sum_{k=1}^{\nu_j-\nu_i+1-r} (-1)^{k-1} L_{\mathbf{g}_{i1}^0} L_{F_0}^{\nu_j-k-r} \gamma_{jr}(H(x)) \bar{\mathbf{g}}_{jk}^0(x) \\
&= \tilde{\mathbf{g}}_{i1}^{s-1}(x) + \sum_{j=1}^{i-1} \sum_{k=1}^{\nu_j-\nu_i+1-s} \sum_{r=s}^{\nu_j-\nu_i+1-k} (-1)^{k-1} L_{\mathbf{g}_{i1}^0} L_{F_0}^{\nu_j-k-r} \gamma_{jr}(H) \bar{\mathbf{g}}_{jk} \\
&= \tilde{\mathbf{g}}_{i1}^{s-1}(x) + \sum_{j=1}^{i-1} (-1)^{\nu_j-\nu_i-s} L_{\mathbf{g}_{i1}^0} L_{F_0}^{\nu_i-1} \gamma_{js}(H(x)) \bar{\mathbf{g}}_{j(\nu_j-\nu_i+1-s)}^0(x) \\
&\quad + \sum_{j=1}^{i-1} \sum_{k=1}^{\nu_j-\nu_i-s} \sum_{r=s}^{\nu_j-\nu_i+1-k} (-1)^{k-1} L_{\mathbf{g}_{i1}^0} L_{F_0}^{\nu_j-k-r} \gamma_{jr}(H(x)) \bar{\mathbf{g}}_{jk}^0(x)
\end{aligned}
$$
$$\tag{8.248}$$

where, for $1 \le s \le \nu_1 - \nu_i + 1$,

$$
\tilde{\mathbf{g}}_{i1}^{s-1}(x) \triangleq \mathbf{g}_{i1}^0(x) + \sum_{j=1}^{i-1} \sum_{r=1}^{s-1} \sum_{k=1}^{\nu_j-\nu_i+1-r} (-1)^{k-1} L_{\mathbf{g}_{i1}^0} L_{F_0}^{\nu_j-k-r} \gamma_{jr}(H(x)) \bar{\mathbf{g}}_{jk}^0(x).
$$
$$\tag{8.249}$$

That is, $\tilde{\mathbf{g}}_{21}^0(x) \triangleq \mathbf{g}_{21}^0(x)$ and, for $1 \le s \le \nu_1 - \nu_i$,

$$
\tilde{\mathbf{g}}_{i1}^s(x) \triangleq \tilde{\mathbf{g}}_{i1}^{s-1}(x) + \sum_{j=1}^{i-1} \sum_{k=1}^{\nu_j-\nu_i+1-s} (-1)^{k-1} L_{\mathbf{g}_{i1}^0} L_{F_0}^{\nu_j-k-s} \gamma_{js}(H(x)) \bar{\mathbf{g}}_{jk}^0(x) \tag{8.250}
$$

and

$$
\bar{\mathbf{g}}_{i1}^0(x) \equiv \tilde{\mathbf{g}}_{i1}^{s-1}(x) + (-1)^{\nu_j-\nu_i-s} L_{\mathbf{g}_{i1}^0} L_{F_0}^{\nu_i-1} \gamma_{js}(H(x)) \bar{\mathbf{g}}_{j(\nu_j-\nu_i+1-s)}^0(x) \\
\mod \Phi_{is}(x) \tag{8.251}
$$

where

$$
\begin{aligned}
\Phi_{is}(x) &\triangleq \mathrm{span}\{\bar{\mathbf{g}}_{jk}^0(x), \ 1 \le j \le i - 1, \ 1 \le k \le \nu_j - \nu_i - s\} \\
&= \mathrm{span}\{\mathbf{g}_{jk}^0(x), \ 1 \le j \le i - 1, \ 1 \le k \le \nu_j - \nu_i - s\}.
\end{aligned}
$$
$$\tag{8.252}$$

Lemma 8.7 *Suppose that system (8.132) is equivalent to a dual Brunovsky NOCF. Then, there exist smooth functions $\gamma_{jr}(y)$, $1 \leq j \leq p - 1$, $1 \leq r \leq \nu_j - \nu_p$ such that for $2 \leq i \leq p$, $1 \leq q \leq i - 1$, and $1 \leq s \leq \nu_j - \nu_i$,*

$$[\bar{\mathbf{g}}_{qv_q}^0(x), \, \tilde{\mathbf{g}}_{i1}^{s-1}(x)] \equiv \sum_{j=1}^{i-1}(-1)^{\nu_j - \nu_i - s + \nu_q} \frac{\partial^2 \gamma_{js}(y)}{\partial y_q \partial y_i}\bigg|_{y = H(x)} \mathbf{g}_{j(\nu_j - \nu_i + 1 - s)}^0(x)$$

$$\mathrm{mod}\ \Phi_{is}(x)$$

$$(8.253)$$

where, for $2 \leq i \leq p$ and $1 \leq s \leq \nu_1 - \nu_i$,

$$\tilde{\mathbf{g}}_{i1}^0(x) \triangleq \mathbf{g}_{i1}^0(x) \tag{8.254}$$

and

$$\tilde{\mathbf{g}}_{i1}^s(x) \triangleq \tilde{\mathbf{g}}_{i1}^{s-1}(x) + \sum_{j=1}^{i-1}\sum_{k=1}^{\nu_j - \nu_i + 1 - s}(-1)^{k-1}L_{\mathbf{g}_{i1}^0}L_{F_0}^{\nu_j - k - s}\gamma_{js}(H(x))\bar{\mathbf{g}}_{jk}^0(x). \tag{8.255}$$

Proof Suppose that system (8.132) is equivalent to dual Brunovsky NOCF. Note that $\xi = T(x)$ and $h(\xi) \triangleq H \circ T^{-1}(\xi) = [\xi_{11} \cdots \xi_{p1}]^T \triangleq \tilde{\xi}_1$. Then, by (2.30), (8.192), (8.193), and (8.224), we have that

$$L_{\tau_{i1}^0}L_{f_0}^{\nu_i - 1}\gamma_{js}(h(\xi)) = \frac{\partial \gamma_{js}(h(\xi))}{\partial \xi_{i1}} = \frac{\partial \gamma_{js}(\tilde{\xi}_1)}{\partial \xi_{i1}}$$

$$\bar{\tau}_{qv_q}^0(\xi) \equiv (-1)^{\nu_q - 1}\frac{\partial}{\partial \xi_{q1}} \ \mathrm{mod\ span}\left\{\frac{\partial}{\partial \xi_{ij}}, \ 1 \leq i \leq p, \ 2 \leq j \leq \nu_i\right\}$$

$$L_{\bar{\tau}_{qv_q}^0}L_{\tau_{i1}^0}L_{f_0}^{\nu_i - 1}\gamma_{js}(h(\xi)) = L_{\bar{\tau}_{qv_q}^0}\frac{\partial \gamma_{js}(\tilde{\xi}_1)}{\partial \xi_{i1}} = (-1)^{\nu_q - 1}\frac{\partial^2 \gamma_{js}(\tilde{\xi}_1)}{\partial \xi_{q1}\partial \xi_{i1}}$$

and

$$L_{\bar{\mathbf{g}}_{qv_q}^0}L_{\mathbf{g}_{i1}^0}L_{F_0}^{\nu_i - 1}\gamma_{js}(H(x)) = L_{\bar{\tau}_{qv_q}^0}L_{\tau_{i1}^0}L_{f_0}^{\nu_i - 1}\gamma_{js}(h(\xi))\bigg|_{\xi = T(x)}$$

$$= (-1)^{\nu_q - 1}\frac{\partial^2 \gamma_{js}(y)}{\partial y_q \partial y_i}\bigg|_{y = H(x)}$$

$$(8.256)$$

where $\bar{\tau}_{ik}^0(\xi) \triangleq T_*\left(\bar{\mathbf{g}}_{ik}^0(x)\right)$, $1 \leq i \leq p$, $1 \leq k \leq \nu_i$ *and for* $1 \leq i \leq p$,

$$\bar{\tau}_{i1}^0(\xi) = \tau_{i1}^0(\xi) + \sum_{j=1}^{i-1}\sum_{k=1}^{\nu_j - \nu_i}\tilde{\ell}_{i,j,k} \circ T^{-1}(\xi)\,\bar{\tau}_{jk}^0(\xi).$$

Thus, we have, by (8.222), (8.251), and (8.256), that for $2 \leq i \leq p$, $1 \leq q \leq i-1$, and $1 \leq s \leq \nu_j - \nu_i$,

$$
0 = [\bar{\mathbf{g}}_{q\nu_q}^0(x), \bar{\mathbf{g}}_{i1}^0(x)] \equiv [\bar{\mathbf{g}}_{q\nu_q}^0(x), \tilde{\mathbf{g}}_{i1}^{s-1}(x)]
$$
$$
+ \sum_{j=1}^{i-1} (-1)^{\nu_j-\nu_i-s} L_{\bar{\mathbf{g}}_{q\nu_q}^0} L_{\mathbf{g}_{i1}^0} L_{F_0}^{\nu_i-1} \gamma_{js}(H(x)) \bar{\mathbf{g}}_{j(\nu_j-\nu_i+1-s)}^0(x) \bmod \Phi_{is}(x)
$$
$$
\equiv [\bar{\mathbf{g}}_{q\nu_q}^0(x), \tilde{\mathbf{g}}_{i1}^{s-1}(x)]
$$
$$
+ \sum_{j=1}^{i-1} (-1)^{\nu_j-\nu_i-s+\nu_q-1} \left. \frac{\partial^2 \gamma_{js}(y)}{\partial y_q \partial y_i} \right|_{y=H(x)} \bar{\mathbf{g}}_{j(\nu_j-\nu_i+1-s)}^0(x) \bmod \Phi_{is}(x)
$$

which implies that (8.253) is satisfied. $\qquad\square$

Theorem 8.10 *System (8.132) is equivalent to a dual Brunovsky NOCF, if and only if*

(i) for $1 \leq i \leq p$,

$$
dL_{F_0}^{\nu_i} H_i(x) \in \text{span}\{dL_{F_0}^{k-1} H_j(x), \ 1 \leq j \leq p, \ 1 \leq k \leq \nu_i\} \tag{8.257}
$$

(ii) there exist smooth functions $\gamma_{js}(y)$ and $\beta_{js}^{q,i}(y)$ for $2 \leq i \leq p$, $1 \leq j \leq i-1$, $1 \leq s \leq \nu_j - \nu_i$, and $1 \leq q \leq i-1$, such that for $2 \leq i \leq p$, $1 \leq j \leq i-1$, $1 \leq s \leq \nu_j - \nu_i$, and $1 \leq q \leq i-1$,

$$
\left[\bar{\mathbf{g}}_{q\nu_q}^0(x), \tilde{\mathbf{g}}_{i1}^{s-1}(x)\right] \equiv \sum_{j=1}^{i-1} (-1)^{\nu_j-\nu_i-s+\nu_q} \beta_{js}^{q,i}(H(x)) \mathbf{g}_{j(\nu_j-\nu_i+1-s)}^0 \tag{8.258}
$$
$$
\bmod \Phi_{is}(x)
$$

and

$$
\frac{\partial^2 \gamma_{js}(y)}{\partial y_q \partial y_i} = \beta_{js}^{q,i}(y) \tag{8.259}
$$

where for $1 \leq i \leq p$ and $1 \leq s \leq \nu_1 - \nu_i$,

$$
\Phi_{is}(x) \triangleq \text{span}\{\mathbf{g}_{jk}^0(x), \ 1 \leq j \leq i-1, \ 1 \leq k \leq \nu_j - \nu_i - s\} \tag{8.260}
$$

$$
\tilde{\mathbf{g}}_{i1}^0(x) \triangleq \mathbf{g}_{i1}^0(x) \tag{8.261}
$$

$$\tilde{\mathbf{g}}_{i1}^{s}(x) \triangleq \tilde{\mathbf{g}}_{i1}^{s-1}(x)$$

$$+ \sum_{j=1}^{i-1} \sum_{k=1}^{\nu_j - \nu_i + 1 - s} (-1)^{k-1} L_{\mathbf{g}_{i1}^0} L_{F_0}^{\nu_j - k - s} \gamma_{js}(H(x)) \tilde{\mathbf{g}}_{jk}^{\nu_1 - \nu_j}(x) \qquad (8.262)$$

and for $1 \le j \le p$ and $1 \le k \le \nu_j$,

$$\bar{\mathbf{g}}_{j1}^{0}(x) \triangleq \tilde{\mathbf{g}}_{j1}^{\nu_1 - \nu_j}(x) ; \quad \bar{\mathbf{g}}_{jk}^{u}(x) \triangleq \mathrm{ad}_{F_u}^{k-1} \bar{\mathbf{g}}_{j1}^{0}(x) \qquad (8.263)$$

(iii) for $1 \le i \le p$, $1 \le j \le p$, $1 \le k \le \nu_i$, and $1 \le r \le \nu_j$,

$$\bar{\mathbf{g}}_{ik}^{u}(x) = \bar{\mathbf{g}}_{ik}^{0}(x) \qquad (8.264)$$

and

$$\left[\bar{\mathbf{g}}_{jr}^{0}(x),\ \bar{\mathbf{g}}_{ik}^{0}(x)\right] = 0. \qquad (8.265)$$

Furthermore, a state coordinates transformation $z = S(x)$ is given by

$$\frac{\partial S(x)}{\partial x} = \left[(-1)^{\nu_1 - 1} \bar{\mathbf{g}}_{1\nu_1}^{0}(x) \cdots \ - \bar{\mathbf{g}}_{12}^{0}(x)\ \bar{\mathbf{g}}_{11}^{0}(x)\right.$$
$$\left. \cdots (-1)^{\nu_p - 1} \bar{\mathbf{g}}_{p\nu_p}^{0}(x) \cdots \ - \bar{\mathbf{g}}_{p2}^{0}(x)\ \bar{\mathbf{g}}_{p1}^{0}(x)\right]^{-1}. \qquad (8.266)$$

Proof Necessity. Suppose that system (8.132) is equivalent to a dual Brunovsky NOCF. Then, it is clear, by Theorem 8.9, that condition (i) is satisfied. Since $\bar{\mathbf{g}}_{i1}^{0}(x) = \tilde{\mathbf{g}}_{i1}^{\nu_1 - \nu_i}(x)$ for $1 \le i \le p$ by (8.248), it is easy to see that conditions (ii) and (iii) are satisfied by Lemma 8.7 and condition (ii) of Theorem 8.9, respectively.

Sufficiency. Suppose that conditions (i)–(iii) of Theorem 8.10 are satisfied. Then, there exist smooth functions $\gamma_{jr}(y)$, $1 \le j \le p - 1$, $1 \le r \le \nu_j - \nu_p$ such that the condition (ii) of Theorem 8.9 are satisfied with $\bar{\mathbf{g}}_{i1}^{0}(x) = \tilde{\mathbf{g}}_{i1}^{\nu_1 - \nu_i}(x)$. Hence, by Theorem 8.9, system (8.132) is state equivalent to a dual Brunovsky NOCF. $\quad\square$

The necessary and sufficient conditions of Theorem 8.10 are still unverifiable. For simple explanation, assume that $\nu_1 > \nu_2$. Condition (ii) of Theorem 8.10 should be considered for $i = 2, 3, \cdots, p - 1$, in sequence. If condition (ii) for $i = 2$ and $s = 1$ is not satisfied, then system (8.132) is not equivalent to a dual Brunovsky NOCF. If condition (ii) for $i = 2$ and $s = 1$ is satisfied, we have that

$$\frac{\partial^2 \gamma_{11}(y)}{\partial y_1 \partial y_2} = \beta_{11}^{1,2}(y)$$

$$\gamma_{11}^2(y) \triangleq \int_0^{y_2} \int_0^{y_1} \beta_{11}^{1,2}(\tilde{y}_1, \tilde{y}_2, y_3, \cdots, y_p) d\tilde{y}_1 d\tilde{y}_2$$

$$\triangleq \int \int \beta_{11}^{1,2}(y) dy_1 dy_2 \tag{8.267}$$

and

$$\gamma_{11}(y) = \gamma_{11}^2(y) + \gamma_{11}(y_1, 0, y_3, \cdots, y_p) + \gamma_{11}(0, y_2, \cdots, y_p)$$

$$- \gamma_{11}(0, 0, y_3, \cdots, y_p) \triangleq \gamma_{11}^2(y) + \hat{\gamma}_{11}^2(y). \tag{8.268}$$

If $\gamma_{11}(y)\big|_{y_1=0} = 0$ and $\gamma_{11}(y)\big|_{y_2=0} = 0$, then $\gamma_{11}(y) = \gamma_{11}^2(y)$ or $\hat{\gamma}_{11}^2(y) = 0$. Thus, $\tilde{g}_{21}^1(x)$ can be obtained by (8.262) and condition (ii) for $i = 2$ and $s = 2$ can be checked. However, since $\hat{\gamma}_{11}^2(y)$ is unknown, $\tilde{g}_{21}^1(x)$ cannot be obtained and thus condition (ii) for $i = 2$ and $s = 2$ cannot be checked unless

$$\sum_{k=1}^{\nu_1 - \nu_2} L_{g_{21}^0} L_{F_0}^{\nu_1 - k - 1} \left(\gamma_{11}(y) - \gamma_{11}^2(y) \right) \bar{g}_{1k}^0(x) = 0$$

or for $1 \le k \le \nu_1 - \nu_2$,

$$L_{g_{21}^0} L_{F_0}^{\nu_1 - k - 1} \left(\gamma_{11}(y) - \gamma_{11}^2(y) \right) = 0.$$

When $p = 2$, $\gamma_{1s}(y) - \gamma_{1s}^2(y) \triangleq \hat{\gamma}_{1s}^2(y) = \gamma_{1s}(y_1, 0) + \gamma_{1s}(0, y_2)$ and it is easy to see that for $1 \le s \le \nu_1 - \nu_2$ and $1 \le k \le \nu_1 - \nu_2 + 1 - s$,

$$L_{g_{21}^0} L_{F_0}^{\nu_1 - k - s} \gamma_{1s}(y_1, 0) = 0.$$

However, it is not always satisfied that for $1 \le s \le \nu_1 - \nu_2$ and $1 \le k \le \nu_1 - \nu_2 + 1 - s$,

$$L_{g_{21}^0} L_{F_0}^{\nu_1 - k - s} \gamma_{1s}(0, y_2) = 0.$$

When $p = 2$, a verifiable sufficient condition can be obtained by using $\gamma_{1k}^2(y)$, $1 \le k \le \nu_1 - \nu_2$ instead of $\gamma_{1k}(y)$, $1 \le k \le \nu_1 - \nu_2$ in (8.262).

Corollary 8.6 *System (8.132) with $p = 2$ is equivalent to a dual Brunovsky NOCF, if*

(i) $dL_{F_0}^{\nu_2} H_2(x) \in \text{span}\{dL_{F_0}^{k-1} H_j(x), \ 1 \le j \le 2, \ 1 \le k \le \nu_2\}$

(ii) *there exist smooth functions $\beta_{1s}^{1,2}(y)$ for $1 \le s \le \nu_1 - \nu_2$ such that for $1 \le s \le \nu_1 - \nu_2$,*

$$\left[g_{1\nu_1}^0(x), \ \tilde{g}_{21}^{s-1}(x) \right] \equiv (-1)^{\nu_2 + s} \beta_{1s}^{1,2}(H(x)) g_{1(\nu_1 - \nu_2 + 1 - s)}^0$$

$$\text{mod } \Phi_{2s}(x) \tag{8.269}$$

where

$$\Phi_{2s}(x) \triangleq \text{span}\{\mathbf{g}_{1k}^0(x), \ 1 \le k \le \nu_1 - \nu_2 - s\} \tag{8.270}$$

$$\gamma_{1s}^2(y_1, y_2) \triangleq \int_0^{y_2} \int_0^{y_1} \beta_{1s}^{1,2}(\tilde{y}_1, \tilde{y}_2) d\tilde{y}_1 d\tilde{y}_2 \tag{8.271}$$

$$\tilde{\mathbf{g}}_{j1}^0(x) \triangleq \mathbf{g}_{j1}^0(x), \ 1 \le j \le 2 \tag{8.272}$$

$$\tilde{\mathbf{g}}_{21}^s(x) \triangleq \tilde{\mathbf{g}}_{21}^{s-1}(x)$$
$$+ \sum_{k=1}^{\nu_1 - \nu_2 + 1 - s} (-1)^{k-1} L_{\mathbf{g}_{21}^0} L_{F_0}^{\nu_1 - k - s} \gamma_{1s}^2(H(x)) \mathbf{g}_{1k}^0(x) \tag{8.273}$$

and for $1 \le j \le 2$ *and* $1 \le k \le \nu_j$,

$$\bar{\mathbf{g}}_{j1}^0(x) \triangleq \tilde{\mathbf{g}}_{j1}^{\nu_1 - \nu_j}(x) \ ; \ \ \bar{\mathbf{g}}_{jk}^u(x) \triangleq \text{ad}_{F_u}^{k-1} \bar{\mathbf{g}}_{j1}^0(x) \tag{8.274}$$

(iii) for $1 \le i \le 2$, $1 \le j \le 2$, $1 \le k \le \nu_i$, *and* $1 \le r \le \nu_j$,

$$\bar{\mathbf{g}}_{ik}^u(x) = \bar{\mathbf{g}}_{ik}^0(x) \tag{8.275}$$

and

$$\left[\bar{\mathbf{g}}_{jr}^0(x), \ \bar{\mathbf{g}}_{ik}^0(x) \right] = 0. \tag{8.276}$$

Furthermore, a state coordinates transformation $z = S(x)$ *is given by*

$$\frac{\partial S(x)}{\partial x} = \left[(-1)^{\nu_1 - 1} \bar{\mathbf{g}}_{1\nu_1}^0(x) \ \cdots \ - \bar{\mathbf{g}}_{12}^0(x) \ \bar{\mathbf{g}}_{11}^0(x) \right.$$
$$\left. (-1)^{\nu_2 - 1} \bar{\mathbf{g}}_{2\nu_2}^0(x) \ \cdots \ - \bar{\mathbf{g}}_{22}^0(x) \ \bar{\mathbf{g}}_{21}^0(x) \right]^{-1}. \tag{8.277}$$

Remark 8.7 If $p = 2$ and $\gamma_{1k}(0, y_2) \ne 0$, then $\tilde{\mathbf{g}}_{21}^s$ in (8.262) cannot be obtained by (8.273). Therefore, the conditions in Corollary 8.6 are not necessary but sufficient unless $\gamma_{1k}(0, y_2) = 0$. (Refer to Example 8.4.3.) For a perfect solution, we need to find $\gamma_{1k}(0, y_2)$. However, $\gamma_{1k}(0, y_2)$ is very difficult and complicated to find, even when $p = 2$.

Suppose that $1 \le k \le p$ and

$$\frac{\partial [Q_1(y) \ \cdots \ Q_k(y)]^{\mathsf{T}}}{\partial [y_1 \ \cdots \ y_k]} = \left(\frac{\partial [Q_1(y) \ \cdots \ Q_k(y)]^{\mathsf{T}}}{\partial [y_1 \ \cdots \ y_k]} \right)^{\mathsf{T}}.$$

Then, by Lemma 2.1, there exists a smooth function $P(y)$ such that for $1 \leq i \leq k$,

$$\frac{\partial P(y)}{\partial y_i} = Q_i(y) \text{ and } P(0, \cdots, 0, y_{k+1}, \cdots, y_p) = 0.$$

We denote $P(y)$ by

$$P(y) = \int [Q_1(y) \cdots Q_k(y)] \, d(y_1, \cdots, y_k).$$

For example, when $p = 4$,

$$\int [y_2 y_3 \quad y_1 y_3 + 2 y_2 y_4] \, d(y_1, y_2) = y_1 y_2 y_3 + y_2^2 y_4.$$

Theorem 8.11 *Suppose that $\sigma_0 = 0$, $\sigma_{\bar{p}} = p$, and*

$$v_1 = \cdots = v_{\sigma_1} > v_{\sigma_1+1} = \cdots = v_{\sigma_2} > \cdots > v_{\sigma_{\bar{p}-1}+1} = \cdots = v_{\sigma_{\bar{p}}}.$$

System (8.132) is equivalent to a dual Brunovsky NOCF, if

(i) for $1 \leq i \leq p$,

$$dL_{F_0}^{v_i} H_i(x) \in \mathrm{span}\{dL_{F_0}^{k-1} H_j(x), \ 1 \leq j \leq p, \ 1 \leq k \leq v_i\} \quad (8.278)$$

(ii) there exist smooth functions $\beta_{js}^{q,i}(y)$ for $\sigma_1 + 1 \leq i \leq p$, $1 \leq j \leq i-1$, $1 \leq s \leq v_j - v_i$, and $1 \leq q \leq i-1$, such that for $2 \leq r \leq \bar{p}$, $1 \leq s \leq v_1 - v_{\sigma_r}$, $1 \leq q \leq \sigma_{r-1}$, and $\sigma_{r-1} + 1 \leq i \leq p$,

$$\left[\bar{\mathbf{g}}_{qv_q}^0(x), \ \tilde{\mathbf{g}}_{i1}^{s-1}(x) \right] \equiv \sum_{j=1}^{i-1} (-1)^{v_j - v_i - s + v_q} \beta_{js}^{q,i}(H(x)) \mathbf{g}_{j(v_j - v_i + 1 - s)}^0 \quad (8.279)$$

$$\mathrm{mod} \ \Phi_{is}(x)$$

where

$$\Phi_{is}(x) \triangleq \mathrm{span}\{\mathbf{g}_{jk}^0(x), \ 1 \leq j \leq i-1, \ 1 \leq k \leq v_j - v_i - s\} \quad (8.280)$$

$$\frac{\partial \left[\beta_{js}^{q,j+1}(y) \cdots \beta_{js}^{q,p}(y) \right]^{\mathsf{T}}}{\partial [y_{j+1} \cdots y_p]^{\mathsf{T}}} = \left(\frac{\partial \left[\beta_{js}^{q,j+1}(y) \cdots \beta_{js}^{q,p}(y) \right]^{\mathsf{T}}}{\partial [y_{j+1} \cdots y_p]^{\mathsf{T}}} \right)^{\mathsf{T}} \quad (8.281)$$

$$\tilde{\beta}_{js}^q(y) \triangleq \int \left[\beta_{js}^{q,j+1}(y) \cdots \beta_{js}^{q,p}(y) \right] d(y_{j+1} \cdots y_p) \tag{8.282}$$

$$\frac{\partial \left[\tilde{\beta}_{js}^1(y) \cdots \tilde{\beta}_{js}^j(y) \right]^\mathsf{T}}{\partial [y_1 \cdots y_j]^\mathsf{T}} = \left(\frac{\partial \left[\tilde{\beta}_{js}^1(y) \cdots \tilde{\beta}_{js}^j(y) \right]^\mathsf{T}}{\partial [y_1 \cdots y_j]^\mathsf{T}} \right)^\mathsf{T} \tag{8.283}$$

$$\gamma_{js}^r(y) \triangleq \int \left[\tilde{\beta}_{js}^1(y) \cdots \tilde{\beta}_{js}^j(y) \right] d(y_1 \cdots y_j) \tag{8.284}$$

$$\tilde{\mathbf{g}}_{j1}^0(x) \triangleq \mathbf{g}_{j1}^0(x), \ 1 \le j \le p \tag{8.285}$$

$$\begin{aligned}
\tilde{\mathbf{g}}_{i1}^s(x) &\triangleq \tilde{\mathbf{g}}_{i1}^{s-1}(x) \\
&+ \sum_{j=1}^{i-1} \sum_{k=1}^{\nu_j - \nu_i + 1 - s} (-1)^{k-1} L_{\mathbf{g}_{i1}^0} L_{F_0}^{\nu_j - k - s} \gamma_{js}^r(H(x)) \tilde{\mathbf{g}}_{jk}^{\nu_1 - \nu_j}(x)
\end{aligned} \tag{8.286}$$

and for $1 \le j \le p$ *and* $1 \le k \le \nu_j$,

$$\mathbf{g}_{j1}^0(x) \triangleq \tilde{\mathbf{g}}_{j1}^{\nu_1 - \nu_j}(x); \quad \mathbf{g}_{jk}^u(x) \triangleq \mathrm{ad}_{F_u}^{k-1} \mathbf{g}_{j1}^0(x) \tag{8.287}$$

(iii) for $1 \le i \le p, \ 1 \le j \le p, \ 1 \le k \le \nu_i$, *and* $1 \le r \le \nu_j$,

$$\bar{\mathbf{g}}_{ik}^u(x) = \bar{\mathbf{g}}_{ik}^0(x) \tag{8.288}$$

and

$$\left[\bar{\mathbf{g}}_{jr}^0(x), \ \bar{\mathbf{g}}_{ik}^0(x) \right] = 0. \tag{8.289}$$

Furthermore, a state coordinates transformation $z = S(x)$ *is given by*

$$\begin{aligned}
\frac{\partial S(x)}{\partial x} = \Big[&(-1)^{\nu_1 - 1} \bar{\mathbf{g}}_{1\nu_1}^0(x) \cdots -\bar{\mathbf{g}}_{12}^0(x) \, \bar{\mathbf{g}}_{11}^0(x) \\
&\cdots (-1)^{\nu_p - 1} \bar{\mathbf{g}}_{p\nu_p}^0(x) \cdots -\bar{\mathbf{g}}_{p2}^0(x) \, \bar{\mathbf{g}}_{p1}^0(x) \Big]^{-1}.
\end{aligned} \tag{8.290}$$

Conditions of Theorem 8.11 are verifiable. In other words, vector fields $\{\tilde{\mathbf{g}}_{i1}^s(x), \ 1 \le i \le p, \ 1 \le s \le \nu_1 - \nu_i\}$ in (8.286) are uniquely determined if conditions of Theorem 8.11 are satisfied. However, they are not necessary but sufficient unless, for $2 \le r \le \bar{p}, 1 \le s \le \nu_1 - \nu_{\sigma_r}, 1 \le q \le \sigma_{r-1},$ and $\sigma_{r-1} + 1 \le i \le p,$

$$L_{\mathbf{g}_{i1}} L_{F_0}^{\nu_j - k - s} \left(\gamma_{js}(H(x)) - \gamma_{js}^s(H(x)) \right) = 0.$$

Example 8.4.1 Consider the following control system: □

$$\dot{x} = \begin{bmatrix} x_2 \\ x_3 + x_1 u_2 (1 + u_2) \\ \alpha_{13}^u(x) \\ x_1^2 + u_2 \end{bmatrix} = F_u(x); \quad y = \begin{bmatrix} x_1 \\ x_4 \end{bmatrix} = H(x) \qquad (8.291)$$

where

$$\alpha_{13}^u(x) = x_3 x_4 + u_2 (2x_4 x_1^2 + x_4 x_1 + x_2) + 4x_1^2 x_2 + x_1 x_4 u_2^2$$
$$+ 2x_1 x_4 (x_1^3 + x_2 x_4) + u_1.$$

Use Corollary 8.6 to show that system (8.291) is state equivalent to a dual Brunovsky NOCF.

Solution By simple calculations, we have, by (8.165), that $(\nu_1, \nu_2) = (3, 1)$ and

$$\xi = T(x) \triangleq \begin{bmatrix} H_1(x) \\ L_{F_0} H_1(x) \\ L_{F_0}^2 H_1(x) \\ H_2(x) \end{bmatrix} = \begin{bmatrix} x_1 \\ x_2 \\ x_3 \\ x_4 \end{bmatrix}.$$

Since $L_{F_0} H_2(x) = x_1^2$ and

$$dL_{F_0} H_2(x) = \begin{bmatrix} 2x_1 & 0 & 0 & 0 \end{bmatrix} = 2x_1 dH_1(x),$$

it is clear that condition (i) of Corollary 8.6 is satisfied. By (8.168) and (8.169), we have that

$$\begin{bmatrix} g_{11}^u(x) & g_{12}^u(x) & g_{13}^u(x) \end{bmatrix} = \begin{bmatrix} 0 & 0 & 1 \\ 0 & -1 & x_4 \\ 1 & -x_4 & 3x_1^2 + 2x_1 x_4^2 + x_4^2 \\ 0 & 0 & 0 \end{bmatrix}; \quad g_{21}^u(x) = \begin{bmatrix} 0 \\ 0 \\ 0 \\ 1 \end{bmatrix}.$$

Since $\tilde{g}_{21}^0(x) \triangleq g_{21}^0(x)$ and $\Phi_{21}(x) = \text{span}\{g_{11}^0(x)\}$, we have, by (8.269) with $s = 1$, that

$$[g_{13}^0(x), \tilde{g}_{21}^0(x)] = \begin{bmatrix} 0 \\ -1 \\ -2x_4(2x_1 + 1) \\ 0 \end{bmatrix} = g_{12}^0(x) \bmod \text{span} \left\{ \begin{bmatrix} 0 \\ 0 \\ 1 \\ 0 \end{bmatrix} \right\}$$

which implies that condition (ii) of Corollary 8.6 is satisfied with $\beta_{11}^{1,2}(y) = 1$ when $s = 1$. Thus, we have, by (8.271), (8.272), and (8.273), that

$$\gamma_{11}^2(y) \triangleq \int_0^{y_2} \int_0^{y_1} \beta_{11}^{1,2}(\tilde{y}_1, \tilde{y}_2) d\tilde{y}_1 d\tilde{y}_2 = y_1 y_2$$

and

$$\tilde{\mathbf{g}}_{21}^1(x) \triangleq \tilde{\mathbf{g}}_{21}^0(x) + L_{\mathbf{g}_{21}^0} L_{F_0} \gamma_{11}^2(H(x)) \mathbf{g}_{11}^0(x) + L_{\mathbf{g}_{21}^0} \gamma_{11}^2(H(x)) \mathbf{g}_{12}^0(x)$$

$$= \mathbf{g}_{21}^0(x) + x_2 \mathbf{g}_{11}^0(x) - x_1 \mathbf{g}_{12}^0(x) = \begin{bmatrix} 0 \\ x_1 \\ x_2 + x_1 x_4 \\ 1 \end{bmatrix}.$$

Also, it is easy to see that $\Phi_{22}(x) = \text{span}\{0\}$ and

$$\left[\mathbf{g}_{13}^0(x), \tilde{\mathbf{g}}_{21}^1(x) \right] = \begin{bmatrix} 0 \\ 0 \\ -4x_1 x_4 \\ 0 \end{bmatrix} = -4x_1 x_4 \mathbf{g}_{11}^0(x) \ \text{mod span} \left\{ \begin{bmatrix} 0 \\ 0 \\ 0 \\ 0 \end{bmatrix} \right\}$$

which implies that condition (ii) of Corollary 8.6 is satisfied with $\beta_{12}^{1,2}(y) = 4y_1 y_2$ when $s = 2(= \nu_1 - \nu_2)$. Thus, we have, by (8.271) and (8.273), that

$$\gamma_{12}^2(y) \triangleq \int_0^{y_2} \int_0^{y_1} \beta_{12}^{1,2}(\tilde{y}_1, \tilde{y}_2) d\tilde{y}_1 d\tilde{y}_2 = y_1^2 y_2^2$$

and

$$\tilde{\mathbf{g}}_{21}^2(x) \triangleq \tilde{\mathbf{g}}_{21}^1(x) + L_{\mathbf{g}_{21}^0} \gamma_{12}^2(H(x)) \mathbf{g}_{11}^0(x)$$

$$= \tilde{\mathbf{g}}_{21}^1(x) + 2x_1^2 x_4 \mathbf{g}_{11}^0(x) = \begin{bmatrix} 0 \\ x_1 \\ x_2 + x_1 x_4 + 2x_1^2 x_4 \\ 1 \end{bmatrix}.$$

Since $\bar{\mathbf{g}}_{1k}^u(x) = \mathbf{g}_{1k}^u(x) = \bar{\mathbf{g}}_{1k}^0(x)$ for $1 \leq k \leq 3(\nu_1)$ and $\bar{\mathbf{g}}_{21}^u(x) \triangleq \tilde{\mathbf{g}}_{21}^2(x) = \bar{\mathbf{g}}_{21}^0(x)$, it is clear that (8.275) is satisfied. It is also easy to see that

$$\left\{ \bar{\mathbf{g}}_{11}^0(x), \bar{\mathbf{g}}_{12}^0(x), \bar{\mathbf{g}}_{13}^0(x), \bar{\mathbf{g}}_{21}^0(x) \right\}$$

is a set of commuting vector fields, which implies that condition (iii) of Corollary 8.6 is satisfied. Hence, by Corollary 8.6, system (8.291) is state equivalent to a dual Brunovsky NOCF with state transformation $z = S(x)$. We have, by (8.277), that

$$\frac{\partial S(x)}{\partial x} = \left[\bar{g}_{13}^0(x) \ -\bar{g}_{12}^0(x) \ \bar{g}_{11}^0(x) \ \bar{g}_{21}^0(x) \right]^{-1}$$

$$= \begin{bmatrix} 1 & 0 & 0 & 0 \\ x_4 & 1 & 0 & x_1 \\ 3x_1^2 + 2x_1x_4^2 + x_4^2 \ x_4 & 1 & x_2 + x_1x_4 + 2x_1^2x_4 \\ 0 & 0 & 0 & 1 \end{bmatrix}^{-1}$$

$$= \begin{bmatrix} 1 & 0 & 0 & 0 \\ -x_4 & 1 & 0 & -x_1 \\ -3x_1^2 - 2x_1x_4^2 & -x_4 & 1 & -x_2 - 2x_1^2x_4 \\ 0 & 0 & 0 & 1 \end{bmatrix}$$

and

$$z = S(x) = \begin{bmatrix} x_1 \\ x_2 - x_1x_4 \\ x_3 - x_2x_4 - x_1^2(x_4^2 + x_1) \\ x_4 \end{bmatrix}.$$

Then it is easy to see that

$$S_*(F_u(x)) = \begin{bmatrix} z_2 \\ z_3 \\ 0 \\ 0 \end{bmatrix} + \begin{bmatrix} z_1 z_4 \\ z_1(u_2^2 + z_1 z_4^2) \\ u_1 \\ z_1^2 + u_2 \end{bmatrix} = \begin{bmatrix} z_2 \\ z_3 \\ 0 \\ 0 \end{bmatrix} + \begin{bmatrix} y_1 y_2 \\ y_1(u_2^2 + y_1 y_2^2) \\ u_1 \\ y_1^2 + u_2 \end{bmatrix}$$

$$= A_o z + \gamma^u(y).$$

$\square$

Example 8.4.2 Consider the following control system:

$$\dot{x} = \begin{bmatrix} x_2 \\ x_3 + x_1 u_3 \\ \alpha_{13}^u(x) \\ x_5 + u_1^2 \\ x_1x_4(x_1 + u_3) + x_2x_4x_6 + x_1x_6(x_5 + u_1^2) + u_2 \\ x_1 + u_3 \end{bmatrix} = F_u(x) \tag{8.292}$$

$$y = \begin{bmatrix} x_1 \\ x_4 \\ x_6 \end{bmatrix} = H(x)$$

where

$$\alpha_{13}^u(x) = 3x_1x_2 + x_1x_5 + x_2x_4 + x_3x_6 + x_1u_1^2 + u_1 + u_3x_2 + x_1x_6u_3.$$

Use Theorem 8.11 to show that system (8.292) is state equivalent to a dual Brunovsky NOCF.

Solution By simple calculations, we have, by (8.165), that $(\nu_1, \nu_2, \nu_3) = (3, 2, 1)$ and

$$\xi = T(x) \triangleq \begin{bmatrix} H_1(x) \\ L_{F_0} H_1(x) \\ L_{F_0}^2 H_1(x) \\ H_2(x) \\ L_{F_0} H_2(x) \\ H_3(x) \end{bmatrix} = \begin{bmatrix} x_1 \\ x_2 \\ x_3 \\ x_4 \\ x_5 \\ x_6 \end{bmatrix}.$$

Since $L_{F_0}^2 H_2(x) = x_1^2 x_4 + x_2 x_4 x_6 + x_1 x_5 x_6$, $L_{F_0} H_3(x) = x_1$,

$$dL_{F_0}^2 H_2(x) = \begin{bmatrix} 2x_1 x_4 + x_5 x_6 & x_4 x_6 & 0 & x_1^2 + x_2 x_6 & x_1 x_6 & x_2 x_4 + x_1 x_5 \end{bmatrix}$$
$$= (2x_1 x_4 + x_5 x_6)dH_1(x) + (x_1^2 + x_2 x_6)dH_2(x) + (x_2 x_4 + x_1 x_5)dH_3(x)$$
$$+ x_4 x_6 dL_{F_0} H_1(x) + x_1 x_6 dL_{F_0} H_2(x),$$

and

$$dL_{F_0} H_3(x) = \begin{bmatrix} 1 & 0 & 0 & 0 & 0 & 0 \end{bmatrix} = dH_1(x),$$

it is clear that condition (i) of Theorem 8.11 is satisfied. By (8.168) and (8.169), we have that

$$\begin{bmatrix} \mathbf{g}_{11}^u(x) & \mathbf{g}_{12}^u(x) & \mathbf{g}_{13}^u(x) \end{bmatrix} = \begin{bmatrix} 0 & 0 & 1 \\ 0 & -1 & x_6 \\ 1 & -x_6 & 2x_1 + x_4 + x_6^2 \\ 0 & 0 & 0 \\ 0 & 0 & x_4 x_6 \\ 0 & 0 & 0 \end{bmatrix}$$

and

$$\begin{bmatrix} \mathbf{g}_{21}^u(x) & \mathbf{g}_{22}^u(x) \end{bmatrix} = \begin{bmatrix} 0 & 0 \\ 0 & 0 \\ 0 & -x_1 \\ 0 & -1 \\ 1 & -x_1 x_6 \\ 0 & 0 \end{bmatrix}; \quad \mathbf{g}_{31}^u(x) = \begin{bmatrix} 0 \\ 0 \\ 0 \\ 0 \\ 0 \\ 1 \end{bmatrix}.$$

Note that $\nu_1 > \nu_2 > \nu_3$, $\sigma_0 = 0$, $(\sigma_1, \sigma_2, \sigma_3) = (1, 2, 3)$, and $\bar{p} = 3$. Since $\tilde{\mathbf{g}}_{21}^0(x) \triangleq \mathbf{g}_{21}^0(x)$, $\tilde{\mathbf{g}}_{31}^0(x) \triangleq \mathbf{g}_{31}^0(x)$, $\Phi_{21}(x) = \text{span}\{0\}$, and $\Phi_{31}(x) = \text{span}\{\mathbf{g}_{11}^0(x)\}$, we have, by (8.279) with $r = 2$ and $s = 1$, that

$$\begin{bmatrix} \mathbf{g}_{13}^0(x), \tilde{\mathbf{g}}_{21}^0(x) \end{bmatrix} = 0 = 0\mathbf{g}_{11}^0(x) \mod \text{span}\{0\}$$
$$= -\beta_{11}^{1,2}(H(x))\mathbf{g}_{11}^0(x) \mod \Phi_{21}(x)$$

and

$$\left[\mathbf{g}_{13}^0(x),\tilde{\mathbf{g}}_{31}^0(x)\right] = \begin{bmatrix} 0 \\ -1 \\ -2x_6 \\ 0 \\ -x_4 \\ 0 \end{bmatrix} = \mathbf{g}_{12}^0(x) - x_4\mathbf{g}_{21}^0(x) \ \text{mod span} \left\{ \begin{bmatrix} 0 \\ 0 \\ 1 \\ 0 \\ 0 \\ 0 \end{bmatrix} \right\}$$

$$= \beta_{11}^{1,3}(H(x))\mathbf{g}_{12}^0(x) - \beta_{21}^{1,3}(H(x))\mathbf{g}_{21}^0(x) \ \text{mod} \ \Phi_{31}(x)$$

which implies that condition (ii) of Theorem 8.11 is satisfied when $r = 2$ and $s = 1$, with

$$\left[\beta_{11}^{1,2}(y) \ \beta_{11}^{1,3}(y)\right] = \begin{bmatrix} 0 & 1 \end{bmatrix}.$$

Thus, we have, by (8.282)–(8.287), that

$$\tilde{\beta}_{11}^1(y) \triangleq \int \left[\beta_{11}^{1,2}(y) \ \beta_{11}^{1,3}(y)\right] d(y_2 \ y_3) = y_3$$

$$\gamma_{11}^2(y) \triangleq \int \tilde{\beta}_{11}^1(y) dy_1 = y_1 y_3$$

$$\tilde{\mathbf{g}}_{21}^1(x) \triangleq \tilde{\mathbf{g}}_{21}^0(x) + L_{\mathbf{g}_{21}^0} L_{F_0} \gamma_{11}^2(H(x))\mathbf{g}_{11}^0(x) = \mathbf{g}_{21}^0(x)$$

and

$$\bar{\mathbf{g}}_{21}^0(x) \triangleq \tilde{\mathbf{g}}_{21}^1(x) = \mathbf{g}_{21}^0(x) \ ; \ \ \bar{\mathbf{g}}_{22}^u(x) \triangleq \text{ad}_{F_u} \bar{\mathbf{g}}_{21}^0(x) = \mathbf{g}_{22}^u(x).$$

Since $\Phi_{31}(x) = \text{span}\{\mathbf{g}_{11}^0(x)\}$, we have, by (8.279) with $r = 3$ and $s = 1$, that

$$\left[\mathbf{g}_{13}^0(x),\tilde{\mathbf{g}}_{31}^0(x)\right] = \begin{bmatrix} 0 \\ -1 \\ -2x_6 \\ 0 \\ -x_4 \\ 0 \end{bmatrix} = \mathbf{g}_{12}^0(x) - x_4\mathbf{g}_{21}^0(x) \ \text{mod span} \left\{ \begin{bmatrix} 0 \\ 0 \\ 1 \\ 0 \\ 0 \\ 0 \end{bmatrix} \right\}$$

$$= \beta_{11}^{1,3}(H(x))\mathbf{g}_{12}^0(x) - \beta_{21}^{1,3}(H(x))\mathbf{g}_{21}^0(x) \ \text{mod} \ \Phi_{31}(x)$$

and

$$\left[\bar{\mathbf{g}}_{22}^0(x), \tilde{\mathbf{g}}_{31}^0(x)\right] = \begin{bmatrix} 0 \\ 0 \\ 0 \\ 0 \\ x_1 \\ 0 \end{bmatrix} = 0\mathbf{g}_{12}^0(x) + x_1\mathbf{g}_{21}^0(x) \mod \mathrm{span} \left\{ \begin{bmatrix} 0 \\ 0 \\ 1 \\ 0 \\ 0 \\ 0 \end{bmatrix} \right\}$$

$$= -\beta_{11}^{2,3}(H(x))\mathbf{g}_{12}^0(x) + \beta_{21}^{2,3}(H(x))\mathbf{g}_{21}^0(x) \mod \Phi_{31}(x)$$

which implies that condition (ii) of Theorem 8.11 is satisfied when $r = 3$ and $s = 1$, with

$$\begin{bmatrix} \beta_{11}^{1,3}(y) \\ \beta_{11}^{2,3}(y) \end{bmatrix} = \begin{bmatrix} 1 \\ 0 \end{bmatrix} \quad \text{and} \quad \begin{bmatrix} \beta_{21}^{1,3}(y) \\ \beta_{21}^{2,3}(y) \end{bmatrix} = \begin{bmatrix} y_2 \\ y_1 \end{bmatrix}.$$

Thus, we have, by (8.282)–(8.286), that

$$\begin{bmatrix} \tilde{\beta}_{11}^1(y) \\ \tilde{\beta}_{11}^2(y) \end{bmatrix} \triangleq \begin{bmatrix} \int \beta_{11}^{1,3}(y)dy_3 \\ \int \beta_{11}^{2,3}(y)dy_3 \end{bmatrix} = \begin{bmatrix} y_3 \\ 0 \end{bmatrix}$$

$$\begin{bmatrix} \tilde{\beta}_{21}^1(y) \\ \tilde{\beta}_{21}^2(y) \end{bmatrix} \triangleq \begin{bmatrix} \int \beta_{21}^{1,3}(y)dy_3 \\ \int \beta_{21}^{2,3}(y)dy_3 \end{bmatrix} = \begin{bmatrix} y_2 y_3 \\ y_1 y_3 \end{bmatrix}$$

$$\gamma_{11}^3(y) \triangleq \int \left[\tilde{\beta}_{11}^1(y) \ \tilde{\beta}_{11}^2(y)\right] d(y_1 \ y_2) = \int \left[y_3 \ 0\right] d(y_1 \ y_2) = y_1 y_3$$

$$\gamma_{21}^3(y) \triangleq \int \left[\tilde{\beta}_{21}^1(y) \ \tilde{\beta}_{21}^2(y)\right] d(y_1 \ y_2) = \int \left[y_2 y_3 \ y_1 y_3\right] d(y_1 \ y_2) = y_1 y_2 y_3$$

and

$$\begin{aligned} \tilde{\mathbf{g}}_{31}^1(x) &\triangleq \tilde{\mathbf{g}}_{31}^0(x) + L_{\mathbf{g}_{31}^0} L_{F_0} \gamma_{11}^3(H(x))\mathbf{g}_{11}^0(x) - L_{\mathbf{g}_{31}^0} \gamma_{11}^3(H(x))\mathbf{g}_{12}^0(x) \\ &\quad + L_{\mathbf{g}_{31}^0} \gamma_{21}^3(H(x))\tilde{\mathbf{g}}_{21}^1(x) \\ &= \mathbf{g}_{31}^0(x) + x_2\mathbf{g}_{11}^0(x) - x_1\mathbf{g}_{12}^0(x) + x_1 x_4 \tilde{\mathbf{g}}_{21}^1(x) \\ &= \begin{bmatrix} 0 \ x_1 \ x_2 + x_1 x_6 \ 0 \ x_1 x_4 \ 1 \end{bmatrix}^\mathsf{T}. \end{aligned}$$

Since $\Phi_{32}(x) = \mathrm{span}\{0\}$, we have, by (8.279) with $r = 3$ and $s = 2$, that

$$\left[\mathbf{g}_{13}^0(x), \tilde{\mathbf{g}}_{31}^1(x)\right] = 0 = 0\mathbf{g}_{11}^0(x) = -\beta_{12}^{1,3}(H(x))\mathbf{g}_{11}^0(x)$$

and

$$\left[\bar{\mathbf{g}}_{22}^0(x), \tilde{\mathbf{g}}_{31}^1(x)\right] = 0 = 0\mathbf{g}_{11}^0(x) = \beta_{12}^{2,3}(H(x))\mathbf{g}_{11}^0(x)$$

which implies that condition (ii) of Theorem 8.11 is satisfied when $r = 3$ and $s = 1$, with

$$\begin{bmatrix} \beta_{12}^{1,3}(y) \\ \beta_{12}^{2,3}(y) \end{bmatrix} = \begin{bmatrix} 0 \\ 0 \end{bmatrix}.$$

Thus, we have, by (8.282)–(8.286), that

$$\begin{bmatrix} \tilde{\beta}_{12}^{1}(y) \\ \tilde{\beta}_{12}^{2}(y) \end{bmatrix} \triangleq \begin{bmatrix} \int \beta_{12}^{1,3}(y)dy_3 \\ \int \beta_{12}^{2,3}(y)dy_3 \end{bmatrix} = \begin{bmatrix} 0 \\ 0 \end{bmatrix}$$

$$\gamma_{12}^{3}(y) \triangleq \int \left[\tilde{\beta}_{12}^{1}(y) \; \tilde{\beta}_{12}^{2}(y) \right] d(y_1 \; y_2) = 0$$

and

$$\bar{\mathbf{g}}_{31}^{0}(x) \triangleq \tilde{\mathbf{g}}_{31}^{2}(x) \triangleq \tilde{\mathbf{g}}_{31}^{1}(x) + L_{\mathbf{g}_{31}^{0}} \gamma_{12}^{3}(H(x))\mathbf{g}_{11}^{0}(x) = \tilde{\mathbf{g}}_{31}^{1}(x).$$

Since $\bar{\mathbf{g}}_{1k}^{u}(x) = \mathbf{g}_{1k}^{u}(x)$ for $1 \le k \le 3$ and

$$\bar{\mathbf{g}}_{ik}^{u}(x) = \bar{\mathbf{g}}_{ik}^{0}(x), \; 1 \le i \le 3, \; 1 \le k \le v_i,$$

it is clear that (8.288) is satisfied. It is also easy to see that

$$\{\bar{\mathbf{g}}_{11}^{0}(x), \bar{\mathbf{g}}_{12}^{0}(x), \mathbf{g}_{13}^{0}(x), \bar{\mathbf{g}}_{21}^{0}(x), \bar{\mathbf{g}}_{22}^{0}(x), \bar{\mathbf{g}}_{31}^{0}(x)\}$$

is a set of commuting vector fields, which implies that condition (iii) of Theorem 8.11 is satisfied. Hence, by Theorem 8.11, system (8.292) is state equivalent to a dual Brunovsky NOCF with state transformation $z = S(x)$. We have, by (8.290), that

$$\frac{\partial S(x)}{\partial x} = \left[\bar{\mathbf{g}}_{13}^{0}(x) \; -\bar{\mathbf{g}}_{12}^{0}(x) \; \bar{\mathbf{g}}_{11}^{0}(x) \; -\bar{\mathbf{g}}_{22}^{0}(x) \; \bar{\mathbf{g}}_{21}^{0}(x) \; \bar{\mathbf{g}}_{31}^{0}(x) \right]^{-1}$$

$$= \begin{bmatrix} 1 & 0 & 0 & 0 & 0 & 0 \\ x_6 & 1 & 0 & 0 & 0 & x_1 \\ 2x_1 + x_4 + x_6^2 & x_6 & 1 & x_1 & 0 & x_2 + x_1x_6 \\ 0 & 0 & 0 & 1 & 0 & 0 \\ x_4x_6 & 0 & 0 & x_1x_6 & 1 & x_1x_4 \\ 0 & 0 & 0 & 0 & 0 & 1 \end{bmatrix}^{-1}$$

$$= \begin{bmatrix} 1 & 0 & 0 & 0 & 0 & 0 \\ -x_6 & 1 & 0 & 0 & 0 & -x_1 \\ -2x_1 - x_4 & -x_6 & 1 & -x_1 & 0 & -x_2 \\ 0 & 0 & 0 & 1 & 0 & 0 \\ -x_4x_6 & 0 & 0 & -x_1x_6 & 1 & -x_1x_4 \\ 0 & 0 & 0 & 0 & 0 & 1 \end{bmatrix}$$

and

$$z = S(x) = \begin{bmatrix} x_1 \\ x_2 - x_1 x_6 \\ x_3 - x_1(x_1 + x_4) - x_2 x_6 \\ x_4 \\ x_5 - x_1 x_4 x_6 \\ x_6 \end{bmatrix}.$$

Then it is easy to see that

$$S_* (F_u(x)) = \begin{bmatrix} z_2 \\ z_3 \\ 0 \\ z_5 \\ 0 \\ 0 \end{bmatrix} + \begin{bmatrix} z_1 z_6 \\ z_1 z_4 \\ u_1 \\ z_1 z_4 z_6 + u_1^2 \\ u_2 \\ z_1 + u_3 \end{bmatrix} = \begin{bmatrix} z_2 \\ z_3 \\ 0 \\ z_5 \\ 0 \\ 0 \end{bmatrix} + \begin{bmatrix} y_1 y_3 \\ y_1 y_2 \\ u_1 \\ y_1 y_2 y_3 + u_1^2 \\ u_2 \\ y_1 + u_3 \end{bmatrix}$$

$$= A_o z + \gamma^u(y).$$

$\square$

Example 8.4.3 Consider the following control system:

$$\dot{x} = \begin{bmatrix} x_2 \\ x_3 \\ x_5^2 + u_1 + x_4 u_2 \\ x_5 \\ u_2 \end{bmatrix} = F_u(x); \quad y = \begin{bmatrix} x_1 \\ x_4 \end{bmatrix} = H(x). \qquad (8.293)$$

(a) Show that system (8.293) does not satisfy the conditions of Corollary 8.6.
(b) Show that system (8.293) is state equivalent to a dual Brunovsky NOCF with
    state transformation

$$z = S(x) = \begin{bmatrix} x_1 \\ x_2 - \frac{1}{2} x_4^2 \\ x_3 - x_4 x_5 \\ x_4 \\ x_5 \end{bmatrix}.$$

Solution (a) By simple calculations, we have, by (8.165), that $(v_1, v_2) = (3, 2)$ and

$$\xi = T(x) \triangleq \begin{bmatrix} H_1(x) \\ L_{F_0} H_1(x) \\ L_{F_0}^2 H_1(x) \\ H_2(x) \\ L_{F_0} H_2(x) \end{bmatrix} = \begin{bmatrix} x_1 \\ x_2 \\ x_3 \\ x_4 \\ x_5 \end{bmatrix}.$$

Since $L_{F_0}^2 H_2(x) = x_5$ and

$$dL_{F_0}^2 H_2(x) = [0\ 0\ 0\ 0\ 1] = dL_{F_0} H_2(x),$$

it is clear that condition (i) of Corollary 8.6 is satisfied. By (8.168) and (8.169), we have that

$$[\mathbf{g}_{11}^u(x)\ \mathbf{g}_{12}^u(x)\ \mathbf{g}_{13}^u(x)] = \begin{bmatrix} 0 & 0 & 1 \\ 0 & -1 & 0 \\ 1 & 0 & 0 \\ 0 & 0 & 0 \\ 0 & 0 & 0 \end{bmatrix}$$

and

$$[\mathbf{g}_{21}^u(x)\ \mathbf{g}_{22}^u(x)] = \begin{bmatrix} 0 & 0 \\ 0 & 0 \\ 0 & -2x_5 \\ 0 & -1 \\ 1 & 0 \end{bmatrix}.$$

Since $\tilde{\mathbf{g}}_{21}^0(x) \triangleq \mathbf{g}_{21}^0(x)$ and $\Phi_{21}(x) = \mathrm{span}\{0\}$, we have, by (8.269) with $s = 1$, that

$$[\mathbf{g}_{13}^0(x), \tilde{\mathbf{g}}_{21}^0(x)] = 0 = 0\mathbf{g}_{11}^0(x) = -\beta_{11}^{1,2}(H(x))\mathbf{g}_{11}^0(x)$$

which implies that condition (ii) of Corollary 8.6 is satisfied with $\beta_{11}^{1,2}(y) = 0$ when $s = 1(= \nu_1 - \nu_2)$. Thus, we have, by (8.271), (8.272), and (8.273), that

$$\gamma_{11}^2(y) \triangleq \int_0^{y_2} \int_0^{y_1} \beta_{11}^{1,2}(\tilde{y}_1, \tilde{y}_2) d\tilde{y}_1 d\tilde{y}_2 = 0$$

and

$$\bar{\mathbf{g}}_{21}^0(x) \triangleq \tilde{\mathbf{g}}_{21}^1(x) \triangleq \tilde{\mathbf{g}}_{21}^0(x) + L_{\mathbf{g}_{21}^0} L_{F_0} \gamma_{11}^2(H(x))\mathbf{g}_{11}^0(x) = \mathbf{g}_{21}^0(x).$$

Since $\bar{\mathbf{g}}_{1k}^u(x) \triangleq \mathbf{g}_{1k}^u(x) = \mathbf{g}_{1k}^0(x)$ for $1 \leq k \leq 3$ and

$$\bar{\mathbf{g}}_{22}^u(x) \triangleq \mathrm{ad}_{F_u} \bar{\mathbf{g}}_{21}^0(x) = \mathrm{ad}_{F_u} \mathbf{g}_{21}^0(x) = \mathbf{g}_{22}^u(x) = \mathbf{g}_{22}^0(x),$$

it is clear that (8.275) is satisfied. However, it is easy to see that

$$[\bar{\mathbf{g}}_{21}^0(x), \bar{\mathbf{g}}_{22}^0(x)] = [\mathbf{g}_{21}^0(x), \mathbf{g}_{22}^0(x)] = \begin{bmatrix} 0 \\ 0 \\ -2 \\ 0 \\ 0 \end{bmatrix} \neq 0$$

which implies that (8.276) is not satisfied. Hence, condition (iii) of Corollary 8.6 is not satisfied.

(b) It is easy to see that

$$S_*(F_u(x)) = \left(\frac{\partial S(x)}{\partial x} F_u(x)\right)_{x=S^{-1}(z)}$$

$$= \begin{bmatrix} 1 & 0 & 0 & 0 & 0 \\ 0 & 1 & 0 & -x_4 & 0 \\ 0 & 0 & 1 & -x_5 & -x_4 \\ 0 & 0 & 0 & 1 & 0 \\ 0 & 0 & 0 & 0 & 1 \end{bmatrix} \begin{bmatrix} x_2 \\ x_3 \\ x_5^2 + u_1 + x_4 u_2 \\ x_5 \\ u_2 \end{bmatrix} \Bigg|_{x=S^{-1}(z)}$$

$$= \begin{bmatrix} x_2 \\ x_3 - x_4 x_5 \\ u_1 \\ x_5 \\ u_2 \end{bmatrix}\Bigg|_{x=S^{-1}(z)} = \begin{bmatrix} z_2 + \frac{1}{2}z_4^2 \\ z_3 \\ u_1 \\ z_5 \\ u_2 \end{bmatrix} = \begin{bmatrix} z_2 \\ z_3 \\ 0 \\ z_5 \\ 0 \end{bmatrix} + \begin{bmatrix} \frac{1}{2}y_2^2 \\ 0 \\ u_1 \\ 0 \\ u_2 \end{bmatrix}.$$

Hence, system (8.293) is state equivalent to a dual Brunovsky NOCF with state transformation $z = S(x)$. In fact, we have, by (8.226), that

$$\left(\frac{\partial S(x)}{\partial x}\right)^{-1} = \begin{bmatrix} 1 & 0 & 0 & 0 & 0 \\ 0 & 1 & 0 & x_4 & 0 \\ 0 & 0 & 1 & x_5 & x_4 \\ 0 & 0 & 0 & 1 & 0 \\ 0 & 0 & 0 & 0 & 1 \end{bmatrix}$$

$$= \begin{bmatrix} \bar{\mathbf{g}}_{13}^0(x) & -\bar{\mathbf{g}}_{12}^0(x) & \bar{\mathbf{g}}_{11}^0(x) & -\bar{\mathbf{g}}_{22}^0(x) & \bar{\mathbf{g}}_{21}^0(x) \end{bmatrix}$$

which implies, together with (8.224), that

$$\bar{\mathbf{g}}_{21}^0(x) = \mathbf{g}_{21}^0(x) + \tilde{\ell}_{2,1,1}(x)\bar{\mathbf{g}}_{11}^0(x) = \mathbf{g}_{21}^0(x) + x_4\bar{\mathbf{g}}_{11}^0(x).$$

In other words, $\gamma_{11}^u(y) = \frac{1}{2}y_2^2 = \gamma_{11}^0(y)$ cannot be found by Corollary 8.6, because $\frac{\partial^2 \gamma_{11}^0(y)}{\partial y_1 \partial y_2} = 0$. (Refer to Remark 8.7.) Therefore, the conditions of Corollary 8.6 are not necessary but sufficient for state equivalence to a dual Brunovsky NOCF. Further investigations on Corollary 8.6 and Theorem 8.11 are needed for the verifiable necessary and sufficient conditions.   □

8.5   Discrete Time Observer Error Linearization

Consider a single output discrete time control system of the form

$$x(t + 1) = F\big(x(t), u(t)\big) \triangleq F_u(x(t))$$
$$y(t) = H(x(t))$$
(8.294)

with $F_0(0) = 0$, $H(0) = 0$, state $x \in \mathbb{R}^n$, input $u \in \mathbb{R}^m$, and output $y \in \mathbb{R}$. By letting $u = 0$ in system (8.294), we obtain the following autonomous system:

$$x(t + 1) = F_0(x(t)) ; \quad y(t) = H(x(t)).$$
(8.295)

Let $F_0^0(x) = x$, $\hat{F}_u^0(x) = x$, and for $k \geq 1$,

$$F_0^k \triangleq F_0^{k-1} \circ F_0(x) \quad \text{and} \quad \hat{F}_u^k(x) \triangleq F_0^{k-1} \circ F_u(x).$$

Definition 8.11   (*state equivalence to a LOCF*)
System (8.294) is said to be state equivalent to a LOCF, if there exists a diffeomorphism $z = S(x) : V_0 \to \mathbb{R}^n$, defined on some neighborhood V_0 of $x = 0$, such that

$$z(t + 1) = Az + \gamma(u) \triangleq S \circ F_u \circ S^{-1}(z)$$
$$y = Cz \triangleq H \circ S^{-1}(z)$$

where the pair (C, A) is observable and $\gamma(u) : \mathbb{R}^m \to \mathbb{R}^n$ is a smooth vector function with $\gamma(0) = 0$.

Definition 8.12   (*state equivalence to a dual Brunovsky NOCF*)
System (8.294) is said to be state equivalent to a dual Brunovsky NOCF, if there exist a diffeomorphism $z = S(x) : V_0 \to \mathbb{R}^n$, defined on some neighborhood V_0 of $x = 0$, such that

$$z(t + 1) = A_o z + \gamma(y, u) \triangleq \bar{f}_u(z)$$
$$y = C_o z \triangleq \bar{h}(z)$$

where $A_o = \begin{bmatrix} O_{(n-1)\times 1} & I_{(n-1)} \\ 0 & O_{1\times(n-1)} \end{bmatrix}$, $C_o = \begin{bmatrix} 1 & O_{1\times(n-1)} \end{bmatrix}$, and $\gamma(y, u) : \mathbb{R} \times \mathbb{R}^m \to \mathbb{R}^n$ is a smooth vector function with $\gamma(0, 0) = 0$.

Definition 8.13   (*state equivalence to a dual Brunovsky NOCF with OT*)
System (8.294) is said to be state equivalent to a dual Brunovsky NOCF with output transformation (OT), if there exist a smooth function $\varphi(y)$ $\left(\left. \dfrac{\partial \varphi(y)}{\partial y} \right|_{y=0} = \right.$

1 and $\varphi(0) = 0\Big)$ and a diffeomorphism $z = S(x) : V_0 \to \mathbb{R}^n$, defined on some neighborhood V_0 of $x = 0$, such that

$$z(t + 1) = A_o z + \gamma(\bar{y}, u) \triangleq \bar{f}_u(z)$$
$$\bar{y} = \varphi(y) = C_o z \triangleq \bar{h}(z)$$

where $\gamma(\bar{y}, u) : \mathbb{R} \times \mathbb{R}^m \to \mathbb{R}^n$ is a smooth vector function with $\gamma(0, 0) = 0$.

State equivalence to a dual Brunovsky NOCF for autonomous system (8.295) can be similarly defined with $u = 0$. If $\bar{f}_u(z) \triangleq S \circ F_u \circ S^{-1}(z) = A_o z + \gamma(z_1, u)$, then it is clear that $\bar{f}_0(z) \triangleq S \circ F_0 \circ S^{-1}(z) = A_o z + \gamma(z_1, 0)$. Thus, we have the following remark.

Remark 8.8 If system (8.294) is state equivalent to a dual Brunovsky NOCF with OT $\bar{y} = \varphi(y)$ and state transformation $z = S(x)$, then system (8.295) is also state equivalent to a dual Brunovsky NOCF with OT $\bar{y} = \varphi(y)$ and state transformation $z = S(x)$. But the converse is not true.

Since observability is invariant under state transformation, we assume the observability rank condition on the neighborhood of the origin. In other words,

$$\dim \text{span}\{ dH(x)|_{x=0}, d(H \circ F_0(x))|_{x=0}, \cdots, d(H \circ F_0^{n-1}(x))|_{x=0}\} = n.$$

Definition 8.14 (*Canonical System*)
The canonical system of system (8.294) is defined by

$$\xi(t + 1) = \begin{bmatrix} \xi_2 + \alpha_1^u(\xi) \\ \vdots \\ \xi_n + \alpha_{n-1}^u(\xi) \\ \alpha_n^u(\xi) \end{bmatrix} \triangleq f_u(\xi) ; \quad y = \xi_1 \triangleq h(\xi) \tag{8.296}$$

where

$$\xi = T(x) \triangleq \begin{bmatrix} H(x) \\ H \circ F_0(x) \\ \vdots \\ H \circ F_0^{n-1}(x) \end{bmatrix} \tag{8.297}$$

$f_u(\xi) \triangleq T \circ F_u \circ T^{-1}(\xi)$, $h(\xi) \triangleq H \circ T^{-1}(\xi)$, $\alpha_i^u(\xi) \triangleq H \circ \hat{F}_u^i \circ T^{-1}(\xi) - H \circ F_0^i \circ T^{-1}(\xi)$, $1 \le i \le n - 1$, and $\alpha_n^u(\xi) \triangleq H \circ \hat{F}_u^n \circ T^{-1}(\xi)$.

Remark 8.9 System (8.294) is state equivalent to a dual Brunovsky NOCF with OT $\varphi(y)$ and state transformation $z = S(x)$, if and only if canonical system (8.296) is state equivalent to a dual Brunovsky NOCF with OT $\varphi(y)$ and state transformation $z = \tilde{S}(\xi)$ ($\triangleq S \circ T^{-1}(\xi)$). Canonical system (8.296) is more convenient to solve the observer problems than system (8.294). Since geometric conditions are coordinate free, any geometric condition in $\xi -$ coordinates (for system (8.296)) can be expressed in $x -$ coordinates (for system (8.294)).

We assume that $F_0(x)$ is a diffeomorphism on a neighborhood of $x = 0$. In other words, $F_0(x)$ has the inverse function $(F_0)^{-1}(\bar{x})$. For system (8.294), we define vector fields $\{\mathbf{g}_1^0(x), \mathbf{g}_2^0(x), \cdots\}$ and $\{\mathbf{g}_1^u(x), \mathbf{g}_2^u(x), \cdots\}$ as follows.

$$L_{\mathbf{g}_1^0(x)}\left(H \circ F_0^{k-1}(x)\right) = \delta_{k,n}, \; 1 \le k \le n$$

$$\left(\text{or } \mathbf{g}_1^0(x) \triangleq \left(\frac{\partial T(x)}{\partial x}\right)^{-1} [0 \; \cdots \; 0 \; 1]^{\mathsf{T}} = T_*^{-1}(\frac{\partial}{\partial \xi_n})\right) \tag{8.298}$$

and for $i \ge 2$,

$$\mathbf{g}_i^0(x) \triangleq (F_0)_*(\mathbf{g}_{i-1}^0) = (F_0^{i-1})_*(\mathbf{g}_1^0)$$

$$\mathbf{g}_1^u(x) \triangleq \mathbf{g}_1^0(x) \; ; \quad \mathbf{g}_i^u(x) \triangleq (F_u)_*(\mathbf{g}_{i-1}^u). \tag{8.299}$$

Then it is easy to see, by Theorem 2.5, (8.298), and (8.299), that for $1 \le i \le n$ and $0 \le k \le n$,

$$L_{\mathbf{g}_i^0(x)}\left(H \circ F_0^k(x)\right) = \begin{cases} 0, & i+k < n \\ 1, & i+k = n. \end{cases} \tag{8.300}$$

If $F_0(x)$ is not invertible, $(F_0)_*(\mathbf{g}_1^0(x))$ might not be a well-defined vector field. Let $\xi = T(x)$ and for $1 \le i \le n$,

$$\mathbf{r}_i^u(\xi) \triangleq T_*(\mathbf{g}_i^u(x)) \; ; \quad \mathbf{r}_i^0(\xi) \triangleq T_*(\mathbf{g}_i^0(x)).$$

Since $f_u(\xi) = T \circ F_u \circ T^{-1}(\xi)$ and $f_0(\xi) = T \circ F_0 \circ T^{-1}(\xi)$, it is easy to see, by mathematical induction, (2.22), (8.298), and (8.299), that

$$\mathbf{r}_1^u(\xi) = \mathbf{r}_1^0(\xi) = \frac{\partial}{\partial \xi_n} \tag{8.301}$$

and for $i \ge 2$,

$$\mathbf{r}_i^u(\xi) \triangleq T_* \circ (F_u)_*(\mathbf{g}_{i-1}^u(x)) = T_* \circ (F_u)_* \circ T_*^{-1}(\mathbf{r}_{i-1}^u(x))$$

$$= (f_u)_*(\mathbf{r}_{i-1}^u(x)) \tag{8.302}$$

$$\mathbf{r}_i^0(\xi) = (f_0)_*(\mathbf{r}_{i-1}^0(x)).$$

The vector fields $\{\mathbf{g}_i^u(x),\ i \geq 1\}$ for system (8.294) are the same as the vector fields $\{\mathbf{r}_i^u(x),\ i \geq 1\}$ for system (8.296). In other words, (8.298) and (8.299) are coordinates free definition. Since

$$f_0(\xi) = T \circ F_0 \circ T^{-1}(\xi) = \left[\xi_2 \ \cdots \ \xi_n \ \alpha_n^0(\xi)\right]^\mathsf{T},$$

it is also easy to see, by (8.301) and (8.302), that for $1 \leq i \leq n$,

$$\mathbf{r}_i^0(\xi) = \begin{bmatrix} O_{(n-i)\times 1} \\ 1 \\ * \\ \vdots \\ * \end{bmatrix} \in \mathrm{span}\left\{\frac{\partial}{\partial \xi_{n+1-i}}, \ \cdots \ , \frac{\partial}{\partial \xi_n}\right\}. \tag{8.303}$$

Theorem 8.12 *System (8.294) is state equivalent to a LOCF, if and only if*

(i)

$$\mathbf{g}_i^u(x) = \mathbf{g}_i^0(x), \quad 2 \leq i \leq n+1$$

(ii)

$$[\mathbf{g}_1^0(x),\ \mathbf{g}_i^0(x)] = 0, \quad 2 \leq i \leq n+1.$$

Furthermore, a state transformation $z = S(x)$ can be obtained by

$$\frac{\partial S(x)}{\partial x} = \left[\ \mathbf{g}_n^0(x) \ \cdots \ \mathbf{g}_2^0(x) \ \mathbf{g}_1^0(x)\right]^{-1}.$$

Proof Proof is omitted. (If $u = 0$, this is the dual of the linearization of control system by state coordinated change that is considered in Sect. 3.2.) $\qquad \square$

Example 8.5.1 Consider the following control system:

$$x(t+1) = \begin{bmatrix} x_2 + (x_1 - x_2^2 + u_1)^2 + u_2^2 \\ x_1 - x_2^2 + u_1 \end{bmatrix} = F_u(x) \tag{8.304}$$

$$y = x_1 - x_2^2 = H(x).$$

Show that the above system is state equivalent to a LOCF without OT and find a state transformation $z = S(x)$ and the LOCF that new state z satisfies.

Solution Since $T(x) \triangleq [H(x) \ L_{F_0}H(x)]^\mathsf{T} = [x_1 - x_2^2 \ x_2]^\mathsf{T}$, it is clear, by (8.298) and (8.299), that

$$\mathbf{g}_1^u(x) \triangleq \mathbf{g}_1^0(x) \triangleq \left(\frac{\partial T(x)}{\partial x}\right)^{-1}\begin{bmatrix}0\\1\end{bmatrix} = \begin{bmatrix}1 & -2x_2\\0 & 1\end{bmatrix}^{-1}\begin{bmatrix}0\\1\end{bmatrix} = \begin{bmatrix}2x_2\\1\end{bmatrix}$$

$$\mathbf{g}_2^u(x) \triangleq (F_u)_*\mathbf{g}_1^u(x) = \begin{bmatrix}1\\0\end{bmatrix}$$

$$\mathbf{g}_3^u(x) \triangleq (F_u)_*\mathbf{g}_2^u(x) = \begin{bmatrix}2x_2\\1\end{bmatrix}$$

which imply that condition (i) and condition (ii) of Theorem 8.12 are satisfied. Hence, system (8.304) is state equivalent to a LOCF with state transformation $z = S(x) = [x_1 - x_2^2 \ \ x_2]^T$ and $\gamma(u) = [u_2^2 \ \ u_1]^T$, where

$$\frac{\partial S(x)}{\partial x} = \left[\, \mathbf{g}_2^0(x) \ \ \mathbf{g}_1^0(x)\,\right]^{-1} = \begin{bmatrix}1 & -2x_2\\0 & 1\end{bmatrix}$$

and

$$z(t+1) = S \circ F_u \circ S^{-1}(z) = \begin{bmatrix}0 & 1\\1 & 0\end{bmatrix}z + \begin{bmatrix}u_2^2\\u_1\end{bmatrix}$$

$$y = H \circ S^{-1}(z) = \begin{bmatrix}1 & 0\end{bmatrix}z.$$

$\square$

Theorem 8.13 *System (8.294) is state equivalent to a dual Brunovsky NOCF with state transformation $z = S(x)$, if and only if*

(i)

$$\mathbf{g}_i^u(x) = \mathbf{g}_i^0(x), \ \ 2 \le i \le n \tag{8.305}$$

(ii)

$$[\mathbf{g}_1^0(x), \ \mathbf{g}_i^0(x)] = 0, \ \ 2 \le i \le n \tag{8.306}$$

(iii)

$$\frac{\partial S(x)}{\partial x} = \left[\, \mathbf{g}_n^0(x) \ \ \cdots \ \ \mathbf{g}_2^0(x) \ \ \mathbf{g}_1^0(x)\,\right]^{-1}. \tag{8.307}$$

Proof Proof is omitted. (Special case of Lemma 8.9 with $\varphi(y) = y$.) $\square$

Example 8.5.2 Consider the following control system:

$$x(t+1) = \begin{bmatrix}x_2 + (x_1 - x_2^2)u^2 + (x_1 - x_2^2 + u)^2\\x_1 - x_2^2 + u\end{bmatrix} = F_u(x) \tag{8.308}$$

$$y = x_1 - x_2^2 = H(x).$$

Show that the above system is state equivalent to a dual Brunovsky NOCF without OT and find a state transformation $z = S(x)$ and the dual Brunovsky NOCF that new state z satisfies.

Solution Since $T(x) \triangleq [H(x) \ L_{F_0} H(x)]^\mathsf{T} = [x_1 - x_2^2 \ x_2]^\mathsf{T}$, it is clear, by (8.298) and (8.299), that

$$\mathbf{g}_1^u(x) \triangleq \mathbf{g}_1^0(x) \triangleq \left(\frac{\partial T(x)}{\partial x} \right)^{-1} \begin{bmatrix} 0 \\ 1 \end{bmatrix} = \begin{bmatrix} 1 & -2x_2 \\ 0 & 1 \end{bmatrix}^{-1} \begin{bmatrix} 0 \\ 1 \end{bmatrix} = \begin{bmatrix} 2x_2 \\ 1 \end{bmatrix}$$

$$\mathbf{g}_2^u(x) \triangleq (F_u)_* \mathbf{g}_1^u(x) = \begin{bmatrix} 1 \\ 0 \end{bmatrix}$$

$$\mathbf{g}_3^u(x) \triangleq (F_u)_* \mathbf{g}_2^u(x) = \begin{bmatrix} 2x_2 + u^2 \\ 1 \end{bmatrix}$$

which imply that $\mathbf{g}_3^u(x) \neq \mathbf{g}_3^0(x)$ and condition (i) of Theorem 8.12 is not satisfied. Therefore, by Theorem 8.12, system (8.308) is not state equivalent to a LOCF. However, since condition (i) and condition (ii) of Theorem 8.13 are satisfied, system (8.308) is state equivalent to a dual Brunovsky NOCF with state transformation $z = S(x) = [x_1 - x_2^2 \ x_2]^\mathsf{T}$ and $\gamma(y, u) = [y^2 u^2 \ y + u]^\mathsf{T}$, where

$$\frac{\partial S(x)}{\partial x} = \left[\mathbf{g}_2^0(x) \ \mathbf{g}_1^0(x) \right]^{-1} = \begin{bmatrix} 1 & -2x_2 \\ 0 & 1 \end{bmatrix}$$

and

$$z(t+1) = S \circ F_u \circ S^{-1}(z) = \begin{bmatrix} 0 & 1 \\ 0 & 0 \end{bmatrix} z + \begin{bmatrix} z_1^2 u^2 \\ z_1 + u \end{bmatrix}$$

$$y = H \circ S^{-1}(z) = \begin{bmatrix} 1 & 0 \end{bmatrix} z.$$

$\square$

Lemma 8.8 *System (8.294) is state equivalent to a dual Brunovsky NOCF with OT $\bar{y} = \varphi(y)$ and state transformation $z = S(x)$, if and only if there exist a diffeomorphism $\bar{y} = \varphi(y)$, smooth functions $\gamma_k^0(\bar{y}) : \mathbb{R} \to \mathbb{R}$, $1 \leq k \leq n$, and smooth functions $\varepsilon_k^u(\bar{y}) : \mathbb{R}^{1+m} \to \mathbb{R}$, $1 \leq k \leq n$ such that for $1 \leq i \leq n$,*

$$S_i(x) = \varphi \circ H \circ F_0^{i-1}(x) - \sum_{k=1}^{i-1} \gamma_k^0 \circ \varphi \circ H \circ F_0^{i-1-k}(x) \tag{8.309}$$

and

$$\varphi \circ H \circ \hat{F}_u^n(x) = \sum_{k=1}^{n-1} \gamma_k^0 \circ \varphi \circ H \circ \hat{F}_u^{n-k}(x) + \gamma_n^u \circ \varphi \circ H(x) \tag{8.310}$$

and

$$S_i \circ F_u(x) - S_i \circ F_0(x) = \varepsilon_i^u \circ \varphi \circ H(x) \tag{8.311}$$

where for $1 \leq i \leq n$,

$$\gamma_i^u(\bar{y}) = \gamma_i^0(\bar{y}) + \varepsilon_i^u(\bar{y}). \tag{8.312}$$

Proof Necessity. Suppose that system (8.294) is state equivalent to a dual Brunovsky NOCF with OT $\bar{y} = \varphi(y)$ and state transformation $z = S(x)$. Then, it is clear that $\varphi \circ H \circ S^{-1}(z) = \bar{h}(z) = z_1$ and $S \circ F_u \circ S^{-1}(z) = \bar{f}_u(z) = A_o z + \gamma^u(z_1)$. Since $\varphi \circ H \circ S^{-1} \circ S(x) = \bar{h} \circ S(x) = z_1 \circ S(x)$, it is clear that $S_1(x) = \varphi \circ H(x)$ and (8.309) is satisfied for $i = 1$. Also, since $S \circ F_u(x) = \bar{f}_u(z) \circ S(x) = A_o S(x) + \gamma^u(S_1(x))$, it is easy to see that for $1 \leq i \leq n - 1$,

$$\begin{aligned} S_{i+1}(x) &= S_i \circ F_u(x) - \gamma_i^u(S_1(x)) \\ &= S_i \circ F_0(x) - \gamma_i^0 \circ \varphi \circ H(x) \end{aligned} \tag{8.313}$$

and

$$S_n \circ F_u(x) = \gamma_n^u(S_1(x)) = \gamma_n^u \circ \varphi \circ H(x). \tag{8.314}$$

Thus, it is easy to see, by mathematical induction, that for $2 \leq i \leq n$,

$$S_i(x) = S_1 \circ F_0^{i-1}(x) - \sum_{k=1}^{i-1} \gamma_k^0 \circ \varphi \circ H \circ F_0^{i-1-k}(x)$$

which implies that (8.309) is also satisfied for $2 \leq i \leq n$. Also, since $S_n(x) = \varphi \circ H \circ F_0^{n-1}(x) - \sum_{k=1}^{n-1} \gamma_k^0 \circ T^1 \circ F_0^{n-1-k}(x)$, we have, by (8.314), that

$$\gamma_n^u \circ \varphi \circ H(x) = \varphi \circ H \circ \hat{F}_u^n(x) - \sum_{k=1}^{n-1} \gamma_k^0 \circ \varphi \circ H \circ \hat{F}_u^{n-k}(x)$$

which implies that (8.310) is satisfied. Finally, it is easy to see, by (8.309), (8.310), (8.312), and (8.313), that for $1 \leq i \leq n$,

$$\begin{aligned} \varepsilon_i^u \circ \varphi \circ H(x) &= \gamma_i^u \circ \varphi \circ H(x) - \gamma_i^0 \circ \varphi \circ H(x) \\ &= S_i \circ F_u(x) - S_{i+1}(x) - \gamma_i^0 \circ \varphi \circ H(x) \\ &= S_i \circ F_u(x) - S_i \circ F_0(x) \end{aligned}$$

which implies that (8.311) is satisfied.

Sufficiency. Suppose that there exist $\varphi(y)$ and $\{\gamma_k^0(\bar{y}), \varepsilon_k^u(\bar{y}) \mid 1 \leq k \leq n\}$ such that (8.309)–(8.312) are satisfied. Then it is easy to see, by (8.309), that $\bar{h}(z) \triangleq \varphi \circ H \circ S^{-1}(z) = z_1$ and for $1 \leq i \leq n - 1$,

$$S_i \circ F_0(x) = \varphi \circ H \circ F_0^i(x) - \sum_{k=1}^{i-1} \gamma_k^0 \circ \varphi \circ H \circ F_0^{i-k}(x)$$

$$= S_{i+1}(x) + \gamma_i^0 \circ \varphi \circ H(x)$$

which implies, together with (8.311) and (8.312), that for $1 \le i \le n - 1$,

$$S_i \circ F_u(x) = S_i \circ F_0(x) + \varepsilon_i^u \circ \varphi \circ H(x)$$

$$= S_{i+1}(x) + \gamma_i^0 \circ \varphi \circ H(x) + \varepsilon_i^u \circ \varphi \circ H(x)$$

$$= S_{i+1}(x) + \gamma_i^u \circ \varphi \circ H(x).$$

Finally, we have, by (8.309) and (8.310), that

$$S_n \circ F_u(x) = \varphi \circ H \circ \hat{F}_u^n(x) - \sum_{k=1}^{n-1} \gamma_k^0 \circ \varphi \circ H \circ \hat{F}_u^{n-k}(x)$$

$$= \gamma_n^u \circ \varphi \circ H(x).$$

Therefore, it is clear that

$$\bar{f}_u(z) \triangleq S \circ F_u \circ S^{-1}(z) = \begin{bmatrix} S_2(x) + \gamma_1^u \circ \varphi \circ H(x) \\ \vdots \\ S_{n-1}(x) + \gamma_{n-1}^u \circ \varphi \circ H(x) \\ \gamma_n^u \circ \varphi \circ H(x) \end{bmatrix}\Bigg|_{x=S^{-1}(z)} \tag{8.315}$$

$$= A_o z + \gamma^u(z_1).$$

Hence, system (8.294) is state equivalent to a dual Brunovsky NOCF with OT $\bar{y} = \varphi(y)$ and state transformation $z = S(x)$. $\qquad\square$

Corollary 8.7 *System (8.294) is state equivalent to a dual Brunovsky NOCF with state transformation $z = S(x)$, if and only if there exist smooth functions $\gamma_k^0(y)$: $\mathbb{R} \to \mathbb{R}$, $1 \le k \le n$ and $\varepsilon_k^u(y) : \mathbb{R}^{1+m} \to \mathbb{R}$, $1 \le k \le n$ such that for $1 \le i \le n$,*

$$S_i(x) = H \circ F_0^{i-1}(x) - \sum_{k=1}^{i-1} \gamma_k^0 \circ H \circ F_0^{i-1-k}(x)$$

$$H \circ \hat{F}_u^n(x) = \sum_{k=1}^{n-1} \gamma_k^0 \circ H \circ \hat{F}_u^{n-k}(x) + \gamma_n^u \circ H(x),$$

and

$$S_i \circ F_u(x) - S_i \circ F_0(x) = \varepsilon_i^u \circ H(x)$$

where for $1 \leq i \leq n$,

$$\gamma_i^u(\bar{y}) = \gamma_i^0(\bar{y}) + \varepsilon_i^u(\bar{y}).$$

Corollary 8.8 *System (8.295) is state equivalent to a dual Brunovsky NOCF with OT* $\bar{y} = \varphi(y)$ *and state transformation* $z = S(x)$, *if and only if there exist a diffeomorphism* $\bar{y} = \varphi(y)$ *and smooth functions* $\gamma_k^0(\bar{y}) : \mathbb{R} \to \mathbb{R}$, $1 \leq k \leq n$ *such that for* $1 \leq i \leq n$,

$$S_i(x) = \varphi \circ H \circ F_0^{i-1}(x) - \sum_{k=1}^{i-1} \gamma_k^0 \circ \varphi \circ H \circ F_0^{i-1-k}(x)$$

and

$$\varphi \circ H \circ \hat{F}_0^n(x) = \sum_{k=1}^{n} \gamma_k^0 \circ \varphi \circ H \circ \hat{F}_0^{n-k}(x).$$

Corollary 8.9 *System (8.295) is state equivalent to a dual Brunovsky NOCF with state transformation* $z = S(x)$, *if and only if there exist smooth functions* $\gamma_k^0(y) : \mathbb{R} \to \mathbb{R}$, $1 \leq k \leq n$ *such that for* $1 \leq i \leq n$,

$$S_i(x) = H \circ F_0^{i-1}(x) - \sum_{k=1}^{i-1} \gamma_k^0 \circ H \circ F_0^{i-1-k}(x)$$

and

$$H \circ \hat{F}_0^n(x) = \sum_{k=1}^{n} \gamma_k^0 \circ H \circ \hat{F}_0^{n-k}(x).$$

Lemma 8.9 *System (8.294) is state equivalent to a dual Brunovsky NOCF with OT* $\bar{y} = \varphi(y)$ *and state transformation* $z = S(x)$, *if and only if there exists a smooth function* $\ell(y)$ $(\ell(0) = 1)$ *such that*

(i)

$$\bar{\mathbf{g}}_i^u(x) = \bar{\mathbf{g}}_i^0(x), \quad 2 \leq i \leq n \tag{8.316}$$

(ii)

$$\left[\bar{\mathbf{g}}_1^0(x), \bar{\mathbf{g}}_i^0(x) \right] = 0, \quad 2 \leq i \leq n \tag{8.317}$$

where

$$\bar{\mathbf{g}}_1^u(x) = \bar{\mathbf{g}}_1^0(x) \triangleq \ell\left(H \circ F_0^{n-1}(x)\right)\mathbf{g}_1^0(x) \tag{8.318}$$

$$\bar{\mathbf{g}}_i^u(x) \triangleq (F_u)_*\left(\bar{\mathbf{g}}_{i-1}^u(x)\right), \quad i \geq 2 \tag{8.319}$$

$$\varphi(y) = \int_0^y \frac{1}{\ell(\bar{y})}d\bar{y} \tag{8.320}$$

$$\frac{\partial S(x)}{\partial x} = \left[\bar{\mathbf{g}}_n^0(x) \cdots \bar{\mathbf{g}}_2^0(x)\,\bar{\mathbf{g}}_1^0(x)\right]^{-1}. \tag{8.321}$$

Proof Necessity. Suppose that system (8.294) is state equivalent to a dual Brunovsky NOCF with OT $\bar{y} = \varphi(y)$ and state transformation $z = S(x)$. Therefore, by Lemma 8.8, there exist a smooth function $\varphi(y)$, smooth functions $\gamma_k^0(\bar{y})$, $1 \leq k \leq n$, and smooth functions $\varepsilon_k^u(\bar{y})$, $1 \leq k \leq n$ such that (8.309)–(8.312) are satisfied. In other words, we have

$$z(t+1) = A_o z + \gamma^u(z_1) \triangleq \bar{f}_u(z)$$
$$\bar{y} = \varphi \circ H \circ S^{-1}(z) = z_1$$

where

$$\bar{f}_u(z) \triangleq S \circ F_u \circ S^{-1}(z) = \begin{bmatrix} z_2 + \gamma_1^u(z_1) \\ \vdots \\ z_n + \gamma_{n-1}^u(z_1) \\ \gamma_n^u(z_1) \end{bmatrix}. \tag{8.322}$$

(See (8.315).) We define vector fields $\{\bar{\psi}_1^u(z), \cdots, \bar{\psi}_n^u(z)\}$ by

$$\bar{\psi}_1^u(z) \triangleq \frac{\partial}{\partial z_n} ; \quad \bar{\psi}_i^u(z) \triangleq (\bar{f}_u)_*\left(\bar{\psi}_{i-1}^u(z)\right), \quad i \geq 2. \tag{8.323}$$

Then, by (8.322), it is clear that

$$\bar{\psi}_i^u(z) = \frac{\partial}{\partial z_{n+1-i}} = \bar{\psi}_i^0(z), \quad 1 \leq i \leq n \tag{8.324}$$

which implies that

$$\left[\bar{\psi}_i^u(z), \bar{\psi}_k^u(z)\right] = 0, \quad 1 \leq i \leq n,\, 1 \leq k \leq n. \tag{8.325}$$

Let $\xi = T(x) \triangleq \left[H(x)\ H \circ F_0(x) \cdots\ H \circ F_0^{n-1}(x)\right]^{\mathsf{T}}$ and $\tilde{S}(\xi) \triangleq S \circ T^{-1}(\xi)$. Then it is clear, by (8.309), that for $1 \leq i \leq n$,

$$\tilde{S}_i(\xi) \triangleq S_i \circ T^{-1}(\xi) = \varphi(\xi_i) - \sum_{k=1}^{i-1} \gamma_k^0 \left(\varphi(\xi_{i-k}) \right)$$

which implies that

$$\frac{\partial \tilde{S}_i(\xi)}{\partial \xi_n} = \begin{cases} 0, & \text{if } 1 \le i \le n-1 \\ \frac{d\varphi(\xi_n)}{d\xi_n}, & \text{if } i = n. \end{cases}$$

Thus, we have, by (8.323), that

$$\tilde{S}_*\left(\frac{\partial}{\partial \xi_n}\right) = \sum_{i=1}^{n} \left. \frac{\partial \tilde{S}_i(\xi)}{\partial \xi_n} \right|_{\xi = \tilde{S}^{-1}(z)} \frac{\partial}{\partial z_i}$$

$$= \left. \frac{d\varphi(\xi_n)}{d\xi_n} \right|_{\xi = \tilde{S}^{-1}(z)} \frac{\partial}{\partial z_n} = \left. \frac{d\varphi(\xi_n)}{d\xi_n} \right|_{\xi = \tilde{S}^{-1}(z)} \bar{\psi}_1^u(z)$$

and

$$\bar{\psi}_1^u(z) = \left. \ell(\xi_n) \right|_{\xi = \tilde{S}^{-1}(z)} \tilde{S}_*\left(\frac{\partial}{\partial \xi_n}\right)$$

where

$$\frac{1}{\ell(\xi_n)} = \frac{d\varphi(\xi_n)}{d\xi_n} \quad \left(\text{or } \varphi(y) = \int_0^y \frac{1}{\ell(\xi_n)} d\xi_n \right).$$

Therefore, we have, by (2.49), that

$$\tilde{S}_*^{-1}(\bar{\psi}_1^u(z)) = \tilde{S}_*^{-1}\left(\left. \ell(\xi_n) \right|_{\xi = \tilde{S}^{-1}(z)} \tilde{S}_*\left(\frac{\partial}{\partial \xi_n}\right) \right) = \ell(\xi_n)\frac{\partial}{\partial \xi_n}.$$

Hence, if we let $\bar{\mathbf{g}}_1^u(x) \triangleq S_*^{-1}(\bar{\psi}_1^u(z))$, we have, by (2.49) and (8.298), that

$$\mathbf{g}_1^u(x) = S_*^{-1}(\bar{\psi}_1^u(z)) = T_*^{-1} \circ \tilde{S}_*^{-1}(\bar{\psi}_1^u(z)) = T_*^{-1}\left(\ell(\xi_n)\frac{\partial}{\partial \xi_n} \right)$$

$$= \ell\left(H \circ F_0^{n-1}(x) \right) T_*^{-1}\left(\frac{\partial}{\partial \xi_n} \right) = \ell\left(H \circ F_0^{n-1}(x) \right) \mathbf{g}_1^0(x)$$

which implies that (8.318) is satisfied. It is easy to show, by mathematical induction, that for $i \ge 2$,

$$\mathbf{g}_i^u(x) = S_*^{-1}\left(\bar{\psi}_i^u(z) \right) \quad \text{or} \quad \bar{\psi}_i^u(z) = S_*\left(\bar{\mathbf{g}}_i^u(x) \right). \tag{8.326}$$

Assume that (8.326) is satisfied for $i = k - 1$ and $k \geq 2$. Since $\bar{f}_u(z) = S \circ F_u \circ S^{-1}(z)$ or $F_u(x) = S^{-1} \circ \bar{f}_u \circ S(x)$, it is easy to see, by (2.22), (8.319), and (8.323), that

$$
\begin{aligned}
\bar{\mathbf{g}}_k^u(x) = (F_u)_* \left(\bar{\mathbf{g}}_{k-1}^u(x) \right) &= S_*^{-1} \circ (\bar{f}_u)_* \circ S_* \left(\bar{\mathbf{g}}_{k-1}^u(x) \right) \\
&= S_*^{-1} \circ (\bar{f}_u)_* \left(\bar{\psi}_{k-1}^u(z) \right) = S_*^{-1} \left(\bar{\psi}_k^u(z) \right)
\end{aligned}
$$

which implies that (8.326) is satisfied for $i \geq 2$. Therefore, condition (i) of Lemma 8.9 is satisfied by (8.324) and (8.326). By (8.316), (8.325), (8.326), and Theorem 2.4, condition (ii) of Lemma 8.9 is also satisfied. Finally, since $S_* \left(\bar{\mathbf{g}}_i^0(x) \right) = \bar{\psi}_i^0(z) = \frac{\partial}{\partial z_{n+1-i}}$, $1 \leq i \leq n$ by (8.324) and (8.326), we have that

$$
\frac{\partial S(x)}{\partial x} \left[\bar{\mathbf{g}}_n^0(x) \cdots \bar{\mathbf{g}}_2^0(x) \, \bar{\mathbf{g}}_1^0(x) \right] = I|_{z=S(x)} = I
$$

which implies that (8.321) holds.

Sufficiency. Suppose that there exists $\beta(y)$ such that (8.316)–(8.320) are satisfied. Then it is easy to see, by (2.28) and (8.317), that for $1 \leq i < k \leq n$,

$$
\begin{aligned}
\left[\bar{\mathbf{g}}_i^0(x), \bar{\mathbf{g}}_k^0(x) \right] &= \left[(F_0)_*^{i-1} (\bar{\mathbf{g}}_1^0(x)), (F_0)_*^{k-1} (\bar{\mathbf{g}}_1^0(x)) \right] \\
&= (F_0)_*^{i-1} \left(\left[\bar{\mathbf{g}}_1^0(x), (F_0)_*^{k-i} (\bar{\mathbf{g}}_1^0(x)) \right] \right) \\
&= (F_0)_*^{i-1} \left(\left[\bar{\mathbf{g}}_1^0(x), \bar{\mathbf{g}}_{k-i+1}^0(x) \right] \right) = 0
\end{aligned}
$$

which implies that $\{ \bar{\mathbf{g}}_1^0(x), \bar{\mathbf{g}}_2^0(x), \cdots, \bar{\mathbf{g}}_n^0(x) \}$ is a set of commuting vector fields. Thus, there exists, by Theorem 2.7, a state transformation $z = S(x)$ such that

$$
S_* \left(\bar{\mathbf{g}}_i^0(x) \right) = \frac{\partial}{\partial z_{n+1-i}}, \quad 1 \leq i \leq n. \tag{8.327}
$$

In fact, $z = S(x)$ can be calculated by (8.321). Now it will be shown that $\bar{h}(z) \triangleq \varphi \circ H \circ S^{-1}(z) = z_1$ and $\bar{f}_u(z) \triangleq S \circ F_u \circ S^{-1}(z) = A_o z + \gamma^u(z_1)$. It is easy to show, by (2.49), (8.318), (8.319), and mathematical induction, that for $1 \leq i \leq n$,

$$
\bar{\mathbf{g}}_i^0(x) = \ell(H \circ F_0^{n-i}(x)) \mathbf{g}_i^0(x). \tag{8.328}
$$

Thus, we have, by Theorem 2.5, (8.300), (8.320), and (8.328), that for $1 \leq i \leq n$,

$$\frac{\partial \bar{h}(z)}{\partial z_{n+1-i}} = L_{S_*(\bar{\mathbf{g}}_i^0)}\left(\varphi \circ H \circ S^{-1}(z)\right) = \left\{L_{\bar{\mathbf{g}}_i^0(x)}\left(\varphi \circ H(x)\right)\right\}\Big|_{x=S^{-1}(z)}$$

$$= \left\{\frac{\partial \varphi(y)}{\partial y}\Big|_{y=H(x)} L_{\bar{\mathbf{g}}_i^0(x)} H(x)\right\}\Big|_{x=S^{-1}(z)}$$

$$= \left\{\frac{\partial \varphi(y)}{\partial y}\Big|_{y=H(x)} \ell(H \circ F_0^{n-i}) L_{\mathbf{g}_i^0(x)} H(x)\right\}\Big|_{x=S^{-1}(z)}$$

$$= \begin{cases} 0, & 1 \le i \le n-1 \\ \left\{\frac{\partial \varphi(y)}{\partial y}\Big|_{y=H(x)} \ell(H)\right\}\Big|_{x=S^{-1}(z)}, & i = n \end{cases}$$

$$= \begin{cases} 0, & 1 \le i \le n-1 \\ 1, & i = n. \end{cases}$$

Therefore, it is clear that $\bar{h}(z) = z_1$. Let

$$\bar{f}_u(z) \triangleq \sum_{k=1}^{n} \bar{f}_{u,k}(z) \frac{\partial}{\partial z_k} = [\bar{f}_{u,1}(z) \quad \cdots \quad \bar{f}_{u,n}(z)]^{\mathsf{T}}. \tag{8.329}$$

Since $\bar{f}_u(z) = S \circ F_u \circ S^{-1}(z)$, it is clear, by (2.22) and (8.319), that for $1 \le i \le n-1$,

$$S_*\left(\bar{\mathbf{g}}_{i+1}^u(x)\right) = S_*\left((F_u)_*(\bar{\mathbf{g}}_i^u(x))\right) = S_* \circ (F_u)_* \circ S_*^{-1}\left(S_*(\bar{\mathbf{g}}_i^u(x))\right) \tag{8.330}$$
$$= (\bar{f}_u)_*\left(S_*(\bar{\mathbf{g}}_i^u(x))\right).$$

Thus, we have, by (8.316), (8.327), (8.329), and (8.330), that for $1 \le i \le n-1$,

$$\frac{\partial}{\partial z_{n-i}} = (\bar{f}_u)_*\left(\frac{\partial}{\partial z_{n+1-i}}\right) = \frac{\partial \bar{f}_u(z)}{\partial z_{n+1-i}}$$
$$= \sum_{k=1}^{n} \frac{\partial \bar{f}_{u,k}(\bar{z})}{\partial \bar{z}_{n+1-i}}\Big|_{\bar{z}=\bar{f}_u^{-1}(z)} \frac{\partial}{\partial z_k}$$

which implies that for $1 \le k \le n$ and $1 \le i \le n-1$,

$$\frac{\partial \bar{f}_{u,k}(z)}{\partial z_{n+1-i}} = \begin{cases} 1, & k = n-i \\ 0, & \text{otherwise} \end{cases}$$

or, for $1 \le k \le n$ and $2 \le i \le n$,

$$\frac{\partial \bar{f}_{u,k}(z)}{\partial z_i} = \begin{cases} 1, & i = k+1 \\ 0, & \text{otherwise.} \end{cases}$$

Hence, $\bar{f}_{u,n}(z) = \gamma_n^u(z_1)$ and $\bar{f}_{u,k}(z) = z_{k+1} + \gamma_k^u(z_1)$, $1 \leq k \leq n-1$, for some functions $\gamma_k^u(z_1)$, $1 \leq k \leq n$. Thus, $\bar{f}_u(z) = A_o z + \gamma^u(z_1)$. $\square$

In order to find the conditions for the problem with OT, we define integer κ ($2 \leq \kappa \leq n+1$) and integer σ ($1 \leq \sigma \leq n$). Note that $L_{g_1^0}(H \circ F_0^{n-1}(x)) = 1 \neq 0$. Let us define integer κ ($2 \leq \kappa \leq n+1$) by the smallest integer such that

$$L_{g_\kappa^0}(H \circ F_0^{n-1}(x)) \neq 0. \tag{8.331}$$

Relation (8.331) will be used in (8.342). Since $L_{g_i^0}(H \circ F_0^{n-1}(x)) = 0$, $2 \leq i \leq \kappa - 1$, it is easy to see, by (2.30), that for $1 \leq i \leq \kappa - 2$,

$$\begin{aligned}
L_{r_i^0}(\alpha_n^0(\xi)) &= L_{r_i^0}(H \circ F_0^n \circ T^{-1}(\xi)) = L_{T_*(r_i^0)}(H \circ F_0^n(x))\big|_{x=T^{-1}(\xi)} \\
&= L_{g_i^0}(H \circ F_0^n(x))\big|_{x=T^{-1}(\xi)} = L_{(F_0)_*(g_i^0)}(H \circ F_0^{n-1}(x))\big|_{x=T^{-1}(\xi)} \\
&= L_{g_{i+1}^0}(H \circ F_0^{n-1}(x))\big|_{x=T^{-1}(\xi)} = 0
\end{aligned}$$

which implies, together with (8.296) and $r_1^0 = \frac{\partial}{\partial \xi_n}$, that

$$\alpha_n^0(\xi) = \alpha_n^0(\xi_1, \cdots, \xi_{n+2-\kappa}, 0, \cdots, 0) \tag{8.332}$$

$$T_*(g_i^0) = r_i^0(\xi) = \frac{\partial}{\partial \xi_{n+1-i}}, \quad 1 \leq i \leq \kappa - 1 \tag{8.333}$$

$$\frac{\partial \alpha_n^0(\xi)}{\partial \xi_{n+2-\kappa}} = L_{r_{\kappa-1}^0}(\alpha_n^0(\xi)) = L_{r_\kappa^0}(\xi_n) \neq 0 \tag{8.334}$$

and

$$[g_1^0(x), g_i^0(x)] = T_*^{-1}([r_1^0(\xi), r_i^0(\xi)]) = 0, \quad 1 \leq i \leq \kappa - 1. \tag{8.335}$$

Let us define σ ($1 \leq \sigma \leq n$) by the largest integer such that

$$H \circ \hat{F}_u^{n-i}(x) = H \circ F_0^{n-i}(x), \quad 1 \leq i \leq \sigma - 1 \tag{8.336}$$

or

$$\alpha_i^u(\xi) = 0, \quad n+1-\sigma \leq i \leq n-1 \tag{8.337}$$

where $\alpha_i^u(\xi) \triangleq H \circ \hat{F}_u^i \circ T^{-1}(\xi) - H \circ F_0^i \circ T^{-1}(\xi)$ for $1 \leq i \leq n-1$. Thus, if $\sigma < n$, we have

$$H \circ \hat{F}_u^{n-\sigma}(x) \neq H \circ F_0^{n-\sigma}(x) \text{ or } \frac{\partial}{\partial u}\left(H \circ \hat{F}_u^{n-\sigma}(x)\right) \neq 0. \tag{8.338}$$

If $\sigma = n$, then it is clear that for $1 \leq i \leq n-1$,

$$H \circ \hat{F}_u^i(x) = H \circ F_0^i(x) \text{ or } \alpha_i^u(\xi) = 0. \tag{8.339}$$

For example, if $H(x) = x_1$ and $F_u(x) = [x_2 + u \;\; x_3 + u^2 \;\; x_1 + u]^\mathsf{T}$, then $\sigma = 1$, because $H \circ \hat{F}_u^{n-1}(x) = x_3 + u^2 \neq H \circ \hat{F}_0^{n-1}(x)$. If $H(x) = x_1$ and $F_u(x) = [x_2 + u \;\; x_3 \;\; x_1 + u]^\mathsf{T}$, then $\sigma = 2$, because $H \circ \hat{F}_u^{n-1}(x) = x_3 = H \circ \hat{F}_0^{n-1}(x)$ and $H \circ \hat{F}_u^{n-2}(x) = x_2 + u \neq H \circ \hat{F}_0^{n-2}(x)$. If $H(x) = x_1$ and $F_u(x) = [x_2 \;\; x_3 \;\; x_1 + u]^\mathsf{T}$, then $\sigma = 3 = n$, because $H \circ \hat{F}_u^{n-1}(x) = x_3 = H \circ \hat{F}_0^{n-1}(x)$ and $H \circ \hat{F}_u^{n-2}(x) = x_2 = H \circ \hat{F}_0^{n-2}(x)$.

Example 8.5.3 Let $\bar{\mathbf{g}}_1^u(x) = \bar{\mathbf{g}}_1^0(x) \triangleq \ell \left(H \circ F_0^{n-1}(x) \right) \mathbf{g}_1^0(x)$ and $\sigma < n$. Show that

(a)

$$\bar{\mathbf{g}}_i^0(x) \triangleq (F_0)_* \left(\bar{\mathbf{g}}_{i-1}^0(x) \right) = \ell \left(H \circ F_0^{n-i}(x) \right) \mathbf{g}_i^0(x), \;\; 1 \leq i \leq n \tag{8.340}$$

(b)

$$\bar{\mathbf{g}}_i^u(x) = \begin{cases} \ell \left(H \circ F_0^{n-i}(x) \right) \mathbf{g}_i^u(x), & 1 \leq i \leq \sigma \\ \ell \left(H \circ F_0^{n-\sigma} \circ F_u^{-1}(x) \right) \mathbf{g}_{\sigma+1}^u(x), & i = \sigma + 1 \end{cases} \tag{8.341}$$

Solution (a) It is easy to see, by (2.22), (2.49), and (8.299), that for $1 \leq i \leq n$,

$$\begin{aligned} \bar{\mathbf{g}}_i^0(x) &= (F_0)_*^{i-1} \left(\bar{\mathbf{g}}_1^0(x) \right) = (F_0)_*^{i-1} \left(\ell \left(H \circ F_0^{n-1}(x) \right) \mathbf{g}_1^0(x) \right) \\ &= \ell \left(H \circ F_0^{n-i}(x) \right) (F_0)_*^{i-1} \left(\mathbf{g}_1^0(x) \right) \\ &= \ell \left(H \circ F_0^{n-i}(x) \right) \mathbf{g}_i^0(x). \end{aligned}$$

(b) (8.341) obviously holds when $i = 1$. Assume that (8.341) is satisfied when $i = k$ and $1 \leq k \leq \sigma - 1$. Then, it is easy to see, by (2.49), (8.299), and (8.336), that

$$\begin{aligned} \bar{\mathbf{g}}_{k+1}^u(x) &\triangleq (F_u)_* \left(\bar{\mathbf{g}}_k^u(x) \right) = (F_u)_* \left(\ell \left(H \circ F_0^{n-k}(x) \right) \mathbf{g}_k^u(x) \right) \\ &= \ell \left(H \circ F_0^{n-k} \circ F_u^{-1}(x) \right) \mathbf{g}_{k+1}^u(x) = \ell \left(H \circ F_0^{n-k-1}(x) \right) \mathbf{g}_{k+1}^u(x) \end{aligned}$$

which implies that (8.341) is satisfied when $i = k + 1$ and $1 \leq k \leq \sigma - 1$. Thus, by mathematical induction, (8.341) is satisfied when $1 \leq i \leq \sigma$. Thus, we have, by (2.49) and (8.299), that

$$\begin{aligned} \bar{\mathbf{g}}_{\sigma+1}^u(x) &\triangleq (F_u)_* \left(\bar{\mathbf{g}}_\sigma^u(x) \right) = (F_u)_* \left(\ell \left(H \circ F_0^{n-\sigma}(x) \right) \mathbf{g}_\sigma^u(x) \right) \\ &= \ell \left(H \circ F_0^{n-\sigma} \circ F_u^{-1}(x) \right) \mathbf{g}_{\sigma+1}^u(x). \end{aligned}$$

$\square$

Theorem 8.14 *Suppose that* $\kappa \le n$. *System (8.294) is state equivalent to a dual Brunovsky NOCF with OT* $\bar{y} = \varphi(y)$ *and state transformation* $z = S(x)$, *if and only if there exists a smooth function* $\beta(y)$, *defined on an open neighborhood of* $y = 0$, *such that*

(i)

$$[\mathbf{g}_1^0(x), \mathbf{g}_\kappa^0(x)] = L_{\mathbf{g}_\kappa^0}(H \circ F_0^{n-1}(x))\beta(H \circ F_0^{n-1}(x))\mathbf{g}_1^0(x) \qquad (8.342)$$

(ii)

$$\bar{\mathbf{g}}_i^u(x) = \bar{\mathbf{g}}_i^0(x), \quad 2 \le i \le n \qquad (8.343)$$

(iii)

$$[\bar{\mathbf{g}}_1^0(x), \ \bar{\mathbf{g}}_i^0(x)] = 0, \quad 2 \le i \le n \qquad (8.344)$$

where

$$\ell(y) \triangleq e^{\int_0^y \beta(\bar{y})d\bar{y}} \qquad (8.345)$$

$$\bar{\mathbf{g}}_1^u(x) = \bar{\mathbf{g}}_1^0(x) \triangleq \ell\left(H \circ F_0^{n-1}(x)\right)\mathbf{g}_1^0(x) \qquad (8.346)$$

$$\bar{\mathbf{g}}_i^u(x) \triangleq (F_u)_*(\bar{\mathbf{g}}_{i-1}^u)(x), \quad i \ge 2 \qquad (8.347)$$

$$\varphi(y) = \int_0^y \frac{1}{\ell(\bar{y})}d\bar{y} \qquad (8.348)$$

$$\frac{\partial S(x)}{\partial x} = \left[\bar{\mathbf{g}}_n^0(x) \cdots \bar{\mathbf{g}}_2^0(x) \ \bar{\mathbf{g}}_1^0(x)\right]^{-1}. \qquad (8.349)$$

Proof Necessity. Suppose that system (8.294) is state equivalent to a dual Brunovsky NOCF with OT $\bar{y} = \varphi(y)$ and state transformation $z = S(x)$. Then, by Lemma 8.9, there exist smooth functions $\ell(y)$ ($\ell(0) = 1$) such that (8.316)–(8.321) are satisfied. Since $2 \le \kappa \le n$, it is clear, by (8.300), that

$$L_{\mathbf{g}_1^0}\ell(H \circ F_0^{n-\kappa}(x)) = \left.\frac{d\ell(y)}{dy}\right|_{y=H \circ F_0^{n-\kappa}(x)} L_{\mathbf{g}_1^0}(H \circ F_0^{n-\kappa}(x)) = 0. \qquad (8.350)$$

Thus, we have, by (2.43), (8.317), (8.340), and (8.350), that

$$0 = [\bar{\mathbf{g}}_1^0(x), \bar{\mathbf{g}}_\kappa^0(x)] = \left[\ell \left(H \circ F_0^{n-1}(x) \right) \mathbf{g}_1^0(x), \ \ell \left(H \circ F_0^{n-\kappa}(x) \right) \mathbf{g}_\kappa^0(x) \right]$$

$$= \ell \left(H \circ F_0^{n-1}(x) \right) \ell \left(H \circ F_0^{n-\kappa}(x) \right) [\mathbf{g}_1^0(x), \ \mathbf{g}_\kappa^0(x)]$$

$$- \ell \left(H \circ F_0^{n-\kappa}(x) \right) L_{\mathbf{g}_\kappa^0} \ell \left(H \circ F_0^{n-1}(x) \right) \mathbf{g}_1^0(x)$$

which implies that

$$[\mathbf{g}_1^0(x), \mathbf{g}_\kappa^0(x)] = \frac{L_{\mathbf{g}_\kappa^0} \ell \left(H \circ F_0^{n-1}(x) \right)}{\ell(H \circ F_0^{n-1}(x))} \mathbf{g}_1^0(x)$$

$$= \frac{1}{\ell(y)} \left. \frac{d\ell(y)}{dy} \right|_{y=H \circ F_0^{n-1}(x)} L_{\mathbf{g}_\kappa^0} \left(H \circ F_0^{n-1}(x) \right) \mathbf{g}_1^0(x).$$

Therefore, (8.342) and (8.345) are satisfied with $\beta(y) = \frac{1}{\ell(y)} \frac{d\ell(y)}{dy} = \frac{d \ln \ell(y)}{dy}$. Condition (ii) and condition (iii) of Theorem 8.14 are obviously satisfied by (8.316) and (8.317).

Sufficiency. It is obvious by Lemma 8.9.   □

Theorem 8.15 *Suppose that $\sigma < n$. System (8.294) is state equivalent to a dual Brunovsky NOCF with OT $\varphi(y)$ and state transformation $z = S(x)$, if and only if there exist smooth scalar functions $\theta_\sigma^u(x)$, $\bar{\beta}^u(x)$, and $\beta(y)$, defined on an open neighborhood of $y = 0$, such that*

(i)

$$\mathbf{g}_i^u(x) = \begin{cases} \mathbf{g}_i^0(x), & 2 \leq i \leq \sigma \\ \theta_\sigma^u(x) \mathbf{g}_i^0(x), & i = \sigma + 1 \end{cases} \tag{8.351}$$

(ii)

$$\bar{\mathbf{g}}_i^u(x) = \bar{\mathbf{g}}_i^0(x), \ \ 2 \leq i \leq n \tag{8.352}$$

(iii)

$$[\bar{\mathbf{g}}_1^0(x), \ \bar{\mathbf{g}}_i^0(x)] = 0, \ \ 2 \leq i \leq n \tag{8.353}$$

where

$$\frac{\partial \left(\theta_\sigma^u \circ F_u(x) \right)}{\partial u} = \bar{\beta}^u(x) \frac{\partial \left(H \circ \hat{F}_u^{n-\sigma}(x) \right)}{\partial u} \tag{8.354}$$

$$\bar{\beta}^0(x) = \beta \left(H \circ F_0^{n-\sigma}(x) \right) \tag{8.355}$$

$$\ell(y) \triangleq e^{\int_0^y \beta(\bar{y}) d\bar{y}} \tag{8.356}$$

$$\bar{\mathbf{g}}_1^u(x) = \bar{\mathbf{g}}_1^0(x) \triangleq \ell\left(H \circ F_0^{n-1}(x)\right)\mathbf{g}_1^0(x) \tag{8.357}$$

$$\bar{\mathbf{g}}_i^u(x) \triangleq (F_u)_*(\bar{\mathbf{g}}_{i-1}^u(x)), \; i \geq 2 \tag{8.358}$$

$$\varphi(y) = \int_0^y \frac{1}{\ell(\bar{y})} d\bar{y} \tag{8.359}$$

$$\frac{\partial S(x)}{\partial x} = \left[\bar{\mathbf{g}}_n^0(x) \cdots \bar{\mathbf{g}}_2^0(x) \; \bar{\mathbf{g}}_1^0(x)\right]^{-1}. \tag{8.360}$$

Proof Necessity. Let $\sigma < n$. Suppose that system (8.294) is state equivalent to a dual Brunovsky NOCF with OT $\bar{y} = \varphi(y)$ and state transformation $z = S(x)$. Then, by Lemma 8.9, there exist smooth functions $\ell(y)$ ($\ell(0) = 1$) such that (8.352), (8.353), and (8.357)–(8.360) are satisfied. It is clear, by (8.340) and (8.341), that

$$\bar{\mathbf{g}}_i^0(x) = \ell\left(H \circ F_0^{n-i}(x)\right)\mathbf{g}_i^0(x), \; 2 \leq i \leq n$$

and

$$\bar{\mathbf{g}}_i^u(x) = \begin{cases} \ell\left(H \circ F_0^{n-i}(x)\right)\mathbf{g}_i^u(x), & 2 \leq i \leq \sigma \\ \ell\left(H \circ F_0^{n-\sigma} \circ F_u^{-1}(x)\right)\mathbf{g}_{\sigma+1}^u(x), & i = \sigma + 1 \end{cases}$$

which imply, together with (8.352), that (8.351) is satisfied with

$$\theta_\sigma^u(x) = \frac{\ell\left(H \circ F_0^{n-\sigma-1}(x)\right)}{\ell\left(H \circ F_0^{n-\sigma} \circ F_u^{-1}(x)\right)} \quad \text{or} \quad \frac{\ell\left(H \circ \hat{F}_u^{n-\sigma}(x)\right)}{\ell\left(H \circ F_0^{n-\sigma}(x)\right)} = \theta_\sigma^u \circ F_u(x).$$

Since

$$\frac{\partial\left(\theta_\sigma^u \circ F_u(x)\right)}{\partial u} = \frac{\left.\frac{d\ell(y)}{dy}\right|_{y=H \circ \hat{F}_u^{n-\sigma}(x)}}{\ell\left(H \circ F_0^{n-\sigma}(x)\right)} \frac{\partial\left(H \circ \hat{F}_u^{n-\sigma}(x)\right)}{\partial u},$$

it is clear that (8.354) is satisfied with

$$\bar{\beta}^u(x) = \frac{\left.\frac{d\ell(y)}{dy}\right|_{y=H \circ \hat{F}_u^{n-\sigma}(x)}}{\ell\left(H \circ F_0^{n-\sigma}(x)\right)}.$$

Since

$$\begin{aligned}\bar{\beta}^0(x) &= \frac{1}{\ell(y)}\left.\frac{d\ell(y)}{dy}\right|_{y=H \circ F_0^{n-\sigma}(x)} = \left.\frac{d}{dy}(\ln \ell(y))\right|_{y=H \circ F_0^{n-\sigma}(x)} \\ &\triangleq \beta\left(H \circ F_0^{n-\sigma}(x)\right),\end{aligned}$$

it is easy to see that (8.355) and (8.356) are satisfied.

Sufficiency. It is obvious by Lemma 8.9. □

If $\kappa < n + 1$, Theorem 8.14 can be used to find whether system (8.294) is state equivalent to a dual Brunovsky NOCF with OT or not. If $\sigma < n$, Theorem 8.15 can be used to find whether system (8.294) is state equivalent to a dual Brunovsky NOCF with OT or not. If $\kappa = n + 1$ and $\sigma = n$, we have, by (8.296), (8.332), and (8.339), that

$$f_u(\xi) \triangleq T \circ F_u \circ T^{-1}(\xi) = \begin{bmatrix} \xi_2 & \cdots & \xi_n & \alpha_n^u(\xi) \end{bmatrix}^{\mathsf{T}} \tag{8.361}$$

and

$$\alpha_n^u(\xi) \triangleq H \circ \hat{F}_u^n \circ T^{-1}(\xi) = \alpha_n^0(\xi_1, 0, \cdots, 0) + \hat{\alpha}_n^u(\xi) \tag{8.362}$$

where $\alpha_n^u(\xi) \triangleq \alpha_n^0(\xi) + \hat{\alpha}_n^u(\xi)$ and $\hat{\alpha}_n^0(\xi) = 0$.

Theorem 8.16 *Suppose that $\kappa = n + 1$ and $\sigma = n$. System (8.294) is state equivalent to a dual Brunovsky NOCF with OT $\varphi(y)$ and state transformation $z = S(x)$, if and only if*

$$\mathbf{g}_i^u(x) = \mathbf{g}_i^0(x), \quad 2 \leq i \leq n. \tag{8.363}$$

Furthermore, $\varphi(y) = y$ and state transformation $z = S(x)$ is given by

$$\frac{\partial S(x)}{\partial x} = \begin{bmatrix} \mathbf{g}_n^0(x) & \cdots & \mathbf{g}_2^0(x) & \mathbf{g}_1^0(x) \end{bmatrix}^{-1}. \tag{8.364}$$

Proof Necessity. Let $\kappa = n + 1$ and $\sigma = n$. Suppose that system (8.294) is state equivalent to a dual Brunovsky NOCF with OT $\bar{y} = \varphi(y)$ and state transformation $z = S(x)$. Then, by Lemma 8.8, it is clear that (8.309)–(8.312) are satisfied. Since $\kappa = n + 1$ and $\sigma = n$, it is easy to see, by (8.310), (8.339), and (8.362), that

$$\varphi \circ \alpha_n^u(\xi) = \varphi \circ H \circ \hat{F}_u^n(x) \circ T^{-1}(\xi)$$

$$= \sum_{k=1}^{n-1} \gamma_k^0 \circ \varphi \circ H \circ \hat{F}_u^{n-k} \circ T^{-1}(\xi) + \gamma_n^u \circ \varphi \circ H \circ T^{-1}(\xi)$$

$$= \sum_{k=1}^{n-1} \gamma_k^0 \circ \varphi \circ H \circ \hat{F}_0^{n-k} \circ T^{-1}(\xi) + \gamma_n^u \circ \varphi \circ H \circ T^{-1}(\xi)$$

$$= \sum_{k=1}^{n-1} \gamma_k^0 \circ \varphi(\xi_{n-k+1}) + \gamma_n^u \circ \varphi(\xi_1)$$

and

$$\varphi \circ \alpha_n^0(\xi_1, 0, \cdots, 0) = \sum_{k=1}^{n-1} \gamma_k^0 \circ \varphi(\xi_{n-k+1}) + \gamma_n^0 \circ \varphi(\xi_1)$$

which imply that $\gamma_k^0(y) = 0$ for $1 \le k \le n - 1$ and

$$\alpha_n^u(\xi) = \varphi^{-1} \circ \gamma_n^u \circ \varphi(\xi_1) \triangleq \tilde{\alpha}_n^u(\xi_1).$$

In other words, we have, by (8.361), that

$$f_u(\xi) \triangleq T \circ F_u \circ T^{-1}(\xi) = \begin{bmatrix} \xi_2 & \cdots & \xi_n & \tilde{\alpha}_n^u(\xi_1) \end{bmatrix}^T.$$

Therefore, it is easy to see that for $1 \le i \le n$,

$$T_*(\mathbf{g}_i^u(x)) = \mathbf{r}_i^u(\xi) \triangleq (f_u)_*^{i-1} \left(\frac{\partial}{\partial \xi_n} \right) = \frac{\partial}{\partial \xi_{n+1-i}}$$

which implies that (8.363) is satisfied.

Sufficiency. Let $\kappa = n + 1$ and $\sigma = n$. Suppose that (8.363) is satisfied. Note, by (8.335), that, if $\kappa = n + 1$, then (8.306) is satisfied. Therefore, by Theorem 8.13, system (8.294) is state equivalent to a dual Brunovsky NOCF with OT $\varphi(y) = y$ (i.e., without OT). $\qquad\square$

Example 8.5.4 Consider the following discrete time control system:

$$x(t + 1) = \begin{bmatrix} x_2 \\ \ln(1 + u + x_1 + x_2^2) \end{bmatrix} = F_u(x) \tag{8.365}$$

$$y = x_1 = H(x).$$

(a) Show that system (8.365) is not state equivalent to a dual Brunovsky NOCF without OT.
(b) Show that $\kappa = 2 \le n$ and $\sigma = 2 = n$.
(c) Use Theorem 8.14 to show that system (8.365) is state equivalent to a dual Brunovsky NOCF with OT. Also find a OT $\bar{y} = \varphi(y)$, a state transformation $z = S(x)$, and the dual Brunovsky NOCF that new state z satisfies.

Solution (a) It is easy to see that $\bar{x} = F_u^{-1}(x) = \begin{bmatrix} e^{x_2} - 1 - u - x_1^2 \\ x_1 \end{bmatrix}$. Since $\xi = T(x) \triangleq \begin{bmatrix} H(x) \\ H \circ F_0(x) \end{bmatrix} = x$, it is clear, by (8.298) and (8.299), that

$$\mathbf{g}_1^u(x) = \mathbf{g}_1^0(x) \triangleq \left(\frac{\partial T(x)}{\partial x} \right)^{-1} \begin{bmatrix} 0 \\ 1 \end{bmatrix} = \begin{bmatrix} 0 \\ 1 \end{bmatrix}$$

and

$$\mathbf{g}_2^u(x) \triangleq (F_u)_*(\mathbf{g}_1^u(x)) = \left. \frac{\partial F_u(\bar{x})}{\partial \bar{x}} \mathbf{g}_1^u(\bar{x}) \right|_{\bar{x} = F_u^{-1}(x)} = \begin{bmatrix} 1 \\ 2x_1 e^{-x_2} \end{bmatrix} = \mathbf{g}_2^0(x)$$

which imply that

$$[g_1^0(x), g_2^0(x)] = \begin{bmatrix} 0 \\ -2x_1 e^{-x_2} \end{bmatrix} = -2x_1 e^{-x_2} g_1^0(x) \neq 0 \qquad (8.366)$$

and condition (ii) of Theorem 8.13 is not satisfied. Therefore, by Theorem 8.13, system (8.365) is not state equivalent to a dual Brunovsky NOCF without OT.

(b) Since $L_{g_2^0}(H \circ F_0(x)) = 2x_1 e^{-x_2} \neq 0$, we have $\kappa = 2 \leq n$ by (8.331). Also, since $H \circ F_u(x) = x_2 = H \circ F_0(x)$, it is clear, by (8.336), that $\sigma = 2 = n$.

(c) It is clear, by (8.366), that condition (i) of Theorem 8.14 is satisfied with $\beta(y) = -1$. From (8.345)–(8.347), we have

$$\ell(y) = e^{\int_0^y \beta(\bar{y}) d\bar{y}} = e^{-y}$$

$$\bar{g}_1^u(x) \triangleq \ell(H \circ F_0(x)) g_1^0(x) = \begin{bmatrix} 0 \\ e^{-x_2} \end{bmatrix}$$

and

$$\bar{g}_2^u(x) \triangleq (F_u)_*(\bar{g}_1^u(x)) = \begin{bmatrix} e^{-x_1} \\ 2x_1 e^{-x_1 - x_2} \end{bmatrix}.$$

Since $\bar{g}_2^u(x) = \bar{g}_2^0(x)$ and $[\bar{g}_1^0(x), \bar{g}_2^0(x)] = 0$, it is clear that condition (ii) and condition (iii) of Theorem 8.14 are satisfied. Hence, by Theorem 8.14, system (8.365) is state equivalent to a dual Brunovsky NOCF with OT

$$\bar{y} = \varphi(y) = \int_0^y \frac{1}{\ell(\bar{y})} d\bar{y} = e^y - 1$$

and state transformation $z = S(x) = \begin{bmatrix} e^{x_1} - 1 \\ e^{x_2} - 1 - x_1^2 \end{bmatrix}$, where

$$\frac{\partial S(x)}{\partial x} = [\bar{g}_2^0(x) \; \bar{g}_1^0(x)]^{-1} = \begin{bmatrix} e^{x_1} & 0 \\ -2x_1 & e^{x_2} \end{bmatrix}.$$

It is easy to see that $\bar{y} = \varphi \circ H \circ S^{-1}(z) = z_1$ and

$$S \circ F_u \circ S^{-1}(z) = \begin{bmatrix} z_2 \\ 0 \end{bmatrix} + \begin{bmatrix} (\ln(1+z_1))^2 \\ \ln(1+z_1) + u \end{bmatrix} = \begin{bmatrix} z_2 \\ 0 \end{bmatrix} + \begin{bmatrix} y^2 \\ y + u \end{bmatrix}.$$

$\square$

Example 8.5.5 Consider the system

$$x(t + 1) = \begin{bmatrix} (1 + x_2)e^{u_2^2} - 1 \\ (1 + x_1)e^{u_1} - 1 \end{bmatrix} = F_u(x)$$

(8.367)

$$y = x_1 = H(x).$$

(a) Show that system (8.367) is not state equivalent to a dual Brunovsky NOCF without OT.
(b) Show that $\kappa = 3 = n + 1$ and $\sigma = 1 < n$.
(c) Use Theorem 8.15 to show that system (8.367) is state equivalent to a dual Brunovsky NOCF with OT. Also find a OT $\bar{y} = \varphi(y)$, a state transformation $z = S(x)$, and the dual Brunovsky NOCF that new state z satisfies.

Solution (a) It is easy to see that $\bar{x} = F_u^{-1}(x) = \begin{bmatrix} (1 + x_2)e^{-u_1} - 1 \\ (1 + x_1)e^{-u_2^2} - 1 \end{bmatrix}$. Since $\xi =$

$$T(x) \triangleq \begin{bmatrix} H(x) \\ H \circ F_0(x) \end{bmatrix} = x, \text{ we have, by (8.298) and (8.299), that}$$

$$\mathbf{g}_1^u(x) = \mathbf{g}_1^0(x) \triangleq \left(\frac{\partial T(x)}{\partial x} \right)^{-1} \begin{bmatrix} 0 \\ 1 \end{bmatrix} = \begin{bmatrix} 0 \\ 1 \end{bmatrix}$$

$$\mathbf{g}_2^u(x) \triangleq (F_u)_* (\mathbf{g}_1^u(x)) = \left. \frac{\partial F_u(\bar{x})}{\partial \bar{x}} \mathbf{g}_1^u(\bar{x}) \right|_{\bar{x} = F_u^{-1}(x)} = \begin{bmatrix} e^{u_2^2} \\ 0 \end{bmatrix}$$

and

$$\mathbf{g}_3^0(x) \triangleq (F_0)_* (\mathbf{g}_2^0(x)) = \left. \frac{\partial F_0(\bar{x})}{\partial \bar{x}} \mathbf{g}_2^0(\bar{x}) \right|_{\bar{x} = F_0^{-1}(x)} = \begin{bmatrix} 0 \\ 1 \end{bmatrix}.$$

Since $\mathbf{g}_2^u(x) \neq \mathbf{g}_2^0(x) = \begin{bmatrix} 1 \\ 0 \end{bmatrix}$, condition (i) of Theorem 8.13 is not satisfied. Therefore, system (8.367) is, by Theorem 8.13, not state equivalent to a dual Brunovsky NOCF without OT.

(b) Since $L_{\mathbf{g}_2^0}(H \circ F_0(x)) = 0$ and $L_{\mathbf{g}_3^0}(H \circ F_0(x)) = 1 + x_2^2 \neq 0$, we have $\kappa = 3 = n + 1$ by (8.331). Also, since $H \circ F_u(x) = (1 + x_2)e^{u_2^2} - 1 \neq H \circ F_0(x)$, it is clear, by (8.336), that $\sigma = 1 < n$.

(c) Note that $\mathbf{g}_1^u(x) = \mathbf{g}_1^0(x)$ and $\mathbf{g}_2^u(x) = e^{u_2^2} \mathbf{g}_2^0(x)$. Thus, it is clear that condition (i) of Theorem 8.15 is satisfied with $\theta_1^u(x) = e^{u_2^2}$. Since

$$\theta_1^u \circ F_u(x) = e^{u_2^2} ; \quad \frac{\partial(\theta_1^u \circ F_u(x))}{\partial u} = \begin{bmatrix} 0 & 2u_2 e^{u_2^2} \end{bmatrix}$$

and

$$H \circ F_u(x) = (1+x_2)e^{u_2^2} - 1 \; ; \quad \frac{\partial(H \circ F_u(x))}{\partial u} = \left[0 \; 2u_2(1+x_2)e^{u_2^2} \right],$$

it is easy to see that (8.354) is satisfied with $\bar{\beta}^u(x) = \frac{1}{1+x_2}$. Also, since $\bar{\beta}^0(x) = \frac{1}{1+x_2}$ and $H \circ F_0^{n-\sigma}(x) = H \circ F_0(x) = x_2$, (8.355) is satisfied with $\beta(y) = \frac{1}{1+y}$. Thus, we have, by (8.356)–(8.358), that

$$\ell(y) = e^{\int_0^y \beta(\bar{y})d\bar{y}} = e^{\ln(1+y)} = 1 + y$$

$$\bar{g}_1^u(x) \triangleq \ell(H \circ F_0)g_1^0(x) = \begin{bmatrix} 0 \\ 1 + x_2 \end{bmatrix}$$

and

$$\bar{g}_2^u(x) \triangleq (F_u)_*(\bar{g}_1^u(x)) = \begin{bmatrix} 1 + x_1 \\ 0 \end{bmatrix} = \bar{g}_2^0(x)$$

which imply that condition (ii) and condition (iii) of Theorem 8.15 are satisfied. Hence, by Theorem 8.15, system (8.367) is state equivalent to a dual Brunovsky NOCF with OT $\bar{y} = \varphi(y) = \int_0^y \frac{1}{\ell(\bar{y})}d\bar{y} = \ln(1+y)$ and state transformation $z = S(x) = \begin{bmatrix} \ln(1+x_1) \\ \ln(1+x_2) \end{bmatrix}$, where

$$\frac{\partial S(x)}{\partial x} = \left[\bar{g}_2^0(x) \; \bar{g}_1^0(x) \right]^{-1} = \begin{bmatrix} \frac{1}{1+x_1} & 0 \\ 0 & \frac{1}{1+x_2} \end{bmatrix}.$$

It is easy to see that $\varphi \circ H \circ S^{-1}(z) = z_1$ and

$$S \circ F_u \circ S^{-1}(z) = \begin{bmatrix} z_2 + u_2^2 \\ z_1 + u_1 \end{bmatrix} = \begin{bmatrix} z_2 \\ 0 \end{bmatrix} + \begin{bmatrix} u_2^2 \\ \ln(1+y) + u_1 \end{bmatrix}.$$

$\square$

Example 8.5.6 Consider the system

$$x(t+1) = \begin{bmatrix} x_2 \\ x_1 + u(1+x_2) \end{bmatrix} = F_u(x)$$

$$y = x_1 = H(x).$$

(8.368)

(a) Show that $\kappa = 3 = n + 1$ and $\sigma = 2 = n$.
(b) Use Theorem 8.16 to show that system (8.368) is not state equivalent to a dual Brunovsky NOCF with OT.

Solution (a) It is easy to see that $\bar{x} = F_u^{-1}(x) = \begin{bmatrix} x_2 - u(1+x_1) \\ x_1 \end{bmatrix}$. Since $\xi =$

$T(x) \triangleq \begin{bmatrix} H(x) \\ H \circ F_0(x) \end{bmatrix} = x$, we have, by (8.298) and (8.299), that

$$\mathbf{g}_1^u(x) = \mathbf{g}_1^0(x) \triangleq \left(\frac{\partial T(x)}{\partial x} \right)^{-1} \begin{bmatrix} 0 \\ 1 \end{bmatrix} = \begin{bmatrix} 0 \\ 1 \end{bmatrix}$$

$$\mathbf{g}_2^u(x) \triangleq (F_u)_* (\mathbf{g}_1^u(x)) = \left. \frac{\partial F_u(\bar{x})}{\partial \bar{x}} \mathbf{g}_1^u(\bar{x}) \right|_{\bar{x}=F_u^{-1}(x)} = \begin{bmatrix} 1 \\ u \end{bmatrix} \neq \mathbf{g}_2^0(x)$$

and

$$\mathbf{g}_3^0(x) \triangleq (F_0)_* (\mathbf{g}_2^0(x)) = \left. \frac{\partial F_0(\bar{x})}{\partial \bar{x}} \mathbf{g}_2^0(\bar{x}) \right|_{\bar{x}=F_0^{-1}(x)} = \begin{bmatrix} 0 \\ 1 \end{bmatrix}.$$

Since $L_{\mathbf{g}_2^0}(H \circ F_0(x)) = 0$ and $L_{\mathbf{g}_3^0}(H \circ F_0(x)) = 1 \neq 0$, we have $\kappa = 3 = n + 1$ by (8.331). Also, since $H \circ F_u(x) = H \circ F_0(x)$, it is clear, by (8.336), that $\sigma = 2 = n$.

(b) Since $\mathbf{g}_2^u(x) \neq \mathbf{g}_2^0(x)$, it is clear that (8.363) is not satisfied. Hence, by Theorem 8.16, system (8.368) is not state equivalent to a dual Brunovsky NOCF with OT.   □

Example 8.5.7 Consider the following discrete time control system:

$$x(t+1) = \begin{bmatrix} x_2(1+u) \\ \ln(1+u+x_1+x_2^2) \end{bmatrix} = F_u(x) \tag{8.369}$$

$$y = x_1 = H(x).$$

(a) Show that $\kappa = 2 \leq n$ and $\sigma = 1 < n$.
(b) Use Theorem 8.14 to show that the above system is not state equivalent to a dual Brunovsky NOCF with OT.
(c) Use Theorem 8.15 to show that the above system is not state equivalent to a dual Brunovsky NOCF with OT.

Solution (a) It is easy to see that $\bar{x} = F_u^{-1}(x) = \begin{bmatrix} e^{x_2} - 1 - u - \frac{x_1^2}{(1+u)^2} \\ \frac{x_1}{1+u} \end{bmatrix}$. Since

$\xi = T(x) \triangleq \begin{bmatrix} H(x) \\ H \circ F_0(x) \end{bmatrix} = x$, it is clear, by (8.298) and (8.299), that

$$\mathbf{g}_1^u(x) = \mathbf{g}_1^0(x) \triangleq \left(\frac{\partial T(x)}{\partial x} \right)^{-1} \begin{bmatrix} 0 \\ 1 \end{bmatrix} = \begin{bmatrix} 0 \\ 1 \end{bmatrix}$$

and

$$g_2^u(x) \triangleq (F_u)_*(g_1^u(x)) = \left. \frac{\partial F_u(\bar{x})}{\partial \bar{x}} g_1^u(\bar{x}) \right|_{\bar{x}=F_u^{-1}(x)} = \begin{bmatrix} 1+u \\ 2x_1 e^{-x_2} \\ \overline{1+u} \end{bmatrix}$$

which imply that

$$g_2^0(x) = \begin{bmatrix} 1 \\ 2x_1 e^{-x_2} \end{bmatrix}.$$

Since $L_{g_2^0}(H \circ F_0(x)) = 2x_1 e^{-x_2} \neq 0$, we have $\kappa = 2$ by (8.331). Also, since $H \circ F_u(x) = x_2(1+u) \neq H \circ F_0(x)$, it is clear, by (8.336), that $\sigma = 1$.

(b) Since $L_{g_2^0}(H \circ F_0(x)) = 2x_1 e^{-x_2}$ and

$$[g_1^0(x), g_2^0(x)] = \begin{bmatrix} 0 \\ -2x_1 e^{-x_2} \end{bmatrix} = -2x_1 e^{-x_2} g_1^0(x),$$

it is clear that (8.342) is satisfied with $\beta(y) = -1$. From (8.345)–(8.347), we have

$$\ell(y) = e^{\int_0^y \beta(\bar{y}) d\bar{y}} = e^{-y}$$

$$\bar{g}_1^u(x) \triangleq \ell(H \circ F_0(x)) g_1^0(x) = \begin{bmatrix} 0 \\ e^{-x_2} \end{bmatrix},$$

and

$$\bar{g}_2^u(x) \triangleq (F_u)_*(\bar{g}_1^u(x)) = \begin{bmatrix} (1+u)e^{\frac{-x_1}{1+u}} \\ 2x_1 e^{-x_2 - \frac{x_1}{1+u}} \\ \overline{1+u} \end{bmatrix}.$$

Since $\bar{g}_2^u(x) \neq \bar{g}_2^0(x)$, condition (ii) of Theorem 8.14 is not satisfied. Hence, by Theorem 8.14, system (8.369) is not state equivalent to a dual Brunovsky NOCF with OT.

(c) Since

$$g_2^0(x) = g_2^u(x)\big|_{u=0} = \begin{bmatrix} 1 \\ 2x_1 e^{-x_2} \end{bmatrix},$$

there does not exist $\theta_1^u(x)$ such that (8.351) is satisfied. Since condition (i) of Theorem 8.15 is not satisfied, system (8.369) is not state equivalent to a dual Brunovsky NOCF with OT.

$\square$

8.6   Discrete Time Dynamic Observer Error Linearization

Consider the following single output control system and autonomous system:

$$x(t + 1) = F\big(x(t), u(t)\big) \triangleq F_u(x(t))$$
$$y(t) = H(x(t)) \tag{8.370}$$

$$x(t + 1) = F\big(x(t), 0\big) \triangleq F_0(x(t))$$
$$y(t) = H(x(t)) \tag{8.371}$$

with $F_0(0) = 0$, $H(0) = 0$, state $x \in \mathbb{R}^n$, input $u \in \mathbb{R}^m$, and output $y \in \mathbb{R}$. Define the restricted dynamic system with index d (called auxiliary dynamics) by

$$\begin{bmatrix} w_1(t + 1) \\ \vdots \\ w_{d-1}(t + 1) \\ w_d(t + 1) \end{bmatrix} = \begin{bmatrix} w_2(t) \\ \vdots \\ w_d(t) \\ y(t) \end{bmatrix} \triangleq p(w(t), y(t)). \tag{8.372}$$

Define the extended system of system (8.370) with index d by

$$x^e(t + 1) \triangleq \begin{bmatrix} w(t + 1) \\ x(t + 1) \end{bmatrix} = \begin{bmatrix} p\big(w(t), H(x(t))\big) \\ F_u(x(t)) \end{bmatrix} \triangleq F_{e,u}(x^e(t))$$
$$y^e(t) = w_1(t) \triangleq H^e(x^e(t)) \tag{8.373}$$

where $x^e \triangleq \begin{bmatrix} w \\ x \end{bmatrix} \in \mathbb{R}^{d+n}$.

Definition 8.15 (*RDOEL with index d*)
System (8.370) is said to be restricted dynamic observer error linearizable (RDOEL) with index d, if there exist a smooth function $\varphi(y)$ $\left(\left. \frac{\partial \varphi(y)}{\partial y} \right|_{y=0} = 1 \text{ and } \varphi(0) = 0 \right)$ and a local extended state transformation $z^e = S^e(w, x) = \begin{bmatrix} \bar{w} \\ z \end{bmatrix} = \begin{bmatrix} \Phi(w) \\ S(w, x) \end{bmatrix}$ such that extended system (8.373) satisfies, in the new states z^e, to a generalized nonlinear observer canonical form (GNOCF) with index d defined by

$$z^e(t + 1) = A_e z^e(t) + \bar{\gamma}^u(z_1^e, \cdots, z_{d+1}^e) \triangleq \bar{f}_u(z^e)$$
$$\bar{y}^e(t) = C_e z^e(t) = z_1^e(t) \triangleq \bar{h}(z^e) \tag{8.374}$$

where $A_e = \begin{bmatrix} O_{(n+d-1)\times 1} & I_{(n+d-1)} \\ 0 & O_{1\times(n+d-1)} \end{bmatrix}$, $C_e = \begin{bmatrix} 1 & O_{1\times(n+d-1)} \end{bmatrix}$,

$$\Phi(w) \triangleq \left[\varphi(w_1) \cdots \varphi(w_d)\right]^{\mathsf{T}}$$

and $\bar{\gamma}^u : \mathbb{R}^{d+1} \times \mathbb{R}^m \to \mathbb{R}^{d+n}$ is a smooth vector function with $\bar{\gamma}_i^u(z_1^e, \cdots, z_{d+1}^e) = 0$ for $1 \leq i \leq d$. In other words,

$$\bar{h}(z^e) \triangleq \varphi \circ H^e \circ (S^e)^{-1}(z^e) = z_1^e \tag{8.375}$$

and

$$\begin{aligned}
\bar{f}_u(z^e) &\triangleq S^e \circ F_{e,u} \circ (S^e)^{-1}(z^e) = A_e z^e + \bar{\gamma}^u(z_1^e, \cdots, z_{d+1}^e) \\
&= A_e z^e + \bar{\gamma}^u(\varphi(w_1), \cdots, \varphi(w_d), \varphi(y)) \\
&\triangleq A_e z^e + \gamma^u(w_1, \cdots, w_d, y).
\end{aligned} \tag{8.376}$$

System (8.370) is said to be RDOEL, if system (8.370) is RDOEL with some index d. If we use a general nonlinear dynamic system $w(t+1) = \bar{p}(w(t), y(t))$ in Definition 8.15 instead of restricted (or linear) dynamic system (8.372), system (8.370) is said to be DOEL with index d.

Let $S^{-1}(\bar{w}, z)$ be the vector function such that $S(\bar{w}, S^{-1}(\bar{w}, z)) = z$ for all $\bar{w} \in \mathbb{R}^d$. In other words,

$$x^e = \begin{bmatrix} w \\ x \end{bmatrix} = (S^e)^{-1}(\bar{w}, z) = \begin{bmatrix} \Phi^{-1}(\bar{w}) \\ S^{-1}(\bar{w}, z) \end{bmatrix}.$$

If system (8.370) is RDOEL with index d, then we can design a state estimator

$$\begin{bmatrix} \bar{w}(t+1) \\ \bar{z}(t+1) \end{bmatrix} = (A_e - L_e C_e) \begin{bmatrix} \bar{w}(t) \\ \bar{z}(t) \end{bmatrix} + \bar{\gamma}^u(\bar{w}(t), y(t)) + L_e \bar{w}_1(t)$$

$$\bar{x}(t) \triangleq S^{-1}(\bar{w}(t), \bar{z}(t))$$

that yields an asymptotically vanishing error, i.e., $\lim_{t \to \infty} \|z^e(t) - \bar{z}^e(t)\| = 0$ or $\lim_{t \to \infty} \|x(t) - \bar{x}(t)\| = 0$, where $\bar{z}^e \triangleq \begin{bmatrix} w \\ \bar{z} \end{bmatrix}$ and $(A_e - L_e C_e)$ is an asymptotically stable $(d+n) \times (d+n)$ matrix. Block diagram for dynamic nonlinear observer can be found in Fig. 8.4.

RDOEL for autonomous system (8.371) can also be similarly defined with $u = 0$. If $\bar{f}_u^e(z^e) \triangleq S^e \circ F_{e,u} \circ (S^e)^{-1}(z^e) = A_e z^e + \gamma^u(w, y)$, then it is clear that $\bar{f}_0^e(z) \triangleq S^e \circ F_0^e \circ (S^e)^{-1}(z^e) = A_e z^e + \gamma^0(w, y)$. Thus, we have the following remark.

Remark 8.10 If system (8.370) is RDOEL with index d and state transformation $z^e = S^e(w, x)$, then system (8.371) is also RDOEL with index d and state transformation $z^e = S^e(w, x)$. But the converse is not true.

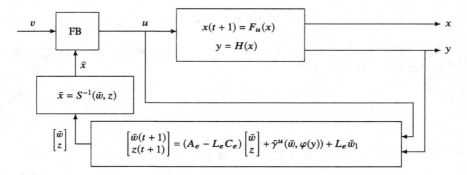

Fig. 8.4 Restricted dynamic nonlinear observer

Definition 8.16 (*state equivalence to a d-GNOCF with OT*)
System (8.370) is said to be state equivalent to a d-GNOCF with output transformation (OT), if there exist a smooth function $\varphi(y)$ $\left(\frac{\partial \varphi(y)}{\partial y} \Big|_{y=0} = 1 \text{ and } \varphi(0) = 0 \right)$
and a local state transformation $z = \tilde{S}(x)$ such that system (8.370) satisfies, in the new states z, a generalized nonlinear observer canonical form (GNOCF) with index d defined by

$$
\begin{aligned}
z(t+1) &= A_o z(t) + \gamma^u \circ \bar{\Phi}^{-1}(z_1, \cdots, z_{d+1}) \triangleq \bar{f}_u(z(t)) \\
\bar{y}(t) &= \varphi(y(t)) = Cz(t) = z_1(t) \triangleq \bar{h}(z(t))
\end{aligned}
\tag{8.377}
$$

where $A_o = \begin{bmatrix} O_{(n-1)\times 1} & I_{(n-1)} \\ 0 & O_{1\times(n-1)} \end{bmatrix}, C = [\, 1 \; O_{1\times(n-1)} \,], \bar{f}_u(z) = \tilde{S} \circ F_u \circ \tilde{S}^{-1}(z), \bar{h}(z) = \varphi \circ H \circ \tilde{S}^{-1}(z)$, $\gamma^u(\bar{z}_1, \cdots, \bar{z}_{d+1}) : \mathbb{R}^{d+1+m} \to \mathbb{R}^n$ is a smooth vector function with $\gamma_i^u = 0$, $1 \le i \le d$, and

$$
\bar{\Phi}^{-1}(z_1, \cdots, z_{d+1}) \triangleq \left[\varphi^{-1}(z_1) \cdots \varphi^{-1}(z_{d+1}) \right]^{\mathsf{T}}.
$$

In other words,

$$
\bar{h}(z) \triangleq \varphi \circ H \circ \tilde{S}^{-1}(z) = Cz(t) = z_1
\tag{8.378}
$$

and

$$
\begin{aligned}
\bar{f}_u(z) &\triangleq \tilde{S} \circ F_u \circ \tilde{S}^{-1}(z) = A_o z + \gamma^u \circ \bar{\Phi}^{-1}(z^1) \\
&\triangleq A_o z + \bar{\gamma}^u(z^1)
\end{aligned}
\tag{8.379}
$$

where $\bar{\gamma}_i^u(z^1) = 0$, $1 \le i \le d$ and

$$
z^1 \triangleq \left[z_1 \cdots z_{d+1} \right]^{\mathsf{T}}.
$$

If system (8.370) is state equivalent to 0-GNOCF with OT, then system (8.370) is said to be state equivalent to a dual Brunovsky NOCF with OT. Also, it is easy to see that system (8.370) is RDOEL with index d, if and only if extended system (8.373) is state equivalent to d-GNOCF with OT.

Since observability is invariant under state transformation, we assume the observability rank condition on the neighborhood of the origin. In other words,

$$\text{rank}\left(\left.\frac{\partial T(x)}{\partial x}\right|_{x=0}\right) = n$$

where

$$T(x) \triangleq \begin{bmatrix} H(x) \\ H \circ F_0(x) \\ \vdots \\ H \circ F_0^{n-1}(x) \end{bmatrix}. \tag{8.380}$$

For extended system (8.373), as in Definition 8.14, the canonical system can also be defined by

$$\xi^e(t+1) = \begin{bmatrix} \xi_2^e + \alpha_1^u(\xi^e) \\ \vdots \\ \xi_n^e + \alpha_{n+d-1}^u(\xi^e) \\ \alpha_{n+d}^u(\xi^e) \end{bmatrix} \triangleq f_u^e(\xi^e); \quad y_a = \xi_1^e = w_1 \triangleq h_E(\xi^e) \tag{8.381}$$

where

$$\xi^e \triangleq \begin{bmatrix} w \\ \xi \end{bmatrix} = T_e(x^e) \triangleq \begin{bmatrix} w \\ T(x) \end{bmatrix} = \begin{bmatrix} w \\ H(x) \\ H \circ F_0(x) \\ \vdots \\ H \circ F_0^{n-1}(x) \end{bmatrix} = \begin{bmatrix} H^e(x^e) \\ H^e \circ F_{e,0}(x^e) \\ \vdots \\ H^e \circ F_{e,0}^{n+d-1}(x^e) \end{bmatrix} \tag{8.382}$$

$f_u^e(\xi^e) \triangleq T_e \circ F_{e,u} \circ T_e^{-1}(\xi^e), \alpha_i^u(\xi^e) \triangleq H^e \circ \hat{F}_{e,u}^i \circ T_e^{-1}(\xi^e) - H^e \circ F_{e,0}^i \circ T_e^{-1}(\xi^e)$, $1 \le i \le n+d-1$, and $\alpha_{n+d}^u(\xi^e) \triangleq H^e \circ \hat{F}_{e,u}^{n+d} \circ T_e^{-1}(\xi^e)$. It is clear that for $1 \le i \le d$,

$$H^e \circ F_{e,u}^i(x^e) = H^e \circ F_{e,0}^i(x^e). \tag{8.383}$$

Lemma 8.10 *System (8.370) is state equivalent to a d-GNOCF with OT $\bar{y} = \varphi(y)$ and state transformation $z = \tilde{S}(x)$, if and only if there exist a smooth function $\varphi(y)$ and smooth functions $\gamma_k^u : \mathbb{R}^{d+1+m} \to \mathbb{R}$, $d + 1 \leq k \leq n$ such that for $1 \leq i \leq n$,*

$$\tilde{S}_i(x) = \varphi \circ H \circ F_0^{i-1}(x) - \sum_{k=d+1}^{i-1} \gamma_k^0 \circ T^1 \circ F_0^{i-1-k}(x) \tag{8.384}$$

$$\varphi \circ H \circ \hat{F}_u^n(x) = \sum_{k=d+1}^{n-1} \gamma_k^0 \circ T^1 \circ \hat{F}_u^{n-k}(x) + \gamma_n^u \circ T^1(x) \tag{8.385}$$

and

$$\tilde{S}_i \circ F_u(x) - \tilde{S}_i \circ F_0(x) = \begin{cases} 0, & 1 \leq i \leq d \\ \varepsilon_i^u \circ T^1(x), & d+1 \leq i \leq n \end{cases} \tag{8.386}$$

where for $d + 1 \leq i \leq n$,

$$\varepsilon_i^u(\xi_1, \cdots, \xi_{d+1}) \triangleq \gamma_i^u(\xi_1, \cdots, \xi_{d+1}) - \gamma_i^0(\xi_1, \cdots, \xi_{d+1}) \tag{8.387}$$

and

$$T^1(x) \triangleq \begin{bmatrix} T_1(x) \\ T_2(x) \\ \vdots \\ T_{d+1}(x) \end{bmatrix} = \begin{bmatrix} H(x) \\ H \circ F_0(x) \\ \vdots \\ H \circ F_0^d(x) \end{bmatrix}. \tag{8.388}$$

Proof Necessity. Suppose that system (8.370) is state equivalent to a d-GNOCF with OT $\bar{y} = \varphi(y)$ and state transformation $z = \tilde{S}(x)$. Then, it is clear, by (8.378) and (8.379), that $\varphi \circ H \circ \tilde{S}^{-1}(z) = \bar{h}(z) = z_1$ and

$$\tilde{S} \circ F_u \circ \tilde{S}^{-1}(z) = \bar{f}_u(z) = A_o z + \gamma^u \circ \bar{\Phi}^{-1}(z^1)$$

where

$$\gamma_i^u(\cdot) = 0, \ 1 \leq i \leq d. \tag{8.389}$$

Let

$$\tilde{S}^1(x) \triangleq [\tilde{S}_1(x), \cdots, \tilde{S}_{d+1}(x)]^\mathsf{T}.$$

Since $\varphi \circ H \circ \tilde{S}^{-1} \circ \tilde{S}(x) = \bar{h} \circ \tilde{S}(x) = z_1 \circ \tilde{S}(x) = \tilde{S}_1(x)$ and

$$\tilde{S} \circ F_u(x) = \bar{f}_u \circ \tilde{S}(x) = A_o \tilde{S}(x) + \gamma^u \circ \bar{\Phi}^{-1} \circ \tilde{S}^1(x), \tag{8.390}$$

it is easy to see, by (8.389), that $\tilde{S}_1(x) = \varphi \circ H(x)$ and

$$\tilde{S}_{i+1}(x) = \tilde{S}_i \circ F_u(x) = \tilde{S}_i \circ F_0(x), \ 1 \le i \le d \tag{8.391}$$

which implies that $\tilde{S}_i(x) = \varphi \circ H \circ F_0^{i-1}(x), \ 1 \le i \le d+1$ and thus (8.384) is satisfied for $1 \le i \le d+1$. In other words,

$$\tilde{S}^1(x) = \bar{\Phi} \circ T^1(x). \tag{8.392}$$

Similarly, we have, (8.390) and (8.392), that for $d+1 \le i \le n-1$,

$$\begin{aligned}
\tilde{S}_{i+1}(x) &= \tilde{S}_i \circ F_u(x) - \gamma_i^u \circ \bar{\Phi}^{-1} \circ \tilde{S}^1(x) = \tilde{S}_i \circ F_u(x) - \gamma_i^u \circ T^1(x) \\
&= \tilde{S}_i \circ F_0(x) - \gamma_i^0 \circ T^1(x)
\end{aligned} \tag{8.393}$$

and

$$\tilde{S}_n \circ F_u(x) = \gamma_n^u \circ \bar{\Phi}^{-1} \circ \tilde{S}^1(x) = \gamma_n^u \circ T^1(x). \tag{8.394}$$

Thus, it is easy to see, by mathematical induction, that for $d+2 \le i \le n$,

$$\tilde{S}_i(x) = \tilde{S}_1 \circ F_0^{i-1}(x) - \sum_{k=d+1}^{i-1} \gamma_k^0 \circ T^1 \circ F_0^{i-1-k}(x)$$

which implies that (8.384) is also satisfied for $d+2 \le i \le n$. Also, since $\tilde{S}_n(x) = \varphi \circ H \circ F_0^{n-1}(x) - \sum_{k=d+1}^{n-1} \gamma_k^0 \circ T^1 \circ F_0^{n-1-k}(x)$, we have, by (8.394), that

$$\gamma_n^u \circ T^1(x) = \varphi \circ H \circ \hat{F}_u^n(x) - \sum_{k=d+1}^{n-1} \gamma_k^0 \circ T^1 \circ \hat{F}_u^{n-k}(x)$$

which implies that (8.385) is satisfied. Finally, it is easy to see, by (8.393) and (8.394), that for $d+1 \le i \le n-1$,

$$\begin{aligned}
\varepsilon_i^u \circ T^1(x) &= \gamma_i^u \circ T^1(x) - \gamma_i^0 \circ T^1(x) = \tilde{S}_i \circ F_u(x) - \tilde{S}_{i+1}(x) - \gamma_i^0 \circ T^1(x) \\
&= \tilde{S}_i \circ F_u(x) - \tilde{S}_i \circ F_0(x)
\end{aligned}$$

and

$$\varepsilon_n^u \circ T^1(x) = \gamma_n^u \circ T^1(x) - \gamma_n^0 \circ T^1(x) = \tilde{S}_n \circ F_u(x) - \tilde{S}_n \circ F_0(x)$$

which imply, together with (8.391), that (8.386) is satisfied.

Sufficiency. Suppose that there exist $\varphi(y)$ and $\{\gamma_k^0, \varepsilon_k(u) \mid d+1 \le k \le n\}$ such that (8.384)–(8.386) are satisfied. Then it is easy to see, by (8.384), that $\bar{h}(z) \triangleq \varphi \circ H \circ \tilde{S}^{-1}(z) = z_1$,

$$\tilde{S}_i \circ F_0(x) = \varphi \circ H \circ F_0^i(x) = \tilde{S}_{i+1}(x), \ 1 \le i \le d \qquad (8.395)$$

and for $d + 1 \le i \le n - 1$,

$$\tilde{S}_i \circ F_0(x) = \varphi \circ H \circ F_0^i(x) - \sum_{k=1}^{i-1} \gamma_k^0 \circ T^1 \circ F_0^{i-k}(x)$$

$$= \tilde{S}_{i+1}(x) + \gamma_i^0 \circ T^1(x)$$

which imply, together with (8.386) and (8.387), that

$$\tilde{S}_i \circ F_u(x) = \tilde{S}_i \circ F_0(x) = \tilde{S}_{i+1}(x), \ 1 \le i \le d$$

and for $d + 1 \le i \le n - 1$,

$$\tilde{S}_i \circ F_u(x) = \tilde{S}_i \circ F_0(x) + \varepsilon_i^u \circ T^1(x)$$

$$= \tilde{S}_{i+1}(x) + \gamma_i^0 \circ T^1(x) + \varepsilon_i^u \circ T^1(x)$$

$$= \tilde{S}_{i+1}(x) + \gamma_i^u \circ T^1(x).$$

Finally, we have, by (8.384) and (8.385), that

$$\tilde{S}_n \circ F_u(x) = \varphi \circ H \circ \hat{F}_u^n(x) - \sum_{k=d+1}^{n-1} \gamma_k^0 \circ T^1 \circ \hat{F}_u^{n-k}(x)$$

$$= \gamma_n^u \circ T^1(x).$$

Note, by (8.395), that

$$\tilde{S}^1(x) = \bar{\Phi} \circ T^1(x) \ \text{ or } \ T^1(x) = \bar{\Phi}^{-1} \circ \tilde{S}^1(x).$$

Therefore, it is clear that

$$\bar{f}_u(z) \triangleq \tilde{S} \circ F_u \circ \tilde{S}^{-1}(z) = \left. \begin{bmatrix} \tilde{S}_2(x) + \gamma_1^u \circ \bar{\Phi}^{-1} \circ \tilde{S}^1(x) \\ \vdots \\ \tilde{S}_{n-1}(x) + \gamma_{n-1}^u \circ \bar{\Phi}^{-1} \circ \tilde{S}^1(x) \\ \gamma_n^u \circ \bar{\Phi}^{-1} \circ \tilde{S}^1(x) \end{bmatrix} \right|_{x = \tilde{S}^{-1}(z)}$$

$$= \begin{bmatrix} z_2 + \gamma_1^u \circ \bar{\Phi}^{-1}(z^1) \\ \vdots \\ z_{n-1} + \gamma_{n-1}^u \circ \bar{\Phi}^{-1}(z^1) \\ \gamma_n^u \circ \bar{\Phi}^{-1}(z^1) \end{bmatrix} = A_o z + \gamma^u \circ \bar{\Phi}^{-1}(z^1)$$

where $\gamma_i^u(\cdot) = 0$, $1 \le i \le d$. Hence, system (8.370) is state equivalent to a d-GNOCF with OT $\bar{y} = \varphi(y)$ and state transformation $z = \tilde{S}(x)$. □

Note, by (8.382) and (8.388), that

$$H^e \circ (\hat{F}_u^e)^i(x^e) = \begin{cases} w_{i+1}, & 0 \le i \le d-1 \\ H \circ \hat{F}_u^{i-d}(x), & i \ge d \end{cases} \tag{8.396}$$

and

$$(T^e)^1 \circ (\hat{F}_u^e)^{d+i}(x^e) = T^1 \circ \hat{F}_u^i(x), \ i \ge 0 \tag{8.397}$$

where $(T^e)^1(x^e) \triangleq [H^e \ H^e \circ (F_0^e) \ \cdots \ H^e \circ (F_0^e)^d]^\mathsf{T} = \begin{bmatrix} w \\ H(x) \end{bmatrix}$.

Lemma 8.11 *The followings are equivalent:*

(i) System (8.370) is RDOEL with index d and

$$H \circ \hat{F}_u^i(x) = H \circ F_0^i(x), \ 1 \le i \le d. \tag{8.398}$$

(ii) System (8.370) is state equivalent to a d-GNOCF with OT $\varphi(y)$ and state transformation $z = \tilde{S}(x)$ and for some $\bar{\varepsilon}_i^u(y)$, $d+1 \le i \le n$,

$$\tilde{S}_i \circ F_u(x) - \tilde{S}_i \circ F_0(x) = \begin{cases} 0, & 1 \le i \le d \\ \bar{\varepsilon}_i^u(H(x)), & d+1 \le i \le n. \end{cases} \tag{8.399}$$

Proof (i) $\Rightarrow$ (ii): Suppose that (8.398) is satisfied and system (8.370) is RDOEL with index d. In other words, extended system (8.373) is state equivalent to a d-GNOCF with OT $\varphi(y)$ and extended state transformation $z^e = S^e(w, x)$. Therefore, it is clear, by Lemma 8.10, that there exist a smooth function $\varphi(y)$ and smooth functions $\hat{\gamma}_k^0 : \mathbb{R}^{d+1} \to \mathbb{R}$ and $\hat{\varepsilon}_k^u : \mathbb{R}^{d+1+m} \to \mathbb{R}$, $d+1 \le k \le n+d$ such that for $1 \le i \le n$,

$$S_{d+i}^e(x^e) = \varphi \circ H^e \circ (F_0^e)^{d+i-1}(x^e) - \sum_{k=d+1}^{d+i-1} \hat{\gamma}_k^0 \circ (T^e)^1 \circ (F_0^e)^{d+i-1-k}(x^e) \tag{8.400}$$

$$\varphi \circ H^e \circ (\hat{F}_u^e)^{n+d}(x^e) = \sum_{k=d+1}^{n+d-1} \hat{\gamma}_k^0 \circ (T^e)^1 \circ (\hat{F}_u^e)^{n+d-k}(x^e) + \hat{\gamma}_{n+d}^u \circ (T^e)^1(x^e) \tag{8.401}$$

and for $1 \leq i \leq n$,

$$S_{d+i}^e \circ \hat{F}_u^e(x^e) - S_{d+i}^e \circ F_0^e(x^e) = \hat{\varepsilon}_{d+i}^u \circ (T^e)^1(x^e). \tag{8.402}$$

Let

$$\tilde{S}_i(x) \triangleq \varphi \circ H \circ F_0^{i-1}(x), \ 1 \leq i \leq d+1 \tag{8.403}$$

and for $d+2 \leq i \leq n$,

$$\tilde{S}_i(x) \triangleq \varphi \circ H \circ F_0^{i-1}(x) - \sum_{k=d+1}^{i-1} \gamma_k^0 \circ T^1 \circ F_0^{i-1-k}(x) \tag{8.404}$$

where

$$\gamma_k^0(\xi_1, \cdots, \xi_{d+1}) \triangleq \hat{\gamma}_k^0(\xi_1, \cdots, \xi_{d+1}), \ d+1 \leq k \leq n-1. \tag{8.405}$$

Also, let for $d+1 \leq i \leq n$,

$$\lambda_i^0(w, \xi_1, \cdots, \xi_d) \triangleq \sum_{k=i}^{d+i-1} \hat{\gamma}_k^0(w_{d+i-k}, \cdots, w_d, \xi_1, \cdots, \xi_{d+i-k}). \tag{8.406}$$

Then we have, by (8.396), (8.397), (8.400), and (8.403)–(8.406), that for $d+1 \leq i \leq n$,

$$\begin{aligned}
S_{d+i}^e(x^e) &= \varphi \circ H \circ F_0^{i-1}(x) - \sum_{k=d+1}^{i-1} \gamma_k^0 \circ T^1 \circ F_0^{i-1-k}(x) \\
&\quad - \sum_{k=i}^{d+i-1} \hat{\gamma}_k^0 \circ (T^e)^1 \circ (F_0^e)^{d+i-1-k}(x^e) \\
&= \tilde{S}_i(x) - \lambda_i^0(w, H(x), \cdots, H \circ F_0^{d-1}(x)).
\end{aligned} \tag{8.407}$$

Thus, it is easy to see, by (8.398), (8.402), (8.403), and (8.407), that for $1 \leq i \leq d$,

$$\tilde{S}_i \circ F_u(x) - \tilde{S}_i \circ F_0(x) = \varphi \circ H \circ \hat{F}_u^i(x) - \varphi \circ H \circ F_0^i(x) = 0$$

and for $d+1 \leq i \leq n$,

$$\begin{aligned}
\tilde{S}_i \circ F_u(x) - \tilde{S}_i \circ F_0(x) &= S_{d+i}^e \circ \hat{F}_u^e(x^e) - S_{d+i}^e \circ F_0^e(x^e) \\
&\quad + \lambda_i^0(p(w, H), H \circ F_u, \cdots, H \circ \hat{F}_u^d) - \lambda_i^0(p(w, H), H \circ F_0, \cdots, H \circ F_0^d) \\
&= \hat{\varepsilon}_{d+i}^u(w, H(x)) = \hat{\varepsilon}_{d+i}^u(0, \cdots, 0, H(x)) \triangleq \bar{\varepsilon}_i^u(H(x))
\end{aligned}$$

which imply that (8.399) is satisfied. Thus, it is also clear that

$$\tilde{S}_i \circ F_u(x) - \tilde{S}_i \circ F_0(x) = \begin{cases} 0, & 1 \le i \le d \\ \varepsilon_i^u \circ T^1(x), & d+1 \le i \le n \end{cases} \tag{8.408}$$

where for $d + 1 \le i \le n$,

$$\varepsilon_i^u(\xi_1, \cdots, \xi_{d+1}) \triangleq \bar{\varepsilon}_i^u(\xi_1).$$

Finally, it is clear, by (8.396)–(8.398), that for $n \le k \le n + d - 1$,

$$(T^e)^1 \circ (\hat{F}_u^e)^{n+d-k}(x^e) = (T^e)^1 \circ (\hat{F}_0^e)^{n+d-k}(x^e). \tag{8.409}$$

Therefore, we have, by (8.396), (8.401), and (8.409), that

$$\begin{aligned}
\varphi \circ H^e \circ (\hat{F}_u^e)^{n+d}(x^e) &= \sum_{k=d+1}^{n-1} \gamma_k^0 \circ T^1 \circ \hat{F}_u^{n-k}(x) + \hat{\gamma}_{n+d}^u \circ (T^e)^1(x^e) \\
&\quad + \sum_{k=n}^{n+d-1} \hat{\gamma}_k^0 \circ (T^e)^1 \circ (\hat{F}_0^e)^{n+d-k}(x^e) \\
&= \sum_{k=d+1}^{n-1} \gamma_k^0 \circ T^1 \circ \hat{F}_u^{n-k}(x) + \hat{\gamma}_{n+d}^u(w_1, \cdots, w_d, H(x)) \\
&\quad + \sum_{k=n}^{n+d-1} \hat{\gamma}_k^0(w_{n+d-k+1}, \cdots, w_d, H(x), \cdots, H \circ \hat{F}_0^{n+d-k}(x)) \\
&\triangleq \sum_{k=d+1}^{n-1} \gamma_k^0 \circ T^1 \circ \hat{F}_u^{n-k}(x) + \tilde{\gamma}_n^u(w_1, \cdots, w_d, \xi_1, \cdots, \xi_{d+1})\Big|_{\xi=T(x)}
\end{aligned}$$

which implies, together with (8.396), that

$$\begin{aligned}
\varphi \circ H \circ \hat{F}_u^n(x) &= \varphi \circ H^e \circ (\hat{F}_u^e)^{n+d}(x^e) = \varphi \circ H^e \circ (\hat{F}_u^e)^{n+d}(x^e)\Big|_{w=0} \\
&= \sum_{k=d+1}^{n-1} \gamma_k^0 \circ T^1 \circ \hat{F}_u^{n-k}(x) + \gamma_n^u \circ T^1(x)
\end{aligned} \tag{8.410}$$

where

$$\begin{aligned}
\tilde{\gamma}_n^u(w_1, \cdots, w_d, \xi_1, \cdots, \xi_{d+1}) &\triangleq \hat{\gamma}_{n+d}^u(w_1, \cdots, w_d, \xi_1) \\
&\quad + \sum_{k=n}^{n+d-1} \hat{\gamma}_k^0(w_{n+d-k+1}, \cdots, w_d, \xi_1, \cdots, \xi_{n+d-k+1})
\end{aligned}$$

and

$$\gamma_n^u(\xi_1, \cdots, \xi_{d+1}) \triangleq \tilde{\gamma}_n^u(O_{d \times 1}, \xi_1, \cdots, \xi_{d+1}).$$

Hence, by (8.403), (8.404), (8.408), (8.410), and Lemma 8.10, system (8.370) is state equivalent to a d-GNOCF with OT $\varphi(y)$ and state transformation $z = \tilde{S}(x)$.

(ii) $\Rightarrow$ (i): Suppose that system (8.370) is state equivalent to a d-GNOCF with OT $\varphi(y)$ and state transformation $z = \tilde{S}(x)$ and that (8.399) is satisfied. Then, by Lemma 8.10 and (8.399), there exist a smooth function $\varphi(y)$ and smooth functions $\gamma_k^u(\xi_1, \cdots, \xi_{d+1})$, $d + 1 \leq k \leq n$ such that for $1 \leq i \leq n$,

$$\tilde{S}_i(x) = \varphi \circ H \circ F_0^{i-1}(x) - \sum_{k=d+1}^{i-1} \gamma_k^0 \circ T^1 \circ F_0^{i-1-k}(x) \tag{8.411}$$

$$\varphi \circ H \circ \hat{F}_u^n(x) = \sum_{k=d+1}^{n-1} \gamma_k^0 \circ T^1 \circ \hat{F}_u^{n-k}(x) + \gamma_n^u \circ T^1(x) \tag{8.412}$$

and

$$\tilde{S}_i \circ F_u(x) - \tilde{S}_i \circ F_0(x) = \begin{cases} 0, & 1 \leq i \leq d \\ \bar{\varepsilon}_i^u \circ H(x), & d + 1 \leq i \leq n \end{cases} \tag{8.413}$$

where for $d + 1 \leq i \leq n$,

$$\bar{\varepsilon}_i^u(\xi_1) \triangleq \gamma_i^u(\xi_1, \cdots, \xi_{d+1}) - \gamma_i^0(\xi_1, \cdots, \xi_{d+1}). \tag{8.414}$$

From (8.411) and (8.413), it is easy to see that $\tilde{S}_i(x) = \varphi \circ H \circ F_0^{i-1}(x)$ and $\varphi \circ H \circ \hat{F}_u^i(x) = \varphi \circ H \circ F_0^i(x)$ for $1 \leq i \leq d$. Since φ is a diffeomorphism and φ^{-1} exists, (8.398) is satisfied. Let

$$\hat{\gamma}_k^0(w, y) \triangleq \begin{cases} \gamma_k^0(w, y), & d + 1 \leq k \leq n \\ 0, & n + 1 \leq n + d \end{cases} \tag{8.415}$$

$$\hat{\varepsilon}_{d+i}^u(w_1) \triangleq \begin{cases} 0, & 1 \leq i \leq d \\ \bar{\varepsilon}_i^u(w_1), & d + 1 \leq i \leq n \end{cases} \tag{8.416}$$

and for $d + 1 \leq k \leq n + d$,

$$\hat{\gamma}_k^u(w, y) \triangleq \hat{\gamma}_k^0(w, y) + \hat{\varepsilon}_k^u(w_1). \tag{8.417}$$

For extended system (8.373), we let, for $1 \leq i \leq d$,

$$S_i^e(x^e) \triangleq \varphi \circ H^e \circ (F_0^e)^{i-1}(x^e) \tag{8.418}$$

for $1 \leq i \leq n - d$,

$$S^e_{d+i}(x^e) \triangleq \varphi \circ H^e \circ (F^e_0)^{d+i-1}(x^e) - \sum_{k=d+1}^{d+i-1} \hat{\gamma}^0_k \circ (T^e)^1 \circ (F^e_0)^{d+i-1-k}(x^e)$$

$$(8.419)$$

and for $n - d + 1 \leq i \leq n$,

$$S^e_{d+i}(x^e) \triangleq \varphi \circ H^e \circ (F^e_0)^{d+i-1}(x^e) - \sum_{k=d+1}^{n} \hat{\gamma}^0_k \circ (T^e)^1 \circ (F^e_0)^{d+i-1-k}(x^e).$$

$$(8.420)$$

Then it is easy to see, by (8.396), (8.397), (8.411), (8.415), (8.419), and (8.420), that for $1 \leq i \leq n$,

$$S^u_{d+i}(x^e) = \varphi \circ H \circ F^{i-1}_0(x) - \sum_{k=d+1}^{i-1} \gamma^0_k \circ T^1 \circ F^{i-1-k}_0(x)$$

$$- \sum_{k=i}^{d+i-1} \hat{\gamma}^0_k \circ (T^e)^1 \circ (F^e_0)^{d+i-1-k}(x^e)$$

$$(8.421)$$

$$= \tilde{S}_i(x) - \lambda^0_i(w, H(x), \cdots, H \circ F^{d-1}_0(x))$$

where for $1 \leq i \leq n$,

$$\lambda^0_i(w, \xi_1, \cdots, \xi_d) \triangleq \sum_{k=i}^{d+i-1} \hat{\gamma}^0_k(w_{d+i-k}, \cdots, w_d, \xi_1, \cdots, \xi_{d+i-k}).$$

Since $S^e_i(x^e) = \varphi \circ H^e \circ (F^e_0)^{i-1}(x^e) = \varphi(w_i)$ for $1 \leq i \leq d$, it is clear, by (8.396), that

$$S^e_i \circ \hat{F}^e_u(x^e) - S^e_i \circ F^e_0(x^e) = 0, \ 1 \leq i \leq d. \tag{8.422}$$

Also, it is easy to see, by (8.398), (8.413), (8.416), and (8.421), that for $1 \leq i \leq n$,

$$S^e_{d+i} \circ \hat{F}^e_u(x^e) - S^e_{d+i} \circ F^e_0(x^e) = \tilde{S}_i \circ F_u(x) - \tilde{S}_i \circ F_0(x)$$

$$- \lambda^0_i(p(w, H), H \circ F_u, \cdots, H \circ \hat{F}^d_u) + \lambda^0_i(p(w, H), H \circ F_0, \cdots, H \circ F^d_0)$$

$$= \begin{cases} 0, & 1 \leq i \leq d \\ \bar{\varepsilon}^u_i(H(x)), & d+1 \leq i \leq n \end{cases} = \hat{\varepsilon}^u_{d+i}(H(x)).$$

$$(8.423)$$

Finally, it is clear, by (8.396), (8.397), (8.412), (8.414), and (8.415), that

$$\varphi \circ H^e \circ (\hat{F}_u^e)^{n+d}(x^e) = \varphi \circ H \circ \hat{F}_u^n(x) = \sum_{k=d+1}^{n-1} \gamma_k^0 \circ T^1 \circ \hat{F}_u^{n-k}(x) + \gamma_n^u \circ T^1(x)$$

$$= \sum_{k=d+1}^{n-1} \gamma_k^0 \circ T^1 \circ \hat{F}_u^{n-k}(x) + \gamma_n^0 \circ T^1(x) + \bar{\varepsilon}_n^u \circ H(x)$$

$$= \sum_{k=d+1}^{n} \hat{\gamma}_k^0 \circ (T^e)^1 \circ (F_u^e)^{n+d-k}(x^e) + \bar{\varepsilon}_n^u \circ H(x)$$

$$= \sum_{k=d+1}^{n+d-1} \hat{\gamma}_k^0 \circ (T^e)^1 \circ (F_u^e)^{n+d-k}(x^e) + \hat{\gamma}_{n+d}^0 \circ (T^e)^1(x^e)$$

(8.424)

where $\hat{\gamma}_{n+d}^0(w, y) \triangleq \bar{\varepsilon}_n^u(y)$. Therefore, by (8.418)–(8.420), (8.422)–(8.424), and Lemma 8.10, extended system (8.373) is state equivalent to a d-GNOCF with OT $\varphi(y)$ and state transformation $z^e = S^e(w, x)$. Hence, system (8.370) is RDOEL with index d. □

Remark 8.11 In the proof of (ii) ⇒ (i) of Lemma 8.11, it has been shown that if system (8.370) is state equivalent to a d-GNOCF with γ^0 and $\varepsilon^u(H(x))$, then extended system (8.373) is state equivalent to a d-GNOCF with $\hat{\gamma}^0 = \begin{bmatrix} \gamma^0 \\ O_{d\times 1} \end{bmatrix}$ and $\hat{\varepsilon}^u(H(x)) = \begin{bmatrix} O_{d\times 1} \\ \varepsilon^u(H(x)) \end{bmatrix}$. In other words, if system (8.370) is state equivalent to a d-GNOCF with OT $\varphi(y)$ and state transformation $z = \tilde{S}(x)$, then system (8.370) is RDOEL with index d and extended state transformation $z^e = S^e(w, x)$, defined by

$$S_i^e(w, x) = \begin{cases} \hat{S}_i(w, H(x), \cdots, H \circ F_0^{n-1-d}(x)), & 1 \le i \le n \\ S_n^e \circ F_0^e(w, x) - \gamma_n^0(w, H(x)), & i = n+1 \\ S_{i-1}^e \circ F_0^e(w, x), & n+2 \le i \le n+d \end{cases}$$

(8.425)

where

$$\hat{S}(\xi) \triangleq \tilde{S} \circ T^{-1}(\xi)$$

$$\gamma_n^0(\xi_1, \cdots, \xi_{d+1}) \triangleq \hat{S}_n \circ T \circ F_0 \circ T^{-1}(\xi)$$
$$= \tilde{S}_n \circ F_0 \circ T^{-1}(\xi).$$

Example 8.6.1 Use (8.411) and (8.418)–(8.420) to show that (8.425) is satisfied.

Solution Solution is omitted. (Problem 8-10.) □

Autonomous system (8.371) satisfies (8.398) and (8.399). Thus, we have the following from Lemma 8.11.

Corollary 8.10 *Autonomous system (8.371) is RDOEL with index d, if and only if autonomous system (8.371) is state equivalent to a d-GNOCF with OT.*

It is obvious, by Lemma 8.10, that autonomous system (8.371) is always state equivalent to a $(n-1)$-GNOCF with OT $\varphi(y) = y$ and $z = \tilde{S}(x) = T(x)$. Assume that (8.398) is satisfied with $d = n$. Then it is clear that $F_u(x) = F_0(x)$ and system (8.370) is the same as autonomous system (8.371). Therefore, system (8.370) is, by Corollary 8.10, RDOEL with index $d = n - 1$. From now on, we will assume that $d \leq n - 1$.

Theorem 8.17 *Suppose that (8.398) is satisfied. System (8.370) is RDOEL with index d and extended state transformation $z^e = S^e(w, x)$, if and only if there exist smooth functions $\ell(y)$ ($\ell(0) = 1$) and $\varepsilon_i^u(y)$, $d + 1 \leq i \leq n$ such that*

$$\bar{g}_i^u(x) = \bar{g}_i^0(x), \quad 1 \leq i \leq n - d \tag{8.426}$$

$$[\bar{g}_1^0(x), \bar{g}_i^0(x)] = 0, \quad 1 \leq i \leq n - d \tag{8.427}$$

and

$$\hat{S}^2 \circ T \circ F_u(x) - \hat{S}^2 \circ T \circ F_0(x) = \begin{bmatrix} \varepsilon_{d+1}^u(H(x)) \\ \vdots \\ \varepsilon_n^u(H(x)) \end{bmatrix} \tag{8.428}$$

where

$$\bar{g}_1^u(x) = \bar{g}_1^0(x) \triangleq \ell(H \circ F_0^{n-1}(x))g_1^0(x) \tag{8.429}$$

$$\bar{g}_i^0(x) \triangleq (F_0)_*(\bar{g}_{i-1}^0(x)); \quad \bar{g}_i^u(x) \triangleq (F_u)_*(\bar{g}_{i-1}^u(x)), \ i \geq 2 \tag{8.430}$$

$$\varphi(y) = \int_0^y \frac{1}{\ell(\bar{y})} d\bar{y} \tag{8.431}$$

$$T_* \left(\bar{g}_i^0 \right) \triangleq \bar{r}_i^0(\xi) = \begin{bmatrix} O_{d \times 1} \\ \hat{r}_i^0(\xi) \end{bmatrix}, \quad 1 \leq i \leq n - d \tag{8.432}$$

$$\hat{S}(\xi) = \begin{bmatrix} \hat{S}^1(\xi) \\ \hat{S}^2(\xi) \end{bmatrix} = \begin{bmatrix} [\varphi(\xi_1) \cdots \varphi(\xi_d)]^{\mathsf{T}} \\ \hat{S}^2(\xi_1, \cdots, \xi_n) \end{bmatrix} \tag{8.433}$$

and

$$\frac{\partial \hat{S}^2(\xi)}{\partial(\xi_{d+1}, \cdots, \xi_n)} = \left[\hat{r}_{n-d}^0(\xi) \cdots \hat{r}_2^0(\xi) \ \hat{r}_1^0(\xi) \right]^{-1}. \tag{8.434}$$

Furthermore, an extended state transformation $z^e = S^e(w, x)$ is given by (8.425).

Proof Necessity. Suppose that system (8.370) is RDOEL with index d and (8.398) is satisfied. Then, by Lemma 8.11, system (8.370) is state equivalent to d-GNOCF

(8.377) with OT $\varphi(y)$ and state transformation $z = \tilde{S}(x)$, and for some $\varepsilon_i^u(y)$, $d + 1 \leq i \leq n$,

$$\tilde{S}_i \circ F_u(x) - \tilde{S}_i \circ F_0(x) = \begin{cases} 0, & 1 \leq i \leq d \\ \varepsilon_i^u(H(x)), & d+1 \leq i \leq n. \end{cases} \tag{8.435}$$

Thus, we have, by (8.384) of Lemma 8.10, that for $1 \leq i \leq n$,

$$\tilde{S}_i(x) = \varphi \circ H \circ F_0^{i-1}(x) - \sum_{k=d+1}^{i-1} \gamma_k^0 \circ T^1 \circ F_0^{i-1-k}(x). \tag{8.436}$$

In other words, we have that $\bar{y} = \bar{h}(z) \triangleq \varphi \circ H \circ \tilde{S}^{-1}(z) = z_1$ and

$$\bar{f}_u(z) \triangleq \tilde{S} \circ F_u \circ \tilde{S}^{-1}(z) = A_o z + \gamma^u \circ \bar{\Phi}^{-1}(z_1, \cdots, z_{d+1}). \tag{8.437}$$

For d-GNOCF system (8.377), we define the following vector fields:

$$\begin{aligned} \bar{\psi}_1^u(z) = \bar{\psi}_1^0(z) &\triangleq \frac{\partial}{\partial z_n} \\ \bar{\psi}_i^u(z) &\triangleq (\bar{f}_u)_*(\bar{\psi}_{i-1}^u(z)) \,; \quad \bar{\psi}_i^0(z) \triangleq (\bar{f}_0)_*(\bar{\psi}_{i-1}^0(z)), \ i \geq 2. \end{aligned} \tag{8.438}$$

Then it is clear, by (8.437), that for $1 \leq i \leq n - d$,

$$\bar{\psi}_i^u(z) = \bar{\psi}_i^0(z) = \frac{\partial}{\partial z_{n+1-i}} \tag{8.439}$$

and

$$[\bar{\psi}_1^0(z), \bar{\psi}_i^0(z)] = 0. \tag{8.440}$$

If we let $\xi = T(x)$ and $z = \tilde{S} \circ T^{-1}(\xi) \triangleq \hat{S}(\xi)$, then we have, by (8.380), (8.388), and (8.436), that

$$\hat{S}_i(\xi) = \varphi(\xi_i), \ 1 \leq i \leq d + 1 \tag{8.441}$$

and for $d + 2 \leq i \leq n$

$$\hat{S}_i(\xi) = \varphi(\xi_i) - \sum_{k=d+1}^{i-1} \gamma_k^0(\xi_{i-k}, \cdots, \xi_{i-k+d}). \tag{8.442}$$

Thus, it is clear that

$$\frac{\partial \hat{S}_i(\xi)}{\partial \xi_n} = \begin{cases} 0, & \text{if } i \leq n - 1 \\ \frac{d\varphi(\xi_n)}{d\xi_n}, & \text{if } i = n \end{cases}$$

which implies, together with (2.22), (8.298), and (8.438), that

$$\tilde{S}_*(\mathbf{g}_1^0(x)) = \hat{S}_* \circ T_*(\mathbf{g}_1^0(x)) = \hat{S}_* \left(\frac{\partial}{\partial \xi_n} \right) = \sum_{i=1}^{n} \frac{\partial \hat{S}_i(\xi)}{\partial \xi_n} \bigg|_{\xi = \hat{S}^{-1}(z)} \frac{\partial}{\partial z_i}$$

$$= \frac{d\varphi(\xi_n)}{d\xi_n} \bigg|_{\xi = \hat{S}^{-1}(z)} \frac{\partial}{\partial z_n} = \frac{d\varphi(\xi_n)}{d\xi_n} \bigg|_{\xi = \hat{S}^{-1}(z)} \bar{\psi}_1^0(z).$$

Thus, if we let

$$\frac{1}{\ell(\xi_n)} = \frac{d\varphi(\xi_n)}{d\xi_n} \quad \left(\text{or} \;\; \varphi(y) = \int_0^y \frac{1}{\ell(\xi_n)} d\xi_n \right)$$

then we have, by (2.49) and (8.380), that $\bar{\psi}_1^0(z) = \ell(\xi_n)|_{\xi = \hat{S}^{-1}(z)} \tilde{S}_*(\mathbf{g}_1^0(x))$ and

$$\tilde{S}_*^{-1}(\bar{\psi}_1^0(z)) = \tilde{S}_*^{-1} \left(\ell(\xi_n)|_{\xi = \hat{S}^{-1}(z)} \tilde{S}_*(\mathbf{g}_1^0(x)) \right)$$

$$= \ell(\xi_n)|_{\xi = T(x)} \mathbf{g}_1^0(x) = \ell(H \circ F_0^{n-1}(x))\mathbf{g}_1^0(x).$$

Hence, if we let $\bar{\mathbf{g}}_1^0(x) \triangleq \tilde{S}_*^{-1}(\bar{\psi}_1^0(z))$, then (8.429) is satisfied. It will be shown, by mathematical induction, that for $1 \leq i \leq n - d$,

$$\bar{\mathbf{g}}_i^u(x) = \tilde{S}_*^{-1}(\bar{\psi}_i^u(z)) = \tilde{S}_*^{-1} \left(\frac{\partial}{\partial z_{n+1-i}} \right) = \bar{\mathbf{g}}_i^0(x). \tag{8.443}$$

Assume that (8.443) is satisfied for $i = k$ and $1 \leq k \leq n - d - 1$. Since $\bar{f}_u(z) = \tilde{S} \circ F_u \circ \tilde{S}^{-1}(z)$ or $F_u(x) = \tilde{S}^{-1} \circ \bar{f}_u \circ \tilde{S}(x)$, it is clear, by (2.22), (8.430), (8.438), and (8.439), that

$$\bar{\mathbf{g}}_{k+1}^u(x) = (F_u)_*(\bar{\mathbf{g}}_k^u(x)) = \tilde{S}_*^{-1} \circ (\bar{f}_u)_* \circ \tilde{S}_*(\bar{\mathbf{g}}_k^u(x)) = \tilde{S}_*^{-1} \circ (\bar{f}_u)_*(\bar{\psi}_k^u(z))$$

$$= \tilde{S}_*^{-1} \left(\bar{\psi}_{k+1}^u(z) \right) = \tilde{S}_*^{-1} \left(\frac{\partial}{\partial z_{n-k}} \right) = \bar{\mathbf{g}}_{k+1}^0(x)$$

which implies, by mathematical induction, that (8.443) is satisfied for $1 \leq i \leq n - d$. Thus, it is easy, by (2.28), (8.440), and (8.443), to see that (8.426) and (8.427) are satisfied. Let $T_*(\bar{\mathbf{g}}_i^0(x)) \triangleq \bar{\mathbf{r}}_i^0(\xi) \triangleq \begin{bmatrix} \tilde{\mathbf{r}}_i^0(\xi) \\ \hat{\mathbf{r}}_i^0(\xi) \end{bmatrix}$ for $1 \leq i \leq n$, where $\tilde{\mathbf{r}}_i^0(\xi)$ is a $d \times 1$ matrix. Now it is clear, by (8.441), that (8.433) is satisfied. Since $\tilde{S} \circ T^{-1}(\xi) = \hat{S}(\xi)$, it is also clear, by (8.443), that for $1 \leq i \leq n - d$,

$$\hat{S}_*(\bar{\mathbf{r}}_i^0(\xi)) = \hat{S}_* \circ T_*(\bar{\mathbf{g}}_i^0(x)) = \tilde{S}_*(\bar{\mathbf{g}}_i^0(x)) = \bar{\psi}_i^0(z) = \frac{\partial}{\partial z_{n+1-i}}$$

and

$$\left[\hat{S}_*(\bar{\mathbf{r}}_{n-d}^0(\xi)) \cdots \hat{S}_*(\bar{\mathbf{r}}_1^0(\xi))\right] = \left[\bar{\psi}_{n-d}^0(z) \cdots \bar{\psi}_1^0(z)\right] = \begin{bmatrix} O_{d \times (n-d)} \\ I_{n-d} \end{bmatrix}$$

which implies, together with (8.433), that

$$\begin{bmatrix} \frac{\partial \hat{S}^1(\xi^1)}{\partial \xi^1} & O_{d \times (n-d)} \\ \frac{\partial \hat{S}^2(\xi)}{\partial \xi^1} & \frac{\partial \hat{S}^2(\xi)}{\partial \xi^2} \end{bmatrix} \begin{bmatrix} \tilde{\mathbf{r}}_{n-d}^0(\xi) \cdots \tilde{\mathbf{r}}_1^0(\xi) \\ \hat{\mathbf{r}}_{n-d}^0(\xi) \cdots \hat{\mathbf{r}}_1^0(\xi) \end{bmatrix} = \begin{bmatrix} O_{d \times (n-d)} \\ I_{n-d} \end{bmatrix}$$

where $\xi^1 \triangleq [\xi_1 \cdots \xi_d]^T$ and $\xi^2 \triangleq [\xi_{d+1} \cdots \xi_n]^T$. Since $\frac{\partial \hat{S}^1(\xi^1)}{\partial \xi^1}$ is nonsingular, it is clear that

$$\left[\tilde{\mathbf{r}}_{n-d}^0(\xi) \cdots \tilde{\mathbf{r}}_1^0(\xi)\right] = O_{d \times (n-d)}$$

and

$$\frac{\partial \hat{S}^2(\xi)}{\partial \xi^2} \left[\hat{\mathbf{r}}_{n-d}^0(\xi) \cdots \hat{\mathbf{r}}_1^0(\xi)\right] = I_{n-d}.$$

In other words, (8.432) and (8.434) are satisfied. If we let $\tilde{S}^2(x) \triangleq \begin{bmatrix} \tilde{S}_{d+1}(x) \\ \vdots \\ \tilde{S}_n(x) \end{bmatrix}$, then

we have $\tilde{S}^2(x) = \hat{S}^2 \circ T(x)$. Thus, it is clear, by (8.435), that (8.428) is satisfied.

Sufficiency. Assume that (8.398) is satisfied. Suppose that there exist $\beta(y)$ and $\varepsilon_i^u(y)$, $d + 1 \leq i \leq n$ such that (8.426)–(8.434) are satisfied. Let $\xi = T(x)$, $f_u(\xi) \triangleq T \circ F_u \circ T^{-1}(\xi)$, and for $1 \leq i \leq n$,

$$\bar{\mathbf{r}}_i^0(\xi) \triangleq T_*(\bar{\mathbf{g}}_i^0(x)) \; ; \quad \bar{\mathbf{r}}_i^u(\xi) \triangleq T_*(\bar{\mathbf{g}}_i^u(x)). \tag{8.444}$$

It will be shown, by mathematical induction, that for $2 \leq i \leq n$,

$$\bar{\mathbf{r}}_i^0(\xi) = (f_0)_*(\bar{\mathbf{r}}_{i-1}^0) \; ; \quad \bar{\mathbf{r}}_i^u(\xi) = (f_u)_*(\bar{\mathbf{r}}_{i-1}^u). \tag{8.445}$$

Assume that (8.445) is satisfied for $i = k$. Since $f_u(\xi) = T \circ F_u \circ T^{-1}(\xi)$ or $F_u(x) = T^{-1} \circ f_u \circ T(x)$, it is clear, by (2.22), (8.430), and (8.444), that

$$\bar{\mathbf{r}}_k^u(\xi) = T_* \circ (F_u)_* \left(\bar{\mathbf{g}}_{k-1}^u(x)\right) = T_* \circ (F_u)_* \circ T_*^{-1} \left(\bar{\mathbf{r}}_{k-1}^u(\xi)\right)$$
$$= (f_u)_* \left(\bar{\mathbf{r}}_{k-1}^u(\xi)\right)$$

which implies, by mathematical induction, that (8.445) is satisfied for $2 \leq i \leq n$. It is also clear, by (2.49), (8.303), (8.426), (8.427), (8.429), and (8.432), that for $1 \leq i \leq n - d$,

$$\bar{\mathbf{r}}_i^u(\xi) = \bar{\mathbf{r}}_i^0(\xi)$$

$$[\bar{\mathbf{r}}_1^0(\xi), \bar{\mathbf{r}}_i^0(\xi)] = T_* \left([\bar{\mathbf{g}}_1^0(x), \bar{\mathbf{g}}_i^0(x)]\right) = 0 \tag{8.446}$$

$$\bar{\mathbf{r}}_i^0(\xi) = T_* \left((F_0)_*^{i-1} \left(\ell(H \circ F_0^{n-1}) \mathbf{g}_1^0(x) \right) \right) = T_* \left(\ell(H \circ F_0^{n-i}) \mathbf{g}_i^0(x) \right)$$
$$= \ell(\xi_{n+1-i}) \mathbf{r}_i^0(\xi) = [O_{1 \times (n-i)} \ \ell(\xi_{n+1-i}) \ * \ \cdots \ *]^\mathsf{T} \tag{8.447}$$

and

$$\left[\hat{\mathbf{r}}_{n-d}^0(\xi) \cdots \hat{\mathbf{r}}_1^0(\xi) \right] = \begin{bmatrix} \ell(\xi_{d+1}) & 0 & \cdots & 0 \\ * & \ell(\xi_{d+2}) & \cdots & 0 \\ * & * & \ddots & \vdots \\ * & * & \cdots & \ell(\xi_n) \end{bmatrix}. \tag{8.448}$$

Note, by (8.446) and (8.447), that $\{\bar{\mathbf{r}}_1^0(\xi), \bar{\mathbf{r}}_2^0(\xi), \cdots, \bar{\mathbf{r}}_{n-d}^0(\xi)\}$ is a set of commuting vector fields such that

$$\text{span} \left\{ \bar{\mathbf{r}}_1^0(\xi), \cdots, \bar{\mathbf{r}}_{n-d}^0(\xi) \right\} = \text{span} \left\{ \frac{\partial}{\partial \xi_{d+1}}, \cdots, \frac{\partial}{\partial \xi_n} \right\}.$$

Therefore, by Corollary 2.1, there exists a state transformation $z = \hat{S}(\xi) = \begin{bmatrix} \hat{S}^1(\xi_1, \cdots, \xi_d) \\ \hat{S}^2(\xi_1, \cdots, \xi_n) \end{bmatrix}$ such that

$$\hat{S}^1(\xi) = \left[\varphi(\xi_1) \cdots \varphi(\xi_d) \right]^\mathsf{T} \tag{8.449}$$

and for $1 \le i \le n - d$,

$$\hat{S}_* \left(\bar{\mathbf{r}}_i^0(\xi) \right) \left(= \hat{S}_* \left(\bar{\mathbf{r}}_i^u(\xi) \right) \right) = \frac{\partial}{\partial z_{n+1-i}} \tag{8.450}$$

which imply, together with (8.432), that

$$\begin{bmatrix} O_{d \times (n-d)} \\ I_{n-d} \end{bmatrix} = \left[\hat{S}_*(\bar{\mathbf{r}}_{n-d}^0(\xi)) \cdots \hat{S}_*(\bar{\mathbf{r}}_1^0(\xi)) \right]$$
$$= \frac{\partial \hat{S}(\xi)}{\partial \xi} \left[\bar{\mathbf{r}}_{n-d}^0(\xi) \cdots \bar{\mathbf{r}}_1^0(\xi) \right] \Big|_{\xi = \hat{S}^{-1}(z)}$$

$$\begin{bmatrix} \frac{\partial \hat{S}^1(\xi)}{\partial \xi^1} & O_{d \times (n-d)} \\ \frac{\partial \hat{S}^2(\xi)}{\partial \xi^1} & \frac{\partial \hat{S}^2(\xi)}{\partial \xi^2} \end{bmatrix} \begin{bmatrix} O_{d \times 1} & \cdots & O_{d \times 1} \\ \hat{\mathbf{r}}_{n-d}^0(\xi) & \cdots & \hat{\mathbf{r}}_1^0(\xi) \end{bmatrix} = \begin{bmatrix} O_{d \times (n-d)} \\ I_{n-d} \end{bmatrix}$$

and

$$\frac{\partial \hat{S}^2(\xi)}{\partial \xi^2} \left[\hat{\mathbf{r}}_{n-d}^0(\xi) \cdots \hat{\mathbf{r}}_1^0(\xi) \right] = I_{n-d}$$

where $\xi^1 \triangleq [\xi_1 \cdots \xi_d]^T$ and $\xi^2 \triangleq [\xi_{d+1} \cdots \xi_n]^T$. Thus, $\hat{S}^2(\xi)$ can be calculated by (8.434). Let $\tilde{S}(x) \triangleq \hat{S} \circ T(x)$. Now we will show that

$$\bar{h}(z) \triangleq \varphi \circ H \circ \tilde{S}^{-1}(z) = \varphi \circ H \circ T^{-1} \circ \hat{S}^{-1}(z) = z_1 \qquad (8.451)$$

and

$$\begin{aligned}
\bar{f}_u(z) &\triangleq \tilde{S} \circ F_u \circ \tilde{S}^{-1}(z) = \hat{S} \circ f_u \circ \hat{S}^{-1}(z) \\
&= A_o z + \bar{\gamma}^u(z_1, \cdots, z_{d+1})
\end{aligned} \qquad (8.452)$$

where $\bar{\gamma}_i(z_1, \cdots, z_{d+1}) = 0$, $1 \le i \le d$. Since $H \circ T^{-1}(\xi) = \xi_1$, it is clear that $\bar{h}(z) = \varphi(\xi_1)|_{\xi = \hat{S}^{-1}(z)} = z_1$ and thus (8.451) is satisfied. We have, by (8.431), (8.434), and (8.448), that

$$\frac{d\hat{S}_1^2(\xi)}{d(\xi_{d+1}, \cdots, \xi_n)} = \left[\frac{1}{\ell(\xi_{d+1})} \ 0 \cdots 0 \right] = \left[\frac{d\varphi(\xi_{d+1})}{d\xi_{d+1}} \ 0 \cdots 0 \right]$$

or

$$\hat{S}_{d+1}(\xi) = \hat{S}_1^2(\xi) = \varphi(\xi_{d+1})$$

which implies, together with (8.449), that for $1 \le i \le d + 1$,

$$z_i = \hat{S}_i(\xi) = \varphi(\xi_i) = \varphi \circ H \circ F_0^{i-1} \circ T^{-1}(\xi). \qquad (8.453)$$

Therefore, if we let

$$\bar{f}_u(z) \triangleq \sum_{k=1}^{n} \bar{f}_{u,k}(z) \frac{\partial}{\partial z_k} = \begin{bmatrix} \bar{f}_{u,1}(z) \\ \vdots \\ \bar{f}_{u,n}(z) \end{bmatrix} \qquad (8.454)$$

then it is clear, by (8.398) and (8.453), that for $1 \le i \le d$,

$$\begin{aligned}
\bar{f}_{u,i}(z) &= \hat{S}_i \circ f_u \circ \hat{S}^{-1}(z) = \varphi \circ H \circ F_0^{i-1} \circ T^{-1} \circ T \circ F_u \circ T^{-1} \circ \hat{S}^{-1}(z) \\
&= \varphi \circ H \circ \hat{F}_u^i \circ T^{-1} \circ \hat{S}^{-1}(z) = \varphi \circ H \circ F_0^i \circ T^{-1} \circ \hat{S}^{-1}(z) = z_{i+1}
\end{aligned} \qquad (8.455)$$

which implies that $\bar{\gamma}_i^u(z_1, \cdots, z_{d+1}) = 0$, $1 \le i \le d$. Since $\bar{f}_u(z) = \hat{S} \circ f_u \circ \hat{S}^{-1}(z)$, it is also easy to show that, for $i \ge 1$,

$$\begin{aligned}
\hat{S}_* \left(\bar{\mathbf{r}}_{i+1}^u(\xi) \right) &= \hat{S}_* \left((f_u)_*(\bar{\mathbf{r}}_i^u(\xi)) \right) = \hat{S}_* \circ (f_u)_* \circ \hat{S}_*^{-1} \left(\hat{S}_*(\bar{\mathbf{r}}_i^u(\xi)) \right) \\
&= (\bar{f}_u)_* \left(\hat{S}_*(\bar{\mathbf{r}}_i^u(\xi)) \right).
\end{aligned} \qquad (8.456)$$

Therefore, we have, by (8.450) and (8.454)–(8.456), that for $1 \le i \le n - d - 1$,

$$\frac{\partial}{\partial z_{n-i}} = (\bar{f}_u)_* \left(\frac{\partial}{\partial z_{n+1-i}} \right) = \sum_{k=1}^{n} \frac{\partial \bar{f}_{u,k}(\bar{z})}{\partial \bar{z}_{n+1-i}} \Bigg|_{\bar{z}=\bar{f}_u^{-1}(z)} \frac{\partial}{\partial z_k}$$

$$= \sum_{k=d+1}^{n} \frac{\partial \bar{f}_{u,k}(\bar{z})}{\partial \bar{z}_{n+1-i}} \Bigg|_{\bar{z}=\bar{f}_u^{-1}(z)} \frac{\partial}{\partial z_k}$$

which implies that, for $d+1 \leq k \leq n$ and $1 \leq i \leq n-d-1$,

$$\frac{\partial \bar{f}_{u,k}(\bar{z})}{\partial \bar{z}_{n+1-i}} \Bigg|_{\bar{z}=\bar{f}_u^{-1}(z)} = \begin{cases} 1, & k = n-i \\ 0, & \text{otherwise} \end{cases}$$

or, for $d+1 \leq k \leq n$ and $d+2 \leq j \leq n$,

$$\frac{\partial \bar{f}_{u,k}(z)}{\partial z_j} = \begin{cases} 1, & j = k+1 \\ 0, & \text{otherwise.} \end{cases}$$

Hence, $\bar{f}_{u,n}(z) = \bar{\gamma}_n^u(z_1, \cdots, z_{d+1})$ and $\bar{f}_{u,k}(z) = z_{k+1} + \bar{\gamma}_k^u(z_1, \cdots, z_{d+1})$, $d + 1 \leq k \leq n-1$, for some functions $\bar{\gamma}_k^u(z_1, \cdots, z_{d+1})$, $d+1 \leq k \leq n$. In other words, (8.452) is satisfied with $\bar{\gamma}_i^u(z_1, \cdots, z_{d+1}) = 0$, $1 \leq i \leq d$ and system (8.370) is, by Definition 8.16, state equivalent to d-GNOCF with OT $\varphi(y)$ and state transformation $z = \tilde{S}(x) \triangleq \hat{S} \circ T(x)$. Finally, it is clear, by (8.398) and (8.428), that (8.399) is satisfied. Hence, by Lemma 8.11, system (8.370) is state equivalent to RDOEL with index d and extended state transformation $z^e = S^e(w, x)$ in (8.425). $\qquad \square$

If we let $\varphi(y) = y$ or $\ell(y) = 1$, the following corollary can be obtained from Theorem 8.17.

Corollary 8.11 *Suppose that (8.398) is satisfied. System (8.370) is RDOEL with index d and OT $\varphi(y) = y$ (i.e., without OT), if and only if there exists smooth functions $\varepsilon_i^u(y)$, $d+1 \leq i \leq n$ such that*

$$\mathbf{g}_i^u(x) = \mathbf{g}_i^0(x), \quad 1 \leq i \leq n-d \tag{8.457}$$

$$[\mathbf{g}_1^0(x), \mathbf{g}_i^0(x)] = 0, \quad 1 \leq i \leq n-d \tag{8.458}$$

$$\hat{S}^2 \circ T \circ F_u(x) - \hat{S}^2 \circ T \circ F_0(x) = [\varepsilon_{d+1}^u(H(x)) \cdots \varepsilon_n^u(H(x))]^{\mathsf{T}} \tag{8.459}$$

where

$$T_* \left(\mathbf{g}_i^0(x) \right) \triangleq \mathbf{r}_i^0(\xi) = \begin{bmatrix} O_{d \times 1} \\ \hat{\mathbf{r}}_i^0(\xi) \end{bmatrix}, \quad 1 \leq i \leq n-d \tag{8.460}$$

$$\hat{S}(\xi) = \begin{bmatrix} \hat{S}^1(\xi) \\ \hat{S}^2(\xi) \end{bmatrix} = \begin{bmatrix} [\xi_1 \cdots \xi_d]^{\mathsf{T}} \\ \hat{S}^2(\xi_1, \cdots, \xi_n) \end{bmatrix} \tag{8.461}$$

and

$$\frac{\partial \hat{S}^2(\xi)}{\partial(\xi_{d+1}, \cdots, \xi_n)} = \left[\hat{\mathbf{r}}_{n-d}^0(\xi) \quad \cdots \quad \hat{\mathbf{r}}_2^0(\xi) \quad \hat{\mathbf{r}}_1^0(\xi) \right]^{-1}. \tag{8.462}$$

Furthermore, an extended state transformation $z^e = S^e(w, x)$ is given by (8.425).

If we let $F_u(x) = F_0(x)$, then (8.398) is satisfied and we can obtain the following corollary for autonomous system (8.371) from Corollary 8.11.

Corollary 8.12 *Autonomous system (8.371) is RDOEL with index d and OT $\varphi(y) = y$ (i.e., without OT), if and only if*

$$[\mathbf{g}_1^0(x), \mathbf{g}_i^0(x)] = 0, \quad 1 \leq i \leq n - d. \tag{8.463}$$

Furthermore, an extended state transformation $z^e = S^e(w, x)$ is given by (8.425), where

$$T_* \left(\mathbf{g}_i^0(x) \right) \triangleq \mathbf{r}_i^0(\xi) = \begin{bmatrix} O_{d \times 1} \\ \hat{\mathbf{r}}_i^0(\xi) \end{bmatrix}, \quad 1 \leq i \leq n - d \tag{8.464}$$

$$\hat{S}(\xi) = \begin{bmatrix} \hat{S}^1(\xi) \\ \hat{S}^2(\xi) \end{bmatrix} = \begin{bmatrix} [\xi_1 \quad \cdots \quad \xi_d]^\mathsf{T} \\ \hat{S}^2(\xi_1, \cdots, \xi_n) \end{bmatrix} \tag{8.465}$$

and

$$\frac{\partial \hat{S}^2(\xi)}{\partial(\xi_{d+1}, \cdots, \xi_n)} = \left[\hat{\mathbf{r}}_{n-d}^0(\xi) \quad \cdots \quad \hat{\mathbf{r}}_2^0(\xi) \quad \hat{\mathbf{r}}_1^0(\xi) \right]^{-1}. \tag{8.466}$$

In order to find whether the conditions of Theorem 8.17 are satisfied, $\ell(y)$ or $\varphi(y)$ should be found. In the following, we further investigate the conditions that $\ell(y)$ or $\beta(y) \triangleq \frac{d \ln \ell(y)}{dy} = \frac{1}{\ell(y)} \frac{d\ell(y)}{dy}$ should satisfy.

Theorem 8.18 *Suppose that (8.398) is satisfied and $\kappa \leq n - d$. System (8.370) is RDOEL with index d and extended state transformation $z^e = S^e(w, x)$, if and only if there exists smooth functions $\beta(y)$, defined on an open neighborhood of $y = 0$, and $\varepsilon_i^u(y)$, $d + 1 \leq i \leq n$ such that*

(i)

$$[\mathbf{g}_1^0(x), \mathbf{g}_\kappa^0(x)] = L_{\mathbf{g}_\kappa^0}(H \circ F_0^{n-1}(x))\beta(H \circ F_0^{n-1}(x))\mathbf{g}_1^0(x) \tag{8.467}$$

(ii)

$$\bar{\mathbf{g}}_i^u(x) = \bar{\mathbf{g}}_i^0(x), \quad 1 \leq i \leq n - d \tag{8.468}$$

(iii)

$$[\bar{\mathbf{g}}_1^0(x), \bar{\mathbf{g}}_i^0(x)] = 0, \quad 1 \leq i \leq n - d \tag{8.469}$$

(iv)

$$\hat{S}^2 \circ T \circ F_u(x) - \hat{S}^2 \circ T \circ F_0(x) = \begin{bmatrix} \varepsilon_{d+1}^u(H(x)) \\ \vdots \\ \varepsilon_n^u(H(x)) \end{bmatrix} \tag{8.470}$$

where

$$\ell(y) = e^{\int_0^y \beta(\bar{y})d\bar{y}} \tag{8.471}$$

$$\bar{\mathbf{g}}_1^u(x) = \bar{\mathbf{g}}_1^0(x) \triangleq \ell(H \circ F_0^{n-1}(x))\mathbf{g}_1^0(x) \tag{8.472}$$

$$\bar{\mathbf{g}}_i^0(x) \triangleq (F_0)_*(\bar{\mathbf{g}}_{i-1}^0(x)) \; ; \quad \bar{\mathbf{g}}_i^u(x) \triangleq (F_u)_*(\bar{\mathbf{g}}_{i-1}^u(x)), \; i \geq 2 \tag{8.473}$$

$$\varphi(y) = \int_0^y \frac{1}{\ell(\bar{y})}d\bar{y} \tag{8.474}$$

$$T_*\left(\bar{\mathbf{g}}_i^0\right) \triangleq \bar{\mathbf{r}}_i^0(\xi) = \begin{bmatrix} O_{d\times 1} \\ \hat{\mathbf{r}}_i^0(\xi) \end{bmatrix}, \; 1 \leq i \leq n-d \tag{8.475}$$

$$\hat{S}(\xi) = \begin{bmatrix} \hat{S}^1(\xi) \\ \hat{S}^2(\xi) \end{bmatrix} = \begin{bmatrix} [\varphi(\xi_1) \; \cdots \; \varphi(\xi_d)]^\mathsf{T} \\ \hat{S}^2(\xi_1, \cdots, \xi_n) \end{bmatrix} \tag{8.476}$$

and

$$\frac{\partial \hat{S}^2(\xi)}{\partial(\xi_{d+1}, \cdots, \xi_n)} = \begin{bmatrix} \hat{\mathbf{r}}_{n-d}^0(\xi) & \cdots & \hat{\mathbf{r}}_2^0(\xi) & \hat{\mathbf{r}}_1^0(\xi) \end{bmatrix}^{-1}. \tag{8.477}$$

Furthermore, an extended state transformation $z^e = S^e(w, x)$ is given by (8.425).

Proof Necessity. Let $\kappa \leq n - d$. Suppose that system (8.370) is RDOEL with index d and (8.398) is satisfied. Then, by Theorem 8.17, there exist smooth functions $\ell(y)$ ($\ell(0) = 1$) and $\varepsilon_i^u(y)$, $d + 1 \leq i \leq n$ such that (8.468)–(8.470) and (8.472)–(8.477) are satisfied. Since $2 \leq \kappa \leq n - d$, it is clear, by (8.300), that

$$L_{\mathbf{g}_1^0}\ell(H \circ F_0^{n-\kappa}(x)) = \frac{d\ell(y)}{dy}\bigg|_{y=H\circ F_0^{n-\kappa}(x)} L_{\mathbf{g}_1^0}(H \circ F_0^{n-\kappa}(x)) = 0. \tag{8.478}$$

Thus, we have, by (2.43), (8.340), (8.469), and (8.478), that

$$0 = [\bar{\mathbf{g}}_1^0(x), \bar{\mathbf{g}}_\kappa^0(x)] = [\ell(H \circ F_0^{n-1}(x))\mathbf{g}_1^0(x), \ell(H \circ F_0^{n-\kappa}(x))\mathbf{g}_\kappa^0(x)]$$
$$= \ell(H \circ F_0^{n-1}(x))\ell(H \circ F_0^{n-\kappa}(x))[\mathbf{g}_1^0(x), \mathbf{g}_\kappa^0(x)]$$
$$- \ell(H \circ F_0^{n-\kappa}(x))L_{\mathbf{g}_\kappa^0}\ell(H \circ F_0^{n-1}(x))\mathbf{g}_1^0(x)$$

which implies that

$$[\mathbf{g}_1^0(x), \mathbf{g}_\kappa^0(x)] = \frac{L_{\mathbf{g}_\kappa^0}\ell\left(H \circ F_0^{n-1}(x)\right)}{\ell(H \circ F_0^{n-1}(x))}\mathbf{g}_1^0(x)$$

$$= \frac{1}{\ell(y)}\frac{d\ell(y)}{dy}\bigg|_{y=H\circ F_0^{n-1}(x)} L_{\mathbf{g}_\kappa^0}\left(H \circ F_0^{n-1}(x)\right)\mathbf{g}_1^0(x).$$

Therefore, (8.467) and (8.471) are satisfied with

$$\beta(y) = \frac{1}{\ell(y)}\frac{d\ell(y)}{dy}\left(= \frac{d\ln\ell(y)}{dy}\right).$$

Sufficiency. It is obvious by Theorem 8.17.   □

If we let $u = 0$, we can obtain the following Corollary Theorem 8.18.

Corollary 8.13 *Let $\kappa \le n - d$. Autonomous system (8.371) is RDOEL with index d and extended state transformation $z^e = S^e(w, x)$, if and only if there exists a scalar function $\beta(y)$, defined on an open neighborhood of $y = 0$, such that*

(i)

$$[\mathbf{g}_1^0(x), \mathbf{g}_\kappa^0(x)] = L_{\mathbf{g}_\kappa^0}(H \circ F_0^{n-1}(x))\beta(H \circ F_0^{n-1}(x))\mathbf{g}_1^0(x)$$

(ii)

$$[\bar{\mathbf{g}}_1^0(x), \bar{\mathbf{g}}_i^0(x)] = 0, \quad 1 \le i \le n - d$$

where

$$\ell(y) = e^{\int_0^y \beta(\bar{y})d\bar{y}}$$

$$\bar{\mathbf{g}}_1^0(x) \triangleq \ell(H \circ F_0^{n-1}(x))\mathbf{g}_1^0(x)$$

and

$$\bar{\mathbf{g}}_i^0(x) \triangleq (F_0)_*(\bar{\mathbf{g}}_{i-1}^0(x)), \quad i \ge 2.$$

Furthermore, an extended state transformation $z^e = S^e(w, x)$ is given by (8.425), where

$$\varphi(y) = \int_0^y \frac{1}{\ell(\bar{y})}d\bar{y}$$

$$T_* \left(\bar{\mathbf{g}}_i^0(x) \right) \triangleq \bar{\mathbf{r}}_i^0(\xi) = \begin{bmatrix} O_{d \times 1} \\ \hat{\mathbf{r}}_i^0(\xi) \end{bmatrix}, \ 1 \le i \le n - d$$

$$\hat{S}(\xi) = \begin{bmatrix} \hat{S}^1(\xi) \\ \hat{S}^2(\xi) \end{bmatrix} = \begin{bmatrix} [\varphi(\xi_1) \ \cdots \ \varphi(\xi_d)]^{\mathsf{T}} \\ \hat{S}^2(\xi_1, \cdots, \xi_n) \end{bmatrix}$$

and

$$\frac{\partial \hat{S}^2(\xi)}{\partial(\xi_{d+1}, \cdots, \xi_n)} = \begin{bmatrix} \hat{\mathbf{r}}_{n-d}^0(\xi) \ \cdots \ \hat{\mathbf{r}}_2^0(\xi) \ \hat{\mathbf{r}}_1^0(\xi) \end{bmatrix}^{-1}.$$

Theorem 8.19 *Suppose that (8.398) is satisfied and $\sigma < n - d$. System (8.370) is RDOEL with index d and extended state transformation $z^e = S^e(w, x)$, if and only if there exist scalar functions $\theta_\sigma^u(x)$, $\bar{\beta}^u(x)$, and $\beta(y)$, defined on an open neighborhood of $y = 0$, such that*

(i)

$$\mathbf{g}_i^u(x) = \begin{cases} \mathbf{g}_i^0(x), & 1 \le i \le \sigma \\ \theta_\sigma^u(x)\mathbf{g}_i^0(x), & i = \sigma + 1 \end{cases} \tag{8.479}$$

(ii)

$$\bar{\mathbf{g}}_i^u(x) = \bar{\mathbf{g}}_i^0(x), \ 1 \le i \le n - d \tag{8.480}$$

(iii)

$$[\bar{\mathbf{g}}_1^0(x), \bar{\mathbf{g}}_i^0(x)] = 0, \ 1 \le i \le n - d \tag{8.481}$$

(iv)

$$\hat{S}^2 \circ T \circ F_u(x) - \hat{S}^2 \circ T \circ F_0(x) = \begin{bmatrix} \varepsilon_{d+1}^u(H(x)) \\ \vdots \\ \varepsilon_n^u(H(x)) \end{bmatrix} \tag{8.482}$$

where

$$\frac{\partial \left(\theta_\sigma^u \circ F_u \right)}{\partial u} = \bar{\beta}^u(x) \frac{\partial \left(H \circ \hat{F}_u^{n-\sigma}(x) \right)}{\partial u} \tag{8.483}$$

$$\bar{\beta}^0(x) = \beta \left(H \circ F_0^{n-\sigma}(x) \right) \tag{8.484}$$

$$\ell(y) = e^{\int_0^y \beta(\bar{y})d\bar{y}} \tag{8.485}$$

$$\bar{\mathbf{g}}_1^u(x) = \bar{\mathbf{g}}_1^0(x) \triangleq \ell(H \circ F_0^{n-1}(x))\mathbf{g}_1^0(x) \tag{8.486}$$

$$\bar{\mathbf{g}}_i^0(x) \triangleq (F_0)_*(\bar{\mathbf{g}}_{i-1}^0(x)) ; \quad \bar{\mathbf{g}}_i^u(x) \triangleq (F_u)_*(\bar{\mathbf{g}}_{i-1}^u(x)), \; i \geq 2 \tag{8.487}$$

$$\varphi(y) = \int_0^y \frac{1}{\ell(\bar{y})} d\bar{y} \tag{8.488}$$

$$T_* \left(\bar{\mathbf{g}}_i^0(x)\right) \triangleq \bar{\mathbf{r}}_i^0(\xi) = \begin{bmatrix} O_{d\times 1} \\ \hat{\mathbf{r}}_i^0(\xi) \end{bmatrix}, \; 1 \leq i \leq n - d \tag{8.489}$$

$$\hat{S}(\xi) = \begin{bmatrix} \hat{S}^1(\xi) \\ \hat{S}^2(\xi) \end{bmatrix} = \begin{bmatrix} [\varphi(\xi_1) \quad \cdots \quad \varphi(\xi_d)]^{\mathsf{T}} \\ \hat{S}^2(\xi_1, \cdots, \xi_n) \end{bmatrix} \tag{8.490}$$

and

$$\frac{\partial \hat{S}^2(\xi)}{\partial(\xi_{d+1}, \cdots, \xi_n)} = \begin{bmatrix} \hat{\mathbf{r}}_{n-d}^0(\xi) & \cdots & \hat{\mathbf{r}}_2^0(\xi) & \hat{\mathbf{r}}_1^0(\xi) \end{bmatrix}^{-1}. \tag{8.491}$$

Furthermore, an extended state transformation $z^e = S^e(w, x)$ is given by (8.425).

Proof Necessity. Let $\sigma < n - d$. Suppose that system (8.370) is RDOEL with index d and (8.398) is satisfied. Then, by Theorem 8.17, there exist smooth functions $\ell(y)$ ($\ell(0) = 1$) and $\varepsilon_i^u(y)$, $d + 1 \leq i \leq n$ such that (8.480)–(8.482) and (8.486)–(8.491) are satisfied. It is clear, by (8.340) and (8.341), that

$$\bar{\mathbf{g}}_i^0(x) = \ell \left(H \circ F_0^{n-i}(x)\right) \mathbf{g}_i^0(x), \; 2 \leq i \leq n$$

and

$$\bar{\mathbf{g}}_i^u(x) = \begin{cases} \ell \left(H \circ F_0^{n-i}(x)\right) \mathbf{g}_i^u(x), & 2 \leq i \leq \sigma \\ \ell \left(H \circ F_0^{n-\sigma} \circ F_u^{-1}(x)\right) \mathbf{g}_{\sigma+1}^u(x), & i = \sigma + 1 \end{cases}$$

which imply, together with (8.480), that (8.479) is satisfied with

$$\theta_\sigma^u(x) = \frac{\ell \left(H \circ F_0^{n-\sigma-1}(x)\right)}{\ell \left(H \circ F_0^{n-\sigma} \circ F_u^{-1}(x)\right)} \quad \text{or} \quad \frac{\ell \left(H \circ \hat{F}_u^{n-\sigma}(x)\right)}{\ell \left(H \circ F_0^{n-\sigma}(x)\right)} = \theta_\sigma^u \circ F_u(x).$$

Since

$$\frac{\partial \left(\theta_\sigma^u \circ F_u\right)}{\partial u} = \frac{\frac{d\ell(y)}{dy}\Big|_{y=H\circ\hat{F}_u^{n-\sigma}(x)}}{\ell \left(H \circ F_0^{n-\sigma}(x)\right)} \frac{\partial \left(H \circ \hat{F}_u^{n-\sigma}(x)\right)}{\partial u},$$

it is clear that (8.483) is satisfied with

$$\bar{\beta}^u(x) = \frac{\frac{d\ell(y)}{dy}\Big|_{y=H\circ\hat{F}_u^{n-\sigma}(x)}}{\ell \left(H \circ F_0^{n-\sigma}(x)\right)}.$$

Since

$$\bar{\beta}^0(x) = \frac{1}{\ell(y)} \frac{d\ell(y)}{dy}\bigg|_{y=H\circ F_0^{n-\sigma}(x)} = \frac{d}{dy}\left(\ln \ell(y)\right)\bigg|_{y=H\circ F_0^{n-\sigma}(x)}$$

$$\triangleq \beta\left(H \circ F_0^{n-\sigma}(x)\right),$$

it is easy to see that (8.484) and (8.485) are satisfied.

Sufficiency. It is obvious by Theorem 8.17.   □

Remark 8.12 Suppose that (8.398) is satisfied and $\sigma \geq n - d$. Then, it is easy to see, by (8.398) and (8.336), that $\sigma = n$ or for $1 \leq i \leq n - 1$,

$$H \circ \hat{F}_u^i(x) = H \circ F_0^i(x) \text{ and } \alpha_i^u(\xi) = 0.$$

Theorem 8.20 *Suppose that (8.398) is satisfied, $\kappa \leq n$, and $\sigma \geq n - d$. Let $F_u(x) \neq F_0(x)$. System (8.370) is RDOEL with index d and extended state transformation $z^e = S^e(w, x)$, if and only if there exist scalar functions $\bar{\beta}^u(x)$ and $\beta(y)$, defined on an open neighborhood of $y = 0$, such that*

(i)

$$\frac{1}{\theta_n^u(\xi)} \frac{\partial \theta_n^u(\xi)}{\partial u} = \bar{\beta}^u(\xi) \frac{\partial \alpha_n^u(\xi)}{\partial u} \tag{8.492}$$

(ii)

$$\bar{\mathbf{g}}_i^u(x) = \bar{\mathbf{g}}_i^0(x), \quad 1 \leq i \leq n - d \tag{8.493}$$

(iii)

$$[\bar{\mathbf{g}}_1^0(x), \bar{\mathbf{g}}_i^0(x)] = 0, \quad 1 \leq i \leq n - d \tag{8.494}$$

(iv)

$$\hat{S}^2 \circ T \circ F_u(x) - \hat{S}^2 \circ T \circ F_0(x) = \begin{bmatrix} \varepsilon_{d+1}^u(H(x)) \\ \vdots \\ \varepsilon_n^u(H(x)) \end{bmatrix} \tag{8.495}$$

where $\alpha_n^u(\xi) \triangleq H \circ \hat{F}_u^n \circ T^{-1}(\xi)$,

$$\theta_n^u(\xi) \triangleq \frac{\partial \alpha_n^u(\xi)}{\partial \xi_{n+2-\kappa}} \tag{8.496}$$

$$\bar{\beta}^u(\xi) = \beta\left(\alpha_n^u(\xi)\right) \tag{8.497}$$

$$\ell(y) = e^{\int_0^y \beta(\bar{y})d\bar{y}} \tag{8.498}$$

$$\bar{\mathbf{g}}_1^u(x) = \bar{\mathbf{g}}_1^0(x) \triangleq \ell(H \circ F_0^{n-1}(x))\mathbf{g}_1^0(x) \qquad (8.499)$$

$$\bar{\mathbf{g}}_i^0(x) \triangleq (F_0)_*(\bar{\mathbf{g}}_{i-1}^0(x)) \; ; \quad \bar{\mathbf{g}}_i^u(x) \triangleq (F_u)_*(\bar{\mathbf{g}}_{i-1}^u(x)), \; i \geq 2 \qquad (8.500)$$

$$\varphi(y) = \int_0^y \frac{1}{\ell(\bar{y})} d\bar{y} \qquad (8.501)$$

$$T_* \left(\bar{\mathbf{g}}_i^0 \right) \triangleq \bar{\mathbf{r}}_i^0(\xi) = \begin{bmatrix} O_{d \times 1} \\ \hat{\mathbf{r}}_i^0(\xi) \end{bmatrix}, \; 1 \leq i \leq n - d \qquad (8.502)$$

$$\hat{S}(\xi) = \begin{bmatrix} \hat{S}^1(\xi) \\ \hat{S}^2(\xi) \end{bmatrix} = \begin{bmatrix} [\varphi(\xi_1) \; \cdots \; \varphi(\xi_d)]^\mathsf{T} \\ \hat{S}^2(\xi_1, \cdots, \xi_n) \end{bmatrix} \qquad (8.503)$$

and

$$\frac{\partial \hat{S}^2(\xi)}{\partial (\xi_{d+1}, \cdots, \xi_n)} = \begin{bmatrix} \hat{\mathbf{r}}_{n-d}^0(\xi) & \cdots & \hat{\mathbf{r}}_2^0(\xi) & \hat{\mathbf{r}}_1^0(\xi) \end{bmatrix}^{-1}. \qquad (8.504)$$

Furthermore, an extended state transformation $z^e = S^e(w, x)$ *is given by (8.425).*

Proof Necessity. Suppose that (8.398) is satisfied, $\kappa \leq n$, and $\sigma \geq n - d$. Then, we have, by Remark 8.12, that $\sigma = n$ or for $1 \leq i \leq n - 1$,

$$H \circ \hat{F}_u^i(x) = H \circ F_0^i(x) \text{ and } \alpha_i^u(\xi) = 0. \qquad (8.505)$$

Suppose that system (8.370) is RDOEL with index d. Then, by Lemma 8.11, system (8.370) is state equivalent to a d-GNOCF with OT $\varphi(y)$ and state transformation $z = \tilde{S}(x)$ and

$$\tilde{S}_i \circ F_u(x) - \tilde{S}_i \circ F_0(x) = \begin{cases} 0, & 1 \leq i \leq d \\ \bar{\varepsilon}_i^u(H(x)), & d + 1 \leq i \leq n \end{cases} \qquad (8.506)$$

for some $\bar{\varepsilon}_i^u(y)$, $d + 1 \leq i \leq n$. Thus, by Lemma 8.10 and (8.506), there exist a smooth function $\varphi(y)$ and smooth functions $\gamma_k^u(\xi_1, \cdots, \xi_{d+1}) : \mathbb{R}^{d+1+m} \to \mathbb{R}, d + 1 \leq k \leq n$ such that for $1 \leq i \leq n$,

$$\tilde{S}_n(x) = \varphi \circ H \circ F_0^{n-1}(x) - \sum_{k=d+1}^{n-1} \gamma_k^0 \circ T^1 \circ F_0^{n-1-k}(x) \qquad (8.507)$$

$$\varphi \circ H \circ \hat{F}_u^n(x) = \sum_{k=d+1}^{n-1} \gamma_k^0 \circ T^1 \circ \hat{F}_u^{n-k}(x) + \gamma_n^u \circ T^1(x)$$

$$= \sum_{k=d+1}^{n-1} \gamma_k^0 \circ T^1 \circ \hat{F}_0^{n-k}(x) + \gamma_n^u \circ T^1(x) \tag{8.508}$$

and

$$\tilde{S}_n \circ F_u(x) - \tilde{S}_n \circ F_0(x) = \bar{\varepsilon}_n^u(H(x)) \tag{8.509}$$

where $\bar{\varepsilon}_n^0(y) = 0$. Since $\tilde{S}_n \circ F_u(x) = \gamma_n^u \circ T^1(x)$ by (8.507) and (8.508), we have, by (8.509), that

$$\gamma_n^u \circ T^1(x) = \gamma_n^0 \circ T^1(x) + \bar{\varepsilon}_n^u(H(x)). \tag{8.510}$$

Thus, it is easy to see, by (8.505), (8.508), and (8.510), that

$$\varphi \circ H \circ \hat{F}_u^n = \sum_{k=d+1}^{n} \gamma_k^0 \circ T^1 \circ F_0^{n-k} + \varepsilon_n^u(H(x))$$

and

$$\varphi \circ \alpha_n^u(\xi) \triangleq \varphi \circ H \circ \hat{F}_u^n \circ T^{-1}(\xi)$$

$$= \sum_{k=d+1}^{n} \gamma_k^0(\xi_{n-k+1}, \cdots, \xi_{n-k+1+d}) + \varepsilon_n^u(\xi_1) \tag{8.511}$$

$$= \varphi \circ \alpha_n^0(\xi) + \varepsilon_n^u(\xi_1).$$

It is clear, by (8.332), (8.334), and (8.511), that $\frac{\partial \alpha_n^u(\xi)}{\partial \xi_{n+2-\kappa}} \neq 0$ and

$$\varphi \circ \alpha_n^u(\xi) = \varphi \circ \alpha_n^0(\xi_1, \cdots, \xi_{n+2-\kappa}, 0, \cdots, 0) + \varepsilon_n^u(\xi_1). \tag{8.512}$$

Let $\bar{\ell}(y) \triangleq \frac{d\varphi(y)}{dy} = \frac{1}{\ell(y)}$. It is easy to see that

$$\frac{1}{\bar{\ell}(y)} \frac{d\bar{\ell}(y)}{dy} = \ell(y) \frac{d\left(\frac{1}{\ell(y)}\right)}{dy} = -\frac{1}{\ell(y)} \frac{d\ell(y)}{dy}.$$

Since $n + 2 - \kappa \geq 2$, we have that

$$0 = \frac{\partial^2 (\varphi \circ \alpha_n^u)}{\partial u \partial \xi_{n+2-\kappa}} = \frac{\partial}{\partial u} \left(\frac{d\varphi(y)}{dy}\Big|_{y=\alpha_n^u(\xi)} \frac{\partial \alpha_n^u(\xi)}{\partial \xi_{n+2-\kappa}} \right) = \frac{\partial}{\partial u} \left(\bar{\ell} \circ \alpha_n^u(\xi) \, \theta_n^u(\xi) \right)$$

$$= \frac{d\bar{\ell}(y)}{dy}\Big|_{y=\alpha_n^u(\xi)} \frac{\partial \alpha_n^u(\xi)}{\partial u} \theta_n^u(\xi) + \bar{\ell}(y)\Big|_{y=\alpha_n^u(\xi)} \frac{\partial \theta_n^u(\xi)}{\partial u}$$

which implies that

$$\frac{1}{\theta_n^u(\xi)}\frac{\partial \theta_n^u(\xi)}{\partial u} = -\left.\frac{1}{\bar{\ell}(y)}\frac{d\bar{\ell}(y)}{dy}\right|_{y=\alpha_n^u(\xi)}\frac{\partial \alpha_n^u(\xi)}{\partial u} = \left.\frac{1}{\ell(y)}\frac{d\ell(y)}{dy}\right|_{y=\alpha_n^u(\xi)}\frac{\partial \alpha_n^u(\xi)}{\partial u}.$$

Therefore, (8.492), (8.497), and (8.498) are satisfied with

$$\bar{\beta}^u(\xi) = \left.\frac{1}{\ell(y)}\frac{d\ell(y)}{dy}\right|_{y=\alpha_n^u(\xi)} = \left.\frac{d\ln\ell(y)}{dy}\right|_{y=\alpha_n^u(\xi)}$$

and

$$\bar{\beta}^u(\xi) = \beta\left(\alpha_n^u(\xi)\right).$$

Sufficiency. It is obvious by Theorem 8.17.   □

Theorem 8.21 *Suppose that (8.398) is satisfied, $\kappa = n + 1$, and $\sigma \geq n - d$. (In other words, $\kappa = n + 1$ and $\sigma = n$.) System (8.370) is RDOEL with index d, if and only if*

$$\mathbf{g}_i^u(x) = \mathbf{g}_i^0(x), \quad 2 \leq i \leq n. \tag{8.513}$$

In other words, system (8.370) is RDOEL with index d, if and only if system (8.370) is state equivalent to a dual Brunovsky NOCF without OT.

Proof Necessity. Let $\kappa = n + 1$ and $\sigma = n$. Suppose that system (8.370) is d-RDOEL. Let $\xi = T(x)$. Then, since $\kappa = n + 1$ and $\sigma = n$, it is easy to see, by (8.505) and (8.512), that

$$\alpha_i^u(\xi) \triangleq H \circ \hat{F}_u^i \circ T^{-1}(\xi) - H \circ F_0^i \circ T^{-1}(\xi) = 0, \ 1 \leq i \leq n - 1$$

and

$$\varphi \circ \alpha_n^u(\xi) = \varphi \circ \alpha_n^0(\xi_1, 0, \cdots, 0) + \varepsilon_n^u(\xi_1) \triangleq \tilde{\gamma}_n^u(\xi_1)$$

which imply, together with (8.296), that

$$f_u(\xi) \triangleq T \circ F_u \circ T^{-1}(\xi) = \begin{bmatrix} \xi_2 \\ \vdots \\ \xi_n \\ \tilde{\alpha}_n^u(\xi_1) \end{bmatrix}$$

where $\tilde{\alpha}_n^u(\xi_1) \triangleq \varphi^{-1} \circ \tilde{\gamma}_n^u(\xi_1) = \alpha_n^u(\xi_1, 0, \cdots, 0)$. Therefore, it is clear, by (8.301) and (8.302), that

$$T_*(\mathbf{g}_1^u(x)) \triangleq \mathbf{r}_1^u(\xi) = \frac{\partial}{\partial \xi_n} = \mathbf{r}_1^0(\xi)$$

and for $2 \leq i \leq n$,

$$T_*(g_i^u(x)) \triangleq r_i^u(\xi) = (f_u)_*(r_{i-1}^u(\xi)) = (f_u)_* \left(\frac{\partial}{\partial \xi_{n+2-i}} \right) = \frac{\partial}{\partial \xi_{n+1-i}}$$

which implies that (8.513) is satisfied.

Sufficiency. Let $\kappa = n+1$ and $\sigma = n$. Suppose that (8.513) is satisfied. Then, by Theorem 8.16, system (8.370) is state equivalent to a dual Brunovsky NOCF with OT $\varphi(y) = y$ (i.e., without OT). Hence, system (8.370) is RDOEL with index d. $\square$

Example 8.6.2 Consider the system

$$x(t+1) = \begin{bmatrix} x_2 \\ x_3 \\ e^{x_1+u+x_2 x_3} - 1 \end{bmatrix} = F_u(x) ; \quad y = x_1 = H(x). \tag{8.514}$$

(a) Show that $\kappa = 2 \leq n$ and $\sigma = 3 = n$.
(b) Use Theorem 8.14 to show that system (8.514) is not state equivalent to a dual Brunovsky NOCF with OT.
(c) Use Theorem 8.18 to show that system (8.514) is RDOEL with index $d = 1$.

Solution (a) It is easy to see that

$$\bar{x} = F_u^{-1}(x) = \begin{bmatrix} \ln(x_3+1) - x_1 x_2 - u \\ x_1 \\ x_2 \end{bmatrix} \quad \text{and} \quad T(x) \triangleq \begin{bmatrix} H(x) \\ H \circ F_0(x) \\ H \circ F_0^2(x) \end{bmatrix} = x.$$

Thus, we have, by (8.298) and (8.299), that

$$g_1^u(x) = g_1^0(x) = T_*^{-1}(\frac{\partial}{\partial \xi_3}) = \frac{\partial}{\partial x_3} = \begin{bmatrix} 0 \\ 0 \\ 1 \end{bmatrix}$$

and

$$g_2^u(x) \triangleq (F_u)_*(g_1^u(x)) = \frac{\partial F_u(\bar{x})}{\partial \bar{x}} g_1^u(\bar{x}) \bigg|_{\bar{x}=F_u^{-1}(x)} = \begin{bmatrix} 0 \\ 1 \\ x_1(x_3+1) \end{bmatrix} = g_2^0(x).$$

Since $L_{g_2^0}(H \circ F_0^2(x)) = L_{g_2^0}(x_3) = x_1(x_3+1) \neq 0$, it is clear, by (8.331), that $\kappa = 2$. Also, since $H \circ \hat{F}_u^{n-1}(x) = x_3 = H \circ F_0^{n-1}(x)$, $H \circ \hat{F}_u^{n-2}(x) = x_2 = H \circ F_0^{n-2}(x)$, and $H \circ \hat{F}_u^{n-3}(x) = x_1 = H \circ F_0^{n-3}(x)$, we have, by (8.336), that $\sigma = 3 = n$.

(b) Since $L_{g_2^0}(H \circ F_0^2(x)) = x_1(x_3+1)$ and

$$[\mathbf{g}_1^0(x), \mathbf{g}_2^0(x)] = \begin{bmatrix} 0 \\ 0 \\ x_1 \end{bmatrix} = L_{\mathbf{g}_2^0}(H \circ F_0^2(x)) \frac{1}{x_3 + 1} \mathbf{g}_1^0(x), \tag{8.515}$$

it is clear that condition (i) of Theorem 8.14 is satisfied with $\beta(y) = \frac{1}{y+1}$. Thus, we have, by (8.345)–(8.347), that

$$\ell(y) \triangleq e^{\int_0^y \beta(\bar{y})d\bar{y}} = e^{\int_0^y \frac{1}{\bar{y}+1}d\bar{y}} = e^{\ln(y+1)} = y + 1 \tag{8.516}$$

$$\bar{\mathbf{g}}_1^u(x) = \bar{\mathbf{g}}_1^0(x) \triangleq \ell(H \circ F_0^2(x))\mathbf{g}_1^0(x) = \ell(x_3) \begin{bmatrix} 0 \\ 0 \\ 1 \end{bmatrix} = \begin{bmatrix} 0 \\ 0 \\ x_3 + 1 \end{bmatrix} \tag{8.517}$$

$$\bar{\mathbf{g}}_2^u(x) \triangleq (F_u)_*(\bar{\mathbf{g}}_1^u(x)) = \begin{bmatrix} 0 \\ (x_2 + 1) \\ x_1(x_2 + 1)(x_3 + 1) \end{bmatrix} \tag{8.518}$$

and

$$\bar{\mathbf{g}}_3^u(x) \triangleq (F_u)_*(\bar{\mathbf{g}}_2^u(x))$$
$$= \begin{bmatrix} (x_1 + 1) \\ (x_1 + 1)(x_2 + 1)(\ln(x_3 + 1) - x_1 x_2 - u) \\ (x_3 + 1)\{x_2 + x_1(x_2 + 1)(\ln(x_3 + 1) - x_1 x_2 - u)\} \end{bmatrix}.$$

Since $\bar{\mathbf{g}}_3^u(x) \neq \bar{\mathbf{g}}_3^0(x)$, condition (ii) of Theorem 8.14 is not satisfied. Therefore, by Theorem 8.14, system (8.514) is not state equivalent to a dual Brunovsky NOCF with OT.

(c) Let $d = 1$. Since $H \circ F_u(x) = x_2 = H \circ F_0(x)$, (8.398) is satisfied. It is clear, by (8.515), that condition (i) of Theorem 8.18 is satisfied with $\beta(y) = \frac{1}{y+1}$. It is also clear, by (8.517) and (8.518), that $\bar{\mathbf{g}}_1^u(x) = \bar{\mathbf{g}}_1^0(x)$ and $\bar{\mathbf{g}}_2^u(x) = \bar{\mathbf{g}}_2^0(x)$. Thus, condition (ii) of Theorem 8.18 is satisfied. Since

$$[\bar{\mathbf{g}}_1^0(x), \bar{\mathbf{g}}_2^0(x)] = \begin{bmatrix} \begin{bmatrix} 0 \\ 0 \\ x_3 + 1 \end{bmatrix}, \begin{bmatrix} 0 \\ (x_2 + 1) \\ x_1(x_2 + 1)(x_3 + 1) \end{bmatrix} \end{bmatrix} = 0,$$

condition (iii) of Theorem 8.18 is satisfied. We have, by (8.474) and (8.516), that

$$\varphi(y) = \int_0^y \frac{1}{\ell(\bar{y})}d\bar{y} = \int_0^y \frac{1}{\bar{y} + 1}d\bar{y} = \ln(y + 1).$$

Since

$$\left[\bar{\mathbf{r}}_2^0(\xi) \ \bar{\mathbf{r}}_1^0(\xi)\right] \triangleq \left[T_*(\bar{\mathbf{g}}_2^0(x)) \ T_*(\bar{\mathbf{g}}_1^0(x))\right]$$

$$= \begin{bmatrix} 0 & 0 \\ (\xi_2 + 1) & 0 \\ \xi_1(\xi_2 + 1)(\xi_3 + 1) & \xi_3 + 1 \end{bmatrix},$$

it is clear, by (8.475) and (8.477), that

$$\left[\hat{\mathbf{r}}_2^0(\xi) \ \hat{\mathbf{r}}_1^0(\xi)\right] \triangleq \begin{bmatrix} (\xi_2 + 1) & 0 \\ \xi_1(\xi_2 + 1)(\xi_3 + 1) & \xi_3 + 1 \end{bmatrix}$$

and

$$\frac{\partial \hat{S}^2(\xi)}{\partial(\xi_2, \xi_3)} = \left[\hat{\mathbf{r}}_2^0(\xi) \ \hat{\mathbf{r}}_1^0(\xi)\right]^{-1} = \begin{bmatrix} \frac{1}{\xi_2 + 1} & 0 \\ -\xi_1 & \frac{1}{\xi_3 + 1} \end{bmatrix}$$

which implies, together with (8.476), that

$$\hat{S}(\xi) = \begin{bmatrix} \hat{S}^1(\xi) \\ \hat{S}^2(\xi) \end{bmatrix} = \begin{bmatrix} \varphi(\xi_1) \\ \hat{S}^2(\xi) \end{bmatrix} = \begin{bmatrix} \ln(\xi_1 + 1) \\ \ln(\xi_2 + 1) \\ \ln(\xi_3 + 1) - \xi_1\xi_2 \end{bmatrix}.$$

Since

$$\hat{S}^2 \circ T \circ F_u(x) - \hat{S}^2 \circ T \circ F_0(x) = \begin{bmatrix} \ln(x_3 + 1) \\ x_1 + u \end{bmatrix} - \begin{bmatrix} \ln(x_3 + 1) \\ x_1 \end{bmatrix} = \begin{bmatrix} 0 \\ u \end{bmatrix},$$

it is clear that condition (iv) of Theorem 8.18 is satisfied. Hence, system (8.514) is, by Theorem 8.18, RDOEL with index $d = 1$ and extended state transformation $z^e = S^e(w, x)$. Finally, the extended state transformation $z^e = S^e(w, x)$ in (8.425) is given by

$$S^e(w, x) = \begin{bmatrix} \hat{S}(w, x_1, x_2) \\ S_3^e \circ F_0^e(w, x) - \gamma_3^0(w, x_1) \end{bmatrix} = \begin{bmatrix} \ln(w_1 + 1) \\ \ln(x_1 + 1) \\ \ln(x_2 + 1) - w_1x_1 \\ \ln(x_3 + 1) - x_1x_2 - w_1 \end{bmatrix}$$

where

$$\gamma_3^0(\xi_1, \xi_2) \triangleq \hat{S}_3 \circ T \circ F_0 \circ T^{-1}(\xi) = \xi_1 + \xi_2\xi_3 - \xi_2\xi_3 = \xi_1$$

and

$$F_u^e(w, x) = \begin{bmatrix} H(x) \\ F_u(x) \end{bmatrix} = \begin{bmatrix} x_1 \\ x_2 \\ x_3 \\ e^{x_1 + u + x_2x_3} - 1 \end{bmatrix}.$$

Since $H^e(w, x) = w$, it is easy to see that $\varphi \circ H^e \circ (S^e)^{-1}(z^e) = z_1^e$ and

$$
S^e \circ F_u^e \circ (S^e)^{-1}(z^e) = \begin{bmatrix} z_2^e \\ z_3^e + (e^{z_1^e} - 1)(e^{z_2^e} - 1) \\ z_4^e + e^{z_1^e} - 1 \\ u \end{bmatrix} = \begin{bmatrix} z_2^e \\ z_3^e \\ z_4^e \\ 0 \end{bmatrix} + \begin{bmatrix} 0 \\ w_1 y \\ w_1 \\ u \end{bmatrix}.
$$

$\square$

Example 8.6.3 Consider the system

$$
x(t+1) = \begin{bmatrix} x_2 \\ x_3 + x_1 u_2^2 \\ u_1 + x_1 - x_2(x_3 + x_1 u_2^2) \end{bmatrix} = F_u(x) ; \quad y = x_1 = H(x). \quad (8.519)
$$

(a) Show that $\kappa = 2$ and $\sigma = 1$.
(b) Use Theorem 8.15 to show that system (8.519) is not state equivalent to a dual Brunovsky NOCF with OT.
(c) Use Theorem 8.19 to show that system (8.519) is RDOEL with index $d = 1$.
(d) Use Theorem 8.14 to show that system (8.519) is not state equivalent to a dual Brunovsky NOCF with OT.
(e) Use Theorem 8.18 to show that system (8.519) is RDOEL with index $d = 1$.

Solution (a) It is easy to see that

$$
\bar{x} = F_u^{-1}(x) = \begin{bmatrix} x_3 - u_1 + x_1 x_2 \\ x_1 \\ x_2 - u_2^2(x_3 - u_1 + x_1 x_2) \end{bmatrix} \quad \text{and} \quad \xi = T(x) = x.
$$

Thus, we have, by (8.298) and (8.299), that

$$
g_1^u(x) = g_1^0(x) = T_*^{-1}\left(\frac{\partial}{\partial \xi_3}\right) = \frac{\partial}{\partial x_3} = \begin{bmatrix} 0 \\ 0 \\ 1 \end{bmatrix}
$$

and

$$
g_2^u(x) \triangleq (F_u)_*(g_1^u(x)) = \frac{\partial F_u(\bar{x})}{\partial \bar{x}} g_1^u(\bar{x}) \Big|_{\bar{x} = F_u^{-1}(x)} = \begin{bmatrix} 0 \\ 1 \\ -x_1 \end{bmatrix} = g_2^0(x).
$$

Since $L_{g_2^0}(H \circ F_0^2(x)) = L_{g_2^0}(x_3) = -x_1 \neq 0$, we have, by (8.331), that $\kappa = 2$. Also, since $H \circ \hat{F}_u^{n-1}(x) = x_3 + x_1 u_2^2 \neq H \circ F_0^{n-1}(x)$, it is clear, by (8.336), that $\sigma = 1$.

(b) Since

$$\mathbf{g}_2^u(x) = \mathbf{g}_2^0(x) = \theta_1^u(x)\mathbf{g}_2^0(x), \tag{8.520}$$

it is clear that condition (i) of Theorem 8.15 is satisfied with $\theta_\sigma^u(x) = 1$. Since

$$\frac{\partial \left(\theta_\sigma^u \circ F_u(x)\right)}{\partial u} = [0 \; 0] \quad \text{and} \quad \frac{\partial \left(H \circ \hat{F}_u^{n-\sigma}(x)\right)}{\partial u} = [0 \; 2x_1 u_2],$$

it is clear that (8.354) and (8.355) are satisfied with

$$\bar{\beta}^u(x) = 0 \quad \text{and} \quad \beta(y) = 0.$$

Thus, we have, by (8.356)–(8.358), that

$$\ell(y) \triangleq e^{\int_0^y \beta(\bar{y})d\bar{y}} = e^0 = 1 \tag{8.521}$$

$$\bar{\mathbf{g}}_1^u(x) = \bar{\mathbf{g}}_1^0(x) \triangleq \ell(H \circ F_0^2(x))\mathbf{g}_1^0 = \begin{bmatrix} 0 \\ 0 \\ 1 \end{bmatrix} \tag{8.522}$$

$$\bar{\mathbf{g}}_2^u(x) \triangleq (F_u)_*(\bar{\mathbf{g}}_1^u(x)) = \mathbf{g}_2^u(x) = \begin{bmatrix} 0 \\ 1 \\ -x_1 \end{bmatrix} \tag{8.523}$$

and

$$\bar{\mathbf{g}}_3^u(x) \triangleq (F_u)_*(\bar{\mathbf{g}}_2^u(x)) = \begin{bmatrix} 1 \\ u_1 - x_3 - x_1 x_2 \\ x_1(x_3 - u_1 + x_1 x_2) - x_2 \end{bmatrix}.$$

Since $\bar{\mathbf{g}}_3^u(x) \neq \bar{\mathbf{g}}_3^0(x)$, condition (ii) of Theorem 8.15 is not satisfied. Therefore, by Theorem 8.15, system (8.519) is not state equivalent to a dual Brunovsky NOCF with OT.

(c) Let $d = 1$. Since $H \circ F_u(x) = x_2 = H \circ F_0(x)$, (8.398) is satisfied. It is clear, by (8.520), that condition (i) of Theorem 8.19 is satisfied with $\theta_\sigma^u(x) = 1$. It is also clear, by (8.522) and (8.523), that $\bar{\mathbf{g}}_1^u(x) = \bar{\mathbf{g}}_1^0(x)$ and $\bar{\mathbf{g}}_2^u(x) = \bar{\mathbf{g}}_2^0(x)$. Thus, condition (ii) of Theorem 8.19 is satisfied. Since

$$[\bar{\mathbf{g}}_1^0(x), \bar{\mathbf{g}}_2^0(x)] = \begin{bmatrix} \begin{bmatrix} 0 \\ 0 \\ 1 \end{bmatrix}, \begin{bmatrix} 0 \\ 1 \\ -x_1 \end{bmatrix} \end{bmatrix} = 0,$$

condition (iii) of Theorem 8.19 is satisfied. We have, by (8.488) and (8.521), that

$$\varphi(y) = \int_0^y \frac{1}{\ell(\bar{y})} d\bar{y} = y.$$

Since

$$\left[\bar{\mathbf{r}}_2^0(\xi) \ \bar{\mathbf{r}}_1^0(\xi)\right] \triangleq \left[T_*(\bar{\mathbf{g}}_2^0(x)) \ T_*(\bar{\mathbf{g}}_1^0(x))\right] = \begin{bmatrix} 0 & 0 \\ 1 & 0 \\ -\xi_1 & 1 \end{bmatrix},$$

it is clear, by (8.489) and (8.491), that

$$\left[\hat{\mathbf{r}}_2^0(\xi) \ \hat{\mathbf{r}}_1^0(\xi)\right] \triangleq \begin{bmatrix} 1 & 0 \\ -\xi_1 & 1 \end{bmatrix}$$

and

$$\frac{\partial \hat{S}^2(\xi)}{\partial(\xi_2, \xi_3)} = \left[\hat{\mathbf{r}}_2^0(\xi) \ \hat{\mathbf{r}}_1^0(\xi)\right]^{-1} = \begin{bmatrix} 1 & 0 \\ \xi_1 & 1 \end{bmatrix}$$

which implies, together with (8.490), that

$$\hat{S}(\xi) = \begin{bmatrix} \hat{S}^1(\xi) \\ \hat{S}^2(\xi) \end{bmatrix} = \begin{bmatrix} \varphi(\xi_1) \\ \hat{S}^2(\xi) \end{bmatrix} = \begin{bmatrix} \xi_1 \\ \xi_2 \\ \xi_3 + \xi_1\xi_2 \end{bmatrix}.$$

Since

$$\hat{S}^2 \circ T \circ F_u(x) - \hat{S}^2 \circ T \circ F_0(x) = \begin{bmatrix} x_3 + x_1 u_2^2 \\ x_1 + u_1 \end{bmatrix} - \begin{bmatrix} x_3 \\ x_1 \end{bmatrix} = \begin{bmatrix} x_1 u_2^2 \\ u_1 \end{bmatrix},$$

it is clear that condition (iv) of Theorem 8.19 is satisfied. Hence, system (8.519) is, by Theorem 8.19, RDOEL with index $d = 1$ and extended state transformation $z^e = S^e(w, x)$. Finally, the extended state transformation $z^e = S^e(w, x)$ in (8.425) is given by

$$S^e(w, x) = \begin{bmatrix} \hat{S}(w, x_1, x_2) \\ S_3^e \circ F_0^e(w, x) - \gamma_3^0(w, x_1) \end{bmatrix} = \begin{bmatrix} w_1 \\ x_1 \\ x_2 + w_1 x_1 \\ x_3 + x_1 x_2 - w_1 \end{bmatrix}$$

where

$$\gamma_3^0(\xi_1, \xi_2) \triangleq \hat{S}_3 \circ T \circ F_0 \circ T^{-1}(\xi) = \xi_1 - \xi_2\xi_3 + \xi_2\xi_3 = \xi_1$$

and

$$F_u^e(w, x) = \begin{bmatrix} H(x) \\ F_u(x) \end{bmatrix} = \begin{bmatrix} x_1 \\ x_2 \\ x_3 + x_1 u_2^2 \\ u_1 + x_1 - x_2(x_3 + x_1 u_2^2) \end{bmatrix}.$$

Since $H^e(w, x) = w$, it is easy to see that $\varphi \circ H^e \circ (S^e)^{-1}(z^e) = z_1^e$ and

$$S^e \circ F_u^e \circ (S^e)^{-1}(z^e) = \begin{bmatrix} z_2^e \\ z_3^e - z_1^e z_2^e \\ z_4^e + z_1^e + z_2^e u_2^2 \\ u_1 \end{bmatrix} = \begin{bmatrix} z_2^e \\ z_3^e \\ z_4^e \\ 0 \end{bmatrix} + \begin{bmatrix} 0 \\ -w_1 y \\ w_1 + y u_2^2 \\ u_1 \end{bmatrix}.$$

(d) Solution is omitted. (Problem 8-11.)
(e) Solution is omitted. (Problem 8-11.)   □

Example 8.6.4 Consider the system

$$x(t + 1) = \begin{bmatrix} x_2 \\ x_3 \\ e^{x_1 x_3 + u_1 + u_2 x_1} - 1 \end{bmatrix} = F_u(x); \quad y = x_1 = H(x). \tag{8.524}$$

(a) Show that $\kappa = 2$ and $\sigma = 3$.
(b) Use Theorem 8.18 to show that system (8.524) is not RDOEL with index $d = 1$.
(c) Use Theorem 8.20 to show that system (8.524) is not RDOEL with index $d = 1$.
(d) Use Theorem 8.20 to show that system (8.524) is RDOEL with index $d = 2$.

Solution (a) It is easy to see that

$$\bar{x} = F_u^{-1}(x) = \begin{bmatrix} \frac{\ln(1+x_3)-u_1}{x_2+u_2} \\ x_1 \\ x_2 \end{bmatrix} \quad \text{and} \quad \xi = T(x) \triangleq \begin{bmatrix} H(x) \\ H \circ F_0(x) \\ H \circ F_0^2(x) \end{bmatrix} = x.$$

Thus, we have, by (8.298) and (8.299), that

$$g_1^u(x) = g_1^0(x) = T_*^{-1}\left(\frac{\partial}{\partial \xi_3}\right) = \frac{\partial}{\partial x_3} = \begin{bmatrix} 0 \\ 0 \\ 1 \end{bmatrix}$$

and

$$g_2^u(x) \triangleq (F_u)_* (g_1^u(x)) = \frac{\partial F_u(\bar{x})}{\partial \bar{x}} g_1^u(\bar{x})\Big|_{\bar{x}=F_u^{-1}(x)} = \begin{bmatrix} 0 \\ 1 \\ \frac{(1+x_3)(\ln(1+x_3)-u_1)}{x_2+u_2} \end{bmatrix}$$

$$\neq \begin{bmatrix} 0 \\ 1 \\ \frac{(1+x_3)\ln(1+x_3)}{x_2} \end{bmatrix} = g_2^0(x).$$

Since $L_{g_2^0}(H \circ F_0^2(x)) = \frac{(1+x_3)\ln(1+x_3)}{x_2} \neq 0$, we have, by (8.331), $\kappa = 2 = n$.
Also, since $H \circ \hat{F}_u^{n-1}(x) = x_3 = H \circ F_0^{n-1}(x)$, $H \circ \hat{F}_u^{n-2}(x) = x_2 = H \circ F_0^{n-2}(x)$, and $H \circ \hat{F}_u^{n-3}(x) = x_1 = H \circ F_0^{n-2}(x)$, it is clear, by (8.336), that $\sigma = 3 = n$.

(b) Let $d = 1$. Then, it is clear that $\kappa \leq n - d$. Since $H \circ F_u(x) = x_2 = H \circ F_0(x)$, (8.398) is satisfied. Note that $L_{g_2^0}(H \circ F_0(x)) = \frac{(1+x_3)\ln(1+x_3)}{x_2}$ and

$$[g_1^0(x), g_2^0(x)] = \begin{bmatrix} 0 \\ \frac{1+\ln(1+x_3)}{x_2} \end{bmatrix} = L_{g_2^0}(H \circ F_0(x)) \frac{1 + \ln(1 + x_3)}{(1 + x_3)\ln(1 + x_3)} g_1^0(x).$$

Since $\frac{1+\ln(1+y)}{(1+y)\ln(1+y)}$ is not defined on a neighborhood of $y = 0$, it is clear that condition (i) of Theorem 8.18 is not satisfied. Hence, by Theorem 8.18, system (8.524) is not RDOEL with index $d = 1$.

(c) Let $d = 1$. Then, it is clear that $\kappa \leq n$ and $\sigma \geq n - d$. Since $H \circ F_u(x) = x_2 = H \circ F_0(x)$, (8.398) is satisfied. Since

$$\alpha_3^u(\xi) \triangleq H \circ \hat{F}_u^3 \circ T^{-1}(\xi) = e^{\xi_1 \xi_3 + u_1 + u_2 \xi_1} - 1$$

$$\theta_3^u(\xi) \triangleq \frac{\partial \alpha_3^u(\xi)}{\partial \xi_{n+2-\kappa}} = \frac{\partial \alpha_3^u(\xi)}{\partial \xi_3} = \xi_1 e^{\xi_1 \xi_3 + u_1 + u_2 \xi_1}$$

$$\frac{1}{\theta_3^u(\xi)} \frac{\partial \theta_3^u(\xi)}{\partial u} = \begin{bmatrix} 1 & \xi_1 \end{bmatrix}$$

and

$$\frac{\partial \alpha_3^u(\xi)}{\partial u} = e^{\xi_1 \xi_3 + u_1 + u_2 \xi_1} \begin{bmatrix} 1 & \xi_1 \end{bmatrix},$$

it is clear that condition (i) of Theorem 8.20 and (8.497) are satisfied with

$$\bar{\beta}^u(\xi) = \frac{1}{e^{\xi_1 \xi_2 + u_1 + u_2 \xi_1}} \quad \text{and} \quad \beta(y) = \frac{1}{1 + y}.$$

Thus, we have, by (8.498)–(8.500), that

$$\ell(y) \triangleq e^{\int_0^y \beta(\bar{y}) d\bar{y}} = e^{\ln(1+y)} = 1 + y \tag{8.525}$$

$$\bar{g}_1^u(x) = \bar{g}_1^0(x) \triangleq \ell(H \circ F_0^2(x)) g_1^0 = \begin{bmatrix} 0 \\ 0 \\ 1 + x_3 \end{bmatrix} \tag{8.526}$$

and

$$\bar{\mathbf{g}}_2^u(x) \triangleq (F_u)_*(\bar{\mathbf{g}}_1^u(x)) = \mathbf{g}_2^u(x) = \begin{bmatrix} 0 \\ 1 + x_2 \\ \frac{(1+x_2)(1+x_3)(\ln(1+x_3)-u_1)}{x_2+u_2} \end{bmatrix}.$$

Since $n - d = 2$ and $\bar{\mathbf{g}}_2^u(x) \neq \bar{\mathbf{g}}_2^0(x)$, it is obvious that condition (ii) of Theorem 8.20 is not satisfied. Hence, by Theorem 8.20, system (8.524) is not RDOEL with index $d = 1$.

(d) Let $d = 2$. Then, it is clear that $\kappa \leq n$ and $\sigma \geq n - d$. Since $H \circ F_u(x) = x_2 = H \circ F_0(x)$ and $H \circ \hat{F}_u^2(x) = x_3 = H \circ F_0^2(x)$, (8.398) is satisfied. It has been shown that condition (i) of Theorem 8.20 is satisfied. Since $n - d = 1$, it is obvious, by (8.526), that condition (ii) and condition (iii) of Theorem 8.20 are satisfied. We have, by (8.501) and (8.525), that

$$\varphi(y) = \int_0^y \frac{1}{\ell(\bar{y})} d\bar{y} = \ln(1 + y).$$

Since

$$\left[\bar{\mathbf{r}}_1^0(\xi)\right] \triangleq \left[T_*(\bar{\mathbf{g}}_1^0(x))\right] = \begin{bmatrix} 0 \\ 0 \\ 1 + \xi_3 \end{bmatrix},$$

it is clear, by (8.502) and (8.504), that

$$\left[\hat{\mathbf{r}}_1^0(\xi)\right] \triangleq \left[1 + \xi_3\right]$$

and

$$\frac{\partial \hat{S}^2(\xi)}{\partial \xi_3} = \left[\hat{\mathbf{r}}_1^0(\xi)\right]^{-1} = \frac{1}{1 + \xi_3}$$

which implies, together with (8.503), that

$$\hat{S}(\xi) = \begin{bmatrix} \hat{S}^1(\xi) \\ \hat{S}^2(\xi) \end{bmatrix} = \begin{bmatrix} \varphi(\xi_1) \\ \varphi(\xi_2) \\ \hat{S}^2(\xi) \end{bmatrix} = \begin{bmatrix} \ln(1 + \xi_1) \\ \ln(1 + \xi_2) \\ \ln(1 + \xi_3) \end{bmatrix}.$$

Since

$$\hat{S}^2 \circ T \circ F_u(x) - \hat{S}^2 \circ T \circ F_0(x) = u_1 + u_2 x_1,$$

it is clear that condition (iv) of Theorem 8.20 is satisfied. Hence, system (8.524) is, by Theorem 8.20, RDOEL with index $d = 1$ and extended state transformation $z^e = S^e(w, x)$. Finally, the extended state transformation $z^e = S^e(w, x)$ in (8.425) is given by

$$S^e(w, x) = \begin{bmatrix} \hat{S}(w, x_1, x_2) \\ S_3^e \circ F_0^e(w, x) - \gamma_3^0(w, x_1) \\ S_4^e \circ F_0^e(w, x) \end{bmatrix} = \begin{bmatrix} \ln(1 + w_1) \\ \ln(1 + w_2) \\ \ln(1 + x_1) \\ \ln(1 + x_2) - w_1 x_1 \\ \ln(1 + x_3) - w_2 x_2 \end{bmatrix}$$

where

$$\gamma_2^0(\xi_1, \xi_2) \triangleq \hat{S}_2 \circ T \circ F_0 \circ T^{-1}(\xi) = \xi_1 \xi_3$$

and

$$F_u^e(w, x) = \begin{bmatrix} w_2 \\ H(x) \\ F_u(x) \end{bmatrix} = \begin{bmatrix} w_2 \\ x_1 \\ x_2 \\ x_3 \\ e^{x_1 x_3 + u_1 + u_2 x_1} - 1 \end{bmatrix}.$$

Since $H^e(w, x) = w$, it is easy to see that $\varphi \circ H^e \circ (S^e)^{-1}(z^e) = z_1^e$ and

$$S^e \circ F_u^e \circ (S^e)^{-1}(z^e) = \begin{bmatrix} z_2^e \\ z_3^e \\ z_4^e + (e^{z_1^e} - 1)(e^{z_3^e} - 1) \\ z_5^e \\ u_1 + u_2(e^{z_3^e} - 1) \end{bmatrix} = \begin{bmatrix} z_2^e \\ z_3^e \\ z_4^e \\ z_5^e \\ 0 \end{bmatrix} + \begin{bmatrix} 0 \\ 0 \\ w_1 y \\ 0 \\ u_1 + u_2 y \end{bmatrix}.$$

□

Remark 8.13 The conditions of Theorem 8.17 without assumption (8.398) are the necessary and sufficient conditions for (ii) of Lemma 8.11. Therefore, it is clear, by Lemma 8.11, that Theorem 8.17–Theorem 8.21 give only sufficient conditions for RDOEL with index d ($d \geq 1$), unless (8.398) is satisfied. For example, if (8.398) is satisfied not with $d = 3$ but with $d = 2$, then necessary and sufficient conditions for RDOEL with index $d = 0$, $d = 1$, and $d = 2$ can be found in the theorems, whereas the conditions for RDOEL with index $d = 3$, $d = 4$, etc., of the Theorems are not necessary but sufficient. (Refer to Example 8.6.5(c) and (d).) The RDOEL problem without assumption (8.398) is much more complicated and remains open.

Example 8.6.5 Consider the system

$$x(t+1) = \begin{bmatrix} x_2 + u \\ x_3 \\ u + x_1 + x_1(x_2 + u)^2 \end{bmatrix} = F_u(x) ; \quad y = x_1 = H(x). \tag{8.527}$$

(a) Show that $\kappa = 3$ and $\sigma = 2$.
(b) Use Theorem 8.15 to show that system (8.527) is not state equivalent to a dual Brunovsky NOCF with OT.
(c) Let $d = 1$. Then $\kappa \le n$ and $\sigma \ge n - d$. Show that system (8.527) does not satisfy the conditions of Theorem 8.20 without assumption (8.398).
(d) Show that system (8.527) is RDOEL with index $d = 1$ and

$$z^e = S_e(w, x) = \begin{bmatrix} w \\ x_1 \\ x_2 \\ x_3 - w_1 x_1^2 \end{bmatrix}.$$

Solution (a) It is easy to see that

$$\bar{x} = F_u^{-1}(x) = \begin{bmatrix} \frac{x_3 - u}{1 + x_1^2} \\ x_1 - u \\ x_2 \end{bmatrix} \text{ and } T(x) \triangleq \begin{bmatrix} H(x) \\ H \circ F_0(x) \\ H \circ F_0^2(x) \end{bmatrix} = x.$$

Thus, we have, by (8.298) and (8.299), that

$$\mathbf{g}_1^u(x) = \mathbf{g}_1^0(x) = T_*^{-1}\left(\frac{\partial}{\partial \xi_3}\right) = \frac{\partial}{\partial x_3} = \begin{bmatrix} 0 \\ 0 \\ 1 \end{bmatrix}$$

$$\mathbf{g}_2^u(x) \triangleq (F_u)_*(\mathbf{g}_1^u(x)) = \frac{\partial F_u(\bar{x})}{\partial \bar{x}} \mathbf{g}_1^u(\bar{x})\Big|_{\bar{x} = F_u^{-1}(x)} = \begin{bmatrix} 0 \\ 1 \\ 0 \end{bmatrix} = \mathbf{g}_2^0(x)$$

and

$$\mathbf{g}_3^u(x) \triangleq (F_u)_*(\mathbf{g}_2^u(x)) = \frac{\partial F_u(\bar{x})}{\partial \bar{x}} \mathbf{g}_2^u(\bar{x})\Big|_{\bar{x} = F_u^{-1}(x)} = \begin{bmatrix} 1 \\ 0 \\ \frac{2x_1(x_3 - u)}{x_1^2 + 1} \end{bmatrix} \ne \mathbf{g}_3^0(x).$$

Since $L_{\mathbf{g}_2^0}(H \circ F_0^2(x)) = 0$ and $L_{\mathbf{g}_3^0}(H \circ F_0^2(x)) = \frac{2x_1x_3}{x_1^2+1} \neq 0$, we have, by (8.331), $\kappa = 3$. Also, since $H \circ \hat{F}_u^{n-1}(x) = x_3 = H \circ F_0^{n-1}(x)$ and $H \circ \hat{F}_u^{n-2}(x) = x_2 + u \neq H \circ F_0^{n-2}(x)$, it is clear, by (8.336), that $\sigma = 2$.

(b) Since

$$\mathrm{rank}\left(\left[\mathbf{g}_3^u(x)\ \mathbf{g}_3^0(x)\right]\right) = \mathrm{rank}\left(\begin{bmatrix} 1 & 0 \\ 0 & 0 \\ \frac{2x_1(x_3-u)}{x_1^2+1} & \frac{2x_1x_3}{x_1^2+1} \end{bmatrix}\right) = 2 \neq 1,$$

there does not exist $\bar{\beta}^u(x)$ such that condition (i) of Theorem 8.15 is satisfied. Hence, by Theorem 8.15, system (8.527) is not state equivalent to a dual Brunovsky NOCF with OT.

(c) Let $d = 1$. Since

$$\alpha_3^u(\xi) \triangleq H \circ \hat{F}_u^3 \circ T^{-1}(\xi) = u + \xi_1 + \xi_1(\xi_2 + u)^2$$

$$\theta_3^u(\xi) \triangleq \frac{\partial \alpha_3^u(\xi)}{\partial \xi_{n+2-\kappa}} = \frac{\partial \alpha_3^u(\xi)}{\partial \xi_2} = 2\xi_1(\xi_2 + u)$$

$$\frac{1}{\theta_3^u(\xi)} \frac{\partial \theta_3^u(\xi)}{\partial u} = \frac{1}{\xi_2 + u}$$

and

$$\frac{\partial \alpha_3^u(\xi)}{\partial u} = 1 + 2\xi_1(\xi_2 + u),$$

it is clear that condition (i) of Theorem 8.20 is satisfied with

$$\bar{\beta}^u(\xi) = \frac{1}{(\xi_2 + u)(1 + 2\xi_1(\xi_2 + u))}.$$

However, there does not exist $\beta(y)$ such that (8.497) is satisfied. Hence, system (8.527) does not satisfy the conditions of Theorem 8.20 without assumption (8.398).

(d) Let $d = 1$. Then it is clear that $H^e(w, x) = w$ and

$$F_u^e(w, x) = \begin{bmatrix} H(x) \\ F_u(x) \end{bmatrix} = \begin{bmatrix} x_1 \\ x_2 + u \\ x_3 \\ u + x_1 + x_1(x_2 + u)^2 \end{bmatrix}.$$

Thus, it is easy to see that $\varphi(y) = y$, $\varphi \circ H^e \circ (S^e)^{-1}(z^e) = z_1^e$ and

$$S^e \circ F_u^e \circ (S^e)^{-1}(z^e) = \begin{bmatrix} z_2^e \\ z_3^e + u \\ z_4^e + z_1^e(z_2^e)^2 \\ z_2^e + u \end{bmatrix} = \begin{bmatrix} z_2^e \\ z_3^e \\ z_4^e \\ 0 \end{bmatrix} + \begin{bmatrix} 0 \\ 0 \\ w_1 y^2 \\ y + u \end{bmatrix}.$$

Hence, by Definition 8.15, system (8.527) is RDOEL with index $d = 1$ and $z^e = S_e(w, x)$. □

Example 8.6.6 Consider the system

$$x(t + 1) = \begin{bmatrix} x_2 \\ x_1 + u(1 + x_2) \end{bmatrix} = F_u(x) \tag{8.528}$$

$$y = x_1 = H(x).$$

It is clear, by Example 8.5.6, that $\kappa = 3 = n + 1$ and $\sigma = 2 = n$. Use Theorem 8.21 to show that system (8.528) is not RDOEL.

Solution It is clear, by Example 8.5.6, that system (8.528) is not state equivalent to a dual Brunovsky NOCF. Therefore, by Theorem 8.21, system (8.528) is not RDOEL.

□

If a system is RDOEL with index d, then it is also RDOEL with index $(d + 1)$. But the converse is not true. It is clear, by Corollary 8.12, that autonomous system (8.371) is RDOEL with index $d = n - 1$. However, it is not true for control system (8.370). (Refer to Example 8.6.6.)

8.7 MATLAB Programs

In this section, the following subfunctions in Appendix C are needed:
adfg, ChCommute, ChConst, ChInverseF, ChZero, Lfh, Lfhk, ObvIndex0, SpanCx, sstarmap

MATLAB program for Theorem 8.1:

```
clear all
syms x1 x2 x3 x4 x5 x6 x7 x8 x9 real
syms u1 u2 u3 u4 u5 real
syms y u real

Fu=[x2+2*x2*(x1-x2^2)+2*x2*u1+u2^2; x1-x2^2+u1];
H=x1-x2^2; m=2; %Ex:8.2.1

% Fu=[x2+2*x2*u+(x1-x2^2)^2*u^2; u];
% H=x1-x2^2; m=1; %Ex:8.2.2

Fu=simplify(Fu)
```

```
H=simplify(H)

n=length(Fu);
x=sym('x',[n,1])
if m>1
  u=sym('u',[m,1])
end

F0=simplify(subs(Fu,u,u-u))
T=x-x;
T(1)=H;
for k=2:n
  T(k)=simplify(Lfh(F0,T(k-1),x));
end
T=simplify(T)

dT=simplify(jacobian(T,x));
idT=simplify(inv(dT));

g0(:,1)=idT(:,n);
gu(:,1)=idT(:,n);
for k=2:n+1
  g0(:,k)=adfg(F0,g0(:,k-1),x);
  gu(:,k)=adfg(Fu,gu(:,k-1),x);
end

g0=simplify(g0)
gu=simplify(gu)

if ChZero(gu-g0)==0
  display('condition (i) of Thm 8.1 is not satisfied.')
  display('System is NOT state equivalent to a LOCF.')
  return
end

if ChCommute(g0,x)==0
  display('condition (ii) of Thm 8.1 is not satisfied.')
  display('System is NOT state equivalent to a LOCF.')
  return
end

display('System is, by Thm 8.1, state equivalent to a LOCF.')

for k=1:n
  idS(:,k)= (-1)^(n-k)*g0(:,n+1-k);
end
idS=simplify(idS)
dS=simplify(inv(idS))
S=Codi(dS,x)

ASu=simplify(dS*Fu)
dAS=simplify(jacobian(ASu,x));
A=simplify(dAS*idS)
```

```
Gammau=ASu-subs(ASu,u,u-u)

return
```

MATLAB program for Theorem 8.2:

```
clear all
syms x1 x2 x3 x4 x5 x6 x7 x8 x9 real
syms u1 u2 u3 u4 u5 real
syms y u real

% Fu=[x2+2*x2*(x1-x2^2)+2*x2*u1+u2^2; x1-x2^2+u1];
% H=x1-x2^2; m=2; %Ex:8.2.1

Fu=[x2+2*x2*u+(x1-x2^2)^2*u^2; u]; H=x1-x2^2; m=1; %Ex:8.2.2

% Fu=[x2; -x2^2+x1^2*exp(-x1)+u]; H=x1; m=1; %Ex:8.2.3

% Fu=[x2; x3; -4*x1*x3-3*x2^2-6*x1^2*x2+u];
% H=x1; m=1; %Ex:8.2.4

% Fu=[x2+x3^2; x3-2*x3*exp(x1)*u; exp(x1)*u];
% H=x1; m=1; %P:8-2(a)

% Fu=[x2+x3^2; x3-2*x3*exp(x1)*u2; exp(x1)*u1];
% H=x1; m=2; %P:8-2(b)

%Fu=[x2+(x1+1)*u2^2;x2*log(x1+1)+x2^2/(1+x1)+(x1+1)*u1+x2*u2^2];
%H=x1; m=2; %P:8-2(c) or P:8-3(a)

% Fu=[x2*exp(x1); x3+x1*u; u ];
% H=x1; m=1; %P:8-2(d) or P:8-3(b)

Fu=simplify(Fu)
H=simplify(H)

n=length(Fu);
x=sym('x',[n,1])
if m>1
  u=sym('u',[m,1])
end

F0=simplify(subs(Fu,u,u-u))

T=x-x;
T(1)=H;
for k=2:n
  T(k)=simplify(Lfh(F0,T(k-1),x));
end
T=simplify(T)

dT=simplify(jacobian(T,x));
idT=simplify(inv(dT));
```

```
gu(:,1)=idT(:,n);
for k=2:n
  gu(:,k)=adfg(Fu,gu(:,k-1),x);
end

gu=simplify(gu)
g0=subs(gu,u,u-u)

if ChZero(gu-g0)==0
  display('condition (i) of Thm 8.2 is not satisfied.')
  display('System is NOT state equivalent to a NOCF.')
  return
end

if ChCommute(g0,x)==0
  display('condition (ii) of Thm 8.2 is not satisfied.')
  display('System is NOT state equivalent to a NOCF.')
  return
end

display('System is, by Thm 8.2, state equivalent to a NOCF')

for k=1:n
  idS(:,k)= (-1)^(n-k)*g0(:,n+1-k);
end
idS=simplify(idS)
dS=simplify(inv(idS))
S=Codi(dS,x)

NFu=simplify(dS*Fu);
Gammau=simplify(NFu-[S(2:n); 0])

return
```

MATLAB program for Theorem 8.3:

```
clear all
syms x1 x2 x3 x4 x5 x6 x7 x8 x9 real
syms y u u1 u2 u3 u4 u5 real

Fu=[x2; -x2^2+x1^2*exp(-x1)+u]; m=1; %Ex:8.2.3

% Fu=[x2; x3; -4*x1*x3-3*x2^2-6*x1^2*x2+u];
% m=1; %Ex:8.2.4 Or Ex:8.3.2

% Fu=[x2; x3; x2^3+u]; m=1; %Ex:8.3.3 or P:8-3

%Fu=[x2+(x1+1)*u2^2;x2*log(x1+1)+x2^2/(1+x1)+(x1+1)*u1+x2*u2^2];
%m=2; %P:8-2(c) or P:8-4(a)

% Fu=[x2*exp(x1); x3+x1*u; u ]; m=1; %P:8-2(d) or P:8-4(b)
```

```
% aa1=(3*x2*x3)/(x1 + 1)-(2*x2^3)/(x1 + 1)^2;
% aa2=-x3*(x1+1)*log(x1+1)+u;
% Fu=[x2; x3; aa1+aa2]; m=1; %P:8-4(c) or P:8-5(c)

% Fu=[x2; x3; x4+u^2; 5*x2*x3+2*x2^2+u]
% m=1; %P:8-4(d) or P:8-5(d)

H=x1
Fu=simplify(Fu)

n=length(Fu);
x=sym('x',[n,1])
if m>1
  u=sym('u',[m,1])
end

F0=simplify(subs(Fu,u,u-u));

T=x-x;
T(1)=H;
for k=2:n
  T(k)=simplify(Lfh(F0,T(k-1),x));
end
T=simplify(T)

dT=simplify(jacobian(T,x));
idT=simplify(inv(dT));

gu(:,1)=idT(:,n);
for k=2:n
  gu(:,k)=adfg(Fu,gu(:,k-1),x);
end

gu=simplify(gu)
g0=subs(gu,u,u-u);

for k=2:n-1
  CC1=adfg(g0(:,1),g0(:,k),x);
  if ChZero(CC1)==0
    display('condition (i) of Thm 8.3 is not satisfied.')
    display('System is NOT state equivalent to a NOCF with OT')
    return
  end
end

if n==2*round(n/2)
  t1=simplify(adfg(g0(:,1), g0(:,n), x))
  [flag1,Cx]=SpanCx(t1,g0(:,1))
  if flag1==0
    display('condition (ii) of Thm 8.3 is not satisfied.')
    display('System is NOT state equivalent to a NOCF with OT')
    return
  end
```

```
  b0=-Cx/2
end

if n~=2*round(n/2)
  t1=simplify(adfg(g0(:,2), g0(:,n), x))
  [flag1,Cx]=SpanCx(t1,g0(:,1:2))
  if flag1==0
    display('condition (ii) of Thm 8.3 is not satisfied.')
    display('System is NOT state equivalent to a NOCF with OT')
    return
  end
  b0=Cx(2)/n
end

bx1=x(2:n)
db0=simplify(jacobian(b0,bx1))
if ChZero(db0)==0
  display('condition (ii) of Thm 8.3 is not satisfied.')
  display('System is NOT state equivalent to a NOCF with OT')
  return
end

beta=b0;
Ibeta=int(b0,x1)
Ibeta0=subs(Ibeta,x1,x1-x1);
ell=exp(Ibeta-Ibeta0)
bgu(:,1)=ell*g0(:,1);

for k1=2:n
  bgu(:,k1)=adfg(Fu,bgu(:,k1-1),x);
end
bgu=simplify(bgu)
bg0=simplify(subs(bgu,u,u-u));

if ChZero(bgu-bg0)==0
  display('condition (iii) of Thm 8.3 is not satisfied.')
  display('System is NOT state equivalent to a NOCF with OT')
  return
end

if ChCommute(bg0,x)==0
  display('condition (iv) of Thm 8.3 is not satisfied.')
  display('System is NOT state equivalent to a NOCF with OT')
  return
end

display('System is, by Thm 8.3, state equi to a NOCF with OT')

varphi=int(1/ell,x1);
varphi=subs(varphi,x1,y);
varphi0=subs(varphi,y,y-y);
varphi=simplify(varphi-varphi0)
```

```
for k=1:n
  idS(:,k)=(-1)^(n-k)*bg0(:,n+1-k);
end
idS=simplify(idS)
dS=simplify(inv(idS))
S=Codi(dS,x)

NFu=simplify(dS*Fu);
Gammaphiu=simplify(NFu-[S(2:n); 0])

return
```

MATLAB program for Theorem 8.5:

```
clear all
syms x1 x2 x3 x4 x5 x6 x7 x8 x9   real
syms w1 w2 w3 w4 w5 w6 w7 w8 w9 real
syms y u u1 u2 u3 u4 u5 real

d=1; Fu=[x2; x3; -4*x1*x3-3*x2^2-6*x1^2*x2+u];
m=1; %Ex:8.2.4 Or Ex:8.3.2

% d=1; Fu=[x2; x3; x2^3+u]; m=1; %Ex:8.3.3

% d=1; Fu=[x2+x1*u^2; x3; 4*x3*x1+u ];
% m=1; %P:8-4(a) or P:8-5(a)

% d=2; Fu=[x2+x3*u^2; x3; x4; 4*x3*x1+u ]; m=1; %P:8-5(b)

% aa1=(3*x2*x3)/(x1+1)-(2*x2^3)/(x1+1)^2;
% aa2=-x3*(x1+1)*log(x1+1)+u;
% d=1; Fu=[x2; x3; aa1+aa2]; m=1; %P:8-4(c) or P:8-5(c)

% d=2; Fu=[x2; x3; x4+u^2; 5*x2*x3+2*x2^2+u];
% m=1; %P:8-4(d) or P:8-5(d)

H=x1
Fu=simplify(Fu)
n=length(Fu);
N=n+d

x=sym('x',[n,1]);
w=sym('w',[d,1]);
xe=[w; x]

P=w-w;
if d>0
  P=[w(2:d); x1];
end

Feu=[P; Fu]
Fe0=subs(Feu,u,u-u);
```

```
Te=xe-xe;
Te(1)=xe(1);
for k=2:N
  Te(k)=simplify(Lfh(Fe0,Te(k-1),xe));
end
Te=simplify(Te)

dTe=simplify(jacobian(Te,xe));
idTe=simplify(inv(dTe));

geu(:,1)=idTe(:,N)
for k=2:n
  geu(:,k)=adfg(Feu,geu(:,k-1),xe);
end

geu=simplify(geu)
ge0=simplify(subs(geu,u,u-u))

CC1=xe-xe;
for k=2:n-1
  CC1(:,k-1)=adfg(ge0(:,1),ge0(:,k),xe);
end
CC1=simplify(CC1)
if ChZero(CC1)==0
  display('condition (i) of Thm 8.5 is not satisfied.')
  display('System is NOT d-RDOEL with')
  display(d)
  return
end

if n==2*round(n/2)
  t1=simplify(adfg(ge0(:,1), ge0(:,n), xe));
  [flag1,Cx]=SpanCx(t1,ge0(:,1))
  if flag1==0
    display('condition (ii) of Thm 8.5 is not satisfied.')
    display('System is NOT d-RDOEL with')
    display(d)
    return
  end
  b0=-Cx/2
end

if n~=2*round(n/2)
  t1=simplify(adfg(ge0(:,2), ge0(:,n), xe));
  [flag1,Cx]=SpanCx(t1,ge0(:,1:2))
  if flag1==0
    display('condition (ii) of Thm 8.5 is not satisfied.')
    display('System is NOT d-RDOEL with')
    display(d)
    return
  end
  b0=Cx(2)/n
end
```

```
bx1=[xe(1:d); xe(d+2:N)]
t2=simplify(jacobian(b0,bx1))
if ChZero(t2)==0
  display('condition (ii) of Thm 8.5 is not satisfied.')
  display('System is NOT d-RDOEL with')
  display(d)
  return
end

beta=b0;
t3=int(b0,x1);
t30=subs(t3,x1,x1-x1);
Bell=exp(t3-t30)
Tg(:,1)=Bell*ge0(:,1);
for k1=2:n
  Tg(:,k1)=adfg(Fe0,Tg(:,k1-1),xe);
end
Tg=simplify(Tg)

for k=1:min(d,n-2)
  if (n+k)==2*round((n+k)/2)
    t1=simplify(adfg(Tg(:,k+1),Tg(:,n),xe))
    [flag1,Cx]=SpanCx(t1,Tg(:,1))
    if flag1==0
      display('condition (iii) of Thm 8.5 is not satisfied.')
      display('System is NOT d-RDOEL with')
      display(d)
      return
    end
    b0=simplify((-1)^(n-1)*Cx/(2*Bell))
  end
  if (n+k)~=2*round((n+k)/2)
    t1=simplify(adfg(Tg(:,k+2), Tg(:,n), xe))
    [flag1,Cx]=SpanCx(t1,Tg(:,1:2))
    if flag1==0
      display('condition (iii) of Thm 8.5 is not satisfied.')
      display('System is NOT d-RDOEL with')
      display(d)
      return
    end
    b0=simplify((-1)^(n-1)*Cx(2)/((n+k)*Bell))
  end
  if ChConst(b0,xe)==0
    display('condition (iii) of Thm 8.5 is not satisfied.')
    display('System is NOT d-RDOEL with')
    display(d)
    return
  end
  beta=[beta; b0]
  Bell=Bell*exp(b0*w(d+1-k))
  Tg(:,1)=Bell*ge0(:,1);
  for k1=2:n
```

```
     Tg(:,k1)=adfg(Fe0,Tg(:,k1-1),xe);
  end
  Tg=simplify(Tg)
end

bgu(:,1)=Tg(:,1);
for k1=2:n
  bgu(:,k1)=adfg(Feu,bgu(:,k1-1),xe);
end
bgu=simplify(bgu)
bg0=simplify(subs(bgu,u,u-u));

CC4=simplify(bgu-bg0)
if ChZero(CC4)==0
  display('condition (iv) of Thm 8.5 is not satisfied.')
  display('System is NOT d-RDOEL with d=')
  d=d
  return
end

if ChCommute(bg0,xe)==0
  display('condition (v) of Thm 8.5 is not satisfied.')
  display('System is NOT d-RDOEL with d=')
  display(d)
  return
end

display('System is, by Thm 8.5, d-RDOEL with')
display(d)

Bell=simplify(Bell)
varphi=int(1/Bell,x1);
varphi0=subs(varphi,xe,xe-xe);
varphi=simplify(varphi-varphi0)

for k=1:n
  idS(:,k)= (-1)^(n-k)*bg0(:,n+1-k);
end
idS=simplify(idS);
idS2=idS((d+1):N, 1:n)
dS2=simplify(inv(idS2))
S2=Codi(dS2,xe(d+1:N));
S2=simplify(S2);
S1=w;
Se=[S1; S2];
Se0=subs(Se,xe,xe-xe);
Se=simplify(Se-Se0)

dSe=jacobian(Se,xe);
NFeu=simplify(dSe*Feu);
Gammaeu=simplify(NFeu-[Se(2:N); 0])

return
```

The following is a MATLAB subfunction program for Theorem 8.6.

```
function [flag,xe,Se,Gammaeu]=dRDOEL(d,Fu,H,x,u,n,m)

syms y x1 w1 w2 w3 w4 w5 w6 w7 w8 w9 real
flag=0;
N=n+d;
w=sym('w',[d,1]);
xe=[w; x];
Se=xe-xe;
Gammaeu=xe-xe;
P=w-w;
if d>0
  P=[w(2:d); H];
end
Feu=[P; Fu];
Fe0=subs(Feu,u,u-u);
Te=xe-xe;
Te(1)=xe(1);
for k=2:N
  Te(k)=simplify(Lfh(Fe0,Te(k-1),xe));
end
Te=simplify(Te);
dTe=simplify(jacobian(Te,xe));
idTe=simplify(inv(dTe));
geu(:,1)=idTe(:,N);
for k=2:n
  geu(:,k)=adfg(Feu,geu(:,k-1),xe);
end
geu=simplify(geu);
ge0=simplify(subs(geu,u,u-u));
for k=2:n-1
  CC1=adfg(ge0(:,1),ge0(:,k),xe);
  if ChZero(CC1)==0
    return
  end
end
if n==2*round(n/2)
  t1=simplify(adfg(ge0(:,1), ge0(:,n), xe));
  [flag1,Cx]=SpanCx(t1,ge0(:,1));
  if flag1==0
    return
  end
  b0=-Cx/2;
end
if n~=2*round(n/2)
  t1=simplify(adfg(ge0(:,2), ge0(:,n), xe));
  [flag1,Cx]=SpanCx(t1,ge0(:,1:2));
  if flag1==0
    return
  end
  b0=Cx(2)/n;
end
bx1=[xe(1:d); xe(d+2:N)] ;
```

```
t2=simplify(jacobian(b0,bx1));
if ChZero(t2)==0
  return
end
t3=int(b0,x1);
t30=subs(t3,x1,x1-x1);
Bell=exp(t3-t30);
Tg(:,1)=Bell*ge0(:,1);
for k1=2:n
  Tg(:,k1)=adfg(Fe0,Tg(:,k1-1),xe);
end
Tg=simplify(Tg);
for k=1:min(d,n-2)
  if (n+k)==2*round((n+k)/2)
    t1=simplify(adfg(Tg(:,k+1),Tg(:,n),xe));
    [flag1,Cx]=SpanCx(t1,Tg(:,1));
    if flag1==0
      return
    end
    b0=simplify((-1)^(n-1)*t1(N)/(2*Bell*Tg(N,1)));
  end
  if (n+k)~=2*round((n+k)/2)
    t1=simplify(adfg(Tg(:,k+2), Tg(:,n), xe));
    [flag1,Cx]=SpanCx(t1,Tg(:,1:2));
    if flag1==0
      return
    end
    b0=simplify((-1)^(n-1)*Cx(2)/((n+k)*Bell));
  end
  if ChConst(b0,xe)==0
    return
  end
  Bell=Bell*exp(b0*w(d+1-k));
  Tg(:,1)=Bell*ge0(:,1);
  for k1=2:n
    Tg(:,k1)=adfg(Fe0,Tg(:,k1-1),xe);
  end
  Tg=simplify(Tg);
end
bgu(:,1)=Tg(:,1);
for k1=2:n
  bgu(:,k1)=adfg(Feu,bgu(:,k1-1),xe);
end
bgu=simplify(bgu);
bg0=simplify(subs(bgu,u,u-u));
CC4=simplify(bgu-bg0);
if ChZero(CC4)==0
  return
end
f ChCommute(bg0,xe)==0
    return
end
Bell=simplify(Bell);
```

```
varphi=int(1/Bell,x1);
varphi0=subs(varphi,xe,xe-xe);
varphi=simplify(varphi-varphi0);
for k=1:n
    idS(:,k)= (-1)^(n-k)*bg0(:,n+1-k);
end
idS=simplify(idS);
idS2=idS((d+1):N, 1:n);
dS2=simplify(inv(idS2));
S2=Codi(dS2,xe(d+1:N));
S2=simplify(S2);
S1=w;
Se=[S1; S2];
Se0=subs(Se,xe,xe-xe);
Se=simplify(Se-Se0);
dSe=jacobian(Se,xe);
NFeu=simplify(dSe*Feu);
Gammaeu=simplify(NFeu-[Se(2:N); 0]);
flag=1;
return
```

The following MATLAB program, that needs subfunction **dRDOEL**, is to check the conditions of Theorem 8.6.

```
clear all
syms x1 x2 x3 x4 x5 x6 x7 x8 x9   real
syms w1 w2 w3 w4 w5 w6 w7 w8 w9 real
syms y u u1 u2 u3 u4 u5 real

% Fu=[x2; x3; -4*x1*x3-3*x2^2-6*x1^2*x2+u];
% m=1; %Ex:8.2.4 or Ex:8.3.2

Fu=[x2; x3; x2^3+u];
m=1; %Ex:8.3.3

% Fu=[x2+x1*u^2; x3; 4*x3*x1+u ];
% m=1; %P:8-4(a) or P:8-5(a)

% Fu=[x2+x3*u^2; x3; x4; 4*x3*x1+u ];
% m=1; %P:8-5(b)

% aa1=(3*x2*x3)/(x1 + 1)-(2*x2^3)/(x1 + 1)^2;
% aa2=-x3*(x1+1)*log(x1+1)+u;
% Fu=[x2; x3; aa1+aa2]; m=1; %P:8-4(c) or P:8-5(c)

% Fu=[x2; x3; x4+u^2; 5*x2*x3+2*x2^2+u]
% m=1; %P:8-4(d) or P:8-5(d)

% Fu=[x2*exp(x1); x3+x1*u; u ];
% m=1; %P:8-2(d) or P:8-4(b)

H=x1
```

```
Fu=simplify(Fu)
n=length(Fu);
x=sym('x',[n,1]);
if m>1
  u=sym('u',[m,1])
end

for k=1:n-1
  d=k-1;
  [flag,xe,Se,Gammaeu]=dRDOEL(d,Fu,H,x,u,n,m);
  if flag==1
    display('System is RDOEL with index')
    display(d)
    display(Se)
    display('and')
    display(Gammaeu)
    return
  end
  if flag==0
    display('System is not RDOEL with index')
    display(d)
  end
end

display('Hence, by Thm 8.6, the system is NOT RDOEL.')

return
```

The following is a MATLAB subfunction program for Theorem 8.11.

```
function [flag,beta]
=betaJS(r,s,ki,q,bg0,Tg,iD,g0,x,y,N,bs,sigma,tx,bx)

flag=0;
beta=x-x;
Phi=x-x;
for k=1:ki-1
  Phi=[Phi g0(:,1:N(k)-N(ki)-s,k)];
end
Phi=Phi(:,2:size(Phi,2));
bPhi=Phi;
for k=1:ki-1
  if N(k)-N(ki)+1-s>0
    bPhi=[bPhi g0(:,N(k)-N(ki)+1-s,k)];
  end
end
cc=simplify(adfg(bg0(:,N(q),q),Tg(:,s,ki),x));
if rank([cc bPhi]) > rank(bPhi)
  return
end
temp1=iD*cc;
for k=1:ki-1
  SS=(-1)^(N(k)-N(ki)-s+N(q));
```

```
  temp2=SS*temp1(bs(k)+N(k)-N(ki)+1-s);
  dtemp2=jacobian(temp2,bx);
  if ChZero(dtemp2)==0
    return
  end
  beta(k)=subs(temp2,tx,y);
end
beta=beta(1:sigma(r-1));
flag=1;

return
```

The following MATLAB program, that needs subfunction **betaJS**, is to check the conditions of Theorem 8.11.

```
clear all
syms x1 x2 x3 x4 x5 x6 x7 x8 x9 x10 x11 x12 real
syms u u1 u2 u3 u4 u5 real
syms y y1 y2 y3 y4 y5 real

aa11=4*x2*x1^2+x3*x4+2*x1*x4*(x1^3+x2*x4)+u1;
aa12=x1*x4*u2^2+(x2+x1*x4+2*x1^2*x4)*u2;
aa1=aa11+aa12;
Fu=[x2; x3+x1*(u2+u2^2); aa1; x1^2+u2];
H=[x1; x4]; m=2; %Ex:8.4.1

% aa1=x1*u1^2+u1+u3*x2+3*x1*x2+x1*x5+2*x2*x4+x3*x6+u3*x1*x6;
% aa2=u2+x1*x4*(u3+x1)+x2*x4*x6+x1*x6*(u1^2+x5);
% Fu=[x2; x3+u3*x1; aa1; u1^2+x5; aa2; u3+x1];
% H=[x1; x4; x6]; m=3; %Ex:8.4.2

% Fu=[x2; x3; x5^2+u1+u2*x4; x5; u2];
% H=[x1; x4]; m=2; %Ex:8.4.3

% aa1=u1+sin(x1)*(u2+x1*x3)+x3*cos(x1)*(x3*u1^2+x2);
% aa2=u2+x1*x3;
% Fu=[x3*u1^2+x2; aa1; aa2];
% H=[x1; x3]; m=2; %P:8.5(a)

% aa1=3*x2^2+2*u2*x2+u1+x3+x6+x1*x3+3*x3*x6+x4*x5+u2*x1*x5;
% aa2=u2+x2;
% Fu=[x2; x3; x4+u2*x1; aa1; x6; aa2];
% H=[x1; x5]; m=2; %P:8.5(b)

% aa1=u1+u3*x4+x4*x5+2*x5*x8+x6*x7; aa3=u3+x5;
% aa2=u2+u3*x1*x4+x1*x4*x5+2*x1*x5*x8+x1*x6*x7+2*x2*x4*x8;
% Fu=[x2; x3; aa1; x5; x6; aa2+2*x2*x5*x7+x3*x4*x7; x8; aa3];
% H=[x1; x4; x7]; m=3; %P:8.5(c)

% aa1=u1+u3*x5+x6*x7; aa2=u2+x2*x8+x1*(x1^2+u4);
% aa3=u3; aa4=x1^2+u4;
% Fu=[x2; u4*x7^2+x3+u2*x1+u3*x4;aa1; x5+u4*x7; x6;aa2;aa3;aa4];
% H=[x1; x4; x7; x8]; m=4; %P:8.5(d)
```

```
% aa1=u1+x1*((x2-x1*x3)*u1^2+x4)+x2*x3;
% aa2=u2+2*x1*x2;
% Fu=[x2; aa1; (x2-x1*x3)*u1^2+x4; aa2];
% H=[x1; x3]; m=2; %P:8.6(a)

% aa1=x2*x3+x1*(x1*u1^2+x4)+u1*(x3+1)
% Fu=[x2; aa1; x1*u1^2+x4; u2+2*x1*x2];
% H=[x1; x3]; m=2; %P:8.6(b)

% aa1=u1+u2*x2+2*x1*x2+x3*x4^2-u2*x1*x4^2+2*u2*x1*x4^3;
% Fu=[x2; x3+2*u2*x1*x4; aa1+2*u2*x2*x4; u2];
% H=[x1; x4]; m=2; %P:8.6(c)

n=length(Fu); p=length(H);
x=sym('x',[n,1]);
if p>1
  y=sym('y',[p,1]);
end
if m>1
  u=sym('u',[m,1]);
end

F0=subs(Fu,u,u-u);
N=ObvIndex0(F0,H,x,n,p);

[bNu,IA]=sort(N,'descend');
IC=zeros(p);
for k=1:p
  IC(k,IA(k))=1;
end
H=IC*H
Fu=simplify(Fu)

N=ObvIndex0(F0,H,x,n,p)

barp=1;
sigma(1)=1;
for k=1:p-1
  if N(k+1)-N(k)==0
    sigma(barp)=sigma(barp)+1;
  else
    sigma=[sigma; sigma(barp)+1];
    barp=barp+1;
  end
end

bs(1)=0;
for k1=1:p
  bs(k1+1)=sum(N(1:k1));
end

T=x-x;
```

```
for k1=1:p
  for k2=1:N(k1)
    T(bs(k1)+k2)=Lfhk(F0,H(k1),x,k2-1);
  end
end
T=simplify(T)

if ChZero(T-x)==0
  display('Solve the problem without MATLAB.')
  return
end

tx=H;
bx=x1-x1;
for k=1:p
  bx=[bx; x(bs(k)+2:bs(k)+N(k))];
end
bx=bx(2:length(bx));

for k=2:p
  C1=Lfhk(F0,H(k),x,N(k));
  dC1=jacobian(C1,x);
  dC2=jacobian(T(1:N(k)),x);
  for k2=2:p
    dC2=[dC2; jacobian(T(bs(k2)+1:bs(k2)+min(N(k),N(k2))),x)];
  end
  if rank([dC1; dC2]) > rank(dC2)
    disp('Condition (i) of Thm 8.11 is not satisfied')
    return
  end
end

dT=simplify(jacobian(T,x));
idT=simplify(inv(dT));
for k1=1:p
  gu(:,1,k1)=idT(:,bs(k1+1));
  for k2=2:N(k1)
    gu(:,k2,k1)=adfg(Fu,gu(:,k2-1,k1),x);
  end
end
gu=simplify(gu)
g0=subs(gu,u,u-u);

for k=1:p
  D(:,bs(k)+1:bs(k)+N(k))=g0(:,1:N(k),k);
end
D=simplify(D);
iD=simplify(inv(D));

for k=1:p
  Tg(:,1,k)=g0(:,1,k);
end
for k1=1:sigma(1)
```

```
  bg0(:,1:N(k1),k1)=g0(:,1:N(k1),k1);
  bgu(:,1:N(k1),k1)=gu(:,1:N(k1),k1);
end

ZeroM=jacobian(x,x)-jacobian(x,x);

for r=2:barp
  for s=1:N(1)-N(sigma(r))
    sg=0;
    for k=1:p
      if N(k) >= N(sigma(r))+s
        sg=sg+1;
      end
    end
    dGam=ZeroM(1:sg,1:sigma(r-1));
    Gam=ZeroM(1:sg,1);
    Beta=ZeroM(1:sigma(r-1),sigma(r-1)+1:p);
    for q=1:sigma(r-1)
      Tbeta=ZeroM(1:sigma(r-1),1);
      for ki=sigma(r-1)+1:p
[fl,beta]=betaJS(r,s,ki,q,bg0,Tg,iD,g0,x,y,N,bs,sigma,tx,bx);
        if fl==0
          disp('Condition (ii) of Thm 8.11 is not satisfied')
          return
        end
        bki=(q-1)*(p-sigma(r-1))+ki-sigma(r-1);
        Beta(1:sg,bki)=beta(1:sg);
      end
    end
    Beta=simplify(Beta);
    for k3=1:sg
      for k4=1:sigma(r-1)
        t4=Beta(k3,(k4-1)*(p-sigma(r-1))+1:k4*(p-sigma(r-1)));
        if ChExact(t4,y(sigma(r-1)+1:p))==0
          return
        end
        Tbeta(k3,k4)=Codi(t4,y(sigma(r-1)+1:p) );
      end
    end
    dGam=simplify(Tbeta(1:sg,:));
    for k3=1:sg
      if ChExact(dGam(k3,:),y(1:sigma(r-1)))==0
        return
      end
      Gam(k3)=Codi(dGam(k3,:),y(1:sigma(r-1)));
    end
    Gam=simplify(Gam(1:sg,:))
    GamH=subs(Gam,y,H);
    for k1=sigma(r-1)+1:sigma(r)
      Tg(:,s+1,k1)=Tg(:,s,k1);
      for k2=1:sigma(r-1)
        for k3=1:N(k2)-N(k1)+1-s
          tempTg1=simplify(Lfhk(F0,GamH(k2),x,N(k2)-k3-s));
```

```
          tempTg2=simplify(Lfh(g0(:,1,k1),tempTg1,x));
          tempTg3=tempTg2*bg0(:,k3,k2);
          Tg(:,s+1,k1)=Tg(:,s+1,k1)+(-1)^(k3-1)*tempTg3;
        end
      end
    end
    for k1=sigma(r-1)+1:sigma(r)
      bg0(:,1,k1)=Tg(:,s+1,k1);
      bgu(:,1,k1)=bg0(:,1,k1);
      for k2=2:N(k1)
        bg0(:,k2,k1)=adfg(F0,bg0(:,k2-1,k1),x);
        bgu(:,k2,k1)=adfg(Fu,bgu(:,k2-1,k1),x);
      end
    end
  end
end
bgu=simplify(bgu)

for k1=1:p
  for k=1:N(k1)
    idS(:,bs(k1)+k)= (-1)^(N(k1)-k)*bgu(:,N(k1)+1-k,k1);
  end
end
idS=simplify(idS)

for k1=1:n
  if ChZero(jacobian(idS(:,k1),u))==0
    disp('Condition (iii-1) of Thm 8.11 is not satisfied')
    return
  end
end

if ChCommute(idS,x)==0
  disp('Condition (iii-2) of Thm 8.11 is not satisfied')
  return
end

disp('System is equivalent to a dual Brunovsky NOCF with z=')

dS=simplify(inv(idS));
S=simplify(Codi(dS,x))

gammau=x-x;
for k1=1:p
  for k=1:N(k1)-1
    gammau(bs(k1)+k)=Lfh(Fu,S(bs(k1)+k),x)-S(bs(k1)+k+1);
  end
  gammau(bs(k1)+N(k1))=Lfh(Fu,S(bs(k1+1)),x);
end
gammau=simplify(gammau)

return
```

MATLAB program for Theorem 8.12:

```
clear all
syms x1 x2 x3 x4 x5 x6 x7 x8 x9 real
syms y u u1 u2 u3 u4 u5 real

Fu=[x2+(x1-x2^2+u1)^2+u2^2; x1-x2^2+u1];
iFu=[x2+(x1-x2^2-u2^2)^2-u1; x1-x2^2-u2^2];
H=x1-x2^2; m=2; %Ex:8.5.1

% Fu=[x2+u^2*(-x2^2+x1)+(-x2^2+u+x1)^2; x1-x2^2+u];
% iFu=[x2+(x1-x2^2-(x2-u)*u^2)^2-u; x1-x2^2-(x2-u)*u^2];
% H=x1-x2^2; m=1; %Ex:8.5.2

Fu=simplify(Fu)
H=simplify(H)
n=length(Fu);
x=sym('x',[n,1])
if m>1
  u=sym('u',[m,1])
end
F0=simplify(subs(Fu,u,u-u));

if ChInverseF(Fu,iFu,x)==0
  display('iFu is not correct.')
  return
end

T=x-x; T(1)=H;
for k=2:n
  T(k)=simplify(subs(T(k-1),x,F0));
end
T=simplify(T)

dT=simplify(jacobian(T,x));
idT=simplify(inv(dT));
gu(:,1)=idT(:,n);
for k=2:n+1
  gu(:,k)=sstarmap(Fu,iFu,gu(:,k-1),x);
end
gu=simplify(gu)
g0=subs(gu,u,u-u)

if ChZero(gu-g0)==0
  display('condition (i) of Thm 8.12 is not satisfied.')
  display('System is NOT state equivalent to a LOCF.')
  return
end

if ChCommute(g0,x)==0
  display('condition (ii) of Thm 8.12 is not satisfied.')
  display('System is NOT state equivalent to a LOCF.')
  return
end
```

```
display('System is, by Thm 8.12, state equiv. to a LOCF with')

for k=1:n
  idS(:,k)= g0(:,n+1-k);
end
idS=simplify(idS);
dS=simplify(inv(idS));
S=Codi(dS,x)

ASu=simplify(subs(S,x,Fu));
dAS=simplify(jacobian(ASu,x));
A=simplify(dAS*idS)
Gammau=ASu-subs(ASu,u,u-u)

return
```

MATLAB program for Theorem 8.13:

```
clear all
syms x1 x2 x3 x4 x5 x6 x7 x8 x9 real
syms y u u1 u2 u3 u4 u5 real

% Fu=[x2+(x1-x2^2+u1)^2+u2^2; x1-x2^2+u1]; H=x1-x2^2; m=2;
% iFu=[x2+(x1-x2^2-u2^2)^2-u1; x1-x2^2-u2^2]; %Ex:8.5.1

Fu=[x2+u^2*(-x2^2+x1)+(-x2^2+u+x1)^2; x1-x2^2+u];
iFu=[x2+(x1-x2^2-(x2-u)*u^2)^2-u; x1-x2^2-(x2-u)*u^2];
H=x1-x2^2; m=1; %Ex:8.5.2

% aa=u+x1+exp(x2)+(x3+u*x1)^2-1; aai=x3-u-exp(x1)-x2^2+1;
% Fu=[x2; x3+u*x1; aa];
% iFu=[aai; x1; x2-u*aai]; H=x1; m=1; %P:8.8(a)

% Fu=[x2; x3; exp(x1+u+x3^2)-1]; H=x1;
% iFu=[ log(x3+1)-u-x2^2; x1; x2 ]; m=1; TYPE=1; %P:8.8(b)

% Fu=[x2; (x3+1)*exp(x1*u)-1; exp(x1+u+(x3+1)*exp(x1*u)-1)-1];
% aai=log(x3+1)-u-x2; iFu=[ aai; x1; (1+x2)*exp(-aai*u)-1];
% H=x1; m=1; TYPE=2; %P:8.8(c)

Fu=simplify(Fu)
H=simplify(H);
n=length(Fu);
x=sym('x',[n,1]);
if m>1
  u=sym('u',[m,1]);
end
F0=simplify(subs(Fu,u,u-u));

if ChInverseF(Fu,iFu,x)==0
  display('iFu is not correct.')
  return
```

```
end

T=x-x;  T(1)=H;
for k=2:n
  T(k)=simplify(subs(T(k-1),x,F0));
end
T=simplify(T)

dT=simplify(jacobian(T,x));
idT=simplify(inv(dT));
gu(:,1)=idT(:,n);
for k=2:n
  gu(:,k)=sstarmap(Fu,iFu,gu(:,k-1),x);
end
gu=simplify(gu)
g0=subs(gu,u,u-u);

if ChZero(gu-g0)==0
  display('condition (i) of Thm 8.13 is not satisfied.')
  display('System is NOT state equivalent to a NOCF.')
  return
end

if ChCommute(g0,x)==0
  display('condition (ii) of Thm 8.13 is not satisfied.')
  display('System is NOT state equivalent to a NOCF.')
  return
end

display('System is, by Thm 8.13, state equiv. to a NOCF.')

for k=1:n
  idS(:,k)= g0(:,n+1-k);
end
idS=simplify(idS);
dS=simplify(inv(idS));
S=Codi(dS,x)

ASu=simplify(subs(S,x,Fu));
gammau=simplify(ASu-[S(2:n); 0])

return
```

The following is a MATLAB subfunction program for Theorems 8.14 and 8.18.

```
function [flag,beta]=BETA_thm814(kappa,g0,x,y,n)

flag=0; beta=y-y;
TEMP1=adfg(g0(:,1),g0(:,kappa),x);
[flag1,TEMP2]=SpanCx(TEMP1,g0(:,1));
if flag1==0
  return
end
```

```
beta=simplify(TEMP2/Lfh(g0(:,kappa),x(n),x));
dbeta=simplify(jacobian(beta,x));
if ChZero(dbeta(1:n-1))==0
  return
end
beta=subs(beta,x(n),y);
flag=1;
return
```

The following is a MATLAB subfunction program for Theorems 8.15 and 8.19.

```
function [flag,beta]=BETA_thm815(sigma,Fu,Tu,gu,g0,x,y,u,n,m)

flag=0; beta=y-y;
if ChZero(gu(:,1:sigma)-g0(:,1:sigma))==0
  return
end
[flag2,theta]=SpanCx(gu(:,sigma+1),g0(:,sigma+1));
if flag2==0
  return
end
temp1=simplify(subs(theta,x,Fu));
tempN=simplify(jacobian(temp1,u));
tempD=simplify(jacobian(Tu(n+1-sigma),u));
for k1=1:m
  if ChZero(tempD(k1))==0
    Betau=simplify(tempN(k1)/tempD(k1));
    break
  end
end
Temp=simplify(tempN-Betau*tempD);
if ChZero(Temp)==0
  return
end
Beta0=simplify(subs(Betau,u,u-u));
dBeta0=simplify(jacobian(Beta0,x));
TdBeta0=[dBeta0(1:n-sigma) dBeta0(n-sigma+2:n)];
if ChZero(TdBeta0)==0
  return
end
beta=subs(Beta0,x(n-sigma+1),y);
flag=1;
return
```

The following MATLAB program, that needs subfunctions **BETA-thm8-5-3** and **BETA-thm8-5-4**, is to check the conditions of Theorems 8.14–8.16.

```
clear all
syms x1 x2 x3 x4 x5 x6 x7 x8 x9  real
syms y u u1 u2 u3 u4 u5 real

Fu=[ x2; log(u + x1+1 +x2^2)]; m=1;
iFu=[ exp(x2)-u-1-x1^2; x1]; TYPE=1; %Ex:8.5.4

% Fu=[ (1+x2)*exp(u2^2)-1; (1+x1)*exp(u1)-1];
% iFu=[ (1+x2)*exp(-u1)-1; (1+x1)*exp(-u2^2)-1];
% m=2; TYPE=2; %Ex:8.5.5

% Fu=[ x2; x1+u*(1+x2)];
% iFu=[ x2-u*(1+x1); x1]; m=1; TYPE=4; %Ex:8.5.6

% Fu=[ x2*(1+u); log(u + x1+1 +x2^2)];
% iFu=[ exp(x2)-u-1-(x1/(1+u))^2; x1/(1+u)];
% m=1; TYPE=1; %Ex:8.5.7(b)

% Fu=[ x2*(1+u); log(u + x1+1 +x2^2)];
% iFu=[ exp(x2)-u-1-(x1/(1+u))^2; x1/(1+u)];
% m=1; TYPE=2; %Ex:8.5.7(c)

% Fu=[ x2; x3; exp(x1+u+x2*x3)- 1];
% iFu=[ log(x3+1)-u-x1*x2; x1; x2 ];
% m=1; TYPE=1; %Ex:8.6.2(b)

% Fu=[ x2; x3+x1*u2^2; u1+x1-x2*(x3+x1*u2^2)];
% iFu=[ x3-u1+x1*x2; x1; x2-u2^2*(x3-u1+x1*x2) ];
% m=2; TYPE=2; %Ex:8.6.3(b)

% Fu=[ x2; x3+x1*u2^2; u1+x1-x2*(x3+x1*u2^2)];
% iFu=[ x3-u1+x1*x2; x1; x2-u2^2*(x3-u1+x1*x2) ];
% m=2; TYPE=1; %Ex:8.6.3(d)

% Fu=[ x2; exp(x1*(x2+u2)+u1)-1 ];
% iFu=[ (log(1+x2)-u1)/(x1+u2) ; x1];
% m=2; TYPE=1; %Ex:8.6.4(b)

% Fu=[ x2+u; x3; u+x1+x1*(u+x2)^2];
% iFu=[ (x3-u)/(1+x1^2) ; x1-u; x2];
% m=1; TYPE=2; %Ex:8.6.5(b)

% Fu=[ x2; log(x1*(x2+u2)+u1+1) ];
% iFu=[ (exp(x2)-1-u1)/(x1+u2) ; x1];
% m=2; TYPE=1; %P:8.13(b)

% Fu=[ x2; x3; exp(x1+u+x3^2)- 1];
% iFu=[ log(x3+1)-u-x2^2; x1; x2 ]; m=1; TYPE=1; %P:8.9(a)

% aa=exp(x1+u+(x3+1)*exp(x1*u)-1)- 1; aai=log(x3+1)-u-x2;
% Fu=[x2; (x3+1)*exp(x1*u)-1; aa];
```

```
% iFu=[aai; x1; (1+x2)*exp(-aai*u)-1]; m=1; TYPE=2; %P:8.9(b)

% Fu=[ x2; x3; x1+u*(1+x2*x3)];
% iFu=[ x3-u*(1+x1*x2); x1; x2 ]; m=1; TYPE=4; %P:8.9(c)

% Fu=[ x2; x3; exp(x1*(x2+u2)+u1)-1]; m=2;
% iFu=[ (log(1+x3)-u1)/(x1+u2) ; x1; x2]; TYPE=1; %P:8.9(d)

n=length(Fu);
x=sym('x',[n,1]);
z=sym('z',[n,1]);
if m>1
  u=sym('u',[m,1])
end

H=x1
Fu=simplify(Fu)
F0=simplify(subs(Fu,u,u-u));

if ChInverseF(Fu,iFu,x)==0
  display('Check inverse function once again.')
  return
end

T=x-x; T(1)=H; Tu=T;
for k=2:n
  T(k)=simplify(subs(T(k-1),x,F0));
  Tu(k)=simplify(subs(T(k-1),x,Fu));
end
T=simplify(T)
Tu=simplify(Tu);
alphaU=x-x;
alphaU(1:n-1)=Tu(2:n)-T(2:n);
alphaU(n)=simplify(subs(T(n),x,Fu))

dT=jacobian(T,x);
idT=inv(dT); gu(:,1)=idT(:,n);
for k=2:n+1
  gu(:,k)=sstarmap(Fu,iFu,gu(:,k-1),x);
end
gu=simplify(gu)
g0=subs(gu,u,u-u);

kappa1=n+1;
for k=2:n
  if ChZero(Lfh(g0(:,k),T(n),x)) == 0
    kappa1=k;
    break
  end
end
kappa=kappa1

sigma1=n;
```

```
for k=1:n-1
  if ChZero(jacobian(Tu(n+1-k),u)) == 0
    sigma1=k;
    break
  end
end
sigma=sigma1

if and(kappa==n+1,sigma==n)
  if ChZero(gu-g0) == 0
    display('condition of Thm 8.16 is not satisfied.')
    display('System is NOT state equiv. to a NOCF with OT.')
    return
  end
end
flag=1;
beta=y-y;

if ChZero(T-x)==0
  display('Find beta(y) without MATLAB.')
  return
end
if kappa<=n
  [flag1,beta1]=BETA_thm814(kappa,g0,x,y,n);
end

if sigma<n
  [flag2,beta2]=BETA_thm815(sigma,Fu,Tu,gu,g0,x,y,u,n,m);
end

if TYPE==1 %Thm 8.14
  flag=flag1; beta=beta1
end
if TYPE==2 %Thm 8.15
  flag=flag2; beta=beta2
end

if flag==0
  display('condition (i) is not satisfied.')
  display('System is NOT state equiv. to a NOCF with OT.')
  return
end

if ChZero(beta)==0
  beta0inv=subs(1/beta,y,y-y);
  if ChZero(beta0inv)==1
    display('condition (i) is not satisfied.')
    display('System is NOT state equiv. to a NOCF with OT.')
    return
  end
end

ibeta=int(beta,y);
```

```
ibeta0=subs(ibeta,y,y-y);
ell = simplify(exp(ibeta-ibeta0))
vphi = simplify(int(1/ell, y));
vphi0 = subs(vphi,y,y-y);
varphi= vphi-vphi0;

bgu(:,1)= subs(ell,y,T(n))*g0(:,1);
for k=2:n
  bgu(:,k)=sstarmap(Fu,iFu,bgu(:,k-1),x);
end
bgu=simplify(bgu)
bg0=simplify(subs(bgu,u,u-u));

if ChZero(bgu-bg0) == 0
  display('condition (ii) is not satisfied.')
  display('System is NOT state equivalent to a NOCF with OT.')
  return
end

for k=1:n-1
  bCC(:,k)=adfg(bg0(:,1),bg0(:,k+1),x);
end
bCC=simplify(bCC);

if ChZero(bCC) == 0
  display('condition (iii) is not satisfied.')
  display('System is NOT state equivalent to a NOCF with OT.')
  return
end

display('System is state equivalent to a NOCF with OT.')

varphi=simplify(varphi)
for k=1:n
  idS(:,k)=bg0(:,n+1-k);
end
idS=simplify(idS);
dS=simplify(inv(idS));
S=Codi(dS,x)

gammaU=x-x;
for k=1:n-1
  gammaU(k)=simplify(subs(S(k),x,Fu)-S(k+1));
end
gammaU(n)=simplify(subs(S(n),x,Fu));
gammaU=simplify(gammaU)

return
```

The following is a MATLAB subfunction program for Theorem 8.20.

```
function [flag,beta]=BETA_thm820(kappa,Fu,iFu,x,y,u,n)

flag=0; beta=y-y;
alphanu=Fu(n);
theta=jacobian(alphanu,x(n+2-kappa));
dtheta=jacobian(theta,u);
dalphanu=jacobian(alphanu,u);
Temp1=simplify((1/theta)*dtheta);
[flag2,Bbetau]=SpanCx(Temp1',dalphanu');
if flag2==0
  return
end
Temp2=jacobian(Bbetau,x);
Temp3=jacobian(alphanu,x);
Temp4=[Temp2; Temp3];
if rank(Temp4)>1
  return
end
xbeta=simplify(subs(Bbetau,x,iFu));
beta=simplify(subs(xbeta,x(n),y));
flag=1;

return
```

The following MATLAB program, that needs subfunctions **BETA-thm8-5-3**, **BETA-thm8-5-4**, and **BETA-thm8-6-4**, is to check the conditions of Theorems 8.18–8.21.

```
clear all
syms x1 x2 x3 x4 x5 x6 x7 x8 x9   real
syms w1 w2 w3 w4 w5 w6 w7 w8 w9 real
syms u y u1 u2 u3 real

d=1; Fu=[x2; x3; exp(x1+u+x2*x3)- 1]; m=1; TYPE=1;
iFu=[ log(x3+1)-u-x1*x2; x1; x2 ]; %Ex:8.6.2(c)

% d=1; Fu=[x2; x3+x1*u2^2; u1+x1-x2*(x3+x1*u2^2)]; m=2; TYPE=2;
% iFu=[ x3-u1+x1*x2; x1; x2-u2^2*(x3-u1+x1*x2) ]; %Ex:8.6.3(c)

% d=1; Fu=[x2; x3+x1*u2^2; u1+x1-x2*(x3+x1*u2^2)]; m=2; TYPE=1;
% iFu=[ x3-u1+x1*x2; x1; x2-u2^2*(x3-u1+x1*x2) ]; %Ex:8.6.3(e)

% d=1; Fu=[x2; x3; exp(x1*(x3+u2)+u1)-1]; m=2; TYPE=1;
% iFu=[(log(1+x3)-u1)/(x2+u2); x1; x2]; %Ex:8.6.4(b)

% d=1; Fu=[x2; x3; exp(x1*(x3+u2)+u1)-1]; m=2; TYPE=3;
% iFu=[(log(1+x3)-u1)/(x2+u2) ; x1; x2]; %Ex:8.6.4(c)

% d=1; Fu=[x2+u; x3; u+x1+x1*(u+x2)^2]; m=1; TYPE=3;
% iFu=[(x3-u)/(1+x1^2); x1-u; x2]; %Ex:8.6.5
```

```
% d=2; Fu=[ x2; x1+u*(1+x2)];
% iFu=[ x2-u*(1+x1); x1]; m=1; TYPE=4; %Ex:8.6.6

% d=1; Fu=[x2; x3+x2*u2^2; u1+x1-x2*(x3+x2*u2^2)]; m=2;
% iFu=[x3-u1+x1*x2; x1; x2-u2^2*x1]; TYPE=1; %P:8.12(b)

% d=1; Fu=[x2; x3+x2*u2^2; u1+x1-x2*(x3+x2*u2^2)]; m=2;
% iFu=[x3-u1+x1*x2; x1; x2-u2^2*x1]; TYPE=2; %P:8.12(c)

% d=1; Fu=[x2; log(x1*(x2+u2)+u1+1)]; m=2; TYPE=3;
% iFu=[(exp(x2)-1-u1)/(x1+u2); x1]; %P:8.13(c)

% d=2; Fu=[x2; x3; x1+u*(1+x2*x3)]; m=1; TYPE=4;
% iFu=[x3-u*(1+x1*x2); x1; x2]; %P:8.14(a)

% d=1; Fu=[x2; x3; exp(x1*(x2+u2)+u1)-1]; m=2; TYPE=3;
% iFu=[(log(1+x3)-u1)/(x1+u2); x1; x2]; %P:8.14(b)

% d=1; Fu=[x2; x3; x1+x1*x3+u]; m=1; TYPE=3;
% iFu=[(x3-u)/(1+x2); x1; x2]; %P:8.14(c)

% d=2; Fu=[x2; x3; x1+x1*x3+u]; m=1; TYPE=3;
% iFu=[(x3-u)/(1+x2); x1; x2]; %P:8.14(c)
n=length(Fu);
x=sym('x',[n,1]); w=sym('w',[d,1]);
if m>1
  u=sym('u',[m,1])
end

H=x1;
Fu=simplify(Fu)
iFu=simplify(iFu);
F0=simplify(subs(Fu,u,u-u));

if ChInverseF(Fu,iFu,x)==0
  display('Check inverse function once again.')
  return
end

T=x-x; T(1)=H; Tu=T;
for k=2:n
  T(k)=simplify(subs(T(k-1),x,F0));
  Tu(k)=simplify(subs(T(k-1),x,Fu));
end
T=simplify(T)
Tu=simplify(Tu);
alphaU=x-x;
alphaU(1:n-1)=Tu(2:n)-T(2:n);
alphaU(n)=simplify(subs(T(n),x,Fu))

if ChZero(alphaU(1:min(d,n-1)))==0
  display('Assumption is not satisfied.')
```

```
    return
end

dT=jacobian(T,x); idT=inv(dT);
gu(:,1)=idT(:,n);
for k=2:n+1
  gu(:,k)=sstarmap(Fu,iFu,gu(:,k-1),x);
end
gu=simplify(gu)
g0=subs(gu,u,u-u);

kappa1=n+1;
for k=2:n
  if ChZero(Lfh(g0(:,k),T(n),x)) == 0
    kappa1=k;
    break
  end
end
kappa=kappa1

sigma1=n;
for k=1:n-1
  if ChZero(jacobian(Tu(n+1-k),u)) == 0
    sigma1=k;
    break
  end
end
sigma=sigma1

if and(kappa==n+1,sigma==n)
  if ChZero(gu-g0) == 0
    display('condition of Thm 8.21 is not satisfied.')
    display('System is NOT RDOEL with')
    d=d
    return
  end
  flag=1; beta=y-y;
end

if ChZero(T-x)==0
  display('Find beta(y) without MATLAB.')
  return
end

if kappa<=n-d
  [flag1,beta1]=BETA_thm814(kappa,g0,x,y,n)
end
if sigma<n-d
  [flag2,beta2]=BETA_thm815(sigma,Fu,Tu,gu,g0,x,y,u,n,m)
end
if and(kappa<n+1,sigma>=n-d)==1
  [flag3,beta3]=BETA_thm820(kappa,Fu,iFu,x,y,u,n)
end
```

```
if TYPE==1
  flag=flag1; beta=beta1
end
if TYPE==2
  flag=flag2; beta=beta2
end
if TYPE==3
  flag=flag3; beta=beta3
end

if flag==0
  display('condition (i) is not satisfied.')
  display('System is NOT RDOEL with')
  d=d
  return
end
if ChZero(beta)==0
  beta0inv=subs(1/beta,y,y-y)
  if ChZero(beta0inv)==1
    display('condition (i) is not satisfied.')
    display('System is NOT RDOEL with')
    d=d
    return
  end
end

ibeta=int(beta,y);
ibeta0=subs(ibeta,y,y-y);
ell = simplify(exp(ibeta-ibeta0))
vphi = simplify(int(1/ell, y));
vphi0 = subs(vphi,y,y-y);
varphi= vphi-vphi0

bgu(:,1)= subs(ell,y,T(n))*g0(:,1);
for k=2:n-d
  bgu(:,k)=sstarmap(Fu,iFu,bgu(:,k-1),x);
end
bgu=simplify(bgu)
bg0=simplify(subs(bgu,u,u-u));

if ChZero(bgu-bg0) == 0
  display('condition (ii) is not satisfied.')
  display('System is NOT RDOEL with')
  d=d
  return
end

if ChCommute(bg0,x) == 0
  display('condition (iii) is not satisfied.')
  display('System is NOT RDOEL with')
  d=d
  return
```

```
end

for k=1:n-d
  idhS2(:,k)=bg0(:,n-d+1-k);
end
idhS2=simplify(idhS2(d+1:n,:))
dhS2=simplify(inv(idhS2))
hS2=Codi(dhS2,x(d+1:n))

TEMP41=subs(hS2,x,T);
TEMP42=subs(TEMP41,x,Fu);
TEMP43=subs(TEMP41,x,F0);
TEMP4=simplify(TEMP42-TEMP43)
if ChZero(jacobian(TEMP4,x(2:n))) == 0
  display('condition (iv) is not satisfied.')
  display('System is NOT RDOEL with')
  d=d
  return
end

display('System is RDOEL with')
d=d

if d>0
  hS1=x(1:d)-x(1:d);
  for k=1:d
    hS1(k)= subs(varphi,y,x(k));
  end
end

if d==0
  hS = hS2
else
  hS = [hS1; hS2]
end

xe=[w; x];
if d>0
  He=xe(1);
  Feu=[w(2:d); H; Fu];
else
  He=H; Feu=Fu;
end
Feu=simplify(Feu)
Fe0=simplify(subs(Feu,u,u-u));

Se=xe-xe;
Se(1:n)=subs(hS,x,xe(1:n));
if d>0
  gammaN0=simplify(subs(hS(n),x,F0))
  Se(n+1)=simplify(subs(Se(n),xe,Fe0)-subs(gammaN0,x,xe(1:n)));
end
for k=n+2:n+d
```

```
  Se(k)=simplify(subs(Se(k-1),xe,Fe0));
end
Se=simplify(Se)

bfeu1=simplify(subs(Se,xe,Feu))
gammaeU=xe-xe;
for k=1:n+d-1
  gammaeU(k)=simplify(subs(Se(k),xe,Feu)-Se(k+1));
end
gammaeU(n+d)=simplify(subs(Se(n+d),xe,Feu))

return
```

8.8 Problems

8-1. Find out whether the following nonlinear control systems are state equivalent
to a dual Brunovsky NOCF or not. If it is state equivalent to a dual Brunovsky
NOCF, find a state transformation $z = S(x)$ and the dual Brunovsky NOCF
that new state z satisfies.

(a)

$$\dot{x} = \begin{bmatrix} x_2 + x_3^2 \\ x_3 - 2x_3 e^{x_1} u \\ e^{x_1} u \end{bmatrix} ; \quad y = x_1$$

(b)

$$\dot{x} = \begin{bmatrix} x_2 + x_3^2 \\ x_3 - 2x_3 e^{x_1} u_2 \\ e^{x_1} u_1 \end{bmatrix} ; \quad y = x_1$$

(c)

$$\dot{x} = \begin{bmatrix} x_2 + (x_1 + 1)u_2^2 \\ x_2 \ln(x_1 + 1) + \frac{x_2^2}{1+x_1} + (x_1 + 1)u_1 + x_2 u_2^2 \end{bmatrix} ; \quad y = x_1$$

(d)

$$\begin{bmatrix} \dot{x}_1 \\ \dot{x}_2 \\ \dot{x}_3 \end{bmatrix} = \begin{bmatrix} x_2 e^{x_1} \\ x_3 + x_1 u \\ u \end{bmatrix} ; \quad y = x_1 .$$

8-2. Show that system (8.131) is not state equivalent to a dual Brunovsky NOCF with OT.

8-3. Find out whether the following nonlinear control systems are state equivalent to a dual Brunovsky NOCF with OT or not. If it is state equivalent to a dual Brunovsky NOCF with OT, find a OT $\bar{y} = \varphi(y)$, a state transformation $z = S(x)$, and the dual Brunovsky NOCF that new state z satisfies.

(a)

$$\dot{x} = \begin{bmatrix} x_2 + (x_1 + 1)u_2^2 \\ x_2 \ln(x_1 + 1) + \frac{x_2^2}{1+x_1} + (x_1 + 1)u_1 + x_2 u_2^2 \end{bmatrix} ; \quad y = x_1$$

(b)

$$\begin{bmatrix} \dot{x}_1 \\ \dot{x}_2 \\ \dot{x}_3 \end{bmatrix} = \begin{bmatrix} x_2 e^{x_1} \\ x_3 + x_1 u \\ u \end{bmatrix} ; \quad y = x_1$$

(c)

$$\dot{x} = \begin{bmatrix} x_2 \\ x_3 \\ u - \frac{2x_2^3}{(x_1+1)^2} + \frac{3x_2 x_3}{x_1+1} - x_3(x_1 + 1) \ln(x_1 + 1) \end{bmatrix} ; \quad y = x_1$$

(d)

$$\dot{x} = \begin{bmatrix} x_2 \\ x_3 \\ x_4 + u^2 \\ 5x_2 x_3 + 2x_2^2 + u \end{bmatrix} ; \quad y = x_1.$$

8-4. Find out whether the following nonlinear control systems are RDEOL or not. If it is RDEOL, find the minimal index d and an extended state transformation $z^e = S^e(w, x)$, and the dual Brunovsky NOCF that new state z^e satisfies.

(a)

$$\dot{x} = \begin{bmatrix} x_2 + x_1 u^2 \\ x_3 \\ 4x_1 x_3 + u \end{bmatrix} ; \quad y = x_1$$

(b)

$$\dot{x} = \begin{bmatrix} x_2 + x_3 u^2 \\ x_3 \\ x_4 \\ 4x_1x_3 + u \end{bmatrix}; \quad y = x_1$$

(c)

$$\dot{x} = \begin{bmatrix} x_2 \\ x_3 \\ u - \frac{2x_2^3}{(x_1+1)^2} + \frac{3x_2x_3}{x_1+1} - x_3(x_1+1)\ln(x_1+1) \end{bmatrix}; \quad y = x_1$$

(d)

$$\dot{x} = \begin{bmatrix} x_2 \\ x_3 \\ x_4 + u^2 \\ 5x_2x_3 + 2x_2^2 + u \end{bmatrix}; \quad y = x_1.$$

8-5. Use Corollary 8.6 or Theorem 8.11 to show that the following nonlinear control systems are state equivalent to a dual Brunovsky NOCF.

(a)

$$\dot{x} = \begin{bmatrix} x_2 + x_3 u_1^2 \\ \sin(x_1)(u_2 + x_1x_3) + x_3\cos(x_1)(x_3u_1^2 + x_2) + u_1 \\ x_1x_3 + u_2 \end{bmatrix}; \quad y = \begin{bmatrix} x_1 \\ x_3 \end{bmatrix}$$

(b)

$$\dot{x} = \begin{bmatrix} x_2 \\ x_3 \\ x_4 + x_1u_2 \\ 3x_2^2 + x_3 + x_6 + x_1x_3 + 3x_3x_6 + x_4x_5 + (2x_2 + x_1x_5)u_2 + u_1 \\ x_6 \\ x_2 + u_2 \end{bmatrix}$$

$$y = \begin{bmatrix} x_1 \\ x_5 \end{bmatrix}$$

(c)

$$\dot{x} = \begin{bmatrix} x_2 \\ x_3 \\ x_4x_5 + 2x_5x_8 + x_6x_7 + u_1 + x_4u_3 \\ x_5 \\ x_6 \\ \alpha_{23}^u(x) \\ x_8 \\ x_5 + u_3 \end{bmatrix} ; \ y = \begin{bmatrix} x_1 \\ x_4 \\ x_7 \end{bmatrix}$$

where

$$\alpha_{23}^u(x) = x_1x_4x_5 + 2x_1x_5x_8 + x_1x_6x_7 + 2x_2x_4x_8 + 2x_2x_5x_7$$
$$+ x_3x_4x_7 + u_2 + x_1x_4u_3.$$

(d)

$$\dot{x} = \begin{bmatrix} x_2 \\ x_3 + x_1u_2 + x_4u_3 + x_7^2u_4 \\ x_6x_7 + u_1 + x_5u_3 \\ x_5 + x_7u_4 \\ x_6 \\ x_2x_8 + x_1(x_1^2 + u_4) + u_2 \\ u_3 \\ x_1^2 + u_4 \end{bmatrix} ; \ y = \begin{bmatrix} x_1 \\ x_4 \\ x_7 \\ x_8 \end{bmatrix}.$$

8-6. Use Corollary 8.5 or Theorem 8.9 to find out whether the following nonlinear control systems are state equivalent to a dual Brunovsky NOCF. If it is state equivalent to a dual Brunovsky NOCF, find a state transformation $z = S(x)$ and the dual Brunovsky NOCF that new state z satisfies.

(a)

$$\dot{x} = \begin{bmatrix} x_2 \\ x_2x_3 + x_1x_4 + u_1 + x_1(x_2 - x_1x_3)u_1^2 \\ x_4 + (x_2 - x_1x_3)u_1^2 \\ 2x_1x_2 + u_2 \end{bmatrix} ; \ y = \begin{bmatrix} x_1 \\ x_3 \end{bmatrix}$$

(b)

$$\dot{x} = \begin{bmatrix} x_2 \\ x_2 x_3 + x_1 x_4 + x_1^2 u_1^2 + (x_3 + 1)u_1 \\ x_1 u_1^2 + x_4 \\ 2x_1 x_2 + u_2 \end{bmatrix} ; \quad y = \begin{bmatrix} x_1 \\ x_3 \end{bmatrix}$$

(c)

$$\dot{x} = \begin{bmatrix} x_2 \\ x_3 + 2x_1 x_4 u_2 \\ 2x_1 x_2 + x_3 x_4^2 + u_1 + (x_2 - x_1 x_4^2 + 2x_1 x_4^3 + 2x_2 x_4)u_2 \\ u_2 \end{bmatrix}$$

$$y = \begin{bmatrix} x_1 \\ x_4 \end{bmatrix}.$$

8-7. Show that (8.300) is satisfied.

8-8. Find out whether the following discrete time nonlinear control systems are state equivalent to a dual Brunovsky NOCF or not. If it is state equivalent to a dual Brunovsky NOCF, find a state transformation $z = S(x)$ and the dual Brunovsky NOCF that new state z satisfies.

(a)

$$x(t+1) = \begin{bmatrix} x_2 \\ x_3 + x_1 u \\ u + x_1 + e^{x_2} + (x_3 + x_1 u)^2 - 1 \end{bmatrix} ; \quad y = x_1$$

(b)

$$x(t+1) = \begin{bmatrix} x_2 \\ x_3 \\ e^{x_1 + x_3^2 + u} - 1 \end{bmatrix} ; \quad y = x_1$$

(c)

$$x(t+1) = \begin{bmatrix} x_2 \\ (x_3 + 1)e^{x_1 u} - 1 \\ e^{x_1 + (x_3 + 1)e^{x_1 u} - 1 + u} - 1 \end{bmatrix} ; \quad y = x_1.$$

8-9. Find out whether the following discrete time nonlinear control systems are state equivalent to a dual Brunovsky NOCF with OT or not. If it is state equivalent to a dual Brunovsky NOCF with OT, find a OT $\bar{y} = \varphi(y)$, a state transformation $z = S(x)$, and the dual Brunovsky NOCF that new state z satisfies.

(a)

$$x(t+1) = \begin{bmatrix} x_2 \\ x_3 \\ e^{x_1+x_3^2+u} - 1 \end{bmatrix}; \quad y = x_1$$

(b)

$$x(t+1) = \begin{bmatrix} x_2 \\ (x_3+1)e^{x_1 u} - 1 \\ e^{x_1+(x_3+1)e^{x_1 u}-1+u} - 1 \end{bmatrix}; \quad y = x_1$$

(c)

$$x(t+1) = \begin{bmatrix} x_2 \\ x_3 \\ x_1 + (x_2 x_3 + 1)u \end{bmatrix}; \quad y = x_1$$

(d)

$$x(t+1) = \begin{bmatrix} x_2 \\ x_3 \\ e^{x_1 x_2 + u_1 + x_1 u_2} - 1 \end{bmatrix}; \quad y = x_1.$$

8-10. Solve Example 8.6.1.
8-11. Solve Example 8.6.3(d) and Example 8.6.3(e).
8-12. Consider the system

$$x(t+1) = \begin{bmatrix} x_2 \\ x_3 + x_2 u_2^2 \\ u_1 + x_1 - x_2(x_3 + x_2 u_2^2) \end{bmatrix}; \quad y = x_1.$$

(a) Show that $\kappa = 2$ and $\sigma = 1$.
(b) Use Theorem 8.18 to show that the above system is not RDOEL with index $d = 1$.
(c) Use Theorem 8.19 to show that the above system is not RDOEL with index $d = 1$.

8-13. Consider the system

$$x(t+1) = \begin{bmatrix} x_2 \\ \ln(x_1 x_2 + u_1 + u_2 x_1 + 1) \end{bmatrix}; \quad y = x_1.$$

(a) Show that $\kappa = 2$ and $\sigma = 2$.
(b) Use Theorem 8.14 to show that the above system is not state equivalent to a dual Brunovsky NOCF with OT.
(c) Use Theorem 8.20 to show that the above system is RDOEL with index $d = 1$.

8-14. Find out whether the following discrete time nonlinear control systems are RDEOL or not. If it is RDEOL, find the minimal index d and an extended state transformation $z^e = S^e(w, x)$, and the dual Brunovsky NOCF that new state z^e satisfies.

(a)

$$x(t+1) = \begin{bmatrix} x_2 \\ x_3 \\ x_1 + (x_2 x_3 + 1)u \end{bmatrix} ; \quad y = x_1$$

(b)

$$x(t+1) = \begin{bmatrix} x_2 \\ x_3 \\ e^{x_1 x_2 + u_1 + x_1 u_2} - 1 \end{bmatrix} ; \quad y = x_1$$

(c)

$$x(t+1) = \begin{bmatrix} x_2 \\ x_3 \\ x_1 + x_1 x_3 + u \end{bmatrix} ; \quad y = x_1 .$$

Chapter 9
Input-Output Decoupling

9.1 Introduction

The problem of input-output decoupling has been introduced briefly in Sect. 1.1. In this chapter, the necessary and sufficient conditions will be studied. Consider the following nonlinear input-output system.

$$\dot{x}(t) = f(x(t)) + g(x(t))u(t)$$
$$y(t) = h(x(t)) \tag{9.1}$$

where $x \in \mathbb{R}^n$, $u \in \mathbb{R}^m$, $y \in \mathbb{R}^m$, and $f(x)$, $g(x)$, and $h(x)$ are analytic functions. Let $y_i^{(j)}(t) \triangleq \frac{d^j}{dt^j} y_i(t)$. Then we have that for $1 \leq i \leq m$ and $j \geq 1$,

$$y_i^{(j)}(t) = Q_i^j(x(t), u(t), \cdots, u^{(j-1)}(t)) \tag{9.2}$$

for some functions Q_i^j, $1 \leq i \leq m$, $j \geq 1$.

Definition 9.1 (*decoupled input-output relation*) System (9.1) is said to have decoupled input-output relationship if output y_i is a function of only input u_i for all i, with changing the order of the inputs.

If the MIMO system's input-output relation is decoupled, the MIMO system has the parallel connection of m SISO systems. Thus, we can control each output without affecting the other outputs. In other words, we have the following equation:

$$y_i^{(j)}(t) = Q_i^j(x(t), u_i(t), \cdots, u_i^{(j-1)}(t)), \ 1 \leq i \leq m. \tag{9.3}$$

If the system does not have the decoupled input-output relation, the nonsingular feedback could obtain the decoupled input-output relation of the closed-loop system, which is called the input-output decoupling.

© The Author(s), under exclusive license to Springer Nature Singapore Pte Ltd. 2022   521
H.-G. Lee, *Linearization of Nonlinear Control Systems*,
https://doi.org/10.1007/978-981-19-3643-2_9

Definition 9.2 (*static input-output decoupling*) System (9.1) is said to be locally static input-output decouplable (on a neighborhood of $x = x_0$), if there exists a non-singular static feedback $u = \alpha(x) + \beta(x)v$ (rank $(\beta(x_0)) = m$) such that the closed-loop system

$$\begin{aligned} \dot{x}(t) &= f(x) + g(x)\alpha(x) + g(x)\beta(x)v(t) \\ &= \tilde{f}(x(t)) + \tilde{g}(x(t))v(t) \\ y(t) &= h(x(t)) \end{aligned} \tag{9.4}$$

has the following decoupled input-output relationship:

$$\begin{bmatrix} y_1^{(\rho_1)} \\ \vdots \\ y_m^{(\rho_m)} \end{bmatrix} = \begin{bmatrix} v_1 \\ \vdots \\ v_m \end{bmatrix} \tag{9.5}$$

where ρ_i is the relative degree of the output y_i.

It is easy to see that the relative degree of system (9.4) is the same as that of system (9.1). (Refer to Problem 5-7.)

Definition 9.3 (*dynamic input-output decoupling*) System (9.1) is said to be locally dynamic input-output decouplable (on a neighborhood of $x = x_0$), if there exists a dynamic feedback

$$u = a(x, z) + b(x, z)w \tag{9.6a}$$

$$\dot{z} = c(x, z) + d(x, z)w, \quad z \in \mathbb{R}^d \tag{9.6b}$$

such that the extended system

$$\begin{aligned} \begin{bmatrix} \dot{x} \\ \dot{z} \end{bmatrix} &= \begin{bmatrix} f(x) + g(x)a(x, z) \\ c(x, z) \end{bmatrix} + \begin{bmatrix} g(x)b(x, z) \\ d(x, z) \end{bmatrix} w \\ &= f_E(x, z) + g_E(x, z)w \\ y(t) &= h(x(t)) \end{aligned} \tag{9.7}$$

has the decoupling matrix $D_E(x, z)$ with rank $(D_E(x_0, 0)) = m$.

In other words, if extended system (9.7) is locally static input-output decouplable, then system (9.1) is said to be locally dynamic input-output decouplable.

9.2 Input-Output Decoupling of the Nonlinear Systems

By Definition 5.6 of the relative degree, we have

$$
\begin{aligned}
y_i^{(\ell)} &= L_f^\ell h_i(x), \ 0 \le \ell \le \rho_i - 1 \\
y_i^{(\rho_i)} &= L_f^{\rho_i} h_i(x) + L_g L_f^{\rho_i - 1} h_i(x) u
\end{aligned}
\tag{9.8}
$$

where ρ_i is the relative degree of the output y_i. Thus, it is clear that

$$
\begin{bmatrix} y_1^{(\rho_1)} \\ \vdots \\ y_m^{(\rho_m)} \end{bmatrix} = \begin{bmatrix} L_f^{\rho_1} h_1(x) \\ \vdots \\ L_f^{\rho_m} h_m(x) \end{bmatrix} + \begin{bmatrix} L_g L_f^{\rho_1 - 1} h_1(x) \\ \vdots \\ L_g L_f^{\rho_m - 1} h_m(x) \end{bmatrix} u
$$
$$
\triangleq E(x) + D(x)u.
\tag{9.9}
$$

If $m \times m$ matrix $D(0)$ is invertible, then the closed-loop system has decoupled input-output relationship

$$
\begin{bmatrix} y_1^{(\rho_1)} \\ \vdots \\ y_m^{(\rho_m)} \end{bmatrix} = \begin{bmatrix} v_1 \\ \vdots \\ v_m \end{bmatrix}
$$

with static feedback

$$
u = -D(x)^{-1} E(x) + D(x)^{-1} v.
\tag{9.10}
$$

Therefore, $D(x)$, defined in (9.9), is called by decoupling matrix.

Theorem 9.1 (conditions for the static IO decoupling problem) *System (9.1) is locally static input-output decouplable (on a neighborhood of $x = x_0$), if and only if*

$$
\mathrm{rank}\,(D(x_0)) = m
\tag{9.11}
$$

where $D(x)$, defined in (9.9), is the decoupling matrix of system (9.1).

Proof Necessity. Suppose that system (9.1) is locally static input-output decouplable. Then, by Definition 9.2 and (9.9), there exists a nonsingular static feedback $u = \alpha(x) + \beta(x)v$ such that

$$
E(x) + D(x)\alpha(x) + D(x)\beta(x)v = v,
$$

which implies that

$$D(x)\beta(x) = I_m.$$

Since $D(x_0)\beta(x_0) = I_m$, it is clear that (9.11) is satisfied.

Sufficiency. Obvious by (9.10) and Definition 2.5.   □

It is clear that system (9.1) is locally static input-output decouplable on a neighborhood of $x = 0$, if and only if

$$\text{rank}\, (D(0)) = m.$$

Example 9.2.1 Show that the following nonlinear system is locally static input-output decouplable. Also, obtain a static feedback for input-output decoupling.

$$\dot{x} = \begin{bmatrix} x_2 \\ x_3 \\ x_4 \\ 0 \end{bmatrix} + \begin{bmatrix} 0 \\ 1 \\ 0 \\ 0 \end{bmatrix} u_1 + \begin{bmatrix} 0 \\ x_4 \\ 0 \\ 1 + x_3 \end{bmatrix} u_2 = f(x) + g_1(x)u_1 + g_2(x)u_2$$

$$\begin{bmatrix} y_1 \\ y_2 \end{bmatrix} = \begin{bmatrix} x_1 \\ x_3 \end{bmatrix} = h(x). \tag{9.12}$$

Solution It is easy, by the definition of the relative degree, to see that $\rho_1 = 2$ and $\rho_2 = 2$. Since

$$\begin{bmatrix} y_1^{(2)} \\ y_2^{(2)} \end{bmatrix} = \begin{bmatrix} x_3 \\ 0 \end{bmatrix} + \begin{bmatrix} 1 & x_4 \\ 0 & 1 + x_3 \end{bmatrix} \begin{bmatrix} u_1 \\ u_2 \end{bmatrix},$$

it is clear that decoupling matrix $D(x)$ is invertible. Therefore, by Theorem 9.1, system (9.12) can be input-output decoupled by static feedback

$$\begin{bmatrix} u_1 \\ u_2 \end{bmatrix} = \begin{bmatrix} 1 & x_4 \\ 0 & 1 + x_3 \end{bmatrix}^{-1} \begin{bmatrix} x_3 \\ 0 \end{bmatrix} + \begin{bmatrix} 1 & x_4 \\ 0 & 1 + x_3 \end{bmatrix}^{-1} \begin{bmatrix} v_1 \\ v_2 \end{bmatrix}$$

$$= \begin{bmatrix} x_3 \\ 0 \end{bmatrix} + \begin{bmatrix} 1 & \frac{-x_4}{1+x_3} \\ 0 & \frac{1}{1+x_3} \end{bmatrix} \begin{bmatrix} v_1 \\ v_2 \end{bmatrix}.$$

                                                                                        □

Example 9.2.2 Show that the following nonlinear system is not locally input-output decouplable by static feedback:

$$\dot{x} = \begin{bmatrix} x_2 \\ x_3 \\ x_4 \\ 0 \end{bmatrix} + \begin{bmatrix} 0 \\ 1 \\ x_4 \\ 0 \end{bmatrix} u_1 + \begin{bmatrix} 0 \\ 0 \\ 0 \\ 1 + x_3 \end{bmatrix} u_2 = f(x) + g_1(x)u_1 + g_2(x)u_2$$

$$\begin{bmatrix} y_1 \\ y_2 \end{bmatrix} = \begin{bmatrix} x_1 \\ x_3 \end{bmatrix} = h(x). \tag{9.13}$$

Solution By simple calculations, it is easy to see that $\rho_1 = 2$, $\rho_2 = 1$, and

$$\begin{bmatrix} y_1^{(2)} \\ y_2^{(1)} \end{bmatrix} = \begin{bmatrix} x_3 \\ x_4 \end{bmatrix} + \begin{bmatrix} 1 & 0 \\ x_4 & 0 \end{bmatrix} \begin{bmatrix} u_1 \\ u_2 \end{bmatrix}. \tag{9.14}$$

Note that decoupling matrix $D(x)$ is not invertible. Therefore, by Theorem 9.1, system (9.13) cannot be locally input-output decoupled by static feedback. $\qquad\square$

In (9.14) of Example 9.2.2, $\begin{bmatrix} y_1^{(2)} \\ y_2^{(1)} \end{bmatrix}$ is a function of input u_1 only, and thus static input-output decoupling is not possible. That is, input u_1 affects the output too early compared to input u_2. We could use integrators to input u_1 and increase the relative degree of the extended closed-loop system until the derivative of the output depends on both of the new inputs simultaneously. In other words, we consider the dynamic feedback

$$u_1 = z_1 \; ; \;\; u_2 = w_2^1$$
$$\dot{z}_1 = w_1^1$$

and the extended closed-loop system

$$\begin{bmatrix} \dot{x}_1 \\ \dot{x}_2 \\ \dot{x}_3 \\ \dot{x}_4 \\ \dot{z}_1 \end{bmatrix} = \begin{bmatrix} x_2 \\ x_3 + z_1 \\ x_4(1 + z_1) \\ 0 \\ 0 \end{bmatrix} + \begin{bmatrix} 0 \\ 0 \\ 0 \\ 0 \\ 1 \end{bmatrix} w_1^1 + \begin{bmatrix} 0 \\ 0 \\ 0 \\ 1 + x_3 \\ 0 \end{bmatrix} w_2^1 \tag{9.15}$$
$$= F^1(x) + G_1^1(x)w_1^1 + G_2^1(x)w_2^1.$$

For extended system (9.15), we have relative degree $(\rho_1^1, \rho_2^1) = (3, 2)$ and

$$\begin{bmatrix} y_1^{(3)} \\ y_2^{(2)} \end{bmatrix} = \begin{bmatrix} x_4(1 + z_1) \\ 0 \end{bmatrix} + \begin{bmatrix} 1 & 0 \\ x_4 & (1 + x_3)(1 + z_1) \end{bmatrix} \begin{bmatrix} w_1^1 \\ w_2^1 \end{bmatrix}$$

which implies that decoupling matrix of extended system (9.15) is nonsingular. Therefore, extended system (9.15) is input-output decouplable by extended state feedback

$$\begin{bmatrix} w_1^! \\ w_2^! \end{bmatrix} = \begin{bmatrix} 1 & 0 \\ x_4 & (1+x_3)(1+z_1) \end{bmatrix}^{-1} \left\{ -\begin{bmatrix} x_4(1+z_1) \\ 0 \end{bmatrix} + \begin{bmatrix} v_1 \\ v_2 \end{bmatrix} \right\}$$

$$= \begin{bmatrix} -x_4(1+z_1) \\ \frac{x_4^2}{1+x_3} \end{bmatrix} + \begin{bmatrix} 1 & 0 \\ \frac{-x_4}{(1+x_3)(1+z_1)} & \frac{1}{(1+x_3)(1+z_1)} \end{bmatrix} \begin{bmatrix} v_1 \\ v_2 \end{bmatrix}.$$

In other words, system (9.13) is input-output decouplable by dynamic feedback

$$\begin{bmatrix} u_1 \\ u_2 \end{bmatrix} = \begin{bmatrix} z_1 \\ \frac{x_4^2}{1+x_3} \end{bmatrix} + \begin{bmatrix} 0 & 0 \\ \frac{-x_4}{(1+x_3)(1+z_1)} & \frac{1}{(1+x_3)(1+z_1)} \end{bmatrix} \begin{bmatrix} v_1 \\ v_2 \end{bmatrix} \qquad (9.16)$$

$$\dot{z}_1 = -x_4(1+z_1) + v_1.$$

□

Example 9.2.3 Show that the following nonlinear system is not locally input-output decouplable by static feedback. Also, find a dynamic feedback to decouple I-O relation.

$$\dot{x} = \begin{bmatrix} x_2 \\ x_3 \\ x_4 \\ 0 \end{bmatrix} + \begin{bmatrix} 0 \\ 1 \\ x_4 \\ 0 \end{bmatrix} u_1 + \begin{bmatrix} 0 \\ 1 \\ x_4 \\ 1+x_3 \end{bmatrix} u_2 = f(x) + g_1(x)u_1 + g_2(x)u_2$$

$$\begin{bmatrix} y_1 \\ y_2 \end{bmatrix} = \begin{bmatrix} x_1 \\ x_3 \end{bmatrix} = h(x). \qquad (9.17)$$

Solution By simple calculations, we have relative degree $(\rho_1, \rho_2) = (2, 1)$ and

$$\begin{bmatrix} y_1^{(2)} \\ y_2^{(1)} \end{bmatrix} = \begin{bmatrix} x_3 \\ x_4 \end{bmatrix} + \begin{bmatrix} 1 & 1 \\ x_4 & x_4 \end{bmatrix} \begin{bmatrix} u_1 \\ u_2 \end{bmatrix}.$$

Since decoupling matrix $D(x)$ is singular, it is clear, by Theorem 9.1, that system (9.17) is not input-output decouplable by static feedback. Unlike Example 9.2.2, the decoupling matrix of the extended system cannot be made nonsingular by applying an integrator to one of the inputs. If we consider static feedback

$$\begin{bmatrix} u_1 \\ u_2 \end{bmatrix} = \begin{bmatrix} 1 & -1 \\ 0 & 1 \end{bmatrix} \begin{bmatrix} \mu_1^1 \\ \mu_2^1 \end{bmatrix} \triangleq L^1(x)\mu^1,$$

we have the following closed-loop system:

$$\dot{x} = \begin{bmatrix} x_2 \\ x_3 \\ x_4 \\ 0 \end{bmatrix} + \begin{bmatrix} 0 \\ 1 \\ x_4 \\ 0 \end{bmatrix} \mu_1^1 + \begin{bmatrix} 0 \\ 0 \\ 0 \\ 1+x_3 \end{bmatrix} \mu_2^1 \qquad (9.18)$$

$$= f^1(x) + g_1^1(x)\mu_1^1 + g_2^1(x)\mu_2^1$$

which is the same as (9.13) with $u = \mu^1$. Therefore, system (9.18) is input-output decouplable by dynamic feedback (9.16). In other words, system (9.17) is input-output decouplable by dynamic feedback

$$
\begin{bmatrix} u_1 \\ u_2 \end{bmatrix} = L^1(x)\mu^1
$$

$$
= \begin{bmatrix} z_1 - \frac{x_4^2}{(1+x_3)} \\ \frac{x_4^2}{(1+x_3)} \end{bmatrix} + \begin{bmatrix} \frac{x_4}{(1+x_3)(1+z_1)} & \frac{-1}{(1+x_3)(1+z_1)} \\ \frac{-x_4}{(1+x_3)(1+z_1)} & \frac{1}{(1+x_3)(1+z_1)} \end{bmatrix} \begin{bmatrix} v_1 \\ v_2 \end{bmatrix}
$$

$$
\dot{z}_1 = -x_4(1 + z_1) + v_1.
$$

$\square$

9.3 Dynamic Input-Output Decoupling

Suppose that input-output decoupling is not possible by static feedback. Then, as shown in Examples 9.2.2 and 9.2.3, we could wait for other inputs to affect the output, by using integrators to some inputs and increasing the relative degree of the extended closed-loop system. We call this process the dynamic extension algorithm.

Lemma 9.1 *If system (9.1) is locally dynamic input-output decouplable (on a neighborhood of $x = x_0$) with dynamic feedback*

$$
u = a(x, z) + b(x, z)w \tag{9.19a}
$$

$$
\dot{z} = c(x, z) + d(x, z)w, \quad z \in \mathbb{R}^d, \tag{9.19b}
$$

then system (9.1) is also locally dynamic input-output decouplable (on a neighborhood of $x = x_0$) with dynamic feedback

$$
u = a(x, 0) + \left. \frac{\partial a(x, z)}{\partial z} \right|_{z=0} z + b(x, 0)w \tag{9.20a}
$$

$$
\dot{z} = c(x, 0) + \left. \frac{\partial c(x, z)}{\partial z} \right|_{z=0} z + d(x, 0)w. \tag{9.20b}
$$

Proof Suppose that system (9.1) is locally dynamic input-output decouplable (on a neighborhood of $x = x_0$) with dynamic feedback (9.19). Then we have, by Definition 9.3, that

$$
\text{rank}\left(\left. \begin{bmatrix} L_{G_E} L_{F_E}^{\rho_1^E - 1} h_1(x) \\ \vdots \\ L_{G_E} L_{F_E}^{\rho_m^E - 1} h_m(x) \end{bmatrix} \right|_{(x,z)=(x_0,0)} \right) = m \tag{9.21}
$$

where $(\rho_1^E, \cdots, \rho_m^E)$ are the relative degrees of the extended system

$$
\dot{x}_E = \begin{bmatrix} f(x) + g(x)a(x, z) \\ c(x, z) \end{bmatrix} + \begin{bmatrix} g(x)b(x, z) \\ d(x, z) \end{bmatrix} w
$$
$$
= F_E(x_E) + G_E(x_E)w.
$$

Consider the extended system of system (9.1) with dynamic feedback (9.20):

$$
\dot{x}_E = \begin{bmatrix} f(x) + g(x)a(x, 0) + g(x) \left.\frac{\partial a(x,z)}{\partial z}\right|_{z=0} z \\ c(x, 0) + \left.\frac{\partial c(x,z)}{\partial z}\right|_{z=0} z \end{bmatrix} + \begin{bmatrix} g(x)b(x, 0) \\ d(x, 0) \end{bmatrix} w
$$
$$
= F_E^0(x_E) + G_E^0(x_E)w.
$$

Note that $F_E(x, 0) = F_E^0(x, 0)$ and

$$
\left.\frac{\partial F_E(x_E)}{\partial x_E}\right|_{z=0} = \begin{bmatrix} \frac{\partial(f(x)+g(x)a(x,0))}{\partial x} & \left. g(x) \frac{\partial a(x,z)}{\partial z}\right|_{z=0} \\ \frac{\partial c(x,0)}{\partial x} & \left.\frac{\partial c(x,z)}{\partial z}\right|_{z=0} \end{bmatrix}
$$

$$
= \begin{bmatrix} \left.\frac{\partial F_E^0(x_E)}{\partial x}\right|_{z=0} & \left.\frac{\partial F_E^0(x_E)}{\partial z}\right|_{z=0} \end{bmatrix} = \left.\frac{\partial F_E^0(x_E)}{\partial x_E}\right|_{z=0}. \tag{9.22}
$$

Thus, it is easy to see, by (2.3) and (9.22), that for $1 \le i \le m$,

$$
\left.\frac{\partial \left(L_{F_E} h_i(x)\right)}{\partial x_E}\right|_{z=0} = \left.\frac{\partial \left(\frac{\partial h_i(x)}{\partial x_E} F_E(x_E)\right)}{\partial x_E}\right|_{z=0}
$$

$$
= \left\{ F_E(x_E)^{\mathsf{T}} \frac{\partial \left(\frac{\partial h_i(x)}{\partial x_E}\right)^{\mathsf{T}}}{\partial x_E} + \frac{\partial h_i(x)}{\partial x_E} \frac{\partial F_E(x_E)}{\partial x_E} \right\}\Bigg|_{z=0}
$$

$$
= F_E^0(x_E)^{\mathsf{T}} \left.\frac{\partial \left(\frac{\partial h_i(x)}{\partial x_E}\right)^{\mathsf{T}}}{\partial x_E}\right|_{z=0} + \left.\frac{\partial h_i(x)}{\partial x_E}\right|_{z=0} \left.\frac{\partial F_E^0(x_E)}{\partial x_E}\right|_{z=0}
$$

$$
= \left.\frac{\partial \left(\frac{\partial h_i(x)}{\partial x_E} F_E^0(x_E)\right)}{\partial x_E}\right|_{z=0} = \left.\frac{\partial \left(L_{F_E^0} h_i(x)\right)}{\partial x_E}\right|_{z=0}.
$$

In this manner, it is easy to show by mathematical induction that for $1 \le i \le m$ and $k \ge 1$,

$$\frac{\partial\left(L_{F_E}^k h_i(x)\right)}{\partial x_E}\bigg|_{z=0} = \frac{\partial\left(L_{F_E^0}^k h_i(x)\right)}{\partial x_E}\bigg|_{z=0}$$

and

$$L_{G_E}L_{F_E}^k h(x)\big|_{z=0} = \frac{\partial\left(L_{F_E}^k h_i(x)\right)}{\partial x_E}\bigg|_{z=0} G_E(x,0)$$

$$= \frac{\partial\left(L_{F_E^0}^k h_i(x)\right)}{\partial x_E}\bigg|_{z=0} G_E(x,0) = L_{G_E^0}L_{F_E^0}^k h(x)\big|_{z=0}$$

which implies, together with (9.21), that

$$\text{rank}\left(\begin{bmatrix} L_{G_E^0}L_{F_E^0}^{\rho_1^E-1}h_1(x) \\ \vdots \\ L_{G_E^0}L_{F_E^0}^{\rho_m^E-1}h_m(x) \end{bmatrix}\bigg|_{(x,z)=(x_0,0)}\right) = m.$$

Hence, system (9.1) is also locally dynamic input-output decouplable (on a neighborhood of $x = x_0$) with dynamic feedback (9.20). $\qquad\square$

Elementary row operations and column operations of the matrix are very useful concepts. The following lemma, which can be proved by using elementary row and column operations of the matrix, plays a key role in the dynamic extension algorithm. (Refer to MATLAB subfunction **Dcolumn**(D, x) in Sect. 9.4.)

Lemma 9.2 *Suppose that system (9.1) has the decoupling matrix $D(x)$ with*

$$\text{rank}\,(D(x)) = r < m.$$

There exist a $m \times m$ permutation matrix R and a $m \times m$ nonsingular matrix $L(x)$ such that

$$RD(x)L(x) = \begin{bmatrix} I_{\bar{r}} & O_{\bar{r}\times(r-\bar{r})} & O_{\bar{r}\times(m-r)} \\ O_{(r-\bar{r})\times\bar{r}} & I_{r-\bar{r}} & O_{(r-\bar{r})\times(m-r)} \\ \hat{D}(x) & O_{(m-r)\times(r-\bar{r})} & O_{(m-r)\times(m-r)} \end{bmatrix} \tag{9.23}$$

and for $1 \le i \le \bar{r}$,

$$\hat{D}^i(x) \neq O_{(m-r)\times 1} \tag{9.24}$$

where $\hat{D}(x) \triangleq \left[\hat{D}^1(x) \cdots \hat{D}^{\bar{r}}(x)\right]$.

Proof We can assume, by changing the order of the outputs, that the first r rows of the matrix $D(x)$ are linearly independent. In other words, there exists a $m \times m$ permutation matrix R_0 such that

$$R_0 D(x) = \begin{bmatrix} \tilde{D}(x) \\ \bar{D}(x) \end{bmatrix} \quad \text{and} \quad \text{rank}\left(\tilde{D}(x)\right) = r.$$

Thus, we can find, by elementary column operations, a $m \times m$ permutation matrix L_0 such that $\tilde{D}^{11}(x)$ is a $r \times r$ nonsingular matrix and

$$\begin{bmatrix} \tilde{D}(x) \\ \bar{D}(x) \end{bmatrix} L_0 = \begin{bmatrix} \tilde{D}^{11}(x) & \tilde{D}^{12}(x) \\ \bar{D}^{21}(x) & \bar{D}^{22}(x) \end{bmatrix}.$$

If we let $L_1(x) = \begin{bmatrix} \tilde{D}^{11}(x)^{-1} & -\tilde{D}^{11}(x)^{-1}\tilde{D}^{12}(x) \\ O & I_{m-r} \end{bmatrix}$, then we have that

$$R_0 D(x) L_0 L_1(x) = \begin{bmatrix} \tilde{D}^{11}(x) & \tilde{D}^{12}(x) \\ \bar{D}^{21}(x) & \bar{D}^{22}(x) \end{bmatrix} L_1(x) = \begin{bmatrix} I_r & O_{r \times (m-r)} \\ \hat{D}^{21}(x) & O_{(m-r) \times (m-r)} \end{bmatrix}$$

where $\hat{D}^{21}(x) \triangleq \bar{D}^{21}(x)\tilde{D}^{11}(x)^{-1}$. It is clear, by elementary column and row operations, that there exist a $r \times r$ permutation matrix $\bar{L}$, a $m \times m$ permutation matrix $L_2 = \begin{bmatrix} \bar{L} & O \\ O & I_{m-r} \end{bmatrix}$, and a $m \times m$ permutation matrix $R_1 = \begin{bmatrix} \bar{L}^{\mathsf{T}} & O \\ O & I_{m-r} \end{bmatrix}$ such that

$$\hat{D}^{21}(x)\bar{L} = \begin{bmatrix} \hat{D}(x) & O_{(m-r) \times (r - \bar{r})} \end{bmatrix}$$

$$R_1 \begin{bmatrix} I_r & O_{r \times (m-r)} \\ \hat{D}^{21}(x) & O_{(m-r) \times (m-r)} \end{bmatrix} L_2 = \begin{bmatrix} \bar{L}^{\mathsf{T}} & O \\ O & I_{m-r} \end{bmatrix} \begin{bmatrix} I_r & O \\ \hat{D}^{21}(x) & O \end{bmatrix} \begin{bmatrix} \bar{L} & O \\ O & I_{m-r} \end{bmatrix}$$

$$= \begin{bmatrix} I_{\bar{r}} & O_{\bar{r} \times (r - \bar{r})} & O_{\bar{r} \times (m-r)} \\ O_{(r - \bar{r}) \times \bar{r}} & I_{r - \bar{r}} & O_{(r - \bar{r}) \times (m-r)} \\ \hat{D}(x) & O_{(m-r) \times (r - \bar{r})} & O_{(m-r) \times (m-r)} \end{bmatrix}$$

and for $1 \leq i \leq \bar{r}$,

$$\hat{D}^i(x) \neq O_{(m-r) \times 1}$$

where $\hat{D}(x) \triangleq \begin{bmatrix} \hat{D}^1(x) \cdots \hat{D}^{\bar{r}}(x) \end{bmatrix}$. In other words, (9.23) and (9.24) are satisfied with

$$R = R_1 R_0 \quad \text{and} \quad L(x) = L_0 L_1(x) L_2.$$

$\square$

9.3.1 Dynamic Extension Algorithm

step 1: For system (9.1), we have

$$\begin{bmatrix} y_1^{(\rho_1^1)} \\ \vdots \\ y_m^{(\rho_m^1)} \end{bmatrix} = E^1(x) + D^1(x)u$$

where $\{\rho_1^1, \cdots, \rho_m^1\}$ is the relative degree of system (9.1). Assume that

$$\text{rank}\left(D^1(x)\right) = r_1.$$

If $r_1 = m$, then the local input-output decoupling is possible by static feedback, and thus, the algorithm terminates. If $r_1 < m$, I-O decoupling by static feedback is not possible. The subspace of the input to apply the integrators should be obtained, as in Examples 9.2.2 and 9.2.3. First, we can assume, by changing the order of the outputs, that the first r_1 rows of the matrix $D^1(x)$ are linearly independent. That is, we have, without loss of generality, that

$$\begin{bmatrix} y_1^{(\rho_1^1)} \\ \vdots \\ y_m^{(\rho_m^1)} \end{bmatrix} = E^1(x) + \begin{bmatrix} \tilde{D}^1(x) \\ \bar{D}^1(x) \end{bmatrix} u$$

where $\text{rank}\left(\tilde{D}^1(x)\right) = r_1$. We can find, by elementary column operations and the output order change, $m \times m$ matrix $L^1(x)$ such that

$$\begin{bmatrix} \tilde{D}^1(x) \\ \bar{D}^1(x) \end{bmatrix} L^1(x) = \begin{bmatrix} I_{\bar{r}_1} & O_{\bar{r}_1 \times (r_1 - \bar{r}_1)} & O_{\bar{r}_1 \times (m - r_1)} \\ O_{(r_1 - \bar{r}_1) \times \bar{r}_1} & I_{r_1 - \bar{r}_1} & O_{(r_1 - \bar{r}_1) \times (m - r_1)} \\ \hat{D}^1(x) & O_{(m - r_1) \times (r_1 - \bar{r}_1)} & O_{(m - r_1) \times (m - r_1)} \end{bmatrix}$$

and for $1 \le j \le m - r_1$ and $1 \le i \le \bar{r}_1$,

$$\hat{D}_j^1(x) \ne O_{1 \times \bar{r}_1} \quad \text{and} \quad \hat{D}^{1i}(x) \ne O_{(m - r_1) \times 1} \tag{9.25}$$

where $\hat{D}^1(x) \triangleq \begin{bmatrix} \hat{D}_1^1(x) \\ \vdots \\ \hat{D}_{m-r_1}^1(x) \end{bmatrix} \triangleq \left[\hat{D}^{11}(x) \cdots \hat{D}^{1\bar{r}_1}(x)\right]$. Thus, we have, with the change of the output's order, that

$$
\begin{bmatrix} y_1^{(\rho_1^1)} \\ \vdots \\ y_m^{(\rho_m^1)} \end{bmatrix} = \hat{E}^1(x) + \begin{bmatrix} I_{\bar{r}_1} & O_{\bar{r}_1 \times (r_1 - \bar{r}_1)} & O_{\bar{r}_1 \times (m - r_1)} \\ O_{(r_1 - \bar{r}_1) \times \bar{r}_1} & I_{r_1 - \bar{r}_1} & O_{(r_1 - \bar{r}_1) \times (m - r_1)} \\ \hat{D}^1(x) & O_{(m - r_1) \times (r_1 - \bar{r}_1)} & O_{(m - r_1) \times (m - r_1)} \end{bmatrix} \mu^1 \qquad (9.26)
$$

with static feedback

$$
u = L^1(x)\mu^1 \triangleq \begin{bmatrix} \tilde{L}^1(x) & \bar{L}^1(x) \end{bmatrix} \begin{bmatrix} \tilde{\mu}^1 \\ \bar{\mu}^1 \end{bmatrix} \qquad (9.27)
$$

where $\tilde{\mu}^1 \in \mathbb{R}^{\bar{r}_1}$, $\bar{\mu}^1 \in \mathbb{R}^{m - \bar{r}_1}$, and $\tilde{L}^1(x)$ and $\bar{L}^1(x)$ are $m \times \bar{r}_1$ and $m \times (m - \bar{r}_1)$ matrices, respectively. Then, with dynamic feedback

$$
u = L^1(x) \begin{bmatrix} z^1 \\ \bar{u}^1 \end{bmatrix} = \tilde{L}^1(x)z^1 + \hat{L}^1(x)u^1 \qquad (9.28a)
$$

$$
\dot{z}^1 = \tilde{u}^1 = \bar{I}_{\bar{r}_1} u^1 \qquad (9.28b)
$$

we have the following extended system:

$$
\Sigma_1 : \quad \begin{bmatrix} \dot{x} \\ \dot{z}^1 \end{bmatrix} = F_E^1(x, z^1) + G_E^1(x, z^1) \begin{bmatrix} \tilde{u}^1 \\ \bar{u}^1 \end{bmatrix} \qquad (9.29)
$$

where $x_E^1 \triangleq [x^{\mathsf{T}} \ (z^1)^{\mathsf{T}}]^{\mathsf{T}}$, $u^1 \triangleq \begin{bmatrix} \tilde{u}^1 \\ \bar{u}^1 \end{bmatrix}$, $\hat{L}^1(x) \triangleq \begin{bmatrix} O & \bar{L}^1(x) \end{bmatrix}$, $\bar{I}_{\bar{r}_1} \triangleq \begin{bmatrix} I_{\bar{r}_1} & O \end{bmatrix}$, and

$$
F_E^1(x_E^1) = \begin{bmatrix} f(x) + g(x)\tilde{L}^1(x)z^1 \\ O \end{bmatrix}; \quad G_E^1(x_E^1) = \begin{bmatrix} O & g(x)\bar{L}^1(x) \\ I_{\bar{r}_1} & O \end{bmatrix}.
$$

step 2: For system (9.29), we have

$$
\begin{bmatrix} y_1^{(\rho_1^2)} \\ \vdots \\ y_m^{(\rho_m^2)} \end{bmatrix} = E^2(x, z^1) + D^2(x, z^1)u^1
$$

where $\{\rho_1^2, \cdots, \rho_m^2\}$ is the relative degree of system (9.29). Assume that

$$
\text{rank}\left(D^2(x, z^1)\right) = r_2.
$$

If $r_2 = m$, then system (9.29) is input-output decouplable by static feedback, and the algorithm terminates. Since $\rho_i^2 = \rho_i^1 + 1$, $1 \le i \le \bar{r}_1$, and $\rho_i^2 = \rho_i^1$, $\bar{r}_1 + 1 \le i \le r_1$, it is clear that $\frac{\partial}{\partial \bar{u}^1}\left(L_{F_E^1 + G_E^1 u^1} \hat{E}_i^1(x)\right) = O_{1 \times \bar{r}_1}$, $1 \le i \le \bar{r}_1$, and thus $r_2 \ge r_1$. If $r_2 < m$, then we can assume, by changing the order of the outputs $y_{r_1 + 1}, \cdots, y_m$, that the first r_2 rows of the matrix $D^2(x)$ are linearly independent. In other words, we have, without loss of generality, that

$$
\begin{bmatrix} y_1^{(\rho_1^2)} \\ \vdots \\ y_m^{(\rho_m^2)} \end{bmatrix} = E^2(x, z^1), + \begin{bmatrix} \tilde{D}^2(x, z^1) \\ \bar{D}^2(x, z^1) \end{bmatrix} u^1
$$

where rank $\left(\tilde{D}^2(x, z^1) \right) = r_2$. We can find, by elementary column operations and the output order change, $m \times m$ matrix $L^2(x, z^1)$ such that

$$
\begin{bmatrix} \tilde{D}^2(x_E^1) \\ \bar{D}^2(x_E^1) \end{bmatrix} L^2(x, z^1) = \begin{bmatrix} I_{\bar{r}_2} & O_{\bar{r}_2 \times (r_2 - \bar{r}_2)} & O_{\bar{r}_2 \times (m - r_2)} \\ O_{(r_2 - \bar{r}_2) \times \bar{r}_2} & I_{r_2 - \bar{r}_2} & O_{(r_2 - \bar{r}_2) \times (m - r_2)} \\ \hat{D}^2(x) & O_{(m - r_2) \times (r_2 - \bar{r}_2)} & O_{(m - r_2) \times (m - r_2)} \end{bmatrix}
$$

and for $1 \le i \le \bar{r}_2$,

$$
\hat{D}^{2i}(x) \ne O_{(m - r_2) \times 1}
$$

where $\hat{D}^2(x) \triangleq \left[\hat{D}^{21}(x) \cdots \hat{D}^{2\bar{r}_2}(x) \right]$. Thus, we have, with the change of the output's order, that

$$
\begin{bmatrix} y_1^{(\rho_1^2)} \\ \vdots \\ y_m^{(\rho_m^2)} \end{bmatrix} = \begin{bmatrix} I_{\bar{r}_2} & O_{\bar{r}_2 \times (r_2 - \bar{r}_2)} & O_{\bar{r}_2 \times (m - r_2)} \\ O_{(r_2 - \bar{r}_2) \times \bar{r}_2} & I_{r_2 - \bar{r}_2} & O_{(r_2 - \bar{r}_2) \times (m - r_2)} \\ \hat{D}^2(x) & O_{(m - r_2) \times (r_2 - \bar{r}_2)} & O_{(m - r_2) \times (m - r_2)} \end{bmatrix} \mu^2
$$
$$
+ \hat{E}^2(x, z^1)
$$

with static feedback

$$
u^1 = L^2(x, z^1) \mu^2 \triangleq \left[\tilde{L}^2(x, z^1) \; \bar{L}^2(x, z^1) \right] \begin{bmatrix} \tilde{\mu}^2 \\ \bar{\mu}^2 \end{bmatrix}
$$

where $\tilde{\mu}^2 \in \mathbb{R}^{\bar{r}_2}$, $\bar{\mu}^2 \in \mathbb{R}^{m - \bar{r}_2}$, and $\tilde{L}^2(x, z^1)$ and $\bar{L}^2(x, z^1)$ are $m \times \bar{r}_2$ and $m \times (m - \bar{r}_2)$ matrices, respectively. Then, with dynamic feedback

$$
u^1 = L^2(x, z^1) \begin{bmatrix} z^2 \\ \bar{u}^2 \end{bmatrix} = \tilde{L}^2(x, z^1) z^2 + \hat{L}^2(x, z^1) u^2 \tag{9.30a}
$$
$$
\dot{z}^2 = \tilde{u}^2 = I_{\bar{r}_2} u^2, \tag{9.30b}
$$

we have the following extended system:

$$\Sigma_2 : \begin{bmatrix} \dot{x} \\ \dot{z}^1 \\ \dot{z}^2 \end{bmatrix} = \dot{x}_E^2 = F_E^2(x, z^1, z^2) + G_E^2(x, z^1, z^2) \begin{bmatrix} \tilde{u}^2 \\ \bar{u}^2 \end{bmatrix} \qquad (9.31)$$

where $x_E^2 \triangleq \begin{bmatrix} x \\ z^1 \\ z^2 \end{bmatrix}$, $u^2 \triangleq \begin{bmatrix} \tilde{u}^2 \\ \bar{u}^2 \end{bmatrix}$, $\hat{L}^2(x, z^1) \triangleq \begin{bmatrix} O & \bar{L}^2(x, z^1) \end{bmatrix}$, $\bar{I}_{\bar{r}_2} \triangleq \begin{bmatrix} I_{\bar{r}_2} & O \end{bmatrix}$, and

$$F_E^2(x_E^2) = \begin{bmatrix} F_E^1(x_E^1) + G_E^1(x_E^1)\hat{L}^2(x_E^1)z^2 \\ O \end{bmatrix}; \quad G_E^2(x_E^2) = \begin{bmatrix} O & G_E^1(x_E^1)\bar{L}^2(x_E^1) \\ I_{\bar{r}_2} & O \end{bmatrix}.$$

step $k (\geq 2)$: At step $(k - 1)$, we obtained the extended closed-loop system

$$\Sigma_{k-1} : \dot{x}_E^{k-1} = F_E^{k-1}(x, z^1, \cdots, z^{k-1}) + G_E^{k-1}(x, z^1, \cdots, z^{k-1})u^{k-1} \qquad (9.32)$$

where $x_E^{k-1} \triangleq \begin{bmatrix} x \\ z^1 \\ \vdots \\ z^{k-1} \end{bmatrix}$ and $u^{k-1} \triangleq \begin{bmatrix} \tilde{u}^{k-1} \\ \bar{u}^{k-1} \end{bmatrix}$. For system (9.32), we have

$$\begin{bmatrix} y_1^{(\rho_1^k)} \\ \vdots \\ y_m^{(\rho_m^k)} \end{bmatrix} = E^k(x_E^{k-1}) + D^k(x_E^{k-1})u^{k-1}$$

where $\{\rho_1^k, \cdots, \rho_m^k\}$ is the relative degree of system (9.32). Assume that

$$\text{rank}\left(D^k(x_E^{k-1})\right) = r_k.$$

If $r_k = m$, then system (9.32) is locally input-output decouplable by static feedback, and the algorithm terminates. If $r_k < m$, then we can assume, by changing the order of the outputs $y_{r_{k-1}+1}, \cdots, y_m$, that the first r_k rows of the matrix $D^k(x)$ are linearly independent. In other words, we have, without loss of generality, that

$$\begin{bmatrix} y_1^{(\rho_1^k)} \\ \vdots \\ y_m^{(\rho_m^k)} \end{bmatrix} = E^k(x_E^{k-1}) + \begin{bmatrix} \tilde{D}^k(x_E^{k-1}) \\ \bar{D}^k(x_E^{k-1}) \end{bmatrix} u^{k-1}$$

where $\text{rank}\left(\tilde{D}^k(x_E^{k-1})\right) = r_k$. We can find, by elementary column operations and the output order change, $m \times m$ matrix $L^k(x_E^{k-1})$ such that

$$\begin{bmatrix} \tilde{D}^k(x_E^{k-1}) \\ \bar{D}^k(x_E^{k-1}) \end{bmatrix} L^k(x_E^{k-1})$$

$$= \begin{bmatrix} I_{\bar{r}_k} & O_{\bar{r}_k \times (r_k - \bar{r}_k)} & O_{\bar{r}_k \times (m - r_k)} \\ O_{(r_k - \bar{r}_k) \times \bar{r}_k} & I_{r_k - \bar{r}_k} & O_{(r_k - \bar{r}_k) \times (m - r_k)} \\ \hat{D}^k(x_E^{k-1}) & O_{(m - r_k) \times (r_k - \bar{r}_k)} & O_{(m - r_k) \times (m - r_k)} \end{bmatrix}$$

and for $1 \le i \le \bar{r}_k$,

$$\hat{D}^{ki}(x_E^{k-1}) \neq O_{(m-r_k) \times 1}$$

where $\hat{D}^k(x_E^{k-1}) \triangleq [\hat{D}^{k1}(x_E^{k-1}) \cdots \hat{D}^{k\bar{r}_k}(x_E^{k-1})]$. Thus, we have, with the change of the output's order, that

$$\begin{bmatrix} y_1^{(\rho_1^k)} \\ \vdots \\ y_m^{(\rho_m^k)} \end{bmatrix} = \begin{bmatrix} I_{\bar{r}_k} & O_{\bar{r}_k \times (r_k - \bar{r}_k)} & O_{\bar{r}_k \times (m - r_k)} \\ O_{(r_k - \bar{r}_k) \times \bar{r}_k} & I_{r_k - \bar{r}_k} & O_{(r_k - \bar{r}_k) \times (m - r_k)} \\ \hat{D}^k(x_E^{k-1}) & O_{(m - r_k) \times (r_k - \bar{r}_k)} & O_{(m - r_k) \times (m - r_k)} \end{bmatrix} \mu^k$$
$$+ \hat{E}^k(x_E^{k-1})$$

with static feedback

$$u^{k-1} = L^k(x_E^{k-1})\mu^k \triangleq [\tilde{L}^k(x_E^{k-1}) \ \bar{L}^k(x_E^{k-1})] \begin{bmatrix} \tilde{\mu}^k \\ \bar{\mu}^k \end{bmatrix}$$

where $\tilde{\mu}^k \in \mathbb{R}^{\bar{r}_k}$, $\bar{\mu}^k \in \mathbb{R}^{m - \bar{r}_k}$, and $\tilde{L}^k(x_E^{k-1})$ and $\bar{L}^k(x_E^{k-1})$ are $m \times \bar{r}_k$ and $m \times (m - \bar{r}_k)$ matrices, respectively. Then, with dynamic feedback

$$u^{k-1} = L^k(x_E^{k-1}) \begin{bmatrix} z^k \\ \bar{u}^k \end{bmatrix} = \tilde{L}^k(x_E^{k-1})z^k + \hat{L}^k(x_E^{k-1})u^k \tag{9.33a}$$

$$\dot{z}^k = \tilde{u}^k = \bar{I}_{\bar{r}_k}u^k, \tag{9.33b}$$

we have the following extended system:

$$\begin{bmatrix} \dot{x}_E^{k-1} \\ \dot{z}^k \end{bmatrix} = \dot{x}_E^k = F_E^k(x_E^k) + G_E^k(x_E^k)u^k \tag{9.34}$$

where $x_E^k \triangleq [x^T \ (z^1)^T \ \cdots \ (z^{k-1})^T \ (z^k)^T]^T = \begin{bmatrix} x_E^{k-1} \\ z^k \end{bmatrix}$, $u^k \triangleq \begin{bmatrix} \tilde{u}^k \\ \bar{u}^k \end{bmatrix}$, $\hat{L}^k(x_E^{k-1}) \triangleq$

$[O \ \bar{L}^k(x_E^{k-1})]$, $\bar{I}_{\bar{r}_k} \triangleq [I_{\bar{r}_k} \ O]$,

$$F_E^k(x_E^k) = \begin{bmatrix} F_E^{k-1}(x_E^{k-1}) + G_E^{k-1}(x_E^{k-1})\tilde{L}^k(x_E^{k-1})z^k \\ O \end{bmatrix},$$

and

$$G_E^k(x_E^k) = \begin{bmatrix} O & G_E^{k-1}(x_E^{k-1})\bar{L}^k(x_E^{k-1}) \\ I_{r_k} & O \end{bmatrix}.$$

If the algorithm does not terminate at a finite step, then let step K be the final step such that $r_1 \leq r_2 \leq \cdots r_{K-1} < r_K = r_{K+1} = r_{K+2} = \cdots$.

The dynamic extension algorithms, which are a little different from the above algorithm, can also be found in [A3, A5] and [G17].

Lemma 9.3 *If system (9.1) is locally dynamic input-output decouplable, then the extended system (9.29) (or Σ_1) is also locally dynamic input-output decouplable.*

Proof Suppose that system (9.1) is locally dynamic input-output decouplable. With static feedback

$$u = L^1(x)\mu^1 \triangleq \begin{bmatrix} \tilde{L}^1(x) & \bar{L}^1(x) \end{bmatrix} \begin{bmatrix} \tilde{\mu}^1 \\ \bar{\mu}^1 \end{bmatrix}$$

in (9.27), we have the following system:

$$\dot{x} = f(x) + \tilde{g}(x)\tilde{\mu}^1 + \bar{g}(x)\bar{\mu}^1 \tag{9.35}$$

where

$$\begin{bmatrix} \tilde{g}(x) & \bar{g}(x) \end{bmatrix} \triangleq g(x)\begin{bmatrix} \tilde{L}^1(x) & \bar{L}^1(x) \end{bmatrix}.$$

Then it is clear that system (9.35) is also locally dynamic input-output decouplable. Thus, there exists a dynamic feedback

$$\begin{bmatrix} \tilde{\mu}^1 \\ \bar{\mu}^1 \end{bmatrix} = \begin{bmatrix} \tilde{a}(x,z) \\ \bar{a}(x,z) \end{bmatrix} + \begin{bmatrix} \tilde{b}(x,z) \\ \bar{b}(x,z) \end{bmatrix} v$$

$$\dot{z} = c(x,z) + d(x,z)v, \quad z \in \mathbb{R}^d \tag{9.36}$$

such that

$$\begin{bmatrix} L_{g_E}L_{f_E}^{\rho_1^E - 1}h_1(x) \\ \vdots \\ L_{g_E}L_{f_E}^{\rho_m^E - 1}h_m(x) \end{bmatrix} = I_m \tag{9.37}$$

where $(\rho_1^E, \cdots, \rho_m^E)$ is the relative degrees of the extended system

$$\begin{bmatrix} \dot{x} \\ \dot{z} \end{bmatrix} = \begin{bmatrix} f(x) + \tilde{g}(x)\tilde{a}(x, z) + \bar{g}(x)\bar{a}(x, z) \\ c(x, z) \end{bmatrix} + \begin{bmatrix} \tilde{g}(x)\tilde{b}(x, z) + \bar{g}(x)\bar{b}(x, z) \\ d(x, z) \end{bmatrix} v$$

$$= f_E(x, z) + g_E(x, z)v.$$

$$(9.38)$$

Assume that there exists i $(1 \le i \le \bar{r}_1)$ such that $\tilde{b}_i(x, z) \ne O_{1 \times m}$. Then it is clear, by (9.26) and (9.37), that $\rho_i^E = \rho_i^1$ and

$$L_{g_E} L_{f_E}^{\rho_i^1 - 1} h_i(x) = L_{\tilde{g}\tilde{b} + \bar{g}\bar{b}} L_f^{\rho_i^1 - 1} h_i(x) = L_{\tilde{g}} L_f^{\rho_i^1 - 1} h_i(x)\tilde{b}(x, z)$$

$$= \bar{e}_i \tilde{b}(x, z) = \tilde{b}_i(x, z) = e_i$$

where $\bar{e}_i$ and e_i are the i-th row of the identity matrix $I_{\bar{r}_1}$ and I_m, respectively. Since $\hat{D}^{1i}(x) \ne O_{(m-r_1) \times 1}$ by (9.25), there exists j $(r_1 + 1 \le j \le m)$ such that j-th component of $\hat{D}^{1i}(x)$ is not zero. Let $\hat{D}_j^1(x) = [\hat{d}_1(x) \cdots \hat{d}_{\bar{r}_1}(x)]$ and $\hat{d}_i(x) \ne 0$. Then it is clear, by (9.26) and (9.37), that $\rho_{j+r_1}^E = \rho_{j+r_1}^1$ and

$$L_{g_E} L_{f_E}^{\rho_{j+r_1}^1 - 1} h_{j+r_1}(x) = L_{\tilde{g}\tilde{b} + \bar{g}\bar{b}} L_f^{\rho_{j+r_1}^1 - 1} h_{j+r_1}(x) = L_{\tilde{g}} L_f^{\rho_{j+r_1}^1 - 1} h_{j+r_1}(x)\tilde{b}(x, z)$$

$$= [\hat{d}_1(x) \cdots \hat{d}_{\bar{r}_1}(x)] \tilde{b}(x, z) = \sum_{k=1}^{\bar{r}_1} \hat{d}_k(x)\tilde{b}_k(x, z) = e_{j+r_1}$$

which implies that $\tilde{b}_i(x, z) = O_{1 \times m}$. ($\tilde{b}_k(x, z) = O_{1 \times m}$ or $\tilde{b}_k(x, z) = e_k$ for $1 \le k \le \bar{r}_1$ and $\hat{d}_i(x) \ne 0$.) It contradicts. Thus, there does not exist i $(1 \le i \le \bar{r}_1)$ such that $\tilde{b}_i(x, z) \ne O_{1 \times m}$. In other words,

$$\tilde{b}(x, z) = 0 \quad \text{or} \quad \tilde{\mu}^1 = \tilde{a}(x, z).$$

$$(9.39)$$

We will show that

$$\text{rank} \left(\left. \frac{\partial \tilde{a}(x, z)}{\partial z} \right|_{z=0} \right) = \bar{r}_1.$$

$$(9.40)$$

Suppose that rank $\left(\left. \frac{\partial \tilde{a}(x,z)}{\partial z} \right|_{z=0} \right) = \hat{r}_1 < \bar{r}_1$. Then, we can find, by elementary row operations, a $\bar{r}_1 \times \bar{r}_1$ matrix $R(x) = \begin{bmatrix} R^1(x) \\ R^2(x) \end{bmatrix}$ such that

$$\begin{bmatrix} R^1(x) \\ R^2(x) \end{bmatrix} \left. \frac{\partial \tilde{a}(x, z)}{\partial z} \right|_{z=0} \triangleq \begin{bmatrix} \left. \frac{\partial \tilde{a}^1(x,z)}{\partial z} \right|_{z=0} \\ \left. \frac{\partial \tilde{a}^2(x,z)}{\partial z} \right|_{z=0} \end{bmatrix} = \begin{bmatrix} O_{(\bar{r}_1 - \hat{r}_1) \times d} \\ \left. \frac{\partial \tilde{a}^2(x,z)}{\partial z} \right|_{z=0} \end{bmatrix}$$

$$(9.41)$$

where $\begin{bmatrix} R^1(x) \\ R^2(x) \end{bmatrix} \tilde{a}(x,z) \triangleq \begin{bmatrix} \tilde{a}^1(x,z) \\ \tilde{a}^2(x,z) \end{bmatrix}$. Let $R(x)\tilde{\mu}^1 \triangleq \begin{bmatrix} \tilde{\mu}^{11} \\ \tilde{\mu}^{12} \end{bmatrix}$ and $\tilde{g}(x)R(x)^{-1} \triangleq$ $\begin{bmatrix} \tilde{g}^1(x) & \tilde{g}^2(x) \end{bmatrix}$, where $\tilde{\mu}^{11} \in \mathbb{R}^{\bar{r}_1 - \hat{r}_1}$ and $\tilde{\mu}^{12} \in \mathbb{R}^{\hat{r}_1}$. Then, it is clear, by (9.39), that the system

$$
\begin{aligned}
\dot{x} &= f(x) + \tilde{g}^1(x)\tilde{\mu}^{11} + \tilde{g}^2(x)\tilde{\mu}^{12} + \bar{g}(x)\bar{\mu}^1 \\
&= f(x) + \tilde{g}^1(x)\tilde{\mu}^{11} + \hat{g}(x)\hat{\mu}
\end{aligned}
\tag{9.42}
$$

is also dynamic input-output decouplable with dynamic feedback

$$
\begin{bmatrix} \tilde{\mu}^{11} \\ \tilde{\mu}^{12} \\ \bar{\mu}^1 \end{bmatrix} = \begin{bmatrix} \tilde{a}^1(x,z) \\ \tilde{a}^2(x,z) \\ \bar{a}(x,z) \end{bmatrix} + \begin{bmatrix} O \\ O \\ \bar{b}(x,z) \end{bmatrix} v
$$
$$
\dot{z} = c(x,z) + d(x,z)v
$$

where $\hat{\mu} \triangleq \begin{bmatrix} \tilde{\mu}^{12} \\ \bar{\mu}^1 \end{bmatrix} \in \mathbb{R}^{m-(\bar{r}_1-\hat{r}_1)}$ and $\hat{g}(x) \triangleq \begin{bmatrix} \tilde{g}^2(x) & \bar{g}(x) \end{bmatrix}$. Therefore, it is clear, by Lemma 9.1 and (9.41), that system (9.42) is also dynamic input-output decouplable with dynamic feedback

$$
\begin{bmatrix} \tilde{\mu}^{11} \\ \tilde{\mu}^{12} \\ \bar{\mu}^1 \end{bmatrix} = \begin{bmatrix} \tilde{a}^1(x,0) \\ \tilde{a}^2(x,0) + \frac{\partial \tilde{a}^2(x,z)}{\partial z}\Big|_{z=0} z \\ \bar{a}(x,0) + \frac{\partial \bar{a}(x,z)}{\partial z}\Big|_{z=0} z \end{bmatrix} + \begin{bmatrix} O \\ O \\ \bar{b}(x,0) \end{bmatrix} v
$$
$$
\dot{z} = c(x,0) + \frac{\partial c(x,z)}{\partial z}\Big|_{z=0} z + d(x,0)v.
$$

In other words, the system

$$
\begin{aligned}
\dot{x} &= f(x) + \tilde{g}^1(x)\tilde{a}^1(x,0) + \hat{g}(x)\hat{\mu} \\
&\triangleq \bar{f}(x) + \hat{g}(x)\hat{\mu}
\end{aligned}
$$

is dynamic input-output decouplable with dynamic feedback

$$
\hat{\mu} = \begin{bmatrix} \tilde{\mu}^{12} \\ \bar{\mu}^1 \end{bmatrix} = \begin{bmatrix} \tilde{a}^2(x,0) + \frac{\partial \tilde{a}^2(x,z)}{\partial z}\Big|_{z=0} z \\ \bar{a}(x,0) + \frac{\partial \bar{a}(x,z)}{\partial z}\Big|_{z=0} z \end{bmatrix} + \begin{bmatrix} O \\ O \\ \bar{b}(x,0) \end{bmatrix} v
$$
$$
\dot{z} = c(x,0) + \frac{\partial c(x,z)}{\partial z}\Big|_{z=0} z + d(x,0)v.
$$

It is possible, only if $m - (\bar{r}_1 - \hat{r}_1) = m$ or $\hat{r}_1 = \bar{r}_1$. Therefore, (9.40) is satisfied and there exists a $(d - \bar{r}_1) \times d$ constant matrix $\bar{S}^0$ such that

$$\text{rank}\left(\begin{bmatrix} \frac{\partial \tilde{a}^1(x,z)}{\partial z}\Big|_{z=0} \\ \bar{S}^0 \end{bmatrix}\right) = d$$

and $\begin{bmatrix} x \\ z^1 \\ \bar{z} \end{bmatrix} = S_E(x,z) \triangleq \begin{bmatrix} x \\ \tilde{a}^1(x,z) \\ \bar{S}^0 z \end{bmatrix}$ or $\begin{bmatrix} x \\ z \end{bmatrix} = S_E^{-1}(x, z^1, \bar{z}) \triangleq \begin{bmatrix} x \\ S_z^{-1}(x, z^1, \bar{z}) \end{bmatrix}$ is

an extended state transformation. The extended system (9.38) satisfies, in $(x, z^1, \bar{z})$-coordinates, the following system:

$$\begin{bmatrix} \dot{x} \\ \dot{z}^1 \\ \dot{\bar{z}} \end{bmatrix} = \begin{bmatrix} f(x) + \tilde{g}^1(x)z^1 + \bar{g}(x)\bar{a}^1(x, z^1, \bar{z}) \\ \bar{c}^{11}(x, z^1, \bar{z}) \\ \bar{c}^{12}(x, z^1, \bar{z}) \end{bmatrix} + \begin{bmatrix} \bar{g}(x)\bar{b}^1(x, z^1, \bar{z}) \\ \bar{d}^{11}(x, z^1, \bar{z}) \\ \bar{d}^{12}(x, z^1, \bar{z}) \end{bmatrix} v$$

$$= (S_E)_*(f_E(x,z)) + (S_E)_*(g_E(x,z))v$$

$$y = h \circ S_E^{-1}(x, z^1, \bar{z}) = h(x)$$

where $\bar{a}^1(x, z^1, \bar{z}) \triangleq \bar{a}(x, S_z^{-1}(x, z^1, \bar{z})), \bar{b}^1(x, z^1, \bar{z}) \triangleq \bar{b}(x, S_z^{-1}(x, z^1, \bar{z}))$, and

$$\begin{bmatrix} O_{n\times 1} \\ \bar{c}^{11}(x, z^1, \bar{z}) \\ \bar{c}^{12}(x, z^1, \bar{z}) \end{bmatrix} \triangleq (S_E)_*\left(\begin{bmatrix} O_{n\times 1} \\ c(x,z) \end{bmatrix}\right); \quad \begin{bmatrix} O_{n\times 1} \\ \bar{d}^{11}(x, z^1, \bar{z}) \\ \bar{d}^{12}(x, z^1, \bar{z}) \end{bmatrix} \triangleq (S_E)_*\left(\begin{bmatrix} O_{n\times 1} \\ d(x,z) \end{bmatrix}\right).$$

In other words, system (9.29) (or Σ_1) is locally input-output decouplable with dynamic feedback

$$\begin{bmatrix} \tilde{u}^1 \\ \bar{u}^1 \end{bmatrix} = \begin{bmatrix} \bar{c}^{11}(x, z^1, \bar{z}) \\ \bar{a}^1(x, z^1, \bar{z}) \end{bmatrix} + \begin{bmatrix} \bar{d}^{11}(x, z^1, \bar{z}) \\ \bar{b}^1(x, z^1, \bar{z}) \end{bmatrix} v$$

$$\dot{\bar{z}} = \bar{c}^{12}(x, z^1, \bar{z}) + \bar{d}^{12}(x, z^1, \bar{z})w, \quad \bar{z} \in \mathbb{R}^{d-\bar{r}_1}.$$

$\square$

Now, using the dynamic extension algorithm, the necessary and sufficient conditions of the dynamic input-output decoupling problem can be found as follows.

Theorem 9.2 (*conditions for the dynamic IO decoupling problem*) *System (9.1) is locally dynamic input-output decouplable, if and only if the Dynamic Extension Algorithm terminates in a finite step or*

$$\text{rank}\left(D^K(x_E^{K-1})\right) \triangleq r_K = m \tag{9.43}$$

where K is the final step of the Dynamic Extension Algorithm for system (9.1).

Proof Necessity. By the Dynamic Extension Algorithm for system (9.1), we have the final extended system

$$\Sigma_{K-1}: \quad \dot{x}_E^{K-1} = F_E^{K-1}(x_E^{K-1}) + G_E^{K-1}(x_E^{K-1})u^{K-1}$$
$$= F_E^{K-1}(x_E^{K-1}) + G_{E,1}^{K-1}(x_E^{K-1})\tilde{u}^{K-1} + G_{E,2}^{K-1}(x_E^{K-1})\bar{u}^{K-1}$$

where

$$x_E^{K-1} \triangleq \begin{bmatrix} x \\ z^1 \\ \vdots \\ z^{K-1} \end{bmatrix} \quad \text{and} \quad u^{K-1} \triangleq \begin{bmatrix} \tilde{u}^{K-1} \\ \bar{u}^{K-1} \end{bmatrix}.$$

Suppose that system (9.1) is locally dynamic input-output decouplable. Then, it is clear, by Lemma 9.3, that the extended system (9.29) is also locally dynamic input-output decouplable. By repeated use of Lemma 9.3, it is easy to see that the final extended system Σ_{K-1} is locally dynamic input-output decouplable. If the Dynamic Extension Algorithm terminates in a finite step, then it is clear that (9.43) is satisfied. Suppose that the Dynamic Extension Algorithm does not terminate in a finite step. Then we have that for $i \geq 0$,

$$L_{G_{E,2}^{K-1}} L_{F_E^{K-1}}^i h(x) = 0$$

which implies that the final extended system Σ_{K-1} is not locally dynamic input-output decouplable and thus system (9.1) is not locally dynamic input-output decouplable.

Sufficiency. Suppose that the Dynamic Extension Algorithm terminates in a finite step K with (9.43). Then it is easy to see, by (9.28), (9.30), and (9.33), that system Σ_{K-1} is the extended system of system (9.1) with the dynamic feedback

$$u = \sum_{i=1}^{K-1} \left(\prod_{j=1}^{i-1} \hat{L}^j(x_E^{j-1}) \right) \tilde{L}^i(x_E^{i-1})z^i + \left(\prod_{j=1}^{K-1} \hat{L}^j(x_E^{j-1}) \right) u^{K-1}$$

and

$$\begin{bmatrix} \dot{z}^1 \\ \vdots \\ \dot{z}^{K-2} \\ \dot{z}^{K-1} \end{bmatrix} = \begin{bmatrix} \bar{I}_{\bar{r}_1} \left\{ \tilde{L}^2 z^2 + \cdots + \hat{L}^2 \cdots \hat{L}^{K-2} \tilde{L}^{K-1} z^{K-1} + \left(\prod_{j=2}^{K-1} \hat{L}^j \right) u^{K-1} \right\} \\ \vdots \\ \bar{I}_{\bar{r}_{K-2}} \left\{ \tilde{L}^{K-1} z^{K-1} + \hat{L}^{K-1} u^{K-1} \right\} \\ \bar{I}_{\bar{r}_{K-1}} u^{K-1} \end{bmatrix}.$$

Since the decoupling matrix of extended system Σ_{K-1} is nonsingular, extended system Σ_{K-1} is locally static input-output decouplable. Hence, system (9.1) is locally dynamic input-output decouplable.   □

Example 9.3.1 Show that the following nonlinear system is locally dynamic input-output decouplable:

$$
\dot{x} = \begin{bmatrix} x_2 \\ x_3 \\ x_4 \\ 0 \\ x_6 \\ x_7 \\ 0 \end{bmatrix} + \begin{bmatrix} 0 \\ 1 \\ x_4 \\ 0 \\ 1 \\ 0 \\ 0 \end{bmatrix} u_1 + \begin{bmatrix} 0 \\ 1 \\ x_4 \\ 1 \\ 1 \\ 0 \\ 0 \end{bmatrix} u_2 + \begin{bmatrix} 0 \\ 0 \\ 0 \\ 0 \\ 0 \\ 0 \\ 1 \end{bmatrix} u_3
$$

$$
\begin{bmatrix} y_1 \\ y_2 \\ y_3 \end{bmatrix} = \begin{bmatrix} x_1 \\ x_3 \\ x_5 \end{bmatrix} = h(x).
$$

(9.44)

Solution step 1: By simple calculations, we have relative degree $(\rho_1^1, \rho_2^1, \rho_3^1) = (2, 1, 1)$ and

$$
\begin{bmatrix} y_1^{(2)} \\ y_2^{(1)} \\ y_3^{(1)} \end{bmatrix} = \begin{bmatrix} x_3 \\ x_4 \\ x_6 \end{bmatrix} + \begin{bmatrix} 1 & 1 & 0 \\ x_4 & x_4 & 0 \\ 1 & 1 & 0 \end{bmatrix} \begin{bmatrix} u_1 \\ u_2 \\ u_3 \end{bmatrix} = E^1(x) + D^1(x)u
$$

where $r_1 \triangleq \operatorname{rank}\left(D^1(x)\right) = 1$. It is easy, by elementary column operations, to find $m \times m$ matrix $L^1(x)$ such that

$$
D^1(x)L^1(x) = \begin{bmatrix} 1 & 1 & 0 \\ x_4 & x_4 & 0 \\ 1 & 1 & 0 \end{bmatrix} \begin{bmatrix} 1 & -1 & 0 \\ 0 & 1 & 0 \\ 0 & 0 & 1 \end{bmatrix} = \begin{bmatrix} 1 & 0 & 0 \\ x_4 & 0 & 0 \\ 1 & 0 & 0 \end{bmatrix}.
$$

Therefore, with

$$
u = L^1(x) \begin{bmatrix} z_1^1 \\ u_2^1 \\ u_3^1 \end{bmatrix} = \begin{bmatrix} 1 & -1 & 0 \\ 0 & 1 & 0 \\ 0 & 0 & 1 \end{bmatrix} \begin{bmatrix} z_1^1 \\ u_2^1 \\ u_3^1 \end{bmatrix}
$$

(9.45)

$$
\dot{z}_1^1 = u_1^1,
$$

we have the following extended system:

$$\begin{bmatrix} \dot{x}_1 \\ \dot{x}_2 \\ \dot{x}_3 \\ \dot{x}_4 \\ \dot{x}_5 \\ \dot{x}_6 \\ \dot{x}_7 \\ \dot{z}_1^1 \end{bmatrix} = \begin{bmatrix} x_2 \\ x_3 + z_1^1 \\ x_4(1 + z_1^1) \\ 0 \\ x_6 + z_1^1 \\ x_7 \\ 0 \\ 0 \end{bmatrix} + \begin{bmatrix} 0 & 0 & 0 \\ 0 & 0 & 0 \\ 0 & 0 & 0 \\ 0 & 1 & 0 \\ 0 & 0 & 0 \\ 0 & 0 & 0 \\ 0 & 0 & 1 \\ 1 & 0 & 0 \end{bmatrix} u^1 \qquad (9.46)$$

where $u^1 \triangleq \begin{bmatrix} \tilde{u}^1 \\ \bar{u}^1 \end{bmatrix}$, $\tilde{u}^1 \triangleq u_1^1$, and $\bar{u}^1 \triangleq \begin{bmatrix} u_2^1 \\ u_3^1 \end{bmatrix}$.

step 2: For system (9.46), we have relative degree $(\rho_1^2, \rho_2^2, \rho_3^2) = (3, 2, 2)$ and

$$\begin{bmatrix} y_1^{(3)} \\ y_2^{(2)} \\ y_3^{(2)} \end{bmatrix} = \begin{bmatrix} x_4(1 + z_1^1) \\ 0 \\ x_7 \end{bmatrix} + \begin{bmatrix} 1 & 0 & 0 \\ x_4 & 1 + z_1^1 & 0 \\ 1 & 0 & 0 \end{bmatrix} \begin{bmatrix} u_1^1 \\ u_2^1 \\ u_3^1 \end{bmatrix}$$

$$= E^2(x, z_1^1) + D^2(x, z_1^1) \begin{bmatrix} \tilde{u}^1 \\ \bar{u}^1 \end{bmatrix}$$

where $r_2 \triangleq \text{rank}\left(D^2(x)\right) = 2$. It is easy, by elementary column operations, to find $m \times m$ matrix $L^2(x)$ such that

$$D^2(x)L^2(x) = \begin{bmatrix} 1 & 0 & 0 \\ x_4 & 1 + z_1^1 & 0 \\ 1 & 0 & 0 \end{bmatrix} \begin{bmatrix} 1 & 0 & 0 \\ \frac{-x_4}{1+z_1^1} & \frac{1}{1+z_1^1} & 0 \\ 0 & 0 & 1 \end{bmatrix} = \begin{bmatrix} 1 & 0 & 0 \\ 0 & 1 & 0 \\ 1 & 0 & 0 \end{bmatrix}.$$

Therefore, with dynamic feedback

$$u^1 = L^2(x) \begin{bmatrix} z_1^2 \\ z_2^2 \\ u_3^2 \end{bmatrix} = \begin{bmatrix} 1 & 0 & 0 \\ \frac{-x_4}{1+z_1^1} & \frac{1}{1+z_1^1} & 0 \\ 0 & 0 & 1 \end{bmatrix} \begin{bmatrix} z_1^2 \\ u_2^2 \\ u_3^2 \end{bmatrix} \qquad (9.47)$$

$$\dot{z}_1^2 = u_1^2,$$

we have the following extended system:

$$
\begin{bmatrix} \dot{x}_1 \\ \dot{x}_2 \\ \dot{x}_3 \\ \dot{x}_4 \\ \dot{x}_5 \\ \dot{x}_6 \\ \dot{x}_7 \\ \dot{z}_1^1 \\ \dot{z}_1^2 \end{bmatrix} = \begin{bmatrix} x_2 \\ x_3 + z_1^1 \\ x_4(1 + z_1^1) \\ \frac{-x_4 z_1^2}{1+z_1^1} \\ x_6 + z_1^1 \\ x_7 \\ 0 \\ z_1^2 \\ 0 \end{bmatrix} + \begin{bmatrix} 0 & 0 & 0 \\ 0 & 0 & 0 \\ 0 & 0 & 0 \\ 0 & \frac{1}{1+z_1^1} & 0 \\ 0 & 0 & 0 \\ 0 & 0 & 0 \\ 0 & 0 & 1 \\ 0 & 0 & 0 \\ 1 & 0 & 0 \end{bmatrix} u^2 \tag{9.48}
$$

where $u^2 \triangleq \begin{bmatrix} \tilde{u}^2 \\ \bar{u}^2 \end{bmatrix}$, $\tilde{u}^2 \triangleq u_1^2$, and $\bar{u}^2 \triangleq \begin{bmatrix} u_2^2 \\ u_3^2 \end{bmatrix}$.

step 3: For system (9.48), we have relative degree $(\rho_1^3, \rho_2^3, \rho_3^3) = (4, 2, 3)$ and

$$
\begin{bmatrix} y_1^{(4)} \\ y_2^{(2)} \\ y_3^{(3)} \end{bmatrix} = \begin{bmatrix} 0 \\ 0 \\ 0 \end{bmatrix} + \begin{bmatrix} 1 & 1 & 0 \\ 0 & 1 & 0 \\ 1 & 0 & 1 \end{bmatrix} \begin{bmatrix} u_1^2 \\ u_2^2 \\ u_3^2 \end{bmatrix}
$$

where $r_3 \triangleq \operatorname{rank}\left(D^3(x)\right) = 3$. Since $r_3 = 3$, then the Dynamic Extension Algorithm terminates at step 3 and thus system (9.48) is input-output decouplable by static feedback

$$
\begin{bmatrix} u_1^2 \\ u_2^2 \\ u_3^2 \end{bmatrix} = \begin{bmatrix} 1 & -1 & 0 \\ 0 & 1 & 0 \\ -1 & 1 & 1 \end{bmatrix} \begin{bmatrix} v_1 \\ v_2 \\ v_3 \end{bmatrix}. \tag{9.49}
$$

It is easy to see, by (9.45), (9.47), (9.48), and (9.49), that

$$
\begin{bmatrix} u_1 \\ u_2 \\ u_3 \end{bmatrix} = \begin{bmatrix} z_1^1 + \frac{x_4 z_1^2}{1+z_1^1} \\ \frac{-x_4 z_1^2}{1+z_1^1} \\ 0 \end{bmatrix} + \begin{bmatrix} 0 & \frac{-1}{1+z_1^1} & 0 \\ 0 & \frac{1}{1+z_1^1} & 0 \\ -1 & 1 & 1 \end{bmatrix} \begin{bmatrix} v_1 \\ v_2 \\ v_3 \end{bmatrix} \tag{9.50}
$$

and

$$
\begin{bmatrix} \dot{z}_1^1 \\ \dot{z}_1^2 \end{bmatrix} = \begin{bmatrix} z_1^2 \\ 0 \end{bmatrix} + \begin{bmatrix} 0 & 0 & 0 \\ 1 & -1 & 0 \end{bmatrix} \begin{bmatrix} v_1 \\ v_2 \\ v_3 \end{bmatrix}. \tag{9.51}
$$

In other words, the extended system

$$
\begin{bmatrix} \dot{x}_1 \\ \dot{x}_2 \\ \dot{x}_3 \\ \dot{x}_4 \\ \dot{x}_5 \\ \dot{x}_6 \\ \dot{x}_7 \\ \dot{z}_1^1 \\ \dot{z}_1^2 \end{bmatrix} = \begin{bmatrix} x_2 \\ x_3 + z_1^1 \\ x_4(1 + z_1^1) \\ \frac{-x_4 z_1^2}{1 + z_1^1} \\ x_6 + z_1^1 \\ x_7 \\ 0 \\ z_1^2 \\ 0 \end{bmatrix} + \begin{bmatrix} 0 & 0 & 0 \\ 0 & 0 & 0 \\ 0 & 0 & 0 \\ 0 & \frac{1}{1+z_1^1} & 0 \\ 0 & 0 & 0 \\ 0 & 0 & 0 \\ -1 & 1 & 1 \\ 0 & 0 & 0 \\ 1 & -1 & 0 \end{bmatrix} v
$$

of system (9.44) with dynamic feedback (9.50) and (9.51) has the following input-output decoupled relation:

$$
\begin{bmatrix} y_1^{(4)} \\ y_2^{(2)} \\ y_3^{(3)} \end{bmatrix} = \begin{bmatrix} v_1 \\ v_2 \\ v_3 \end{bmatrix}.
$$

□

Example 9.3.2 Show that the following nonlinear system is not locally dynamic input-output decouplable:

$$
\begin{bmatrix} \dot{x}_1 \\ \dot{x}_2 \\ \dot{x}_3 \end{bmatrix} = \begin{bmatrix} x_3^2 \\ 0 \\ 0 \end{bmatrix} + \begin{bmatrix} 1 \\ 1 \\ 1 \end{bmatrix} u_1 + \begin{bmatrix} 1 \\ 0 \\ 1 \end{bmatrix} u_2
$$

$$
\begin{bmatrix} y_1 \\ y_2 \end{bmatrix} = \begin{bmatrix} x_1 \\ x_3 \end{bmatrix} = h(x). \tag{9.52}
$$

Solution step 1: By simple calculations, we have relative degree $(\rho_1^1, \rho_2^1) = (1, 1)$ and

$$
\begin{bmatrix} y_1^{(1)} \\ y_2^{(1)} \end{bmatrix} = \begin{bmatrix} x_3^2 \\ 0 \end{bmatrix} + \begin{bmatrix} 1 & 1 \\ 1 & 1 \end{bmatrix} \begin{bmatrix} u_1 \\ u_2 \end{bmatrix} = E^1(x) + D^1(x)u
$$

where $r_1 \triangleq \operatorname{rank}\left(D^1(x)\right) = 1$. It is easy, by elementary column operations, to find $m \times m$ matrix $L^1(x)$ such that

$$
D^1(x)L^1(x) = \begin{bmatrix} 1 & 1 \\ 1 & 1 \end{bmatrix} \begin{bmatrix} 1 & -1 \\ 0 & 1 \end{bmatrix} = \begin{bmatrix} 1 & 0 \\ 1 & 0 \end{bmatrix}.
$$

Therefore, with

$$u = L^1(x) \begin{bmatrix} z_1^1 \\ u_2^1 \end{bmatrix} = \begin{bmatrix} 1 & -1 \\ 0 & 1 \end{bmatrix} \begin{bmatrix} z_1^1 \\ u_2^1 \end{bmatrix}$$

$$\dot{z}_1^1 = u_1^1,$$

we have the following extended system:

$$\begin{bmatrix} \dot{x}_1 \\ \dot{x}_2 \\ \dot{x}_3 \\ \dot{z}_1^1 \end{bmatrix} = \begin{bmatrix} x_3^2 + z_1^1 \\ z_1^1 \\ z_1^1 \\ 0 \end{bmatrix} + \begin{bmatrix} 0 & 0 \\ 0 & -1 \\ 0 & 0 \\ 1 & 0 \end{bmatrix} u^1 \tag{9.53}$$

where $u^1 \triangleq \begin{bmatrix} \tilde{u}^1 \\ \bar{u}^1 \end{bmatrix}$, $\tilde{u}^1 \triangleq u_1^1$, and $\bar{u}^1 \triangleq u_2^1$.

step 2: For system (9.53), we have relative degree $(\rho_1^2, \rho_2^2) = (2, 2)$ and

$$\begin{bmatrix} y_1^{(2)} \\ y_2^{(2)} \end{bmatrix} = \begin{bmatrix} 2x_3 z_1^1 \\ 0 \end{bmatrix} + \begin{bmatrix} 1 & 0 \\ 1 & 0 \end{bmatrix} \begin{bmatrix} u_1^1 \\ u_2^1 \end{bmatrix}$$

$$= E^2(x, z_1^1) + D^2(x, z_1^1) \begin{bmatrix} \tilde{u}^1 \\ \bar{u}^1 \end{bmatrix}$$

where $r_2 \triangleq \text{rank}\left(D^2(x)\right) = 1$. It is easy, by elementary column operations, to find $m \times m$ matrix $L^2(x)$ such that

$$D^2(x)L^2(x) = \begin{bmatrix} 1 & 0 \\ 1 & 0 \end{bmatrix} \begin{bmatrix} 1 & 0 \\ 0 & 1 \end{bmatrix} = \begin{bmatrix} 1 & 0 \\ 1 & 0 \end{bmatrix}.$$

Therefore, with dynamic feedback

$$u^1 = L^2(x) \begin{bmatrix} z_1^2 \\ u_2^2 \end{bmatrix} = \begin{bmatrix} 1 & 0 \\ 0 & 1 \end{bmatrix} \begin{bmatrix} z_1^2 \\ u_2^2 \end{bmatrix}$$

$$\dot{z}_1^2 = u_1^2,$$

we have the following extended system:

$$\begin{bmatrix} \dot{x}_1 \\ \dot{x}_2 \\ \dot{x}_3 \\ \dot{z}_1^1 \\ \dot{z}_1^2 \end{bmatrix} = \begin{bmatrix} x_3^2 + z_1^1 \\ z_1^1 \\ z_1^1 \\ z_1^2 \\ 0 \end{bmatrix} + \begin{bmatrix} 0 & 0 \\ 0 & -1 \\ 0 & 0 \\ 0 & 0 \\ 1 & 0 \end{bmatrix} u^2$$

where $u^2 \triangleq \begin{bmatrix} \tilde{u}^2 \\ \bar{u}^2 \end{bmatrix}$, $\tilde{u}^2 \triangleq u_1^2$, and $\bar{u}^2 \triangleq u_2^2$. In this manner, it is easy to see that the Dynamic Extension Algorithm does not terminate at the final step with $r_1 = r_2 = r_3 = \cdots$. Thus, we have the final step $K = 1$ and $r_K = \text{rank}\left(D^1(x)\right) = 1 \neq m$. Hence, by Theorem 9.2, system (9.52) is not dynamic input-output decouplable. $\square$

9.4   MATLAB Programs

In this section, the following subfunctions in Appendix C are needed:
   CharacterNum, ChZero, Decoupling-M, Lfh, Lfhk, RelativeDegree, Yreorder

MATLAB program for Theorem 9.1:

```
clear all
syms x1 x2 x3 x4 x5 x6 x7 x8 x9 real

f=[x2; x3; x4; 0]; g=[0 0; 1 x4; 0 0; 0 1+x3];
h=[x1; x3]; %Ex:9.2.1

% f=[x2; x3; x4; 0];
% g=[0 0; 1 0; x4 0; 0 1+x3]; h=[x1; x3]; %Ex:9.2.2

% f=[x2; x3; x4; 0];
% g=[0 0; 1 1; x4 x4; 0 1+x3]; h=[x1; x3]; %Ex:9.2.3

% g=[0 0 0; 1 1 0; x4 x4 0; 0 1 0; 1 1 0; 0 0 0; 0 0 1];
% f=[x2; x3; x4; 0; x6; x7; 0]; h=[x1; x3; x5]; %Ex:9.3.1

% f=[x3^2; 0; 0]; g=[1 1; 1 x1-x1; 1 1];
% h=[x1; x3]; %Ex:9.3.2

% f=[x2; 0; x1^2]; g=[0 0; 1 1; 0 1+x2^2];
% h=[x1; x3]; %P:9-2(a)

% f=[x2; 0; x1^2]; g=[0 0; 1 1; 0 1+x2^2];
% h=[x1+x3^2; x3]; %P:9-2(b)

f=simplify(f)
g=simplify(g)
h=simplify(h)
[n,m]=size(g); x=sym('x',[n,1]);
if length(h) ~= m
  return
end

rho=RelativeDegree(f,g,h,x)
[E,D]=Decoupling_M(f,g,h,x,rho)
```

```
if rank(D)<m
  display('By Thm 9.1, NOT locally static decoupable.')
  return
end

display('By Thm 9.1, locally static decoupable with')

beta=simplify(inv(D))
alpha=simplify(-beta*E)

return
```

The following is a MATLAB subfunction program for Theorem 9.2. (Refer to Lemma 9.2.)

```
function [R,L,br]=Dcolumn(D)

m=size(D,2); u=sym('u',[m,1]);
L0=jacobian(u,u); L1=L0; bL=L0;
R0=RowReorder(D);
D=R0*D;
r=rank(D);
L0trans=RowReorder(D');
L0=L0trans';
tD=D*L0;
L1(1:r,1:r)=simplify(inv(tD(1:r,1:r)));
L1(1:r,(r+1):m)=-L1(1:r,1:r)*tD(1:r,(r+1):m);
L=simplify(L0*L1);
tD=D*L;
t1=bL(:,1)-bL(:,1); t2=t1;
for k1=1:r
  if ChZero(tD(r+1:m,k1))==0
    t1=[t1 bL(:,k1)];
  else
    t2=[t2 bL(:,k1)];
  end
end
br=size(t1,2)-1;
bL(:,1:r)=[t1(:,2:br+1) t2(:,2:r-br+1)];
L=L*bL;
R=bL'*R0;
```

The following is a MATLAB subfunction program for Theorem 9.2.

```
function [flag,L,br,R,he,fe,ge,xe]=DEAstep(k,f,g,h,x)

flag=0; [n,m]=size(g);
z=sym('z',[n,n]); u=sym('u',[m,1]);
rho=RelativeDegree(f,g,h,x);
[E,D]=Decoupling_M(f,g,h,x,rho);
if rank(D)==m
  R=jacobian(u,u); L=R;
  br=0; flag=1;
  he=h; fe=f; ge=g, xe=x;
  return
end
[R,L,br]=Dcolumn(D,x)
tL=L(:,1:br);
bL=L(:,br+1:m);
he=R*h;
zz=z(1:br,k);
xe=[x; zz];
fe=[f+g*tL*zz; zz-zz];
teye=jacobian(xe,xe);
G11=teye(1:n,1:br)-teye(1:n,1:br);
G12=g*bL;
G21=teye(1:br,1:br);
G22=teye(1:br,1:m-br)-teye(1:br,1:m-br);
ge=[G11 G12; G21 G22];
```

The following is a MATLAB subfunction program for Theorem 9.2.

```
function [Kf,LL,Br,R,he,fe,ge,xe]=DEA(f,g,h,x)

[n,m]=size(g); u=sym('u',[m,1]);
R=jacobian(u,u); LL=R;
for k=1:n+1
  Kf=k
  [flag,L,br,R1,h,f,g,x]=DEAstep(k,f,g,h,x);
  R=R1*R;
  fe=f; ge=g; he=h; xe=x;
  LL=[LL L]
  Br(k)=br;
  if flag==1
    return
  end
end
```

MATLAB program for Theorem 9.2:

```
clear all
syms x1 x2 x3 x4 x5 x6 x7 x8 x9 real

% f=[x2; x3; x4; 0]; g=[0 0; 1 x4; 0 0; 0 1+x3];
% h=[x1; x3]; %Ex:9.2.1

% f=[x2; x3; x4; 0];
% g=[ 0 0; 1 0; x4 0; 0 1+x3]; h=[x1; x3]; %Ex:9.2.2

% f=[x2; x3; x4; 0];
% g=[ 0 0; 1 1; x4 x4; 0 1+x3]; h=[x1; x3]; %Ex:9.2.3

f=[x2; x3; x4; 0; x6; x7; 0];
g=[ 0 0 0; 1 1 0; x4 x4 0; 0 1 0; 1 1 0; 0 0 0; 0 0 1];
h=[x1; x3; x5]; %Ex:9.3.1
%h=[x5; x1; x3];

% f=[x3^2; 0; 0]; g=[1 1; 1 x1-x1; 1 1];
% h=[x1; x3]; %Ex:9.3.2

% f=[x2; 0; x1^2]; g=[0 0; 1 1; 0 1+x2^2];
% h=[x1+x3^2; x3]; %P:9-2(b)

% f=[0; 0; x3]; g=[1 1; 1 x1-x1; 1 1];
% h=[x1; x3]; %P:9-2(c)

% g=[x1-x1 0 0; 1 0 0; 0 0 0; 0 1 0; 1 0 0; 0 1 0; 0 0 1];
% f=[x2; x3; x4; 0; x6; x7; 0]; h=[x1; x3; x5]; %P:9-2(d)

% g=[1 x1 x1; 1 x1 x1; 0 1 x1; 1 x1 x1; 0 1 x1; 0 0 1];
% f=[0; x3; 0; x5; x6; x1^2]; h=[x1; x2; x4]; %P:9-2(e)

f=simplify(f)
g=simplify(g)
h=simplify(h)
[n,m]=size(g); x=sym('x',[n,1]); u=sym('u',[m,1]);
w=sym('w',[m,1]); v=sym('v',[m,1]); z=sym('z',[n,n]);

rho=RelativeDegree(f,g,h,x)
[E1,D1]=Decoupling_M(f,g,h,x,rho)

if rank(D1)==m
  display('System is static decoupable with.')
  u=inv(D1)*(-E1+v)
  return
end

[Kf,LL,Br,R,he,fe,ge,xe]=DEA(f,g,h,x)

display('DEA terminates at')
STEP=Kf
```

```
if Kf > n
  display('By Thm 9.2, NOT locally dynamic decoupable.')
  return
end

display('By Thm 9.2, locally dynamic decoupable.')
Im=jacobian(u,u); Zerom=Im-Im;
LL=LL(:,m+1:length(LL));
aa=LL(:,1:Br(1))*z(1:Br(1),1);
bu=[Zerom(:,1:Br(1)) LL(:,Br(1)+1:m)]*u;
abu=aa+bu;

for k=2:Kf-1
  taa=LL(:,(k-1)*m+1:(k-1)*m+Br(k))*z(1:Br(k),k);
  tbu=[Zerom(:,1:Br(k)) LL(:,(k-1)*m+Br(k)+1:k*m)]*u;
  tabu=taa+tbu;
  abu=simplify(subs(abu,u,tabu));
end

abw=simplify(subs(abu,u,w));
fegew=fe+ge*w;
cdw=fegew(n+1:length(fegew));

rhoE=RelativeDegree(fe,ge,he,xe)
[E,D]=Decoupling_M(fe,ge,he,xe,rhoE)

W=simplify(inv(D)*(v-E))

fegev=fe+ge*W;
Ge=jacobian(fegev,v)
Fe=simplify(subs(fegev,v,v-v))
abv=subs(abw,w,W);
cdv=fegev(n+1:length(fegev));
aa=simplify(subs(abv,v,v-v))
bb=jacobian(abv,v)
cc=simplify(subs(cdv,v,v-v))
dd=jacobian(cdv,v)

rhoE=RelativeDegree(Fe,Ge,he,xe)
[Ee,De]=Decoupling_M(Fe,Ge,he,xe,rhoE)

return
```

9.5 Problems

9-1. Find out whether the following nonlinear systems are locally input-output decouplable by static feedback. If static input-output decoupling is not possible, then find out whether it is locally input-output decouplable by dynamic feedback.

(a)

$$\dot{x} = \begin{bmatrix} x_2 \\ 0 \\ x_1^2 \end{bmatrix} + \begin{bmatrix} 0 \\ 1 \\ 0 \end{bmatrix} u_1 + \begin{bmatrix} 0 \\ 1 \\ 1 + x_2^2 \end{bmatrix} u_2 ; \quad \begin{bmatrix} y_1 \\ y_2 \end{bmatrix} = \begin{bmatrix} x_1 \\ x_3 \end{bmatrix}.$$

(b)

$$\begin{bmatrix} \dot{x}_1 \\ \dot{x}_2 \\ \dot{x}_3 \end{bmatrix} = \begin{bmatrix} x_2 \\ 0 \\ x_1^2 \end{bmatrix} + \begin{bmatrix} 0 & 0 \\ 1 & 1 \\ 0 & 1 + x_2^2 \end{bmatrix} u ; \quad \begin{bmatrix} y_1 \\ y_2 \end{bmatrix} = \begin{bmatrix} x_1 + x_3^2 \\ x_3 \end{bmatrix}.$$

(c)

$$\begin{bmatrix} \dot{x}_1 \\ \dot{x}_2 \\ \dot{x}_3 \end{bmatrix} = \begin{bmatrix} 0 & 0 & 1 \\ 0 & 0 & 0 \\ 0 & 0 & 0 \end{bmatrix} x + \begin{bmatrix} 1 & 1 \\ 1 & 0 \\ 1 & 1 \end{bmatrix} u ; \quad y = \begin{bmatrix} 1 & 0 & 0 \\ 0 & 0 & 1 \end{bmatrix} x.$$

(d)

$$\begin{bmatrix} \dot{x}_1 \\ \dot{x}_2 \\ \dot{x}_3 \\ \dot{x}_4 \\ \dot{x}_5 \\ \dot{x}_6 \\ \dot{x}_7 \end{bmatrix} = \begin{bmatrix} x_2 \\ x_3 \\ x_4 \\ 0 \\ x_6 \\ x_7 \\ 0 \end{bmatrix} + \begin{bmatrix} 0 & 0 & 0 \\ 1 & 0 & 0 \\ 0 & 0 & 0 \\ 0 & 1 & 0 \\ 1 & 0 & 0 \\ 0 & 1 & 0 \\ 0 & 0 & 1 \end{bmatrix} \begin{bmatrix} u_1 \\ u_2 \\ u_3 \end{bmatrix} ; \quad \begin{bmatrix} y_1 \\ y_2 \\ y_3 \end{bmatrix} = \begin{bmatrix} x_1 \\ x_3 \\ x_5 \end{bmatrix} = h(x).$$

(e)

$$\begin{bmatrix} \dot{x}_1 \\ \dot{x}_2 \\ \dot{x}_3 \\ \dot{x}_4 \\ \dot{x}_5 \\ \dot{x}_6 \end{bmatrix} = \begin{bmatrix} 0 \\ x_3 \\ 0 \\ x_5 \\ x_6 \\ x_1^2 \end{bmatrix} + \begin{bmatrix} 1 & x_1 & x_1 \\ 1 & x_1 & x_1 \\ 0 & 1 & x_1 \\ 1 & x_1 & x_1 \\ 0 & 1 & x_1 \\ 0 & 0 & 1 \end{bmatrix} \begin{bmatrix} u_1 \\ u_2 \\ u_3 \end{bmatrix} ; \quad \begin{bmatrix} y_1 \\ y_2 \\ y_3 \end{bmatrix} = \begin{bmatrix} x_1 \\ x_2 \\ x_4 \end{bmatrix} = h(x).$$

9-2. For continuous nonlinear control systems, define the local input-output decoupling problem by restricted dynamic feedback, and find the necessary and sufficient conditions for this problem by defining the corresponding dynamic extension algorithm.

9-3. For the discrete time nonlinear control systems, define the static and dynamic input-output decoupling problems. In other words, find the discrete versions of Definitions 9.2 and 9.3.

9-4. Find out a nonsingular feedback for the input-output decoupling of the following nonlinear discrete time system.

$$\begin{bmatrix} x_1(t+1) \\ x_2(t+1) \\ x_3(t+1) \end{bmatrix} = \begin{bmatrix} x_2(t) \\ x_1(t)^2 + u_1(t) + u_2(t)^2 \\ x_3(t)^2 + u_2(t) \end{bmatrix}$$

$$\begin{bmatrix} y_1(t) \\ y_2(t) \end{bmatrix} = \begin{bmatrix} x_1(t) \\ x_3(t) \end{bmatrix}.$$

9-5. Find out the necessary and sufficient conditions for local static input-output decoupling problems of the discrete time nonlinear control systems. In other words, obtain the discrete version of Theorem 9.1.

9-6. Find out the discrete version of Theorem 9.2.

9-7. Find out the discrete version of Problem 9.2.

Appendix A
Basics of Topology

A.1 Topology of Real Numbers

Definition A.1 (*open interval, neighborhood, open set, limit point, closure, and closed set*)

(a) Open interval (a, b) is the set $\{x \in \mathbb{R} \mid a < x < b\}$.

(b) A neighborhood of point $p \in \mathbb{R}$ is an open interval U such that $p \in U$.

(c) A set $E\,(\subset \mathbb{R})$ is said to be open, if, for every point $p \in E$, there exists a neighborhood $U \subset E$ of p.

(d) A point p is said to be a limit point of set E, if every neighborhood U of p contains at least one point of E different from p itself.

(e) $\bar{E}\left(\triangleq E \cup E'\right)$ is said to be the closure of set E, where E' is the set of all limit points of E.

(f) A set $E\,(\subset \mathbb{R})$ is said to be closed, if every limit point of E belongs to E or $\bar{E} = E$.

Example A.1.1 (a) $E = (0, 2)$ is an open set. But, E is not a closed set.

(b) $E = [0, 2] \triangleq \{x \in \mathbb{R} \mid 0 \leq x \leq 2\}$ is a closed set. But, E is not an open set.

(c) $E = (0, 1] \triangleq \{x \in \mathbb{R} \mid 0 < x \leq 1\}$ is neither open nor closed.

(d) $E = \mathbb{R}$ is both open and closed.

(e) $E = \phi$ is both open and closed.

□

As shown in the example above, open and closed sets are not the opposite. In other words, it should not be considered a closed set unless it is an open set.

Example A.1.2 Prove the following:

(a) If the set E is a closed set, the complement E^c is an open set and vice versa.

(b) If $\{G_\alpha \mid \alpha \in A\}$ is a collection of open sets, then $\bigcup_{\alpha \in A} G_\alpha$ is also an open set.

© The Editor(s) (if applicable) and The Author(s), under exclusive license to Springer Nature Singapore Pte Ltd. 2022
H.-G. Lee, *Linearization of Nonlinear Control Systems*,
https://doi.org/10.1007/978-981-19-3643-2

(c) If $\{G_1, \cdots, G_n\}$ is a finite collection of open sets, then $\bigcap\limits_{i=1}^{n} G_i$ is also an open
   set.

$\square$

The union of infinite open sets is an open set. However, the intersection of infinite
open sets may not be an open set. See the following example.

Example A.1.3 Show that $\bigcap\limits_{\ell=1}^{n} (0, 1 + \frac{1}{\ell})$ is an open set. Also, show that $\bigcap\limits_{\ell=1}^{\infty} (0, 1 + \frac{1}{\ell})$
is not an open set. $\square$

A.2   General Topology

Definition A.2 (*topological space*)
A topological space is an ordered pair (X, τ), where X is a set and τ is a collection
of subsets of X, satisfying the following axioms:

(i) $\phi, X \in \tau$.
(ii) (closed under arbitrary union)

$$A_\lambda \in \tau \text{ for } \lambda \in \Lambda \;\; \Rightarrow \;\; \bigcup_{\lambda \in \Lambda} A_\lambda \in \tau.$$

(iii) (closed under finite intersection)

$$A, B \in \tau \;\; \Rightarrow \;\; A \cap B \in \tau.$$

Example A.2.1 (indiscrete topology)
For some set X, let $\tau = \{\phi, X\}$. Prove that (X, τ) is a topological space. $\square$

Example A.2.2 (discrete topology)
For some set X, let τ be the collection of all subsets of X. Prove that (X, τ) is a
topological space. $\square$

Example A.2.3 Let $X = \{1, 2, 3\}$.

(a) Prove that (X, τ) is a topological space when $\tau = \{\phi, \{1\}, \{2, 3\}, X\}$.
(b) Prove that (X, τ) is not a topological space when $\tau = \{\phi, \{1, 2\}, \{2, 3\}, X\}$.

$\square$

Definition A.3 (coarser topology and finer topology)
Let τ_1 and τ_2 be two topologies on a set X such that $\tau_1 \subset \tau_2$.

(a) τ_1 is said to be a coarser (or weaker) topology than τ_2.
(b) τ_2 is said to be a finer (or stronger) topology than τ_1.

The coarsest topology on X is the indiscrete topology in Example A.2.1. The finest topology on X is the discrete topology in Example A.2.2.

Definition A.4 (*open set*)
Let (X, τ) be a topological space. A is said to be an open set, if $A \in \tau$.

Definition A.5 (*closed set*)
Let (X, τ) be a topological space. A is said to be a closed set, if $X - A \in \tau$. In other words, A is said to be a closed set, if the complement of A is an open set.

Definition A.6 (*Hausdorff topological space*)
(X, τ) is said to be a Hausdorff topological space, if, for all $x_1 \in X$ and $x_2 \in X$ ($x_1 \neq x_2$), there exist open sets U_1 and U_2 such that $x_1 \in U_1$, $x_2 \in U_2$, and $U_1 \cap U_2 = \phi$.

Definition A.7 (*basis*)
A basis B for a topological space (X, τ) is a collection of open sets such that every open set can be written as a union of elements of B.

For example, the collection of all open intervals in the real line forms a basis for the usual (or standard) topology on the real numbers. However, a basis is not unique. For example, the collection of all open intervals with rational endpoints is also a basis for the usual real topology.

Definition A.8 (*countable set*)
Let $\mathbb{N} = \{1, 2, 3, ...\}$ be the set of the natural numbers. A set X is said to be countable, if there exists an injective (or $1 - 1$) function $f : X \rightarrow \mathbb{N}$. X is said to be an uncountable set, if X is not countable.

Example A.2.4 (a) Show that $X = \{a, b, c\}$ is countable.
(b) Show that $\mathbb{Z}$, the set of all integers, is countable.
(c) Show that $\mathbb{Q}$, the set of all rational numbers, is countable.
(d) Show that $\mathbb{R}$, the set of all real numbers, is uncountable.

$\square$

Definition A.9 (*second countable*)
A topological space (X, τ) is said to be second countable, if τ has a countable basis.

Definition A.10 (*closure, interior, boundary, and dense subset*)
Let $A \subset X$, where (X, τ) is a topological space.

(a) $\bar{A}$, the closure of A, is defined by the intersection of all closed subsets containing A.
(b) $\text{Int}(A)$, the interior of A, is defined by the union of all open sets contained in A.
(c) ∂A, the boundary of A, is defined by $\partial A = \bar{A} - \text{Int}(A)$.
(d) A is said to be a dense subset of X, if $\bar{A} = X$.

Definition A.11 (*neighborhood*)
Let (X, τ) be a topological space. A set U is said to be a neighborhood of $x (\in X)$, if $x \in U$ and $U \in \tau$. In other words, a neighborhood of $x (\in X)$ is an open set U such that $x \in U$.

Let (X, τ) and (Y, σ) be topological spaces. Suppose that $f : X \to Y$ is a function. We define that for a subset B of Y,

$$f^{-1}(B) \triangleq \{x \in X \mid f(x) \in B\}.$$

Note that $f^{-1}(B) \subset X$.

Example A.2.5 Show the following:

(a) $f^{-1}\left(\bigcup_{\alpha \in A} B_\alpha\right) = \bigcup_{\alpha \in A} f^{-1}(B_\alpha).$

(b) $f^{-1}\left(\bigcap_{\alpha \in A} B_\alpha\right) = \bigcap_{\alpha \in A} f^{-1}(B_\alpha).$

(c) $f^{-1}(Y - B) = X - f^{-1}(B).$

(d) $f\left(\bigcup_{\alpha \in A} A_\alpha\right) = \bigcup_{\alpha \in A} f(A_\alpha).$

(e) $f\left(\bigcap_{\alpha \in A} A_\alpha\right) \subset \bigcap_{\alpha \in A} f(A_\alpha).$

□

Definition A.12 (*continuous function*)
Let (X, τ) and (Y, σ) be topological spaces. Suppose that $f : X \to Y$ is a function.

(a) f is said to be continuous at point x_0, if, for every neighborhood V of $f(x_0)$, there exists a neighborhood U of $x_0 (\in X)$ such that $f(U) \subset V$.
(b) f is said to be continuous, if $f^{-1}(V) \in \tau$ for all $V \in \sigma$.

Definition A.13 (*metric*)
A metric (or distance function) on a set X is a function $d : X \times X \to [0, \infty)$ satisfying the following axioms:

(a) $d(x, y) = d(y, x), \ \forall x, y \in X.$
(b) $x = y \ \Leftrightarrow \ d(x, y) = 0.$
(c) $d(x, z) \leq d(x, y) + d(y, z), \ \forall x, y, z \in X.$

Definition A.14 (*metric space*)
A metric space is an ordered pair (X, d) where M is a set and d is a metric on X.

The topology of $\mathbb{R}$, described in Appendix A.1, can be simply defined as follows.

Definition A.15 (*usual topology of* $\mathbb{R}$)
If τ, the usual topology of $\mathbb{R}$, is defined by the collection of arbitrary union of open intervals, then $(\mathbb{R}, \tau)$ is a topological space.

For the metric space $(\mathbb{R}^n, d)$, the open ball $B_q(x)$ with the center x and the radius q is defined by

$$B_q(x) \triangleq \left\{ y \in \mathbb{R}^n \mid d(x, y) < q \right\}.$$

Example A.2.6 Let $B = \left\{ B_q(x) \mid x \in \mathbb{Q} \text{ and } q \in \mathbb{Q} \right\}$ where $\mathbb{Q}$ is the set of all rational numbers. Show that B is a countable basis for the usual topological space $(\mathbb{R}, \tau)$. Thus, the usual topological space $(\mathbb{R}, \tau)$ is second countable.   □

Definition A.16 (*product topology*)
Suppose that (S_1, τ_1) and (S_2, τ_2) are topological spaces. For the Cartesian product $S_1 \times S_2$, the product topology τ is defined by

$$\tau = \{U_1 \times U_2 \mid U_1 \in \tau_1 \ \& \ U_2 \in \tau_2\}.$$

Definition A.17 (*usual topology of* $\mathbb{R}^n$)
Suppose that B is a basis for the topological space $(\mathbb{R}^n, \tau)$, where B is the collection of all open balls of $\mathbb{R}^n$. τ is said to be the usual topology of $\mathbb{R}^n$.

Definition A.18 (*subset topology*)
Suppose that (S, τ) is a topological space and $S_1 \subset S$. If we define subset topology τ_1 by

$$\tau_1 = \{U \cap S_1 \mid U \in \tau\},$$

then (S_1, τ_1) is also a topological space.

Usually, we use the subset topology when we define the topology for a subset of some topological space. If a function from a topological space to another topological space is defined, it may be used to define the following topological space.

Definition A.19 (*induced topology*)
Suppose that $F : S_1 \to S_2$ is a continuous function, where (S_1, τ_1) and (S_2, τ_2) are topological spaces. If we define induced topology τ by

$$\tau = \{F(U) \mid U \in \tau_1\},$$

then $(F(S_1), \tau)$ is also a topological space.

From the above definitions, it can be seen that two different topologies (subset topology and induced topology) can be defined for $F(S_1) (\subset S_2)$ as needed.

Appendix B
Manifold and Vector Field

B.1 Manifold

Definition B.1 (*topological manifold*)
A topological space (M, τ) is said to be an n-dimensional topological manifold, if the following conditions are satisfied:

(i) (M, τ) is Hausdorff.
(ii) (M, τ) is second countable.
(iii) For every $x(\in M)$, there exists a neighborhood U of x such that U is homeomorphic to an open set of $\mathbb{R}^n$.

In other words, an n-dimensional topological manifold is a locally second countable Hausdorff Euclidean space $\mathbb{R}^n$. An n-dimensional topological manifold is locally homeomorphic to an open set of $\mathbb{R}^n$, even though the universal set M may not be homeomorphic to an open set of $\mathbb{R}^n$.

Example B.1.1 Show that the unit circle is a one-dimensional topological manifold. □

Example B.1.2 Show that a sphere is a two-dimensional topological manifold. □

Example B.1.3 Let τ be the usual topology of $\mathbb{R}^n$. Suppose that M is open subset of $\mathbb{R}^n$. Show that (M, τ_1) is an n-dimensional topological manifold where τ_1 is the subset topology. □

Let (M, τ) be an n-dimensional topological manifold. Suppose that U_λ is an open subset of M. An ordered pair $(U_\lambda, \varphi_\lambda)$ is said to be a chart (or coordinate chart), if φ_λ is a homeomorphism from U_λ to an open subset of $\mathbb{R}^n$. An atlas for a topological space M is a collection $\{(U_\lambda, \varphi_\lambda) \mid \lambda \in \Lambda\}$ such that $\bigcup_{\lambda \in \Lambda} U_\lambda = M$.

© The Editor(s) (if applicable) and The Author(s), under exclusive license to Springer Nature Singapore Pte Ltd. 2022
H.-G. Lee, *Linearization of Nonlinear Control Systems*,
https://doi.org/10.1007/978-981-19-3643-2

Definition B.2 (*C^∞-compatible*)

Let $(U_\alpha, \varphi_\alpha)$ and (U_β, φ_β) be charts of an atlas such that $U_\alpha \cap U_\beta \neq \phi$. $(U_\alpha, \varphi_\alpha)$ and (U_β, φ_β) are said to be C^∞-compatible, if transition map $\varphi_\beta \circ \varphi_\alpha^{-1} : \varphi_\alpha(U_\alpha \cap U_\beta) \to \varphi_\beta(U_\alpha \cap U_\beta)$ is a smooth diffeomorphism.

Definition B.3 (*differentiable structure*)

An atlas $\mathcal{U} = \{(U_\lambda, \varphi_\lambda) \mid \lambda \in \Lambda\}$ is said to be a differentiable structure, if the following conditions are satisfied:

  (i) $(U_\alpha, \varphi_\alpha)$ and (U_β, φ_β) are C^∞-compatible for all $\alpha(\in \Lambda)$ and $\beta(\in \Lambda)$.

  (ii) If (V, ψ) is C^∞-compatible with every $(U_\alpha, \varphi_\alpha) \in \mathcal{U}$, then $(V, \psi) \in \mathcal{U}$.

Definition B.4 (*smooth manifold*)

A smooth manifold is a topological manifold equipped with a differentiable structure.

That is, for a smooth manifold, coordinate transformation and differentiation can be used. In the literature on linearization of nonlinear systems, an n-dimensional smooth manifold can be interpreted as a locally Euclidean space $\mathbb{R}^n$. If you are not familiar with the differential geometry, you do not have to pay much attention to the terminology of a smooth manifold.

B.2   Vector Space and Algebra

Definition B.5 (*vector space or linear space*)

A set V is said to be a vector space over field $\mathbb{R}$, if there exist two operations, $+ : V \times V \to V$ (vector addition or addition) and $\cdot : \mathbb{R} \times V \to V$ (scalar multiplication), which satisfy the following eight axioms:

  (i) (commutativity of addition)

$$x + y = y + x, \ \forall x, y \in V.$$

  (ii) (associativity of addition)

$$x + (y + z) = (x + y) + z, \ \forall x, y, z \in V.$$

  (iii) There exists an element $O_V \in V$, called additive identity, such that

$$x + O_V = O_V + x = x, \ \forall x \in V.$$

  (iv) For every $x \in V$, there exists an element $-x \in V$, called the additive inverse of x, such that

$$x + (-x) = O_V.$$

(v)

$$r_1 \cdot (r_2 \cdot x) = (r_1 r_2) \cdot x, \ \forall r_1, r_2 \in \mathbb{R}, \ \forall x \in V.$$

(vi) (distributivity of scalar multiplication with respect to vector addition)

$$r \cdot (x_1 + x_2) = r \cdot x_1 + r \cdot x_2, \ \forall r \in \mathbb{R}, \ \forall x_1, x_2 \in V.$$

(vii) (distributivity of scalar multiplication with respect to field addition)

$$(r_1 + r_2) \cdot x = r_1 \cdot x + r_2 \cdot x, \ \forall r_1, r_2 \in \mathbb{R}, \ \forall x \in V.$$

(viii)

$$1 \cdot x = x, \ \forall x \in V.$$

The vector space has only vector addition $(+)$ and scalar multiplication $(\cdot)$. If vector multiplication $(\odot)$ is further defined in the vector space, it becomes an algebra.

Definition B.6 (*algebra*)
A vector space A over field $\mathbb{R}$ is said to be an algebra over field $\mathbb{R}$, if there exists a vector multiplication (or product) $\odot : A \times A \to A$ which satisfies the following:

(i)

$$f \odot (g + h) = f \odot g + f \odot h, \ \forall f, g, h \in A$$
$$(f + g) \odot h = f \odot h + g \odot h, \ \forall f, g, h \in A.$$

(ii)

$$\alpha \cdot (f \odot g) = (\alpha \cdot f) \odot g = f \odot (\alpha \cdot g), \ \forall \alpha \in \mathbb{R}, \ \forall f, g \in A.$$

Definition B.7 (*algebra with identity, associative algebra, and commutative algebra*)

(a) An algebra A is said to be an algebra with identity, if there exists an identity element $1 \in A$ for vector multiplication $\odot$ such that $f \odot 1 = 1 \odot f = f$, $\forall f \in A$.

(b) An algebra A is said to be an associative algebra, if vector multiplication $\odot$ satisfies the following associative law:

$$f \odot (g \odot h) = (f \odot g) \odot h, \ \forall f, g, h \in A.$$

(c) An algebra A is said to be a commutative algebra, if vector multiplication $\odot$ satisfies the following commutative law:

$$f \odot g = g \odot f, \ \forall f, g \in A.$$

Definition B.8 (*Lie algebra*)
Let $f \odot g \triangleq [f, g]$. An algebra A is said to be a Lie algebra, if vector multiplication $\odot$ satisfies the following axioms:

(i) (bilinear)

$$[r_1 f + r_2 g, h] = r_1[f, h] + r_2[g, h], \ \forall r_1, r_2 \in \mathbb{R}, \ \forall f, g, h \in A$$
$$[h, r_1 f + r_2 g] = r_1[h, f] + r_2[h, g], \ \forall r_1, r_2 \in \mathbb{R}, \ \forall f, g, h \in A.$$

(ii) (anticommutative or skew-commutative)

$$[f, g] = -[g, f], \ \forall f, g \in A.$$

(iii) (Jacobi identity)

$$[f, [g, h]] + [g, [h, f]] + [h, [f, g]] = 0, \ \forall f, g, h \in A.$$

The bracket operation that satisfies the three properties of Definition B.8 is called the Lie bracket.

B.3 Vector Field on Manifold

Let M be a smooth manifold of dimension n. A real-valued function $f : M \to \mathbb{R}$ is said to belong to $C^\infty(M)$, if for every coordinate chart $\varphi : U \to \mathbb{R}^n$, the map $f \circ \varphi^{-1} : \varphi(U) \, (\subset \mathbb{R}^n) \to \mathbb{R}$ is C^∞ (or smooth).

Definition B.9 (*tangent vector*)
A tangent vector at a point $p \in M$ is a function $v : C^\infty(M) \to \mathbb{R}$ which satisfies the following two conditions:

(i) (linearity)

$$v(\alpha h_1 + \beta h_2) = \alpha v(h_1) + \beta v(h_2), \ \ \forall h_1, h_2 \in C^\infty(M), \ \ \forall \alpha, \beta \in \mathbb{R}.$$

(ii) (Leibniz rule)

$$v(h_1 h_2) = h_2(p)v(h_1) + h_1(p)v(h_2), \quad \forall h_1, h_2 \in C^\infty(M).$$

Definition B.10 (*tangent space*)
We define the tangent space $T_p(M)$ at p as the set of all tangent vectors at p.

If we define vector addition and scalar multiplication on $T_p(M)$ by

$$(X_p + Y_p)(f) = X_p(f) + Y_p(f), \quad \forall f \in C^\infty(M)$$

$$(\alpha \cdot X_p)(f) = \alpha X_p(f), \quad \forall \alpha \in \mathbb{R}, \ \forall f \in C^\infty(M),$$

then $T_p(M)$ is a vector space over field $\mathbb{R}$.

Definition B.11 (*tangent bundle*)
The tangent bundle of a differentiable manifold M is a manifold $T(M)$ which assembles all the tangent vectors in M. In other words,

$$T(M) = \bigcup_{p \in M} \{p\} \times T_p(M) = \{(p, v) \mid p \in M, \ v \in T_p(M)\}.$$

Define the projection map $\pi : TM \to M$ by $\pi(p, v) = p$. A smooth assignment of a tangent vector to each point of a manifold is called a vector field.

Definition B.12 (*vector field*)
A vector field is a function $X : M \to T(M)$ such that $X(p) \in \{p\} \times T_p(M)$.

If we define $X(f)$ for all $f \in C^\infty(M)$ by

$$X(f)(p) = X_p(f), \quad \forall p \in M,$$

then a smooth vector field X on a manifold M is a linear map $X : C^\infty(M) \to C^\infty(M)$ such that

$$X(fg) = fX(g) + X(f)g, \quad \forall f, g \in C^\infty(M).$$

If X is a smooth vector field, then $X(f)$ is a smooth function on M. Note that $T_p(M)$ is a vector space over field $\mathbb{R}$. If we define vector addition and scalar multiplication on the set of all smooth vector fields on M by

$$(X + Y)_p = X_p + Y_p, \quad \forall p \in M$$

$$(\alpha \cdot X)_p = \alpha \cdot X_p, \quad \forall \alpha \in \mathbb{R}, \ \forall p \in M,$$

then the set of all smooth vector fields on M is also a vector space over field $\mathbb{R}$. The Lie bracket, [X, Y], of two smooth vector fields X and Y is the smooth vector field [X, Y] such that

$$(f) \triangleq X\left(Y(f)\right) - Y\left(X(f)\right), \quad \forall f \in C^{\infty}(M). \tag{B.1}$$

It is not difficult to see that the Lie bracket operation, defined by (B.1), satisfies the axioms of Definition B.8. In other words, for all smooth vector fields X, Y, and Z on M, the following conditions are satisfied:

(i) (bilinear)

$$[\alpha_1 X + \alpha_2 Y, Z] = \alpha_1[X, Z] + \alpha_2[Y, Z], \quad \forall \alpha_1, \alpha_2 \in \mathbb{R}$$

$$[Z, \alpha_1 X + \alpha_2 Y] = \alpha_1[Z, X] + \alpha_2[Z, Y], \quad \forall \alpha_1, \alpha_2 \in \mathbb{R}.$$

(ii) (anticommutativity or skew-commutative)

$$[X, Y] = -[Y, X].$$

(iii) (Jacobi identity)

$$[X, [Y, Z]] + [Y, [Z, X]] + [Z, [X, Y]] = 0.$$

Therefore, the set of all smooth vector fields on M equipped with the bracket operation is a Lie algebra.

Definition B.13 (*differential map*)
Let M and N be smooth manifolds. Suppose that $S : M \rightarrow N$ is a smooth map between smooth manifolds. Given some $p \in M$, the differential map $S_* : T_pM \rightarrow T_{S(p)}N$ of S at p is a linear map from the tangent space of M at p to the tangent space of N at $S(p)$ such that

$$S_*(X_p)(f) = X_p(f \circ S), \quad \forall f \in C^{\infty}(N).$$

Definition B.14 (*smooth distribution*)
Suppose that D_p is a k-dimensional subspace of $T_p(M)$ for all $p \in M$. D is said to be a k-dimensional smooth distribution on M, if there exist a neighborhood U of p and a set of smooth vector fields $\{X_1, \cdots, X_k\}$ on U for any $p \in M$, such that for any $q \in U$,

$$D_q = \text{span}\{X_1, \cdots, X_k\}$$

$$\triangleq \left\{ \sum_{i=1}^{k} c_i X_i \; \middle| \; c_i \in C^\infty(U), \; 1 \le i \le k \right\}.$$

$\{X_1, \cdots, X_k\}$ is a local basis of distribution D.

Definition B.15 (*involutive distribution*)
Smooth distribution D is said to be involutive, if $[X, Y] \in D$ for any smooth vector fields $X \in D$ and $Y \in D$. In other words, smooth distribution D is said to be involutive, if D is closed under bracket operation.

Definition B.16 (*completely integrable distribution*)
Suppose that D is a k-dimensional smooth distribution on M. Distribution D is said to be completely integrable, if each point has a chart (U, φ) such that $\left\{ \varphi_*^{-1}\left(\frac{\partial}{\partial z_1}\right), \cdots, \varphi_*^{-1}\left(\frac{\partial}{\partial z_k}\right) \right\}$ is a local basis on U.

Theorem B.1 (*Frobenius Theorem*)
A distribution D is completely integrable, if and only if D is involutive.

Appendix C
MATLAB Subfunctions

In this section, MATLAB Subfunctions used in this book are introduced. The readers who are expert in programming can improve the programs.

$\mathbf{adfg}(f, g, x) : \text{out}=\text{ad}_f g(x) = [f(x), g(x)]$

```
function out=adfg(f,g,x)

dg=jacobian(g,x);
df=jacobian(f,x);
out=simplify(dg*f-df*g);
```

$\mathbf{adfgk}(f, g, x, k) : \text{out}=\text{ad}_f^k g(x)$

```
function out=adfgk(f,g,x,k)

out=g;
for k1=1:k
  out=simplify(adfg(f,out,x));
end
```

$\mathbf{adfgM}(f, \left[g_1(x) \cdots g_m(x) \right], x) : \text{out}=\left[\text{ad}_f g_1(x) \cdots \text{ad}_f g_m(x) \right]$

```
function out=adfgM(f,G,x)

out=G-G;
for k1=1:size(G,2)
  out(:,k1)=adfg(f,G(:,k1),x);
end
out=simplify(out);
```

© The Editor(s) (if applicable) and The Author(s), under exclusive license to Springer Nature Singapore Pte Ltd. 2022
H.-G. Lee, *Linearization of Nonlinear Control Systems*,
https://doi.org/10.1007/978-981-19-3643-2

adfgkM(f, g, x, k) : out=$\left[\mathrm{ad}_f^k g_1(x) \cdots \mathrm{ad}_f^k g_m(x)\right]$

```
function out=adfgkM(f,G,x,k)

out=G;
for k1=1:k
  out=simplify(adfgM(f,out,x));
end
```

CharacterNum$(f(x), g(x), h(x), x)$: characteristic number
out $= \rho$

```
function rho=CharacterNum(f,g,h,x)

[n,m]=size(g);
for k=1:n
  t1=simplify(Lfh(g,Lfhk(f,h,x,k-1),x));
  if ChZero(t1)==0
    rho=k;
    return
  end
end
rho=n+1;
```

ChCommute$(T(x), x)$: out $= \begin{cases} 1, & \text{if } [T_i(x), T_j(x)] = 0, \ 1 \leq i, j \leq s \\ 0, & \text{otherwise} \end{cases}$

```
function out=ChCommute(T,x)

out=0;
s=size(T,2);
for k1=1:s
  for k2=k1:s
  cc=adfg(T(:,k1),T(:,k2),x);
  if ChZero(cc)==0
    return
  end
  end
 end
out=1;
```

ChConst$(M(x), x)$: out $= \begin{cases} 1, & \text{if } M(x) \text{ is a constant matrix} \\ 0, & \text{otherwise} \end{cases}$

```
function out=ChConst(M,x)

out=0;
s=size(M);
for k1=1:s(1)
  for k2=1:s(2)
    t1=jacobian(M(k1,k2),x);
    if ChZero(t1) == 0
      return
    end
  end
end
out=1;
```

$$\textbf{ChExact}(\omega(x), x) : \text{out} = \begin{cases} 1, & \text{if } \omega(x) \text{ is exact one form} \\ 0, & \text{otherwise} \end{cases}$$

```
function out=ChExact(omega,x)

out=0;
Tomega=omega';
dTomega=jacobian(Tomega,x);
t1=dTomega'-dTomega;
if ChZero(t1)==0
  return
end
out=1;
```

$$\textbf{ChInverseF}(F_u(x), G_u(x), x) :$$
$$\text{out} = \begin{cases} 1, & \text{if } F_u \circ G_u(x) = x \text{ and } G_u \circ F_u(x) = x \\ 0, & \text{otherwise} \end{cases}$$

```
function out=ChInverseF(F,iF,x)

out=0;
cc1=simplify(subs(F,x,iF));
if ChZero(cc1-x)==0
  return
end
cc2=simplify(subs(iF,x,F));
if ChZero(cc2-x)==0
  return
end
out=1;
```

$$\textbf{ChInvolutive}(D(x), x) :$$

$$\text{out} = \begin{cases} 1, & \text{if distribution } \text{span}\{d_1(x), \cdots, d_s(x)\} \text{ is involutive} \\ 0, & \text{otherwise} \end{cases}$$

```
function out=ChInvolutive(D,x)

out=0;
s=size(D,2);
r=rank(D);
for k1=1:s
  for k2=k1+1:s
    t1=adfg(D(:,k1),D(:,k2),x);
    t2=[D t1];
    if rank(t2)>r
      return
    end
  end
end
out=1;
```

$$\textbf{ChZero}(M(x)) : \text{out} = \begin{cases} 1, & \text{if } M(x) = O \\ 0, & \text{if } M(x) \neq O \end{cases}$$

```
function out=ChZero(M)

out=0;
s=size(M);
for k1=1:s(1)
  for k2=1:s(2)
    if M(k1,k2) ~= M(1,1)-M(1,1)
      return
    end
  end
end
out=1;
```

$$\textbf{Codi}(\frac{\partial S(x)}{\partial x}, x) : \text{out}=S(x)$$

```
function out=Codi(dS,x)

s1=size(dS,1);
out=dS(:,1)-dS(:,1);
for m=1:s1
  out(m)=dS(1,1)-dS(1,1);
  for k=1:length(x)
    dbeta(m,k)=simplify(dS(m,k)-diff(out(m),x(k)));
    beta(m,k)= simplify(int(dbeta(m,k),x(k)));
    out(m)=simplify(out(m)+beta(m,k));
  end
end
out=simplify(out-subs(out,x,x-x));
```

CXexact$(\omega(x), x)$:

$$\text{out1} = \begin{cases} 1, & \text{if } \frac{-1}{\omega_K(0)} \mathbf{q}_K(x) \text{ is exact} \\ 0, & \text{otherwise} \end{cases}$$

out2=$c(x)$ such that $c(x)\omega(x)$ is exact one form when out1= 1.

One form $\frac{-1}{\omega_K(0)} \mathbf{q}_K(x)$ is defined in (4.24).

```
function [flag,CX]=CXexact(omega,x)

flag=0;
CX=x(1)-x(1)+1;
n=size(x,1);
if ChExact(omega,x)==1
  flag=1;
  return
end
domega=jacobian(omega',x);
bQ=domega-domega';
omega0=subs(omega,x,x-x);
for k1=1:n
  if ChZero(omega0(k1))==0
    K=k1;
    break
  end
end
dLCX=(1/omega(K))*bQ(:,K)';
if ChExact(dLCX,x)==0
  return
end
LCX=Codi(dLCX,x);
CX=exp(LCX);
if ChExact(CX*omega,x)==0
  return
end
flag=1;
```

Decoupling-M$(f(x), g(x), h(x), x, \rho)$: Decoupling matrix

$$\text{out1}= E(x) = \begin{bmatrix} L_f^{\rho_1} h_1(x) \\ \vdots \\ L_f^{\rho_q} h_q(x) \end{bmatrix} \quad ; \quad \text{out2}= D(x) = \begin{bmatrix} L_g L_f^{\rho_1} h_1(x) \\ \vdots \\ L_g L_f^{\rho_q} h_q(x) \end{bmatrix}$$

```
function [E,D]=Decoupling_M(f,g,h,x,rho)

t1=h-h;
for k=1:length(h)
  t1(k)=Lfhk(f,h(k),x,rho(k)-1);
end
t1=simplify(t1);
D=simplify(Lfh(g,t1,x));
E=simplify(Lfh(f,t1,x));
```

Delta$(f(x), g(x), x)$:
$$\text{out}=\left[g_1(x) \cdots g_m(x) \cdots \text{ad}_f^{n-1} g_1(x) \cdots \text{ad}_f^{n-1} g_m(x)\right]$$

```
function D=Delta(f,g,x)

[n,m]=size(g);
D=g;
for k=1:n-1
  D=[D adfgM(f,D(:,(k-1)*m+1:k*m),x)];
end
```

DeltaDT$(F_u(x), x, u)$:
$$\text{out}=\left[\frac{\partial F_u(x)}{\partial u} \cdots \frac{\partial \hat{F}_u^{\max(\kappa)}(x)}{\partial u}\right]$$

```
function D=DeltaDT(F,x,u)

n=size(F,1);
m=size(u,1);
xu=[x; u];
D=simplify(jacobian(F,u));
for k=2:n
  t1=HatF(F,x,u,k);
  t2=simplify(jacobian(t1,u));
  newD=[D t2];
  if rank(subs(newD,xu,xu-xu)) == rank(subs(D,xu,xu-xu))
    return
  end
  D=newD;
end
```

HatF$(F_u(x), x, u, k)$: out=$\hat{F}^k(x, u) \triangleq F_0^{k-1} \circ F_u(x)$

```
function out=HatF(F,x,u,k)

out = x;
for n=1:k
  out=simplify(subs(out,u,u-u));
  out=simplify(subs(out,x,F));
end
out = simplify(out);
```

ker-sF$(\mathcal{F}(w), w, n)$: out=ker $\mathcal{F}_*$

```
function out=ker_sF(sF,w,n)

Y=jacobian(sF,w);
t1=Y(:,1:n);
t2=Y(:,n+1);
t3=simplify(inv(t1)*t2);
out=[-t3; 1];
```

ker-sF-M$(\mathcal{F}(w), w, \tilde{U}, \kappa, n, m)$: out=ker $\mathcal{F}_*$

```
function out=ker_sF_M(sF,w,tW,ka,n,m)

bW=transp(w(1,1:ka(1)));
for k1=2:m
  bW=[bW; transp(w(k1,1:ka(k1)))];
end
for k1=1:m
  bW=[bW; w(k1,ka(k1)+1)];
end
Y=jacobian(sF,bW);
t1=Y(:,1:n);
t2=Y(:,(n+1):(n+m));
t3=simplify(-inv(t1)*t2);
bkersF=[t3; eye(m)];
iP=jacobian(tW,bW);
out=iP*bkersF;
```

Kindex$(f(x), g(x), x)$: Kronecker indices
out1=$\kappa = \begin{bmatrix} \kappa_1 \cdots \kappa_m \end{bmatrix}^\mathsf{T}$ on a nbhd of $x = x_0$
out2=$\begin{bmatrix} g_1(x) \cdots g_m(x) \cdots \mathrm{ad}_f^{\max(\kappa)-1} g_1(x) \cdots \mathrm{ad}_f^{\max(\kappa)-1} g_m(x) \end{bmatrix}$

```
function [kappa,D]=Kindex(f,g,x)

[n,m]=size(g);
D1=Delta(f,g,x);
kappa=zeros(m,1);
DD=x-x;
for k1=1:n
  for k2=1:m
    t1 =[DD D1(:,m*(k1-1)+k2)];
    if rank(t1)>rank(DD)
      kappa(k2)=kappa(k2)+1;
      DD=t1;
      if rank(DD)==rank(D1)
        D=D1(:,1:m*max(kappa))
        return
      end
    end
  end
end
```

Kindex0$(f(x), g(x), x)$: Kronecker indices
out1=$\kappa = \begin{bmatrix} \kappa_1 \cdots \kappa_m \end{bmatrix}^\mathsf{T}$ on a nbhd of $x = 0$
out2=$\begin{bmatrix} g_1(x) \cdots g_m(x) \cdots \mathrm{ad}_f^{\max(\kappa)-1} g_1(x) \cdots \mathrm{ad}_f^{\max(\kappa)-1} g_m(x) \end{bmatrix}$

```
function [kappa,D]=Kindex0(f,g,x)

[n,m]=size(g);
D1=Delta(f,g,x);
D0=subs(D1,x,x-x);
kappa=zeros(m,1);
DD=x-x; for k1=1:n
  for k2=1:m
    t1 =[DD D0(:,m*(k1-1)+k2)];
    if rank(t1)>rank(DD)
      kappa(k2)=kappa(k2)+1;
      DD=t1;
      if rank(DD)==rank(D0)
        D=D1(:,1:m*max(kappa));
        return
      end
    end
  end
end
```

KindexDT0($F_u(x)$, x, u) : Kronecker indices of discrete time system
out=$\kappa = \begin{bmatrix} \kappa_1 \cdots \kappa_m \end{bmatrix}^{\mathsf{T}}$ on a nbhd of $x = 0$

```
function kappa=KindexDT0(F,x,u)

n=size(F,1);
m=size(u,1);
xu=[x; u];
D=DeltaDT(F,x,u);
tD=subs(D,xu,xu-xu);
kappa=zeros(m,1);
N=size(D,2)/m;
DD=x-x; for
k1=1:N
  for k2=1:m
    t1 =[DD tD(:,m*(k1-1)+k2)];
    if rank(t1)>rank(DD)
      kappa(k2)=kappa(k2)+1;
      DD=t1;
      if rank(DD)==rank(tD)
        return
      end
    end
  end
end
```

Lfh$(f(x), h(x), x)$: out=$L_f h(x)$

```
function out=Lfh(f,h,x)

dh=jacobian(h,x);
out=simplify(dh*f);
```

Lfhk$(f(x), h(x), x, k)$: out=$L_f^k h(x)$

```
function out=Lfhk(f,h,x,k)

out = h;
for n=1:k
    out=simplify(Lfh(f,out,x));
end
out = simplify(out);
```

ObvIndex0$(F0(x), H(x), x, n, p)$: observability indices
out=$v = \begin{bmatrix} v_1 & \cdots & v_p \end{bmatrix}^\mathsf{T}$ on a nbhd of $x = 0$

```
function nu=ObvIndex0(F0,H,x,n,p)

TM=H;
for k=1:n-1
  TM=[TM; Lfh(F0,TM((k-1)*p+1:k*p),x)];
end
dTM=simplify(jacobian(TM,x));
dTM0=subs(dTM,x,x-x);
nu=zeros(p,1);
DD=dTM0(1,:)-dTM0(1,:);
for k1=1:n
  for k2=1:p
    t1 =[DD; dTM0(p*(k1-1)+k2,:)];
    if rank(t1)>rank(DD)
      nu(k2)=nu(k2)+1;
      DD=t1;
      if rank(DD)==rank(dTM0)
        return
      end
    end
  end
end
end
```

Psi-sF$(F_u(x), x, u, w, n)$:
out1 $= \Psi(w_1, \cdots, w_n) \triangleq F_{w_1} \circ F_{w_2} \circ \cdots \circ F_{w_n}(0)$
out2 $= \mathcal{F}(w_1, \cdots, w_{n+1}) \triangleq F_{w_1} \circ \cdots \circ F_{w_n} \circ F_{w_{n+1}}(0)$

```
function [Psi,sF]=Psi_sF(F,x,u,w,n)

FF(:,1)=subs(F,[x;u],[x-x;w(n+1)]);
for k=1:n
  FF(:,k+1)=subs(F,[x;u],[FF(:,k);w(n+1-k)]);
end
sF=simplify(FF(:,n+1));
Psi=simplify(subs(sF,w(n+1),w(n+1)-w(n+1)));
```

Psi-sF-M$(F_u(x), x, u, w, m, \kappa)$:

$\text{out1} = \Psi(w_1, \cdots, w_n) \triangleq F_{w^1} \circ \cdots \circ F_{w^{\kappa_{\max}}}(0)|_{w_j^i=0, \, i \geq \kappa_j+1}$

$\text{out2} = \mathcal{F}(\tilde{U}) \triangleq F_{w^1} \circ \cdots \circ F_{w^{\kappa_{\max}}} \circ F_{w^{\kappa_{\max}+1}}(0)|_{w_j^i=0, \, i \geq \kappa_j+2}$

$\text{out3} = U = \begin{bmatrix} w_1^1 \ w_2^1 \ w_1^2 \ w_1^3 \end{bmatrix}^{\mathsf{T}}$, when $(\kappa_1, \kappa_2) = (3, 1)$

$\text{out4} = \tilde{U} = \begin{bmatrix} w_1^1 \ w_2^1 \ w_1^2 \ w_2^2 \ w_1^3 \ w_1^4 \end{bmatrix}^{\mathsf{T}}$, when $(\kappa_1, \kappa_2) = (3, 1)$

$\text{out5} = \begin{bmatrix} w_1^1 \ w_1^2 \ w_1^3 \ w_1^4 \\ w_2^1 \ w_2^2 \ 0 \ 0 \end{bmatrix}^{\mathsf{T}}$, when $(\kappa_1, \kappa_2) = (3, 1)$

```
function [Psi,sF,W,tW,U]=Psi_sF_M(F,x,u,w,m,ka)

kamax=max(ka);
FF(:,1)=subs(F,[x;u],[x-x;w(:,kamax+1)]);
for k=1:kamax
  FF(:,k+1)=subs(F,[x;u],[FF(:,k);w(:,kamax+1-k)]);
end
BW=w-w;
BtW=w-w;
for k1=1:m
  BW(k1,1:ka(k1))=w(k1,1:ka(k1));
  BtW(k1,1:ka(k1)+1)=w(k1,1:ka(k1)+1);
end
sF=simplify(subs(FF(:,kamax+1),w,BtW));
Psi=simplify(subs(FF(:,kamax+1),w,BW));
W=w(1,1)-w(1,1);
tW=w(1,1)-w(1,1);
for k1=1:kamax+1
  for k2=1:m
    if ChZero(BW(k2,k1))==0
      W=[W; BW(k2,k1)];
    end
    if ChZero(BtW(k2,k1))==0
      tW=[tW; BtW(k2,k1)];
    end
  end
end
W=simplify(W(2:length(W)));
tW=simplify(tW(2:length(tW)));
U=simplify(BtW(:,1:kamax+1));
```

RelativeDegree$(f(x), g(x), h(x), x)$: out=$(\rho_1, \cdots, \rho_q)$

```
function rho=RelativeDegree(f,g,h,x)

for k=1:length(h)
  rho(k)=CharacterNum(f,g,h(k),x);
end
```

RowReorder$(D(x))$: out=R_0

where R_0 is a row permutation matrix such that $R_0 D(x) = \begin{bmatrix} R_1 \\ R_2 \end{bmatrix} D(x) = \begin{bmatrix} \tilde{D}(x) \\ \bar{D}(x) \end{bmatrix}$

and rank $\left(\tilde{D}(x) \right) =$ rank $(D(x))$.

```
function R=RowReorder(D)

q=size(D,1); r=rank(D); R=eye(q);
id=0;
for k1=1:q
  if rank(D(1:k1,:))>rank(D(1:k1-1,:))
     id=[id k1];
  end
end
id=id(2:r+1);
t1=R(1,:)-R(1,:);
t2=t1;
for k=1:q
  if k==id(1)
     t1=[t1; R(k,:)];
     id=[id(2:r) 0];
  else
     t2=[t2; R(k,:)];
  end
end
R=[t1(2:r+1,:); t2(2:q-r+1,:)];
```

S1$(f(x), g(x), x)$:

out1 $= \begin{cases} 1, & \text{if } \frac{-1}{\omega_K(0)} \mathbf{q}_K(x) \text{ is exact} \\ 0, & \text{otherwise} \end{cases}$

out2=$S_1(x)$ in Lemma 4.1 when out1= 1.

If out1= 0, then $S_1(x)$ should be found without MATLAB function **S1**.

One form $\frac{-1}{\omega_K(0)} \mathbf{q}_K(x)$ is defined in (4.24).

```
function [flag,S1]=S1(f,g,x)

flag=0;
S1=g(1)-g(1);
[n,m]=size(g);
T(:,1)=g;
for k=2:n
  T(:,k)=adfg(f,T(:,k-1),x);
end
T=simplify(T);
BD=T(:,1:n);
BD(:,n)=subs(BD(:,n),x,x-x);
iBD=inv(BD);
Eyen=jacobian(x,x);
omega=(-1)^(n-1)*Eyen(n,:)*iBD
[flagCX,CX]=CXexact(omega,x)
if flagCX==0
  return
end
flag=1;
dS1=CX*omega;
S1=Codi(dS1,x);
```

S1M$(f(x), g(x), x, \kappa)$:

$$\text{out1} = \begin{cases} 1, & \text{if } z = S(x) \text{ can be found by S1M} \\ 0, & \text{if } z = S(x) \text{ should be found by hand calculation} \end{cases}$$

$$\text{out2} = \begin{bmatrix} S_{11}(x) \\ \vdots \\ S_{m1}(x) \end{bmatrix} \text{ in Lemma 4.3 when out1} = 1.$$

```
function [flag,S1]=S1M(f,g,x,kappa)

flag=0;
S1=g(1,:)'-g(1,:)';
[n,m]=size(g);
for k1=1:m
  for k2=1:kappa(k1)
    T(:,k2,k1)=adfgk(f,g(:,k1),x,k2-1);
  end
end
T=simplify(T);
for k=1:m
  BD=g(:,1)-g(:,1);
  tt2=x(1)-x(1);
  for k1=1:m
    BD=[ BD T(:,1:min(kappa(k),kappa(k1)),k1) ];
    t2=x(1:min(kappa(k),kappa(k1)));
    tt2=[tt2 (t2-t2)' ];
```

```
    if kappa(k1) >= kappa(k)
      BD(:,size(BD,2))=subs(BD(:,size(BD,2)),x,x-x);
      if k1==k
        tt2(size(tt2,2))= (-1)^(kappa(k1)-1);
      end
    end
    BD=simplify(BD);
  end
  BD=simplify(BD(:,2:size(BD,2)));
  tt2=simplify(tt2(:,2:size(tt2,2)));
  ipBD=simplify(inv(BD'*BD));
  omega=simplify(tt2*ipBD*BD');
  [flagCX,CX]=CXexact(omega,x);
  if flagCX==0
    return
  end
  dS1=CX*omega;
  S1(k)=Codi(dS1,x);
end
flag=1;
S1=simplify(S1);
```

SpanCx$(X(x), Y(x))$:

$$\text{flag} = \begin{cases} 1, & \text{if rank}\,([X\ Y]) = \text{rank}(Y) = k \\ 0, & \text{otherwise} \end{cases}$$

out=$c(x)$

where $X(x) = Y(x)c(x) = \sum_{i=1}^{k} c_i(x)Y_i(x)$ and $Y(x)$ is a $n \times k$ matrix.

```
function [flag,out]=SpanCx(X,Y)

flag=0;
[n,K]=size(Y)
out=(Y(1,1:K)-Y(1,1:K))'
Z=[X Y] if rank(Y) <
K
  return
end
if rank([X Y]) > rank(Y)
  return
end
Z0=Z(1,:)-Z(1,:);
for k1=1:n
  if rank([Z0; Z(k1,:)]) > rank(Z0)
    Z0=[Z0; Z(k1,:)];
  end
end
Z0=Z0(2:size(Z0,1),:)
out=simplify(inv(Z0(:,2:K+1))*Z0(:,1))
flag=1;
```

sstarmap$(S(x), S^{-1}(z), f(x), x)$:

$$\text{out}=S_*(f(x)) \triangleq \left.\frac{\partial S(x)}{\partial x}f(x)\right|_{x=S^{-1}(z)}$$

```
function out=sstarmap(s,invs,f,x)

ds=jacobian(s,x);
out2=ds*f;
out = subs(out2,x,invs);
out=simplify(out);
```

TauFG$(f(x), g(x), x, \kappa)$:

$$\text{out}=\left[g_1(x) \cdots \text{ad}_f^{\kappa_1-1} g_1(x) \cdots g_m(x) \cdots \text{ad}_f^{\kappa_m-1} g_m(x) \right]$$

```
function out=TauFG(f,g,x,kappa)

[n,m]=size(g);

T=x-x;
for k1=1:m
  for k2=1:kappa(k1)
    T=[T adfgk(f,g(:,k1),x,k2-1)];
  end
end
out=simplify(T(:,2:size(T,2)));
```

transp$(M(x))$: out=$M(x)^\mathsf{T}$

```
function out=transp(M)

s=size(M);
out=M'-M';
for k1=1:s(1)
  out(:,k1)=M(k1,:);
end
out=simplify(out);
```

References

Text Book

[A1] W.A. Boothby, *An Introduction to Differentiable Manifolds and Riemannian Geometry* (Academic, New York, 1975)
[A2] C.T. Chen, *Introduction to Linear System Theory* (HRW, inc., 1970)
[A3] A. Isidori, *Nonlinear Control Systems*, 3rd ed. (London Ltd., Springer, 1995)
[A4] R. Marino, P. Tomei, *Nonlinear Control Design* (Prentice Hall Europe, 1995)
[A5] H. Nijmeijer, A. van der Schaft, *Nonlinear Dynamical Control Systems* (New York Inc., Springer, 1990)
[A6] W. Rudin, *Principle of Mathematical Analysis*, 3rd ed. (McGraw-Hill, Inc., 1976)
[A7] M. Spivak, *A Comprehensive Introduction to Differential Geometry*, vol. I (Publish or Perish, Boston, 1970)
[A8] M. Vidyasagar, *Nonlinear Systems Analysis* (Prentice-Hall Inc., 1978)

Linerization by state transformation

[B1] L.R. Hunt, M. Luksic, R. Su, Exact linearization of input-output systems. Int. J. Control **43**, 247–255 (1986)
[B2] A.J. Krener, On the equivalence of control systems and the linearization of nonlinear systems. SIAM J. Control **11**, 670–679 (1973)
[B3] H.J. Sussmann, Lie brackets, real analyticity and geometric control, in *Differential Geometric Control Theory*, ed. by R.W. Brockett, et al. (Birkhauser, Boston, 1983), pp. 1–116

© The Editor(s) (if applicable) and The Author(s), under exclusive license to Springer Nature Singapore Pte Ltd. 2022
H.-G. Lee, *Linearization of Nonlinear Control Systems*,
https://doi.org/10.1007/978-981-19-3643-2

Static Feedback Linerization

[C1] A. Arapostathis, B. Jakubczyk, H.G. Lee, S.I. Marcus, E.D. Sontag, The effect of sampling on linear equivalence and feedback linearization. Syst. Control Lett. **13**, 373–381 (1989)

[C2] W.M. Boothby, Some comments on global linearization of nonlinear system. Syst. Control Lett. **4**, 147 (1984)

[C3] R.W. Brockett, *Feedback Invariants for Nonlinear Systems* (IFAC Congress, Helsinki, 1978)

[C4] L.R. Hunt, R. Su, G. Meyer, Design for multi-input nonlinear system, in *Differential Geometric Control Theory*, ed. by R.W. Brockett et al. (Birkhauser, Boston, 1983), pp. 268–293

[C5] I.-J. Ha, S.-J. Lee, Input-output linearization with state equivalence and decoupling. IEEE Trans. Autom. Control **39**(11), 2269–2274 (1994)

[C6] B. Jakubczyk, W. Respondek, On the linearization of control systems. Bull. Acad. Polon. Sci. Ser. Math. Astron. Phys. **28**, 517–522 (1980)

[C7] A.J. Krener, Linearization and bilinearization of control systems, in *Proceedings of the 1974 Allerton Conference on Circuits and System Theory* (1974)

[C8] A.J. Krener, W. Respondek, Nonlinear observers with linearizable error dynamics. SIAM J. Control Optim. **23**, 197–216 (1985)

[C9] R. Su, On the linear equivalents of nonlinear systems. Syst. Control Lett. **2**, 48–52 (1982)

[C10] A.J. van der Schaft, Linearization and input-output decoupling for general nonlinear systems. Syst. Control Lett. **5**, 27–33 (1984)

Linearization of the Discrete Time

[D1] J.P. Barbot, S. Monaco, D. Normand-Cyrot, Discrete-time approximated linearization of SISO systems under output feedback. IEEE Trans. Autom. Control **44**(9), 1729–1733 (1999)

[D2] J.W. Grizzle, Feedback linearization of discrete time systems, in *Lecture Notes in Control and Information Sciences*, vol. 83 (Springer, 1986), pp. 273–281

[D3] J.W. Grizzle, P.V. Kokotovic, Feedback linearization of sampled-data systems. IEEE Trans. Autom. Control **33**, 857–859 (1988)

[D4] B. Jakubczyk, Feedback linearization of discrete time systems. Syst. Control Lett. **9**, 411–416 (1987)

[D5] H.G. Lee, A. Arapostathis, S.I. Marcus, Linearization of discrete-time systems. Int. J. Control **45**, 1803–1822 (1987)

[D6] H.G. Lee, S.I. Marcus, Approximate and local linearizability of nonlinear discrete time systems. Int. J. Control **44**, 1103–1124 (1986)

[D7] K. Nam, Linearization of discrete time nonlinear systems and a canonical structure. IEEE Trans. Autom. Control **34**(1), 119–122 (1989)

Dynamic Feedback Linerization

[E1] E. Aranda-Bricaire, C.H. Moog, J.B. Pomet, A linear algebraic framework for dynamic feedback linearization. IEEE Trans. Autom. Control **40**, 127–132 (1995)

[E2] B. Charlet, J. Levine, R. Marino, On dynamic feedback linearization. Syst. Control Lett. **13**, 143–151 (1989)

[E3] B. Charlet, J. Levine, R. Marino, New sufficient conditions for dynamic feedback linearization, *Geometric Methods in Nonlinear Control Theory*, pp. 39–45 (1989)

[E4] B. Charlet, J. Levine, R. Marino, Sufficient conditions for dynamic state feedback linearization. SIAM J. Control Optim. **29**, 38–57 (1991)

[E5] M. Fliess, J. Levine, P. Martin, P. Rouchon, Flatness and defect of non-linear system: introductory theory and examples. Int. J. Control **61**(6), 1327–1361 (1995)

[E6] M. Fliess, J. Levine, P. Martin, P. Rouchon, On differentially flat nonlinear systems, in *Proceedings of the 2nd IFAC NOLCOS*, Bordeaux, pp. 408–412 (1992)

[E7] R.B. Gardner, W.F. Shadwick, The GS algorithm for exact linearization to Brunovsky normal form. IEEE Trans. Autom. Control **37**, 224–230 (1992)

[E8] M. Guay, P.J. Mclellan, D.W. Bacon, A condition for dynamic feedback linearization of control-affine nonlinear systems. Int. J. Control **68**(1), 87–106 (1997)

[E9] L.R. Hunt, R. Su, G. Meyer, Design for multi-input nonlinear system, in *Differential Geometric Control Theory*, ed. by R.W. Brockett et al. (Birkhauser, Boston, 1983), pp. 268–293

[E10] B. Jakubczyk, W. Respondek, On the linearization of control systems. Bull. Acad. Polon. Sci. Ser. Math. Astron. Phys. **28**, 517–522 (1980)

[E11] B. Jakubczyk, Remarks on equivalence and linearization of nonlinear systems, in *Proceedings of the 2nd IFAC NOLCOS*, Bordeaux, pp. 393–397 (1992)

[E12] H. Lee, A. Arapostathis, S.I. Marcus, Linearization of discrete-time systems via restricted dynamic feedback. IEEE Trans. Autom. Control **48**(9), 1646–1650 (2003)

[E13] H.G. Lee, Y.M. Kim, H.T. Jeon, On the linearization via a restricted class of dynamic feedback. IEEE Trans. Autom. Control **45**(7), 1385–1391 (2000)

[E14] P. Rouchon, Necessary condition and genericity of dynamic feedback linearization. J. Math. Syst. Estim. Control **4**, 1–14 (1994)

[E15] W.F. Shadwick, Absolute equivalence and dynamic feedback linearization. Syst. Control Lett. **15**, 35–39 (1990)

[E16] W.M. Sluis, A necessary condition for dynamic feedback linearization. Syst. Control Lett. **21**, 277–283 (1993)

[E17] W.M. Sluis, D.M. Tilbury, A bound on the number of integrators needed to linearize a control system. Syst. Control Lett. **29**, 43–50 (1996)

Applications of Feedback Linearization

[F1] E. Freund, Fast nonlinear control with arbitrary pole-placement for industrial robots and manipulators, in *Brady, Hollerbach, Johnson, Lozano-Perez, and Mason*. ed. by R. Motion (MIT Press, Cambridge, MA, 1982)

[F2] I. Ha, A. Tugcu, N. Boustany, Feedback linearizing control of vehicle longitudinal acceleration. IEEE Trans. Automat. Control **34**, 689–698 (1989)

[F3] R. Hunt, R. Su, G. Meyer, Application of nonlinear transformations to automatic flight control. Automatica **20**, 103–107 (1984)

[F4] M. Ilic-Spong, F. Mak, A new class of fast nonlinear voltage controllers and their impact on improved transmission capacity, Proc. Am. Control Conf. (1989)

[F5] M. Ilic-Spong, R. Marino, S. Persada, D. Taylor, Feedback linearizing controls of switched reluctance motors. IEEE Trans. Automat. Control **32**, 371–374 (1987)

[F6] D.I. Kim, I. Ha, M.S. Ko, Control of induction motors via feedback linearization with input-output decoupling. Int. J. Control **51**(4), 863–883 (1990)

[F7] T. Tarn, A. Bejczy, A. Isidori, Y. Chen, Nonlinear feedback in robot arm control in Proceedings of the 23rd IEEE Conference on Decision and Control, Las Vegas, NV, pp. 736–751 (1984)

[F8] T.J. Tarn, A.K. Bejczy, and X. Yun, Coordinated control of two robot arms, in *Proceedings of 25th IEEE Conference on Decision and Control*, Athens, Greece, pp. 1193–1202 (1986)

Input-Output Linearization and Decoupling

[G1]   R.W. Brockett, Nonlinear systems and differential geometry. Proc. IEEE **64**(1), 61–72 (1976)

[G2]   E. Freund, Decoupling and pole assignment in nonlinear systems. Electron. Lett. **9**, 373–374 (1973)

[G3]   E. Freund, The structure of decoupled non-linear systems. Int. J. Control **21**, 443–450 (1975)

[G4]   J.W. Grizzle, Local input-output decoupling of discrete-time nonlinear systems. Int. J. Control **43**(5), 1517–1530 (1986)

[G5]   J.W. Grizzle, A. Isidori, Block noninteracting control with stability via static state feedback. Math. Control Sig. Syst. **2**, 315–341 (1989)

[G6]   I.J. Ha, Canonical forms of decouplable and controllable linear systems. Int. J. Control **56**(3), 691–701 (1992)

[G7]   I.J. Ha, E.G. Gilert, A complete characterization of decoupling control laws for a general class of nonlinear systems. IEEE Trans. Autom. Control **31**, 823–830 (1986)

[G8]   R.M. Hirschorn, Invertibility of multivariable nonlinear control systems. IEEE Trans. Autom. Control **24**(6), 855–865 (1979)

[G9]   H.J.C. Huijberts, H. Nijmeijer, L.L.M. van der Wegen, Minimality of dynamic input-output decoupling for nonlinear systems. Syst. Control Lett. **18**, 435–443 (1992)

[G10] A. Isidori, J.W. Grizzle, Fixed modes and nonlinear noninteracting control with stability. IEEE Trans. Autom. Control **33**, 907–914 (1988)

[G11] A. Isidori, A.J. Krener, C. Gori-Giogi, S. Manaco, Nonlinear decoupling via feedback: a differential geometric approach. IEEE Trans. Autom. Control **26**, 331–345 (1981)

[G12] A. Isidori, A.J. Krener, C. Gori-Giorgi, S. Monaco, Locally (f, g) invariant distributions. Syst. Control Lett. **1**(1), 12–15 (1981)

[G13] A. Isidori, A. Ruberti, On the synthesis of linear input-output responses for nonlinear systems. Syst. Control Lett. **4**, 17–22 (1984)

[G14] H.G. Lee, S.I. Marcus, On input-output linearization of discrete time nonlinear systems. Syst. Control Lett. **8**, 249–259 (1987)

[G15] S. Monaco, D. Normand-Cyrot, Some remarks on the invertibility of nonlinear discrete-the systems, *American Control Conference*, pp. 324–327 (1983)

[G16] H. Nijmeijer, W. Respondek, Decoupling via dynamic compensation for nonlinear control systems, *IEEE Conference on Decision and Control*, pp. 192–197 (1986)

[G17] H. Nijmeijer, W. Respondek, Dynamic input-output decoupling of nonlinear control systems. IEEE Trans. Autom. Control **33**, 1065–1070 (1988)

[G18] H. Nijmeijer, Local (dynamic) input-output decoupling of discrete time nonlinear systems. IMA J. Math. Control Inf. **4**, 237–250 (1987)

[G19] W.A. Porter, Diagonalization and inverses of nonlinear system. Int. J. Control **11**, 67–76 (1970)

[G20] W. Respondek, Dynamic input-output linearization and decoupling of nonlinear systems, in *Proceedings of the 2nd European Control conference*, Groningen, pp. 1523–1527 (1993)

[G21] L.M. Silverman, Inversion of multivariable linear systems. IEEE Trans. Autom. Control **14**, 270–276 (1969)

[G22] S.N. Singh, Decoupling of invertible nonlinear systems with state feedback and precompensation. IEEE Trans. Autom. Control **25**(6), 1237–1239 (1980)

[G23] S.N. Singh, A modified algorithm for invertibility in nonlinear systems. IEEE Trans. Autom. Control **26**(2), 595–598 (1981)

Observer Error Linearization

[H1] J. Back, J.H. Seo, An algorithm for system immersion into nonlinear observer form. Automatica **42**, 321–328 (2006)

[H2] J. Back, K.T. Yu, J.H. Seo, Dynamic observer error linearization. Automatica **42**, 2195–2200 (2006)

[H3] G. Besancon, On output transformations for state linearization up to output injection. IEEE Trans. Autom. Control **44**(10), 1975–1981 (1999)

[H4] D. Bestle, M. Zeitz, Canonical form observer design for non-linear time-variable systems. Int. J. Control **38**(2), 419–431 (1983)

[H5] C. Califano, S. Monaco, D. Normand-Cyrot, On the observer design in discrete-time. Syst. Control Lett. **49**(4), 255–265 (2003)

[H6] C. Califano, S. Monaco, D. Normand-Cyrot, Canonical observer forms for multi-output systems up to coordinate and output transformations in discrete-time. Automatica **45**, 2483–2490 (2009)

[H7] C. Califano, C.H. Moog, The observer error linearization problem via dynamic compensation. IEEE Trans. Autom. Control **59**, 2502–2508 (2014)

[H8] S.T. Chung, J.W. Grizzle, Sampled-data observer error linearization. Automatica **26**(6), 997–1007 (1990)

[H9] H. Hong, H. Kang, H.G. Lee, Discrete generalized nonlinear observer canonical form. J. Electr. Eng. Technol. **15**, 1357–1365 (2020)

[H10] M. Hou, A.C. Pugh, Observer with linear error dynamics for nonlinear multi-output systems. Syst. Control Lett. **24**(4), 291–300 (1999)

[H11] H.J.C. Huijberts, On existence of extended observers for nonlinear discrete-time systems. *Lecture Notes in Control Information Science*, vol. 244 (Springer, 1999), pp. 73–92

[H12] H.J.C. Huijberts, T. Lilge, H. Nijmeijer, Nonlinear discrete-time synchronization via extended observers. Int. J. Bifur. Chaos **11**, 1997–2006 (2001)

[H13] P. Jouan, Immersion of nonlinear systems into linear systems modulo output injection. SIAM J. Control Optim. **41**(6), 1756–1778 (2003)

[H14] H. Keller, Non-linear observer design by transformation into a generalized observer canonical form. Int. J. Control **46**(6), 1915–1930 (1987)

[H15] A.J. Krener, A. Isidori, Linearization by output injection and nonlinear observers. Syst. Control Lett. **3**(1), 47–52 (1983)

[H16] A.J. Krener, W. Respondek, Nonlinear observers with linearizable error dynamics. SIAM J. Control Optim. **23**(2), 197–216 (1985)

[H17] H.G. Lee, Dynamic observer of general nonlinear control systems. J. Electr. Eng. Technol. **16**(6), 3275–3287 (2021)

[H18] H.G. Lee, Verifiable conditions for multi-output observer error linearizability. IEEE Trans. Autom. Control **62**(9), 4876–4883 (2017)

[H19] H.G. Lee, Verifiable conditions for discrete-time multioutput observer error linearizability. IEEE Trans. Autom. Control **64**(4), 1632–1639 (2019)

[H20] H.G. Lee, A. Arapostathis, S.I. Marcus, Necessary and sufficient conditions for state equivalence to a nonlinear discrete-time observer canonical form. IEEE Trans. Autom. Control **53**(11), 2701–2707 (2008)

[H21] H.G. Lee, K.D. Kim, H.T. Jeon, Restricted dynamic observer error linearizability. Automatica **53**, 171–178 (2015)

[H22] W. Lee, K. Nam, Observer design for autonomous discrete-time nonlinear systems. Syst. Control Lett. **17**(1), 49–58 (1991)

[H23] T. Lilge, On observer design for nonlinear discrete-time systems. Eur. J. Control **4**, 306–319 (1998)

[H24] W. Lin, C.I. Byrnes, Remarks on linearization of discrete-time autonomous systems and nonlinear observer design. Syst. Control Lett. **25**(1), 31–40 (1995)

[H25] D.G. Luenberger, Canonical forms for linear multivariable systems. IEEE Trans. Autom. Control **12**(3), 290–293 (1967)

[H26] D. Noh, N.H. Jo, J.H. Seo, Nonlinear observer design by dynamic observer error linearization. IEEE Trans. Autom. Control **49**(10), 1746–1750 (2004)

[H27] X.H. Xia, W.B. Gao, Nonlinear observer design by observer error linearization. SIAM J. Control Optim. **27**(1), 199–216 (1989)

Index

A
Additive identity, 560
Additive inverse, 560
Algebra, 560, 561
Algebra with identity, 561
Annihilator, 53, 76
Anticommutativity, 20, 562
Anticommutativity or skew-commutative, 564
Approximate linearization, 4, 155
Associative algebra, 561
Associative law, 561
Associativity, 560
Auxiliary dynamics, 350, 440

B
Basis, 555
Bilinear, 20, 562, 564
Binary relation, 60, 76
Boundary, 555
Brunovsky canonical form, 65, 67, 100, 102, 166, 199, 225, 265, 266, 308

C
Canonical system, 352, 377, 443
Chain rule, 11, 25, 32, 62
Characteristic number, 167, 309
Chart, 559
C^∞-compatible, 560
Closed, 553
Closed-loop system, 87, 113, 116, 522
Closed set, 555
Closure, 553, 555

Coarser topology, 555
Commutative algebra, 562
Commutative law, 562
Commutativity, 560
Complement, 553, 555
Completely integrable distribution, 48, 565
Continuous at point x_0, 556
Continuous function, 556
Controllability condition, 105
Controllable, 65, 158
Coordinate free, 27–29, 335, 378, 417
Countable set, 555
Covector field, 39

D
Decoupled input-output relation, 521
Decoupling matrix, 185, 523–526
Dense subset, 555
Diffeomorphism, 12, 13, 560
Difference equation, 266
Differentiable structure, 560
Differential map, 21, 23, 564
Differential one form, 40
Discrete topology, 554
Distribution, 46, 53, 76, 105, 114, 279, 280, 297, 298
Distributivity, 561
Dual Brunovsky canonical form, 333
Dual Brunovsky NOCF, 333, 337, 338, 340, 341, 346, 370, 373, 379, 382, 383, 385, 391, 396–399, 401, 403, 415, 419, 420, 422, 423, 430
Dynamic Extension Algorithm, 543

© The Editor(s) (if applicable) and The Author(s), under exclusive license to Springer Nature Singapore Pte Ltd. 2022
H.-G. Lee, *Linearization of Nonlinear Control Systems*,
https://doi.org/10.1007/978-981-19-3643-2

587

Printed in the United States
by Baker & Taylor Publisher Services